Climate Change Management

Series Editor

Walter Leal Filho, International Climate Change Information and Research Programme, Hamburg University of Applied Sciences, Hamburg, Germany

More information about this series at http://www.springer.com/series/8740

Yazidhi Bamutaze • Samuel Kyamanywa
Bal Ram Singh • Gorettie Nabanoga • Rattan Lal
Editors

Agriculture and Ecosystem Resilience in Sub Saharan Africa

Livelihood Pathways Under Changing Climate

Springer

Editors
Yazidhi Bamutaze
Department of Geography, Geo-Informatics
Makerere University
Kampala, Uganda

Samuel Kyamanywa
Department of Agricultural Production
Makerere University
Kampala, Uganda

Bal Ram Singh
Department of Environmental Science, Soil
Science Building
Norwegian University of Life Science
Ås, Norway

Gorettie Nabanoga
Department of Agricultural Production
Makerere University
Kampala, Uganda

Rattan Lal
The Ohio State University
Columbus, OH, USA

ISSN 1610-2002 ISSN 1610-2010 (electronic)
Climate Change Management
ISBN 978-3-030-12973-6 ISBN 978-3-030-12974-3 (eBook)
https://doi.org/10.1007/978-3-030-12974-3

This Springer imprint is published by the registered company Springer Nature Switzerland AG
The registered company address is: Gewerbestrasse 11, 6330 Cham, Switzerland

Foreword

The United Nations Global Development Agenda 2030 and specifically the Sustainable Development Goals (SDGs) aver, among other goals, a global development trajectory that leaves no one behind. This is a daunting challenge in light of the contemporary global processes and structures, but very achievable if we optimize and maximize the enormous resources and opportunities at our disposal and unlock the existing potentials. To realize this, we need an ecosystem of committed and dedicated actors to confront the contemporary challenges at all scales, in varied geographies, societies and contexts. We must undertake a correct diagnosis and have the right prescription at the right place and right time.

Climate change remains a contemporary challenge in Sub Saharan Africa (SSA) due to its multiplier effects. The IPCC special report on global warming reaffirms the threat posed by the 1.5 °C change in temperature in on SSA on populations, ecosystems, agriculture productivity, biodiversity, and health, among others. In SSA, livelihoods are largely ecosystem dependent and agricultural productions systems are heavily rainfed. Overall, climate change is compunding the structural and nonstructual issues constraining optimal development.

What can science contribute to the Global Development Agenda 2030 and in realizing the aspirations for SSA development? The main domain of science is to create new knowledge through accurate research of integrity that can assist in a positive transformation of society. The consumption of knowledge created by practitioners and policy-makers can yield evidence-based policy formulation and interventions that can catalyze best practices that better protect the environment, secure societies against adverse hazards, and propel innovations and technologies that boost economies.

Recognizing the value of agriculture in SSA, Environment and Natural Resources (ENR), and the challenge of climate change, NORAD under the NORHED program supported multiple transformative research projects implemented in a partnership arrangement between Norway and a cohort of institutions in low- and middle-income countries (LIMIC). Two institutions were outstanding in the number of granted project: Makerere University among the LIMIC countries and Norwegian University of Life Sciences (NMBU) among the Norwegian institutions.

Consequently, a substantial number of research projects were conducted in SSA culminating into new knowledge. However, the research under NORHED projects can only give a snapshot of what is happening in the region owing to its geographical extent and the attendant physical and social heterogeneity. In order to capture a wider spectrum, more articles focusing on SSA were obtained from other contributors beyond those on the NORHED projects.

This book entitled *Agricultural and Ecosystems Resilience in Sub-Saharan Africa* tackles both the natural science and social science issues under conditions of changing climate and increasing climate variability.

We convey tributes to the successful authors, editors, and reviewers of the chapters in this book. We have no doubt that the knowledge presented here is a crucial piece in the ingredients required for transformative development and ultimately realizing the Global Development Agenda 2030, particularly in the SSA context.

Bernard Nawangwe
Vice-Chancellor, Makerere University, Kampala, Uganda

Mari Sundli Tveit
Rector, Norwegian University of Life Sciences (NMBU), Ås, Norway

Preface

A consensus has emerged globally that climate change poses the greatest threat to development and stability. These threats are more severe in regions whose economies are climate sensitive and whose livelihoods are heavily ecosystem and agricultural dependent. A compendium of metrics strikingly highlights the sub-Saharan Africa (SSA) as a hotspot region where a significant proportion of livelihoods are heavily ecosystem dependent, highly vulnerable to climate variability and change, and have low adaptive capacities. Cognizant of the fact that livelihoods and economies in SSA are strongly climate sensitive, it is not possible to continuously experience these complex situations without triggering conflict and instability under the "business as usual" scenario. While SSA national economies are depicting noticeable promise with impressive annual growth rates, there is an increasing concern on declining ecosystems and natural resources therein and agricultural productivity due to the rapidly increasing population. These challenges underpin the need for generating new evidence and decision-making support tools to create an enabling environment for poorer smallholder farmers to engage in sustainable agricultural practices.

A huge demand exists for scientific knowledge to guide interventions and policy decision-making for better management of risks linked to climate change and increasing variability in SSA. However, there are also huge knowledge gaps, and contributions from SSA to the global scientific pool remain dismally low on aspects of ecosystem and livelihood resilience under changing climate. It is this lacuna that largely constitutes the niche and necessity for the publication of this book. The book draws contributions by about 90 scientists, from around the world but with greater proportions from the SSA region. These contributors, with keen interest in the SSA region, have expertise in both natural (biophysical) and social science (the human dimension) disciplines. A total of 35 diverse chapters are distributed under a range of subthemes including the following: transformative agriculture, water, agriculture and ecosystem interactions, landscape processes and human security, climate risk management, and ICT for ecosystem and human resilience. The book addresses issues at micro, meso, and macro levels using various analytical lenses. The conceptualization, thematic focus, and contributions in the book are envisioned to lead to a better development trajectory in SSA.

The timing of the book and its relevancy coincide with contemporary global, continental, and national development processes and efforts with an eye on the SSA region. At the apex is the Global Development Agenda (2030) with the attendant 17 Sustainable Development Goals (SDGs) agreed by the United Nations General Assembly in 2015, which are largely focused on reducing risk and building resilience of the ever-increasing world population. Under the SDGs, climate change adaptation and mitigation are addressed directly under SDG 13 (Climate Action). The African Development Agenda (2063) aims to accelerate growth and development in order to attain socioeconomic transformation over the next 50 years. Nationally, a series of development plans and visions have been formulated by most SSA countries. It is precisely in this context that the book makes a contribution to the development aspirations by tackling impediments via better understanding of landscape processes and policy options but also through provisions of some geographic-specific solutions or ICT-based innovations. It thus responds to the increasing recognition that science must provide smart solutions to obviate the current challenges and improve the futuristic situations for positive transformation of SSA.

We convey our tribute and great appreciation to all the authors for their scientific contributions and dedication to the book. They brought a diversity of thoughts and insights which enriched the book content and, increased its relevancy in multiple domains. We are also thankful to all the reviewers who assured the quality of the manuscripts presented in the book. A dedicated team of persons, including Paul Mukwaya, Twaha Basamba, Settumba Mukasa, and Frank Kansiime, reinforced the editorial team, and we very much appreciate their contributions.

Kampala, Uganda — Yazidhi Bamutaze
Kampala, Uganda — Samuel Kyamanywa
Ås, Norway — Bal Ram Singh
Kampala, Uganda — Gorettie Nabanoga
Columbus, OH, USA — Rattan Lal

Contents

About the Editors

Yazidhi Bamutaze, PhD is working as Associate Professor and is currently the Head of the Department of Geography, Geo-Informatics, and Climatic Sciences, College of Agricultural and Environmental Sciences, Makerere University. He holds a PhD in Geography from Makerere University; an MSc in Geo-Information Science and Earth Observation from the ITC, the Netherlands; and a Bachelor's degree in Geography from Makerere University. He has over 21 years of proven professional experience, leadership, and training grounded largely in Africa but also with collaborations in Europe, North America, and Asia. His line of research expertise primarily lies in geomorphology, land degradation, natural hazards, and disaster risk management coupling applications of geographic information science and remote sensing. He has coauthored over 40 scientific publications and co-edited a series of books. He has contributed toward the development of active learning pedagogy, training programs at Makerere University and other universities in Africa, and mentoring of the next generation of scientists. He has chaired a range of international scientific meetings and currently serves on the executive of the African Association of Remote Sensing of the Environment (AARSE), International Geographical Union (IGU) Commission on African Studies, and International Association for Landscape Ecology (IALE-Africa). As a major dissertation adviser to graduate students, he has successfully directed 20 students, 5 at PhD level. Many of his former students have attained responsible positions in academia, industry, and government. He has also served on over 10 PhD and 30 Master's committees.

Samuel Kyamanywa obtained his PhD in Agri-cultural Entomology from Makerere University under the African Regional Postgraduate Programme in Insect Science at the ICIPE in Nairobi, Kenya, and a BSc in Agriculture from Makerere University. He has over 30 years of teaching and research expertise in agricultural entomology and pest management. He was Principal of the College of Agricultural and Environmental Science, Dean of the Faculty of Agriculture, and Head of the Crop Science Department at Makerere University. He was Deputy Vice-Chancellor at Kyambogo University and a Scientist at the International Centre of Insect Physiology and Ecology in Nairobi, Kenya. He was Chairman of the Agricultural Chemicals Technical Committee of the Uganda Agricultural Chemicals Board, Ministry of Agriculture, Animal Industry and Fisheries (MAAIF). He was Member of the Advisory Board of the Center for Sustainable Rural Livelihoods, Iowa State University. He has served as (i) Council Member of African Rural University and (ii) Chairman of the Advisory Committee of Uganda National Coffee Resources Research Institute. Professor Kyamanywa has been Principal Investigator for 18 research projects, including 8 regional projects. He has supervised 66 MSc and 10 PhD students. He has over 73 publications in international journals, 68 papers in conference proceedings, 3 chapters in 3 textbooks, and 1 handbook.

Bal Ram Singh, PhD is a Professor Emeritus at the Norwegian University of Life Sciences. He earned his PhD degree from G. B. Pant University of Agriculture and Technology, India. His program focuses on bioavailability and mobility of heavy metals in the soil and plant system, fertility management, and agricultural sustainability in soils of the tropics and on carbon sequestration in soils. He has served as Chairman of the program board "Soils and Plants" of the Research Council of Norway and as Deputy Head of the Department, in addition to many national and international committees. He chaired the Cost Action FA0905 (EU) on "Mineral Improved Crop Production for Healthy Food and Feed," in which >200 scientists from 31 countries participated. He has supervised 76 graduate students and 16 visiting fellows/scientists from 20 countries and published 430 articles, of which 240 are in peer-reviewed journals and books.

Professor Singh is a Fellow of the ASA (2004) and SSSA (2005) and Recipient of the International Award in Soil Science (SSSA) in 2011. He is currently Chair of Division 3 of the International Union of Soil Sciences, President of the Norwegian Society of Soil Science, and Member of the Geomedicine committee – Food, Environment and Health of the Norwegian Academy of Science and Letters.

Gorettie Nabanoga, PhD is an Associate Professor at Makerere University. She attained her PhD degree in Social Sciences from Wageningen University of Agriculture, the Netherlands. She holds a Master's degree in Natural Resources Management from the Agricultural University of Norway and a Bachelor of Science in Forestry from Makerere University. She is currently the Deputy Principal of the College of Agricultural and Environmental Sciences at Makerere University. Her teaching and research focus on gender and natural resources management with emphasis on gendered rights to land and natural resources. She has vast experience in participatory approaches to community development and local livelihoods improvement, indigenous knowledge, and traditional resources management practices. She is a Gender Specialist and has conducted research mainly in socioeconomic and institutional changes in communities close to natural forests, institutional capacity building for sustainable natural resource use, and management for local livelihoods with special interest in engendering the processes. She is also a Collaborator in several research projects as the gender focal person and leads the engendering of teaching, research, and outreach activities. She has facilitated/spearheaded the mainstreaming and streamlining of gender in several graduate training and research programs, which are highly diverse in terms of the participants, facilitators, and environment. She coordinated an international capacity building program in which over 200 Master's students from 20 countries were trained. She has supervised over 70 students and published over 40 articles, of which 25 are in peer-reviewed journals. Associate Professor Nabanoga is a Member of the UFA (1994) and a Fellow of the SCB (2006) and UNAS (2017).

Rattan Lal, PhD is a Distinguished University Professor of Soil Science and Director of the Carbon Management and Sequestration Center, The Ohio State University, and an Adjunct Professor of the University of Iceland. He received his BS from the Punjab Agricultural University, Ludhiana; MS from the Indian Agricultural Research Institute, New Delhi; and PhD from the Ohio State University. He served as Senior Research Fellow at the University of Sydney, Australia (1968–1969); Soil Physicist at the IITA, Ibadan, Nigeria (1970–1987); and Professor of Soil Science at Oregon State University (1987–to date). He has authored/coauthored 919 refereed journal articles and 535 book chapters, has written 22 books, and has edited/co-edited 70 books. He is included in the Thomson Reuters list of the World's Most Influential Scientific Minds (2014, 2015) and among the most cited scientists (2014–2018). He received the Honoris Causa degree from six universities in Europe, the USA, and Asia: the Medal of Honor from UIMP, Santander, Spain (2018); and the Distinguished Service Medal of the IUSS (2018) and is Fellow of the five professional societies. Dr. Lal mentored 111 graduate students and 174 visiting scholars from around the world. He was President of the World Association of Soil and Water Conservation (1987–1990), the International Soil and Tillage Research Organization (1988–1991), and the Soil Science Society of America (2006–2008) and is the current President of the International Union of Soil Sciences (2017–2018). He is Laureate of the GCHERA World Agriculture Prize 2018 and Glinka Soil Prize 2018.

Part I
Transformative Agriculture: Science and Policy Interfaces

Agricultural Food Crop Production and Management Challenges Under Variable Climatic Conditions in Rungwe District, Tanzania

Brown Gwambene, Emma T. Liwenga, and Claude G. Mung'ong'o

Abstract This study explores the dynamics and challenges for agricultural food crop production from the perspective of climate variability in Tanzania. It describes the variations in agricultural crop production as a result of differences in the types of farming systems, socio-economic situation and implications of climate variability across agroecological zones. The technological aspects and policy options for developing adaptation options and minimising adverse effects on agricultural production under changed climate are also presented. The study is based on both primary and secondary data, collected through household surveys, key informant interviews, focus group discussions and field observations. The qualitative data were subjected to content and trend analysis, whereas the quantitative data were analysed using Microsoft Excel and SPSS software. The results indicate the decline in food crop production and increased vulnerability to food insecurity among the smallholder and subsistence farmers. The main reasons for the change in production include climate change and variability (manifested by the fluctuating rainfall, drought and changes in temperature), land exhaustion and low adaptive capacity. Addressing the main challenges to increasing agricultural productivity on the existing farm lots to meet the growing demands for food and to offset the climate variability-induced yield losses was recommended. This includes putting in place appropriate policies, strategies, investment plans and collective actions to support the local farmers and enhance their resilience to the changing environment. This will need to enhance productivity and market efficiency among the farming community to increase the uptake of the recommended packages for agricultural crop production.

Keywords Agriculture · Crop production · Food security · Climate variability

---

B. Gwambene (✉)
Department of Geography, Marian University College, Bagamoyo, Tanzania

E. T. Liwenga · C. G. Mung'ong'o
Institute of Resource Assessment, University of Dar es Salaam, Dar es Salaam, Tanzania

Y. Bamutaze et al. (eds.), *Agriculture and Ecosystem Resilience in Sub Saharan Africa*, Climate Change Management,
https://doi.org/10.1007/978-3-030-12974-3_1

1 Introduction

Climate change and variability are being experienced in terms of gradual increase in temperature, increased variability in annual rainfall regimes and greater prevalence of extreme events such as droughts and floods. Seasonal shifts and changes in temperature and precipitation patterns have severe impact on agricultural production and other livelihood activities (Pew 2006; Majule et al. 2009; Obiora 2013). In fact, annual variability in climate already contributes to food shortages and poverty in places where exposure is high and adaptive capacity is low. Moreover, decline in agricultural production in farming communities results in low harvests that leads to the need for communities to buy rather than produce the food they use. However, such a strategy only manages to make communities vulnerable to rising food prices apart from increases production costs (Low 2005; IUCN 2007).

The occurrence and distribution of rainfall, numbers of wet and dry days, minimum and maximum daily temperatures and the strength of the wind have undergone changes (Paul et al. 2009; Gwambene 2011) in the Southern Highlands. These changes have had significant impacts on agricultural crop production and increased challenges of food crop production. For example, changes in patterns and distribution of rainfall and temperature and land use have significantly altered land productivity and hydrology of crop-producing areas (Gwimbi and Mundoga 2010). Moreover, fluctuation in rainfall and temperature has also triggered the emergence of new pests and diseases that wreak havoc on crops and livestock growth and development (Orindi and Murray 2005). As such, understanding the impacts of climate variability on agriculture is vital in formulating adaptation strategies and options to minimise the adverse effects and, hence, enhance food crop production and ensure sustainable livelihoods.

In literature, food security is described as a complex topic at the intersection of multidiscipline (FAO 2006). The cause of food shortages in many African countries is associated with climatic risks (especially, drought and floods), poor technology, political instability and the AIDS pandemic effects (FAO 2006). The impact of climate variability on food production and food security has impacted millions of people, who are driven from their homes and are unable to work in their fields (Zilberman et al. 2012). Also, impacted people are here cut off from the markets; they are unable to sell their products and get supplies of seed, fertilizer and credit. Climate variability constitutes a major factor that influenced food production as it contributes to other factors. Thus, the challenge of food security requires a substantial investment at all levels with more efforts being exerted at the local level.

The climate is one of the key factors of agricultural production with significant impacts on production. It is reflected in the context of essential inputs (water, solar radiation and temperature) that are necessary to support plant and animal growth (Antle 2009). The variability in temperature and rainfall distribution and patterns is one of the major threats to rural livelihoods and food security in the developing world that depends on climate-sensitive activities (Orindi and Murray 2005; Kangalawe et al. 2007; FAO 2009; Leliveld et al. 2013). In Tanzania, the predicted increase in temperature is up to 1.5 °C in the next 20 years and up to 4.3 °C by 2080 with changes in rainfall amount; the intensity and distribution projected will increase challenges on food crop production (URT 2007a; Zilberman et al. 2012; Hoffmann and Smith 2013). The staple food crop yields are predicted to decrease by 20% by 2075 and by as much as 80% in the Central Tanzania (Paavola 2008; Hoffmann and Smith 2013). Although the poor generally, are the most vulnerable to the impact of climate variability, the interlinkages between potential risks under the climate change, livelihood systems, including the farming systems; and challenges in agricultural crop production are not well communicated.

Other factors influencing the farmers' decisions on production include transportation costs, access to labour, land, crop varieties, use of inputs, quality of products, market sources and land tenure (Gwambene 2011). In addition, farming practices/technology, weeds and a wide array of public policies increase vulnerability to the loss of crops. Such factors are affected by individual attitudes towards risk and change, adaptive capacity, and external factors constraining or supporting adaptation and mitigation measures (URT 2007b; Tumbo et al. 2008; Antle 2009; Gwambene 2012; Kangalawe 2012; Leliveld et al. 2013). In fact, the complexity of interacting factors over time induced a spatial organisation of agriculture that tends to be economically efficient in the face of various constraints on the system.

Human activities, institutional arrangements and natural stressors form an important component for the sustainability of livelihood, food production and ecological resources (Spangenberg 2002). This study investigated the challenges and prospects of agricultural crop production in the context of climate variability and changing environment. It assessed the interaction between the context of adaptation measures and transforming institutions. Such transformations need to enhance and sustain livelihood strategies that consider the existing knowledge, practices and coping strategies in order to foster effective adaptation measures.

2 Data and Methodology

The study was undertaken in three villages of Rungwe District across the agroecological zone, namely, Mbeye 1 in the highlands, Kikota in the midlands and Kapulampunguti in lowlands. The district is located in the Southern highlands in Mbeya region and is bordered to the North and West by Mbeya District, to the East by Njombe Region, to the Southeast by Kyela District and to the Southwest by Ileje District (URT 2009). The district lies between latitudes 8°30' and 9°30' south and longitudes 33° and 34°5' east, with an altitude ranging from 770 to 2265 metres above sea level (masl) (Gwambene 2011). Rungwe District has good road networks of 1033.3 km. There is a tarmac road network from the Mbeya City to Rungwe, Kyela and Malawi (URT 2010; Gwambene 2012). The area was selected due to its importance in agricultural food crop production.

To acquire accurate and detailed information, a mixed method research design that combines both qualitative and quantitative data was used in the assessment of the agricultural food crop production challenges under climate variability. The qualitative method established the knowledge and experiences of smallholder farmers on the implications of climate variability on agricultural crop production, spatial and temporal changes in agricultural production and response strategies to climate variability and other stressors. The quantitative method, on the other hand, provided the percentages and other descriptive statistical information. Based on the nature of the study and its objectives, the quantitative method was employed at a household level whereby a total of 147 (9% of the total household in all study areas) households were interviewed. Qualitative methods involved key informant interviews (15 persons from the district, wards and village leaders), focus group discussions (3 groups comprising 12 to18 people per group) and field observations. Key informant interviews and FGDs were guided by structured and semi-structured checklists. To enhance the smooth running of the study, secondary information was reviewed from various documents.

The data collected were edited, coded, tabulated, compiled, processed and analysed using different techniques. Quantitative data from the household survey and meteorological stations were compiled and analysed using the Statistical Package for Social Sciences (SPSS) version 20 and Microsoft Excel software, whereas qualitative data were subjected to content and trend analysis (Ashley and Boyd 2006; ACAPS 2015). Chi-square (X^2) test was used to test the climate variability (rainfall and temperature) and agricultural crop production-related parameters across the agroecological zones. Cross-tabulation allows a comparison of different villages and across agroecological zones. Rainfall and temperature data from meteorological stations were analysed using Microsoft Office Excel to establish their patterns and trends.

3 Results and Discussion

3.1 Climate Variability Perceptions and Trends

Meteorological data were analysed in comparison with farmer's observations and perceived changes. Farmers' responses indicate that they were well aware of the general climate in their locations, its variability and the probabilistic nature of that variability, including its impacts on crop production. In relation to long-term changes in the climate, farmer's observations on rainfall patterns were supported and correlated well with trends in meteorological rainfall data. Their perception also corresponded well with the situation reported by other farmers from elsewhere in the country and across sub-Saharan Africa (Mbilinyi et al. 2013). Also, the farmers' ability to synthesise the knowledge gained through their observation was affected by compounding interactions between climate and other factors such as the decline in soil fertility, land use change and land fragmentations resulting from population pressure that determine the overall influence of climate on crop production.

Perceiving involves the process of using the senses to acquire information on the surrounding environment or situation. This information is obtained after closely watching and scrutinising the environment and noting the changes (Liwenga 2003). The short- and long-term farmers' perceptions of weather and climate variability indicated their ability to realise the different trends. In this study, the perceived trends conform to the actual weather observations examined from rainfall data obtained from the Katumba Wakulima Tea Company and Rungwe District Agricultural Department in the study area. Table 1 shows the changing patterns of climate variability as perceived by various farmers in the three zones of the study area.

The analysis across the study area indicated changes in some major climate variables/aspects. The main climate variables reported by respondents across villages include rainfall patterns (96.1%), intra-seasonal dry spells (94.3%), strong winds

Table 1 Perceived trends of climate variability and change in the study area over the 30 past years

Aspect of change	Trends of change	Kapulampunguti		Kikota		Mbeye 1		Pooled values	
		n	%	*N*	%	*N*	%	*N*	%
Drought	Increasing	55	98.2	53	88.3	25	80.6	133	89.0
	Decreasing	0	0	1	1.7	1	3.2	2	1.6
	No change	0	0	5	8.3	5	16.1	10	8.1
	Not sure	0	0	1	1.7	0	0	1	0.6
High temperature	Increasing	46	82.1	48	80.0	20	64.5	114	75.5
	Decreasing	3	5.4	2	3.3	5	16.1	10	8.3
	No change	4	7.1	8	13.3	6	19.4	18	13.3
	Not sure	2	3.6	2	3.3	0	0	4	2.3
Low temperature	Increasing	22	39.3	26	43.3	10	32.3	58	38.3
	Decreasing	3	5.4	15	25.0	9	29.0	27	19.8
	No change	11	19.6	7	11.7	8	25.8	26	19.0
	Not sure	3	5.4	0	0	0	0	3	1.8
	Fluctuating	0	0	1	1.7	1	3.2	2	1.6
Heavy rainfall	Increasing	1	1.8	0	0	2	6.5	3	2.8
	Decreasing	53	94.6	59	98.3	25	80.6	137	91.2
	Fluctuating	1	1.8	0	0	1	3.2	2	1.7
	No change	0	0	0	0	2	6.5	2	2.2
Strong wind	Increasing	11	19.6	17	28.3	5	16.1	33	21.3
	Decreasing	7	12.5	4	6.7	2	6.5	13	8.6
	No change	19	33.9	30	50.0	10	32.3	59	38.7
	Fluctuating	2	3.6	1	1.7	0	0	3	1.8
	Not sure	3	5.4	0	0	2	6.5	5	4.0
Pest and diseases	Increasing	27	48.2	51	85.0	24	77.4	102	70.2
	Decreasing	6	10.7	0	0	0	0	6	3.6
	No change	6	10.7	5	8.3	2	6.5	13	8.5
	Fluctuating	3	5.4	0	0	0	0	3	1.8
	Not sure	6	10.7	3	5.0	1	3.2	10	6.3

(70.3%), high temperature (64.5%) and seasonal droughts (77.6%). Other changing aspects reported by the respondents included the occurrence of plant diseases and insect pests (91.9%), livestock diseases (87.3%) and human diseases (85.5%).

Farmers' observations on changes in the amount of rainfall, patterns and distribution were confirmed by the meteorological data as indicated in Fig. 1. The figure confirms higher variation in rainfall in the same month for different periods. These changes have a negative effect on rainfall predictability during the growing season as reported by farmers in all the study villages.

Figure 1 shows a general decrease in rainfall amount and high fluctuation during the growing season, which had a negative effect on the planting dates. The results indicate a decline in rainfall amounts and high fluctuations. The main fluctuation was observed in 2004, 2008 and 2011 as rainfall continued to decline to a greater extent. Almost all the seasons in the period experienced a fluctuation in rainfall, with high rainfall in December, March and April. The effects of such temporal variation observed in maize production, for example, are on the variation in planting date, weeding and harvesting dates. Maize is harvested before the long rain in the midland area to avoid rooting due to high and long rain; late planting affects the harvesting period, which may lead to crop loss.

Results from FGDs and key informant interviews also supported the view that changes in climate in the area, especially in temperature, rainfall and wind, had a great influence on farming activities in the area. The findings from the historical trend line and seasonal calendar indicated that the onset of rainfall had changed and was becoming unpredictable. The respondents reported that they used to have rainfall in the highlands from September to May, with short dry spells of 1–2 weeks between December and March. The fog was in May and June, whereas the cold season was from May to July, with frost falling between June and August. Currently, there is a significant change. For example, no rains fall in September, as was

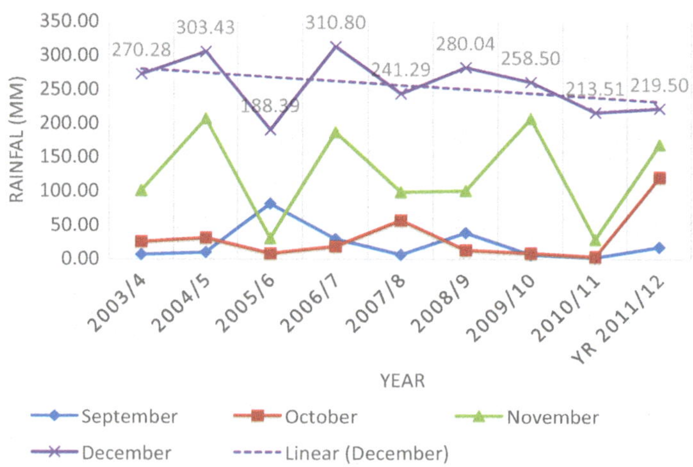

Fig. 1 Rainfall amount, trends and variation in Rungwe District (September to December)

traditionally the case. Rainfall could start in early October or as late as January. The normal distribution patterns had also changed as had the onset of cold months and the occurrence of frost.

Traditionally, in the midlands the rainfall used to be from November to August, the dry spell in February, whereas the dry season used to be September and early October, and the cold season was late May, June and July. The windy season was between October and December and sometimes in February. Currently, the rain is largely unpredictable with the cold season from June to early July, sometimes extending into August. Likewise, in the lowlands rainfall was reported to fluctuate and was becoming more unpredictable. Temperatures have also increased. In the meantime, the frequency of floods has also decreased compared to the past 10–20 years. The farmers' perception of variation in rainfall patterns, especially, from August to December, is also supported by the meteorological data as indicated in Fig. 1. The variation in rainfall patterns affects the growing season in the area through alterations of the growing calendar. In fact, this was reported to be one of the main challenges of farming practices, especially for annual food crop production.

More than 75% of the farmers interviewed in all the three zones were aware of the significant changes in rainfall and other aspects of weather over 5–10 years. The results show that 65% of farmers interviewed in all surveyed villages believed long-term temperatures were rising. An overwhelming 96% of the farmers also believed that precipitation was declining with more fluctuations and pronounced changes in the timing. Moreover, farmers perceived the increase of frequency of droughts as compared to the past 20 years. The farmers' observation on rainfall conditions is confirmed by existing rainfall data as presented in Fig. 2.

The result of the precipitation observed in the area indicates a decrease in the amount of rainfall especially in April whereby, the Lowest rainfall was experienced in April 2006/2007. Such a decline in rainfall affected agricultural production in the study area. In the lowlands, for example, low rainfall means lower rice production since paddy fields need enough water for its growth.

Farmers in all the zones were able to remember extreme events that posed a challenge to their livelihoods. Through the use of resource mapping (providing information on natural resources surrounding the village), historical trend line (providing an overview of significant historical events of the area being studied) and seasonal calendar (indicating the annual variations and the different activities carried out in different seasons) techniques, farmers described climate-related event that had significant impacts in their locality (Liwenga 2003). The climate-related events they remembered in the highlands (Mbeye 1Village) include hail in 2008 and heavy rainfall in 1995 and 1998 which devastated maize and round potatoes, hence resulting in food shortage in this area. In addition the drought of 2010 (there were no rain up to December) caused crops to wither, hence eroding the farmers' capital. In the midland farmers remembered the hail and winds of 2011 that resulted in decrepiting/ dilapidating of bananas, the extreme cold of 2010 that destroyed beans, heavy rainfall of 1997 (El Niño) that destroyed crops and houses and the drought of 1973 that caused failure in maize germination. In the lowland farmers remembered the

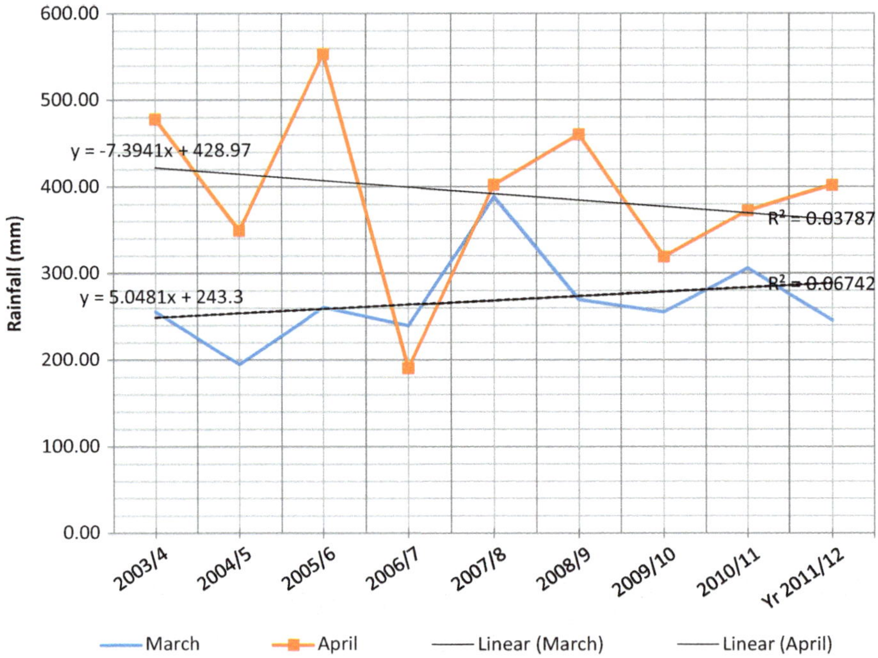

Fig. 2 Rainfall variability in March and April in Rungwe District

floods of 1975, 1979, 1997/1998 and 2009, armyworms in 2006, the drought of 2010/2011, the late onset of rainfall in 2012 and the earthquakes of 2010/2011. These events were reported to have affected their agricultural production and farmers' livelihood in general and challenged the farming systems in all the zones.

3.2 Local Perceptions of Other Stressors with Regard to Climate Variability

Farmers during FGDs also recognised the effects of the multiple stressors in all the three zones. Accordingly, many farmers still indicated that most of the stressors emanated from the current climate variability and changes. It was reported that farmers in all zones linked the challenges to climate variability; for example, farmers have always had to contend with the unpredictability of weather. The favourable weather conditions that include many aspects, such as the optimal range of temperatures, sunshine and moisture, together support the production of good harvests. It is important, therefore, to understand climate variability and the magnitude of changes for improvement of adaptation efforts in agricultural production to become a reality.

The results from FGDs and key informant interviews revealed that in the past 20 years, farming activities were regarded as a stress-free activity due to the favourable production environment. Today, farming as an occupation is one of the most vulnerable activities in terms of harvest loss and crop failure due to hazardous situations and different physical, biological, chemical and sociological factors. The negative effects of different factors such as crop failure, loss of income and health deterioration on the part of individual farmers create stressful situations. The most critical stressors that affect small farmers in a study area include climate variability in form of unpredictability, erratic rains, frost, floods, drought, strong winds and temperature. According to Ramesh and Madhavi (2009), stress is a perceived discrepancy between activity demands and the ability of an individual to perform when the consequences of failure are thought to be important. This situation can arise from the interaction of people and their activities and can be characterised by changes in people.

Other stressors in all the study areas were expressed in terms of challenges and constraints in agricultural production. The mostly mentioned stressors included land exhaustion, high prices for input and unreliable markets, especially for bananas, which lack a standard measurement. However, many of the smallholder farmers singled out increased climate variability (in terms of unreliable rainfall and drought) and land exhaustion as the major causes of their present predicament. These factors pushed farmers into producing by minimising risk rather than maximising production. The reasons behind the changing agricultural crop production mentioned have been presented in Table 2.

The findings indicate greater risk for agricultural production and food security posed by land exhaustion (71.4%), the use of inorganic fertilisers (65.2%) and variability and extreme weather events (72.2%). As in most of sub-Saharan Africa, the largest falls in agricultural crop production and food insecurity in the study area were attributed to abnormally low or too much precipitation and land degradation (IUCN 2007; FAO 2009; Gwambene 2011; Obiora 2013). The direction of future rainfall and temperature variability and changes reflect the magnitude of drought and warming relative to historical variability.

Continuous land cultivation, which led to land exhaustion, is a major factor for change in production in all the study zones. The use of chemical fertilizers, on the

Table 2 Reason for change in agricultural crop production

	Kapulampunguti %		Kikota %		Mbeye 1%		
Reason for change	Maize	Rice	Maize	Rice	Maize	Round potatoes	Pooled values %
Land exhaustion	16.1	9.0	15.0	1.7	22.6	7.0	71.4
Use of fertilizers	20.0	0.0	0.0	3.3	25.8	16.1	65.2
Drought	17.9	19.6	0.0	0.0	3.2	1.0	41.7
Climate change	12.5	8.9	5.0	0.0	3.2	1.0	30.6
The use of fake seeds		0.0	10.0	0.0	0.0	0.0	10.0
Shortage of land	5.0	0.0	0.0	0.0	0.0	0.0	5.0

other hand, was reported in the middle and highland zones. These reasons were linked to climate variability as the main driving factor. For farmers, extreme rainfall disrupts plant growth and destroys crops, whereas changes in temperature alter cropping patterns and suitability of the crops to certain agroecological zones and increase attacks on crops by diseases and insect pests (URT 2010; Mbilinyi et al. 2013). Frequent droughts and floods erode assets and crops, leaving people more vulnerable to climate variability. These changes account for shifts in food crop production.

3.3 Perceived Climate Variability Implications on Food Crop Production

The nature of agriculture and farming practices in a particular location is influenced by the long-term mean climate conditions coupled with the experience/knowledge and infrastructure/technology of local farming (Gornall et al. 2012; Zilberman et al. 2012). Changes in the mean climate away from current conditions might need adjustments to the current practices so as to maintain productivity, and in some situations, it can change the optimum type of farming. The farming systems and crop type and variety which are known to be appropriate for communities under the normal climate are particularly vulnerable to climate variability. Heavy, fluctuating and unpredictability of rainfall and changes in temperature have increased vulnerability and posed a challenge to agricultural crop production and food security.

The sensitivity of agriculture to climate variability and weather extremes poses more challenges to smallholder farmers in terms of increased potential for droughts, floods and heat, and the enduring changes in climate increased vulnerability in agricultural production. Such change affects productivity through decreasing water supply and soil moisture and makes it less feasible to continue crop production or planting some crop varieties in certain area (IPCC 2007; Dai 2011). Higher temperatures and fluctuating rainfall reduce yields of desirable crops and encourage weeds and pest proliferation (Gwimbi and Mundoga, 2010). Changes in precipitation patterns increase the likelihood of short-run crop failures and long-run production declines (IPCC 2007; Ngusaru 2007). Such changes may result in additional price increases for the most important agricultural crops (Orindi and Murray 2005). In fact, further changes in rainfall patterns are likely to affect the amount of water needed for crop production.

The results revealed that changes in rainfall patterns and increased temperature shortened the growing season and forced farmers to switch to weather-tolerant crops that are expensive though highly productive. Frequent droughts and floods were reported to erode assets and capital, which left people more vulnerable to extreme events. In all the three zones, changes in aspects of climate such as minimum and maximum daily temperature, occurrence and distribution of rainfall, decrease in the number of wet and increase in dry spells period were reported.

During FGDs and key informant interviews in the midlands, it emerged that due to late onset of rains, farmers planted maize and had to use irrigation to add moisture for seed germination. For example, an elderly man in the Kikota village (midland zone) said:

> We used to have rainfall in August and September that provided moisture for growing maize, but now the situation has changed. We are not sure of rain. Because of late coming of rainfall and lack of soil moisture, we grow maize and must water it to increase moisture for germination. The situation is becoming more complicated because in our area, we cannot wait for rains to come because a delay in planting maize can result in rotting of crops due to the high rainfall in March and April.

Changing temperature and rainfall patterns attributed to land use changes have significantly altered agricultural crop production in the study areas. The statistical test on the main food production and land degradation, as a cause of drought, indicated the significant relationship and a negative strong correlation. For example, the statistical test on the maize and rice has an X^2 of 6.00, df 4 with $p = 0.199$, and X^2 of 10.00, df 9 with $p = 0.350$, whereas the correlations (interval by interval – Pearson's r) have value of −0.663; Asymp. Std. Error[a] = 0.229; Approx. $T^b = -0.885$; and $p = 0.539$ and value of −0.271; Asymp. Std. Error[a] = 0.261; Approx. $T^b = -0.487$; and $p = 0.660$. Similar results were also revealed in other studies (see, e.g., Lobell and Burke 2008; Gwimbi and Mundoga 2010). Variations in rainfall patterns triggered the emergence of new pests and diseases that affect crop production, livestock growth and development. During the FGDs the reported effects of climate change include withering of crops, decline in economy and recurrent food shortages in smallholder farmers' households.

The impacts of the extreme events can destabilise agricultural production, particularly when their frequencies increased. These changes are anticipated to cause tremendous changes in food production. Alterations in temperature and moisture will affect the crops that are sensitive to changes in climate conditions. In addition, climate variability influences the populations of beneficial organisms and alters their effective roles in agricultural ecosystems. These results are also supported in the literature (Stern 2006a, b; IPCC 2007; Ngusaru 2007; Paul et al. 2009; World Agroforestry Centre 2009; Gwimbi and Mundoga 2010; Mubaya 2010). Climate variability and change favour the occurrence of diseases and insect pests due to both increased temperature and change in rainfall pattern. In such instances, farmers use more agrochemicals and disease-resistant varieties, which increase production costs. For example, more pesticides are being used due to the warmer weather, and this has increased the number of destructive pests. This results in additional cost of pesticides, which reduce profit. The use of excessive pesticides also results in soil and water pollution, especially in the highland zone where more agrochemicals are used for round potato farming as also described in Nyunza and Mwakaje (2012).

In all the three zones, planning for agricultural activities is becoming more challenging due to climate variability and population increase that reduced crop production and increased pressure on resources. In the past, farmers used to harvest more crops as land was available and fertile; and the climate was good with rains coming on time. The rainy season in the midland area, for example, was known to start from mid-October through May, but now the rainfall pattern has become unpredictable and unreliable and can start in November or December. The FGDs and key informant interviews revealed that remarkable changes in rainfall had been observed in the last two decades. In this regard, understanding the effects of climate variability on agricultural production is a vital component in formulating strategies and adaptation options.

3.4 Agricultural Food Crop Production Trends

Agricultural crop production is not static as it evolves over time with the management practices and a combination of products and production factors applied by the farm households (Fresco 1986). These factors include any approaches adopted for the purpose of growing crops and using them for food or fibre. Farmers use different approaches to actuate the production of plants and to manipulate agricultural production systems. On the whole, the production of crops has significantly declined in the study area. The harvest of the main crops grown in the area has significantly changed. The changes in productivity and changes in crop types and varieties were mainly associated with climate variability, land exhaustion, shortage of land and poor agronomic practices. Also, the varieties used in the past were no longer suitable in the area due to changes in the growing seasons, soil moisture content and soil fertility. Many farmers in the past used local varieties, but now they have adopted improved varieties due to the poor performance of the local varieties as a result of environmental change and climate variability. Table 3 shows the perceived production trends on main food crops.

The results indicate a decline in main food crop per unit land, but which increased through extensification of area, except for the production of round-potatoes under

Table 3 Perceived production and trend on main food crops across the zone

S/N	Main crop produced	Lowland		Midland		Highland		Productivity Trends (General)
		Past (kg/ha^2)	Current (kg/ ha^2)	Past (kg/ha^2)	Current (kg/ha^2)	Past (kg/ha^2)	Current (kg/ha^2)	
1	Maize	X	X	X	X	449	1483	Increased
2	Rice	4942	1235	NA	NA	NA	NA	Decreased
3	Banana	X	X	X	X	NA	NA	Decreased
4a	Round potato (improved practices)	NA	NA	X	X	5436	11,861	Increased
4a	Round potato (normal practices)	NA	NA	X	X	5436	1235	Decreased

Key: X = Use without the conventional measurement 1 ac^2 = 0.4046863 ha^2

improved practices that has increased per unit area. Mbilinyi et al. (2013) and Tilumanywa (2013) also reported the decline in main agricultural crop production in the area. The increase in round-potato production under improved practices was associated with the increase of the production knowledge and use of inputs such as industrial fertilisers and chemicals. However, due to the high production costs, only a few farmers managed to produce; and for those who used the normal practices, production had generally decreased. Such a decrease, coupled with increased production cost, had led some farmers to abandon its production. For example, 19.4% of the respondents in the highlands had abandoned production of round potatoes. The farmers associated the decline observed in yields of round potatoes with inadequate and unreliable rainfall, prolonged drought (especially in the lowlands), pests and diseases and land exhaustion, as illustrated in Table 4.

Climate variability and land exhaustion were cited as causes of the decline in yields for paddy rice and maize in all zones. Inappropriate maize seeds used in the area were reported to affect production in the lowlands and midlands, while the high price of inputs did the same in the midlands and the highlands. Other reasons for the decline of production in highlands included heavy rainfall, low temperature, frost, wilting diseases, low capital for buying agricultural inputs and pests. Along the same lines, MAFAP (2013) suggest that most of the increased production was due to area increase and not yields, as the yield remained below average in East Africa. The production has increased due to improved production knowledge, the use of inputs and a booming market of round potatoes.

The main challenges for food crop production in the lowland zone included land shortage, climate variability, land exhaustion, the shortened growing season, late-onset rainfall, pests, the high price of inputs (fertilisers), poor marketing for simsim

Table 4 Main reasons behind change in production of main food crops

Main crop	Reason for change in crop production		
	Lowland	Midland	Highland
Rice	Climate variability and land exhaustion	Not produced	Not produced
Maize	Climate variability, land exhaustion and type of seeds used are not appropriate in the area	Climate variability, soil exhaustion, high price of inputs and knowledge on seeds	Land exhaustion, heavy rainfall, inputs, cold weather and production have increased due to increase of production knowledge
Bananas	Not much change in banana production	Knowledge of improved cultivation increased the production	Not produced
Round potatoes	Not produced	The crop is not grown much in the area	Frost in June to August as usual pattern, erratic rainfall, wilt disease, low capital for buying agricultural inputs and pest. The production has increased due to increased knowledge, the use of inputs and a booming market

and lack of production knowledge. In the highlands the main challenges included climate variability, especially too much rain and late onset of rain, wilt disease, low capital for buying agricultural inputs and pesticides that increased production costs. High production costs in the highlands resulted in abandoning round potato production by many low-income local people. Soil exhaustion, high production cost, climate variability and pests are among the challenges in the midlands. For maize production, fluctuating rainfall during the growing season resulted in watering when planting seeds to increase soil moisture due to delays in the onset of the rainy season. The challenges mentioned for banana production include unreliable market and lack of proper measurement (conventional measuring unit) for the product.

Assessments of general changes that have been taking place in the study villages since independence in 1961 show a number of changes that had taken place in each village. Some of the changes are directly or indirectly linked to climate change and variability. The reported changes include decreased yields due to erratic rainfall leading to expansion of off-farm land and use of fertilisers to increase yield; inclusion of nonfarm livelihood activities; increased thefts of farm produce and livestock; change in traditional gender roles (activities such as weaving mats and casual labour for piecework which are now carried out by both men and women); and abandonment of some of the local crop varieties in favour of early-maturing and disease-resistant crop varieties such as cassava, maize, groundnuts and sweet potatoes. In general, most of the changes associated with climate variability are specific to a particular zone, which suggests that similar climatic events may impose differential effects on people's livelihoods in different sites, despite having similar climatic characteristics or lying in the same geographical area.

The major changes occurred over time in agricultural crop production, especially in farming systems, land and labour allocation, management practices, type of crops, marketing, use of inputs, total production output, capital and public supports (TNB 2009; URT 2015). However, smallholder farmers dominate the agricultural sector with average farm sizes of between 0.2 and 2.0 hectares, depending on the location (URT 2015). Household crop production for own consumption has been the primary and the main purpose in many smallholder farmers' households. Likewise sale of agricultural products accounts for about 70% of rural household income (URT 2006). Tanzania's national data records indicate an increase in agricultural food crop production (URT 2007b). For example, food crop production has grown at a rate of 3%, which is about the rate of population growth and accounts for about 65% of agricultural GDP, with cash crops accounting for only about 10% (URT 2006; URT 2007a). It was argued that the yields of the main food crops were mostly stagnant for the last 20 years and agricultural productivity gains have been based more on the expansion of cultivated land, which is one of the major drivers of land degradation in the country (Gwambene 2012; Kangalawe 2012; Leliveld et al. 2013; URT 2015). Agricultural intensification considered to be driven by population growth and its associated increase in density (Paavola 2008; Sokoni 2013).

Agricultural food crop production has increased over the last two decades, mainly for maize, paddy, sugarcane and meat production that almost doubled over this period. Although the contribution of agriculture to Tanzania's GDP has fallen from 27% in 1998 to 24.7% in 2012–2013 and agricultural growth has not exceeded 4–5% per year since 1998, the sector still accounts for over 70% of the country's employment today (URT 2015; Bowman and Zilberman 2013). On the whole, the production of cash crops has declined, mostly due the fall of market price under the liberalized economy. For example, the continuing disappearance of pyrethrum, formerly the main dominant cash crop, denotes a major change in crop composition in the Uporoto highland farming systems (Sokoni 2013). It indicates that the production of round potatoes has replaced a greater proportion (36.2%) of the pyrethrum plots and maize (15.4%), whereas other crops and trees occupy the rest. According to Sokoni (2013), most of the households stopped pyrethrum production between 1991 and 1995 due to decrease in market price.

3.5 *Food Security in the Climate Variability Perspective*

In the context of changing climate, farmers had to contend with the challenge of increasing agricultural crop production and income for purchasing food (due to poor harvests) and other basic needs. The results indicate that more frequent and severe extreme climatic events, especially droughts and heavy rainfall, pose challenges to agricultural production and food security in the study area. It was found that during food shortages, most of the households reduced the number of meals and portions and types of food they ate per day and ate the available types of food or switched to affordable food types. Table 5 shows the number of meal and type of

Table 5 Number and main meals taken per day during food shortages

Season	Number of meal	Mbeye 1 (%)	Kikota (%)	Kapulampunguti (%)	Pooled values (%)
During bad year (food shortage)	1	10	8	54	24
	2	36	32	39	36
	3	12	10	7	10
During good year (food available)	1	10		4	5
	2	26	43	43	37
	3	65	47	52	55
	4			2	1
Main meals					
Whatsoever available		48.2	3.3	19.4	24
Maize		30.4	28.3	32.3	30
Bananas (green)		30.4	33.3	0.0	21
Rice		7.1	0.0	0.0	2
Cassava		5.4	5.0	0.0	3
Tea		1.8	8.3	6.5	6
Beer tubers		1.8	3.3	0.0	2
Round potatoes		0.0	0.0	12.9	4

food taken during food shortages. Such situations affect the health and nutritional status of the villagers.

The food shortage in the study area is experienced mainly from December to March in the lowlands, from December to February in the midlands and January and February in the highlands. Unlike in the central zone of Tanzania (Lamboll et al. 2011), in Rungwe District, only few farmers experience food shortages. These food shortages are mostly experienced in the lowlands where there is an increase in the frequency of drought. During the FGDs, it was reported that some households did not reduce the number of meals taken per day, but they reduced the portions and changed the type of food taken. It was further revealed that in many cases food shortage for most households was due to the shortage of the main staple foods such as maize in the highlands and the midlands and rice in the lowlands during the periods of scarcity (lack or shortage of such staple foods means food shortage).

It was further reported during FGDs and key informant interviews that rainfall fluctuation has largely affected agricultural crop production. In fact, rainfall fluctuation was perceived to be the main factor for production, as it modifies other weather elements. Climate-related causes of food shortages reported included changes in temperature, strong winds, too much rain, floods, crop pests and diseases and livestock diseases. Other causes are lack of social networks, lack of labour, low soil fertility, low household income/lack of assets, loss of livestock, overselling of crops and laziness, as indicated in Table 6.

Years of low rainfall lead to people harvesting insufficient food. In addition, livestock tend to have no feeds due to drought, and water supply also remains a

Table 6 Main causes of food shortages

Causes of food shortages	Kapulampunguti		Kikota		Mbeye 1		Pooled values	
	N	%	*N*	%	*N*	%	*N*	%
Drought	52	92.9	30	50.0	16	51.6	98	64.8
Low temperature	35	62.5	25	41.7	8	25.8	68	43.3
High temperatures	21	37.5	24	40.0	9	29.0	54	35.5
Shortage of Labour	15	26.8	9	15.0	4	12.9	28	18.2
Strong winds	19	33.9	9	15.0	0	0.0	28	16.3
Floods	9	16.1	4	6.7	7	22.6	20	15.1
Lack of social networks	12	21.4	9	15.0	2	6.5	23	14.3
Too much rain	18	32.1	4	6.7	0	0.0	22	12.9
Low household income	20	35.7	0	0.0	0	0.0	20	11.9
Laziness	13	21.7	0	0.0	3	9.7	16	10.5
Low soil fertility	12	21.4	5	8.3	0	0.0	17	9.9
Crop pests and diseases	8	14.3	5	8.3	2	6.5	15	9.7
Loss of livestock	10	17.9	0	0.0	0	0.0	10	6.0
Overselling of crops	9	16.1	1	1.7	0	0.0	10	5.9
Livestock diseases	7	12.5	0	0.0	0	0.0	7	4.2

problem. Although everyone in the community is vulnerable to climate change and variability, the poor are the most vulnerable because they have very limited assets to rely on and/or livestock to sell. The most affected in this regard include women, children and the aged, due to their inability to migrate and look for alternative sources of livelihood. The source of income includes casual labour, remittances, petty trade, selling livestock and engagement in other off-farm activities. During FGDs and key informant interviews, it was reported that better-off households in terms of income were less vulnerable to food shortage, as food security is partly linked to the people's ability to buy food. However, climate variability has a cost implication for inflation because of food price hike (FAO 2008; IFAD 2008). The increased prices in turn have serious implications of food security for low-income groups, as food constitutes a greater percentage of the household expenditure.

The study found that low soil fertility and inadequate nutrient replenishment resulted in low crop yield per unit land in all the three zones. The low soil fertility was due to inappropriate soil management and adverse impact of climate variability. The loss of land productivity, and the resultant food shortage affects the livelihoods of smallholder farmers. In the highlands, contour bands had been knocked down and removed from the sloping lands as noted during field observation and FGDs. Similarly, the findings from studies by Sokoni (2008) and Tilumanywa (2013) indicate a decline in soil fertility on the slopes of the highlands. Furthermore, Yemefack (2015) reported that soil fertility depletion is one of the main constraints to sustaining agriculture in most of sub-Saharan Africa that causes low per capita food production. As such, soil fertility management and enhancing productivity through appropriate soil management play a major role in achieving household food security and increasing agricultural production.

The subsistence nature of agricultural production, with food security being a main concern among smallholder farmers, increased vulnerability to climate variability. This compounded other non-climatic factors, contributing to household food shortage. For example, in the study area, other causes of food shortage include *overselling* of food *crops and* laziness. Smallholder farmers are compelled to *oversell* their produce during harvest time, so as to meet other important needs. L*aziness, which* is inherent in the culture of some local people involved in productive activities, also contributed to food shortages. Such factors were more pronounced in the lowlands than in the other two zones because of the short growing season that limits agricultural activities.

3.6 Crop Management Challenges in Smallholder Farming

The analysis of this study revealed the main vulnerability factors in the agricultural sector as including the decrease in agricultural crop production for different crops. These factors are exacerbated by the unpredictability of seasonality, climate variability, soil erosion and environmental degradation. Fluctuating rainfall, uncertainty in cropping patterns and prolonged dry spells, together with shifting characteristics of agroecological zones, increased the vulnerability of agricultural crop production in the area. Similar results have also been discussed in other studies (Rosegrant et al. 2008; Tumbo et al. 2008; Majule et al. 2009; World Agroforestry Centre 2009; Mubaya 2010; URT 2015), which have confirmed the system's vulnerability to climate change impact and adaptation measures.

During FGDs, farmers reported that maize and rice were among the most affected crops by climate variability, especially in the lowlands and the midlands. In the highlands round potatoes were reported to be the most affected. Due to erratic and fluctuating rainfall in the lowlands and the midlands, moisture has decreased and posed a significant threat to rice production, which requires more moisture. In the highlands there is sufficient moisture for maize production; however, frost and heavy rainfall with high intensity in a short time affect round potato production. The production and yield of these main food crops are decreasing, as was revealed during the FGDs and key informant interviews. The decrease was associated with an increase in weeds that compete with crops for moisture, nutrients and light in the highland area and climate change and variability. It was further reported that ecological changes (especially, climatic conditions) that favour weeds, pests and diseases, together with climate variability, shortage of land and exhaustion are expected to increase vulnerability in crop production. Due to temperature rises and frequent occurrences of drought, a decline in food crop of 33% at the national level is predicted (Bisanda et al. 1998). The main challenge has to do with building the capacity of smallholder farming households to adapt to the changing climate.

In addition to rainfall and temperature, strong wind was reported to affect banana production in the midlands. For example, increased wind strength devastated weaker crops, as revealed during FDG. Strong winds affected crops and trees, especially bananas during the onset of the short rainy season in the midlands. Climate change in the area, especially in temperature, rainfall and wind had a great influence on farming activities.

FGDs and key informant interview respondents reported climate variability as a critical factor for formulating strategies and adaptation options to minimise adverse consequences in all the three zones. For example, floods in the lowlands result in loss of life, crops and animals, outbreaks of pests and diseases, destruction of fish and wildlife habitat, displacement of people and environmental degradation. Tumbo et al. (2008) and Tadesse et al. (2013) report a similar finding on the impact of floods on agriculture and livelihoods in lowland areas. Interviews with village leaders and FGDs held revealed that droughts cause crop failures, water scarcity, drying of water resources, food shortage, loss of human and animal lives, loss of biodiversity and environmental degradation. Strong winds also damage structures and destroy crops, especially bananas in the midlands. This finding correlates with observations on the adaptation to the impacts of climate change in Tanzania (IUCN 2007; Lamboll et al. 2011; Nyunza and Mwakaje 2012).

A general shift in the characteristics of agroclimatic zones and change in the cropping patterns require farmers to adhere to improved agricultural practices and better crop management. Because of less and erratic rainfall received in the lowlands and midlands, there was a need for irrigation and introduction of drought-resistant crop varieties to compensate for moisture losses due to increased evapotranspiration. The major adaptation options needed included multiple crop diversification, intercropping, mixed crop-livestock systems and switching from rain-fed to irrigation agriculture. In addition, in the lowlands and midlands, fluctuating rainfall calls for irrigation to buffer crops and agricultural production from negative impacts during dry periods. In this regard, there was a need to have efficient and effective irrigation systems, including good strategies for rainwater harvesting for agricultural production enhancement.

The analysis revealed the information and awareness gap on climate and current agricultural production strategies among smallholder farmers. These include the packaging and dissemination of information on inputs, technologies and mainstreaming of the adaptation options in planning and development activities at different levels. Findings from the study confirmed poor access to information on climate, production and credit, which is crucial in enhancing farmers' awareness and decision-making in adaptation and planning. Such results are in line with those of studies by Tumbo et al. (2008) and Liwenga et al. (2009) who described the need/power of information in improving livelihood diversification in smallholder farmers. For instance, information on extension services is essential in decision-making and taking up adaptation measures. The Tanzania Agriculture Policy of 2013 underscores the importance of information on agricultural production and the need to involve the media in information dissemination and public education.

3.7 Policy and Agricultural Food Crop Production Changes

Generally, changes in policy and registration have contributed to the production and development of the agricultural sector. Indeed, policies and regulations affect the profitability and the evolution of different farming systems by influencing farmer's decisions on what crops to grow or how much land to farm (Bowman and Zilberman 2013). For instance, Tanzania's agricultural policy of 2007 describes food crop production in Tanzania as ranging from the major staples (maize, rice and wheat), drought-resistant crops (sorghum, millet and cassava) and other sub-staples (round potatoes, sweet potatoes, bananas and plantains) (URT 2007b). However, the market liberalisation policy has drastically changed the production dynamics of food crops, especially maize. The policy dynamic in food crop production in Tanzania has evolved through different phases with a profound impact on the agricultural sector (URT 2008). For example, the primary aim of food crop production is to satisfy domestic demand and the secondary objective is to facilitate the entrance of some of these crops into the export market on a regular basis (URT 2015). However, the policies remained fragmented and failed to put in place an orderly marketing system which guaranteed reliable markets for producing crops in all localities taking into account portability and other associated risks.

The study results from discussions with key informants indicated that most of the farmers in the study villages had high expectation of receiving input from the government. Low awareness of policy and strategies is one of the challenges to efficient and enhanced crop production. In this regard, the study revealed that many of the farmers were not aware of the government policy and strategies. The low awareness of policies and strategies was attributed to lack of awareness creation and lack of enforcement and implementation strategies and plans at the local level. In this regard, the study suggests that policy interventions need to provide guidance for awareness creation on policy issues, improving land management and agricultural crop production. Moreover, line ministry in charge of the sector needs to initiate and lead agricultural policy implementation mechanisms at a grass-roots level and support the supervision, sector planning, standard setting, quality assurance, sector monitoring and guidance (Bategeka et al. 2013). After all, policy regulating the sector should provide greater support to agricultural production and improve farmers' livelihood to have the desired impact. However, the applicability of such a policy and attendant strategies remained generally weak at the individual and local level where most farmers operate.

Through the major initiative to enhance and improve the agriculture sector in Tanzania, agricultural transformation for food self-sufficiency and export was identified as a core priority, with a focus on high-value crops including horticulture and spices (OECD 2013). In terms of agricultural investment, the most notable programme is the Agriculture First ("Kilimo Kwanza") policy initiatives launched in 2009 with the objective of fostering a green revolution and transforming agriculture into a modern sector. Southern Agricultural Growth Corridor of Tanzania (SAGCOT), an international private-public partnership (PPP), is another major initiative aimed at enhancing investment in agriculture that aims to catalyse large volumes of private investment to develop commercially viable agriculture and increase productivity in the southern corridor. Also, the Tanzania Agriculture and Food Security Investment Plan (TAFSIP) was launched in 2011 in the context of the African Union's Comprehensive African Agriculture Development Programme (CAADP); however, it has not yet to be fully implemented (2013). In addition, these initiatives have limited or inadequate implementation strategies for improving agricultural food production among smallholder farmers under the changing environment.

These studies illustrate that agricultural policies implemented varyingly influence farmers' income across policies. In Tanzania, the country's agriculture policy of 2013 identified poor market performance and inefficient processing plants as causes of the reduction in the farm-gate prices of food crops, without necessarily reducing consumer prices. In addition to increasing farmers' income, most of the agricultural policies carried out also induced adjustments. For example, farm subsidies altered farmers' incentives on demand inter-sectorally, and in terms of allocation and ownership factors, Bategeka et al. (2013) contended.

As a result, policy changes have left farmers to treat farming merely as a survival business. Inadequate levels of entrepreneurial skills and inability to cope with the market dynamics, especially among smallholder farmers, increased their vulnerability to the changing environment and policy. Many smallholder farmers produce crops which they have been producing traditionally and, subsequently, continue searching for the markets for these products, even when environmental conditions and the market require improved or entirely different products (URT 2008). Awareness of policy and strategies and possession of entrepreneurial and marketing skills are relatively low among farmers and key stakeholders (including traders and processors). Subsequently, farming activities are becoming more vulnerable and costly now than ever before. Yet proper production skills, awareness of environmental changes and proper utilisation of opportunities are prerequisites for survival and growth. Developing agricultural production, entrepreneurial and management skills and creating an enabling environment are essencial options for identification and utilisation of opportunities in the agriculture sector and for improvement of agricultural crop production and enhancement of food security.

The multiplicity of agencies and the weak co-ordination mechanisms in the sector, complicated by political risks involved, constitute institutional factors affecting crop production. The Tanzania Agriculture Policy of 2013 acknowledges the poor income status of households engaging in farming, livestock-keeping, fishing and forestry. However, the mechanism for spearheading measures to reduce the poverty identified at the local level remains a crucial factor in improving agricultural production. The agricultural policy measures analysed in many studies point out the need to support farmers' income through controlling provider price production and factor subsidies as well as coupled and decoupled payments (Alston and James 2002; Guyomard et al. 2004; Ciaian and Swinnen 2006; Tumbo et al. 2008). However, the implementation mechanism and strategies lack important factors that could reduce production costs to increase profitability among farming communities.

Since 2011, the Tanzania government has developed policies, strategies and registrations for addressing the persistent challenges. In this regard, the active involvement of the private sector in agriculture has been identified as one of the priority sectors in the drive towards a modernist, commercial, highly productive and profitable sector. Specifically, the Agricultural Sector Development Strategy (ASDS), adopted in 2005 and implemented through the Agricultural Sector Development Programme (ASDP), provides a framework for implementing agricultural policy. It is also important to note that policies that do not specifically target agriculture, such as labour and immigration or water policies, have a significant effect on the costs of agricultural production. In the short run, such regulations may have a negative impact on farmers' welfare when it comes to complying with regulations and reducing the costs of production. There was also a need to stimulate innovation and adoption of new technologies (Bowman and Zilberman 2013; OECD 2013). Through policy, a stable and growing food supply should be viewed as a comprehensive package and should be extended to include investment in rural infrastructure and agricultural services, such as seasonal weather forecasting, in addition to providing a supportive policy environment.

The Tanzania Agriculture Policy of 2013 suggests that public expenditure should focus on marketing, storage and processing, as well as providing an enabling environment. The taxes, levies, fees and high-energy tariffs that are applied on the macro- and micro-economic levels are identified as a policy challenge that negatively affects the sector's profitability. The policy also recommends a continuous review and reforms of the taxable and nontax system, so as to increase profitability and investment in the sector. This study established a need for increasing production knowledge and reducing production cost at the local and national levels. At the local level, the focus should be on building farmers' capacity in improving production knowledge and using locally available resources that could reduce production costs and bring about more benefits. Under the liberal economy, the reduction of production cost and postharvest losses are important factors in the production process.

At the household institution, the typical quandary is that women, who produce food, do not own the land due to patriarchal values based on entrenched traditions and stereotyping. The males, who predominantly own the land, emphasise the production of 'cash crop', which results in limited attention being paid to agricultural production for household and national food security (Majule et al. 2007; Tumbo et al. 2008; Lamboll et al. 2011). In this regard, the study found that cash crops claim a larger share of household land and a greater share of family labour and other household productive assets.

The availability of services was one of the important factors in agricultural production reported by respondents in all zones during key informant interviews and FGDs. The services include irrigation and transport infrastructure, market services, storage facilities, extension services, health services, sanitation and education. In this regard, the government should continue ensuring there is reliable access by the smallholder farmer to relevant public goods, services and infrastructure, in addition to promoting the sustainable use of natural resources. The supply and demand in the market may also play a role because farmers would adjust their activities to meet the needs of consumers, which can have an impact on agricultural production.

In addition, farmers should address their main challenges to increasing agricultural productivity on the existing farm lots, both to meet the growing demands for food and to offset the climate variability-induced yield losses. Innovation is a key factor in boosting agricultural production. For instance, creative and well-informed farmers experienced an increase in production as it was noted in field observations and also confirmed during discussions with key informants.. It is therefore important to invest in developing new farming techniques and approaches to farming. Doing so would help to enhance resilience and the adaptive capacity of farmers to the impact of climate variability. However, this orientation needs a clear strategy and an implementation plan to guide the process.

4 Conclusions

The study revealed the high vulnerability of the agricultural sector to climate variability and weather extremes, such as droughts, floods and severe storms. It indicated that the changing rainfall patterns affect the amount of water available for crop production and pose a significant threat to agricultural production and food security and affect the ecology and livelihoods of millions of people who depend on small-scale production. Such effects increase the burden and the vulnerability of the poor and jeopardise poverty alleviation strategies. The yield of the most important agricultural crops is expected to decline and result in a price increase for crops. Increasing the adaptation capacity of smallholder farmers is, therefore, an important strategic goal in coping with climate change and variability, as doing so can reduce the vulnerability to climate variability and extreme climatic events.

The positive opportunities arising from climate change and its variability in agriculture should be maximised to reduce the threat to sustainable agriculture development. Such opportunities include shifting from traditional crops, farming of high-value crops, double cropping and irrigation opportunity that are likely to increase the number of growing seasons per year. More information and efforts are needed to strengthen the linkages between different stakeholders for sharing and scaling up successful strategies and opportunities to mitigate climate change and ensure sustainable environment management. The information includes addressing challenges and utilising opportunities arising from agricultural production.

Raising awareness on the problem and potential benefits of taking action is imperative in determining the decisions behind adaptation measures, especially with regard to changes in climate attributes (temperature and precipitation). In this study, it was observed that the adoption of soil management measures in the midlands was induced by the farmers' awareness and perceptions of the problems of soil erosion in the area. It was noted that the adaptation measures to reduce the impact and take advantage of the opportunities associated with climate variability were adopted more by farmers who were aware of changes in the local climate than those who were not.

To address climate impacts in the future, policies and strategies that reduce vulnerability were suggested as being the most important interventions. Appropriate policy, strategies, investment and collective actions are required to support the local smallholder farmers; these should include improving coping and adaptation strategies that would help to reduce the negative consequences of projected changes. The effective application of these recommended strategies would represent a significant contribution towards increasing production, food security and the achievement of the targeted goals of reducing hunger and increasing production. Thus, there is a need to enhance co-operation among relevant stakeholders and finding proper ways and techniques for ensuring the effective and efficient dissemination of information and knowledge to all key stakeholders.

References

ACAPS (2015) Qualitative and quantitative research techniques for humanitarian needs assessment: an introductory brief. http://tinyurl.com/q4vywcv. Accessed 01 Nov 2015

Alston JM, James JS (2002) The Incidence of Agricultural Policy. In: Gardner BL, Rausser GC (eds) Handbook of Agricultural Economics, vol 2B. Elsevier, Amsterdam, pp 1689–1749

Antle JM (2009) Agriculture and the food system: adaptation to climate change. An initiative of the climate policy program at RFF report. http://tinyurl.com/qbxqu8w. Accessed 23 June 2011

Ashley P, Boyd WE (2006) Quantitative and qualitative approaches to research in environmental management. Aust J Environ Manag 13(2):70–78. http://preview.tinyurl.com/hdfvo2v. Accessed on 03 Nov 2013

Bategeka L, Kiiza J, Kasirye I (2013) Institutional constraints to agriculture development in Uganda. http://tinyurl.com/nj3rmeu. Accessed 16 July 2013

Bisanda SW, Mwangi H, Verkuijl A, Moshi J, Anandajayasekeram P (1998) Adoption of maize production technologies in the Southern Highlands of Tanzania. Mexico, D.F. International Maize and Wheat Improvement Center (CIMMYT), the United Republic of Tanzania, and the Southern Africa Centre for Cooperation in Agricultural Research (SACCAR)

Bowman MS, Zilberman D (2013) Economic factors affecting diversified farming systems. Ecol Soc 18(1):33. https://doi.org/10.5751/ES-05574-180133

Ciaian P, Swinnen JFM (2006) Land market imperfections and agricultural policy impacts in the new EU Member States: a partial equilibrium analysis. Am J Agric Econ 88(4):799–815

Dai A (2011) Drought under global warming: a review. Adv Rev 2:45–65. Wiley

FAO (2006) Agriculture and Consumer Protection Department. Rome, Italy Available from http://www.fao.org/ag/magazine/0110sp.htm. Accessed 15 May 2015

FAO (2008) Climate Change, Water and Food Security: technical background document from the expert consultation held on 26 to 28 February 2008. FAO, Rome

FAO (2009) Global agriculture towards 2050. Rome http://tinyurl.com/63trq5u. Accessed 16 July 2013

Fresco LO (1986) Cassava in shifting cultivation: a systems approach to agricultural technology development in Africa. Royal Tropical Institute, Amsterdam

Gornall J, Betts R, Burke E, Clark R, Camp J, Willett K, Wiltshire A (2012) The implications of climate change for agricultural productivity in the early twenty-first century. The Royal Society, Partner of Research 4Life, PERii-INASP, http://tinyurl.com/pwj6kw4. Accessed 28 Oct 2012

Guyomard H, Le Mouël C, Gohin A (2004) Impacts of alternative agricultural income support schemes on multiple policy goals. Eur Rev Agric Econ 31(2):125–148

Gwambene B (2011) Climate Change and variability adaptation strategies: implication of adaptation strategies on land resources in Rungwe District, Tanzania. LAP Lambert Academic Publishing, Saarbrucken

Gwambene B (2012) Land utilization for agriculture and environment management, soil management in Rungwe, Tanzania. LAP LAMBERT Academic Publishing, Saarbrucken Deutschland

Gwimbi P, Mundoga T (2010) Impact of climate change on cotton production under rainfed conditions: case of Gokwe. J Sustain Dev Afr 12(8, 2010ISSN):1520–5509. Clarion University of Pennsylvania, Clarion, Pennsylvania

Hoffmann MP, Smith DL (2013) Feeding our great cities: climate change and opportunities for agriculture in Eastern Canada and the North eastern US; Cornell University, Ithaca, NY. US& McGill University, Montreal. http://preview.tinyurl.com/jp8rn4r. Accessed 12 July 2007

IFAD (2008) Challenges and opportunities for smallholder farmers in the context of climate change and new demands on agriculture. Proceedings of the Governing Council Round Tables (3) in Conjunction with the Thirty-first Session of IFAD's Governing Council. Workshop Report: The Royal Society, London. Accessed 30 Mar 2006

IPCC (2007) Climate change 2007 - impacts, adaptation and vulnerability. Contribution of Working Group II to the fourth assessment report of the Intergovernmental Panel on climate change, 2007. Cambridge University Press, Cambridge, United Kingdom and New York, NY, USA. Glossary available online: http://www.ipcc.ch/pdf/glossary/ar4-wg2.pdf. Accessed 6 Apr 2013

IUCN (2007) Climate change and development recognizing the role of forest and water resources in climate change adaptation. Project document prepared for: The Government of Finland, Ministry for Foreign Affairs in collaboration with: The Governments of Mozambique, Tanzania and Zambia and IUCN – the World Conservation Union

Kangalawe R (2012) Food security and health in the southern highlands of Tanzania: a multidisciplinary approach to evaluate the impact of climate change and other stress factors. Afr J Environ Sci Technol 6(1):50–66, http://tinyurl.com/mszj2z3. Accessed 23 Jan 2015

Kangalawe RYM, Liwenga ET, Majule AE, Madulu NF (2007) The dynamics of the farming systems, food security and poverty alleviation strategies in semi-arid areas of Sukumaland, Tanzania. Research Report Submitted to REPOA, Dar es Salaam

Lamboll R, Nelson V, Nathaniels N (2011) Emerging approaches for responding to climate change in African agricultural advisory services: challenges, opportunities and recommendations for an AFAAS climate change response strategy. AFAAS, Kampala, Uganda and FARA, Accra, Ghana. Available from http://tinyurl.com/krxsk29. Accessed 15 July 2013

Leliveld A, Dietz T, Foeken D, Klaver W (2013) Agricultural dynamics and food security trends in Uganda. Published on behalf of the Developmental Regimes in Africa Project by the Overseas Development Institute, 203 Blackfriars Road, London SE1 8NJ, UK. http://preview.tinyurl.com/hp3nrk8. Accessed 12 Oct 2016

Liwenga ET (2003) Food insecurity and coping strategies in semiarid areas: the case of Mvumi in Central Tanzania. PhD Dissertation No. 11. Stockholm Studies in Human Geography, Stockholm University, Sweden

Liwenga ET, Kangalawe RYM, Sokoni C (2009) Climate change/variability and implications on agricultural production and livelihoods in the southern highlands of Tanzania. Regional Conference on strengthening local agricultural innovations to adapt to climate change 24–26th August 2009. Dar es Salaam, Tanzania

Lobell DB, Burke MB (2008) Why are agricultural impacts of climate change so uncertain? The importance of temperature relative to precipitation. Environ Res Lett 3:1–8. http://tinyurl.com/ndnhpgl (Accessed July 15, 2013)

Low PS (2005) Climate Change and Africa. Cambridge University, New York

Majule AE, Gibson R, Chiwatakwenda A (2007) Climate change in low potential areas of Tanzania: local perceptions, vulnerability, current adaptations and future strategies in Sanjaranda village, Manyoni, Singida. CCAA-Tanzania, Malawi Working Paper No. 2

Majule AE, Mbonile MJ, Campbell DJ (2009) Ecological gradients as a framework for analysis of land use change in East Africa. Afr J Ecol 47:55–61

Mbilinyi A, Saibul GO, Kazi V (2013) Impact of climate change to small scale farmers: voices of farmers in village communities in Tanzania. Economic and Social Research Foundation, ESRF Discussion Paper No. 47.http://preview.tinyurl.com/jp8rn4r. Accessed 23 July 2015

Mubaya CP (2010) Farmers strategies towards climate variability and change in Zimbabwe and Zambia, PhD Thesis, University of the Free State

Ngusaru A (2007) Climate change impacts in Tanzania. http://tinyurl.com/p3n8sc7. Accessed 01 Aug 2007

Nyunza G, Mwakaje AEG (2012) Analysis of round potato marketing in Tanzania: the case of Rungwe District, Tanzania. Int J Bus Soc Sci 3(23):86–97

Obiora CJ (2013) Climate change and sustainable agriculture in Nigeria: challenges and opportunities. J Renewable Agric 1(3):53–57. Available from http://manuscript.sciknow.org/uploads/jra/pub/jra_1368044786.pdf. Accessed 3 Nov 2013

OECD (2013) Overview of progress and policy challenges in Tanzania, in OECD Investment Policy Reviews: Tanzania 2013. OECD Publishing, Paris. http://preview.tinyurl.com/hpsy7qj. Accessed on 29 Sept 2016

Orindi VA, Murray LA (2005) Adaptation to climate change in East Africa: a strategic approach, iied. Gatekeeper Series 117: International Institute for Environment and Development

Paavola J (2008) Livelihoods, vulnerability and adaptation to climate change in Morogoro, Tanzania. Environ Sci Policy 11(7):642–654. https://doi.org/10.1016/jenvsci200806002

Paul H, Ernsting A, Semino S, Gura S, Lorch A (2009) Agriculture and climate change: real problems, false solutions. Report published for the conference of the Parties, COP15, of the United Nations Framework, Convention on Climate Change in Copenhagen. EcoNexus, Oxford, UK. Available online at www.econexus.info

Pew C (2006) http://tinyurl.com/nbxf2cbRetrieved on October 12, 2006. Accessed 22 June 2007

Ramesh AS, Madhavi C (2009) Occupational stress among farming people. J Agric Sci 4(3):115–125. http://tinyurl.com/px2f75k. Accessed 4 Feb 2013

Rosegrant MW, Ewing M, Yohe G, Burton I, Huq S, Valmonte- Santos R (2008) Climate change and agriculture: threats and opportunities. Eschborn, Deutsche Gesellschaftfür Technische Zusammenarbeit (GTZ) GmbH Climate Protection Programme for Developing Countries

Sokoni C (2008) Commercialisation of smallholder production in Tanzania: implications for sustainable resources management. Geogr J 174:158–161. http://tinyurl.com/jlv7p6o (Accessed November 29, 2015)

Sokoni CH (2013) Analysis of agricultural change using field allocation to crops technique: a case of Uporoto Highlands, Southern Tanzania. J Geograph Assoc Tanzan 34:63–75

Spangenberg JH (2002) Institutional sustainability indicators: an analysis of the Institutions in Agenda 21 and a draft set of indicators for monitoring their inactivity. Sustain Dev 10:103–115

Stern N (2006a) Stern review on the economics of climate change. Executive Summary. HM treasury. http://tinyurl.com/ordbcor and Annex 7.g: Emissions from agriculture sector http://tinyurl.com/nqnudvo. Accessed 01 Nov 2012

Stern N (2006b) Stern review on the economics of climate change. Her Majesty's Treasury and the Cabinet Office, London. available from http://preview.tinyurl.com/vgzxv. Accessed 01 Nov 2012

Tadesse T, Dechassa N, Bayu W, Gebeyehu S (2013) Effects of farmyard manure and inorganic fertilizer application on soil physico-chemical properties and nutrient balance in rain-fed lowland rice ecosystem. Am J Plant Sci 4:309–316

Tilumanywa VT (2013) Land use and livelihood changes in the mount Rungwe ecosystem, Tanzania. Stellenbosch University

TNBC (Tanzania National Business Council) (2009) Kilimo Kwanza, towards Tanzania Green Revolution, Dar es Salaam

Tumbo M, Nsemwa L, Gwambene B (2008) Stakeholder consultation survey report for Rungwe and Kyela District in Mbeya Region. Strengthening local agricultural innovation systems in less favoured and more favoured areas of Tanzania and Malawi to adapt to the challenges and opportunities arising from climate change and variability. Institute of Resource Assessment, University of Dar es Salaam, Tanzania

URT (2006) Agricultural Sector Development Programme (ASDP), Support through Basket Fund. Government Programme Document

URT (2007a) National Adaptation Programme of Action (NAPA). Vice President's Office, Division of Environment, Dar es Salaam. Http://www.unitar.org/ccp/napaworkshops.htm. Accessed 23 June 2007

URT (2007b) Agricultural and livestock policy. Ministry Of Agriculture and Co-Operative Development, Dar es Salaam

URT (2008) Agricultural marketing policy. Ministry for Industry, Trade and Marketing, Dar es Salaam

URT (2009) Investment potential and opportunities in agriculture (Crop sub-sector). Ministry of agriculture food security and cooperatives. http://tinyurl.com/zr9ymyn (Accessed September 01, 2014)

URT (2010) Rungwe District Investment Profile, 2010. Rungwe Council: Government Printers

URT (2015) Tanzania climate smart agriculture program 2015–2025. Ministry of Agriculture, Food Security and Cooperatives and Vice President's Office

World Agroforestry Centre (2009) Mainstreaming climate change into agricultural education: challenges and perspectives. ICRAF Working Paper no. 82, World Agroforestry Centre. Nairobi, Kenya

Yemefack M (2015) Preserving soil fertility and ensuring food security. International Institute of Tropical Agriculture. Conference Paper. Yaoundé Available from: http://tinyurl.com/gpqnc2t. Accessed 29 Nov 2015

Zilberman D, Zhao J, Heiman A (2012) Adoption vs. adaptation, with emphasis on climate change. Ann Rev Resour Econ 4

Constraints to Agricultural Transformation in Yumbe District, Uganda

Nabalegwa Muhamud Wambede, Asaba Joyfred, and Alule Rhombe Jimmy

Abstract This book chapter analysed the constraints to agricultural transformation, focussing on Yumbe district in Uganda. Since the 1980s, many resources have been devoted to agricultural sector by the government and donor community with the aim of transforming it from predominantly subsistence to one where farmers produce for the market. This was expected to take place with increased farmers' adoption of modern farming technologies. Scientists developed a number of technologies with a hope that once they are introduced to the farmers,they would be readily adopted. This has not been the case. The uptake of the modern technologies has not been impressive. Even where adoption has taken place, it is on a piecemeal basis. Therefore there is a need to examine the factors that have led to the low rate of technology uptake and hence limited agricultural transformation in the country. The data collection methods used in this study included questionnaires, interviews, direct observation and document review. The study covered the social and economic conditions of the farming communities that are hypothesised to be limiting the uptake of technologies that would lead to agricultural modernisation. The study analysed the relationship between the socio-economic factors and adoption to modern farming technologies. The chi-square test was used to establish whether there were significant differences in education level, membership to farmer organisations/ farmer groups, family size, etc. between adopters and non-adopters of modern farming technologies and hence agricultural transformation. Multiple regression was used to establish the most important factors influencing agricultural transformation in the area. The findings indicated that the most important factors influencing farmers' adoption of modern farming technologies and hence agricultural transformation are membership to agricultural organisation and cost of inputs. It can be concluded that the government and other organisations promoting agricultural

N. M. Wambede (✉) · A. Joyfred
Kyambogo University, Kampala, Uganda

A. R. Jimmy
Gulu University, Gulu, Uganda

Y. Bamutaze et al. (eds.), *Agriculture and Ecosystem Resilience in Sub Saharan Africa*, Climate Change Management,
https://doi.org/10.1007/978-3-030-12974-3_2

modernisation should encourage farmers to form groups where they learn modern methods. It should also provide subsidies to farm input so that they are affordable to the farmers.

Keywords Agricultural transformation · Constraints · Socio-economic factors · Farming technologies · Agricultural modernisation

1 Introduction

Agricultural modernisation and transformation according to Grove (1993) involves many technological, institutional and management changes aimed at increasing productivity per unit area, per animal and per person. As noted by Waugh (2002), all over the world, agricultural modernisation is better developed in the developed countries (Europe and North America) than the developing countries. He points out that this variation is largely attributed to differences in rate of adoption of modern farming technologies such as agrochemicals, high-yielding varieties (HYVs), soil conservation measures and mechanisation. The farmers in developed countries are responsive to innovations leading to high agricultural productivity (OECD 1994). Fellman et al. (1999) observe that, between 1965 and 1995, world cereal production rose by more than 90%. Over three quarters of that increase was due to increases in yields rather than expansions in croplands. Contrastingly, the farmers in the developing world are so married to traditional technologies such as shifting cultivation, rudimental tools and local crop varieties. Any increase in yield can only be realised with expansion in land.

In Africa, the rate of adoption of modern farming technologies is low (Waugh 2002). The low rate of adoption of modern farming technologies is responsible for low agricultural productivity (Lowe 1986). Yet Lowe has noted that innovative farming can only solve the agricultural problems of Africa and this requires innovative farmers. It needs an audience of farmers who are receptive to new ideas and experimental findings and those that are capable of putting them into effect. However, not all farmers are receptive to new ideas. As noted by Arnon (1981), there are two types of communities and groups in developing countries: those who do not require much persuasion to adopt new techniques and those in which tradition and customs are conducive to resistance to change in spite of all forms of persuasion and inducements. Unfortunately, the latter group form the majority in Africa; they can refuse to increase their production and incomes even if opportunities are available. Even for the former group, where technological development such as new strains of seeds, cross-breeding of animals, improved machinery and irrigation may extend the area of optimal condition and the limits of production (Green Revolution), due to lack of capital and expertise, the farmers are rarely able to take advantage of these advances (Waugh 2002). However, Lowe (1986) points out that within Africa some countries tend to embrace modern farming technologies much more than others as is the case in Kenya, South Africa and Egypt.

In Uganda, the current modernisation of agriculture is closely linked to the Green Revolution, which swept and continues to sweep across both the developed and the developing world (Waugh 2002). As pointed out by Enger and Smith (2004), the Green Revolution introduced modern farming technologies such as new plant varieties, farming methods and effective use of agrochemicals, which resulted in increased agricultural productivity. The high-yielding varieties (HYVs) not only matured faster than the native varieties but were also less vulnerable to local climatic conditions and diseases and more adaptable to a wider range of environmental conditions (Africa Farming 2002). Despite all these technologies, the farmers were only able to realise 13–33% of the yields obtained at the research stations (NEMA 1998). This implies that earlier attempts to modernise agriculture did not make any headway.

In response to declining agricultural productivity, the Plan for Modernisation of Agriculture (PMA) was started as an initiative of Ugandan government with a mission of 'eradicating poverty by transforming subsistence agriculture to commercial agriculture' by 2017 (PMA 2007). The Plan for Modernizing Agriculture (PMA)'s mission was broken down into four objectives, namely,

- Increase incomes and improve the quality of life of poor subsistence farmers through increased productivity and increased share of marketed production.
- Improve household food security through the market rather than emphasising self-sufficiency.
- Provide gainful employment through the secondary benefits of PMA implementation such as agro-processing factories and services.
- Promote sustainable use and management of natural resources by developing a land use and management policy and promotion of environmentally friendly technologies.

To ensure that the PMA objectives are achieved, the government came up with eight strategies. These include:

- Making poverty eradication the overriding objective of agricultural development
- Extending decentralisation to lower levels of government for efficient service delivery
- Removing direct government involvement in commercial aspects of agriculture and promoting the role of the private sector
- Supporting the dissemination and adoption of productivity-enhancing technologies
- Guaranteeing food security through the market and improved incomes
- Ensuring that all interventions are gender focused
- Involving and empowering local governments in the planning and budget process to enable them influence public policy
- Ensuring coordination of the multi-sectoral interventions to remove any constraints to agricultural modernisation

With such strategies, it was expected that the PMA objectives would be a success story (PMA 2007).

To achieve this mission, the National Agricultural Advisory Services (NAADS) was formed in 2000 with a mission that targeted 'increased farmer access to information, knowledge and technology through effective, efficient, sustainable and decentralised extension with increasing private sector involvement in line with government policy'. Thus, the plan to transform agriculture was about enabling farmers adopt agricultural innovations. Ideally, as stated in the plan's mission, these innovations aimed at transforming agriculture from being predominantly subsistence to commercial through the dissemination and adoption of productivity-enhancing technologies relevant to the farmers (PMA 2003). The plan assumed that once the technologies are made available, they would be adopted. The adoption of these technologies is a prerequisite in modernising agriculture in Uganda.

According to the innovation diffusion model, adoption is a process and can only be successful if the farmers are ready to adopt them (Roger's 1983). He points that an adopter of an innovation must first learn about the innovation, be persuaded as to the advantages of the innovation, decide to adopt, implement the innovation or reject the decision to adopt the innovation. This according to him depends on four important factors such as distance to the origin, the level of education, the nature of transport and the mode of communication. As a result, fewer farmers adopt innovations at its inception, and the largest proportion adopts at later date when the innovation is almost obsolete as reflected on the sigmoid diffusion curve.

It appears that the rate of adoption of these technologies is rather slow and does not occur homogeneously throughout the agricultural sector in the country. There are severe differences in speed and level of dispersion of these farming technologies. For instance, 2.2% of the farmers used improved seeds in Western Uganda compared to 11.9% in Eastern Uganda, 7.6% in Northern Uganda and 5.5% in Central Uganda UBOS (2007). The situation in Yumbe District was exacerbated by the socio-economic and physical setting characterised by unreliable rainfall, declining soil fertility, low levels of income (poverty level of 60.7%), low levels of literacy (59%) and small farm sizes made complex by the traditional land tenure system and rapidly rising population with a growth rate of 7.9% (Vision 2035, 2008). In some parts of the district, the land has reached a point where the available land for cultivation is unable to support the population. This has created a situation of continuous exodus to sparsely populated areas. In some cases, the land area for agriculture has been reduced causing a radical change in agricultural productivity (Yumbe District Profile 2004). Because of such a setting, the cost of inputs is far beyond the means of the farmers, membership to local organisations and extension services are very ineffective. These in turn negatively affects adoption of modern farming technologies under the Plan for Modernisation of Agriculture.

1.1 The Study Problem

The high population growth in Uganda of 3.0% per annum has exerted pressure on land resources especially for agriculture (UBOS 2015). This has resulted into accelerated soil erosion, reduced soil fertility, land fragmentation and low crop yields leading to persistently low agricultural productivity (NEMA 2008). This concern led to the establishment of the PMA with the aim of transforming agriculture from subsistence to commercial. Under PMA, modern farming technologies were developed, such as high-yielding varieties (HYVs), use of agrochemicals, improved tillage systems and soil and water conservation measures which were to be adopted by the farmers to increase agricultural yields.

However, according to PMA (2007) and World Bank Report (2008), the plan registered some serious failures across the country. The NAADS audit report 2007 revealed that most of the technologies introduced showed low level of adoption despite the fact that the same technologies were successful elsewhere (New Vision, May 17, 2010). For example, adoption rates were high in the districts of Kayunga, Luweero, Hoima, Kabale, Kisoro and Kabarole and low in the districts of Lira, Kanungu, Iganga, Bulisa and Karamoja region and in other parts of country. Despite the low level of adoption, most farmers adopted these technologies on piecemeal basis rather than as a package leading to low productivity (PMA 2007). The situation was compounded further where the technologies adopted were often abandoned in case of slight problems. As such, adoption rates in the country were generally low even when some of the technologies were offered free (NEMA 2008). This study therefore seeks to establish the reasons for the low rate of adoption since it is a serious constraint to agricultural transformation in the county as a whole and Yumbe District in particular.

1.2 The General Objective and the Specific Objectives

The general objective of the study was to gain knowledge about the factors constraining agriculture transformation in Uganda, and the specific objectives were to:

(i) Establish the rate of adoption of modern farming technologies in the study area.
(ii) Examine the socio-economic factors constraining adoption of modern technologies towards agricultural transformation in Yumbe District, in Uganda.
(iii) Identify the most dominant factors constraining agricultural transformation in the study area.

2 Methodology

The study was carried out in Yumbe District located in Northwestern Uganda. It is bordered by the Republic of South Sudan in the North, Moyo *District* in the East, Koboko District in the Northwest and Arua District in the South. Apart from the hilly relief along the border with South Sudan, the area slopes gently from the west towards the east with slopes in the range of 30–120 (NEMA 1996). The area has a mean annual rainfall ranging from 1250 to 1500 mm categorised by NEMA, as a Climatic Region Zone III – Western Uganda (NEMA 1996) with a dry season, which runs from December to March, and a short dry spell from May to June. According to Koppen Climate Classification, the mean annual temperature is 24.4 °C (Atlas of Uganda); maximum temperatures in the study area occur during the day of 26–31 °C.

According to the FAO classification system, the soils of the area of study are mainly ferralitic in nature and fall under soils of fair productivity (Atlas of Uganda 1967; NEMA 1996). The soil is largely made up of sand, and as a result, the proportion of land with sandy soils is greater than 60% (Wortman and Eledu 1999). The ferralitic soils cover more than three quarters of the total area. A narrow belt in the east has ferrisols (sandy loam) and vertisols (Yumbe District Profile 2004). The area has a moderate population density with an average of 209 persons per Km^2 and very high population growth rate of 7.2% (UBOS 2015). Subsistence agriculture is the major economic activity employing 94% of the population.

The research design that was adopted for this study was cross-sectional in that the findings were to apply to situation as it was on the ground. However, efforts were made to understand the situation before the implementation of PMA from existing documents. The main source of this data was written records from agricultural departments at the district and the sub-county levels. Where some of the data could not be obtained at the district and sub-county, it was collected from farmers who had practiced agriculture for not less than10 years and were able to recall events before the inception of the PMA. This provided a base line data for this research. The second part included collecting data from a cross section of respondents after the introduction of the PMA. The categories of respondents involved peasant farmers, LC officials and agricultural officers.

The researcher predetermined the area of the study, Kuru sub-county in Yumbe District. The reasons behind this choice were that the sub-county had an interesting demographic characteristic; one portion had dense population and the other had a sparse population. This facilitated a comparative analysis between farm sizes in densely populated areas and sparsely populated areas. Secondly, over 90% of the households in the sub-county depended entirely on agriculture.

Though over 90% of the 5828 households in the area practiced agriculture, those purely peasant farmers were about 4987 (UBOS 2014). Two hundred fifty respondents were selected covering 235 household peasant farmers and the leaders. A random sampling procedure and a fixed percentage were used to identify targeted households from each parish to cater for uneven population distribution. A total of 250 respondents were selected including 235 household peasant farmers; others included local council official (LC1), agricultural officers and sub-county and district leaders. However, due to high rate of illiteracy, the researcher felt most of the households including the less educated would not be in the position to interpret the questionnaires well. As a result, the interview guide was the most convenient research instrument for data collection. This in turn had a multiplier effect on time factor, costs and the workforce. For this reason, the researcher felt the number 250 (5%) of the total respondents was manageable. To cater for gender balance, at least five of the total respondents investigated in each parish were from female-headed households.

The major interest of the study was the socio-economic factors that influence the farmer's decision and consequent adoption of modern farming technologies. The degree of adoption of modern farming technologies was rated based on percentage of households adopting particular farming technologies. The percentage of farmers adopting a particular technology was rated as high (>70%), medium (50–69%), low (30–49%) and poor (<30%).

The data collected was analysed at both bivariate and multivariate levels. This applies when each of the subjects studied has measurements on two variables (Everilt 1998). Thus in order to determine the influence of each of the factors influencing farmers' adoption of modern farming technologies, the chi-square (X^2) was used to test the level of significance as a result of differences between the expected and the observed. At multivariate level, qualitative response regression model was used because adoption studies are qualitative in nature. As noted earlier, the bivariate data analysis best applies to statistical data where independent variable (regressor) takes effect on the dependent variable (regressand). However, labour economics research suggests that the farmers' decision to adopt is a function of several factors such as education, income, farm size, land tenure and so on (Gujarati 2003). Thus, to determine the combined effect of the stated independent variables on the adoption of modern farming technologies and to isolate the most significant factors require the logit model. In this case, the dependent variable which reflects farmers' response to modern technology is qualitative and can only take two values, that is, adoption or none adoption. In other words, the dependent variable is a binary or dichotomous variable (Gujarati 2003).

3 Results and Discussions

3.1 *The Rate of Adoption of Modern Farming Technologies*

It should be noted that given the present population pressure on land resulting from the high population growth rate, restricting farming in areas with severe limitations would be impractical, unrealistic and would obviously be met with a lot of resistance from the people whose livelihood depends entirely on farming. Thus, the future for this region lies in ensuring that farmers adopt appropriate farming technologies and combine them with traditional farming practices to increase agricultural productivity.

The data on adoption of modern technologies is shown in Table 1.

Results of the rate of adoption indicate that despite the level of household awareness among of modern farming technologies, adoption rates remained low. From data analysis, 41.2% of the households had adopted modern farming technologies, while 58.8% never adopted. Of the adopters, 40.4% adopted on piecemeal basis, while 0.8% adopted on package basis. The most popular and widely adopted modern

Table 1 Adoption of the various farming technologies in Yumbe District

Type of technology	Households	Adopters	Non-adopters	Total
HYC/animal breeds	Number:	103	147	250
	Percentage:	41.2	58.8	100.0
Agrochemicals	Number:	61	189	250
	Percentage:	24.4	75.6	100.0
Improved tillage systems	Number:	00	250	250
	Percentage:	00.0	100.0	100.0
Soil and water conservation	Number:	02	248	250
	Percentage:	0.8	99.2	100.0

Source: Field Data Analysis

farming technology was improved seeds and animal breeds with adoption rate of 41.2%. The 41.2% adoption rate was attributed to over 60% free provision of some technologies by either the government or the non-governmental organisations. The rates of adoption for the various technologies were 45% for maize varieties such as *Longe 1–6, High breed A and B*; 33% for cassava varieties such as *NASE 1–4* and *ME*; 9% for groundnut varieties such as *Red beauty* and *Serena*; *17% for* beans varieties such as *NABE 1–10c, K131/K132*; 12% for sorghum variety such as *Sekendo*; and *41% for* rice variety such as *NARIC 1–3.* Improved livestock breeds included *Boer* goats, *Mubende* goats with adoption rate of 3% and the *Friesian* cattle with adoption rate of 0.4%.

Nearly 59% (58.8%) of the households never adopted HYC varieties and improved animal breeds; 24.4% of the households adopted agrochemicals. The most adopted agrochemical was pesticides since the government freely supplied them, while chemical fertiliser was the least adopted agrochemical because it was not free. The most unpopular modern technologies were soil and water conservation techniques and improved tillage practices. The likely reason was high costs associated them and lack of adequate information about these technologies. As a result, most households chose to use family labour or hired labour using traditional tools. Even then, there was practically little evidence that soil erosion control measures were adopted in areas with steep slopes.

Thus despite the traditional technologies being too low in productive capacity to meet the demands of the ever-rising population amidst dwindling resources (Ojungu 1992), they dominated the agricultural scene in nearly all the households except the 0.8% who adopted these technologies as a package. In a similar study, Adesina et al. (2006) found that, compared to other regions in the world, African agriculture uses predominantly traditional technologies and is heavily undercapitalised. This is also in line with the work of Tarrant (1980), who found that the low level of adoption of farming technologies in developing countries was attributed to most agricultural research being crop oriented rather than concentrating on the system of traditional agriculture within which the crop is produced. This proves right Nyerere's comment about *Ujamaa* in White (1973) that we have to think in terms of what is available or can be made available cheaply if agriculture has to be modernised.

3.2 Analysis of the Socio-economic Factors Affecting Adoption of Modern Farming Technologies by Farmers

As stated in the objectives, this study was meant to examine the factors influencing farmers' decision to adopt modern farming technologies under the Government Plan to Modern Agriculture in Kuru sub-county, Yumbe District. The factors that were examined in this research included farm size, level of education, level of income, land tenure, cost of input, extension services, membership to farmer organisations, access to credit, slope of the farm, soil depth and rainfall reliability.

The socio-economic factors in this research study included membership to farmer organisations, level of education, extension services, land tenure system, farm size, household income and so on. They were tested against adoption of modern farming technologies as follows.

3.3 Membership to Farmer Organisations

The diffusion of information on modern farming technologies may be influenced by farmer's access to farmer organisations. An analysis of membership to farmer organisations in Table 2 revealed that 38.0% of the households belonged to at least one local farmer organisation, while 62.0% did not belong to any farmer organisation. Those who belonged to farmer groups were mainly male households. This was partly attributed to the fact that majority of the respondents were male household heads. Cross-tabulation of membership to farmer organisations and adoption indicates that, among the adopters, 67% were members of farmer organisation, while 33% were non-members of farmer organisation. On the other hand, among the non-adopters, 18% were members of farmer organisation, while 82% were non-members of farmer organisations. From the above analysis, it can be deduced that most of the adopters belonged to at least one farmer organisation. These groups helped the farmers to access loans and agricultural inputs. As a result, the contribution of membership to farmer organisation plays an important role in adoption of agricultural innovations.

Table 2 Relationship between types of membership and adoption

	Membership to farmer organisation		
Adoption	Members	Non-members	Total
Adopters (number)	69	34	103
Percentage	67%	33%	100%
Non-adopters (number)	26	121	147
Percentage	18%	82%	100%
Total	**95**	**155**	**250**
Percentage	**38%**	**62%**	**100%**

Source: Field Data Analysis

Table 3 Chi-square for membership to farmer organisations and adoption

Chi-square tests	Value	Df	Asymp. Sig. (2-sided)
Pearson chi-square	56.762	1	0.000
Continuity correction	54.796	1	0.000
Likelihood ratio	58.160	1	0.000
Linear-by-linear association	56.535	1	0.000
No. of valid cases	250		

As such, a chi-square test was run to ascertain whether there was a significant difference in membership to farmer organisations between adopters and non-adopters of new farming technologies.

The chi-square analysis (Table 3) indicated that there was significant difference in membership to farmer organisations between adopters and non-adopters in the adoption of modern farming technics. Since the level of significance of 0.000 is less than the required significance level of 0.05(5%), membership to farmer organisations significantly influences the adoption of modern farming technologies among households. This is in agreement with UNEP (2006) recommendation that membership to local farmer organisations is an important factor influencing farmers decision to choose and adopt modern farming technologies. This could be due to the advantage of social capital that is associated with membership to a formal grouping. Thus, the way forward to increased adoption of these technologies will be to encourage active farmer organisations because through these groups it is easy to deal with the household farmers.

3.4 Level of Education and Adoption of Modern Farming Technologies

The education levels in the study area were typical of most parts of rural Northern Uganda. Table 4 shows that 19.6% of the households had no education at all, 57.2% had primary education, 18.8% had secondary education, and 4.4% had tertiary education. The male household heads registered higher education levels compared to their female counterparts. It was also found that majority of the educated ranged between 20 and 50 years. This appears to be related to increasing number of educational institutions after independence (1962).

Analysis of the relationship between education and adoption revealed that the adopters were more educated than the non-adopters. Among the adopters, 17% had no education, 49% had primary education, 27% had secondary education, and 7% had tertiary education. For the non-adopters, 21% had no education, 63% had primary education, 13% had secondary education and 3% had tertiary education. The results showed that adoption rate more than doubled with increase in education background, for instance, 27% adopters against 13% non-adopters for farmers with

Table 4 Adoption of modern farming technologies and education

	Level of education				
Adoption	No education	Primary	Secondary	Tertiary	Total
Adopters (number)	18	50	28	7	103
Percentage	17%	49%	27%	7%	100%
Non-adopters (number)	31	93	19	4	147
Percentage	21%	63%	13%	3%	100%
Total	**49**	**143**	**47**	**11**	**250**
Percentage	**19.6%**	**57.2%**	**18.8%**	**4.4%**	**100%**

Source: Field Data Analysis

Table 5 Chi-square for level of education and adoption in Kuru sub-county

Chi-square tests	Value	Df	Asymp. Sig. (2-sided)
Pearson chi-square	12.530	3	0.006
Likelihood ratio	12.388	3	0.006
Linear-by-linear association	7.817	1	0.005
No. of valid cases	250		

secondary education background and 7% adopters versus 3% non-adopters for farmers with tertiary education background. Even though post-primary education reduces labour intensity as noted previously, it increased the rate of adoption of agricultural innovations and other inputs in production. This seems to be related to high receptivity for new ideas. However, the 17% of adopters with no educational background seem to suggest that people may be aware of certain innovations but without any formal education.

A chi-square test was run to establish whether there is a statistically significant difference in education between adopters and non-adopters of modern farming technologies, and the results are presented in Table 5.

The chi-square test results showed a significant difference in the level of education between adopters and non-adopters of modern farming technologies since the significance level of 0.006 is less than the 0.05 (5%) required for this study. Exposure to education was found to increase farmer's ability to adopt agricultural innovations because it influenced the degree to which one would understand government policies and amass a range of alternative strategies for survival (Ojungu 1992). The results agree with the argument put forward by Rogers (1983) that education enhances awareness of a new technology and increases the farmer's capacity to adopt such a technology.

3.5 *Extension Services*

Extension services provided are believed to improve the ability of households to adopt modern farming technologies. As such, in the analysis of the influence of extension services in adoption of modern farming technologies, it was found out

Table 6 Access to extension services and adoption

	Access to extension services			
Adoption	No access	Inadequate	Adequate	Total
Adopters (number)	47	53	3	103
Percentage	46%	51%	3%	100%
Non-adopters (number)	108	38	1	147
Percentage	73%	26%	1%	100%
Total	**155**	**91**	**4**	**250**
Percentage	**62%**	**36%**	**2%**	**100%**

Table 7 Chi-square for access to extension services and adoption

Chi-square tests	Value	Df	Asymp. Sig. (2-sided)
Pearson chi-square	16.945	2	0.000
Likelihood ratio	16.957	2	0.000
Linear-by-linear association	16.812	1	0.000
No. of valid cases	250		

that 62.0% of the households had no access to extension services, 36% had inadequate extension services, while only 2% reported having adequate extension services as shown on Table 6. This categorisation was based on World Bank Report (2008) that a farmer should access extension services at least three times a season in order to get good results.

As seen in Table 6, adopters were more associated with the level of extension services be it adequate or inadequate. Among adopters, 46% had no access to extension services, 51% had inadequate access, while 3% had adequate access to extension services. Among the non-adopters, 73% had no access, 26% had inadequate access, while 1% had adequate access to extension services.

A chi-square test was run to establish whether there is a statistically significant difference in access to extension services between adopters and non-adopters of modern farming technologies; the results are presented in Table 7.

The results of the chi-square in Table 7 revealed a significant difference in access to extension services between adopters and non-adopters of modern farming technologies. The results show that access to extension services significantly influences the adoption of modern farming technologies since the significance level of 0.000 is much lower than 0.05 (5%) the required level for this study.

It showed that households who accessed extension services adopted modern farming technologies much more easily than those who never accessed extension services. However, the positive association of PMA with high crop yields per acre may have less to do with extension services but rather the changing farming patterns. For example, majority of the farmers complained that most of the training was done at the sub-county headquarters instead of visiting them on their farms. This had a negative impact on adoption of modern farming technologies. As noted in OECD (1991), sustainable agriculture is not a return to low technology practices but rather is highly dependent on incorporating field experiences, ecological data and information on technological advances and new equipment into farm management. This can only be achieved through intensive extension services. The chi-square findings here agree with the research carried by Nkonya et al. (1997), on factors affecting adoption of improved maize variety and fertiliser in Northern Tanzania where there was a significant relationship between extension services and adoption, though Khaukha (1992) found otherwise. It is important to note that adoption of modern farming technologies can be improved in Kuru sub-county if extension services are stepped up.

3.6 *Land Tenure (Ownership)*

Security of tenure is believed to play an important role in influencing adoption of agricultural innovations. On analysing the data on land tenureship, four types of ownership were identified. In this study 87.6% of the respondents were found to have inherited land, 5.2% borrowed land, 7.2% bought land, while no one had leased their land (Table 8). Theoretically, leased, inherited and bought land is secure in terms of use than borrowed land with tenure insecurity. Theoretically, leased, inherited and bought land is secure in terms of use than borrowed land with tenure insecurity. However, the research findings showed otherwise. In relation to adoption of modern farming technologies, among the adopters, 84.5% had inherited land, 5.8% had borrowed land, while 9.7% had bought land. Among the non-adopters, 89.8% had inherited land, 4.8% had borrowed land, and 5.4% had bought land.

A chi-square test was run to establish whether there is a statistically significant difference in land tenure system between adopters and non-adopters of modern farming technologies; the results are presented in Table 9.

The chi-square results in Table 9 reveal that there is no significant difference in land tenure between adopters and non-adopters in the adoption of modern farming technologies. Hence, the null hypothesis is taken to be correct, and therefore land tenure was dropped from the list of factors influencing adoption of modern farming technologies in Yumbe District. The results here disagree with the findings of a

Table 8 Relationship between land tenure and adoption

	Land tenure				
Adoption	Inherited	Borrowed	Bought	Lease	Total
Adopters (number)	87	06	10	00	103
Percentage	84.5%	5.8%	9.7%	00	100
None Adopters (number)	132	07	08	00	147
Percentage	89.8%	4.8%	5.4%	00	100%
Total	**219**	**13**	**18**	**00**	**250**
Percentage	**87.6**	**5.2**	**7.2**	**00**	**100**

Source: Field Data Analysis

Table 9 Chi-square for land tenure and adoption in Kuru sub-county

Chi-square tests	Value	Df	Asymp. Sig. (two-sided)
Pearson chi-square	17.680	3	0.115
Likelihood ratio	17.655	3	0.115
Linear-by-linear association	9.817	1	0.110
No. of valid cases	250		

study carried by Ayodele and Eshalomi (1998) in Southwestern Nigeria on adoption of high-yielding palm varieties that revealed that land tenure system had a significant bearing on its adoption. The non-significance of land tenure in influencing adoption is that majority of the adopters and non-adopters live on customarily owned land with no security of tenure.

3.7 *Farm Size*

An analysis of the farm size owned by each household indicated that household land holdings were generally small basing on the acreage. Due to the wide spread practice of land inheritance, land in Kuru sub-county was highly fragmented into small portions as land is shared between the father and sons over generations. On average, each household owns three pieces of farm plots. The average farm size was 4.3 acres (1.72 ha).

A survey of the farm sizes in Kuru sub-county, Yumbe District, showed that 8.8% of the households owned less than 1 acre of land, 67.6% owned between 1 and 5 acres, while 23.6% owned more than 5 acres (>2 ha.) of land. The households that owned more than 5 acres of land were mostly located in sparsely populated areas in the east, while most of the households in densely populated areas in the central and the western parts of the sub-county had farm sizes ranging between 1 and 3 acres. Thus, majority of the household heads who owned land were male compared to female. This seems to be attributed to the traditional land tenure system were men own land.

A bivariate analysis was carried to establish the relationship between farm size and adoption of modern farming technologies. As shown in Table 10, among the non-adopters of modern farming technologies, 9% had land below 1 acre, 67% had land of 1 to 5 acres, and 25% had land above 5 acres. Of the non-adopters, 9% had land of below 1 acre, 68% had land of 1 to 5 acres, and 23% had land above 5 acres.

A chi-square test was run to ascertain whether there was a significant difference in land sizes between adopters and non-adopters (Table 11).

Since the level of significance value of 0.070 is greater than the required value of 0.05 (5%), there is no significant difference in land size between adopters and non-adopters. Therefore, farm size did not significantly influence farmers' adoption decision. These results relates to the findings of Buyukcolak (1978), who studied adoption of improved wheat varieties in Turkey and found no significant relationship

Table 10 Relationship between household farm size and adoption in Kuru sub-county

	Household farm size			
Adoption	<1 acre	1–5 acres	>5 acres	Total
Adopters (number)	9	69	25	103
Percentage	9%	67%	24%	100%
Non-adopters (number)	13	100	34	147
Percentage	9%	68%	23%	100%
Total	**22**	**169**	**59**	**250**
Percentage	**8.8%**	**67.6%**	**23.6%**	**100.0%**

Source: Field Data Analysis

Table 11 Chi-square for farm size and adoption

Chi-square tests	Value	df	Asymp. Sig. (2-sided)
Pearson chi-square	8.830	2	0.070
Likelihood ratio	8.600	2	0.070
Linear-by-linear association	5.817	1	0.005
No. of valid cases	250		

between farm size and adoption, although a similar research by Nabalegwa and Sudarmadji (1996) on the role of socio-economic factors in soil conservation implementation and management indicated a positive relationship between sizes of the farm owned by the farmers and their response to soil conservation practices.

3.8 *Household Income*

The household income from agricultural produce is believed to be an important factor in influencing farmers' choice in adopting a given technology. The higher the income, the higher the probability of adoption of the planned technology. As revealed by the data presented in Table 12, majority of the peasant farmers in Kuru sub-county Yumbe District were poverty-stricken. Of the households in Kuru sub-county, 41.6% reported getting monthly income of less than or equal to 25,000 shillings, 32.4% reported earning between 25,000 and 50,000 shillings, 23.6% earned between 50,000 and 100,000 and only 2.4% earned more than 100,000 shillings.

As observed from results in Table 12, among the adopters, 27% of the households had monthly income of less than 25,000 Shs. (US $ <10), 38% had monthly income between 25,000 and 50,000 (US $ 20–40), 31% had monthly income ranging from 50,000 to 100,000 Shs. (US $ 40–80), while 4% earned over 100,000 Shs. (US $ >20) monthly. Among the non-adopters, 52% of the households had monthly income of less than 25,000 Shs. (US $ <10), 29% had monthly income between 25,000 and 50,000 (US $ 20–40), 18% had monthly income ranging from 50,000 to 100,000 Shs. (US. $ 40–80), while 1% earned over 100,000 Shs. (US. $ >20) monthly.

Table 12 Relationship between household's monthly income and adoption

	Household monthly income				
Adoption	≤25,000	25,001–50,000	50,000–100,000	>100,000	Total
Adopters (number)	28	39	32	4	103
Percentage	27%	38%	31%	4%	100%
Non-adopters (number)	76	42	27	2	147
Percentage	52%	29%	18%	1%	100%
Total	**104**	**81**	**59**	**6**	**250**
Percentage	**41.6%**	**32.4%**	**23.6%**	**2.4%**	**100%**

Source: Field Data Analysis

Table 13 Chi-square for level of income (monthly) and adoption in Kuru sub-county

Chi-square tests	Value	Df	Asymp. Sig. (2-sided)
Pearson chi-square	16.110	3	0.001
Likelihood ratio	16.447	3	0.001
Linear-by-linear association	14.748	1	0.000
No. of valid cases	250		

The chi-square was used to test whether there was a significant difference in household's monthly income between adopters and non-adopters of modern farming technologies. The results are presented in Table 13.

The chi-square results in Table 13 show that there is a significant difference in income between adopters and non-adopters. Since the level of significance value of 0.001 is less than 0.05 used in the study, the level of household income influences adoption of modern technologies. This was attributed to the *power* of money to buy the available technologies. The results agree with Fellman et al. (1999), who observe that the green revolution is not a success story in Africa because the farmers lack capital for 'high-input-high-yield' agricultural innovations it has created. Bhalla (1996), in a related study on application of fertilisers in India, found that lack of income was a major constraint that limited 48% of small farmers from adoption of modern farming technologies. Thus, it is no wonder that attempts to modernise agriculture in Kuru sub-county have met with strong resistance from the farmers due to widespread poverty of 60.7%.

3.9 *The Cost of Inputs/Technology*

An analysis of the cost of inputs indicated that their prices were generally high. The widespread poverty coupled with the inaccessibility to credit meant that majority of the households were unable to afford them. From the analysis, only 14.4% of the households reported that the costs of the inputs were affordable, 36.8% reported that

Table 14 Relationship between cost of inputs and adoption

	Cost of inputs			
Adoption	Affordable	High	Very high	Total
Adopters (number)	34	44	25	103
Percentage	33	43	24	100
Non-adopters (number)	2	48	97	147
Percentage	1	33	66	100
Total	**36**	**92**	**122**	**250**
Percentage	**14.4**	**36.8**	**48.8**	**100.0**

Source: Field Data Analysis

Table 15 Chi-square for cost of input/innovation and adoption

Chi-square tests	Value	Df	Asymp. Sig. (2-sided)
Pearson chi-square	65.392	2	0.000
Likelihood ratio	72.232	2	0.000
Linear-by-linear association	63.206	1	0.000
No. of valid cases	250		

the costs were high, while the majority, 48.8%, indicated very high costs. As a result, nearly 86.0% of the households reported that the agricultural inputs/equipment were difficult to afford due to the high prices as shown on Table 14.

The results show that among the adopters, 33% reported that the inputs were cheap and therefore affordable, 43% reported that the cost of inputs was high, and 24% reported that the cost was very high. Among non-adopters, 2% reported that the inputs were cheap and therefore affordable, 33% reported that the cost of inputs was high, and 66% reported that the cost was very high. As such, affordability of the agricultural inputs seems to influence adoption.

In this study, the cost of inputs/innovations was assumed to significantly affect adoption of modern farming technologies.

As such, a chi-square test was run to ascertain whether there was statistically significant difference in farmers' perception of cost of inputs between adopters and non-adopters. The results are presented in Table 15.

The results in Table 15 show that there is a significant difference in farmer's perception of affordability of inputs between adopters and non-adopters since the level of significance value of 0.000 is less than 0.05 used in the study. As put by Alamgir and Arora (1991), modern technologies always come at a cost. The lower the cost, the higher the rate of adoption. Hence, the result here is consistent with the findings of a survey carried by GoK, Economic Survey (1993), on use of fertilisers. It showed that the low level of fertiliser usage in Kenya was due to high-input prices. The low rate of adoption of these technologies in Kuru sub-county could partly be attributed to high costs of these innovations, which the poor farmers are unable to afford.

Table 16 Relationship between access to credit and adoption

	Access to credit		
Adoption	Accessed credit	Not accessed credit	Total
Adopters (number)	06	97	103
Percentage	6%	94%	100%
Non-adopters (number)	10	137	147
Percentage	7%	93%	100%
Total	**16**	**234**	**250**
Percentage	**6.4**	**93.6**	**100%**

Source: Field Data Analysis

Table 17 Chi-square for access to credit and adoption in Kuru sub-county

Chi-square tests	Value	df	Asymp. Sig. (2-sided)
Pearson chi-square	46.660	1	0.730
Likelihood ratio	46.951	1	0.740
Linear-by-linear association	7.621	1	0.700
No. of valid cases	250		

3.10 Access to Credit

Majority of poor households in developing countries are rural based and therefore lack access to credit, which may influence adoption of modern farming technologies. This in turn may affect agricultural productivity. In the area of study, it was found out that 6.4% households had received credit at least once in the last 3 years, while 93.6% households never accessed any form of credit (Table 16).

Analysis of the data in Table 16 revealed that among the adopters, 6% received credit, while 94% never received credit. Among non-adopters, 7% accessed credit, while 93% never accessed credit. From the above analysis, it can be seen that access to credit did not significantly influence adoption of modern farming technologies.

The chi-square test was run to establish the relationship between access to credit and adoption of modern farming technologies. The results are presented in Table 17.

The chi-square significance value of 0.730 is far much greater than the required 0.05 as in Table 17. This shows that access to credit did not significantly influence adoption behaviour of households. As shown in the cross-tabulation, more than a half of the households who accessed credit facilities never adopted modern farming technologies. This was due to lack of interest to invest in agriculture due to several risks involved. Most farmers (62.5%) who accessed agricultural credit invested it in nonagricultural businesses. This result is consistent with the findings of Pender (2004) and Nkonya et al. (2004), who found out that credit constraints are not a major impediment to adoption of improved land management practices and that access to credit may promote less intensive land management by facilitating more remunerative nonfarm activities. However, the results of the finding strongly disagreed with the view of Upton and Kim (1999), who notes that provision of credit

Table 18 The logistic model summary

−2 log likelihood	Cox and Snell R-square	Nagelkerke R-square
33.525	0.496	0.664

Table 19 Regression model of the dependent variable (adoption of modern farming methods) and independent variables

Dependent variable	Standardised coefficients	*t*-ratio	Sig.
Adoption of modern farming methods	**Beta**		
Independent variable			
Farm size	0.030	0.628	0.530
Level of education	0.115	2.336	0.020*
Level of income(monthly)	0.112	2.244	0.026*
Land tenure system	0.007	0.143	0.886
Cost of input/innovation	−0.393	−7.801	0.000*
Extension services	0.081	1.592	0.113
Membership to any organisation	−0.364	−7.247	0.000*
Access to credit	0.039	0.793	0.429
Constant		4.431	0.000

Note * means that those independent variables are significant at 5% level
Source: Computer Statistical Analysis

facilities accelerates the necessary investments and hence the adoption of the technology. His finding relates to that of Ruttan and Thirtle (1987), who identified credit as a major factor affecting adoption of hybrid rice technologies in Thailand. Thus, in implementing adoption of agricultural innovations in Kuru sub-county, other factors need to be given more consideration than access to credit as it had proved to be insignificant.

3.11 The Most Important Factors Influencing Farmers Adoption of Modern Farming Technologies

To ascertain the most influential factor in the adoption of modern farming technologies, a logistic regression was run as seen in Table 19. Identification of the most significant variables was important as it was meant to focus adoption efforts to the most influential factors in determining farmers' response to modern farming technologies.

From the model, the Nagelkerke R-square shows that 66.4% of the variance in the dependent variable can be explained by the independent variables used in this study, an indication that the independent variables considered have a great combined effect on farmer's decision to adopt modern farming technologies. The other 33.6% variance in the dependent variables can be explained by other factors not considered in this research (Table 18).

In order to establish the variables that have the greatest influence on the farmer's response towards adoption of modern farming technologies, coefficients and significance values for each variable were used. The different standardised coefficients and levels of significance occurring were computed. Since the relationship between two variables is reflected in the magnitude of the coefficient and significance values, variables with larger coefficient and smaller significance values were said to be more significant in influencing adoption than those with smaller coefficient values and larger significance values. A positive coefficient value indicates a negative relationship betwen the variable and adoption while a negative coefficient value indicates a negative relationship. A coefficient value of zero indicates no relationship. When the coefficient value was zero, the factor was equal to one, which left the odds unchanged.

The results from the model show that four variables, namely, level of education, level of income (monthly), cost of input/innovation and membership to farmer organisation, were significant at 5% level in determining which household/farmer should adopt the modern farming technologies. However, surprisingly, results in the table show that extension services were not significant in determining adoption of the modern farming technologies, as was the case with chi-square test. Other factors such as farm size, land tenure system and access to credit were not significant at 5% level in determining adoption.

The result revealed that the cost of input/innovation had the greatest influence on adoption of modern farming technologies than any other variables. The significance value of 0.000 indicated a perfect relationship between adoption and cost of input/innovation. This was followed by membership to any farmer organisation, level of education and level of income. Since the rest of the levels of significance values are more than 5% (0.05), these factors did not significantly influence adoption of modern farming technologies in Kuru sub-county, Yumbe District.

3.12 Discussion of Econometric Model Estimations

The education level of the head of the household was found to be an important determinant of adoption of modern farming technologies. Similar findings by Nkonya and Norman (1997) revealed that adoption of improved maize seed and fertiliser in Northern Tanzania was positively related to the level of education. Another finding by Nabalegwa and Sudarmadji (1998) revealed that adoption of soil conservation measures is positively related to education level among farmers. This is supported further by findings of Nyirenda et al. (2001), where education was positively associated with adoption of agroforestry technologies. The above studies based their arguments on the fact that formal and informal training can potentially increase the rate of adoption by directly increasing awareness and imparting skills and knowledge of the new technology. Contrary to the results of this research, Feder et al. (1985), in a review of empirical work on adoption, did not find education a key explanatory factor.

The uptake of modern farming technologies is often influenced by farmer's level of income. This study yielded findings similar to that of Bhalla (1979), who found level of income as a major factor in the application of fertiliser by small-scale farmers in India. This is supported by Fellman (1999), who attributes the failure of the Green Revolution in Africa to low incomes due to widespread poverty. Important to note is that income level plays an important role in adoption of technologies because it enhances the necessary investments and encourages the farmer to take risks and enables the farmer an accurate understanding of the cost benefits accruing from the technology. However, Khaukha (1992) found that farmer's income level was not significant in explaining the demand for pesticides among cotton farmers in Iganga district.

Farmer's membership to local organisations was found to influence adoption decision of modern farming technologies in Yumbe district. The success of previous farmer organisations led to high expectations because it made such groups a prime target of the government in modernising agriculture. Through these organisations, the farmers easily accessed credit, inputs and advisory services. This is consistent with the findings of Ntege-Nanyeenya, (1998), who found that membership to an extension group significantly affected the probability of adopting hybrid maize variety (*longel*). However, similar findings by Buyinza and Naagula (2009) revealed that membership to local farmer organisations had no significant relationship in adoption of modern farming technologies. As a result, the high expectations unfortunately ended up with less impact on the farmer's welfare. Hence, majority of women and male farmers with busy schedules were reluctant to join such organisations.

The cost of the input/technology had a positive impact on adoption of modern farming technologies. The lower the cost, the higher the rate of adoption. This is in line with research findings by GoK (1993), which revealed that the low level of fertiliser usage in Kenya is partly due to high-input costs. In a similar study, Semgalawe (1998) identified cost of the input/technology as an important factor determining investment levels required in influencing adoption of soil conservation practices in North Eastern Mountains of Tanzania. Thus, effective adoption of modern farming technologies can be achieved in Yumbe district if income level among farmers is improved.

The factors that influence the adoption of modern farming technologies in Yumbe district include level of education, level of income, cost of input/innovation and membership to farmer organisation and to a lesser extent extension services. Yumbe District is concerned about the dwindling status of their land in terms of size and productivity and is struggling to adopt modern farming technologies such as HYVs, agrochemicals, improved tillage methods and soil/water conservation methods to mitigate against the declining agricultural productivity. To promote greater adoption of such technologies, particular attention should be put on the appropriateness of the technology and the socio-economic and physical factors.

3.13 Conclusions

Basing on the results from this research about technological appropriateness and factors that influence farmer's response to modern farming technologies under the Plan to Modernise Agriculture in Yumbe District, the following conclusions can be made.

It is evident that majority of the farmers in Yumbe District have not adopted modern farming methods. HYC varieties and improved animal breeds were the most widely adopted in the study area. None of the farmers practiced improved village. Most of the adopters except 0.8% were found to be utilising a mix of traditional and modern farming technologies. Farms with moderm technologies registered faily higher productivity compared to those with traditional methods.

Majority of the households argued that the existing modern farming technologies were inappropriate because of the socio-economic and physical implications. However, most of the households attached technological inappropriateness to economic factors more especially the cost attribute other than socio-cultural and physical environment.

The diffusion of these technologies from the sub-county headquarters formed a pattern similar to the findings of Hägerstrand in Sweden in the 1950s and 1960s. The pattern was determined by the distance from the sub-county headquarter. Households close to the sub-county headquarters adopted these technologies much more easily than those further away. Similarly, households close to demonstration plots easily adopted these technologies compared to those far away. However, it was found out that most of the demonstration plots existed by name (signpost) but were missing in action.

The factors that were considered in this study as having an influence on the farmers' response on modernisation of agriculture programmes have significant influence. Therefore, these variables are suitable in predicting and making conclusions about the farmer's response to adopt modern farming technologies in a bid to modernise agriculture in Yumbe District. Other factors outside the researched variables must be considered so that accurate predictions are made.

Among the socio-economic and physical factors which are considered in this research including farm size, level of education, level of income, land tenure system, cost of input/technology, extension services, membership to any farmer organisation, access to credit, slope of the farm, soil depth and rainfall reliability, it was found out that the factors of membership to farmer organisations and cost of input/technology had the greatest influence on the farmers response to agricultural modernisation programmes. These factors cannot adequately be used with accuracy in determining and developing policies on adoption of modern farming technologies in Yumbe District. However, in combination with other factors, good results are inevitable.

References

Adesina JO, Graham Y, Olukoshi A (2006) Africa and development challenges in the new millennium: the NEPAD debate. CODESRIA, London/New York
Alamgir M, Arora P (1991) Providing food security for all (IFAD series on rural poverty). ITDG Publishing. ISBN 10: 1853391174 ISBN 13: 9781853391170
Arnon (1981) Modernisation of agriculture in developing countries: resources, potentials and problems. Willey, New York
Atlas of Uganda (1967) Department of lands surveys, Kampala, Uganda
Ayodele T, Eshalomi MO (1998) African case study: palm oil and economic development in Nigeria and Ghana. Lagos. Initiative for Public Policy Analysis (IPPA)
Buyinza M, Naagula A (2009) Predictors of agro-forestry technology adoption and land conservation strategies in the highlands of southwestern Uganda. Online J Earth Sci 3(2):46–55
Buyukcolak U (1978) Transfer of technology: factors influencing small farmers in the adoption of modern agricultural related to high yielding varieties of wheat production in Turkey, Dissertation Abstracts International, A (1979):38
Enger DE, Smith FB (2004) Environmental science: a study of interrelationships, 9th edn. McGraw-Hill Higher Education, Boston
Everilt BS (1998) The Cambridge dictionary of statistics. Cambridge University Press, London
Feder G, Just RE, Zilberman D (1985, January) Adoption of agricultural innovations in developing countries: a survey. Econ Dev Cult Chang 33(2):255–298. https://doi.org/10.1086/451461
Fellmann DJGA (1999) Human geography- landscapes of human activities. McGraw-Hill Higher Education, Boston
Fellmann DJ, Getis A, Judith G (1999) Human geography- landscapes of human activities. McGraw-Hill Higher Education, Boston
Grove AT (1993) The changing geography of Africa, Second edn. Oxford University Press, London
GoK (1993) Government of Kenya Economic Survey 1993. Kenya National Bureau of Statistics. Nairobi, Kenya
Gujarati DN (2003) Basic econometrics, 4th edn. McGraw- Hill Higher Education, Burr Ridge- Boston
Khaukha R (1992) An economic analysis of demand and supply of agricultural inputs at farm level. In: Mukama PC (ed) Economic benefit and willingness to pay for improved rangelands in Uganda. Unpub MSC Thesis, Makerere University
Lowe RG (1986) Agricultural revolution in Africa. Macmillan Publishers, London/Basingstoke
Muhamud NW (1996) Direction of soil conservation as a way of conserving the capability of the environment in Sedayu Sub-County-Bantul District, Yogyakarta, Indonesia
Muhamud NW, Sudarmadji (1998, December) The role of socio- economic factors in soil conservation implementation and management. Indones J Geogr 30(76):13–19
NEMA (National Environment Management Authority) (1996) The state of environment report for Uganda. Kampala, Uganda
NEMA (National Environment Management Authority) (1998) The state of environment report for Uganda. Kampala, Uganda
NEMA (National Environment Management Authority) (2008) The state of environment report for Uganda 2008. Kampala, Uganda
Nkonya ES, Norman D (1997) Factors affecting adoption of improved maize seeds and fertilizer in Northern Tanzania. J Agric Econ 48:1
Nkonya E, Schroeder, Norman D (1997) Factors affecting adoption of improved maize seeds and fertilizer in northern Tanzania. J Agric Econ 48:1–12
Nkonya E, Pender J, Kato E, Muwonge J (2004) Community natural resource management and its linkage to poverty in Uganda. International Food Policy Research Institute, Washington D.C Mimeo
Nyirenda M, Kanyama-Phiri G, Bohringer A, Haule C (2001) Economic performance of improved fallow agroforestry technology for smallholder maize production in central Malawi. In: African

Crop Science Conference Proceedings, vol 5. African Crop Science Society, Kampala, Uganda, pp 638–687
OECD (1991) The state of the environment. OECD, Paris
OECD (Organisation for Economic Cooperation and Development) (1994) Towards sustainable agricultural production: cleaner technology. Paris
Omara Ojungu PH (1992) Resource management in developing countries. Longman Group, London
Pender J (2004) Development pathways for hillsides and highlands: some lessons from central America and east Africa. Food Policy 29:339–367. https://doi.org/10.1016/j.foodpol.2004.07.005
PMA, (Plan for Modernisation of Agriculture), Annual Report for Financial Year 2006/2007
PMA, (Plan for Modernisation of Agriculture), PMA Progress Report, June 2003
Semgalawe ZM, (1998) House adoption behaviour and agricultural sustainability in the North Eastern Mountains of Tanzania: the case of soil conservation in the North Pare and West Usambara Mountains. PhD Thesis, Wageningen Agricultural University, The Netherlands
Rogers (1983) Diffusion of innovations, 3rd edn. Free Press, New York
Tarrant JR (1980) Food policies. Wiley, Chichester/New York/Brisbane/Toronto
Thirtle CG, Ruttan VW (1987) The role of demand and supply in the generation and diffusion of technical change. Routledge
Waugh D (2002) Geography: an integrated approach, 3rd edn. Nelson Thornes
UBOS (Uganda Bureau of Statistics) (2007) Agricultural sector: gender statistics profile. Kampala, Uganda
UBOS (Uganda Bureau of Statistics) (2014) Uganda national housing population vensus, provisional results. Kampala, Uganda
UBOS (Uganda Bureau of Statistics) (2015) Uganda national housing population census. Kampala, Uganda
UNEP (United Nations Environment Programme) (2006) Africa environment outlook 2. Nairobi, Kenya
Upton DM, Kim B (1999) Alternative methods of learning and process improvement in manufacturing. J Oper Manag 16:1–20
White RG (1973) Africa: studies for east African students. Heinemann Educational Books, Nairobi
World Bank (2008) World development report- agriculture for development. Washington D.C
Wortmann CS, Eledu CA (1999) Ugandas agro- ecological zones: a guide for planners and policy makers. Centro International de Agricultura Tropical, Kampala
Yumbe District Local Government (2004) The Yumbe district state of the environment 2004. Yumbe, Uganda

Indigenous Grasses for Rehabilitating Degraded African Drylands

Kevin Z. Mganga, Dickson M. Nyariki, Nashon K. R. Musimba, and Agnes W. Mwang'ombe

Abstract Drylands provide an important livelihood stream to its inhabitants across the globe through a range of products and ecosystem services. However, these fragile ecosystems are threatened and believed to experience various degrees of land degradation. Estimates of the landmass affected by land degradation in the global drylands range from 10% to 20%, a percentage that is increasing at an annual global rate of 12 million ha of soil lost from desertification and drought. African drylands are especially highly susceptible to severe degradation because of their poor soil structure aggravated by scarce vegetation cover. Causes of degradation in these environments are both natural and anthropogenic in nature. Change in vegetation cover, decline in soil fertility, biodiversity loss and soil erosion demonstrate degradation in African drylands. Grass reseeding using indigenous species is one of the promising sustainable land management strategies to combat degradation in the drylands. Reseeding programmes are aimed at improving vegetation cover and biomass, and they conserve the soil to an extent not possible by grazing and land management alone. Indigenous drought-tolerant grasses notably African foxtail grass (*Cenchrus ciliaris*), bush rye grass (*Enteropogon macrostachyus*) and Maasai lovegrass (*Eragrostis superba*) have produced promising rehabilitation outcomes. Previous studies in African drylands have demonstrated the potential of such indigenous forage grasses in improving both vegetation cover (plant frequency and

K. Z. Mganga (✉)
Department of Agricultural Sciences, South Eastern Kenya University, Kitui, Kenya

Department of Forest Sciences, University of Helsinki, Helsinki, Finland
e-mail: kmganga@seku.ac.ke

D. M. Nyariki
Department of Agricultural Sciences, South Eastern Kenya University, Kitui, Kenya

School of Business and Economics, Murangá University of Technology, Murangá, Kenya

N. K. R. Musimba
Department of Agricultural Sciences, South Eastern Kenya University, Kitui, Kenya

A. W. Mwang'ombe
Department of Plant Science and Crop Protection, University of Nairobi, Nairobi, Kenya

Y. Bamutaze et al. (eds.), *Agriculture and Ecosystem Resilience in Sub Saharan Africa*, Climate Change Management,
https://doi.org/10.1007/978-3-030-12974-3_3

densities, basal cover) and soil hydrological properties (increased infiltration capacity, reduced runoff and sediment production) as indicators of rehabilitation success. Despite their comparative and widespread success, natural and anthropogenic challenges persist. This makes reseeding programmes a risky and often expensive venture, especially for the resource-poor pastoral communities in African drylands. Despite the risks, grass reseeding using indigenous pastures remains a viable sustainable land management option to combat degradation in African drylands. However, to ensure its continued success in the long term, multifaceted approaches and strategies that will integrate land and water management and seed systems suitable for African drylands need to be developed, strengthened and promoted.

Keywords Degradation · Reseeding · Desertification · Vegetation cover · Infiltration capacity · Runoff

1 Introduction

Land degradation is a major environmental problem of global significance. It is described as the persistent reduction in the bundle of goods and services provided to humans by the ecosystem under consideration (Verstraete et al. 2009). Degradation can also be defined as the loss of both biological and economic productivity, which includes deterioration of soil, vegetation and water. These processes are attributed to climatic variations, human activities or a combination of both factors (Mganga et al. 2018). Land degradation occurs in the forms of impoverishment and depletion of vegetative cover, loss of biophysical and economic productivity through exposure of the soil surface to wind and water erosion, salinisation and water logging leading to deterioration of physical, chemical and biological soil properties.

Global drylands, comprising hyperarid, arid and semiarid lands with very low (<0.65) ratio of long-term mean annual precipitation to potential evapotranspiration ratios, are considered the most threatened by land degradation. These xeric environments are estimated to occupy approximately 38% of the earth's surface (Verstraete et al. 2009). In these areas, land degradation, of which desert encroachment is only a small part, is widespread and therefore a very important subject of inquiry. Additionally, wetland enclosures in drylands are affected by desertification through deforestation and subsequent erosion of soils and loss of nutrients. Semiarid to weakly arid lands of Africa are particularly vulnerable as they have fragile soils, localised high population densities and generally a low-input form of agriculture (Mganga et al. 2015). Africa is particularly threatened because land degradation affects about 73% of the continent (Mganga et al. 2018; Gisladottir and Stocking 2005). For example, in Kenya, it has been estimated that approximately 30–40% of the arid and semiarid lands (ASALs) are quickly degrading and that another 2% has completely been lost through this process. High rates of soil loss up to 50 tonnes per hectare annually from degraded grazing lands in semiarid areas are common (Nyangito et al. 2009). According to Nyangito et al. (2009), grazing contributes about 34.5% of the total soil degradation.

Although some land degradation is attributable to natural causes (climate change, droughts and localised weather events), a substantial part can be attributed to human activities. Drylands are characterised by limited availability of arable land, limited and highly variable rainfall and scarcity of water resources. Demands placed on land and water resources by rapidly expanding populations, through agricultural intensification, urbanisation and industrialisation, have combined to intensively exploit these natural resources. Moreover, resource use strategies are largely exploitative leading to land degradation. The principal causes of degradation in the African drylands can generally be grouped into two (2) categories, namely:

1. Removal of vegetation through cutting and uprooting trees and shrubs, ploughing previously uncultivated and marginal land for annual cropping and overgrazing natural rangelands
2. Agricultural intensification using inappropriate cultural practices that degrade soil fertility and encourage erosion and overuse of irrigation that leads to poor drainage, water logging, rising water tables and salinisation

Overgrazing and deforestation remove the vital soil cover, exposing the soil to erosion. Overcultivation, which refers to the cropping of the land without replenishing the plant nutrients, exhausts the soil, destroying its structure and fertility. Reduction of land productive potential through degradation is not only linked to destructive human activities but also associated with human-induced global warming and climate change on a global scale. The destruction of forests that are important carbon sinks contributes to higher carbon dioxide (CO_2) concentration in the atmosphere, leading to global warming.

Grass reseeding has been used successfully as a means of rehabilitating degraded drylands in East Africa (Mganga et al. 2010, 2015; Opiyo et al. 2011; Nyangito et al. 2009), thus reversing the degradation process. Indigenous grass species are more successful in reseeding degraded drylands. As a rule of thumb, the best grasses for a reseeding programme are those that are native and found on ecological sites similar to those to be reseeded. This chapter discusses the potential of three important indigenous forage grasses in African drylands, namely, African foxtail grass (*Cenchrus ciliaris*), bush rye grass (*Enteropogon macrostachyus*) and Maasai lovegrass (*Eragrostis superba*), for rehabilitating arid and semiarid rangelands and improving soil hydrological properties, thus combating desertification and reducing soil erosion, respectively.

2 Reseeding: Common Indigenous Grasses for Reseeding

2.1 Reseeding

The primary purpose of dryland reseeding programmes is to improve existing ground cover and biomass to an extent or manner not possible by grazing management alone. This can be accomplished by oversowing into existing vegetation with a superior species, establishing a completely new pasture, with or without the aid of irrigation and reseeding a denuded land. Grass reseeding programmes are not new in the ASALs of Kenya. In the pre- and postcolonial eras of the 1950s and 1960s, respectively, a number of reseeding techniques were developed and introduced in Machakos, Taita-Taveta, Baringo and Kitui and Kajiado counties. These reseeding programmes were aimed at rehabilitating degraded drylands for improved pasture and livestock production. It is worth noting that reseeding programmes can sometimes be costly and risky, especially in arid and semiarid ecosystems. In many ASALs, the cost of rehabilitating a degraded pastureland may far exceed the potential returns from livestock production. Furthermore, rehabilitation techniques are often unreliable in environments where precipitation is unpredictable, especially in dryland climates characterised by large spatial and temporal variations in rainfall. Moisture conditions suitable for active growth are usually short-lived and unpredictable in many instances.

Traditional methods of reseeding degraded ASALs are expensive and often unsuccessful, due to the high rates of seed and seedling mortality and predation. As a rule of thumb, seeding should not be attempted in areas with less than 300 mm of average annual rainfall because they are apt to fail. When a seed stock is healthy, only two environmental factors will stop it from germinating and establishing in the semiarid drylands, namely, soil type and moisture. Apart from soils and rainfall, other factors including human interventions (burning and grazing) and individual species physiological and morphological differences affect germination and subsequent growth. Successful reseeding has however been achieved in low rainfall areas, e.g. Thar Desert in India, where the rainfall ranges between 100 and 500 mm annually, 90% of which is received between July and September (Chauhan 2003). This is only achievable with water management. Water harvesting techniques such as pitting, contour furrows and trenches are important for the creation of both microcatchments and macrocatchments. In semiarid drylands in Kenya, carve half-moon-shaped microcatchments in checkerboard fashion are constructed along slopes of land, situated in such a way as to capture and retain rainwater. Half-moons are used to rehabilitate degraded land for crop cultivation and fodder production; conserve water and fertile soil; improve soil fertility, with the addition of compost; expand agricultural land; and improve the infiltration and stock of water in the soil. Such techniques are used in reducing runoff, thus ensuring that the grass seeds get enough water for a prolonged period of time, thus improving their chances of germination and subsequent establishment.

Dryland reseeding requires soil disturbance. Minimal soil disturbance helps in replenishing deficient plant species by allowing seedling root penetration through provision of conditions suitable for germination, emergence and subsequent establishment (Mganga et al. 2015). Different methods are used to prepare denuded lands for reseeding. These methods include seedbed preparation and soil disturbance and involve some form of rainwater harvesting through creation of microcatchments. This ensures that all available moisture is utilised effectively to increase water penetration and slow runoff. Low-cost techniques for the rehabilitation of drylands are more sustainable. Soil disturbance by the use of an ox-drawn plough and hand hoes to create microcatchments to trap enough moisture for seed germination is among the most economical practices for resource-poor farmers in the drylands. Field preparation methods and techniques are defined by such factors as the size of area to be reclaimed, degree of degradation, soil types, rainfall, amount and type of invasive species, presence of wildlife and the financial and human resources available.

2.2 Indigenous Grasses for Reseeding

Over the years, indigenous grasses in African drylands have evolved adaptive mechanisms for survival in harsh arid and semiarid environments. This has made them the preferred choice in many dryland reseeding programmes in Africa (Opiyo et al. 2011; Mganga et al. 2010). They are widespread in Africa and occur mostly in savannas and woodland vegetation types (Fig. 1). Furthermore, perennial grasses are preferable to all other plants, except in ecoclimate zone VI, where rainfall is mostly too low to support perennials and where annual grasses are favoured. Perennial grasses have the ability to survive dry seasons and regenerate with each rain to produce fresh growth from the original rootstock. Moreover, perennial grasses have good self-seeding ability. Thus, with proper management, they can establish and spread quickly to give good cover. Although perennials may produce seed every season, they leave for a few to several years.

Generally, grasses selected for reseeding programmes in African drylands should be drought tolerant to survive and self-perpetuate; provide a good quantity of herbage of fair or good grazing value; and produce adequate amount of viable seed, which can be easily harvested and established (Opiyo et al. 2011; Mganga et al. 2010, 2015). In addition to these characteristics, tolerance to grazing and ability to establish fast during spells of favourable climatic conditions are very important traits in choosing grass species for reseeding (Jordan 1957). Previous reseeding programmes in arid and semiarid environments in Kenya (Nyangito et al. 2009; Mganga et al. 2010, 2015; Opiyo et al. 2011) have demonstrated rehabilitation success using African foxtail grass (*Cenchrus ciliaris*), horsetail grass (*Chloris roxburghiana*), star grass (*Cynodon dactylon*), bush rye grass (*Enteropogon macrostachyus*) and Maasai lovegrass (*Eragrostis superba*).

African Foxtail Grass (*Cenchrus ciliaris* L.)

Cenchrus ciliaris (Photo Plate 1) is native to tropical and subtropical Africa and one of the most drought-tolerant of perennial grasses. It is an extremely variable species, tufted (sometimes shortly rhizomatous) and persistent perennial which occurs in a wide variety of types, some of which have become reputed cultivars (Marshall et al. 2012). Numerous cultivars have been created in order to improve productivity and vigour in extreme conditions of drought, disease and frequent fires. The panicle is erect or nodding with a bur-like fascicle comprising a single spikelet or cluster

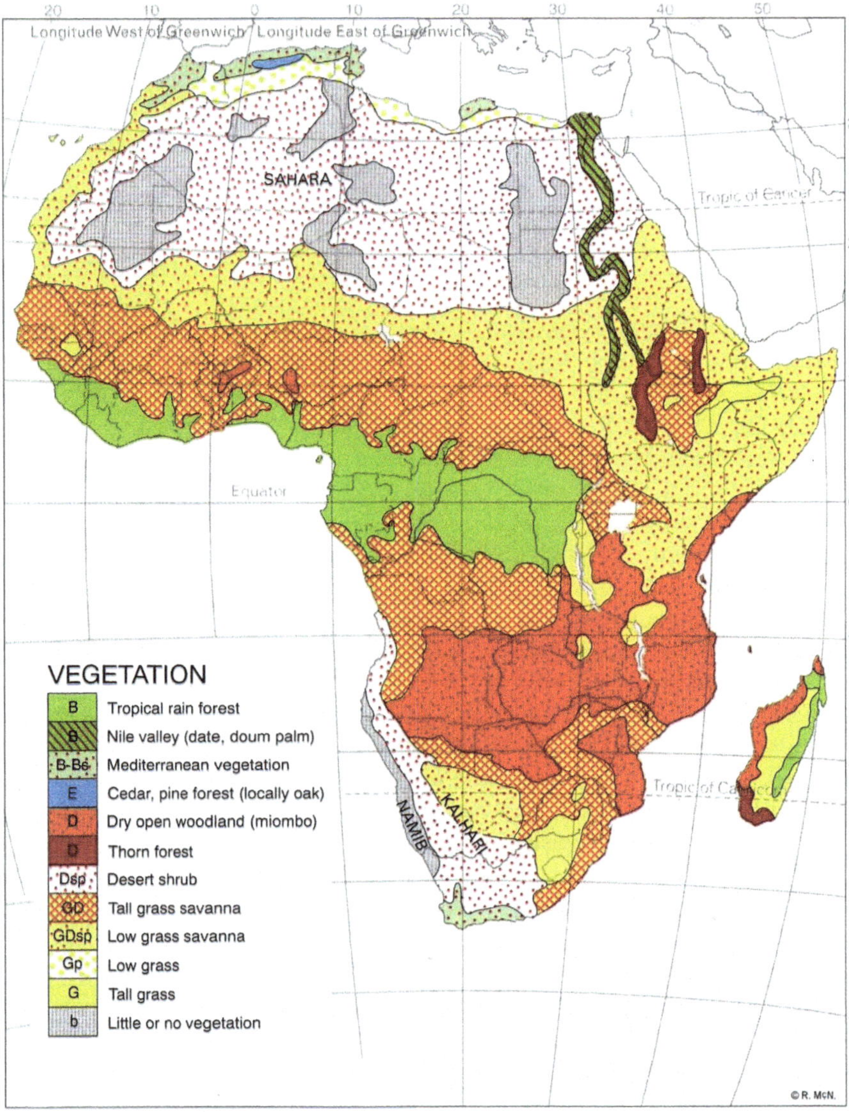

Fig. 1 Map of Africa showing different vegetation types. Indigenous grasses occurring primarily in dry open woodland and grassland savannas. (Source: Küchler and Zonneveld (1988))

spikelets, surrounded by an involucre of bristles of various lengths. Bristles are barbed and hairy, giving the fascicle an adhesive quality, thus dispersing profusely by clinging to animal fur and by wind as well (Marshall et al. 2012). *Cenchrus ciliaris* free-seeding nature can make it a major pasture component by simply spreading into relatively undisturbed native pasture from nucleus areas established by

Photo Plate 1 African foxtail grass (*Cenchrus ciliaris*)

traditional methods. *Cenchrus ciliaris* has a deep strong fibrous root system that exceeds 2 m. It is a particularly aggressive grass, by virtue of its extensive root system competing with associated species for water and nutrients. It also appears to be allelopathic (Mganga et al. 2015).

It occurs in the wild on sandy soils, but is also well adapted to deep, freely draining sandy loams, loams, clay loams and red earth soils. *Cenchrus ciliaris* has been recommended for reseeding areas receiving an annual rainfall of 350–900 mm. Whole seeds of this grass species have been shown to result in better grass stands than when hulled seeds are used (Opiyo et al. 2011). Seeds of *C. ciliaris* germinate better after predrying for 10 days at 40 °C than prechilling for the same period at 5 °C (Hussey and Bashaw 1996). Degraded pastures in the drylands are reseeded with *C. ciliaris* to enhance productivity, prolong grazing period and increase carrying capacity. Furthermore, it is a palatable and nutritious grass species and highly nutritious forage for livestock and recovers well from grazing (Marshall et al. 2012). It produces reasonable quality hay when cut in the early-flowering stage, yielding up to 2500 kg/ha per cut with a protein content of 6–10% of dry matter (Koech et al. 2016).

Bush Rye Grass (*Enteropogon macrostachyus* A. Rich. Munro ex Benth.)

Enteropogon macrostachyus (Photo Plate 2) is a native species to Africa. It is a widely distributed perennial grass species very common in arid and semiarid areas where it grows in bush, in forest edges and to a lesser extent in open grassland (Opiyo et al. 2011). It is a tufted perennial suitable for reseeding rock slopes or bushland. It has proved an excellent grass for reseeding African drylands under moderately dry conditions. This species has been tried with moderate success for reseeding denuded pastoral land in Kenya (Mganga et al. 2015) under annual rainfall of 550–800 mm. *Enteropogon macrostachyus* is a good grass for arid and semi-arid ecosystems because it is drought resistant.

This species has erected culms of 30–100 cm high. Leaf sheaths are without a keel. Surface of the sheath and outer margins are glabrous. Leaf blades are narrow and flat, approximately 10–60 cm long and 1.5–10 mm wide. Depending on the environmental conditions, *E. macrostachyus* may be leafy or stemmy. Although stemmy, it is drought resistant and provides useful grazing for grazing herbivores.

Photo Plate 2 Bush rye grass (*Enteropogon macrostachyus*)

Previous studies (e.g. Koech et al. 2016) have shown *E. macrostachyus* to have a crude protein and crude fibre percentages of 5.2% and 28%, respectively.

In tropical Africa, *E. macrostachyus* occurs naturally in open rangelands and rocky outcrops in semiarid environments at an altitude ranging between 300 and 1600 m above sea level. The species occurs in areas receiving around 600 mm of rainfall per annum. It is a very good seeder, and seed can be collected rapidly by cutting the seedheads or stripping the heads by hand. It lends itself easily to mechanical harvesting. The seeds germinate readily and grow vigorously. It is palatable; thus, its reintroduction in degenerated swards is of obvious value to grazers (Opiyo et al. 2011).

Maasai Lovegrass (*Eragrostis superba* Peyr.)

The grass species derives its scientific name from both Greek and Latin words. *Eragrostis* comes from the Greek words *Eros* meaning love, possibly referring to the heart-shaped spikelets, and *Agrostis* meaning grass. The species name *superba* is the Latin for splendid, probably descriptive of the spikelets. *Eragrostis superba* (Photo Plate 3) occurs naturally in South Africa and northwards throughout East Africa to Sudan. It is widespread in the semiarid areas of East Africa and is very common in various vegetation types mainly grassland and savannah types and rocky and sandy areas throughout its ecological range.

Eragrostis superba is a tufted perennial 20–120 cm high with a high shoot/root ratio (Opiyo et al. 2011) which is a disadvantage during drought periods but is advantaged by having deep root system which goes as far as 2.2 m with 73% of the roots limited to the upper 0.4 m from the soil surface, which enable the grass to make full use of light showers of rain (Opiyo et al. 2011; Mganga et al. 2010, 2015). It is a moderate tiller with a poor regrowth ability compared to *C. ciliaris* and *C. roxburghiana*. *Eragrostis superba* grows in disturbed places and thus has been used successfully for reseeding denuded and pastoral drylands and soil erosion control in African drylands (Nyangito et al. 2009; Mganga et al. 2010, 2015; Opiyo et al. 2011) because it has excellent establishment characteristics. It has been successful in the semiarid areas of East Africa particularly in ecoclimatic zone VI where mean annual rainfall ranges between 500 and 900 mm. *Eragrostis superba* occurs at an altitude of 0–2000 m above sea level. *Eragrostis superba*, along with *C. ciliaris*, has been the basis of the seed mixtures used for large-scale reseeding programmes in ASALs in Kenya.

Eragrostis superba is a quick growing species and shows green vegetative growth almost throughout the year and has good drought tolerance. It is very palatable when young. Depending on the soil types and other environmental conditions, chemical and digestibility analyses indicate that the grass has between 5% and 12% crude protein (%CP) in the dry matter at an early-flowering stage with 25–35% crude fibre (%CF) (Koech et al. 2016). Wasonga et al. (2003) demonstrated that pastoral communities in Kenya have identified *E. superba* as one of the grass species suitable for fattening and improving the condition of their livestock herd.

Photo Plate 3 Maasai lovegrass (*Eragrostis superba*)

The grass species grows very easily from seeds which are in the form of a small, plump grain, which are particularly susceptible to insect damage. The grass seeds can be collected easily from open grassland or at roadsides by stripping the ripe panicles. Mature spikelets, each with numerous florets, detach easily with the caryopses enclosed. The grass can be established in gravely, sandy, loamy or clay soils. However, it does best in sandy soils but occurs also on clay loams and clays. The species has a high tolerance to salinity and alkalinity and can also be found on termite mounds common in the African drylands.

3 Ecological Indicator I: Vegetation Characteristics

Vegetation attributes, e.g. biomass yields, plant frequency, plant density, basal cover and seed production, have been used in past ecological study to assess the success of reseeding programmes in dryland environments. Table 1 below highlights some of the vegetation attributes of indigenous grass species used for rehabilitating degraded dryland environments in Africa.

Table 1 Vegetation attributes of some indigenous grass species used for reseeding programmes

Grass species	Biomass yield (kg/ha)	Plant frequency (%)	Plant density (plants m^{-2})	Basal cover (%)	Seed production (kg/ha)
C. ciliaris	1026.6 ± 55a	44.4 ± 19.63a	7 ± 5.23a	30 ± 26.44a	145 ± 13a
E. macrostachyus	744.0 ± 28b	72.3 ± 25.43b	36 ± 2.1b	54 ± 19.29b	56 ± 25b
E. superba	896.5 ± 44c	38.7 ± 9.82a	5 ± 6.25a	23 ± 15.44a	119 ± 17c

Source: Modified from Mganga et al. (2010)
Column means followed by different letters are significantly different at $P < 0.05$

Differences in plant frequency, plant densities and basal cover among the grass species are explained by germination success and seedling emergence rates and subsequent establishment of the seedlings. Nature, morphology and size of the seeds determine germination percentage and thus percentage frequency, plant density and basal cover (Mganga et al. 2010, 2015). Seed size has an effect on seedling emergence. Vegetation attributes highlighted in Table 1 above clearly demonstrate that *E. macrostachyus* has the best rehabilitation results among the three species. Higher frequency of *E. macrostachyus* is attributed to its bigger seed size compared to *C. ciliaris* and *E. superba*. Additionally, *E. macrostachyus* dormancy mechanism involves only the integument and thus explains its rapid germination relative to the other two grass species. Seed morphology of *C. ciliaris*, characterised by the hairy bristle coat, aides germination by maintaining a high humidity within the fascicle and thereby helps reduce water loss from the caryopsis, thus enhancing germination (Sharif-Zadeh and Murdoch 2001) as compared to those of *E. superba*. In addition, these fascicles are known to contain more than one caryopsis (Daehler and Georgen 2005).

4 Ecological Indicator II: Soil Hydrological Properties

Soil hydrological measurements, e.g. infiltration capacity and runoff, have been used in previous ecological studies (Nyangito et al. 2009; Mganga et al. 2010) to determine the effect of increased vegetation cover as an indicator of rehabilitation success, in reducing soil erosion. Tables 2 and 3 show the effect stubble heights of different indigenous grass species used to rehabilitate degraded drylands on soil water infiltration capacity and runoff.

All the grass species show a general increase in the infiltration capacity with an increase in stubble height. *C. ciliaris* maintained the highest infiltration capacity of 3.02 cm and 3.72 cm at stubble heights of 20 cm and 40 cm, respectively, compared to the other grasses. *Enteropogon macrostachyus* had 2.79 cm and 3.48 cm, and *Eragrostis superba* recorded 2.11 cm and 2.99 cm within the same stubble height range, and both were ranked second and third, respectively.

Table 2 Infiltration capacity (cm) in relation to grass stubble heights (cm)

	Infiltration capacity (cm)		
Stubble height (cm)	*Cenchrus ciliaris*	*Enteropogon macrostachyus*	*Eragrostis superba*
0	2.07a	2.07a	2.07a
20	3.02b	2.79c	2.11d
40	3.72c	3.48d	2.99b

Source: Modified from Mganga et al. (2010)
Column and row mean followed by different letters are significantly different at $P < 0.05$

Table 3 Runoff (cm) in relation to grass stubble heights (cm)

	Runoff (cm)		
Stubble height (cm)	*Cenchrus ciliaris*	*Enteropogon macrostachyus*	*Eragrostis superb*
0	1.88a	1.88a	1.88a
20	0.93b	1.16c	1.84a
40	0.23c	0.47d	0.96b

Source: Modified from Mganga et al. (2010)
Column and row mean followed by different letters are significantly different at $P < 0.05$

Indigenous grasses with higher and lower infiltration capacities gave lower and higher runoffs, respectively. *Cenchrus ciliaris* yielded the lowest runoff of 0.93 and 0.23 cm at 20 and 40 cm grass stubble heights, respectively. *Enteropogon macrostachyus* yielded 1.16 and 0.47 cm, while *E. superba* recorded runoffs of 1.84 and 0.96 cm, respectively, at the same range of grass stubble heights.

Previous studies have demonstrated that perennial vegetation can increase infiltration (Seobi et al. 2005). In their study in a typical semiarid environment in Kenya, Nyangito et al. 2009 also observed higher infiltration capacity in rehabilitated sites dominated by *E. macrostachyus* compared to those dominated by *E. superba*. Differences in infiltration capacity between the different indigenous grasses are attributed to their growth and morphological characteristics. *Cenchrus ciliaris* is densely leafed with branching culms arranged in a funnel shape. It is also relatively broad leafed. These characteristic presents a greater surface area for collecting water and rain drops that is directed more into its rhizosphere. *Enteropogon macrostachyus*, though narrow leafed, tends to be leafy than stemmy especially at its base and, therefore, closely compares with *C. ciliaris* in trapping and directing rainwater. In contrast, *E. superba* has a higher stem-to-leaf ratio and thus is less effective in concentrating rainwater into their rhizosphere.

Consequently, these outcomes strongly suggest a general decline in sediment production, and index of sheet erosion, with an increase in grass stubble height. This is attributed to the reduction of the force of water drops hitting and destabilising the soil surface. Generally, vegetation cover intercepts rainfall kinetic energy and thereby decreases the mobilisation of soil particles. Taller grass stubble heights trap more rainwater drops and funnel it down its crown, thus concentrating more water around the rhizosphere compared to shorter stubble heights. Larger leaf blades also reduce the force of the water drops directly hitting the ground. This improves infiltration capacity and reduces runoff, thus less sediment production (soil erosion).

5 Challenges of Rehabilitating Degraded Drylands Using Reseeding

Rainfall in the African drylands is usually low, erratic and unpredictable in both space and time. High variability in rainfall amounts and distribution is a common characteristic of semiarid lands often leading to soil moisture deficits (Ekaya et al. 2001). Low amounts of rainfall and drought period characteristic of the African drylands are the main factors which contribute immensely to rangeland rehabilitation failures. Low amounts of rainfall reflected by the soil moisture deficits hinder seed germination. Often reseeded drylands experience poor establishment. When the seed bank is healthy, the main environmental factors that will stop seeds from germinating and establishing in the drylands are soil type and moisture.

Conversely, during heavy rains, drylands are also very susceptible to flash floods leading to massive soil erosion and consequently destruction of reseeded areas. High-intensity rainfall that lasts for a very short period of time often causes flash floods in dryland environments in Africa. Coupled with the soils characterised by high levels of salinity, poor drainage, soil erosion, soil compaction, soil crusting and low soil fertility (Nyangito et al. 2009; Opiyo et al. 2011) expose the sown grass seeds to agents of erosion notably water and wind. Floodwaters often transport the sown grass seeds as it flows along the gradient. Furthermore, heavy storms also destroy the microcatchments created to trap rainwater for the grass seedlings. Soil crusting common in dry environments in Africa further compounds the problem and hinders grass seedling emergence. This combination of climatic and edaphic factors leads to poor rates of germination and thus poor rates of establishment.

Livestock keeping is the main source of livelihood among pastoral communities inhabiting the African drylands. Grazing animals are mainly kept under free-range/herding system in defined household grazing areas (Nyangito et al. 2009). This form of grazing system also contributes immensely to rehabilitation failures. Free-grazing animals often cause destruction to young grass seedlings often by trampling on newly established grass stands hindering their subsequent development to maturity and seed setting. Destruction of seedlings before seed setting leads to continual deprivation of the seed bank in the soil which in the long run leads to bare patches.

Fencing is the only way of excluding domestic and wild herbivores from reseeded areas. However, the use of long-lasting fences, e.g. chain-link and barbed wire, is often expensive for the pastoral and agropastoral communities. Consequently, they are forced to settle for cheap and local materials notably branches and poles of *Acacia* and *Commiphora* species to keep free-roaming herbivores. Such materials only provide a short-term solution, since they are easily destroyed by livestock. Continued exposure of the fence to rainfall further weakens and destroys the fence. Moreover, notorious herders easily remove the fences to allow their animals graze the newly established seedlings and later replace it. To ensure successful establishment, newly established grass stands should be protected from grazing animals for at least two growing seasons. However, this is hardly possible in pastoral areas.

Seedbed preparation and sowing for reseeding programmes in dryland Africa are mainly done prior to the onset of the rainy season. The grass seeds are sown at a shallow depth and covered with minimal amounts of soil. This guarantees that the sown seeds get enough moisture for germination and consequent establishment. Dry planting ensures that the first showers of rain find the sown seeds in the ground. However, this period also coincides with scarcity of food for rodents, birds and insects. Rats, squirrels and other small rodents and insects often feed on the sown seeds. Termites also transport the seeds for long distances and store them in their food stores. Damage by termites is a general concern in African drylands (Mugerwa 2015) as they also destroy vegetation, crops and farm structures. Additionally, *Quelea quelea* birds common in African drylands often invade the rehabilitation plots in hundreds and feed on the sown grass seeds, especially those of *E. superba* which is mostly preferred by pastoralists because of its role in improving milk production (Wasonga et al. 2003). These factors contribute to poor rates of initial germination in some patches.

Inadequate supply of good quality and quantities of indigenous grass seeds, e.g. *E. macrostachyus*, *E. superba*, *C. ciliaris* and *C. roxburghiana*, in the formal markets limits their accessibility for use in reseeding programmes. Land users' needs for herbage seeds are rarely met by formal seed production and marketing systems in sub-Saharan Africa. This is partly attributed to degradation of natural vegetation where grasses are steadily being replaced by more woody vegetation type in the rangelands. Consequently, there is a shortage of supply of grass seeds commonly harvested in open-grazing areas. Proper seed technology is therefore necessary at harvesting, processing and storage, especially where dryland communities are expected to undertake pasture seed multiplication to meet the current shortage.

Poor postharvest storage methods, skills and facilities to store harvested seeds and use of poor quality seeds also contribute to poor establishment in reseeded drylands. Due to lack of infrastructure and inadequate knowledge and skills of handling grass seeds, pastoralists inhabiting drylands store seeds harvested in the wild in gunny bags and place them in grass-thatched granaries which are sometimes not well done or poorly maintained. Quality of indigenous grass seeds is highly variable and poor in tropical Africa as a result of poor harvesting and storage technology. Dry seeds, especially those of rangeland pasture grasses, are known to be highly hygroscopic, and exposure of dry seeds to moisture has been reported to worsen seed dormancy and often leads to fungal infection (Mganga et al. 2015). However, the ability to withstand moisture conditions during storage varies between species. During heavy storms, traditional granaries allow some rainwater to penetrate and thus contaminate the stored grass seeds. This leads to spoilage. Additionally, some of the grass seeds used to reseed denuded patches in the dryland are dormant and often not viable. The seeds used can either be too old or fresh and immature yet to break dormancy.

Competition for limited resources in drylands soils between weeds and reseeded grasses especially during the early stages of development can be very severe. Soils in the semiarid environments in Africa may contain 1.5 million seeds per hectare on a landscape scale (Witkowski and Garner 2000). This far exceeds the recommended sowing rates of indigenous pasture species usually at 30–500 viable seeds/m^2. Some of the common weeds found in established reseeded drylands in semiarid Kenya include Sodom apple (*Solanum incanum*), milkweed (*Lactuca capensis*), Kitui morning glory (*Ipomoea kituensis*), crab grass (*Digitaria scalarum*), weeping lovegrass (*Eragrostis curvula*), Jimson weed (*Datura stramonium*), coat buttons (*Tridax procumbens*), *Barleria taitensis*, wandering jew (*Commelina bengalensis*) and natal grass (*Rhynchelytrum repens*). *Ipomoea kituensis* is a very common and notorious weed in reseeded pastures in semiarid lands in Kenya. It spreads very fast and thus colonises a wide area within a very short period, making it difficult to eradicate. Its creeping nature engulfs newly established grass seedlings suppressing their growth and development by depriving them of sunlight necessary for normal photosynthetic function. Furthermore, due to its aggressive nature, *I. kituensis* often outcompetes the grass seedlings for soil nutrients and water.

6 Conclusions

Land degradation is an environmental problem of great importance in the African drylands. Both human and climatic factors contribute significantly to accelerated degradation in the dryland environments. Grass reseeding, using indigenous perennial grasses, is a viable option to combat degradation and halt the desertification process in African drylands. Previous studies have shown that reseeded areas improve vegetation attributes such as cover, biodiversity and density. These consequently enhance soil hydrological properties notably increased water infiltration and decreased runoff, thereby significantly reducing soil loss. Despite its widespread success, grass reseeding programmes in Africa continue to face multifaceted challenges that hinder their full potential in rehabilitating degraded drylands. This necessitates multi-stakeholder collaborations and interaction between land users, researchers, local and national governments and other non-governmental organisations and institutions to develop and introduce multifaceted sustainable land management technologies to combat land degradation in dryland environments. Such initiatives in African drylands, a hotspot for successful land restoration projects due to innovations in technology and social engineering, will go a long way to supporting sustainable rural livelihoods and promoting environmental management.

References

Chauhan SS (2003) Desertification control and management of land degradation in the Thar desert of India. Environmentalist 23:219–227

Daehler C, Georgen EM (2005) Experimental restoration of an indigenous Hawaiian grassland after invasion of buffel grass (*Cenchrus ciliaris*). Restor Ecol 13:380–389

Ekaya WN, Kinyamario JI, Karue CN (2001) Abiotic and herbaceous vegetational characteristics of arid rangeland in Kenya. Afr J Range Forage Sci 18:125–129

Gisladottir G, Stocking M (2005) Land degradation control and its global environmental benefits. Land Degrad Dev 16(2):99–112
Hussey AM, Bashaw EC (1996) Performance of buffel grass germplasm with improved winter survival. Agron J 88:944–946
Jordan SM (1957) Reclamation and pasture management in the semiarid areas of Kitui district, Kenya. East Afr Agric For J 25:18–22
Koech OK, Kinuthia RN, Karuku GN, Mureithi SM, Wanjogu R (2016) Field curing methods and storage duration affect the quality of hay from six rangeland grass species in Kenya. Ecol Process 5(3). https://doi.org/10.1186/s13717-016-0048-2
Küchler AW, Zonneveld IS (1988) Handbook of vegetation science. Kluwer Academic Publishers, Dordrecht
Marshall VM, Lewis MM, Ostendorf B (2012) Buffel grass (*Cenchrus ciliaris*) as an invader and threat to biodiversity in arid environments: a review. J Arid Environ 78:1–12
Mganga KZ, Musimba NKR, Nyangito MM, Nyariki DM, Mwang'ombe AW (2010) Improving hydrological properties of degraded soils in semi-arid Kenya. Afr J Plant Sci Technol 4:138–144
Mganga KZ, Musimba NKR, Nyariki DM, Nyangito MM, Mwangombe AW (2015) The choice of grass species to combat desertification in semi-arid Kenyan rangelands is greatly influenced by their forage value for livestock. Grass Forage Sci 70:161–167
Mganga KZ, Nyariki DM, Musimba NKR, Amwata DA (2018) Determinants and rates of land degradation: application of stationary time-series model to data from a semi-arid environment in Kenya. J Arid Land 10(1):1–11
Mugerwa S (2015) Magnitude of the termite problem and its potential anthropogenic causes in Nakasongola district of Uganda. Grassl Sci 61(2):75–82
Nyangito MM, Musimba NKR, Nyariki DM (2009) Hydrological properties of grazed perennial swards in semi-arid southeastern Kenya. Afr J Environ Sci Technol 3:26–33
Opiyo FEO, Ekaya WN, Nyariki DM, Mureithi SM (2011) Seedbed preparation influence on morphometric characteristics of perennial grasses of a semi-arid rangeland in Kenya. African J Plant Sci 5:460–468
Seobi T, Anderson SH, Udawatta RP, Gantzer CJ (2005) Influence of grass and agroforestry buffer strips on soil hydraulic properties for an Albaqualf. Soil Sci Soc Am J 69:893–901
Sharif-Zadeh F, Murdoch AJ (2001) The effects of temperature and moisture on after-ripening Cenchrus ciliaris seeds. J Arid Environ 49:823–831
Verstraete MM, Scholes RJ, Smith MS (2009) Climate and desertification: looking at an old problem through new lenses. Front Ecol Environ 7(8):421–428
Wasonga VO, Ngugi RK, Kitalyi A (2003) Traditional range condition and trend assessment: lessons from Pokot and Il Chamus pastoralists of Kenya. Anthropologist 5(2):79–88
Witkowski ETF, Garner RD (2000) Spatial distribution of soil seed banks of three African savanna woody species at two contrasting sites. Plant Ecol 149(1):91–106

Adoption of Recommended Maize Production Practices and Productivity Among Farmers in Morogoro District, Tanzania

Consolatha J. Gahanga and Justin K. Urassa

Abstract The chapter is based on a study that aimed at assessing the adoption of recommended maize production practices and productivity between farmers who were members of Farmer Field Schools (FFSs) and those who were not. The study was conducted in Morogoro District, Tanzania. Specifically, the study aimed at identifying FFS recommended maize production technologies; determining socio-economic factors influencing farmers' adoption of the technologies; and comparing maize productivity and income between households involved in FFS and those that are not. Lastly, it determined the contribution of maize sales to the household incomes of the two groups. The study adopted a cross-sectional research design whereby data was collected from 166 individuals through household surveys, focus group discussions and key informant interviews. Quantitative data was analysed using the Statistical Package for Social Science (SPSS), whereby descriptive statistics such as frequencies and percentages were determined. A logistic regression model was used to determine the association of socio-economic factors and the adoption of FFS technologies. Study findings show that age, education, household income and farm size significantly influenced the adoption of recommended FFS practices. Results also show that farmers who participated in the FFS had a higher maize productivity and maize sales were the main source of income in the study area. Thus, extension agents need to do more to encourage more farmers to join FFS so as to get access to improved maize technologies which will enable them to raise their maize productivity and ultimately their income and general living standards.

Keywords Farmer field schools · Adoption · Maize · Productivity · Tanzania

---

C. J. Gahanga
Ilala Municipal Council, Dar es Salaam, Tanzania

J. K. Urassa (✉)
Department of Policy Planning and Management, College of Social Sciences and Humanities, Sokoine University of Agriculture, Morogoro, Tanzania
e-mail: urassa@sua.ac.tz

Y. Bamutaze et al. (eds.), *Agriculture and Ecosystem Resilience in Sub Saharan Africa*, Climate Change Management,
https://doi.org/10.1007/978-3-030-12974-3_4

1 Background

Over the past decades, the level of produced food has decreased dramatically in Africa, particularly in sub-Saharan Africa (SSA), resulting in general deterioration of the population's living standard (Ojo and Jibowa 2008; FAO 2009). Generally, the decline has resulted in increasing rural poverty, rising food prices, widespread famines and increasing food imports. According to Kydd et al. (2004), as cited by Urassa (2015), the rate of growth in agricultural production in SSA over the last 30 years has been lower than those of other regions. The report also indicates that in SSA and other parts of the developing world, farmers have relatively low rates of adoption of modern crop technologies associated with increased crop yields, hence poor living standards (Peterman et al. 2010).

In Africa, some of the factors hindering agricultural development include, inter alia, inadequate investment in agriculture, limited access to credit by smallholder farmers, high cost and unavailability of inputs, inadequate use of modern technologies, inefficient agricultural input markets and the absence of a conducive policy environment. In addition, the use of improved agricultural technologies in Africa is very low and has remained largely static over the last 25 years, especially among smallholder food crop and livestock production systems (Tsusaka and Otsuka 2013).

Agriculture in Tanzania plays an important role in the overall economy as it accounts for the larger portion of rural employment, food security and provision of industrial raw materials for other sectors in the country (Ministry of Agriculture, Food Security and Cooperatives [MAFC] 2008). For example, in the year 2011, the agricultural sector contributed approximately 51% of foreign exchange, 75% of total employment and 27.1% to the country's gross domestic product (GDP) (URT 2011). However, Tanzania's agriculture, like that of other SSA countries, is still characterized by a number of challenges including ineffective extension services, inadequate use of improved seeds and fertilizers and unreliable rainfall, which usually result in low crop productivity. According to FAO (2014), crop production in Tanzania is mainly rain-fed and dominated by smallholder farmers, cultivating small farms ranging from 0.9 to 3.0 hectares. The main staple foods cultivated in Tanzania include maize, rice, banana and cassava. In addition, close to three quarters of the crop area is cultivated by hand hoe; the remainder is cultivated by ox plough (20%) and tractor (10%). Further to the above, agricultural labour and land productivity are decreasing mainly due to the use of poor technology and heavy dependence on unreliable and irregular weather conditions (FAO 2014). According to literature (Keenja 2001; Liberio 2012), about 93% of Tanzania's smallholder farmers cultivate less than two hectares. Additionally, use of chemical fertilizers and improved seeds is very low. For example, the 2010/2011 National Panel Survey (NPS) as cited by the World Bank (2012) showed that only 16.5% and 16.8% of rural households use chemical fertilizers and improved seeds, respectively.

The importance of agricultural extension services in raising agricultural productivity of smallholder farmers cannot be overstated; evidence from literature Okoth

(2013) shows that access to these services is paramount for increased crop yields. Extension services and the promotion of input use, for example, use of chemical fertilizers, were able to raise smallholder maize production in northern Tanzania (SAA 2007 cited by Urassa 2015). Generally, extension services aim at transferring specific knowledge to producers, such as the transfer of technology, improvement of management practices or transfer of knowledge and capacities (Cerdán-Infantes et al. 2008). Several extension approaches, such as the Transfer of Technology (TOT), Training and Visit (T&V), Frontal Approach (FA), Commodity Extension Approach (CEA), Integrated Agricultural Development (IAD) and Improvement and Transformation (IT), have been used in the past to disseminate adoption of recommended crop production practices with the ultimate aim of improving agricultural productivity. However, despite the varying approaches, literature shows that agricultural development has not produced the intended results (URT 2004).

As pointed above, use of ineffective extension approaches and low adoption of improved recommended technologies by Tanzanian smallholder farmers have been among the reasons for low agricultural productivity (Michelle 2005). Apart from the literature showing many factors associated with low agricultural productivity, ineffective extension approaches could nonetheless be the major problem in Tanzania. In order to address the shortfalls, the government through the MAFC adopted the farmer field school (FFS) extension approach. The FFS approach was deemed more participatory than the other extension approaches like the Training and Visit (T&V). The FFS approach is being promoted country wide through the ASDP (Agricultural Sector Development Programme) and is being scaled out in all districts. By using the FFS approach, farmers could be empowered, and hence, extension services could become more demand-driven because farmers are enabled to establish their problems and identify needs in agricultural production (Ngeno 2003).

The FFS approach has been implemented and adopted worldwide (Braun et al. 2006 as cited by Larsen and Lilleør 2014). Since its emergence, the FFS approach has become an innovative, participatory and interactive model for educating farmers in more than 78 countries throughout Asia and extended to several countries in Africa and Latin America (Bunyata et al. 2011). The FFS approach uses experiential learning and a group approach to facilitate farmers in making decisions, solving problems and learning new techniques (Duveskog and Frii-Hansen 2013; Davis et al. 2010). FFS has brought a shift from focusing on a single constraint of one crop (IPM for rice-based systems) to emphasizing multiple aspects of crop production and management, such as cropping systems, non-crop/forest (livestock production, etc.) and natural resource management (soil fertility, water conservation) (van den Berg and Jiggins 2007). In SSA, the approach started to be used in the mid-1990s on rice irrigation schemes and other crops in Burkina Faso, Ghana, South Africa, Mali, Nigeria and East Africa, Tanzania included (FAO Undated). According to Nathaniels (2005), the FFS approach has led to increased agricultural productivity, hence improvement of people's livelihoods.

In Tanzania, the FFS approach was adopted in 1998. Its practice started in the Southern Highlands, namely, Iringa, Mbeya, Rukwa and Ruvuma under the Highlands Extension and Rural Financial Services Project (SHERFS) financed by the International Fund for Agricultural Development (IFAD). The main mission of the project was to improve farmers' inclusion in defining and solving their farming problems. In addition to its low cost of implementation (Khisa and Heinemann 2006), other benefits include directly influencing the use of agricultural innovations and increased household farm productivity, and eventually livelihoods were targeted. Nevertheless, the ultimate aim of FFS was to build farmers' capacity in analysing their production systems, identifying their main constraints, testing possible solutions and eventually identifying and adopting the practices that are most suitable to their farming system (Muhammad and Muhamad 2012). Once the competence is acquired, it helps to develop favourable attitudes towards improved practices and thereby motivates an individual to take certain action in accepting an innovation.

In Morogoro Region the FFS was adopted in 1999 in Mvomero District; the pilot study involved Mkindo village and later expanded to Kilosa, Kilombero, Ifakara and Ulanga districts. There are various FFS classes under operation dealing with various improved practices in crops and livestock with the aim of increasing productivity at a minimum cost. In 2001, FFS groups were established and trained on recommended practices in maize productivity in Morogoro District. The introduced FFS practices were expected to promote the adoption of recommended maize production and productivity in rural farming households in Tanzania and in particular Morogoro District. Over the years, farmers in the district have been using different maize growing systems which are characterized by traditional farming practices that lead to low yields.

Despite efforts made by the government of Tanzania to promote extension services to facilitate adoption of agricultural technologies among the farmers, the actual adoption rate of the recommended technologies such as improved varieties, fertilizers, weed control and pest and diseases control technologies is still quite limited. As a result maize productivity among the farmers has been decreasing. Available data shows that the average maize yield per hectare in Tanzania has declined from over 3000 kg (i.e. 3 metric tons (MT)) in 2001 to 1500 kg (i.e. 1.5 MT) in 2010 (World Bank 2012). This implies that if special attention is not paid to reverse the situation, the country may face severe food insecurity and negative outcomes from rural poverty alleviation by the government.

Morogoro is one of Tanzania's main maize-producing regions. However, the region has also experienced the problem of decreasing maize productivity. Empirical findings show that yields are dramatically low (i.e. less than 1 ton/ha), and the use of recommended technologies such as fertilizers is still very limited (Agriculture Council of Tanzania 2010). Maize is one of the most important food crops in Tanzania; it accounts for about 75% of the total cereal consumption, making it one of the strategic crops for food security in the country (Msuya et al. 2008). The crop provides about 60% of dietary calories to the Tanzanian population (Kaliba et al. 1998). The crop is grown in almost every region, and Tanzania has the potential to double productivity through adoption of recommended technologies. Therefore, the study on which the chapter is based was conducted to determine the contribution of FFS practice to maize production practices and productivity in Morogoro District.

Specifically, the chapter identifies recommended maize production practices by farmer field schools (FFSs), determines socio-economic factors influencing farmers' adoption of the FFS recommended technologies in Morogoro District and compares productivity between households involved in the FFS and those not. Lastly, it assesses the contribution of maize sales to households' income in Morogoro District based on participation in FFS. The chapter also tests two hypotheses: first "there is no significant difference in maize productivity between household involved in FFS and those that are not" and second "maize productivity does not have a significant impact on households' income in Morogoro District". Generally, the chapter provides in-depth information to all stakeholders, namely, farmers, researchers, extension staff and policy makers, on the level of adoption of recommended maize production practices and factors influencing their adoption. Lastly, the chapter presents recommendations which if put into practice could lead to increases in farming households' maize productivity, food security and income earnings in Morogoro District and other areas with similar characteristics.

1.1 Conceptual Framework

The study's conceptual framework (Fig. 1) describes the relationship between the FFS approach and the adoption of recommended maize production practices/technologies and productivity among farmers in the study area. The assumptions are farmers' adoption of the recommended maize production practices and technologies is influenced by their participation in FFS and other factors, such as institutional factors, i.e. policies, extension services, financial and market access and access to information related to agricultural production. Several researchers argue that the access to these services is a precursor for increased crop yields (Kaliba et al. 2000; World Bank 2007; Urassa 2010; Mattee et al. 2015). Generally, extension services aim at transferring specific knowledge to producers, such as technology, and improvement of management practices or knowledge and capacities (Mattee et al. 2015). To justify the importance of extension in raising agricultural productivity, SAA (2007) as cited by Urassa (2015) and the World Bank (2007) argues that reliable extension services and promotion of input use led to a rise in smallholder's maize yields of between 4.5 tons/ha and 5.1 tons/ha, compared to the national average yield of around 1.3 tons/ha, due to appropriate use of chemical fertilizers in the 2000 project. In addition, according to Kaliba et al. (2000) and Urassa (2010), availability of extension services was one of the most important factors that influenced the adoption of improved maize seeds and use of inorganic fertilizer in maize production. According to Ayinde et al. (2010), access to agricultural extension services in Nigeria at a level of one or two visits per agricultural year led to rise in farmer's crop productivity by 15% on average, other factors remaining constant.

Other factors included in the conceptual framework are socio-economic factors such as age, education levels, household head's sex, household size, number of farm practices adopted and income, which are assumed to influence a household's crop

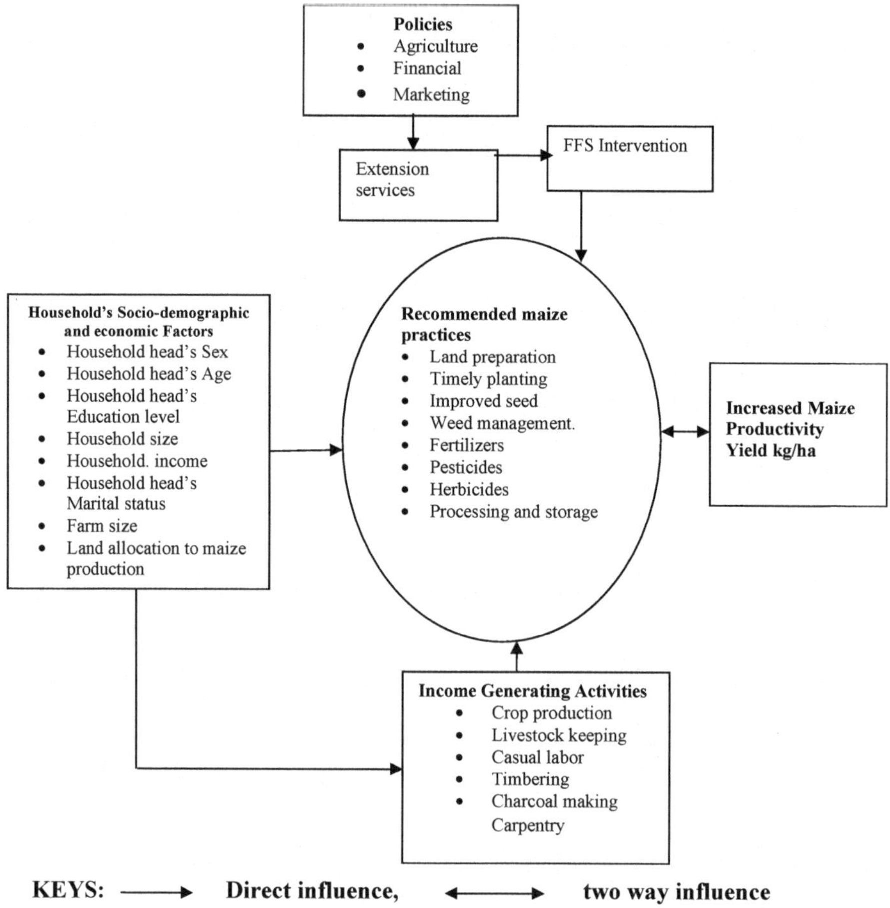

Fig. 1 Conceptual framework in adoption of technologies using FFS approach. Keys: ⟶ direct influence, ⟷ two way influence

output and yield. Education levels of household heads could also be very important in making decisions related to production of general commodities and agricultural products. For example, Salasya et al. (2007) argue that households with more education or other forms of human capital stand a better chance of accessing nonfarm income and/or credit and could thus be more able to purchase inputs. Salasya et al. (2007) further point out that such households may also be more aware of the benefits of modern technologies and are more efficient in their farming practices. The influence of age on crop production has been reported to influence adoption of new agricultural technologies (Nanai 1993; Adesina and Baidu-Forson 1995; Akadugu et al. 2012). Younger household heads are reported to be more flexible and eager to practise new innovations than older farmers.

The influence of a household head's sex on agricultural productivity of households is well documented. For example, Murage et al. (2012) point out that female-headed households in western Kenya used significantly less labour and draft power in farming. Murage et al. (2012) also argue that, in instances where hired labour is costly to monitor, households with a greater endowment of labour are better placed to farm their land more intensively and to conduct critical operations at the right time compared to households dependent on hired labour or with labour shortage. Generally, it is expected that large-sized rural households would be more able to supply the labour required for their crop production, basically due to abundance of their own labour. On the other hand, the World Bank (2007) noted that inadequate transport infrastructure and services in rural areas were associated with reduced crop production in sub-Saharan Africa.

1.2 Conceptualization of Key Concepts

Adoption

The definitions of adoption vary widely across studies. Due to that, the definition of adoption needs to be agreed upon. Rogers (1962) as cited by Feder et al. (1985) defines the adoption process as the mental process an individual passes through from first hearing about an innovation to final adoption. The above suggests that the adoption of new technologies, skills or innovations or ways of doing things does not happen automatically. Further to the above, adoption can vary between the individual (farm level) and the community at large (aggregate adoption). Final adoption at the individual farmer's level is defined as the degree of use of a new technology in a long-run equilibrium when the farmer has full information about the new technology and its potential. Corresponding to that definition, Loevinsohn et al. (2013) further define adoption as an integration of new technologies into existing practices, usually proceeded by a period of trying and some degree of adaptation. Taking all the above definitions into account, the chapter defines adoption of recommended technologies as a farmer's decision to make full use of recommended technologies as the best course of action available.

Farmer Field School (FFS)

Farmer field schools (FFSs) are a popular education and extension approach currently used worldwide. FFSs are schools without walls that bring together farmers to share existing knowledge and experience (Kokate et al. 2009). The approach is hinged on Kolb's theory and adult learning theories (Duveskog and Frii-Hansen 2013). In its operation the farmer's field becomes a learning site, whereby most of the materials such as plants, pests, soil particles and real problems are found. It is a season-long training programme conducted in the field (Anandajayasekeram et al. 2007). Any new "language" learned in the course of the study can be applied directly to real objects, and local names can be used and agreed upon. In these schools, farmers are encouraged to experiment, analyse, discover and evaluate and are equipped to address challenges that make appropriate changes in their recommended production practices and finally adopt developed farming methods that suit their ecosystem. FFS works with small groups; for example, a big group of about 20–25 farmers is subdivided into smaller groups of about 5 members to allow better participation by individuals in field observations, analysis, discussion and presentations. On the basis of group size, the research findings show that bigger groups of more than 30 farmers cause inactive participation compared to small groups (Pontius et al. 2002).

Productivity

According to OECD (2001), productivity is defined as a ratio of a volume measure of output to a volume measure of input use. The objectives of productivity measurement include to trace technological change; to identify changes in efficiency; to identify real cost savings in the production process; to express production in physical units; and to determine living standards through per capita income and crop productivity on the basis of crop yield (output per unit of land used, i.e. kg/ha). It is also a measure of land productivity, which does not include other important inputs, such as labour, and other forms of capital, including purchased inputs such as seeds and fertilizers. Agricultural productivity is defined in several ways throughout the literature, i.e. as general output per unit of input, farm yield by crop or total output per hectare (ha) and output per worker (Schneider and Gugerty 2011). However, in this study maize productivity refers to a total output per ha. Generally, many factors affect Tanzania's maize productivity. According to Urassa (2010), these include those related to the access to inputs (e.g. fertilizers, seeds, pesticides) and others to price variation, climate, crop husbandry practices and farm households' socio-economic characteristics. In addition, access to extension services has also been reported by SAA (2007) as cited by Urassa (2010).

2 Research Methodology

2.1 Description of Study Area

The study was conducted in Morogoro District, one of the six local government administrative authorities (LGAs) of Morogoro Region, Tanzania. The district lies between latitudes 6° 50′ and 06° 54′ south and longitudes 37° 40′ and 37° 54′ east of the Greenwich. The district occupies an area of 19, 056 square kilometres. The district is bordered by Tanga Region to the north, to the east by Coast Region, to the south by Kilombero District and to the west by Kilosa District. Administratively, the district is divided into 4 divisions, 24 wards and 214 villages. Morogoro District was selected purposively because maize is a major food crop for most of households followed by other crops such as beans, rice, sorghum, cassava and banana (Morogoro District Council 2012). Moreover, empirical evidence shows Morogoro District has the largest area (137,106 ha) under maize production (URT 2003).

Climate and Soil

Morogoro District has a bimodal rainfall: short rains start in October and end in January, while long rains last from mid-February to May, with average of 600–2000 mm per annum. The average annual temperature is 29 °C. The temperature may become lower and sometimes reaches 15 °C in May up to July in the Uluguru Mountains. The type of soil available in the area is sandy loam to sand clay loams; in some areas it includes clay and heavy clay soils.

Agroecological Zones

Geographically, Morogoro District is divided into three agroecological zones: highland and mountain zone, low mountain zone/Miombo woodland and savannah and river basin/lowland zone. The highland and mountain zone receives more than 2000 mm rainfall per annum and has an average altitude between 1200 m and 2000 m above the sea level and covers about 25% of the total area of the district. This zone is the source of main rivers in the district such as Ruvu and Wami Rivers. The low mountain zone rainfall ranges from 1000 to 2000 mm per annum. The area has an altitude between 600 m and 1200 m above sea level and covers about 20% of the total area, while the low land zone covers about 55% of the whole area and has an altitude ranging from 400 to 600 m above sea level with rainfall between 600 mm and 1200 mm per annum.

Economic Activities

The major economic activities in Morogoro District include agricultural, i.e. crop production and livestock keeping. The main food crops and cash crops grown in the district include banana, rain-fed upland rice, fruits, coffee, beans, spices, vegetables, maize, cassava, sorghum and simsim. The lowland areas have fertile soils deposited from highlands by floods during heavy rainfall, and the area is suitable for rice production. Other land uses include livestock keeping (pigs, goat, cattle, ducks and chicken) and fishing, wildlife conservation and forest reserves. The bulk of Morogoro District's economic activities are based on agricultural crop production and pastoralist livestock keeping in the southern part. More than 85% of the population engaged in agriculture produce maize, beans, cassava, sorghum, rice, fruits, coffee and cotton. The animal husbandry includes cattle, goat, sheep, pigs, chicken and ducks (Fig. 2).

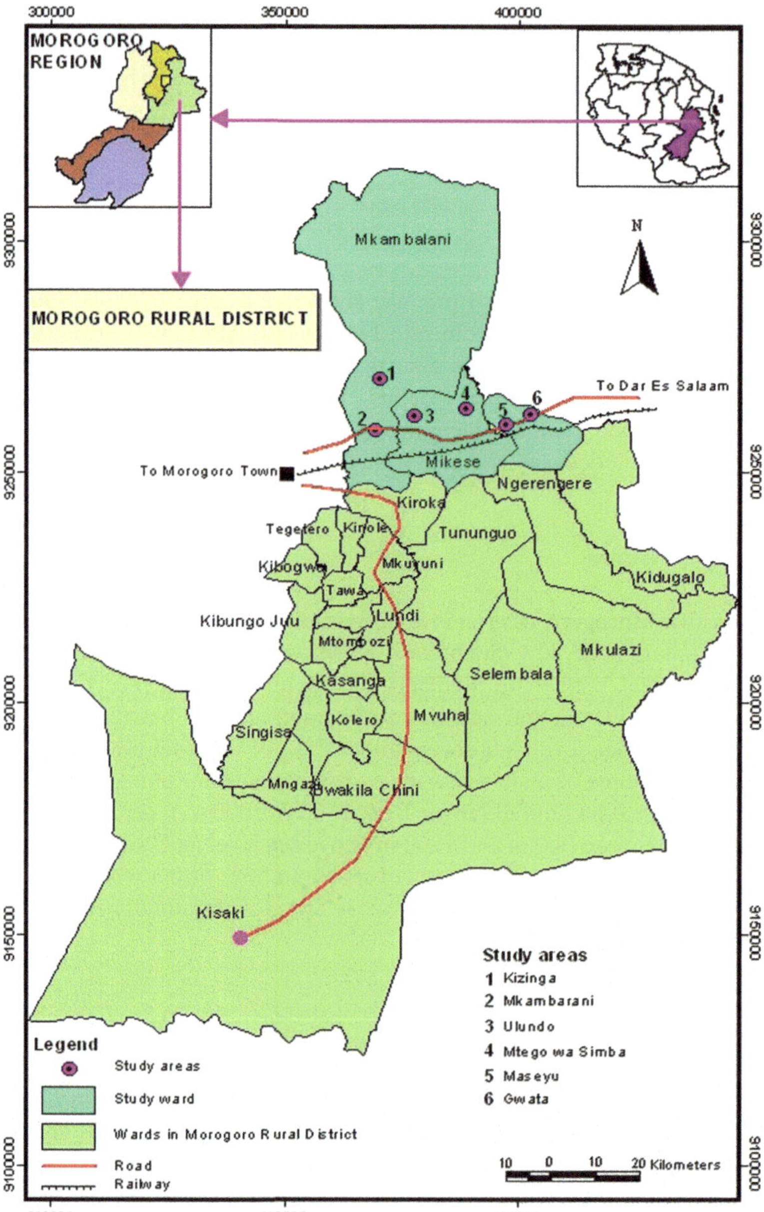

Fig. 2 The map of Morogoro Rural District

2.2 Research Design

This study adopted a cross-sectional research design whereby data was collected at a single point and time (Bob and Liz 2010). The design was chosen because it is cost-effective, it is less time-consuming and a lot of information can be collected in a relatively short period of time. The cross-sectional technique is recommended for descriptive studies and, more specifically, for studies that analyse the relationships between and among variables.

2.3 Sampling Technique and Sample Size

The study's population included all farmers in Morogoro District engaged in small-scale maize production. However, to get a representative sample of the above, a multistage sampling technique was adopted whereby purposive sampling was used to select the division of Mikese and three wards of Mkambarani, Mikese and Gwata based on availability of FFS and non-FFS maize growers. The reason for using the multistage sampling was to increase research precision by ensuring that key subjects are represented in the sample (Garson 2012). Thereafter, stratified random sampling technique was used to select six villages, two from each ward. These villages (strata) were Kizinga and Mkambarani (Mkambarani ward), Mtego wa Simba and Ulundo (Mikese ward) and Gwata and Maseyu (Gwata ward). In each village, sampling frames for FFS and non-FFS were used to randomly select 20 smallholder maize farmers (10 FFS and 10 non-FFS). For the sampled household, the head or his representative was interviewed. Therefore, the study covered a total of 166 individuals, of which 10 were key informants (3 ward executive officers, 6 village executive officers and 1 extension agent) and 36 were focus group participants.

2.4 Data Collection

Quantitative Data

A structured questionnaire with both closed and open-ended questions was used to obtain quantitative information on stated objectives from maize farmers. Before the actual interview, the questionnaire was pretested to check its reliability and validity. After the pretesting, the questionnaire was revised to obtain the final version. Thereafter, the modified version was used to solicit primary information from farmers.

Qualitative Data

For qualitative data, focus group discussions (FGDs) and key informant interview were conducted. Generally, FGDs offer the researcher a means of obtaining a better understanding (insight) of a wide range of views that people have about a specific issue as well as how they interact and discuss the issue (Matthew and Ross 2010). Based on the above, FGDs were conducted using an FGD guide. To constitute the FGDs, FFS leaders were asked to gather individuals into groups of six people (men and women). These representative groups were not involved in the individual interviews. A total of 36 farmers, 6 from each village (men and women) were chosen based on their experience with FFS and maize production. The FGDs were useful as they allowed freedom of expression and maximum participation in respect to knowledge, experience, opinion and feelings.

Key informant interviews were also used to get primary data. The key informant interviews were used to collect detailed additional information from knowledgeable and informed people on the subject matter. The key informants included agricultural extension officers, DALDO (District Agriculture and Livestock Development Officer), FFS leaders, WEOs (ward extension officers) and VEOs (village extension officers). A key informant checklist was used to collect relevant information from the above to complement the other forms of data collected.

2.5 *Data Analysis*

Descriptive statistics including means, frequencies and percentages were determined in connection to objectives 1 and 4 of the study using SPSS. In addition, inferential analysis, i.e. independent sample t-test, and a binary logistic regression model were used to compare and determine the means and association between the independent variables and the dependent variables, respectively. The independent samples t-test was used to compare the mean score on maize productivity trends between members of farmer field school (FFS) and non-FFS members. However, before analysis, the test (dependent variable) assumption for normality was tested. The binary regression model was used to determine some of the socio-economic factors affecting adoption of recommended technologies in FFS. The dependent variable "adoption of recommended FFS technologies" was binary with a value of 1 if the farmer was FFS technology adopter or 0 for FFS non-adopters. The formation of the model was influenced by a number of working hypotheses. It was hypothesized that a farmer's decision to adopt or reject recommended FFS maize practices at any time is influenced by combined (simultaneous) effects of a number of socio-economic factors related to the farmer's objectives and constraints. The variables shown in Table 1 were hypothesized to influence the adoption of recommended FFS maize practices by maize farming households. According to literature (Adesina et al. 2000; Chaves and Riley 2002), decision-making process by farmers involved in adopting FFS technologies can quantitatively be analysed using the logistic regression modelling approach.

Table 1 Description of independent variables and the expected relationship with the dependent variable

Description	Type	Description	Expected sign
Dependent variable (Yi)	Dummy	0 = non-adopter, 1 = adopter	
Explanatory variables			
Age of head of household	Continuous	Age of household head in years	+/−
Household head's sex	Dummy	0 = female, 1 = male	+
Household head's marital status	Dummy	0 = single, 1 = married	+
Household head's education	Dummy	0 = no formal education, 1 = formal education	+
Household size	Continuous	Number of people	+
Household's annual income	Continuous	Annual income from selling crops	+
Land under cultivation	Continuous	Amount of land under cultivation in ha (hectares)	+

The likelihood of the farmer being an adopter of FFS recommended maize production practices was predicted using the following equation:

$$Lg\left(P/1-P\right)=\beta_o+\beta_1x_1+\beta_2x_2+\beta_3x_3+\ldots\ldots\ldots\ldots\ldots\ldots\beta_nx_n+i$$

where:

P = decision for farmers to adopt recommended FFS technologies
$1-P$ = decision for farmers not to adopt recommended FFS technologies
X_1-X_7 = explanatory socio-economic predictor variables
X_1 = age of head of household
X_2 = sex of head of household
X_3 = marital status of head of household
X_4 = education qualification of head of household
X_5 = household size
X_6 = income (annual estimation of income from crop produces)
X_7 = land size (number of acres cultivated)

3 Results and Discussion

3.1 *Socio-economic Characteristics of Respondents and Their Participation in FFS*

An understanding of various socio-economic characteristics of the surveyed respondents is important. In the study on which the chapter is based, socio-economic characteristics, i.e. age, sex, marital status, education level, occupation, household income and farm size, were considered in order to understand the respondent's status (Table 2). Generally, findings in Table 2 show that the average age of the respondents was 47 years. According to the findings, the majority of FFS and non-FFS participants were in the age group ranging from 36 to 59 (Table 2). The above suggests that the majority of the respondents were economically active. A study by Uaiene et al. (2009) indicates that younger household heads are more flexible and active, hence more likely to adopt new technologies and practices.

Findings in Table 2 also show that more than a half of the FFS participants were females. However, for the non-FFS, male participants were the majority. Generally, results indicate that the majority of FFS maize producers were females. During the focus group discussions (FGDs), women participants reported that male farmers tend not to participate in FFSs because they do not want to waste time on activities such as the agroecosystem analysis (splitting into groups and working with pictures on what is going on in the fields). Moreover, the findings are supported by a study by Leavens and Anderson (2011) and Nyomora et al. (2012), who found that women play an essential role in agricultural production.

Table 2 Socio-economic characteristics of respondents (n = 120)

Characteristic	Category	Non-FFS participants (n = 60)	FFS participants (n = 60)	Overall % FFS&NFFS (n = 120)
Respondent's age	15–35	16 (26.6)	10 (16.6)	26 (21.7)
	36–59	35 (58.3)	36 (60.0)	71(59.2)
	60 and above	9 (15.0)	14 (23.4)	23(19.1)
Respondent's sex	Male	32 (53.3)	28 (46.7)	60(50)
	Female	28 (46.7)	32 (53.3)	60(50)
Respondent's marital status	Single	19 (31.6)	14 (23.4)	33(27.5)
	Married	41 (68.4)	46 (76.6)	87(72.5)
Respondent's education level	No formal education	2 (3.4)	1 (1.6)	3(2.5)
	Primary	52 (86.7)	43 (71.6)	95(79.2)
	Secondary	3 (5.5)	6 (10)	9(7.5)
	Tertiary	3 (5.0)	10 (16.8)	13 (10.8)
Respondent's occupation	Crop producer	57 (95.2)	55 (91.6)	112(93.33)
	Livestock keeper	1 (1.6)	1 (1.6)	2(1.7)
	Self-employed off farm	1 (1.6)	0 (0.0)	1(0.8)
	Causal labour on farm	1 (1.6)	4 (6.8)	5(4.2)
Household's farm size	Small (<0.4 ha)	21 (35.0)	15 (25.0)	36(30)
	Medium (0.4–1 ha)	29 (48.4)	36 (60.0)	65(54.2)
	Large (>1 ha)	10 (16.6)	9 (15)	19(15.8)
Average household income	<200,000	27 (45.0)	6 (10)	33(27.5)
	300,000–500,000	23 (38.4)	23 (38.4)	46(38.33)
	600,000–1,000,000	6 (10)	24 (40.0)	30(25)
	More than 1000 000	4 (6.6)	7 (11.6)	11(9.2)

NB Numbers in parenthesis indicate percentage

Table 2 further shows that more than three quarters of the interviewed farmers were married, compared to just over two thirds of the non-FFS participants. This indicates married individuals were more likely to be members of a farmers' group relative to unmarried individuals. Similar findings have also been reported by Adong et al. (2013) in a study conducted in Uganda. Other literature (Mwaura 2014; Mwanzia 2014) shows that a household head's marital status does influence farmer's decision to join FFS group.

The role of education in adoption of recommended technologies has been reported in literature (Uaiene et al. 2009). This is because educated farmers have a better opportunity to acquire and process information on new technologies (Salasya et al. 2007). Study findings (Table 2) show that the majority of both FFS and non-FFS farmers had attained primary education. The study observed further that relatively more FFS participants had attained secondary and tertiary education, compared to the non-FFS participants. This suggests that education is an important determinant for a farmer's decision to participate in FFS in the study area.

Occupation is a characteristic that can determine farmer's prospect in adoption of FFS recommended technologies. Findings in Table 2 show that the majority of respondents were full-time crop farmers with a slight difference observed between FFS and non-FFS participants. These findings are similar to those reported in literature (Livingston et al. 2011; FAO 2012) that agriculture is the main occupation which provides employment to most (80%) of economically active farming households in sub-Saharan Africa. According to NBS (2013), about three quarters (74%) of mainland Tanzanians depend on agriculture as their main occupation.

As regards respondents' farm size (defined as total farm size (ha) owned or rented by a household head), the study findings (Table 2) show that more than half of the interviewed households had medium land size ranging from 0.4 to 1 ha. However, under a third of all the respondents had less than 0.4 ha. Generally, this indicates that the majority of interviewed farmers had small farm sizes compared to the national average of 2.47 ha per household (Minot et al. 2010 cited by Derksen-Schrock et al. 2011). According to Tabi et al. (2010), increasing farm size can lead to less adoption of improved farming technologies as the decision to adopt is not contingent upon having large acres of land.

The findings in (Table 2) also show that the majority of surveyed households had an average annual income of 500,000.00 Tanzanian shillings (Tsh) and below. Nonetheless, more FFS participants earned incomes above 500,000.00 Tsh relative to the non-FFS participants. The study's observation conforms to what has been reported in literature whereby many agricultural technology adoption studies have associated the impact of technology with income (Karanja et al. 2003; Mojo et al. 2007; Mendola 2003; Becerril and Abdulai 2009; Simtowe et al. 2010). Generally, results from the study show that the majority of the surveyed households had low incomes compared to Tanzanian's per capita income of Tsh. 869,436.3 and Tsh 1,025,038 Ts for 2011 and 2012, respectively (URT 2013).

3.2 *Recommended Farmer Field Schools Maize Production Practices in the Study Area*

Findings in Table 3 summarize the major recommended maize farming practices and the proportion of FFS and non-FFS participants using them in the study area. The results show that more than two fifths of the FFS participants were practising early land preparation compared to about a quarter of the non-FFS participants. This difference may be due to the fact that the FFS farmers are involved in the learning process, which might have enhanced their competence. In most cases, land preparation depends on the onset of the main rains; hence farmers undertake land preparation from December to end of January.

Apart from early land preparation, about two fifths of the FFS participants practise timely planting and proper spacing compared to just over a quarter of the non-FFS participants. Generally, literature (Goodbody 1990; Katinila et al. 1998) shows that planting early is the most important single factor among others for increased yields. This is because in the area, the main rains take 2–3 months; and once a farmer delays planting, the crops are likely to suffer from water stress, thus affecting maize yields. The findings are also supported by the FGD participants whereby it was reported that it is generally recommended that planting be done at the onset of the main rains. However, dry planting is barely practised before the expected rain comes. In dry planting, seeds are placed in dry soil 1–2 weeks before the start of the rain season. The above conforms to what was observed in the key informant interviews as shown in the quote below:

> Normally the FFS participants have been following what they have been taught and have been implementing this on their plots. For example, planting date and use of proper spacing between seeds. This knowledge helps them to increase their maize yields compared to non-FFS participants. This is because when maize is planted late, the critical growth period at flowering will coincide with drought and this will make the crop become stressed and lead to poor grain filling... (Mikese ward extension officer, Morogoro District, 19/11/2014)

In case of weed management, just under a half of the FFS participants were practising timely weed management compared to just over a tenth of the non-FFS participants (Table 3). Generally, early weeding is critical for obtaining a good maize

Table 3 Adoption of FFS recommended maize production practices (n = 120)

	FFS participants (n = 60)		Non-FFS participants (n = 60)	
Types of practices	No (%)	Yes (%)	No (%)	Yes (%)
Early land preparation	5(4.2)	55 (45.8)	28(23.5)	32(26.5)
Timely planting and spacing	11(9.2)	49(40.8)	27(22.5)	33(27.5)
Timely weed management	5(4.2)	55 (45.8)	44(36.6)	16(13.4)
Application of fertilizers	6(5.0)	54(45.0)	47(39.2)	13(10.8)
Pest and natural management	2(1.7)	58(48.3)	48(40.0)	12(10.0)
Processing and storage management	9(7.5)	51(42.5)	50(41.6)	10(8.4)

NB The parenthesis indicates percentages

yield. A study by Zvonko and Gordana (2009) reported that the most critical period of weed competition in maize is during the first 4–6 weeks after seedling emergence. In addition, weeds tend to thrive better than the maize crop under marginal environmental conditions, including moisture stress (drought stress). It is on the above basis that weed management helps to minimize yield losses and also increase the quality and quantity of the maize crop (Zvonko and Gordana 2009).

Weed management practices are one of the key areas of FFS as indicated in the quote below:

> In FFS we were trained on the importance of early weed management which can be done manually or through the application of herbicides for the best result of maize yields because the maize field may be infested with broad-leaf or narrow-leaf weeds, or with a mixture of both. Generally, if crops are kept weed-free during the early stages of plant growth, yields will not be affected significantly. (A 56-year-old FGD participant from Kizinga Village, Morogoro District, 23/11/2014)

The level of fertilizer application in maize production was another recommended practice observed by the study. Table 3 shows that over two fifths of FFS participants in the study area had adopted and were applying fertilizer in their maize production. In addition, the table shows that over a third of the non-FFS in the study do not apply fertilizer, and just over a tenth did use fertilizers. The results show that adoption of fertilizer as a recommended maize production practice is higher among farmers who participated in FFS than those who did not and vice versa. These results conform to findings reported by Haefele et al. (2003) that fertilizers are very important inputs for optimum maize yield and productivity. As regards the nonuse of fertilizers, the main reasons given were soils in the area had enough fertility; high fertilizer prices; long distance to fertilizer sale points; and the of use fertilizers destroys the soil. Based on the above reasons, one may see the importance of the FFS. Generally, the FFS approach as an extension model does give farmers a forum of sharing their experiences and knowledge on fertilizer use through usual field observation. This is because in the study area, the adoption of recommended use of fertilizer is high among the FFS participants than among non-FFS participants.

Study findings (Table 3) on maize processing and storage management show that over two fifths of the FFS members were practising improved methods compared to under a tenth of the non-FFS participants. Generally, postharvest losses are recognized as one of the critical constraints upon food security among many poor farmers across Africa (Owusu et al. 2011). It is on this basis that the recommended FFS processes and storage management systems are important practices, which could help farmers improve their food security. In Tanzania, farming households lose about 30% of the produce at postharvest (Rugumamu 2009). Normally, processing and storage losses are encountered during drying and grading of cobs, shelling of cobs, drying the grains, winnowing of grains, application of pesticides and packing grains at household level and in storage for later use.

During the household survey, participants pointed out that in FFS, both local and improved maize storage practices were used to avoid postharvest losses. They reported that with close supervision of an extension agent, participants were able to improve their traditional processing and storage practices. In the study area, respondents reported that maize harvest stretches from June to July and the activity is done by picking maize using hand which is thereafter transported to homesteads for drying and storage. Drying maize on plastic sheets or local mats is a common practice for farmers who favour keeping maize off the ground during the drying process. FFS participants reported that more often in the last weeks before harvesting, they bend down the upper part of maize holding the ear, a practice that also prevents the kernels from becoming soaked when it rains. They further reported that the majority of farmers store unshelled maize on cribs (see Plate 1), while shelled maize is stored in sacks (bags). FFS participants reported that they avoid drying harvested maize on the ground because the grain that is in contact with the ground absorbs moisture in addition to picking up dirt and pests.

3.3 Socio-economic Factors Influencing Adoption of Recommended Maize Farming Practices

This section presents the results of a binary logistic regression on the influence of socio-economic factors influencing adoption of recommended maize practices. The socio-economic characteristics included in the model were age, sex, marital status, education, income and land and household size. According to Pallant (2011), a binary logistic regression is used to predict a categorical variable. With a categorical dependent variable, discriminant function analysis is usually employed if all of the

Plate 1 An improved crib within the homestead at Maseyu Village in Morogoro Rural District

predictors are continuous and nicely distributed; logic analysis is usually employed if all of the predictors are categorical; and logistic regression is often chosen if the predictor variables are a mix of continuous and categorical variables. It is on the above basis that the logistic regression model was performed to determine the association of the socio-economic factors on adoption of the recommended FFS practices.

The model with all predictors was statistically significant χ^2 ($p \leq 0.05$) implying that the model was able to predict the respondents' decision to adopt farmers field schools recommended maize practices. According Pallant (2011), the Cox and Snell R Square and the Nagelkerke R Square values provide an indication of the amount of variation in the dependent variable explained by the model (from a minimum value of 0 to a maximum of approximately 1). Cox and Snell R Square suggests that 0.330 of the variation in the dependent variable was explained by the variables included in the logistic regression model. The Nagelkerke R Square was 0.450, which means that the independent variables included in the model explained 45% of the chances adopting recommended maize technologies occur. Results in Table 4 show the importance of each socio-economic variable and its contribution to the adoption of the farmer field school recommended maize production practices.

Age and Adoption of FFS Recommended Practices

Age of household head was hypothesized to have a positive influence on adoption of recommended FFS maize practices (Table 1). According to literature (Wekesa et al. 2003; Ebewore 2013; Beshir and Wegary 2014), age has a direct bearing on a farmer's approach, openness or conservativeness and level of exposure to new technologies. The results obtained from the study (Table 4) show that age was significantly ($p < 0.037$) related to adoption of the recommended FFS maize practices with a Wald Statistic of 4.341, respectively. These results conform to those by Kariyisa and Dewi (2013) and Ebojei et al. (2012) who have reported that age was significantly related to adoption of IPM (integrated pest management)

Table 4 Socio-economic factors influencing adoption of recommended maize farming practices ($n = 120$)

Predictor	B	S.E.	Wald	Df	Sig.	Exp (B)
Household head's age	4.448	2.135	4.341	1	0.037**	85.449
Household head's education level	5.403	2.509	4.638	1	0.031 **	222.167
Household size	−0.775	1.543	0.252	1	0.616	0.461
Household's average annual income	2.539	0.602	17.808	1	0.000 ***	12.669
Household farm size	−1.514	0.773	3.831	1	0.050 **	0.220
Household head's sex	0.158	0.510	0.096	1	0.757	1.171
Household head's marital status	−0.841	0.579	2.109	1	0.146	0.431
Constant	−20.793	5.239	15.752	1	0.000	0.000

NB ** Significant at $p = 0.05$ while ***is significant at $p = 0.001$

technologies under FFS in Indonesia and the latter age was significantly related to adoption of recommended hybrid maize in Nigeria. Generally, younger farmers are more interested in trying out new agricultural technologies because of their risk-taking character, and hence, they try to maximize their maize productivity by adopting recommended FFS maize practices.

Marital Status and Adoption of FFS Recommended Maize Practices

Household head's marital status was hypothesized to have a positive influence on adoption of recommended FFS maize practices (Table 1). This is based on the fact that in marriage, wives can play a double role as housewives on one side and, on the other side, as the recipients of innovation and have a chance to apply the recommended technologies with the head of the family (husband). Moreover, in Tanzania and many other parts of SSA, women play an important role in agricultural production (Mmasa 2013). Contrary to the expectation, results in Table 4 show that marital status was not significantly related to adoption of recommended FFS maize practice. This could be perhaps due to married couples having different perspectives on the recommended FFS practices in the study area. These results are consistent with observations by Fadare et al. (2014) who found out that marital status was not a significant factor influencing adoption of maize technologies in Nigeria. However, Adong et al. (2013) found out that marital status had a positive relationship with adoption of recommended technologies in Uganda.

Farmers' Education and Adoption of FFS Recommended Maize Production Practices

Education level is an important factor in adoption of new technologies. In this study education level was hypothesized to have a positive influence on the adoption of recommended FFS maize practices. A study by Mutuah (2013) showed that literacy level of farmers affects the degree or level of acceptance and subsequent adoption of recommended farming technologies. The results (Table 4) obtained from the study show that education was significantly ($p \leq 0.05$) and positively related to adoption of FFS recommended maize production practices. These results conform to those by Fisher and Mazunda (2011), Ebojei et al. (2012) and Beshir and Wegary (2014) who observed that education level positively influenced adoption of recommended hybrid maize production in Malawi, Nigeria and Ethiopia, respectively.

Occupation and Adoption of FFS Recommended Maize Production Practices

In areas where households indulge in varying income-earning activities such as employed work, business, mining and agricultural production, adoption of agricultural technologies could only be confined to those who are preoccupied with the activity. Study results show occupation was not significantly related to a farmer's decision to participate in FFS. The reason could be due to both FFS and non-FFS depending solely on agriculture as their main source of income.

Farm Sizes and Adoption of FFS Recommended Maize Production Practices

Farm size was considered a factor in farmers' adoption of technologies through the FFS approach (Table 1). According to Foster and Rosenzweig (2010), large farmers or those with large land holdings have been reported to have greater likelihood of adopting recommended practices as they are more intensively managed. The results obtained from the study (Table 4) show that farm size was positively and significantly ($p \leq 0.05$) related to adoption of FFS recommended maize production practices. These results conform to Akadugu et al. (2012), who observed that farm size had a positive relationship to adoption of recommended modern technologies in Ghana.

Household Income and Adoption of FFS Recommended Maize Production Practices

Income of the household head was hypothesized to have a positive influence on adoption of FFS recommended maize production practices (Table 1). Generally, a household with sufficient income from selling crops or other sources is likely to adopt recommended FFS maize practices compared to low-income households. The results (Table 4) obtained from the study are in agreement with the hypothesis, whereby income was positively and significantly ($p \leq 0.000$) related to adoption of the FFS recommended maize production practices. The observed results conform to those obtained by Ebojei et al. (2012), who observed a positive relationship between adoption of hybrid maize production and farmer's income in Nigeria.

Household Size and Adoption of FFS Recommended Maize Production Practices

Household size was hypothesized to have a positive influence on adoption of recommended FFS maize practices (Table 1). It is believed that adequate and a readily accessible labour pool can affect the adoption decision and amount of technology to be adopted (Weyori et al. 2012). In contrast, the results in Table 4 show that household size was not significantly ($p \leq 0.05$) related to adoption of recommended FFS maize production practices. This could be due to the fact that not all household members are necessarily an active labour force in the study area. These results are consistent with the observation by Salasya et al. (2007), who found out that household size was not significantly related to adoption of recommended stress-tolerant maize hybrid (WH 502) in western Kenya. Nevertheless, Alene et al. (2000) observed that household size significantly influenced the adoption of recommended improved maize varieties in the central highlands of Ethiopia. These contrasting findings could be due to varying traditions and culture on household labour in different parts of Africa.

Household Head's Sex and Adoption of FFS Recommended Maize Production Practices

A household head's sex was hypothesized to influence adoption of FFS recommended maize production practices (Table 1). For example, it was expected that male-headed households would have a greater chance of adopting FFS recommended maize production practices. Generally, male-headed households in developing countries have been reported to have higher access to resources and information compared to female-headed households and therefore a greater capacity to adopt technologies (Kaliba et al. 2000). For this reason, male heads were assigned a priori positive sign. In contrast to the priori expectation, results in Table 4 show a household head's sex was not significantly related to adoption of FFS recommended maize production practices. The findings conform to those by Murage et al. (2012) who found that gender/sex was not significantly related to adoption of recommended IPM technologies in western Kenya. These results could be due to the fact that men were reluctant to participate in FFS compared to their female counterparts.

3.4 A Comparison of Maize Production and Productivity Between FFS and Non-FFS Participants

An independent sample t-test was conducted to compare maize production between FFS participants and non-FFS participants in three consecutive seasons of 2011/2012, 2012/2013 and 2013/2014. The results in Table 5 show that the mean score of maize production by FFS adopters of recommended maize practices was significantly ($p \leq 0.05$) higher than that of non-FFS farmers. The results suggest that the two groups differ in maize production in the consecutive seasons. These results suggest an expected yield in the study area is higher with the adoption of recommended FFS maize production practices than without. Apart from the yield variability between the two groups, results also show that FFS recommended maize production practices are suitable for enhancing yields. These results are consistent with observation by Nyangena and Juma (2014), who found that there were differences in maize yields between adopters and non-adopters of recommended maize improved farm technologies from 2004 to 2007 in Kenya.

Table 5 A comparison of maize production between FFS and non-FFS ($n = 120$)

Seasons		N	Mean in ton/ha	t-value	Sig. (2 tailed)
2011/2012	Non-FFS	60	0.38	−21.19	0.000
	FFS	60	1.24	−21.19	0.000
2012/2013	Non-FFS	60	0.39	−23.02	0.000
	FFS	60	1.26	−23.02	0.000
2013/2014	Non-FFS	60	0.40	−30.70	0.000
	FFS	60	1.46	−30.70	0.000

Table 6 Reason for high maize yields among the FFS adopters ($n = 120$)

Reasons	Adopters ($n = 60$)	Non-adopters ($n = 60$)
Easy access to credit	1 (1.7)	8 (13.4)
Access to information and inputs	50 (83.3)	9 (15)
Farming knowledge	9 (15.0)	43 (71.6)

NB Number in parentheses denotes percent of respondents

Table 6 presents the reasons for high maize yields among the FFS adopters of recommended practices. An understanding of the reasons for the differences in production between FFS and non-FFS adopters is critical for farmers to join FFS in the study area. Many reasons have been reported by various researchers that expose reasons for differences. For example, a study in Nigeria by Bamire et al. (2010) observed that maize yield increased among the conservation farming (CF) adopting farmers due to early planting, timely management of weed and use of recommended inputs relative to others.

Three reasons for the difference in maize yield between FFS participants and non-FFS participants were established. These reasons were easy access to information and inputs, access to farming knowledge and access to credit. Across the surveyed villages, access to information and inputs was a major reason, which was reported by most (82.4%) FFS participants; this was followed by easy access to farming knowledge (71.6%) reported by non-adopters. The above was supported by the FGDs and key informant interviews as shown below:

> It is easy for a farmer to access farming information and inputs while he/she is in the farmer field school compared to non-FFS members, this has been an advantage for participating in the FFS. Thus, the FFS participants have been able to increase their farm productivity. (An extension officer, Mikese ward, Morogoro District, 29/11/2014)

On the other hand, non-FFS farmers support these results, as they believed that the FFS cohort obtains high maize yield because of the farming knowledge they obtain through the FFS in the study area as shown in the quote below.

> I agree that participation in FFS has helped farmers in our village in a number of ways especially with the farming education they have acquired. Generally, farmers who follow and properly apply the knowledge and skills offered tend to get higher farming outputs. My advice is that the government should create enough awareness to the farmers on importance

of FFS. In addition, a better environment should be created for farmers to learn the practices in order to improve maize production which is our main source of food and income in this village… (A 45-year-old non-FFS participant in Maseyu Village in Morogoro District, 18/11/2014)

3.5 Contribution of Maize Sales to Household Income

Income is widely used as a welfare measure because it is strongly correlated with the capacity to acquire many things that are associated with an improved standard of living, such as food, clothing, shelter, health care, education and recreation (Morris et al. 1999). Literature indicates that the majority of rural farming households in Tanzania earn most of their income from crop sales (Covarrubias et al. 2012). Study findings (Table 7) from the study show that maize was the main source of income accounting for 47% of the total income of the interviewed household. In addition, sales of maize had a positive and significant relationship ($p < 0.001$) with household income. The observation suggests that maize plays an important role in stabilizing the socio-economic capability of households by providing a reliable income. The study's observations conform to that by Mathenge et al. (2012) who found sales of maize have positive relationship to household income in Kenya.

During the FGDs and key informant interviews, respondents mentioned maize as a leading crop for earning income. Other crops contributing to their household income include beans, rice, sorghum, cassava and banana. These results are consistent with observation by Ellis and Mdoe (2003) who reported that maize was an important source of rural households' income in Morogoro. Apart from maize the present study observed that sales of livestock accounted for (31%) of the total income of the interviewed households. Livestock keeping is another important source of income of the people in Morogoro District. Moreover, petty business recorded a small contribution (1.3%) of the income source of the people in the study area. These results were supported by the FDGs and key informant interviews as shown in the quotes below:

In our area, maize ranks as the most important crop that contributes a greater percent of our income, followed by cash crops such as sunflower and sesame. (A 40-year-old male, FGD participant Mkambarani Village, Morogoro District, 20/11/2014)

Table 7 Sources and household income in the study area ($n = 120$)

Source of income	Amount in Tshs.	Percentage (%)	Average HH income in Tshs.
Maize	46,060,000	47.0	761,322
Cash crops (*sunflower, sesame,* etc.)	17,705,000	18.0	147,542
Livestock	30,230,000	31.0	251,917
Causal labour	2,660,000	2.7	22,167
Petty business	1,645,000	1.3	13,708

> Most households in our village depend on maize sales to get their daily bread. However, currently the change in price of a bag from the former 50 000 Tshs to Tsh.30 000, is discouraging as it is relative lower compared to expenses that we incur in farming. (A female FGD participant aged 30 years, Mtego wa Simba Village, Morogoro Rural District, 25/11/2014)
>
> In this area for a long time maize is both a staple food and a cash crop. Therefore, if yields decline, farmers' livelihoods are also affected. Apart from maize, farmers depend on sales of livestock (local chicken, goats, cattle and pigs etc.) and other crops such as beans, paddy, cassava, sorghum, sunflower and sesame.... (Extension officer, Gwata wards, Morogoro Rural District, 30/11/2014)

4 Conclusions

The chapter assessed the adoption of recommended FFS technologies and productivity between farmers who are members of FFS and those who are not. Generally, it has been shown that mean maize yields between those who adopted FFS recommended FFS maize production practices and those who did not differed significantly. As expected maize yields in the study area were higher with the adoption of FFS recommended maize production practices than without. The chapter has also shown that four major socio-economic variables, i.e. household head's age, education, household income and farm size, significantly influenced adoption of FFS recommended maize production practices. In addition, maize sales have been observed to be major sources of household income in the study area. Based on the above, increased maize productivity (i.e. yield/ha) through the adoption of FFS recommended practices can stabilize households' economic capabilities through provision of a reliable income. The study concludes that adoption of FFS recommended maize production practices is imperative if maize productivity of smallholder farmers in Morogoro District is to be raised. Therefore, the chapter recommends that policy makers should work in collaboration with other agricultural development partners in the Morogoro District to mobilize and support extension services through farmer field schools. In addition, there is a need for promotion of the formation of microfinance institutions such as SACCOs (Savings and Credit Cooperative Societies); these can easily increase farmer's access to affordable credit and hence hasten access to improved maize production technologies.

References

ACT (Agriculture Council of Tanzania) (2010) Value chain analysis of rice and maize in selected districts in Tanzania: Volume I: introduction, context analysis and recommended way forward. Match Maker Associated Limited, Dar es Salaam, Tanzania, p 89

Adesina AA, Baidu-Forson J (1995) Farmers' perceptions and adoption of new agricultural technology: evidence from analysis in Burkina Faso and Guinea, West Africa. Agric Econ 13(1):1–9

Adesina AA, Mbila D, Nkamleu GB, Endaman D (2000) Economic analysis of the determinants of adoption of alley farming by farmers in the forest zone of South west Cameroon. Agric Ecosyst Environ 80:255–265

Adong A, Mwaura F, Okoboi G (2013) What factors determine membership of farmer group in Uganda? evidence from the Uganda census of agriculture 2008/9. J Sustain Dev 6(4):1–19
Akadugu MA, Guo E, Dadzie SK (2012) Adoption of modern agricultural production technologies by farm households in Ghana: what factors influence their decisions. J Biol Agric Healthc 2(3):1–13
Alene AD, Poonyth D, Hassan RM (2000) Determinants of adoption and intensity of use of improved maize varieties in the central highlands of Ethiopia: a tobit analysis. Agrekon 39(4):1–11
Anandajayasekeram P, Davis K, Workneh S (2007) Farmer field schools: an alternative to existing extension systems? experience from eastern and southern Africa. J Int Agric Exten Educ 14(1):81–93
Ayinde OE, Adewumi MO, Olatunji GB, Babalola OA (2010) Determinants of adoption of downy mildew resistant maize by small-scale farmers in Kwara State, Nigeria. Afr J Agric Res 3(7):465–468
Bamire AS, Abdoulaye T, Sanogo D, Langyintuo A (2010) Characterization of maize producing households in the dry savanna of Nigeria. CIMMYT, Nigeria, p 88
Becerril J, Abdulai A (2009) The impact of improved maize varieties on poverty in Mexico: a propensity score-approach. World Dev 38:1024–1035
Beshir B, Wegary D (2014) Determinants of smallholder farmers' hybrid maize adoption in the drought prone central Rift Valley of Ethiopia. Afr J of Agric Res 9(17):1334–1343
Bob M, Liz R (2010) Research methods. A Practical Guide for the Social Sciences, Rotolito Lombarda, Italy, p 121
Braun A, Jiggins J, Roling N, van den Berg H, Sunijders P (2006) A global survey and review of farmer field schools experience. International Livestock Research Institute 35(4):661–676
Bunyata DK, Mureithi JG, Onyango CA, Ngesa FU (2011) Farmer field school effectiveness for soil and crop management technologies in Kenya. J Int Agric Exten Educ 13(3):47–63
Cerdán-Infantes P, Maffioli A, Ubfa D (2008) The impact of agricultural extension services: the case of grape production in Argentina. http://www20.iadb.org/intal/catalogo/PE/2009/03159.pdf. [28/08/2017]
Chaves B, Riley J (2002) Determination of factors influencing integrated pest management adoption in coffee berry borer in Colombian farms. Agric Ecosyst Environ 87:159–177
Covarrubias K, Nsiima L, Zezza A (2012) Livestock and livelihoods in rural Tanzania: a descriptive analysis of the 2009 National Panel Survey. International Livestock Research Institute, Nairobi, Kenya, p 57
Davis K, Nkonya E, Kato E, Mekonnen DA, Odendo M, Miiro R, Nkuba J (2010) Impact of farmer field schools on agricultural productivity and poverty in East Africa. International Food Policy Research Institute, Nairobi, Kenya, p 56
Berg H van den, Jiggins J (2007) Farmer field schools reap long-term rewards. Pesticides News 78 December 2007
Derksen-Schrock K, Anderson CL, Gugerty MK (2011) Tanzania: agricultural sector overview Evans School Policy Analysis and Research (EPAR), EPAR Brief No. 133
Duveskog D, Frii-Hansen E (2013) Farmer field schools; a platform for transformative learning in rural Africa. In: Mezirow J, Taylor E (eds) Transformative learning in action: hand book of practice. Jossey-Bass Press, Uppsala
Ebewore SO (2013) Cocoa farmer field graduate farmers perceived benefits of ffs training in Ondo state, Nigeria. Asian J Agric Sci 5(4):74–82
Ebojei CO, Ayinde TB, Akogwu GO (2012) Socio-economic factors influencing the adoption of hybrid maize in Giwa local government area of Kaduna State, Nigeria. J Agric Sci 7(1):23–32
Ellis F, Mdoe N (2003) Livelihoods and rural poverty reduction in Tanzania. World Dev 31(8):1367–1384
Fadare AO, Akerele D, Toritseju B (2014) Factors affecting adoption decisions of maize farmers in Nigeria. Int J Food Agric Econ 2(3):45–54
FAO (2009) High level expert forum report - how to feed the world in 2050. 00153 Rome, Italy

FAO (2012) "Climate change, smallholders and traditional farming communities" demonstrates how smallholder farmers have developed strategies to increase their resilience to external shocks while maintaining ecosystem goods and services. Food and Agriculture Organization, Rome, Italy, p 24

FAO (Undated) Farmer Field Schools and Empowerment: Community Empowerment, Social Inclusion and Gender Equality Experience from Jordan and Tunisia. http://www.fao.org/filead-min/user_upload/rne/docs/FFS_Empowerment.pdf. [29/08/2017]

FAO (Food and Agriculture Organization of the United Nations) (2014) Tanzania Country Programming Framework January 2014 – June 2016. www.fao.org/3/a-bp609e.pdf. [12/09/2018]

Feder G, Just RE, Zilberman D (1985) Adoption of agricultural innovations in developing countries: a survey. Econ Dev Cult Chang 33(3):255–297

Fisher M, Mazunda J (2011) Could low adoption of modern maize varieties in Malawi be explained by farmers' interest in diverse seed characteristics?. International Food Policy Research Institute, Nairobi, Kenya, p 4

Foster AD, Rosenzweig MR (2010) Microeconomics of technology adoption. Annu Rev Econ 2010:2

Garson DG (2012) General linear models: univariate GLM, ANOVA/ANCOVA, repeated measures, Vol 6(7). Statistical Associates Publishers, Asheboro, pp 21–24

Goodbody S (1990) Maize, time of planting: a review of recent work in Tanzania. In: Moshi AJ and Ransom JK (eds) Maize research in Tanzania proceedings of the first Tanzania National Maize Workshop, Dar es Salaam, Tanzania, pp 2–10

Haefele SM, Wopereis MCS, Ndiaye MK, Barro SE, Ould IM (2003) Internal nutrient efficiencies fertilizer recovery rates and indigenous nutrient supply of irrigated lowland rice in Sahelian West Africa. J Field Crop Res 80:19–32

Kaliba ARM, Verkuijl H, Mwangi W, Mwilawa AJT, Anandajayasekeram P, Moshi AJ (1998) Adoption of Maize Production Technologies in Central Tanzania. Mexico, D.F.: International Maize and Wheat Improvement Center (CIMMYT), the United Republic of Tanzania, and the Southern Africa Centre for Cooperation in Agricultural Research (SACCAR).

Kaliba RMA, Verkuijl H, Mwangi W (2000) Factors affecting adoption of improved maize seeds and use of inorganic fertilizer for maize production in the intermediate and lowland zones of Tanzania. J Agric Appl Econ 32(1):35–47

Karanja DD, Renkow M, Crawford EW (2003) Welfare effects of maize technologies in marginal and high potential regions of Kenya. Agric Econ 28:193–216

Kariyisa K, Dewi YA (2013) Analysis of factors affecting adoption of integrated crop management farmer field school (Icm-Ffs) in swampy areas. Int J Food Agric Econ 1(2):29–38

Katinila N, Verkuijl H, Mwangi W, Anandajayasekeram P, Moshi AJ (1998) Adoption of maize production technologies in Southern Tanzania. International Maize and Wheat Improvement. Center (CIMMYT), the United Republic of Tanzania, and the Southern Africa and the Southern Africa Centre for Cooperation in Agricultural Research (SACCAR), Mexico, D.F.

Keenja C (2001) Food security in Tanzania. The way forward. The Eighth Sokoine Memorial Lecture. Sokoine University of Agriculture, Morogoro, Tanzania, p 75

Khisa GS, Heinemann E (2006) Chapter 5: Farmer Empowerment through Farmer Field Schools. https://www.cabdirect.org/cabdirect/FullTextPDF/2009/20093203632. [28/08/2017]

Kokate KD, Kharde PB, Patil SS, Deshmukh BA (2009) Farmers led. Extension experience and road ahead. Indian Res J Ext Educ 9(2):18–21

Larsen AF, Lilleør HB (2014) Beyond the field: the impact of farmer field schools on food security and poverty alleviation. World Dev 64:843–859

Leavens MK, Anderson CL (2011) Gender and agriculture in Tanzania EPAR Brief No. 134. Evans School Policy Analysis and Research (EPAR), University of Washington

Liberio J (2012) Factors contributing to adoption of sunflower farming innovations in Mlali ward, Mvomero District, Morogoro Region. Dissertation for Award of MSc. Degree at Sokoine University of Agriculture, Morogoro, Tanzania, p 91

Livingston G, Schonberger S, Delaney S (2011) Sub-Saharan Africa: the state of smallholders in agriculture. International Fund for Agricultural Development, Rome, p 36
Loevinsohn M, Sumberg J, Diagne A, Whitfield S (2013) Under what circumstances and conditions does adoption of technology result in increased agricultural productivity? Institute of Development studies, Brighton, UK. https://opendocs.ids.ac.uk/opendocs/bitstreamhandle/123456789/3208/Productivity%20systematic%20review%20report%203.pdf;jsessionid=FD90D6085FDDDB2250324C881A89FFF6?sequence=1. [28/08/2017]
MAFC (Ministry of Agriculture, Food Security and Cooperatives) (2008) Agricultural sector review and public. Expenditure review 2008/2009. Dar es Salaam, Tanzania, p 45
Mathenge MK, Smale M, Olwande J (2012) The impact of maize hybrids on income, poverty, and inequality among smallholder farmers in Kenya. Michigan State University, Department of Agricultural, Food, and Resource Economics and Department of Economics
Mattee AZ, Mussa KR, Mwaseba DL, Mahonge CP, Nsenga JV (2015) Factors in smallholder farmers' vulnerability to climate change impacts in the Uluguru Mountains, Morogoro, Tanzania. In: Lal R, Singh B, Mwaseba D, Kraybill D, Hansen D, Eik L (eds) Sustainable intensification to advance food security and enhance climate resilience in Africa. Springer, Cham, pp 185–195
Matthew B, Ross L (2010) Research methods, a practical guide for sciences. Perason Education Limited 2010, Rotolito Lombarda, Italy, p 490
Mendola M (2003) Agricultural technology and poverty reduction: a micro-level analysis of causal effects (November 2003). Centro Studi Luca D'Agliano Development Studies Working Paper No. 179
Michelle KJ (2005) Technology adoption in West Africa: adoption and disadoption of soy beans on the Togo-Benin Border. Dissertation for Award of MSc Degree at North Carolina State University, Releigh, p 175
Mmasa JJ (2013) Women participation in agriculture in Tanzania – challenges and policy recommendations - Tanzania Country Level Knowledge Network - CLKnet Policy Brief No 8: 2013 CLKnet - Policy Brief No: 8
Mojo S, Norton GW, Alwang J, Rhinehart I, Deom CM (2007) Peanut research and poverty reduction: impacts of variety improvement to control peanut viruses in Uganda. Am J Agric Econ 89:448–460
Morogoro District Council (2012) Morogoro District planning report, Morogoro
Morris ML, Tripp R, Dankyi AA (1999) Adoption and impacts of improved maize production technology: a case study of the Ghana grains development project. Economics program paper 99-01. CIMMYT, Mexico, D.F.
Msuya EE, Hisano S, Nairu T (2008) Explaining productivity variation among smallholder maize farmers in Tanzania. Munich Personal RePEc Archive. https://mpra.ub.uni-muenchen.de/14626/1/Explaining_Productivity_Variation_among_Smallholder_Maize_[10/04/2018]
Muhammad A, Muhamad SH (2012) Analysis of farmer field school (ffs) as a potential source of advanced technology dissemination. OIDA Int J Sustain Dev 3(1):65–70
Murage AW, Obare G, Chianu J, Amudavi DM, Midega CAO, Pickett JA, Khan ZR (2012) The effectiveness of dissemination pathway on adoption of "push-pull" technology in Western Kenya. Q J Int Agric 51(1):51–71
Mutuah BP (2013) Factors influencing adoption of recommended cassava production practices by farmers in Bwari and Kuje Area Councils, Abuja, Federal Capital Territory. A Thesis submitted to the School of Postgraduate Studies, Ahmadu Bello University, Zaria, in partial fulfillment of the requirement for the award of degree of Masters of Science in Agricultural Extension. Department of Agricultural Economics and Rural Sociology, Faculty of Agriculture, Ahmadu Bello University, Zaria July, 2013. http://hdl.handle.net/123456789/5913. [29/08/2017]
Mwanzia KJ (2014) Influence of farmer groups capacity on their ability to advocate for government agricultural services in Khwisero Sub-County Kenya. Dissertation for Award of MA Rural Development at University of Nairobi, Kenya, p 91
Mwaura F (2014) Effect of farmer group membership on agricultural technology adoption and crop productivity in Uganda. Afr Crop Sci J 22:1–11

Nanai NAK (1993) Peasant participation in community development projects: its implication in laying a strategy for participatory extension. Dissertation for Award of MSc Degree at Sokoine University of Agriculture, Morogoro, Tanzania. p 138

Nathaniels QRN (2005) Cowpea, Farmer Field Schools and Farmer-to-Farmer Extension: network paper no. 148. AgREN, Benin, p 15

NBS (National Bureau of Statistics) (2013) 2011/12 Household budget survey: key findings

Ngeno JK (2003) Farmer Field Schools. The Kenyan Experience Report of the Farmer Field School Stake holder's Forum held on the 27th March 2003 at International Livestock Research Institute, Nairobi, Kenya, pp 1–10

Nyangena W, Juma OM (2014) Impact of improved farm technologies on yields: the case of improved maize varieties and inorganic fertilizer in Kenya. Environment for development, discussion paper series EfD DP 14-02

Nyomora A, Kanyeka Z, Nduguru A (2012) Supporting Tanzania's cocoa farmers. Research report no. 3. REPOA, Dar es Salaam, Tanzania, p 43

OECD (2001) Measuring productivity: measurement of aggregate and industry-level. OECD Manual. [www.SourceOECD.org] site visited 14/06/2014

Ojo MA, Jibowa AA (2008) Socio-economic characteristics influencing role performance of rural community power actors in agricultural extension delivery system in Osun State, Nigeria. J Agric Rural Dev 3375(2):27–40

Okoth JO (2013) Tillage and variety effects on soil moisture content, biological nitrogen fixation and soybean (*Glycine Max* L.) yield in Western Kenya. MSc. Thesis, Egerton University, p 45

Owusu V, Abdulai A, Abdul-rahman S (2011) Non-farm work and food security among farm households in Northern Ghana. Food Policy 36:108–118

Pallant J (2011) SPSS Survival manual: a step by step guide to data analysis using SPSS 4th edition. Allen & Unwin, Australia. Printed in China at Everbest Printing Co. p 345

Peterman A, Behrman J, Quisumbing A (2010) A review of empirical evidence on gender differences in non land agricultural inputs, technology, and services in developing countries. Discussion paper 00975. International Food Policy Research Institute, Nairobi, Kenya

Pontius J, Dilts R, Bartlett A (2002) Ten years of IPM. Food and Agriculture Organisation Community Programme, Jakarta

Rugumamu CP (2009) Assessment of post-harvest technologies and gender relations in maize loss reduction in Pangawe village Eastern Tanzania. Tanzania J Sci 35:1–10

Salasya B, Mwangi W, Mwabu D, Diallo A (2007) Factors influencing adoption of stress-tolerant maize hybrid (WH 502) in western Kenya. Afr J Agric Res 2(10):544–551

Schneider K, Gugerty MK (2011) Agricultural productivity and poverty reduction: linkages and pathways. Evans School Rev 1(1):56–74

Simtowe F, Asfaw S, Diagne A, Shiferaw B (2010) Determinants of agricultural adoption: the case of improved groundnut varieties in Malawi. Paper presented at Joint 3rd African Association of Agricultural Economists and 48th Agricultural Economists Association of South Africa Conference, Cape Town, South Africa, 19–23 Sept 2010

Tabi AJ, Vabi MB, Malaa DK (2010) Adoption of maize and cassava production technologies in the Forest-savanna zone of Cameroon: implication for poverty reduction. World Appl Sci J 11(2):196–209

Tsusaka T, Otsuka K (2013) The impact of technological changes on crop yields in sub-Saharan Africa, 1967–2004. In: Otsuka K, Larson D (eds) An African green revolution: finding ways to boost productivity on small farms. Springer Publisher, Dordrecht, Netherlands, pp 1–23

Uaiene R, Arndt C, Masters W (2009) Determinants of agricultural technology adoption in Mozambique. Discussion paper no. 67

Urassa JK (2010) Rural household livelihoods, crop production and well-being after a period of trade reforms: a case study of Rukwa, Tanzania. Doctoral thesis, University of Sussex, Sussex

Urassa JK (2015) Factors influencing maize crop production at household levels: a case of Rukwa region in the southern highlands of Tanzania. Afr J Agric Res 10(10):1097–1106, March 2015

URT (2003) Tanzania agriculture sample census, country report, p 288. www.kilimo.go.tz/agricultural%20statistics/Agric%20census/large%20scale. [10/3/2013]

URT (2004) Ministry of Agriculture and Food and Ministry of Regional Administration and Local Government. Agricultural Extension Reform in Tanzania. A Vision and Strategy outline to year 2010 Main Report. Task Force on Agricultural Extension Reform, Dar es Salaam, Tanzania, p 53

URT (2011) Agricultural sector development programme performance report. Dar es Salaam

URT (2013) The economic survey 2012. Ministry of Finance, Dar es Salaam, Tanzania

Wekesa E, Mwangi W, Verkuijl H, Danda K, De Groote H (2003) Adoption of maize production technologies in the Coastal Lowlands of Kenya. [http://repository.cimmyt.org:8080/xmlui/bitstream/handle/10883/915/448348.PDF?sequence=4&isAllowed=y]. [2/08/2014]

Weyori AE, Mulubrhan A, Waibel H (2012) Agricultural innovations systems and adoption decision: findings from a study of Ghanaian Plantation Sector. Institute of Development and Agricultural Economics, Leibniz University Hannover, Germany, p 30

World Bank (2007) World development report 2008 agriculture for development. Int Bank Reconstr Dev 4(7):39–51

World Bank (2012) Agribusiness indicators: Tanzania. World Bank, Washington, DC. © World Bank. https://openknowledge.worldbank.org/handle/10986/26561. License: CC BY 3.0 IGO. [13/09/2018]

Zvonko P, Gordana G (2009) The use of herbicides for weed control in direct wet-seeded rice (*Oryza sativa* L.). J Plant Prot Sci 45(3):113–188

Nitrate-Nitrogen Pollution and Attenuation Upstream of the Okavango Delta in Angola and Namibia

Andrea Vushe

Abstract Development plans in the Okavango River Basin in Angola and Namibia will abstract 18.13% of mean annual flow for urban water supply and irrigate 108,992 hectares of infertile Kalahari sandy soils requiring high fertilization rates. This alludes to high leaching and hence nutrient pollution of the river system. The loads of nitrates leached from Namibia's existing 4,000-hectare irrigated area into the river's reaches are unknown. Hence, a unique simulation model was developed by combining a gross nitrogen balance, river nitrate transfer and CROPWAT 8.0 models, for estimating quantities of nitrates in agricultural leachate and urban effluent, nitrates transported into the river's mainstream and river reach nitrate attenuation and transfer and to predict future mainstream nitrate levels. The model showed that 7,000 m^3/ha/year was leaching 23% and 18% of applied N fertilizer in wheat and maize crops, respectively, which had negligible impact on river nitrate levels. Irrigating at most 15,659 hectares in Namibia will increase mode nitrate levels from the long-term 0.5 mg/L to 0.6 mg/L. Okavango Delta inflow nitrate levels will increase above 1.0 mg/L if 17.49% of the mean annual flow is abstracted for irrigation and 60 million m^3/year is abstracted for Namibian urban water supply.

Keywords Fertilizer · Irrigated commercial farms · Nitrate transport and attenuation · Kalahari sandy soils · Namibia · Okavango River

1 Introduction

Intensive use of pesticides and chemical fertilizers inputs in agriculture is polluting the environment and hence threatening human health. Agricultural pollutants of greatest concern include pesticides, nitrates in groundwater, trace metallic elements and emerging pollutants. Worldwide, the most common chemical contaminant found in groundwater aquifers is nitrate from farming (Mateo-Sagasta et al. 2018).

A. Vushe (✉)
Namibia University of Science and Technology, Windhoek, Namibia
e-mail: avushe@nust.na

Y. Bamutaze et al. (eds.), *Agriculture and Ecosystem Resilience in Sub Saharan Africa*, Climate Change Management,
https://doi.org/10.1007/978-3-030-12974-3_5

Severe environmental degradation due to nutrient pollution from areas practicing fertilizer applications and little animal manure recycling in arable farms is now evident in many Chinese rivers and lakes (Strokal et al. 2016). Lake Victoria in East Africa and Lake Chivero in Zimbabwe have eutrophication problems due to nutrient pollution from urban and agricultural sources (Kanda et al. 2017; Moyo 1997). Generally, nutrient transport from farms and contamination of mainstreams through groundwater develop gradually for several years, taking 20–30 years before it becomes apparent (Covert 2014), because agricultural pollutants are more difficult to measure and control (FAO 2005), compared to effluent discharges from urban sewers. Developing strategies for management of river nutrient pollution requires understanding nutrient pollutant loading and transport in the river systems, hence the need for appropriate models.

Globally, 56% of water abstracted for various uses is released into the environment as urban and industrial wastewater or irrigation return flow, and the latter is usually the largest volume (Mateo-Sagasta et al. 2018). In the near-pristine transboundary Okavango River Basin (ORB), (OKACOM 2009), with low urbanization and industrial development, expanding irrigation schemes were identified as the main sources of nutrient pollution (Trewby 2003). The urban population and irrigated agriculture are increasing (NPCN 2012; Vushe et al. 2016), and urban effluent and irrigation return flows are unknown, and moreover they are releasing unknown nutrient loads into the river. The fate of the applied N fertilizers, rates of N leaching from irrigated fields and impact of nutrients on the Okavango River (OR) water quality are not well understood. OKACOM (2009) identified large-scale irrigated agriculture as the most potent human activity to cause the greatest impact on the Okavango River's natural systems.

Annually, the ORB-irrigated farms of Namibia produce two crops (maize and wheat), and 400 kg/ha of nitrogen fertilizer is applied annually (Trewby 2003; Vushe et al. 2014) and sometimes over 470 kg N/ha/year (Vushe et al. 2016). Namibian commercial farm general target to harvest at least 10 t/ha of maize grain, and hence the nitrogen fertilizer application rate is at least 250 kg/ha, which is about 1.5 times more than the average amounts recommended in South Africa, for a minimum maize grain yield of 8 t/ha (du Plessis 2003) and USA's Corn Belt (Sawyer 2007; Alley et al. 2009). Mmualefe and Torto (2011) observed that pesticides were present in sediment and water samples in the mouth of the Okavango Delta (Namibia/Botswana border), which implied existence of pollution from human activities upstream of the Okavango Delta. Duda and El-Ashry (2000) predicted that the basin will have serious water shortages by 2025. The planned large-scale intensive agriculture on Kalahari sandy soils may increase pollution from the agricultural activities and human settlements. This can exacerbate the predicted water shortage challenges, especially if compounded by agrochemical pollution and subsequent growth of pernicious aquatic plants in the river water.

Degradation of water resources has been the most globally visible impact of overapplication of nitrogen fertilizers which often gets serious attention when drinking water supplies are contaminated (Lord and Anthony 2002; Schroeder et al. 2004). Surface and groundwater pollution associated with nitrate leaching has become a concern in intensive agricultural areas (Eneji et al. 2013), where nutrient loads are causing algae blooms, hypoxia and toxicity, which lead to loss of fish and other aquatic lives (Heisler et al. 2008; Cohen et al. 2007). Typical examples in the region are pollution effects of Manyame River and reservoirs in Zimbabwe (Moyo 1997) and water quality deterioration of Goreangab and Swakoppoort reservoirs that are urban water sources of central towns of Namibia (Lehmann and Leon 2010). In the event of higher water abstractions and more nutrient loads from increasing anthropogenic activities, the Okavango River may fail to adequately attenuate the nitrate pollutant loads. Hence this study sought to understand the nitrate loading and the river's nitrate attenuation capacity upstream of the delta, which may help in making decisions for the management of nitrate pollution.

2 Study Area

The Okavango River (OR) flow is generated predominantly in Angola, where sparsely populated communities are practicing shifting cultivation on Kalahari sandy soils, the predominant soil of the ORB. The shifting cultivation is slowly giving way to more settled agricultural practices that are changing the savanna miombo woodlands and grasslands (Mendelsohn and el Obeid 2003; Kgathi et al. 2006). In the Angolan headwaters, mean annual rainfall is over 1200 mm/year, in Namibia at Rundu mean annual rainfall is 577 mm, and the Okavango Delta receives less than 450 mm/year. Average annual inflow into the Okavango Delta in Botswana is 9,350 million m^3/year, where up to 98% of the inflow is lost as evapotranspiration and infiltration in the Kalahari sandy soils (Vushe et al. 2014; Wolski et al. 2010; Farquharson et al. 1990). As shown in Fig. 1, the Cubango River which enters and forms the Angola/Namibia boundary at Katwitwi and Cuito Rivers are the two main tributaries of the OR, which exits Namibia at Kwetze. The Cuito River contributes about 45% of the 9,350 Mm^3 mean annual discharge into the inland delta, measured at Mohembo flow gauging station in Botswana. The mean annual discharge rates of the Cubango tributary measured at Rundu is 5,200 Mm^3. Namibia and Botswana contribute almost no surface water into the delta, on the entire length of 400 km, because of semiarid climatic conditions (el Obeid and Mendelsohn 2001; OKACOM 2010).

In 2017, Namibia was the only country using the Okavango River for over 4000 ha irrigated commercial agriculture, whereby 75% of the current irrigated area is located upstream of the Cuito confluence (18° 00′ 39″ S and 20° 52′ 19″E). Generally, the Kalahari sandy soils have high infiltration rates, inherent low fertility and low nutrient and water retention. Average fertilizer application rates are about 195 and 280 kg N/ha for wheat and maize, respectively (Vushe et al. 2016), which were higher than rates recommended by Yang et al. (2017), for an intensive wheat/maize rotation that obtained high crop yields with low environmental risks. Therefore, high nutrient leaching rates are expected to occur in Kalahari sandy soils, which can be a serious threat to the ecological health of the OR (OKACOM 2009;

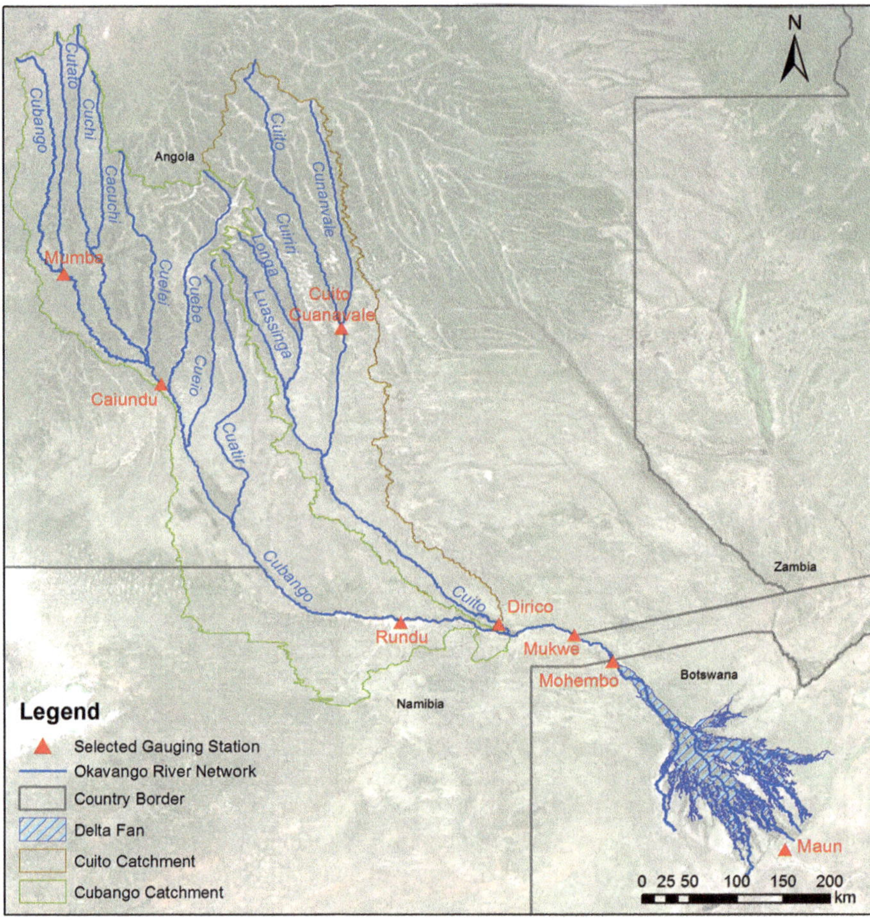

Fig. 1 Okavango River Basin including the main tributaries, flow gauging stations and the Okavango Delta with its entrance at Mohembo gauging station and outlet at Maun (Data source: NIWR, ORI; background, MODIS 16-day BRDF-adjusted NBAR data (MCD43A4) courtesy of EOSDIS-NASA) in Steudel et al. 2013

MET 2000). Runoff generated in the irrigated fields drains into numerous local drainage sinkholes; hence subsurface flows in the Kalahari sandy soils and underlying calcrete and silicate formations mainly carry the runoff and leachate to the mainstream. Groundwater flow direction is chiefly towards the river, and any recharge of river water into the Kalahari aquifer occurs during floods; hence groundwater interactions can transport the nitrate-polluted runoff and leachate into the mainstream (Vushe et al. 2016). Wastewater from waste stabilization ponds and pit latrines and sceptic tanks is also released indirectly by infiltration into the OR riparian zones. Therefore, OKACOM (2009) has sensible fears that increasing polluted return flows from irrigated agriculture and urban wastewater are likely to cause significant hydrochemical pollution and damage to the ecological health of the Okavango River.

2.1 *Irrigated Agriculture Development and Urbanization Trends and Driver Upstream of the Okavango Delta*

The most densely populated part of the ORB is in the Kavango Regions of Namibia (Trewby 2003), and 68% of the population of the Kavango Regions in Namibia are living within a locus of 10 km from the Okavango River. From 1991 to 2001, the Kavango Region experienced a 90% population increase to 222,500, probably because of the Angolan Civil War between 1975 and 2002 that could have forced emigration from large rural parts of southern Angolan ORB and resettlements, probably in northern towns of the ORB in Angola and across the river in the Namibian part of the ORB (Kgathi et al. 2006). Rundu, the largest town in the ORB, abstracted about 6.0 Mm^3 of water per year (OKACOM and FAO 2014) to supply a 63,431 population that increased by 72% between 2001 and 2011 (NPCM 2012) and will probably count 100,000 people by 2025 (Domptail and Mundy 2014).

Between 1993 and 2012, human activities along the river increased, and large-scale commercial irrigated agriculture increased from 300 to 2,600 ha (Vushe et al. 2014). In 2017, the area had increased above 4,000 ha, and water allocation for irrigation was 15,000 m^3/ha/year. The Namibian government has plans for expanding irrigation in the ORB, and in a high development scenario, 15,659 ha would be irrigated in Namibia, abstracting 2.18% of the total average flow per year (OKACOM 2009). Angola's long armed conflict restricted development in the ORB (OKACOM 2009). The Angolan ORB had a population of less than 370,000 that practices shift cultivation on about 740,000 ha of dryland (Mendelsohn and el Obeid 2003), which currently has low impact on nutrient water quality (Vushe et al. 2014). Angola plans to develop over 200,000 ha and abstract 7000 m^3/ha/year (OKACOM 2017; FAO 1997). Botswana has 6,060 ha irrigation potential, but she had no plans for major irrigation development around the delta (FAO 1997; Masamba and OKACOM 2009). The semiarid climate of the ORB in southern Angola and Namibia receives erratic summer rainfall, which is often inadequate for crop production (Vushe et al. 2016); hence food security challenges for the increasing population can be the main driver of irrigation development in the ORB.

2.2 *Historic Okavango River Nitrate-Nitrogen Pollution and Attenuation*

In the hydrological year 2011/2012, river water quality measurements showed that water flowing from Angola had excellent quality according to Namibian Drinking Water Standards. The water had higher total nitrogen and total phosphorus concentrations, which were attenuated and passed on to the Okavango Delta (in Botswana), with a lower nutrient load (Vushe et al. 2014). Although nitrates were leached from the irrigated field and transported into the uncultivated riparian zones of the Okavango River (Vushe et al. 2016), overall human activities had low impact on the water quality of the river which was adequately attenuating the nutrient load that it received. Mode nitrate levels measured by NamWater since 2001 were below 0.5 mg/L (for over 80% of the measurements), with temporal fluctuations in both low-flow and high-flow periods, for example, a few spikes in nitrate concentration reached 1.5 mg/L (Vushe et al. 2014).

The OR system which is characterized by relatively steady flows in a river channel that meanders through a vegetation-covered flood plain, naturally attenuates the nutrient load it gets in Angola and Namibia, and hence inflows into the delta had lower nutrient levels (Vushe et al. 2014). The seasonally inundated flood plains of the Okavango River, often with a rich mix of pasture vegetation and reeds that include papyrus, phragmites (common reed), cattail reeds and grasses, provided grazing for livestock and wildlife especially during dry seasons (Bethune 2009; OKACOM 2011). If the flood plain/wetland plants are harvested and/or browsed by animals regularly, then active growth and high N removal rates in the flood plain may be maintained for long periods of time.

3 Determinants of Nitrate-Nitrogen Load from Urban and Agricultural

Nitrogen load from urban and agricultural sources transported into rivers depends on nitrogen concentration of return flows, and much of the nitrogen pollution from human contributions enter drainage systems, primarily as nitrates (Nee 2015). Although natural sources of nitrates in groundwater exist, concentrations greater than 1 mg/L nitrate-N are generally a sign of fertilizer or sewage effluent contamination (Brodie and Mitchell 2006). The main sources of nitrogen pollution into the Okavango River were identified as return flows from irrigated farms and urban settlements.

Nitrate loads from urban return flows are dependent on urban water demand, return to sewer and biological nitrogen removal efficiencies of wastewater treatment systems. Nitrate-nitrogen load from agricultural return flows depends on agricultural management, especially irrigation water demand, fertilization and leaching rates. Nitrogen leaching on most soils is intrinsically linked with soil water dynamics, especially nitrates are highly water soluble and are assumed to move with the wetting front (Zotarelli et al. 2007).

Global studies partitioned the total N (especially fertilizers) used by the farmers into the following: 35% was utilized by the crops, 35% was lost to the atmosphere annually, and the rest (30%) was lost to drainage or natural river systems (Islam et al. 2008). Nitrate pollution load from agricultural fields can be estimated at a river basin scale using a gross nitrogen balance (GNB), a mass balance of nitrogen input and outputs. Inputs are wet and dry atmospheric deposition, mineral fertilizers applied, biological fixation, nitrogen credit in irrigation water, soil organic matter, manure and crop residues. Outputs are crop uptake (crop residue and harvest), gaseous losses (denitrification and volatilization), microbial uptake and leaching (Buss et al. 2005).

Nitrogen balances are appreciated for being objective, transparent and readily verified agro-environmental indicators (Lord and Anthony 2002) and can be assessed or computed on a farm as well as at a national scale (Oenema et al. 2003; Watson and Atkinson 1999). The agro-environmental indicators derived at representative agricultural enterprises can also be up-scaled regionally (OECD and EUROSTAT 2007). Despite some inaccuracies in predicting observable nitrate levels under short time steps, the GNB is a statistically reliable method as a groundwater nitrate pollution predictor (Wick et al. 2012; Wijayanti et al. 2017), especially for annual time steps and at large spatial scales. Hence in the simulation model for nitrate pollution, transport and attenuation in the Okavango River, the GNBs are used to estimate the average annual load of nitrates leached from the irrigated farms in the river basin. Hence, farm records of fertigation schedules and rainfall were used in the GNB model (Wijayanti et al. 2017) in combination with a CROPWAT 8.0 model to estimate the water used by the crop and leachate load. Data on nitrate quality of river water was obtained from measurements done on field campaigns that took grab river water samples in hydrological years 2001/2002 and 2011/2012 and historical records (between 2001 and 2017) kept by NamWater's water treatment plants that abstract water from Okavango River reaches in Namibia.

3.1 Biological Fixation as an N Input in Cereal-Cropped Fields

One natural source that inputs nitrogen into all cropped fields is the several types of free-living nitrogen-fixing bacteria that grow within the rhizosphere of host tropical grasses, including cereal crops such as wheat, maize, oats and barley (Wagner 2012). Nitrogen fixing by non-symbiotic, free-living soil bacteria only averages about 3 kg/ha/year within a range of 1–5 kg per year (Schlesinger and Bernhardt 2013). In the simulation model, as per estimates by OECD and EUROSTAT (2007), the quantity of nitrogen fixed in the soil by non-symbiotic bacteria in a cultivated field was estimated as the average (3 kg N/ha/year), with a calibration factor ranging from 0.3 to 1.7.

3.2 Atmospheric Deposition of Nitrogen in Fields

The wet and dry atmospheric depositions of nitrogen compounds on the irrigated farms were estimated as the product of cultivated agricultural area and the nitrogen deposition rates (OECD and EUROSTAT 2007). Average wet atmospheric deposition estimates were 7.0 kg per ha per year with a calibration factor ranging 0.4–1.6, and dry depositions estimates were 4.75 kg/ha/year with a calibration factor ranging 0.5–2. The summer rainy season is about 6 months long; hence it was assumed that the summer crop (maize) got half of the wet and half of the dry depositions, while the winter wheat crop obtained half of the dry deposition, with no wet deposition because the winter is a dry season.

3.3 Nitrogen Credit in Irrigation Water and Soil Organic Matter

The average amount of N contributed by irrigation water was estimated as the product of annual average TN concentration and irrigation water applied (Patterson 2003). The Okavango River's long-term average TN level of 2 mg/L was used (Vushe et al. 2014). Rundu town was identified as the only urban settlement with significant contribution of wastewater return flows to the Okavango River. The town has three wastewater treatment plants, and all are wastewater stabilization ponds (WSPs). The efficiency of the Rundu ponds was unknown; hence, the nitrate levels in the WSP effluent were assumed to be the mean of range 5.1–14.4 mg/L (Gad and Abdalla 2017); therefore the calibration factor ranged between 0.51 and 1.44.

Soil organic matter mainly from crop residues could provide some nitrogen nutrients; therefore organic matter content of the soil was measured. It was assumed that for each percent of organic matter measured in the soil, 22.42 kg N/ha could be converted into mineralized N every year which was divided equally between the maize and wheat crops (O'Leary et al. 2002; Roy et al. 2006; Angus 2001).

3.4 Agricultural Crops and Fertigation Scheduling

Maize and wheat are the main crops in Namibia's green schemes, mainly for national food security, and other horticultural crops are produced at a smaller scale. Average N fertilizer application in Namibian irrigation schemes were about 288 and 195 kg/ha (Vushe et al. 2016) and measured average soil residual N after harvesting maize and wheat crops were about 35.7 and 4.5 kg/ha, respectively. Farming practice in Angola's dryland farms was mainly slash-and-burn clearance of woodlands for shifting cultivation of pearl millet, maize, sorghum and manioc (cassava) crops. Two crops were planted per rainy season, and the peasant farmers shifted to new lands after 4 to 6 years with fallow periods of about 7 years (Mendelsohn and el Obeid 2004); hence no fertilizers were applied in Angolan farms.

In irrigated farms infield fertilizer application, less uptake and gaseous losses were assumed to be equal to the sum of residual fertilizer and leaching losses. Nitrogen leaching losses depended on nitrogen source and application rates, crop removal capacity and water displacement below the active root zone. Excessive irrigation and/or nitrogen application rates combined with intense rainfall on excessively drained sandy soils with low water-holding capacity greatly enhance the potential risk of nitrogen leaching. Leached nitrate load from cereal-cropped fields was estimated using the GNB equation. The GNB was used to estimate the mass of excess nitrates that were available to the crop in each cropping season, which was computed as total nitrogen inputs into the field minus total nitrogen outputs, according to the methods stated by (OECD and EUROSTAT 2007). A fraction of the excess nitrates was leached from the root zone, and hence the GNB was used to estimate the leached nitrates in each cropping season.

3.5 Sources and Sinks of Nitrogen Nutrients in the Cropped Fields

Six main sinks of N in the irrigated field, riparian zones between the field and Okavango River's mainstream were uptake by the wheat and maize crops; gaseous losses due to volatization, nitrification and denitrification; microbial sequestration; leaching and uptake by riparian savannah vegetation; and seepage into the river channel. Generally, plants take up 50 percent of the nitrogen fertilizer applied to agricultural land (FAO and IFA 2001), and N losses due to volatilization and leaching may range from 0 to 56 kg Ha^{-1} $year^{-1}$ (DeBruin and Butzen 2016).

3.6 Gaseous N Loss Estimates in the Field

Okavango farmers applied nitrogen fertilizers in form of urea, a compound, calcium nitrate, sulphate of ammonium or ammonium nitrate. For a more accurate GNB equation, estimates of N loss from applied fertilizers due to conversion of solid N into N gases were made. Nitrates in the soil and groundwater can be lost to the atmosphere during microbial denitrification processes. This is a sequential, stepwise reduction of nitrate to nitrogen gases which involves the following processes: nitrate to nitrite to nitrogen oxide (NO) to nitrous oxide (N_2O) to nitrogen gas (Smith et al. 2004). NO loss rates in the range of 0.003–11% of applied N fertilizer were obtained on different types of soils with different plants (Yan et al. 2003; Huang and Li 2014). The annual NO emissions in some Chinese uplands were within a range of 0.29–50.46 kg N/ha/year. Global NO emissions from uplands ranged from 0.2 to 23 kg N/ha/year, whereas annual NO emissions from Chinese uplands in 2012 were greater than the global average probably due to a higher N fertilization rate of 380 kg N/ha/year (Huang and Li 2014); hence higher fertilization rates can cause higher NO emissions. NO emissions are highest when the soil is dry and are almost equal to N_2O emissions when soil moisture is at field capacity. N_2O emission rates increase with temperature and soil moisture content, and when the soil is moisture saturated, N_2 is emitted (Signor and Cerri 2013).

Therefore, in this study NO gas losses were estimated at about 10% of N in the soil with a calibration range of 0.003–1.1, and NO loss limits ranged from 0.29 to 50.46 kg N/ha per cropping season. N_2O emissions were computed according to IPCC (2006) methodologies for use in national GHG inventories which stated an emission factor of 16 kg N_2O-N/ha/year for tropical climates with an uncertainty range of 5–48 kg N_2O-N/ha/year. Therefore, N_2O losses were estimated as 8 kg N_2O-N/ha for each cropping season with a calibration factor ranging from 0.3 to 3.0.

Volatilization of NH_3 from urea and other N fertilizers is controlled by many diverse soil properties and environmental conditions that make losses difficult to predict in the field (Jones et al. 2007). NH_3 volatilization rates were 6.2–20.6 kg N/ha, and N losses were 2.5–8.9% during one maize-growing season (Han et al. 2014). Application of urea alone can cause cumulative NH_3-N losses to be 7.2–9.7% of the applied urea. Generally, NH_3 and NO losses can be 10% of applied N artificial fertilizers if rates are less than 180 kg/ha (IPCC 1996). Therefore, the average NH_3 volatization loss rate was estimated as 8.45% for urea, with a calibration range from 0.9 to 1.2, and 5.7% NH_3 loss for other N fertilizers, with a calibration range from 0.4 to 1.6.

3.7 *Leached Water and Nitrate Leachate Estimation*

The GNB equation was used to estimate nitrates leached from irrigated fields in Namibia, and a water balance was used to estimate seasonal quantities of leached water. Historic and recent farm records of crop management including fertigation schedules were obtained from the Okavango-irrigated commercial farms between the years 2014 and 2017. Residual nitrates after each crop harvest were measured at Mashare commercial farm, located at 17°53′S, 20°11′E, about 40 km east of Rundu town. Mashare commercial farm records on crop management practices were compared with records from other irrigated commercial farms. The agronomic practices and fertigation schedules for all farms were similar because they all had target yields of 10 t/ha and 6 t/ha for the maize and wheat crops respectively. Basing on the Namibian government's food security policy, the agronomic practices were assumed to remain the same in the future (OKACOM 2009). Long-term rainfall data was collected from weather stations at Rundu Weather Station, Mashare Research Centre. The Food and Agricultural Organization (FAO) CROPWAT 8.0 model was used to estimate the seasonal crop evapotranspiration per hectare and effective monthly rainfall. Input data into the CROPWAT 8.0 model included climatic data (monthly averages) and information related to the soil and the crops.

Centre pivot sprinkler irrigation technology was the main irrigation system, and the irrigation efficiency was assumed to be 83%, which was the average of range 75–90% for low- and high-pressure centre pivots (Bauder et al. 2014). Water available to the crop was calculated as the sum of effective rainfall and net irrigation water applied. The average leached water was the difference in available water and crop water demand or evapotranspiration. The annual irrigation return flow into the Okavango River was estimated as a ratio of leached water to the irrigation water abstracted each year. The GNB for the irrigated farms was used to partition the total N nutrients input into the amount utilized by each crop, lost as gases and the N amount leached beyond the root zone similar to a procedure used by Cui et al. (2009).

The nitrate leachate load transported from the farms was calculated as the product of nitrates leached and the leached water quantity. The leached water was estimated as the sum of irrigation water and effective rainfall minus crop water demand. The nitrate load entering the river's mainstream was estimated as a product of loss fractions in series (of dry woodland and seasonally inundated flood plain) riparian zones and the nitrate leachate load transported from the farms.

3.8 Estimating Wheat and Maize Crop Nitrogen Uptake

Generally, wheat grain yield and N uptake (in grain and straw) response to N application rate fit a linear plus plateau model (Liu et al. 2016; Cui et al. 2009). The ratio of wheat grain yield to N uptake rate (YNR) equal to 0.052 t grain/kg N, obtained in a temperate climate, was assumed to be inappropriately high, compared to Australian commercial farming conditions. Studies in Australia obtained wheat crop N uptake of about 50 kg N ha^{-1}, and the average yield was 1.9 t/ha (Carson and Phillips 2016; Angus 2001); hence the YNR was 0.038 t grain/kg N. Australian conditions were assumed to be similar to Namibian conditions. The highest ratio of harvested grain yield to gross soil N available to the crop was 0.036 t/kg N (FAO 2006), and this was defined as the uptake efficiency of N (UEN) in this study.

At the economically optimum N rate, maize grain yield and N uptake response to soil nitrate-N content could be simulated by a linear plus plateau model (Cui et al. 2008; Sawyer 2007). According to Cui et al. (2008), for a maximum grain yield, the maize crop can take about 186 kg N/ha from the soil. Connor et al. (2011) stated that, in Australia, the maize's N fertilizer use efficiency was highest (and N losses were smallest) when application rates were below 224 kg N/ha since little inorganic N remained in the soil after harvest and the maximum yield for maize was obtained at this application rate. Each metric tonne of maize grain removed 15.0–18.0 kg of nitrogen from the soil (du Plessis 2003) which implied that the highest maize's YNR was 0.067 t/kg N.

Residual nitrogen was measured in soil samples after each harvest, and gaseous N losses were estimated as per IPCC (2006) and IPCC (1996) methods and global estimates. The leached nitrates were estimated using the GNB equation:

$$\mathrm{GAN} = \mathrm{GNL} + \mathrm{LN} + \frac{\mathrm{HGY}}{\mathrm{YNR}} + \mathrm{RN} \tag{1}$$

where GAN is the gross available nitrogen, GNL is the gaseous N losses, LN is the leached nitrogen, HGY is the harvested grain yield, RN is the residual nitrogen in the soil root zone and YNR is the harvested grain yield to N uptake ratio and, for

South African maize and wheat varieties in Namibian agronomic conditions, YNR was estimated as 0.067 and 0.038, respectively, while for temperate condition wheat YNR was found to be 0.052.

The product of crop's YNR and the farmer's target yield were used to estimate the N demand or uptake under optimum and suboptimum conditions. It was assumed that the YNR depended on the crop (type and variety) and readily available soil N. To achieving a target yield, a farmer applied a specific amount of N fertilizer; hence the crop's nitrogen uptake efficiency UEN was assumed to depend mainly on the amount of gross available nitrogen (GAN). The GAN was the sum of applied N fertilizer, atmospheric N deposition, biological N fixation, soil organic matter N releases, N from irrigation water and residual N from previous crop. Therefore, it was also assumed that under optimum conditions the optimum yield harvested was equal to the target grain yield (TGN) which could be estimated by the equation:

$$\mathrm{TGN} = \mathrm{UEN} * \mathrm{GAN} \tag{2}$$

where UEN is the nitrogen utilization efficiency of the crop under optimum conditions. The gross unutilized N (gaseous N loss + leached N + residual N) was assumed to be at a minimum, and N uptake for luxuriant vegetative growth would be negligible.

It was assumed that the leachate was transported mainly by subsurface runoff into the riparian zones of the river, because the sandy soil has high infiltration rates, since erosion and direct runoff into the river's mainstream is minimal even on 5% slopes, but ORB average slope was less than 0.07% (OKACOM 2009; MET 2000). As leachate upwells in the riparian zone and mainstream and the nitrates are transported downriver, sequestration and denitrification attenuate the nitrate load. Sequestration of the nitrates by plants and microbes is a short-term process, but the long-term nitrogen removal is played by the denitrifying bacteria in the hyporheic zone, the interface between groundwater and surface water where exchange of nutrients occurs (Nee 2015). In-stream denitrification happens primarily within the oxygen-depleted hyporheic zone and normally removes about half (50%) of the original leached N input (Seitzinger et al. 2002; Boyer et al. 2006).

Modelling N attenuation in river systems with vegetation must account for the intrinsic linkage of N processes to the C:N ratio of the environments and microbial activities which can be affected by many environmental factors. Balances between N inputs and outputs lead to ecosystem N retentions or releases, with associated carbon storages or releases depending on C:N ratios in the riparian zones, hyporheic zones and stream water and the soils under vegetative ecosystems. Leaching and denitrification fluxes are not always well understood even biogeochemically, and in many models the ultimate fate of the N after it leaves the boundaries of the model generally is not specified (Nevison et al. 2016). For example, in agricultural lands a fraction of N lost to the atmosphere through volatization and denitrification could be a part of the local N cycle, due to possible dry and/or wet N redepositions. Since levels of C:N ratios affect retention and release of N, hence terrestrial ecosystem models utilizing carbon-nitrogen interactions can be the most appropriate prognostic tools for modelling the export of N in river basins to coastal zones. Most current models of river N export are primarily diagnostic rather than prognostic (Wollheim et al. 2008; Nevison et al. 2016). In this study, the prognostic tool was a simulation model of the OR nitrate levels based on mass balance relationship, water abstraction and return flows with dissolved nitrogen (mainly as nitrates), as well as estimation of nitrate concentration attenuation by denitrification along the length of the river. Denitrification loss within the OR mainstream was parameterized a function of hydrological properties of the river.

The attenuation capacity of riparian zones and the hyporheic zones determined how much of the leached nitrates are transferred by the river downstream. The longitudinal attenuation and transfer of nitrates in the river's mainstream was estimated using equations by Wollheim et al. (2008). Mean annual TN removal in the water body within each river reach was modelled as a function of upstream plus local nutrient inputs, hydrological characteristics and biological activity. It was assumed that first-order benthic processes dominated mean annual nitrogen removal (Seitzinger et al. 2002; Boyer et al. 2006). Biological nitrogen removal activity or uptake velocity was adjusted for local conditions (Wollheim et al. 2008). Many studies seem to show that unamended denitrification (U_{DEN}) in natural riverbeds could be ranging from 0.2 to 75 mg N/m^2/h.

3.9 Hydrological Properties and Denitrification in the River Mainstream

Denitrification loss in flowing water can be parameterized as a function of hydrological properties such as reach length, flow rate and velocity. Fractional loss is greatest in small, slow-moving streams, where long residence times and shallow depths allow for frequent contact of dissolved nitrogen with sediments (Wollheim et al. 2008; Nevison et al. 2016). Although river N export estimates are model based, they are useful as reference values in cases where observations are not available (Wollheim et al. 2008). Therefore, the model provided in Wollheim et al. (2008) was appropriate for estimating annual rate of nitrate and total nitrogen attenuation and transport or export of rivers like the Okavango River where nitrate attenuation capacity was unknown (Trewby 2003). Mean annual TN removal in the water body within each river reach can be modelled as a function of upstream plus local nutrient inputs, hydrological characteristics and biological activity (Wollheim et al. 2008).

Total nitrogen (TN) flux exported from each river reach i (kg/year) can be estimated based on a modification of the model of (Green et al. 2004; Wollheim et al. 2008), determined as in Eq. (3):

$$\mathrm{TN}_i = \left[\mathrm{LocalPoint}_i + \left(\mathrm{LocalNonPoint}_i \times \mathrm{TE}_{\mathrm{localriver}_i}\right) + \mathrm{UpstreamIn}_i\right] \times \mathrm{TE}_{\mathrm{largeriver}_i} \times \mathrm{TE}_{\mathrm{lake}_i} \quad (3)$$

where LocalPoint = all point sources in reach or aquatic system i (kg/year), LocalNonPoint = all nonpoint N inputs from tributaries into the reach or aquatic system in i (kg/year), UpstreamIn = TN inputs from upstream reach immediately upstream from *i* (kg/year) and TE = transfer efficiency associated with the local tributary rivers, large river channels, lakes and/or reservoirs. TE = 1 – *R*, where *R* was the proportion of inputs removed by the water body or river reach, and *R* and TE were unitless.

Transfer efficiency (TE) for each reach (water body) was a function of hydrologic conditions and biological activity. For N over annual timescales, benthic processes were assumed to dominate N removal via denitrification or sequestration and were represented as:

$$\mathrm{TE} = \exp\left(V_f / H_L\right) \tag{4}$$

where V_f represented biological removal activity (N uptake velocity), or mass transfer coefficient of the nutrient (m/year), and H_L represented hydrological conditions in the form of the hydraulic load (m/year).

According to Lansdown et al. (2015), annual removal rate of nitrates highly depends on the river's benthic or denitrifying active surface area. Benthic processes were assumed to dominate, and, hence, biological and hydrological factors were separated as in Eq. (4). This Eq. (4) was used to scale biological activity across water bodies of widely varying size, especially in the USA (Wollheim et al. 2006; Wollheim et al. 2008). H_L can be calculated in several related ways as follows:

$$H_L = \frac{dv}{L} = \frac{d}{t} = \frac{Q}{A_{sw}} = \frac{Q}{wL} \tag{5}$$

where *d* is mean water depth (m), *v* is mean water velocity (m/year), *w* is mean channel width (m), *L* is channel length (m), *t* is residence time (year), *Q* is discharge (m^3/year) and A_{sw} is surface area of the water body (m^2).

For each reach, the amount of nitrogen removed $TE_{riverreach}$ was determined using Eq. (4) based on river flow or discharge *Q* and benthic surface area (wL) of the reach. The width of the reach was assumed to be the mean annual channel width (*w*). Since the removal rate of nitrates (mg/year) highly depend on the river's benthic or denitrifying active surface area (Lansdown et al. 2015), therefore, the denitrifying active surface area was estimated as the product of the reach length and the width of the river, which was related to annual flow rate (*Q*). According to Wollheim et al. (2008), the *Q* fields rely solely on accumulated water balance model (WBM) – computed runoff – and the mean annual channel widths (*w*) are defined by the equation:

$$w = 8.3Q^{0.52} \tag{6}$$

Equation (6) was based on the analysis of US stream gages and was assumed valid globally. Hence using the river's mean annual flow rates (*Q*), only w is required to calculate H_L using Eq. (5).

3.10 Biological Processes and Mainstream Nitrate Concentration

It was assumed that first-order benthic processes dominated mean annual nitrogen removal, and the dissolved nitrogen pollutants were nitrates. Biological nitrogen removal activity or uptake velocity was estimated using the equation:

$$V_f = \mathrm{U}_{\mathrm{DEN}} / C_n \tag{7}$$

where $\mathrm{U}_{\mathrm{DEN}}$ was the areal process rate (mg-TN/m^2/year) and C_n was nutrient concentration in the water column (mg-TN/m^3). V_f was applied assuming local mean annual water temperature control variability (Wollheim et al. 2008). The reference TN uptake velocity (TN-V_{fref}) = 35 m/year for a reference temperature (T_{ref}) of 20 °C. A value of 35 m/year was typical for denitrification based on measurements in lake and river systems (Howarth et al. 2006; Pina-Ochoa and Alvarez-Cobelas 2006; Wollheim et al. 2008). For local reaches and climatic conditions, the V_f was determined assuming V_f changes by a factor of 2 for every 10 °C change in temperature (i.e. V_f doubles for every 10 °C increase in temperature (Donner et al. 2002; Seitzinger 1988; Wollheim et al. 2008). According to Wollheim et al. (2008), V_f ranges from 8 to70 m/year between 0 and 30 °C. For the Okavango River water, the V_f was determined using the long-term average temperature of 25 °C.

4 Simulation Modelling of Nitrate Levels in the Okavango

4.1 Projected Changes in Water Demands and Predicting Nitrate Pollution from ORB Riparian Settlements and Irrigated Agriculture

The hydrological year 2015/2016 scenario was used for testing the impact of most recent developments and two future scenarios were used for prediction of future nitrate loads due to human developments upstream of the Okavango Delta. The future scenarios were as follows: (i) expected expansion in irrigation area in Namibia from about 4,000 ha (in 2017) to 15,659 ha and urban population in Rundu of 100,000 and (ii) the development of over 115,000 ha irrigation schemes in Angola and Namibia and interbasin transfer of 60 Mm^3/year to central and coastal towns of Namibia.

4.2 Model Calibration and Validation

Simulations of river water nitrate levels were done for the hydrological years 2001/2002, for calibration of the simulation model. The input data was obtained from records of Rundu population, irrigated area, crops, climatic data and long-term average annual river flow at Rundu and Mukwe (as per recommendations by Legates and McCabe (1999)); and the long-term mode nitrate levels were measured in mainstream water grab samples taken between Katwitwi and Kwetze (Steudel et al. 2013; NamWater 2013, 2017; Trewby 2003). Long-term mode values of nitrate concentration were assumed to be the most probable or representative river nitrate levels (U.S. Department of Energy 1982; Martin 2013), since river nitrate levels were assumed to have a log normal frequency distribution in a free-flowing river that naturally remove nitrogen (O'Gorman 2004).

The main calibration factors were areal denitrification rate (U_{DEN}), crop target yield, grain yield to nitrate ratios (YNR), return to sewer percentage, average nitrate level in WSP effluent and riparian and hyporheic zone (in series) attenuated percentage. The selected values of calibration factors gave the lowest mean absolute percentage error (MAPE) and a relative efficiency index of agreement (d_{rel}) lying between 0 and 1, for no correlation and perfect fit, respectively (Hauduc et al. 2011; Krause et al. 2005). The calibrated model had a MAPE of 7.48% and d_{rel} efficiency of 0.75.

Agricultural management data and mode nitrate levels measured in the hydrological year 2011/2012 (NamWater 2013; Vushe et al. 2014) were used for model validation. The main Namibian contributors of nitrates were assumed to be the 63,431 people of Rundu town and the 2,600 ha under irrigation. The population of urban centres and villages had changed, but the changes were assumed to have negligible impact; hence Rundu town's 2011 population census value was used. On validation, the model slightly underestimated nitrate levels in all reaches although the nitrate levels approximated 0.5 mg/L. The MAPE reduced to 3.6%, while the d_{rel} efficiency remained constant at 0.75, which was similar to the d_{rel} value obtained in the calibration; hence the model was good enough for prediction of nitrate levels in the Okavango River mainstream.

4.3 Scenario 0: Irrigated Area in Namibia in 2016 and No Irrigated Area in Angola

Namibia irrigated mainly wheat and maize crops on 3,445 ha along the Cubango tributary and 555 ha downstream of the Cubango-Cuito confluence at 15,000 m^3/ha/year and about 6 Mm^3/year for Rundu urban water supply for a population of over 63,431. It was assumed that in 2016 Angola was using a negligible amount of Okavango water for industrial purposes and urban water supply. Figure 2 shows the simulated and measured nitrate levels in hydrological year 2015/2016, when 4,000 ha was irrigated in Namibia and there was no irrigation in Angola.

The simulation model almost perfectly estimated observed nitrate levels because mean absolute percentage error (MAPE) was 3.15% and a relative efficiency index of agreement (d_{rel}) was 0.75 (Hauduc et al. 2011; Krause et al. 2005).

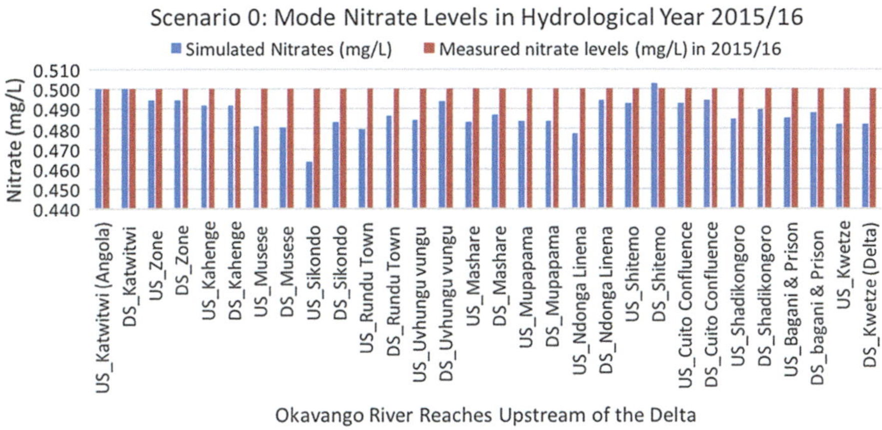

Fig. 2 Simulated and measured nitrate levels in hydrological year 2015/2016 4,000 ha irrigated in Namibia and no irrigation in Angola (Scenario 0)

4.4 *Scenario 1: High Irrigation Development (15,659 ha) in Namibia and No Irrigation in Angola*

In a high development scenario, OKACOM (2009) stated that 15,659 ha would be irrigated in Namibia abstracting 2.18% of the total average flow per year. The irrigation upstream of the Cuito-Cubango confluence would be 4,545 ha, which will abstract 37.27% of the river's low flow rate, while 11,114 ha could be irrigated downstream of the Cuito-Cubango confluence, and 15,000 m^3/ha/year will be allocated mainly for irrigating wheat and maize crops. Hence in this scenario, Namibia will fully develop irrigated wheat and maize crop production, while Angola and Botswana use a negligible amount of Okavango water for irrigation. The mode nitrate levels of 0.5 mg/L at Katwitwi and Cuito confluence were used as long-term inflow nitrate levels from unpolluted parts of the ORB. Figure 3 shows the simulated results for the Scenario 1, whereby Namibia irrigates 15,659 ha, and there is no irrigation in Angola.

As shown in Fig. 3, the simulation model indicated that mode nitrate levels will increase to 0.61 mg/L in water flowing into the delta. The model predicted that increasing Rundu population from 63,431 to 100,000 and reducing effluent nitrate levels from 10.0 mg/L to 1.0 mg/L had negligible effect on river nitrate levels.

4.5 *Scenario 2: High Irrigation and Urban Water Demand (1694.88 Mm^3/year) as a Worst-Case Scenario*

According to FAO (1997), Angola has a potential for irrigation of 200,000 ha at 7,000 m^3/ha/year which means full development will abstract 15% of mean annual flow and Botswana has potential for 6,060 ha using 6,000 m^3/ha/year, hence abstract

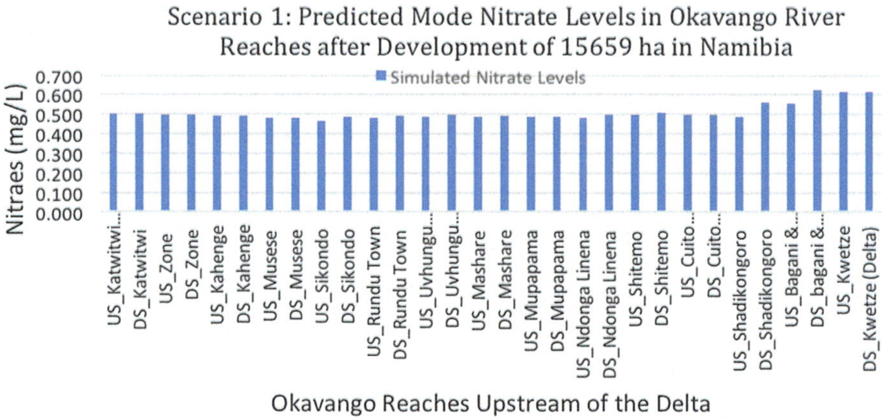

Fig. 3 Simulated results of high irrigation development in Namibia and no irrigation in Angola (Scenario 1)

36.36 Mm^3/year. Namibia would irrigate 2,000 ha/year using 5,500 m^3/ha/year for as well as transfer 60 Mm^3/year of water from the Okavango River to the central and western coastal towns of the country, hence abstract 71 Mm^3/year. In 2017, Namibia was irrigating over 4,000 ha using 15,000 m^3/ha/year, and government plans and approved water permits showed that 5,960 ha will be irrigated on the Cubango tributary. OKACOM (2009) stated that Namibia will irrigate 15,659 ha in high irrigation development scenario. Therefore, Scenario 2 is this study's highest water demand scenario in the ORB which was described as follows:

Namibia will fully develop its Okavango green schemes and irrigate 15,659 ha, abstracting 234.885 Mm^3/year; Rundu urban population will grow to 100,000, as per projection by Domptail and Mundy (2014). At least 60 million m^3/year will be abstracted downstream of the Cubango-Cuito confluence for urban water supply to the central and coastal towns of Namibia. It was assumed that Angola will locate a bulk of its irrigated area in the driest parts of the ORB, that is, close to Namibia and just across the river boundary. Hence, Angolan irrigation water requirements will be like Namibian water demands of 15,000 m^3/ha/year; hence Angola could irrigate 93,333 ha in this scenario. At most 45% and 10% of the area will be irrigated by the Cuito tributary, and only 10% of the irrigated area will be located downstream of the Cubango-Cuito confluence because of possible limitations in available land. Total irrigated land in Angola and Namibia will be 108,992 ha abstracting 17.49% of the mean annual flow, while total water abstraction upstream of the Okavango Delta will be 18.13% of the mean annual flow. It was assumed that all return flows from irrigation schemes will flow back into the Okavango River. The mode nitrate levels of 0.5 mg/L at Katwitwi and Cuito confluence were used as long-term inflow nitrate levels from unpolluted parts of the ORB. Figure 4 shows the results of simulated river nitrate levels in the scenario of high-water demand for irrigation in Angola and Namibia and urban water supply in Namibia.

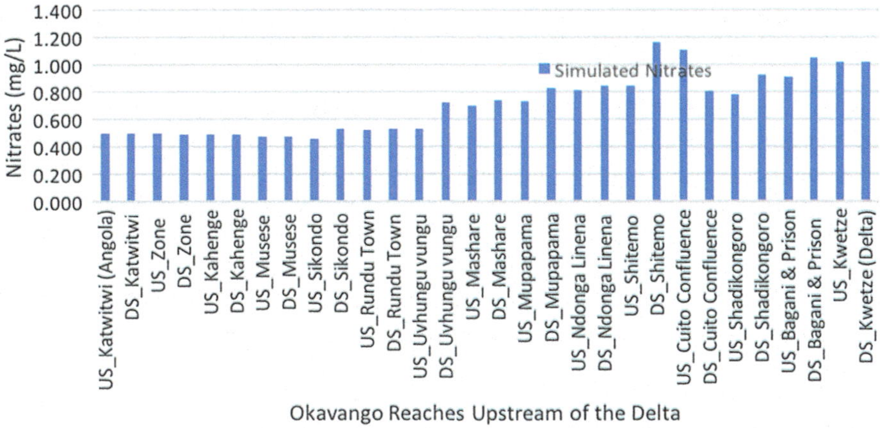

Fig. 4 Simulated river nitrate levels for Scenario 2 (high irrigation development in Angola and Namibia plus water supply to central and coastal towns of Namibia)

The simulated results showed (in Fig. 4) that irrigating 46,545 ha in the catchment of the Cubango tributary can increase mode nitrate levels from 0.5 to 1.161 mg/L downstream of Shitemo (DS-Shitemo). The mode nitrate levels can be attenuated to 0.802 mg/L at the Cubango-Cuito confluence (DS_Cuito Confluence) due to the dilution effect of the Cuito tributary inflows. Then the mode nitrate levels can increase to 1.046 mg/L due to agricultural return flow pollution downstream of the confluence but can be attenuated to 1.017 mg/L before entering the delta. According to Kincheloe et al. (1979), nitrate tolerance of fish eggs and fish fry had mortality rates of 10% at 1.1 mg/L. Mabaya et al. (2016) classified nitrate concentration in freshwater as follows: below 1 mg/L as clean, from 1 to 3 mg/L as tolerable. Therefore, increasing irrigation area from the current 4,000 ha to about 108,992 ha can double mode nitrate levels in water flowing into the Okavango Delta. This can degrade the water quality classification from clean to tolerable for freshwater fish. Increasing Rundu population from 63,431 in 2011 to 100,000 and effluent nitrate levels from 1.0 mg/L to 10.0 mg/L may have negligible effect on mode nitrate levels. Increasing the reach length of the Cuito tributary with irrigation Schemes (42,000 ha in Angola) from 200 km to 330 km had negligible effect on nitrate attenuation capacity of the reach.

5 Comparison with Other Studies

The OR nitrate simulation model showed that 2015/2016 nitrate pollution by urban and agricultural return flows was adequately attenuated in riparian zones and in each successive river reach; hence OR nitrate levels were constant at just under

0.5 mg/L. This level was similar to constant long-term nitrate levels measured in water sampled from each river reach. Also, Vushe et al. (2014) stated that the OR system attenuated nitrate pollution and kept mode mainstream nitrate levels constant at 0.5 mg/L. Matějıček et al. (2003) simulated a river basin's nitrate pollution and transport and found out that the agricultural and urban areas act as sources of pollution, but forested zones of the river basin with less pollution operate as sinks. The nitrate levels were decreased downstream, if the proportion of the forest inside contribution zones increases. Anthony et al. (2009) used an EveNFlow model for estimation of monthly river flow and nitrate-N concentrations in river systems polluted by intensive agriculture, and results compare favourably with hydrological and total nitrate load data collected over 10 years at monitoring sites on rivers getting contributions from both point and diffuse sources.

The ORB nitrate pollution and transfer simulation model can be used for predicting annual mode river nitrate concentrations but inaccurately predicts observable nitrate levels in shorter time steps. The river's U_{DEN} rate can increase in response to increase in nitrate load up to a threshold because attenuation rates of some biological uptake, cation sorption and microbial immobilization are not persistent, but ecosystems eventually reach homeostasis with respect to N metabolism (Cohen et al. 2007); hence the model can only successfully use a long-term constant U_{DEN} rate. The use of a nitrate load-sensitive U_{DEN} function will probably enhance the prediction accuracy and the model's sensitivity. The low U_{DEN} value (4.83 mg nitrate-N m^2/h) obtained in calibration of the model indicated that currently the OR mainstream had low nitrate pollution load, since low values of U_{DEN} were obtained in pristine rivers (Seitzinger 1988; Wollheim et al. 2008). Measured nitrate levels were less than 1.0 mg/L, which probably indicated high attenuation in dry and wetland riparian zones or low impact of human activities on the N in mainstream water. Hickey and Martin (2009) stated that pristine rivers with limited or no catchment development had nitrate-N concentrations, often less than1.0 mg/L. The water entering and leaving the Namibian reaches of the OR had relatively low nitrate-N load, of 0.5 mg/L, which implied low net nitrate pollution by human activity upstream of the delta.

5.1 Wheat and Maize Nitrogen Demands, Irrigation Water and the Gross Nitrate Mass Balance (GNB) in the Simulation Model

The respective maize and wheat N uptakes estimated as 54.5% and 74.4% of applied fertilizer were higher than the general estimate by FAO (2001) that plants take up 50% of the nitrogen fertilizer applied to agricultural land. The maize uptake rate was slightly lower than the optimum rate (58%) given by Liang and Mackenzie (1994), which indicated that the simulation model's GNB estimates for maize cropping seasons were within the expected ranges. Probably, further studies, on the variation of readily available soil N with YNR and luxuriant N uptake in the ORB environment, may enhance the accuracy of estimates of the uptake responses of different crop varieties to N fertilization rates and hence estimate the leached N more accurately for each crop variety.

The model estimated that about 51.0 and 44.9 kg N ha^{-1} (17.7% and 23.1% of applied fertilizer) were leached from maize and wheat crops, respectively. The total annual leached N was 288% higher than estimates by Yousefi et al. (2017) where 33.35 kg N/ha was leached annually from Iranian irrigated fields (548 ha, wheat; 572 ha, barley; 291 ha, maize; 286 ha, alfafa; and 423 ha, tomatoes) after applying 386 kg N/ha on sandy loams and silt loam soils. Probably the increase in total fertilization rate by 25% and the maize/wheat crop rotation, which indicated different agronomic practices, contributed to higher leaching rates in the ORB. Despite the erratic wet spells which increased leaching potential as farmers applied more nitrogen fertilizer (Vushe et al. 2016), the model showed that the summer maize crop lost about 5 kg N/ha more to leaching compared to the wheat crop losses, although an additional 61 kg N/ha was available to the maize crop. The small difference in leached N amounts was attributed to maize's higher N uptake efficiency and/or higher YNR values of 0.067 and 0.038 t/kg N for maize and wheat, respectively.

The annual N fertilization rate of 482 kg/ha was higher than 208–230 kg N/ha rates recommended by Liu et al. (2016) for cost-effective winter wheat-summer maize rotation with least environmental impact in China. The wheat yield of 5.5 t/ha was obtained in the ORB farm under lower fertilization rates of 195 kg N/ha for wheat (Vushe et al. 2016) compared to the recommendations by Liu et al. (2016). This might indicate that ORB wheat crop production met some international best practices, while improvements were required in maize crop production.

The model estimated that 24% of irrigation water and rain was an unintentional leaching fraction, and this was slightly higher than findings in wheat crop experiments done in Saudi Arabia, where the deep percolation water loss ranged from 3.24% to 23.04%. Traditional flood irrigation in Northwest China had loss to deep percolation ranging from 29.5% to 34.0% of irrigation water (675–745 mm) in wheat fields and 28.5% of 765 mm in maize fields (Wang et al. 2012). Generally, flood irrigation is more inefficient compared to centre pivot sprinkler irrigation; hence loss of water to leaching was expected to be lower in the Namibian irrigated farms. The estimates of leached water by the ORB simulation model were relatively accurate since they were comparable to findings in other agroclimates.

The model estimated that mean sums of effective rainfall and net irrigation water demand were about 12,500 m^3/ha/year and the leached water as 7,000 m^3/ha/year, and hence the annual return flows were about 47% of the allocated irrigation water. Deep percolation water loss from the summer maize crop was about 71% of the annual return flows, which seem to agree with the suggestion by Vushe et al. (2016) that the excess soil moisture during long wet spells in summer was the main transporter of nutrients from the irrigated fields. Therefore, meticulous fertigation scheduling during the summer wet spells may help to reduce N losses to leaching and transport from the field into the river.

6 Sustainability Options Under Increasing Urbanization and Irrigated Agriculture Upstream of the Okavango Delta

The model showed that full-scale development of irrigation in Angola and Namibia and urban water supply to central and coastal towns of Namibia can double mode nitrate levels in Okavango Delta inflows. Also, improving Rundu urban effluent nitrate quality from 10 mg/L down to 1.0 mg/L can slightly reduce river mainstream nitrate levels from 0.532 to 0.522 mg/L downstream of Rundu. Therefore intensive efforts should be directed towards managing pollution from irrigation systems since they are likely to have the largest impact on river water quality.

Irrigation schemes are always accompanied by increased human settlements and urbanization, which always increase water resource pollution; hence for economic and environmental sustainability, population densities close to the river should be as low as possible. One solution can be sparsely locating large irrigation schemes and human settlements. Further studies may be required to determine the minimum separation between neighbouring schemes and/or settlements and the minimum width of the uncultivated riparian zone with natural vegetation for attenuation of pollutants and aesthetic reasons.

The irrigation schemes may affect the livelihoods of local rural communities. Most Okavango River Basin communities are peasant crop and livestock producers who may fail to adapt to new challenges that may include a water resource polluted with agrochemicals (Mmualefe and Torto 2011; Vushe et al. 2016) and other emerging pollutants. Population growth increases competition for limited natural resources, and hence irrigation development may increase the competition and conflicts, because irrigation schemes may fail to socio-economically empower all the communities. The schemes may deprive some communities of ecosystem services they are currently enjoying, e.g. pastoralists can lose some traditional pastures. Therefore, water and land allocation to irrigation schemes and associated settlements should minimize negative impacts on livelihoods, but probably ensure some benefit sharing and/or economic cooperation for uplifting livelihoods of disadvantaged surrounding communities.

Also, resilience to the above-mentioned challenges may be enhanced by integration and active participation of communal peasant farmers into the national or global economy by commercializing indigenous drought-resistant rainfed field crops and livestock production systems. This can be achieved through agricultural extension and integration of communal livestock production with irrigated agriculture, e.g. some crop residues may be used by surrounding communities for livestock supplementary feeding and/or fattening, which may help in offsetting the loss of pastures. The manure can be sold to irrigation schemes and hence used to improve the condition of the infertile Kalahari sandy soils. This interdependence and economic relationship may reduce possible conflicts between irrigators and livestock farmers.

An excellent water quality of the unpolluted ORB under the high-water demand or full development scenario should be one major indicator of sustainable utilization of the Okavango water. Maintaining the prevailing excellent drinking water quality standards (Namibian) would probably require enforcing pollution reduction measures, minimizing water demand and minimizing polluted return flows by employing irrigation water demand management measures (e.g. environmentally friendly and prudent fertigation scheduling and crop protection) together with irrigator education and awareness campaigns, as well as promoting agro-economic efficiency and effectiveness, and appropriate and innovative modernization of communal agricultural systems.

7 Conclusions

In 2017, Namibia was the only country using Okavango water for commercial irrigated crop production on about 4,000 ha which had negligible impact on nitrate levels of the Okavango River. The OR simulation model estimated that mode nitrate levels in delta inflows will slightly increase from the long-term 0.5 to 0.6 mg/L, if the irrigated area is increased to 15,659 ha. If 18.13% of the mean annual flow is abstracted for irrigation of the projected 108,992 ha in Angola and Namibia, and 60 million m^3/year is abstracted for water supply to central and coastal urban centres of Namibia, the model predicted that mode nitrate concentration in the Okavango Delta inflows will be greater than 1.0 mg/L which is double long-term mode nitrate levels. This will be a water quality degradation, and classification will drop from clean to tolerable for freshwater fish.

Acknowledgements The author is very grateful to the BMZ (Germany Government) for financial support under The Future Okavango Research Project, Management of Mashare Commercial Farm for giving assistance, agronomic data and permission to do field tests. The data, information and support from Namibia Water Corporation, managers of several other irrigated commercial farms in the ORB, Namibia University of Science and Technology and Namibian government departments is also appreciated.

References

Alley MM, Martz J, Marvin E, Davis PH, Hammons JL (2009) Nitrogen and phosphorous fertilization of corn: Virginia Cooperative Extension, Virginia Tech, and Virginia State University. Available at: http://pubs.ext.vt.edu/424/424-027/424-027.html. Accessed 22 Sept 2017

Anthony SG, Sigram M, Collins AL, Fawcett LE (2009) Modelling nitrate river water quality for policy support. JRBM [online] 7(3):259–275. https://doi.org/10.1080/15715124.2009.9635388. Accessed 21 July 2017

Angus JF (2001) Nitrogen supply and demand in Australian agriculture. Aust J Exp Agric [Online] 41(3):277–288. https://doi.org/10.1071/EA00141. Accessed 18 May 2016

Bauder TA, Waskom RM, Andales A (2014) Nitrogen and irrigation management. Colorado State University Extension. Extension 3/99. Available at: www.ext.colostate.edu. Accessed 12 May 2016

Bethune S (2009) Environmental protection and sustainable management of the Okavango River basin (EPSMO). Transboundary Diagnostic Analysis Basin Ecosystems report. Available from: http://www.okacom.org/site-documents/tda-background-reports/transboundary-diagnostic-analysis-reports-for-the-namibia-section-of-the-cubango-okavango-basin/report-transboundary-diagnostic-analysis-basin-ecosystems-report. Accessed 16 June 2012

Brodie J, Mitchell A (2006) Sediments and nutrients in north Queensland tropical streams: changes with agricultural development and pristine condition status. CRC Reef Research Centre Technical Report No. 62, CRC Reef Research Centre, Townsville, Australia. Available at: http://www.reef.crc.org.au/. Accessed 23 June 2016

Boyer EW, Howarth RW, Galloway JN, Dentener FJ, Green PA, Vorosmarty CJ (2006) Riverine nitrogen export from the continents to the coasts. Global Biogeochem Cycles [online] 20:GB1S91. https://doi.org/10.1029/2005GB002537. Accessed 24 Aug 2016

Buss SR, Rivett MO, Morgan P, Bemment CD (2005) Attenuation of nitrate in the sub-surface environment. Science Report SC030155/SR2. Geography, Earth and Environmental Sciences, University of Birmingham, Edgbaston, Birmingham B15 2TT. Available at: https://www.gov.uk/government/uploads/system/uploads/attachment_data/file/291473/scho0605bjcs-e-e.pdf. Accessed 2 June 2017

Carson J, Phillips L (2016) Soil nitrogen supply. (The University of Western Australia) and Lori Phillips (The Department of Environment and Primary Industries – Victoria). Available at: http://soilquality.org.au/factsheets/soil-nitrogen-supply. Accessed 21 Oct 2017

Cohen JM, Lamsal S, Kohrnak LV (2007) Sources, transport and transformations of Nitrate-N in the Florida environment. St Johns River Management District Special Pubication-SP10 [online]. http://www.sjrwmd.com/technicalreports/pdfs/SP/SJ2007-SP10.pdf. Accessed 30 May 2017

Connor JD, Loomis SR, Cassman GK (2011) Crop ecology: productivity and management in agricultural systems. 2nd Ed. [ebook] Cambridge University Press. Available at: https://books.google.com.na/books?id=O2eh7vyvuscC&pg. Accessed 12 July 2017

Covert DMN (2014) Irrigation management, chapter 5, nitrate movement and loss under irrigated crop production. Why nitrate loss is a problem: University of Nebraska, Plant Soil Sci. eLibrary. Available at: http://croptechnology.unl.edu/pages/informationmodule.php?idinformationmodule. Accessed 06 July 2017

Cui Z, Zhang F, Miao Y, Sun O, Li F, Chen X, Li J, Ye Y, Yang Z, Zhang O, Liu C (2008) Soil nitrate-N levels required for high yield maize production in the North China Plain. Nutr Cycl Agroecosys [Online] 82:187–196. Available at: https://doi.org/10.1007/s10705-008-9180-4. Accessed 16 July 2017

Cui Z, Zhang F, Dou Z, Miao Y, Sun O, Chen X, Li J, Ye Y, Yang Z, Zhang O, Liu C, Huang S (2009) Regional evaluation of critical nitrogen concentrations in winter wheat production of the north china plain. Agron J [Online] 101(1):159–166. Available at: https://doi.org/10.2134/agronj2008.0102. Accessed 16 July 2017

DeBruin J, Butzen S (2016) Nitrogen uptake in corn. DuPont Pioneer. Available at: https://www.pioneer.com/home/site/us/agronomy/library/n-uptake-corn. Accessed 27 May 2016

Domptail S, Mundy O (2014) VISIONS 2030 for the Okavango river basin inspirations from 27 stakeholders from all levels. Learning from one another to better anticipate the future [ebook]. Available at: http://www.future-okavango.org/downloads/vision2030-booklet.pdf. Accessed 2 Jan 2017

Donner S, Coe M, Lenters J, Twine T, Foley J (2002) Modeling the impact of hydrologic changes on nitrate transport in the Mississippi, Global Biogeochem Cycles [online] 16(3):1043. https://doi.org/10.1029/2001GB001396. Accessed 16 Jan 2017

du Plessis J (2003) Maize production, Directorate agricultural information services. [pdf] Department of Agriculture in cooperation with ARC-Grain Crops Institute Publication. p.2. Available at: http://www.arc.agric.za/arc-gci/Library/Maize-infopak.pdf. Accessed 31 Jan 2017

Duda AM, El-Ashry M (2000) Addressing the global water and environment crises through integrated approaches to the management of land, water and ecological resources. Water Int [Online] 25(1):115–126. Available at: https://doi.org/10.1080/02508060008686803. Accessed 20 Jan 2016

Eneji AE, Islam R, An P, Amalu UC (2013) Nitrate retention and physiological adjustment of maize to soil amendment with superabsorbent polymers. J Clean Prod [online] 52:474–480. https://doi.org/10.1016/j.jclepro.2013.02.027. Accessed 11 Sept 2017

el Obeid S, Mendelsohn J (2001) A preliminary profile of the Kavango Region in Namibia. Namibia Nature Foundation, Windhoek. Available at: https://www.nnf.org.na/publications/a-preliminary-profile-of-the-kavango-region/29/26.html. Accessed 25 Jan 2016

FAO (Food and Agriculture Organization of the United Nations) (1997) Irrigation potential in Africa: a basin approach. Chapter 6: Review of existing information on irrigation potential. The Okavango basin. FAO Land and Water Bulletin 4. http://www.fao.org/docrep/W4347E/w4347e0p.htm. Accessed 19 Jan 2018

FAO (Food and Agriculture Organization of the United Nations) (2006) Plant nutrition for food security: a guide for integrated nutrient management. FAO Fertilizer and Plant Nutrition Bulletin 16. Chapter 8. Nutrient management guidelines for some major field crops. http://www.fao.org/docrep/010/a0443e/a0443e00.htm. Accessed 19 May 2017

FAO (Food and Agriculture Organization) (2005) Control of water pollution from agriculture – FAO irrigation and drainage paper 55. Chapter 1: Introduction to agricultural water pollution. Available at: http://www.fao.org/docrep/w2598e/w2598e04.htm. Accessed 24 July 2017

FAO and IFA (Food and Agriculture Organization) and (International Fertilizer Association) (2001) Global estimates of gaseous emissions of NH3, NO and N_2O from agricultural land. [pdf] pp 11–17. Available at: http://nue.okstate.edu/Ammonia_Loss/globest.pdf. Accessed 25 June 2016

Farquharson FAK, Gash JHC, Roberts JM (1990) Report on a review of evaporation estimation for the Okavango Delta. [pdf] Institute of Hydrology Wallingford Oxon, OXIO8BB, UK. https://nora.nerc.ac.uk/id/eprint/14937/1/N014937CR.pdf. Accessed 23 Jan 2018

Gad AAM, Abdalla MAA (2017) Fate of heavy metals and nutrients in waste stabilization. [pdf] J Eng Sci 45:1–16. http://www.aun.edu.eg/journal_files/515_J_9505.pdf. Accessed 25 June 2016

Green PA, Vörösmarty CJ, Meybeck M, Galloway JN, Peterson BJ, Boyer EW (2004) Pre-industrial and contemporary fluxes of nitrogen through rivers: a global assessment based on typology, Biogeochemistry [online] 68: 71–105. https://doi.org/10.1023/B:BIOG.0000025742.82155.92. Accessed 14 Sept 2017

Han K, Zhou C, Wang L (2014) Reducing ammonia volatilization from maize fields with separation of nitrogen fertilizer and water in an alternating furrow irrigation system. J Integr Agric [Online] 13(5):1099–1112. Available at: https://doi.org/10.1016/S2095-3119(13)60493-1. Accessed 20 Aug 2016

Hauduc H, Neumann MB, Muschalla D, Gamerith V, Gillot S, Vanrolleghem PA (2011) Towards quantitative quality criteria to evaluate simulation results in wastewater treatment– A critical review. [ebook] 8th IWA Symposium on Systems Analysis and Integrated Assessment. https://books.google.com.na/books?isbn=186872963X. Accessed 23 Jan 2018

Heisler J, Glibert P, Burkholder J, Anderson D, Cochlan W, Dennison W, Gobler C, Dortch Q, Heil C, Humphries E, Lewitus A, Magnien R, Marshall H, Sellner K, Stockwell D, Stoecker D, Suddleson M (2008) Eutrophication and harmful algal blooms: a scientific consensus. Harmful Algae [online] 8(1):3–13. https://doi.org/10.1016/j.hal.2008.08.006. Accessed 4 Feb 2017

Hickey CW, Martin ML (2009) A review of nitrate toxicity to freshwater aquatic species. Report No. R09/57 prepared for Environment Canterbury, New Zealand [online]. Available at ISBN 978-1-86837-997-1. Accessed 23 Nov 2017

Howarth RW et al (2006) The influence of climate on average nitrogen export from large watersheds in the northeastern United States, Biogeochemistry [online] 79: 163–186. https://doi.org/10.1007/s10533-006-9010-1. Accessed 15 Sept 2017

Huang Y, Li D (2014) Soil nitric oxide emissions from terrestrial ecosystems in China: a synthesis of modelling and measurements. Scientific Reports 2014. [Online] 4:7406

IPCC (Intergovernmental Panel on Climate Change) (2006) Guidelines for national greenhouse gas inventories. Chapter 11: N_2O emissions from managed soils, and CO_2 emissions from Lime and Urea application. [pdf] Available at: http://www.ipcc-nggip.iges.or.jp/public/2006gl/pdf/4_Volume4/V4_11_Ch11_N2O&CO2.pdf. Accessed 16 July 2017

IPCC (Intergovernmental Panel on Climate Change / Organization for Economic Cooperation and Development) (1996) Revised IPCC guidelines for national greenhouse gas inventories. Reference Manual (Volume 3). Agriculture Available at: https://www.ipcc-nggip.iges.or.jp/public/gl/invs6c.html. Accessed 16 July 2017

Islam MS, Ullahb SM, Khanb TH, Huqb SMI (2008) Retention of nitrate and phosphate in soil and their subsequent uptake by plants. Bangladesh J Sci Ind Res [Online] 43(1):67–76. Accessed 17 Apr 2017

Jones AC, Koenig RT, Ellsworth JW, Brown BD, Jackson GD (2007) Management of urea fertiliser to minimize volatilization. [pdf] The U.S. Department of Agriculture (USDA), Washington State University Extension and the Montana State University Extension. Available at: www.extension.uidaho.edu/nutrient/pdf/smallgrain/New/Management%2520of%2520Urea%2520fertilizer%2520. Accessed 11 June 2017

Kanda EK, Kipkorir EC, Kosgei J (2017) Modelling of nitrates in River Nzoia using MIKE 11. Water Pract Tech [Online] 12(1):217–223. Available at: https://doi.org/10.2166/wpt.2017.026. Accessed 04 Sept 2018

Kgathi DL, Kniveton D, Ringrose S, Turton AR, Vanderpost CHM, Lundqvist J, Seely M (2006) The Okavango; a river supporting its people, environment and economic development. J Hydrol [Online] 331:3–17. Available at: https://ubrisa.ub.bw/bitstream/handle/10311/154/kgathi_j_hydrology_2006.pdf? Accessed 05 Sept 2018

Kincheloe JW, Wedemeyer GA, Koch DL (1979) Tolerance of developing salmonid eggs and fry to nitrate exposure. Bull Environ Contam Toxicol [Online] 23:75–578. Available at: https://doi.org/10.1007/BF01770006. Accessed 14 June 2017

Krause P, Boyle DP, Bäse F (2005) Comparison of different efficiency criteria for hydrological model assessment. In: Advances in geosciences, [Online], vol 5, pp 89–97, Available at: SRef-ID: 1680-7359/adgeo/2005-5-89 Accessed 14 June 2017

Lansdown K, Heppell CM, Trimmer MA, Binley A, Heathwaite AL, Byrne P, Zhang H (2015) The interplay between transport and reaction rates as controls on nitrate attenuation in permeable, streambed sediments. J Geophys Res [Online] 120(6):1093–1109. https://doi.org/10.1002/2014JG002874. Accessed 19 Jan 2017

Legates RD, McCabe Jr GJ (1999) Evaluating the use of "goodness-of-fit" measures in hydrologic and hydroclimatic model validation. Water Resour Res [Online] 35(1):233–241. Paper number 1998WR900018. Available at: https://doi.org/10.1029/1998WR900018. Accessed 10 Feb 2017

Lehmann F, Leon F (2010) Water quality of the Swakoppoort Dam, Namibia. [pdf] Master's Thesis Freiburg im Breisgau, Germany. Available at: http://www.hydrology.uni-freiburg.de/abschluss/Lehmann_F_2010_MA.pdf. Accessed 05 Sept 2018

Liang BC, MacKenzie AF (1994) Corn yield, nitrogen uptake and nitrogen use efficiency as influenced by nitrogen fertlization. Can J Soil Sci [online] 74(2): 235–240. https://doi.org/10.4141/cjss94-032; https://doi.org/10.1029/1998WR900018. Accessed 24 Feb 2019

Liu H, Wang Z, Yu R, Li F, Li K, Cao H, Yang N, Li M, Dai J, Zan Y, Li Q, Xue C, He G, Huang D, Huang M, Liu J, Qiu W, Zhao H, Mao H (2016) Optimal nitrogen input for higher efficiency and lower environmental impacts of winter wheat production in China. Agric Ecosyst Environ [Online] 224:1–11. Available at: https://doi.org/10.1016/j.agee.2016.03.022. Accessed 10 Feb 2017

Lord I, Anthony SG (2002) Agricultural nitrogen balance and water quality in the UK. Soil Use Manage [Online] 18(4):363–369. Available at: https://onlinelibrary.wiley.com/doi/10.1111/j.1475-2743.2002.tb00253.x/pdf. Accessed 11 May 2017

Mabaya G, Koichi U, Masayuki F (2016) Stochastic optimal control of agrochemical pollutant loads in reservoirs for irrigation. J Clean Prod [Online] 146:37–46. https://doi.org/10.1016/j.jclepro.2016.05.108. Accessed 18 June 2017

Martin EJ (2013) Physics for radiation protection. Chapter 13.7, Log Normal Data Distributions. 3rd Edition. [ebook] Wiley. Available at: https://books.google.com.na/books?isbn=3527667083. Accessed 04 Nov 2017

Masamba RLW, OKACOM (2009) Transboundary Diagnostic Analysis of the Botswana Portion of the Okavango River Basin. Output 3: Irrigation Development. Available at: http://www.okacom.org/site-documents/tda-background-reports/transboundary-diagnostic-analysis-reports-for-the-botswana-portion-of-the-cubango-okavango-basin/transboundary-diagnostic-analysis-of-the-botswana-portion-of-the-okavango-river-basin-output-3-irrigation-development. Accessed 03 Oct 2017

Matějıček L, Benešová L, Tonika J (2003) Ecological modelling of nitrate pollution in small river basins by spreadsheets and GIS author links open overlay panel. Ecol Modell [Online] 170(2):245–263. Available at: https://doi.org/10.1016/S0304-3800(03)00232-1. Accessed 05 Sept 2018

Mateo-Sagasta J, Marjani ZS, Turral H (2018) More people, more food, worse water? a global review of water pollution from agriculture. Water Land and Ecosystems research program of the CGIAR Colombo, 2018. [pdf] Food and Agriculture Organization of the United Nations Rome, 2018 and the International Water Management Institute Publication. Available at: www.fao.org/3/ca0146en/CA0146EN.pdf. Accessed 05 Sept 2018

Mendelsohn J, el Obeid S (2003) Sand and water: a profile of the Kavango region in Namibia. Struik, Cape Town

Mendelsohn J, el Obeid S (2004) Okavango River: the flow of a lifeline. [ebook]. Struik, Cape Town. Available at: http://www.worldcat.org/title/okavango-river-the-flow-of-a-lifeline/oclc/55607774. Accessed 07 Oct 2016

MET (Ministry of Environment and Tourism) (2000) Natural resources mapping of the Kavango Region, Final Report. Available at: http://www.the-eis.com/data/literature_OK/Interconsult.pdf. Accessed 07 Oct 2016

Mmualefe LC, Torto N (2011) Review Water quality in the Okavango Delta. Water SA, [Online] vol. 37, p 3. Available at: https://doi.org/10.4314/wsa.v37i3.68492. Accessed 17 Jan 2018

Moyo NA (1997) Lake Chivero: a polluted lake, University of Zimbabwe Publications, Harare

NamWater Namibia Water Corporation (2013) Rundu Water Supply Records, Unpublished

NamWater Namibia Water Corporation (2017) Rundu Water Supply Records, Unpublished

Nee B (2015) A survey of groundwater, macrophytes, and habitat types below the Maple River Dam. Available at: https://deepblue.lib.umich.edu/bitstream/handle/2027. Accessed 05 May 2017

Nevison CP, Riddick S, Ward D (2016) Denitrification, leaching, and river nitrogen export in the Community Earth System Model. J Adv Model Earth Syst [Online] 8:272–291. Available at: https://doi.org/10.1002/2015MS000573. Accessed 05 May 2017

NPCM (National Planning Commission, Namibia) (2012) Republic of Namibia 2011 Population and Housing Census Preliminary Report. [pdf] Available at: www.npc.gov.na/. Accessed 25 May 2012

O'Gorman WT (2004) Applied adaptive statistical methods: tests of significance and confidence intervals. American Staistical Association and the Society for Industrial and Applied Mathematics. [ebook]. Available at: https://books.google.com.na/books?isbn=0898718430 Accessed 30 Sept 2017

O'Leary M, Rehm G, Schmitt M (2002) Understanding nitrogen in soils. Available at: http://www.extension.umn.edu/agriculture/nutrient-management/nitrogen/understanding-nitrogen-in-soils/. Accessed 30 May 2014

OECD and EUROSTAT (Organisation for Economic Co-operation and Development and Statistical Office of the European Communities) (2007) Gross Nitrogen Balance Handbook. Available at: www.oecd.org/tad/env/indicators. Accessed 03 Oct 2017

Oenema O, Kros H, de Vries W (2003) Approaches and uncertainties in nutrient budgets: implications for nutrient management and environmental policies. Eur J Agron [Online] 20:3–16. Available at: https://doi.org/10.1016/S1161-0301(03)00067-4. Accessed 05 May 2017

OKACOM (The Permanent Okavango River Basin Water Commission) (2009) Technical Report on Irrigation Development in the Namibia Section of the Okavango River Basin. Available at: http://iwlearn.net/resolveuid/5f52b9baa53084d2e83ecf3c13b5056c. Accessed 10 June 2017

OKACOM (The Permanent Okavango River Basin Water Commission) (2010) Environmental Protection and Sustainable management of the Okavango River basin (ESPSMO). Available at: http://www.okacom.org/okacoms-work/partners-and-projects/projects/okacom-projects/epsmo-1. Accessed 18 Oct 2017

OKACOM (The Permanent Okavango River Basin Water Commission) (2011) Cubango-Okavango River Basin Transboundary Diagnostic Analysis 2011. Maun Botswana [pdf] Available at: http://www.okacom.org/search?SearchableText=Transboundary+Diagnostic+Analysis%2C+2011. Accessed 12 July 2017

OKACOM (2017) Programme for Transboundary Water Management in the Cubango – Okavango River Basin. CRIS number: RSO/FED/039-558. Available at: https://ec.europa.eu/transparency/regdoc/rep/3/2017/EN/C-2017-7381-F1-EN-ANNEX-3-PART-1.PDF. Accessed 14 Sept 2018

OKACOM and FAO (2014) Synthesis Report. Cubango-Okavango River Basin Water Audit (CORBWA) Project [pdf] Available at: www.fao.org/3/a-i3743e.pdf. Accessed 24 Sept 2017

Patterson RA (2003) Nitrogen in Wastewater and its Role in Constraining On-Site Planning in Future Directions for On-site Systems: Best Management Practice. Proceedings of On-site '03 Conference University of New England, Armidale. Published by Lanfax Laboratories Armidale. p.313-320. Available at: http://www.lanfaxlabs.com.au/papers/P54-03-nitrogen.PDF. Accessed 12 June 2017

Pina-Ochoa E, Alvarez-Cobelas M (2006) Denitrification in aquatic environments: a cross-system analysis. Biogeochemistry 81:111–130. https://doi.org/10.1007/s10533-006-9033-7. Accessed 14 Sept 2017

Roy RN, Finck A, Blair GJ, Tandon HL (2006) Plant nutrition for food security: a guide for integrated nutrient management. FAO Fertilizer and Plant Nutrition Bulletin 16. [pdf] Available at: http://www.fao.org/docrep/010/a0443e/a0443e00.htm. Accessed 19 May 2017

Sawyer J (2007) Crop advantage series, nitrogen fertilization for continuous and rotated corn. [pdf] Available at: http://www.agronext.iastate.edu/soilfertility/info/07CAScornsheet.pdf. Accessed 21 May 2017

Schlesinger HW, Bernhardt SE (2013) Biogeochemistry: an analysis of global change. 3rd Edition; Chapter 12, The Global cycles of Nitrogen and Phosphorus, http://books.google.com.na/books?id=sy. Accessed 28 May 2017

Schroeder JJ, Scholefield D, Cabral F, Hofman G (2004) The effect of nutrient losses from agriculture on ground and surface water quality: the position of science in developing indicators for regulation. Environ Sci Pol [online] 7:15–23. https://doi.org/10.1016/j.envsci.2003.10.006. Accessed 14 July 2017

Seitzinger PS (1988) Denitrification in freshwater and coastal marine ecosystems: ecological and geochemical significance. Limnol Oceanogr [Online] 33(4,2):702–724. Available at: https://onlinelibrary.wiley.com/doi/10.4319/lo.1988.33.4part2.0702/full. Accessed 12 Feb 2017

Seitzinger SP, Styles RV, Boyer EW, Alexander RB, Billen G, Howarth RW, Mayer B, Van Breemen N (2002) Nitrogen retention in rivers: model development and application to watersheds in the northeastern USA. Biogeochemistry [Online] 57(58):199–237. Available at: https://doi.org/10.1007/978-94-017-3405-9_6. Accessed 11 Feb 2017

Signor D, Cerri CEP (2013) Nitrous oxide emissions in agricultural soils: a review. Pesq Agropec Trop Goiânia [Online] 43(3):322–338. Available at: https://doi.org/10.1590/S1983-40632013000300014. Accessed 19 Feb 2016

Smith RL, Böhlke JK, Garabedian SP, Revesz KM, Yoshinari T (2004) Assessing denitrification in groundwater using natural gradient tracer tests with 15N: in situ measurement of a sequential multistep reaction. Water Resour Res [Online] 40:(W07101). Available at: https://doi.org/10.1029/2003WR002919. Accessed 13 Mar 2016

Steudel T, Göhmann H, Mosimanyana E, Quintino M, Flügel WA, Helmschrot J (2013) Okavango Basin – Hydrology, Biodiversity and Ecology 5. Available at: www.biodiversity-plants.de/biodivers_ecol/publishing/b-e.00238. Accessed 08 Apr 2017

Strokal M, Ma L, Bai Z, Luan S, Kroeze C, Oenema O, Velthof G, Zhang F (2016) Alarming nutrient pollution of Chinese rivers as a result of agricultural transitions. Environ Res Lett 11:024014. Available at: https://doi.org/10.1088/1748-9326/11/2/024014. Accessed 07 Sept 2018

Trewby F (2003) The effect of land use/ land cover change on the water quality of the Okavango River, Namibia. [pdf] M.A. Dissertation, Department of Geoscience, University of Arkansas, USA. Available at: http://www.the-eis.com/data/literature/Fiona%20Trewby%20thesis%20on%20Okavango%20water%20quality.pdf. Accessed 18 Apr 2017

U.S. Department of Energy (1982) Site characterisation report for the basalt waste isolation project. Basalt Waste Isolation Project, Hanford Site Characterization Report, [ebook] Volume 2. Available at: https://books.google.com.na/books?id=eQoxAQAAMAAJ&pg. [Acessed 04 Nov 2017]

Vushe A, Haimene EP, Mashauri D (2014) Namibian land use changes and nutrient water quality of the Okavango River. J Agr Env Sci [Online] 3(2):219–239. Available from: http://jaesnet.com/journals/jaes/Vol_3_No_2_June_2014/15.pdf. Accessed 06 Nov 2016

Vushe A, von Landsberg L, Groengroeft A, Mashauri AD (2016) Nitrate leaching in irrigated inorganic agriculture: a case study of Mashare commercial farm in Namibia, Okavango River Basin. Afr J Agric Res [Online] 11(31):2871–2881. Available at: https://doi.org/10.5897/AJAR2016.11016. Accessed 06 Nov 2016

Wagner SC (2012) Biological nitrogen fixation. Nature Education Knowledge [Online] 3(10):15. Available at: https://sites.google.com/site/biologyhust/topic-1/biologicalnitrogenfixation? Accessed 04 Oct 2017

Wang S, Li H, Yang Y, Wang H, Yang Y, Jia Y (2012) Using DSSAT model to assess spring wheat and maize water use in the arid oasis of northwest China. JFAE [online] 10(1):911–918. https://doi.org/10.1234/4.2012.2825Water. Accessed 29 Mar 2016

Watson CA, Atkinson D (1999) Using nitrogen budgets to indicate nitrogen use efficiency and losses from whole farm systems: a comparison of three methodological approaches. Nutr Cycl Agroecosys [Online] 53(3):259–267. Available at: https://link.springer.com/content/pdf/10.1023/A:1009793120577.pdf. Accessed 04 Sept 2017

Wick K, Heumesser C, Schmid E (2012) Groundwater nitrate contamination: factors and indicators. J Environ Manag [Online] 111(3):178–186. Available at: https://doi.org/10.1016/j.jenvman.2012.06.030. Accessed 04 Sept 2017

Wijayanti Y, Budihardjo K, Sakamoto Y, Setyandito O (2017) Topsoil N-budget model in orchard farming to evaluate groundwater nitrate contamination. In: The International Conference on Eco Engineering Development 2017 (ICEED 2017) [Online], IOP Conf. Series: Earth and Environmental Science 109 (2017) 012034, IOP Publishing. https://doi.org/10.1088/1755-1315/109/1/012034. Accessed 08 Sept 2018

Wollheim WM, Vorosmarty CJ, Peterson BJ, Seitzinger SP, Hopkinson CS (2006) Relationship between river size and nutrient removal. Geophys Res Lett [online] 33:L06410. https://doi.org/10.1029/2006GL025845. Accessed 16 Sept 2017

Wollheim WM, Vörösmarty J, Charles BAF, Green P, Harrison J, Linder E, Peterson JB, Seitzinger PS, Syvitski PMJ (2008) Global N removal by freshwater aquatic systems using a spatially distributed, within-basin approach. Global Biogeochem Cy [Online] 22:GB2026. https://doi.org/10.1029/2007GB002963. Accessed 14 Sept 2017

Wolski P, Ramberg L, Magole L, Mazvimavi D (2010) Evolution of river basin management in the Okavango system, Southern Africa. Handbook of Catchment Management. 1st Edition. [ebook] Backwell Publishing Ltd. John Wiley & Sons. Available at: https://onlinelibrary.wiley.com/doi/10.1002/9781444307672.ch21. Accessed 06 Sept 2017

Yan XY, Akimoto H, Ohara T (2003) Estimation of nitrous oxide, nitric oxide and ammonia emissions from croplands in East, Southeast and South Asia. Global Change Biol [Online] 9:1080–1096. Available at: https://eurekamag.com/pdf/003/003755747.pdf. Accessed 06 Sept 2017

Yang X, Yongli L, Ding Y, Yin X, Raza S, Tong Y (2017) Optimising nitrogen fertilisation: a key to improving nitrogenuse efficiency and minimising nitrate leaching losses in an intensive wheat/maize rotation (2008–2014). Field Crop Res [online] 206:1–10. https://doi.org/10.1016/j.fcr.2017.02.016. Accessed 10 Sept 2018

Yousefi M, Soltani J, RahimiKhob A, Banihabib ME, Soltani E (2017) Assessment of nitrogen leaching of cropping pattern by soil nitrogen balance equation (Case Study: Varamin Irrigation and Drainage Network). Mod Appl Sci [Online] 11(4):30–38. Available at: https://doi.org/10.5539/mas.v11n4p30. Accessed 17 Sept 2017

Zotarelli L, Scholberg JM, Dukes MD, Mvfioz-Carpena R (2007) Monitoring of nitrate leaching in sandy soils: comparison of three methods. J Environ Qual [Online] 36:953–962. Available at: https://abe.ufl.edu/faculty/carpena/files/pdf/publications/refereed_articles/MonitoringOfNitrateLeaching.pdf

Biochar Application to Soil for Increased Resilience of Agroecosystems to Climate Change in Eastern and Southern Africa

Alfred Obia, Vegard Martinsen, Gerard Cornelissen, Trond Børresen, Andreas Botnen Smebye, Jose Luis Munera-Echeverri, and Jan Mulder

Abstract With the current unreliable rainfall pattern, which is expected to worsen due to climate change, agricultural production might become more challenging especially among resource-poor farmers in Eastern and Southern Africa. This calls for adaptation of farming systems to overcome this emerging challenge. Biochar, a product of biomass pyrolysis, with long-term evidence from Amazonia, might contribute to a climate-resilient farming system. This is due to its positive effects on soil chemical and physical properties resulting in increased crop yields, which has been experimentally demonstrated largely within the last two decades. In acidic low cation exchange capacity (CEC) soils, biochar derived from corncob at 5% application rate, for example, increased pH by ≥1 unit and CEC by ≥2 $cmol_c\ kg^{-1}$ in addition to direct nutrient supply. Increased CEC may be linked to the observed increase in soil organic carbon content (biochar carbon/sequestered carbon) due to biochar addition. Sequestration of carbon due to biochar has been reported to be stronger in soils that have low pH and low carbon contents, with greater effects from biochars

A. Obia (✉)
Norwegian Geotechnical Institute (NGI), Department of Environmental Engineering, Oslo, Norway

Department of Agronomy, Faculty of Agriculture and Environment, Gulu University, Gulu, Uganda

V. Martinsen (✉) · T. Børresen · J. L. Munera-Echeverri · J. Mulder
Faculty of Environmental Sciences and Natural Resource Management (MINA), Norwegian University of Life Sciences (NMBU), Aas, Norway
e-mail: vegard.martinsen@nmbu.no

G. Cornelissen
Norwegian Geotechnical Institute (NGI), Department of Environmental Engineering, Oslo, Norway

Faculty of Environmental Sciences and Natural Resource Management (MINA), Norwegian University of Life Sciences (NMBU), Aas, Norway

A. B. Smebye
Norwegian Geotechnical Institute (NGI), Department of Environmental Engineering, Oslo, Norway

Y. Bamutaze et al. (eds.), *Agriculture and Ecosystem Resilience in Sub Saharan Africa*, Climate Change Management,
https://doi.org/10.1007/978-3-030-12974-3_6

produced from woody materials or those produced at high temperature. Such soils with low pH and carbon contents are common in tropical areas. Another effect of biochar at field-relevant doses of ≤5% is the improvement of physical properties such as increased aggregate stability by up to 7%, increased aggregate mean weight diameter by 8–13%, increased soil water contents and reduced soil temperature fluctuations. Similar changes to soil properties have been found to increase yield by 10% based on global dataset, but up to fourfold increases have also been reported for acidic low CEC soils. One key challenge to implementation of biochar technology for increased yields is probably the unavailability of large quantity of biochar often in tens of Mg ha^{-1} that is required for field application. Conservation agriculture (CA) with its three principles of minimum tillage, residue retention and crop diversification may partly address this key challenge of biochar implementation. This can be achieved by applying biochar only in the tilled part of land in a minimum tillage operation, which can be only 10% of the land surface. In this way, biochar may increase the positive attributes of CA such as increased rainwater harvesting and crop yields. Further solution to the quantity of biochar material for field application is the application of limited quantity of biochar in combination with traditional amendment such as manures either in mixtures or after co-composting. In this way, biochar is expected to deliver its benefits to manure resulting in improved soil properties and increased yields. Overall, incorporation of biochar in CA-based farming system alone or in combination with manures might contribute to building of climate-resilient agroecosystem.

Keywords Biochar · Conservation agriculture · Tropical soils · Climate resilient farming systems · Crop yields

1 Introduction

Biochar is a type of charcoal produced from various biomass feedstocks such as crop residues, manure and waste from agro-industries under low or no oxygen availability, a process called pyrolysis for the purpose of application to soil. Recent interest in biochar research stems mainly from work that has been conducted in Amazonia. A number of areas in Amazonia have black-earth-like "man-made" soils (Anthrosols) containing charcoal (Glaser et al. 2001). These man-made soils are also called Terra Preta or Amazonian Dark Earth. Terra Preta soils are characterized by greater content of soil organic carbon and higher pH, cation exchange capacity (CEC), nutrient availability and water holding capacity compared to surrounding soils with similar texture and mineralogy, resulting into an overall enhanced soil fertility and productivity (Glaser et al. 2001; Lehmann et al. 2003; Novotny et al. 2009). Biochar with evidence from Amazonia has been found to last in the soil for hundreds to thousands of years. Depending on the proportion of stable and degradable carbon in biochar (Spokas 2010; Lal 2016), biochar may lock up carbon in the soil for a longer timescale not comparable to any carbon sequestration approach such as afforestation currently in practice (Lehmann 2007). This soil productivity improvement and carbon sequestration potentials are important given the "low-tech" nature of the overall biochar technology (Renner 2007).

So far few studies have assessed the potential of biochar in Southern Africa (Gwenzi et al. 2015). However, the observed benefits of biochar in Amazonia, which date up to over 1000 years, can be expected in Eastern and Southern Africa given the similar geographical placement in the tropics. Tropical soils are generally low in soil organic matter, pH, CEC, nutrients and water holding capacity largely due to advanced weathering status. Biochar, generally having greater pH, CEC, nutrients and water holding capacity (Martinsen et al. 2015; Revell et al. 2012), can therefore be used to improve tropical soils and hence resilience of agroecosystems to climate change. This improved resilience to climate change will definitely contribute to increase in crop yields (Jeffery et al. 2011). Now and in the future, the importance of this lies in the ever-challenging need to feed the rapidly growing human population.

Feedstock for biochar production can be from any biomass, but this should be restricted to agricultural and forest/wood product wastes and not cutting of trees for biochar production. The choice of feedstock depends on local availability and costs of acquisition (Gwenzi et al. 2015) and can include manures and agricultural waste such as corncobs (*Zea mays*), groundnut (*Arachis hypogaea*) shells and rice (*Oryza sativa. L.*) husks. However, for large-scale implementation, crops with large amounts of biomass wastes such as pigeon pea (*Cajanus cajan*) may be preferred. Production of biochar can take different forms such as traditional earth mounts, drum/Adam retort kiln, flame curtain/Kon-Tiki technique and a number of advanced techniques that may not be relevant to small-scale rural farmers in Eastern and Southern Africa. Traditional earth mount is unhealthy as it emits large amounts of toxic gases to the air, while retort kilns may be costly and require technical know-how for its construction and operation (Sparrevik et al. 2013, 2015). Flame curtain kiln, which can be a conical hole in the ground (Fig. 1) or metallic, is cheap and low-tech (Schmidt and Taylor 2014). It has been found to produce large quantity of high-quality biochar in a short time with low emission of toxic gases (Cornelissen et al. 2016; Pandit et al. 2017). The low emissions during production of biochar with such a kiln is however outweighed by the carbon sequestration if the biochar is used as a soil amendment, leaving a neutral environmental impact (Smebye et al. 2017). Flame curtain kiln is perhaps the best technique for biochar production in rural areas much as production procedure is non-standardized with no control on production temperature and oxygen concentration, which may result in variation in char quality.

The objective of this chapter is to assess the potential of biochar application to soils for increased resilience of agroecosystems to climate change in Eastern and Southern Africa. We review studies on effects of biochar and biochar formulations on soil physical and chemical properties and crop yield under different management practices including conservation farming.

Fig. 1 Pyrolysis in flame curtain kiln in Zambia. (Photo: Left, corncob by A. Obia; right, pigeon pea by G. Cornelissen)

2 Vulnerability of Eastern and Southern Africa Agriculture to Climate Change

The economies of countries in Eastern and Southern African countries are predominantly based on agriculture. In these countries, agricultural growth rate has generally remained low at less than 1% per year in the last 50 years (Chauvin et al. 2012). Yet population growth rate for a number of these countries far exceeds 1%. For example, Zambia had a population of only 2.3 million people in 1950, which had increased to 17.1 million by 2017 and is expected to further grow to 41 million people by 2050 (United Nations Department of Economic and Social Affairs Population Division 2017). The increasing population presents a challenge of food insecurity, which is likely to remain into a distant future unless drastic measures are carried out.

The food insecurity has been and will be compounded by land degradation and climate change, which are interlinked (Lal 2013). Agricultural-based degradation such as soil erosion, compaction, soil acidification and pollution is widespread. In addition, most of the lands in Eastern and Southern Africa are characterized as drylands, where evapotranspiration exceeds precipitation in some parts of the year, making the lands inherently vulnerable to degradation by erosion. Common drylands are dry sub-humid, semi-arid and arid areas. Besides being dry during part of the year with significantly reduced vegetation cover, other socio-economic factors are also driving people to continuously remove vegetation cover to provide energy for cooking. This is due to either lack of clean energy sources or clean energy sources are simply too expensive for many households.

With climate change, real or merely perceived, agricultural droughts are increasingly becoming more common due to soil degradation (Cornelis et al. 2013) and will necessitate adaptation of farming systems to reduce impact on productivity. Such adaptation may involve the use of long-established technologies such as incorporation of manures and other agronomic practices such as conservation agriculture (Farooq and Siddique 2015). Meanwhile, completely degraded agricultural areas may require restoration using afforestation with appropriate vegetation type. In addition to traditional soil amendments such as poultry manure and farm yard manure, biochar, due to its long resident time in soil, may offer sustained benefits for improved soil productivity. Among the benefits that can reduce vulnerability to climate change are the improved soil water retention and chemical properties, including direct addition of nutrients (Obia et al. 2016; Martinsen et al. 2014).

3 Effect of Biochar to Soil Physical Properties

Soil is composed of solid matter (organic and inorganic) and pore space, which can be filled with water and/or air. The proportion and composition of the solid matter and pore space determine the physical characteristics. Characteristics that are important for crop production are structural/mechanical, hydraulic and thermal/soil temperature. These soil characteristics are interdependent such that the effect of biochar on one may influence the other.

In Eastern and Southern Africa, studies exist showing that biochar can improve soil structure (Obia et al. 2016; Fungo et al. 2017), especially when applied in combination with fertilizer. Such improvement in soil structure may be attributed directly to biochar or indirectly to increased biomass production and microbial activities due to both biochar and fertilizer. Increased biomass production may be related to increased root exudates that promote aggregation. The changes in soil structure after biochar application have been assessed in terms of soil aggregation (aggregate size distribution and aggregate stability), porosity and pore-size distribution, bulk density and penetration resistance. Biochar produced from corncob has been found to increase aggregate stability by 2.5–7.0% per percent biochar, added in a sandy loam Acrisol in Zambia (Obia et al. 2016), while aggregate mean weight diameter increased by 8–13% in a similar soil type (Ultisol ≈ Acrisol) in Kenya only when biochar was combined with other amendments (Fungo et al. 2017). In a sandy loam Acrisol in Zambia, porosity significantly increased by 1.2% per percent biochar, and the increase was mainly associated with increase in large pores with sizes >100 μm. Increase in porosity coupled with the light weight of biochar relative to soil reduced the bulk density of the soil after biochar application. In sandy Arenosol, biochar increased porosity due to its high porosity and reduced bulk density due to weight dilution (Obia et al. 2016). The reduction in bulk density was in the range of 0.02–0.06 g cm^{-3} per percent biochar added. In another study in Zambia (Obia et al. 2017), biochar reduced soil penetration resistance in a loamy sand Acrisol by 2.1–2.9 N cm^{-2} per percent biochar added, while no effect was observed in sand. The reduction in penetration resistance was found to be due to the initiation of structural development (i.e. aggregation).

Soil structure, a dynamic soil property, is a key determinant of soil porosity and pore structure. Soil pores provide space through which fluids (water and air) are transported and stored in soil. Pore size distribution determines the proportion between water and air in soil, because large pores normally drain fast and fill with air. The effect of biochar on water transport commonly measured in terms of hydraulic conductivity has shown mixed results depending on soil type/texture, biochar type and particle size (Obia et al. 2017). In Zambia, a reduction in saturated hydraulic conductivity by 0.17 cm h^{-1} per percent biochar added was observed in sandy loam Acrisol, while no effect was observed in sandy Arenosol after application of biochar derived from corncob (Obia et al. 2017). The reduction in hydraulic conductivity was likely due to clogging of water-conducting macro-pores and introduction of micro-pores by biochar, which need to be filled first before water transport. The reduction in conductivity was not related to the water-repellent nature of the biochar, because the hydrophobic biochar did not affect the water repellency of the non-repellent soil after 1 year of experiment. Water repellency of biochar may be due to surface coating by semi-volatile organic compounds (Yi et al. 2015) or due to alkyl groups on BC surfaces (Kinney et al. 2012). However, such hydrophobic compounds can be rapidly lost from biochar upon mixing with soil to percolating water as suggested by Yi et al. (2015). In addition, in these coarse-textured Zambian soils, biochar generally increased water retention resulting in greater plant available water (Cornelissen et al. 2013; Martinsen et al. 2014; Obia et al. 2016). In clayey soil of Ethiopian highlands, an increase in hydraulic conductivity was observed after application of woody chars (Bayabil et al. 2015). Increased hydraulic conductivity resulted in reduced water retention at low tensions but an overall no effect on plant available water in this clayey soil.

Under field conditions in Zambia in a sandy loam soil, biochar derived from pigeon pea feedstock increased maximum soil water content by up to ~80% under conservation farming similar to the effect of corncob biochar resulting in reduced soil temperature fluctuation by 0.09 °C per percent increase in volumetric moisture content (Obia et al. 2018, unpublished data) and therefore lower evaporation. This effect of biochar is similar to the observation by Zhang et al. (2013) where biochar reduced temperature fluctuations by moderating extreme high and low temperatures in the North China Plain. In the North China Plain, reduction in temperature fluctuation was due to reduced thermal conductivity in addition to increased reflectance of near-ultraviolet and blue-light wavelengths and decreased reflectance in infrared wavelength range. In Zambia where soil/biochar mixtures were covered with soil, reduced temperature could be due to increased soil heat capacity caused by greater water contents. Moderation of extreme soil temperature coupled with greater water content can be an important component of climate resilience of agro-ecosystem in a tropical setting and can reduce the negative effect of high temperature and unreliable rainfall on seed germination, root activity, crop growth and eventually yields.

4 Effect of Biochar on Soil Chemical Properties

Given the low inherent fertility of tropical soils associated with low pH, low CEC, low organic carbon and low nutrient availability, biochar is a promising amendment that can ameliorate these soils. However, the effect of biochar depends on feedstock, pyrolysis condition and soil type (Martinsen et al. 2015; Cornelissen et al. 2018). The pH of biochar has been reported to be mostly in the alkaline range but can vary from six to over ten (Atkinson et al. 2010; Martinsen et al. 2015). Such generally high pH of biochar can increase pH of tropical soils as has been found for some acidic soils of Zambia, where pH increased by ≥1 unit (Cornelissen et al. 2013; Martinsen et al. 2014). From these same studies, the CEC was also significantly increased by ≥2 $cmol_c\ kg^{-1}$ owing to the low CEC of the soils and high CEC of the biochar derived from corncob and wood. The positive effect of biochar on soil pH may, however, be short term, as reported in a recent study from a highly acidic Indonesian Ultisol by Cornelissen et al. (2018). Elevating the pH in such highly acidic soils will also increase the Ca/Al ratio, limiting the availability of toxic aluminium to plants. Results of the effect of biochars produced from rice (*Oryza sativa. L.*) husk and cacao (*Theobroma cacao. L.*) shell on soil chemical properties and maize (*Zea mays*) yields revealed an initial positive effect of biochar addition, but there was a fading effectiveness after multiple planting seasons. This was believed to be due to leaching of biochar-associated alkalinity and led to the conclusion that biochar should be applied approximately every third season in order to maintain positive effects on yields (Cornelissen et al. 2018).

Biochar contains plant nutrients such as potassium, calcium, phosphorus and some nitrogen (N). However, N in biochar with a high C:N ratio is only to a little extent available to plants, and negative effects of addition of such biochar on crop yields may be observed due to microbial immobilization of N (Gwenzi et al. 2015). Assessing the effect of addition of biochar derived from maize cob (5 $Mg\ ha^{-1}$) to sandy or loamy sands (Arenosols) in Zambia, Martinsen et al. (2014) found significantly greater levels of exchangeable potassium in the soil, resulting in significantly higher concentrations of potassium in maize stover. However, the fertilizing effect of biochar is most likely of short duration, since a significant proportion of base cations will be leached when exposed to precipitation and soil water. Recently, Munera-Echeverri et al. (2018) found that the amount of base cations removed during three washings with water and acid prior to determination of CEC was about three (biochar derived from rice husk) to ten times (biochar derived from pigeon pea) greater than the amount of exchangeable cations. A large proportion (14%, 47%, 66% and 71% for biochar derived from pigeon pea, caca shell, corncob and rice husk, respectively) of the cations removed during the initial washings (i.e. not cations associated with negatively charged binding sites) was due to dissolution of salts (Munera-Echeverri et al. 2018). Therefore, direct addition of plant nutrients is one key benefit of application of biochar, though this effect may be short lived. Raising the pH of soil from acidic range to near neutral can increase nutrient availability, especially of phosphorus which is normally fixed to iron oxide in the acidic weathered soils of tropical areas with pH < 5 and hence not fully available to plants. With enhanced CEC of soil, nutrient retention of base cations can increase, reducing loss of base cations from the root zone to deeper subsoil with percolating water.

With increased nutrient content of the soil and availability of farming systems incorporating biochar, better crop growth can be expected. Well-established crops can tolerate the impact of dry spells, hence improving resilience of the farming system to climate change.

5 Effect of Biochar on Crop Yields

To the farmer, improvement in soil quality due to biochar should translate to an increase in crop yield, as indicated above. The effects of biochar are not positive in all systems, but it may increase yields in low-nutrient, acidic soils in the tropics (Jeffery et al. 2017). In a meta-analysis of a global dataset, Jeffery et al. (2011) showed an overall increase in crop yield of 10% with greatest yield increase in sandy acidic soils (Fig. 2). Acidic soils are very widespread in tropical areas, and hence biochar may have great potentials for increasing productivity of these soils. In an acidic sandy Arenosol in Western Zambia, maize yield increased by 350–450%, whereas in the red sandy clay loam Ultisol of Central Zambia, a non-significant increase in maize yield of 142% and 131% for biochar derived from corncob and wood, respectively, was observed (Cornelissen et al. 2013). From this

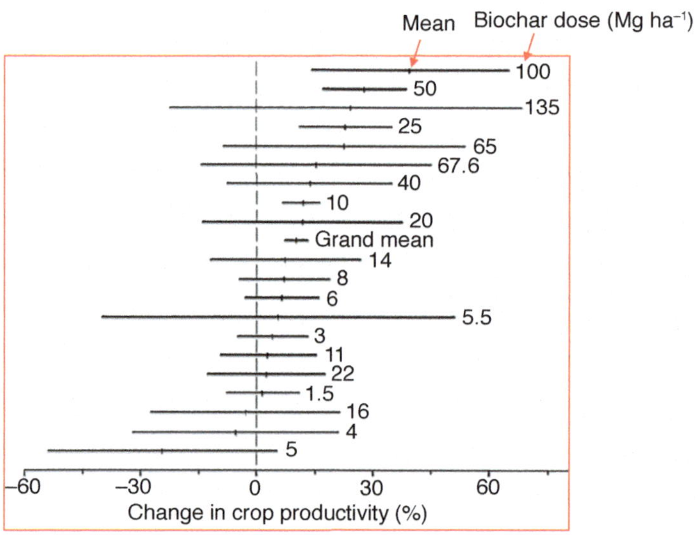

Fig. 2 Result from a meta-analysis of data from various studies showing changes in yield after application of various doses of different biochars (Jeffery et al. 2011)

same study, biochar had no effect on crop yield in "good" soils with loamy texture and near-neutral pH. In a related study by Martinsen et al. (2014), an increase in maize yield of 232% was observed in a fertilized sandy Arenosol with biochar compared to a fertilized control without biochar. The benefit of biochar in increasing yields in very sandy soil has been associated with strong increase in plant available water of the otherwise inherently low plant available water. Therefore, the two main mechanisms for increased yields in tropical soils are pH and plant available water, in addition to CEC particularly of very sandy soils with low CEC.

6 Biochar and Soil Carbon Sequestration

Carbon sequestration as a result of different management practices is due to soil organic carbon sequestered from atmospheric CO_2 through plants, plant residues and other organic solids (Olson et al. 2014). Biochar may improve soil fertility leading to increased biomass production and thus increased carbon sequestration. In addition, depending on the proportion of stable and degradable carbon in biochar (Spokas 2010; Lal 2016), biochar may lock up carbon in the soil for a longer timescale, not comparable to any carbon sequestration approach such as afforestation currently in practice (Lehmann 2007). The carbon content of biochar varies considerably from as low as 17–88% (Atkinson et al. 2010). The low carbon biochar is from non-woody materials such as poultry wastes, whereas high carbon biochars are generally those produced from woody materials. Pyrolysis temperature and retention time during pyrolysis also affect carbon content with higher temperature and longer retention time producing biochars with greater carbon contents. Of the carbon content of biochar, only a small amount of <1% is easily degradable (Luo et al. 2011), leaving large amount of recalcitrant materials, which may be responsible for the long resident time of biochars in soils (potentially hundreds to thousands of years).

Despite low amounts of labile carbon in biochars, it may still affect the decomposition of native soil organic matter similar to what has long been observed upon addition of fresh organic materials (Bingeman et al. 1953). Addition of fresh organic materials to soil has been found to increase the decomposition of native soil organic matter, also called priming effects. Some biochars such as those derived from sewage/paper sludge, *Miscanthus* spp., rye grass (*Lolium perenne*), manure, etc. have been found to increase the decomposition of native soil organic matter (Naisse et al. 2015; Cely et al. 2014; Singh and Cowie 2014; Luo et al. 2011). On the other hand, other biochars such as those produced from wood (mixed), barley (*Hordeum vulgare*) straw, oak (*Quercus*), oil seed etc. have been found to decrease decomposition (Rittl et al. 2015; Cely et al. 2014; Bruun and EL-Zehery et al. 2012; Jones et al. 2011; Zimmerman et al. 2011). A number of reasons seem to explain the differences in the observations including soil texture, pH, soil carbon content, land use, biochar feedstock, pyrolysis temperature and retention time. Greater positive priming effect upon addition of biochar to soil has been found in coarse-textured soils with low pH. Similarly, greater positive priming by biochar addition has also been found in land uses that reduce soil pH and carbon content, e.g. bare fallow or arable land compared to grasslands. Biochar derived from woody materials as well as high temperature pyrolysis and longer retention time results in lower priming effects or even negative priming (i.e. build-up of non-pyrogenic carbon) possibly due to low amounts of labile carbon it carries. In the long term, biochar could as well store native soil organic matter by sorption and physical protection (Zimmerman et al. 2011).

Application of biochar derived from woody materials pyrolysed at high temperature ≥ 400 °C with longer retention time in the pyrolysis unit of more than 1 hour may stabilize soil organic carbon with stronger effects in fine-textured soil with pH of >4 and greater carbon content of >3% (Cely et al. 2014; Cross and Sohi 2011; Zimmerman et al. 2011). One reason for this could be the low amount of labile carbon on these types of biochars to stimulate priming of native soil organic carbon. Overall, this may be critical in a tropical setting where increasing soil carbon content is challenging.

In Zambia, application of biochar made from corncob to a coarse-textured soil increased soil carbon content from ~0.7% up to >3% (depending on biochar dose) with much of the carbon coming from biochar after 2 years (Obia et al. 2016). Studies on effect of biochar on native soil organic carbon in Eastern and Southern Africa are lacking, but biochar's effect may be similar to results from similar soils elsewhere. Increase in soil organic carbon after biochar application would be an ideal outcome given the importance of organic carbon/matter in soil. Soil organic matter is known to increase soil aggregation, nutrient retention (CEC), water retention and overall soil quality improvement. In effect, it can improve the resilience of farming system to disturbances, both man-made, e.g. tillage, and natural ones, e.g. droughts. In addition, build-up of soil organic matter is positive in terms of climate change mitigation (i.e. long-term storage of carbon).

7 Incorporation of Biochar in Climate-Resilient Farming System: Conservation Agriculture

Large-scale implementation of biochar has come under scrutiny in recent years due to potentially large quantity of biochar reaching several tens of tonnes per hectare that are not normally available. This is attributable to shortage of feedstock. The current widespread promotion of conservation agriculture (CA) in Eastern and Southern Africa presents an opportunity where the amount of biochar required to cause significant effects is drastically reduced. The quantity of biochar may be reduced to cover only 10% of the land surface (Cornelissen et al. 2013). This is because CA involves minimum tillage where only a small area of the surface is tilled and biochar is added or concentrated only to the tilled area instead of application over the entire area. In addition to minimum tillage, the two other principles of CA are residue retention and crop diversification or rotation. The above principles may contribute to the climate resilience of a farming system. The climate resilience may be attributed to the water harvesting capacity of CA minimum tillage where water over a wider area is funnelled to the tilled spots (Cornelis et al. 2013; Obia et al. unpublished data). This then increases water availability for crops in the event of reduced rainfall/dry spell. Minimum tillage also prevents exposure of soil organic matter to decomposition, hence maintaining the functions of organic matter in the soil (Corbeels et al. 2018). However, it is believed that climatic and edaphic conditions, combined with management practices such as seeding system, degree of residue retention, fertilizer addition, weeding and crop rotation, determine whether CA has positive, negative or no effect on yields and soil fertility. Under on-farm conditions in Zambia, CA was found to have negligible effects on soil carbon contents and stocks (Martinsen et al. 2017). Other benefits may include reduced chances of soil degradation due to erosion. Residue retention in CA may further enhance soil water availability due to improved soil water infiltration and reduce water evaporation. Residue cover also moderates soil temperature, which is often high in tropical settings reducing water evaporation from the surface. Such temperature moderation also helps in seed germination and stimulating biological activities. Inclusion of crop diversification such as cover crops may have effects similar to residue retention in addition to other benefits such as pest and disease control. The above formed the basis where CA has been considered as a climate-resilient farming system that increases yields (Thierfelder and Wall 2009; Thierfelder et al. 2018). On the other hand, CA minimum tillage is commonly associated with soil compaction over time and weed management challenges, all of which may reduce crop yields. Inclusion of biochar, given its potential benefits in CA systems, can enhance resilience to climate change and increase yields as has been demonstrated in Zambia (Cornelissen et al. 2013). We therefore suggest that among resource-poor farmers in Eastern and Southern Africa, biochar and CA should be implemented as a single intervention. This may then form a sustainable and profitable low-input agriculture among poor smallholder farmers.

Other solutions to limited amounts of biochar due to feedstock shortage, in addition to CA, have been suggested. These include cultivation of crops that have large quantities of biomass wastes such as pigeon peas to provide feedstock and incorporation of limited amounts of biochar in a formulation such as co-composting.

8 Production and Application of Biochar Formulations in Agriculture

There are two reasons why biochar formulations have been considered necessary for field application in agriculture. Firstly, formulations are a way to reduce the amount of biochar in field application while still delivering its benefits in improving soil quality and crop yields. Secondly, the delivery of good biochar properties into the traditional amendment such as manures or removing negative attributes of manures. Several studies have reported two forms of formulations: biochar-manure mixtures and biochar-manure co-compost (Dias et al. 2010; Kammann et al. 2015; Qayyum et al. 2017; Yuan et al. 2017). Given the easily decomposable nature of manures, its effect is expected to fade within a few seasons unlike biochar. The contribution of manures to the effect of biochar-manure mixture may also fade over a short time. In an experiment on Vertisols in West Timor, Indonesia, the positive effect of 5 Mg ha^{-1} biochar and 5 Mg ha^{-1} compost mixture dropped in the second year of the experiment though not significantly (Fig. 3).

A number of attributes of biochar have been identified and can be delivered to manure in co-composting process. In a comparison of biochar, coffee husk and saw dust, Dias et al. (2010) found that biochar promoted humification more than the other bulking materials during composting and further reduced nitrogen losses in mature compost. Increased humification due to biochar may result in better stabilization of native soil organic matter as well as organic matter in manure during co-composting as has been reported by Qayyum et al. (2017). Biochar-manure co-composting has been found to significantly reduced soil CO_2 and N_2O climate gas emissions compared to manure compost (Yuan et al. 2017). In some instances, biochar-manure co-composting has been found to cause yield increases even in soil/

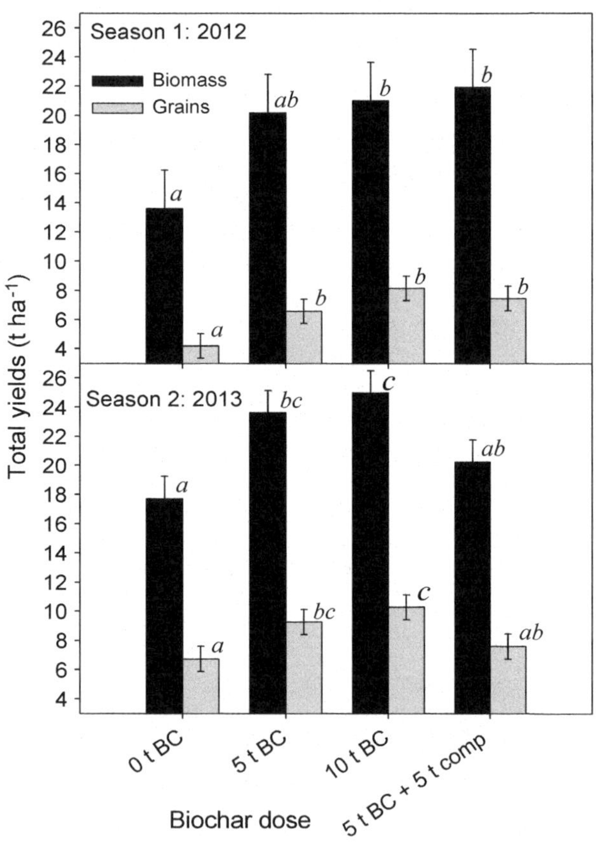

Fig. 3 Maize yields from field plots amended with BC and BC + compost in Vertisols, West Timor. 5 and 10 Mg (t) BC correspond to 2.5% and 5% BC, respectively. Means followed by the same letter are not significantly different. Error bars are the SEs, $n = 6$. (Adapted from Obia et al. 2018)

regions where biochar has been found to have either no or limited effects such as in temperate soil (Kammann et al. 2015; Qayyum et al. 2017). The enhanced benefits of co-composted biochar may be attributed to inherent properties of biochars such as large surface area, porosity, CEC and pH.

9 Conclusions

Building climate-resilient farming systems may remain part of the greater efforts in the coming years to increase agricultural production in the face of climate change. This will require practices that allow sustainable use of resources. Biochar, as one of the possible resources with positive effects on soil quality, can be concentrated into the minimum tillage planting stations of CA. Biochar can then increase the positive attributes of CA in soils such as improvement of soil hydraulic properties (e.g. water infiltration and retention) contributing to increase in the productivity of agricultural lands. In addition, biochar can directly add nutrients to soils and increase the nutrient retention ability of the soil (CEC) and soil pH. Soil water infiltration and retention, nutrient level, CEC and pH are among the soil properties that are generally below optimal range for agricultural production in tropical soils largely due to the advance weathering status. These soil properties can be improved directly or indirectly through application of biochar. Due to long resident time of biochar in soils as demonstrated in Amazonia, some of its effects in soil are expected to last for hundreds to thousands of years alongside the associated carbon sequestration potentials. Therefore, biochar technology and CA can complement each other to increase productivity and should be implemented as a single intervention. In particular, both biochar and CA will improve the resilience of agroecosystem in Eastern and Southern Africa, which is one of the most vulnerable regions to climate change.

References

Atkinson CJ, Fitzgerald JD, Hipps NA (2010) Potential mechanisms for achieving agricultural benefits from biochar application to temperate soils: a review. Plant Soil 337:1–18

Bayabil HK, Stoof CR, Lehmann JC, Yitaferu B, Steenhuis TS (2015) Assessing the potential of biochar and charcoal to improve soil hydraulic properties in the humid Ethiopian Highlands: the Anjeni watershed. Geoderma 243–244:115–123

Bingeman CW, Varner JE, Martin WP (1953) The effect of the addition of organic materials on the decomposition of an organic soil. Soil Science Society of America Proceedings 29:692–696

Bruun S, EL-Zehery T (2012) Biochar effect on the mineralization of soil organic matter. Pesquisa Agropecuária Brasileira 47(5):665–671

Chauvin ND, Mulangu F, Porto G (2012) Food production and consumption trends in sub-Saharan Africa: prospects for the transformation of the agricultural sector. Working paper 2012–011, UNDP. Regional Bureau for Africa

Cely P, Tarquis AM, Paz-Ferreiro J, Méndez A, Gascó G (2014) Factors driving the carbon mineralization priming effect in a sandy loam soil amended with different types of biochar. Solid Earth 5:585–594

Corbeels M, Cardinael R, Naudin K, Guibert H, Torquebiau E (2018) The 4 per 1000 goal and soil carbon storage under agroforestry and conservation agriculture systems in sub-Saharan Africa. Soil Tillage Res. In press. https://doi.org/10.1016/j.still.2018.02.015

Cornelis WM, Araya T, Wildermeersch J, Mloza-Banda MK, Waweru G, Obia A, Verbist K, Boever D (2013) Building resilience against drought: the soil-water perspective. In: De Boever M, Khlosi M, Delbecque N, De Pue J, Ryken N, Verdoodt A, Cornelis WM, Gabriels D (eds) Desertification and land degradation: processes and mitigation. Belgium, Ghent, UNESCO Chair of Eremology, Ghent University, pp 1–15

Cornelissen G, Martinsen V, Shitumbanuma V, Alling V, Breedveld GD, Rutherford DW, Sparrevik M, Hale SE, Obia A, Mulder J (2013) Biochar effect on maize yield and soil characteristics in five conservation farming sites in Zambia. Agronomy 3:256–274. https://doi.org/10.3390/agronomy3020256

Cornelissen G, Pandit NR, Taylor P, Pandit BH, Sparrevik M, Schmidt HP (2016) Emissions and char quality of flame-curtain "Kon Tiki" kilns for farmer-scale charcoal/biochar production. PLoS One 11(5):e0154617. https://doi.org/10.1371/journal.pone.0154617

Cornelissen G, Nurida NL, Hale SE, Martinsen V, Silvani L, Mulder J (2018) Fading positive effect of biochar on crop yield and soil acidity during five growth seasons in an Indonesian Ultisol. Sci Total Environ 634:561–568

Cross A, Sohi SP (2011) The priming potential of biochar products in relation to labile carbon contents and soil organic matter status. Soil Biol. Biochem. 43, 2127–2134.

Dias BO, Silva CA, Higashikawa FS, Roig A, Sánchez-Monedero MA (2010) Use of biochar as bulking agent for the composting of poultry manure: effect on organic matter degradation and humification. Bioresour Technol 101(4):1239–1246

Farooq M, Siddique MKH (2015) Conservation agriculture: concepts, brief history, and impacts on agricultural systems. In: Farooq M, Siddique MKH (eds) Conservation agriculture. Springer International Publishing, Cham, pp 3–17

Fungo B, Lehmann J, Kalbitz K, Thiongo M, Okeyo I, Tenywa M, Neufeldt H (2017) Aggregate size distribution in a biochar-amended tropical Ultisol under conventional hand-hoe tillage. Soil Tillage Res 165:190–197

Glaser B, Haumaier L, Guggenberger G, Zech W (2001) The 'Terra Preta' phenomenon: a model for sustainable agriculture in the humid tropics. Naturwissenschaften 88:37–41. https://doi.org/10.1007/s001140000193

Gwenzi W, Chaukura N, Mukome FND, Machado S, Nyamasoka B (2015) Biochar production and applications in sub-Saharan Africa: opportunities, constraints, risks and uncertainties. J Environ Manag 150:250–261

Jeffery S, Verheijen FGA, van der Velde M, Bastos AC (2011) A quantitative review of the effects of biochar application to soils on crop productivity using meta-analysis. Agric Ecosyst Environ 144:175–187

Jeffery S, Abalos D, Prodana M, Bastos AC, van Groenigen JW, Hungate BA, Verheijen F (2017) Biochar boosts tropical but not temperate crop yields. Environ Res Lett 12:053001

Jones DL, Murphy DV, Khalid M, Ahmad W, Edwards-Jones G, DeLuca TH (2011) Short-term biochar-induced increase in soil CO^2 release is both biotically and abiotically mediated. Soil Biol. Biochem. 43:1723–1731

Kammann CI, Schmidt HP, Messerschmidt N, Linsel S, Steffens D, Müller C, Koyro HW, Conte P, Joseph S (2015) Plant growth improvement mediated by nitrate capture in co-composted biochar. Sci Rep 5:11080. https://doi.org/10.1038/srep11080

Kinney TJ, Masiello CA, Dugan B, Hockaday WC, Dean MR, Zygourakis K, Barnes RT (2012) Hydrologic properties of biochars produced at different temperatures. Biomass Bioenergy 41:34–43

Lal R (2013) Food security in a changing climate. Ecohydrol Hydrobiol 13:8–21

Lal R (2016) Biochar and soil carbon sequestration. In: Guo M, He Z, Uchimiya SM (eds) Agricultural and environmental applications of biochar: advances and barriers. Soil Science Society of America, Inc., Madison, WI, pp 175–198

Lehmann J (2007) A handful of carbon. Nature 447:143–144
Lehmann J, da Silva JP Jr, Steiner C, Nehls T, Zech W, Glaser B (2003) Nutrient availability and leaching in an archaeological Anthrosol and a Ferralsol of the Central Amazon basin: fertilizer, manure and charcoal amendments. Plant Soil 249:343–357
Luo Y, Durenkamp M, De Nobili M, Lin Q, Brookes P C (2011) Short term soil priming effects and the mineralisation of biochar following its incorporation to soils of different pH. Soil Biol. Biochem. 43:2304–2314
Martinsen V, Mulder J, Shitumbanuma V, Sparrevik M, Børresen T, Cornelissen G (2014) Farmer-led maize biochar trials: effect on crop yield and soil nutrients under conservation farming. J Plant Nutr Soil Sci 177:681–695
Martinsen V, Alling V, Nurida NL, Mulder J, Hale SE, Ritz C, Rutherford DW, Heikens A, Breedveld GD, Cornelissen G (2015) pH effects of the addition of three biochars to acidic Indonesian mineral soils. Soil Sci Plant Nutr:1–14. https://doi.org/10.1080/00380768.2015.1052985
Martinsen V, Shitumbanuma V, Mulder J, Ritz C, Cornelissen G (2017) Effects of hand-hoe tilled conservation farming on soil quality and carbon stocks under on-farm conditions in Zambia. Agric Ecosyst Environ 241:168–178
Munera-Echeverri JL, Martinsen V, Strand LT, Zivanovic V, Cornelissen G, Mulder J (2018) Cation exchange capacity of biochar: an urgent method modification. Sci Total Environ 642:190–197
Naisse C, Girardin C, Davasse B, Chabbi A, Rumpel C (2015) Effect of biochar addition on C mineralisation and soil organic matter priming in two subsoil horizons. J Soils Sediments 15:825
Novotny EH, Hayes MHB, Madari BE, Bonagamba TJ, de Azevedo ER, de Souza AA, Song G, Nogueira CM, Mangrich AS (2009) Lessons from the Terra Preta de Índios of the Amazon region for the utilisation of charcoal for soil amendment. J Braz Chem Soc 20(6):1003–1010
Obia A, Mulder J, Martinsen V, Cornelissen G, Børresen T (2016) In situ effects of biochar on aggregation, water retention and porosity in light-textured tropical soils. Soil Tillage Res 155:35–44
Obia A, Børresen T, Martinsen V, Cornelissen G, Mulder J (2017) Effect of biochar on crust formation, penetration resistance and hydraulic properties of two coarse-textured tropical soils. Soil Tillage Res 170:114–121
Obia A, Mulder J, Hale SE, Nurida NL, Cornelissen G (2018) The potential of biochar in improving drainage, aeration and maize yields in heavy clay soils. PLoS One 13(5):e0196794. https://doi.org/10.1371/journal.pone.0196794
Olson KR, Al-Kaisi MM, Lal R, Lowery B (2014) Experimental consideration, treatments, and methods in determining soil organic carbon sequestration rates. Soil Sci Soc Am J 78:348–360
Pandit NR, Mulder J, Hale SE, Schmidt HP, Cornelissen G (2017) Biochar from "Kon Tiki" flame curtain and other kilns: effects of nutrient enrichment and kiln type on crop yield and soil chemistry. PLoS One 12(4):e0176378. https://doi.org/10.1371/journal.pone.0176378
Qayyum MF, Liaquat F, Rehman RA, Gul M, Hye MZ, Rizwan M, Rehaman MZ (2017) Effects of co-composting of farm manure and biochar on plant growth and carbon mineralization in an alkaline soil. Environ Sci Pollut Res 24:26060–26068. https://doi.org/10.1007/s11356-017-0227-4
Renner R (2007) Technology solutions. Rethinking biochar. Environ Sci Technol 41:5932–5933
Revell KT, Maguire RO, Agblevor FA (2012) Influence of poultry litter biochar on soil properties and plant growth. Soil Sci 177(6):404–408
Rittl TF, Novotny EH, Balieiro FC, Hoffland E, Alves BJ, Kuyper TW (2015) Negative priming of native soil organic carbon mineralization by oilseed biochars of contrasting quality. Eur J Soil Sci 66:714–721
Schmidt H -P, Taylor P (2014) Kon-Tiki flame cap pyrolysis for the democratization of biochar production. Ithaka J. www.ithaka-journal.net
Singh BP, Cowie AL (2014) Long-Term Influence of Biochar on Native Organic Carbon Mineralisation in a Low-Carbon Clayey Soil. Sci Rep 4:1–9
Smebye AB, Sparrevik M, Schmidt HP, Cornelissen G (2017) Life-cycle assessment of biochar production systems in tropical rural areas: comparing flame curtain kilns to other production methods. Biomass Bioenergy 101:35–43

Sparrevik M, Field JL, Martinsen V, Breedveld GD, Cornelissen G (2013) Life cycle assessment to evaluate the environmental impact of biochar implementation in conservation agriculture in Zambia. Environ Sci Technol. https://doi.org/10.1021/es302720k

Sparrevik M, Adam C, Martinsen V, Cornelissen G (2015) Emissions of gases and particles from charcoal/biochar production in rural areas using medium-sized traditional and improved "retort" kilns. Biomass Bioenergy 72:65–73

Spokas KA (2010) Review of the stability of biochar in soils: predictability of O:C molar ratios. Carbon Manag 1:289–303

Thierfelder C, Wall PC (2009) Effects of conservation agriculture techniques on infiltration and soil water content in Zambia and Zimbabwe. Soil Tillage Res 105:217–227

Thierfelder C, Baudron F, Setimela P, Nyagumbo I, Mupangwa W, Mhlanga B, Lee N, Gérard B (2018) Complementary practices supporting conservation agriculture in southern Africa. a review. Agron Sustain Dev 38:16

United Nations, Department of Economic and Social Affairs, Population Division (2017) World population prospects: the 2017 revision, key findings and advance tables. Working paper no. ESA/P/WP/248

Yi S, Witt B, Chiu P, Guo M, Imhoff P (2015) The origin and reversible nature of poultry litter biochar hydrophobicity. J Environ Qual 44:963–971

Yuan Y, Chen H, Yuan W, Williams D, Walkere JT, Shi W (2017) Is biochar-manure co-compost a better solution for soil health improvement and N_2O emissions mitigation? Soil Biol Biochem 113:14–25

Zhang Q, Wang Y, Wu Y, Wang X, Du Z, Liu X, Song J (2013) Effects of biochar amendment on soil thermal conductivity, reflectance, and temperature. Soil Sci Soc Am J 77:1478–1487. https://doi.org/10.2136/sssaj2012.0180

Zimmerman AR, Gao B, Ahn MY (2011) Positive and negative carbon mineralization priming effects among a variety of biochar-amended soils. Soil Biol Biochem 43:1169–1179

The Efficacy of the Soil Conservation Technologies Adopted in Mountain Agro-Ecosystems in Uganda

Nabalegwa Wambede Muhamud and Asaba Joyfred

Abstract This study examined the efficiency of soil conservation practices, under current cropping systems in Kasese district, Mt. Rwenzori. A longitudinal transect design was used in this study. Transects of 100 m wide were drawn on 6 selected ridges, within which 102 peasant farms with different conservation technologies were selected. The severity of erosion on the selected farms, assessed from the areal extent of physical erosion indicators, was used to determine the efficiency of the soil conservation technologies in controlling erosion. The results indicate that terracing, found on 57.8% of the selected peasant farms, was the most dominant soil conservation practice on the Rwenzori Mountain slopes, followed by trash bunds, water diversion channels, water collection ditches, and mulching. It was, however, evident that despite the adoption of soil conservation practices by some farmers, soil erosion remained high, especially on farms where one conservation practice was solerly used. It is therefore recommended that a more comprehensive soil conservation strategy, involving the use of agronomic soil management and mechanical methods of soil conservation, be emphasized in the Rwenzori highlands.

Keywords Erosion · Conservation technologies · Adoption and efficiency

1 Introduction

Soil has been of particular interest to man since the beginning of organized agriculture many thousands of years ago. No single resource is more important to the achievement of sustainable agriculture than the soil, which contains the nutrients and stores water essential for plant growth (Brader 1994). Indeed the quality of soil largely determines the level of agricultural development, food production, and the quality of life for a large part of the population on the earth's surface. Soil erosion by water is commonly recognized as one of the main reasons of land degradation

N. W. Muhamud (✉) · A. Joyfred
Kyambogo University, Kampala, Uganda

Y. Bamutaze et al. (eds.), *Agriculture and Ecosystem Resilience in Sub Saharan Africa*, Climate Change Management,
https://doi.org/10.1007/978-3-030-12974-3_7

worldwide (Beskow et al. 2009). Soil erosion rates, caused by water, are highest in agrosystems located in hilly or mountainous regions of Asia, Africa, and Southern America especially in less developed countries (Widomski 2011). The noticeable site effects of soil erosion such as rills and gullies together with changes in soil fertility and water storage capacity drastically reduce agricultural or forestry productivity (Kuhlman et al. 2010). In view of the losses caused by water erosion, the development and adoption of efficient soil conservation techniques are fundamental (Miranda et al. 2012). According to Valentin et al. (2005), efficient soil erosion control should reduce the destructive processes occurring in top soil and improve infiltration of surface water into soil profile, thus improving the water balance of eroded basins and increasing the amount of water available to plants. It will also limit soil fertility deterioration caused by soil composition changes and removal of nutrients and organic matter from soil.

The history of conservation in Uganda shows that efforts geared toward soil conservation have often failed, while the problem of soil erosion has continued to escalate. Although the severity of soil erosion problems in the cultivated parts of the highlands was realized as early as the 1920s and consequently conservation measures enforced there after (GOU 1949), most of the conservation works that were introduced are today almost none existent (Nakileza 1992). Nsubuga (1994) notes that initial soil conservation programs in Uganda failed due to lack of responsibility for soil conservation, lack of expertise in the subject area, and dependence on legislation and enforcement other than education for policy administration. In short, structural weaknesses failed the initial soil conservation programs established for this nation.

In Busongora County in particular, the problem of soil has reached alarming levels and threatens the survival of peasant farmers unless urgent mitigation measures are undertaken. The county's total population that has grown from 280,290 in 2004 to 338,009 people in 2014 (UBOS 2014) is concentrated in a fixed stretch of land between the Mount Rwenzori National Park and Queen Elizabeth National Park. This has exerted population pressure on the fragile mountainous land, leading to overcultivation of the lower mountain slopes and encroachment on the steeper slopes and forest lands, resulting in accelerated soil erosion, reduced soil productivity, low crop yields, food shortages, and poverty. Currently, efforts are being made to address the problem of soil degradation. The Government of Uganda through relevant agencies such as the ministries of agriculture, lands, water and environment is continuously encouraging farmers to adopt soil conservation practices so as to improve land productivity and increase their agricultural yields. Unfortunately, in the Rwenzori region, these efforts are generally poorly funded, with peasant farmers being left to fund conservation initiatives on their farms, including those that yield communal benefits, such as water ways, cut-off drains, water diversion channels, soak pits and terraces. As a result, farmers' response to government's soil conservation efforts in the region has been considerably slow. Obrien (1991) indicates that, for farmers to adopt a particular soil conservation practice, they must be convinced that it will produce the desired results in form of increased yields, in an appropriate time frame and they must have access to appropriate resources and skills.

In Busongora County however, the efficiency of the adopted soil conservation technologies is unknown. Thus, this absence of local data on the performance of the adopted soil conservation techniques constitutes a knowledge gap and highlights the need to evaluate the efficiency of these conservation technologies in the region in order to ascertain the most appropriate and technically efficient techniques in erosion control. Such knowledge is vital in making the necessary recommendations as to what soil conservation practices should be given agent attention by those designing conservation programs to mitigate the problem of soil erosion. As such, this study evaluated the efficacy of the adopted soil conservation technologies in Busongora County on Mount Rwenzori.

2 Study Area

Busongora County is located in the eastern part of Kasese district. It is bordered by Bukonzo County in the west, Kamwenge District in the east, Kabarole District in the north and Bushenyi District in the south (Fig. 1). The county's physical landscape is characterized by a low flat land in the northeast and south and a mountainous land rising to over 7500 ft in the northwest. The county is part of the Rwenzori region, which was subjected to the complex internal processes of faulting, folding, and vulcanicity responsible for its striking relief drainage and geology. Nakimera (2001), citing Tanner (1971) and Maasha (1975), notes that the Rwenzori Mountain is considered to be a recently raised horst within the rift valley and comprises of the undifferentiated gneisses of the gneissic-granulitic complex overlaid uncomfortably in places occupied by the younger Precambrian rocks of the Toro Super group. The rise of the mountain and the formation of the rift valley were a result of a series of successive tectonic episodes reported to have started in the Miocene. During the major tectonic episode at about 8 million years ago, the floor of the rift valley which demarcates the northeastern and southern parts of Busongora was downthrown. Later, around about 2.6 to 2.3 million years ago, another major tectonic event resulted in the emergence and rising of the Rwenzori to considerable altitudes and in the formation of the present rift valley lakes of Edward in the southwest, George in the southeast, and the Kazinga Channel in the southern part of Busongora.

Important to note is that the sharp contrast in relief between the mountainous northwest and the lowland south and northeast of Busongora has had a bearing on land use practices and on the severity of erosion. As such, the erosion risk is higher on the mountainous northwest, a phenomenon that calls for mitigation measures in the form of soil conservation, if land has to be used sustainably.

Busongora County, as part of the larger Rwenzori Mountain, experiences montane climate. Due to its location near the equator, the intertropical convergence zone (ITCZ) lies close to the mountain range for most of the year and supplements it with heavy orographic rainfall. Available data indicates that Busongora County experiences two long wet seasons, that is, March to May and August to December, with mean monthly rainfall values of 104 mm and 83.6 mm, respectively, and two short

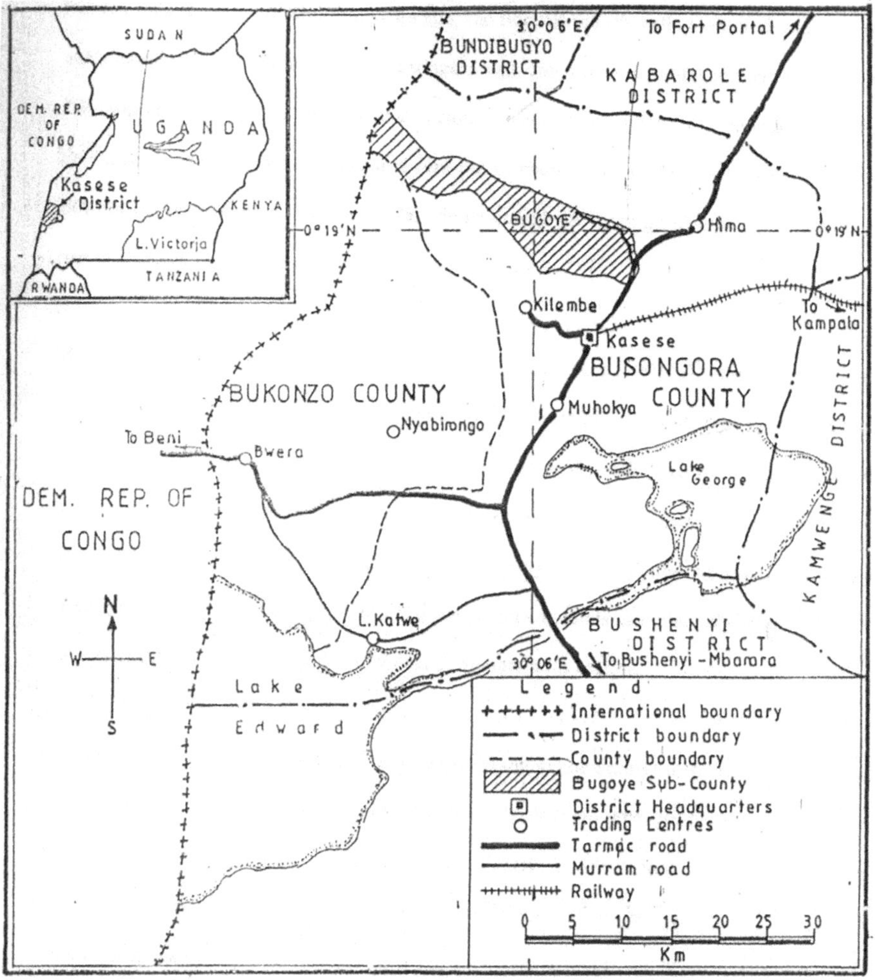

Fig. 1 Map of Kasese district showing the location of Bugoye sub-county Busongora County Kasese district. (Source: Kasese district map of 2002 population and housing census)

dry seasons of January to February and June to July, with mean monthly rainfall values of 29.6 mm and 37.1 mm, respectively. January is the driest month with a mean rainfall value of 28.3 mm, while November is the wettest month with mean rainfall of 117.6 mm (Department of Meteorology 2000–2006). It is important to note that the more frequent rains over this mountainous area often cause rapid mass movements which are readily accelerated by human use of the mountain resources (Cook and Doornkamp 1990). The lowland northeastern and southern parts of Busongora are however relatively dry due to their generally low relief, characteristic of the East African rift valley floor. In this study, the high intensity and frequency of rainfall in this mountainous region were presumed to be related to the high rates of erosion over the Busongora upland peasant farms, thus the need for soil conservation.

Like any other mountainous region, soils in Busongora County are distributed largely according to topography and climate. Because of the effect of slope gradient, steep slopes have thin soils (lithosols), except at their base where alluvial fans develop. In line with this, Nakimera (2001) notes that higher elevations tend to have thin soils due to the fact that biochemical changes by soil organisms are sensitive to temperature changes as well as moisture. Areas of low temperatures, and subsequently poor vegetation, have thin soils, implying that soil-forming processes are more active at lower than higher elevations. Resource use over both the steep and gentle slopes in the form of cultivation, grazing, logging, and road construction increases the dangers of soil erosion, hence the need for soil conservation. This study explores techniques that should be adopted by the Busongora upland peasant farmers to ensure that the fertility of soil, the only resource base for agricultural development, is enhanced and sustained.

3 The Study Methods

Bugoye sub-county, which is one of the most densely populated sub-counties in Busongora County, was purposively chosen as a representative sample for purposes of assessing the efficiency of the present soil conservation technologies in controlling erosion. The area also represents some of the most fragile and erosion prone slopes in Busongora and, as such, requires urgent attention. To assess the efficiency of the existing soil conservation technologies in controlling erosion, three out of the five parishes in the sub-county, that is, Kibirizi, Muhambo, and Bugoye, were purposively chosen for the study (Fig. 2). Six (6) ridges, namely, Nyakajoro, Muramba, Bulindiguru, Kibirizi, Kyikubangali, and Ndugutu, were then selected from the three parishes taking two ridges from each parish to represent the rest of the sub-county. The ridges selected were those with relatively higher levels of agricultural activity and soil conservation. Each of the six ridges was then divided into three slope segments, namely, the upper slope, middle slope, and lower slopes, at significant breakpoints along slope profiles, in line with Speight's (1990) morphological type classification. A longitudinal slope transect, 100 m wide, was drawn along each of the six ridges. The site of the transect was chosen in such a way that the most cultivated part of the slope with the highest number of farms under conservation was selected. A total of 102 farm plots with conservation practices were then chosen from the transects on the 6 ridges in such a way as to represent the slope facets and a range of slope angles. On the sampled farm plots, measurement of the various variables, namely, slope gradient, the height, width, and spacing of terrace risers and trash bunds, depth of water diversion channels, water collection ditches, and mulch depth, was made to determine their appropriateness to local farm conditions. The severity of erosion estimated from the measured dimensions of rills and

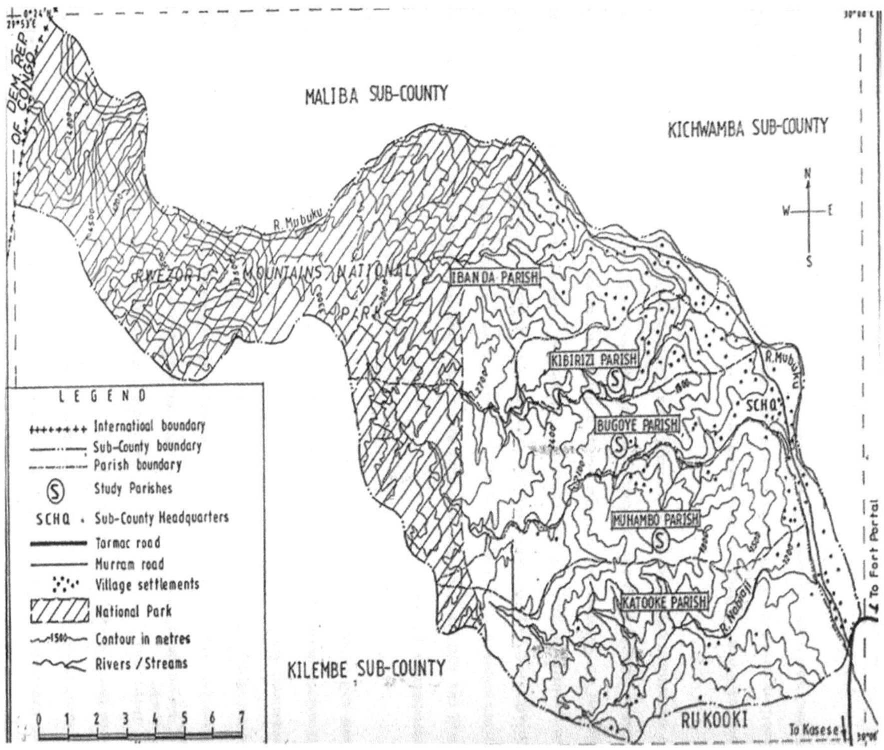

Fig. 2 Map of Bugoye sub-county showing the location of sampled parishes in Bugoye sub-county. (Source: UBOS 2002)

observations of other visual indicators of erosion, such as gullies, rain splash pedestals, and exposed plant roots, was used as a measure of the efficiency of the current soil conservation technologies. Based on field measurements of rill erosion and observations of other physical erosion indicators, erosion on farm plots was characterized as slight, moderate, and severe, according to the guidelines established by the Land and Water Development Department, FAO (2000) shown in Table 1. As such, soil conservation was said to be inefficient over farm plots with moderate to severe erosion and efficient over farm plots with slight or no erosion.

4 Results and Discussion

4.1 The Efficiency of Terraces on the Upland Peasant Farms

To assess the appropriateness and efficiency of the terracing conservation practice, measurements of terrace height and spacing in relation to farm plot slope angle were made over the 59 terraced farm plots randomly distributed over the 6 ridges.

Table 1 Checklist for characterization of erosion severity based on the physical indicators of erosion

Erosion severity characterization	Physical indicators of erosion
No erosion	No evidence of visual erosion indicators
Slight erosion	Average rill depth below 0.1 m
	Little topsoil lost, no rain splash pedestals, no plant roots exposed, and no subsoil exposed
	Areal extent of rill erosion is 5% of the plot
Moderate erosion	Average rill depth of 0.1–0.2 m
	Evidence of topsoil removal but no subsoil exposed, exposure of plant roots in topsoil in some parts
	Areal extent rill erosion is between 6 and 25% of the plot
Severe erosion	Average rill depth of 0.21–0.3 m
	Severe loss of top soil, existence of splash pedestals, exposure of plant roots, subsoil exposed
	Areal extent rill erosion is over 25% of the plot

Adopted from: FAO (2000) guidelines and reference material on integrated soil and nutrient management. pp 141–146

The standard formula which emphasizes the importance of slope in the spacing of terraces was used to calculate the standard terrace spacing values for the selected farm plots. The measured spacing values were then compared with the calculated standard values. Basing on the standard results obtained, terraces were characterized as standard (S), above standard (AS), and below standard (BS). Measurements of terrace heights and spacing are summarized in Table 2.

In terms of terrace spacing, the results summarized in Table 2 revealed that 52 farm plots, which is 88.1% of the 59 terraced farm plots chosen over the 6 transects, had overspaced terraces and were, therefore, characterized as above standard; only 5 farm plots, 8.5%, had terraces whose average spacing was characterized as standard; and two farm plots, 3.4%, had under-spaced terraces. The reason given by farmers for the widely spaced terraces was that they save the badly needed and scarce arable land that would otherwise be used up by close terraces. The farmers also revealed that having close terraces would mean constructing more of the structures per farm plot, which would be so demanding in terms of labor and as such prove costly for poor farmers.

It should also be noted that terrace heights and spacing did not show any systematic change, from the lower through the middle to the upper slopes. In many instances, farm plots with same slope angles registered differences in terrace heights and spacing. This confirms the fact that farmers did not follow any standard guidelines in constructing terraces over their farms. The same view is echoed by Bagoora (1997) who observed that contour bunds in Rukiga were arranged arbitrarily, instead of being laid accurately along the contours, an indication of lack of the technical knowledge necessary for designing appropriate soil conservation structures.

Table 2 Characterization of terrace risers by spacing

Slope transact	Mean farm plot angle %			Number of terraces			Mean terrace risers' height			Mean terrace risers' spacing			Characterization by spacing			Total number of terraces	Total number of farm plots
Ridge	LS	MS	US	LS	MS	US	LS	MS	US	LS	MS	US	S	AB	BS		
Nyakajoro	15.8	46.9	53.5	7	12	5	0.63	1	1.2	16.5	15.5	14.1	1	9	0	24	10
Muramba	17.3	31.3	57.2	8	19	6	0.73	1	0.9	18.1	15	12.9	2	13	0	32	15
Bulindiguru	13.9	34.8	53.6	7	10	8	0.53	1	0.7	16.2	16.6	17.7	1	10	0	25	11
Kibirizi	21	29.1	39.8	4	7	4	0.55	1	0.7	15.6	16	17.9	0	5	2	15	7
Kyikubangali	12.6	41.2	57	4	8	5	0.55	63	0.75	17	19.4	13.6	1	6	0	17	7
Ndugutu	13.8	35.6	56.9	4	11	6	0.65	1	0.9	17	15.4	16.2	0	9	0	21	9
Total													**5**	**52**	**2**	**134**	**59**
Percentage													**8.5**	**88**	**3.4**		

LS Lower slope, *MS* middle slope, *US* upper slope, *S* Standard slope, *BS* below standard, *AS* above standard

Soil conservation is meant to reduce the current erosion rate to tolerable levels and as such maintain land productivity and reduce sedimentation of stream channels downslope. The most important observed function of terrace risers on the upland arable farms of Bugoye sub- county was the trapping of runoff and sediment originating from the immediate upslope farm plots, which would have otherwise caused serious sediment pollution in the streams at the bottom of the slope. In the area of study, successive terraces over upland farms combine to result in a net reduction in sediment deposited in river valleys, thus reducing the danger of sedimentation of streams, which would otherwise have affected water quality and led to serious flooding downstream over the lowland areas of Bugoye. Indeed over the terraced farm plots, runoff and eroded soil were observed to accumulate at the upper terrace ends, thereby facilitating water infiltration where the accumulated soils were deep enough. As such terraces on the upland farms of Bugoye sub-county play an important role in alleviating off-site erosion effects. The contribution of terraces in trapping sediment over the Bugoye upland farms, however, should not be overemphasized. From the observations made and measurements taken, most of the terraces were found to be both overspaced and undersized. As observed, the wide spacing of the terraces compromises their role in reducing slope length, while their low heights compromise their ability to adequately trap sediment for a long time, so that the terraces are easily overrun by high-velocity runoff water from heavy rainstorms, making them unable to control erosion over upland peasant farms. Worse still, the wide spacing of the terraces allows runoff to recollect in between the terraces within the farm plots, thus gaining velocity and becoming very erosive before it meets the next terrace. As such, it is not uncommon to find large and deep rills in between two widely spaced terraces on a terraced farm plot.

In addition, observations and field measurements revealed higher terrace heights on the middle slopes and some sections of the upper slopes with deeper soils, while the steeper upper slopes with shallow stony soils had shorter undersized terraces. The construction and maintenance of sizable terraces over steep slopes is a big problem and is indeed a challenge to soil conservation in mountainous areas. Fitz (1983), in explaining the effect of topography on soil formation, elaborates on the difficulty conservationists are likely to face in trying to maintain sizable terraces on steeper slopes by arguing that there is a tendency for material to remain in place on flat lands and gently sloping sites and for the pedo-unit to be thick, but as slope angle increases, so does the erosion hazard, resulting in thin soils on strongly sloping grounds. The thin, stony soils obviously restrict terraces to only shorter heights making the practice inappropriate and inefficient over steeper mountain slopes. Given this observation, it is important to note that though terraces appear to be reasonably satisfactory as a conventional conservation measure, they are not necessarily suited to all steeply sloping areas.

Bagoora (1997) in analyzing the efficacy of terraces in the Rukiga highlands observed a decline in terrace heights with increase in slope and notes that while terrace heights on the Rukiga upland peasant farms increased from lower to middle slopes, they declined on the upper slopes. He attributes the low heights of terraces on steeper slopes to slope processes and argues that on the lower slopes, transfer of soil downslope due to upslope cultivation is less dynamic, but it increases on the middle slope due to the existence of reasonably deep soils. As such the loss of soil below the terrace increases the lower terrace walls to reasonable heights. But on the upper slopes where soil loss is even more dynamic, the shallow soil depth limits the growth of terrace walls. Temple (1972) cited by Morgan (1986) found terraces unsuitable for controlling erosion in the Uluguru Mountains in Tanzania because the soil was found to be too thin, so that their construction exposed the infertile subsoil thereby curtailing plant growth.

It was also observed that where the poorly maintained terraces were continuously overrun by high-velocity runoff, the upper terrace ends were filled up with sediment, while the lower ends were modified into steep slopes due to continuous erosion. This over steeping of the lower terrace ends increases the velocity of runoff, causing severe rill and sheet erosion, creating deep rills in case of sufficient soil depth and massive wash of soil up to bedrock or subsoil in case of shallow soil, thus causing severe crop destruction (Plate 1).

The presence of poorly maintained terraces has indeed aggravated erosion problems on some Bugoye farm plots. Plate 1 above shows a poorly maintained modified bench terrace with overgrown grass on a beans farm plot. As seen in the photograph, the lower terrace end has been over-steepened by both cultivation and high-velocity runoff, and as such the steep slope created accelerates the speed of runoff increasing its erosive power leading to the development of large and deep rills and thus massive loss of soil and destruction of crops.

The results of this study generally revealed the fact that sizeable terraces can hardly be constructed and maintained on the steeper farm plots of Bugoye. As such,

Plate 1 Severe rill erosion over a beans farm plot caused by high-velocity runoff flowing over the steep lower terrace end on the Kibirizi middle slope

conservationists should focus on combining terraces with other conservation techniques, such as cutoff drains, diversion channels, mulching, and soil cover crops, to ensure higher infiltration levels and a reduction in the amount of runoff; otherwise terraces alone will remain an inefficient method of controlling erosion.

To assess the extent of erosion on the terraced farms, a detailed analysis of erosion severity based on the physical indicators of erosion was done over 1 square meter micro plots randomly demarcated over selected terraced farm plots. The results of the measurements of rill depth, width, and length over the three farm plots taken from each slope transect are given in Table 3.

Rill erosion was observed to be the most dominant erosion type on the upland peasant farms of Bugoye. The dominance of rill erosion in the area of study is in itself an indicator of severe erosion levels. On the significance of erosion type in determining erosion severity, Spomer et al. (1973) note that, since rills and gullies remove larger volumes of soil per unit area than does sheet wash, the dominance of rill erosion indicates severe erosion rates. As such, the rill dimensions of depth, width, and length were measured over micro plots drawn on randomly chosen farm plots and used to classify erosion as slight, moderate, and severe.

Generally the results in Table 3 show mean rill depth values ranging from 0.10 to 0.21 m, mean rill width values between 0.11 and 0.21 m, and mean rill length values of 0.65–0.88 m. Muramba and Bulindiguru registered the highest erosion rates with Muramba registering a mean rill depth and width of 0.21 m and 0.20 m, a mean rill length of 0.88 m, and an average rill number of 3.3 rills per square meter. Bulindiguru registered 0.20 m and 0.19 m as mean rill depth and width, 0.76 m as mean rill length, and an average rill number of 3.3 rills per square meter. Ndugutu had a mean rill depth and width of 0.19 m and 0.16 m, respectively, a mean rill length of 0.83 m, and an average rill number of 3.3 rills per square meter.

These results reveal generally moderate to severe erosion rates over all the selected farm plots. The mean rill depth values for all the selected farm plots were found to be over 0.1 m, indicating that erosion rates were all above the standard value for slight erosion and thus a threat to the growth of crops. In addition to being deep, the rills were also found to be considerably large. The implication of these rills with high depth, width, and length values is that they act as drainage channels

Table 3 Measurements of rill dimensions on selected terraced farm plots

Slope transect (name of ridge)	Average farm plot slope (%) Lower slope	Middle slope	Upper slope	Mean rill number per Sq.m	Mean rill depth (m)	Mean rill width (m)	Mean rill length (m)
Nyakojoro	16.5	40.2	51.8	3.7	0.14	0.12	0.65
Muramba	17.0	47.6	50.0	3.3	0.21	0.21	0.88
Bulindiguru	14.5	34.7	59.0	3.3	0.20	0.19	0.76
Kibirizi	23.2	37.0	41.0	3.3	0.10	0.15	0.72
Kyikubangali	12.9	43.0	55.2	3.7	0.18	0.11	0.70
Ndugutu	14.0	38.2	53.6	3.3	0.19	0.16	0.83

Plate 2 Measurements of the dimensions of the rills over a micro plot drawn on a freshly cultivated severely eroded cassava farm plot on the Muramba upper slope. The terrace is far below and was not captured

facilitating the transportation of large amounts of sediment leading to massive loss of soil from farmland and therefore severe erosion rates (Plate 2). Indeed as Morgan (1986) observes, over areas affected by rill erosion, the total sediment eroded is a reflection of the density and size of rills.

Terraces, especially those on the upper slopes, were found to be a failed venture for the Bugoye farmers. Plate 2 shows the researcher measuring the rill dimensions to estimate erosion severity on a farm plot; and as seen in the photograph, high-velocity runoff was found to have formed deep and wide rills on land portions in between terraces. As already noted, farmers in Bugoye were found to locate terraces at considerable distance away from one another. Such wide terrace spacing allows runoff to accumulate in between the terraces, thereby creating deep rills and destroying crops. In the photograph the rills formed by high-velocity runoff were found to have swept the newly sowed seeds downslope causing a big loss to the farmer.

It is also important to note that while the measured rill dimensions were higher over Muramba, Ndugutu, and Bulindiguru, they were slightly lower over the three other transects of Kyikubangali, Nyakajoro, and Kibirizi. Kyikubangali registered 0.18 m and 0.11 m as mean rill depth and width; its mean rill length was 0.70 m, while its average rill number was 3.7 rills. Nyakajoro had 0.14 m and 0.12 m as mean rill depth and width and 0.65 m as mean rill length and an average of 3.7 rills. Kibirizi registered an average rill number of 3.3 rills per square meter, a mean rill length of 0.72 m and 0.10 m and 0.15 m as mean rill depth and width, respectively. Even then, these values are still indicative of high erosion rates. In fact the estimated real coverage of erosion over all the 18 farm plots selected for detailed study was found to be over 10%, thus indicating moderate to severe erosion in the area of study.

Analysis of erosion severity along the six selected ridges in relation to slope gradient revealed a strong relationship between slope gradient and rate of erosion. As already noted, rill depths were found to be lower over the upper steeper slopes and higher on farms over the middle and lower slope facets with gentle slopes. But despite these lower rill depths over the higher slopes, their rill widths were found to be larger and rill floors exposed the parent rock an indicator of severe erosion. The existence of shallow rills over steeper slopes is well explained by Fitz (1983), who argues that over steeper slopes, soils are shallow and stony due to higher rates of erosion, while those over less steeply sloping surfaces are deep with a high percentage of clay and other resistant minerals. It therefore follows that the occurrence of shallow rills over the upper slopes of Bugoye is a reflection of the dominance of shallow soils, which restrict rill formation only to shallow depths. In fact erosion was found to be more severe over farms located on the upper slope segments with steep slopes, than on the middle and lower slopes mainly because, on steep slopes, loss of even smaller amounts of soil through the shallow rills was found to leave behind a bare rock, making crop growth difficult. The exposure of the parent rock by rills on higher middle and upper slopes, as observed in the area of study, is an indicator of the effect of erosion on plants, which need considerable soil depth to grow, and accounts for the low crop yields, occasional food shortages, and poverty in Rwenzori region and Busongora in particular. This underscores the urgent need to intensify conservation measures over these steep slopes, if life in this mountainous part of Uganda is to be sustained.

In addition, the results revealed the importance of cropping systems in erosion control. In Bugoye, erosion was found to be higher on farm plots with a single annual crop, such as cassava, beans, and groundnuts, but lower over plots with mixed crops as cassava groundnuts. Although not captured in micro plot analysis, field surveys revealed that erosion rates were even much lower over plots with a combination of annuals and perennials as on the banana-beans farm plots. The practice offers considerable protection to soil from rain droplet effect and checks on speed of runoff.

While rill erosion was found to be the most dominant type of erosion over the six ridges, it was found to occur together with sheet and rain splash erosion. Indeed as noted by Morgan (1986), part of the material transported by rills is derived from inter-rill areas and is moved into the rills by overland flow or rain splash. Over steeper slopes where soils were too thin to allow the formation of deep rills, sheet erosion dominated, exposing rocky surfaces. Although not captured within the transects, gully erosion was occasionally found to occur in the farm boundaries due to poor disposal of runoff from diversion channels and terraces and in areas where footpaths released runoff onto farm plots downslope.

It is, therefore, true to say that the terraces over the upland peasant farms of Bugoye only play the role of simply trapping sediment from upslope, thereby leading to the accumulation of soil at the upper terrace ends and reducing the amount of sediment delivered to streams and farm plots downslope. In fact observations and measurements of the physical indicators of erosion on the terraced farms in the area revealed that despite the existence of terraces, on-site point erosion in the form of rill and sheet erosion still remained high. Terraces in Bugoye were, however, found to be capable of trapping more soils over the lower slope segments than on the upper slopes; but even here, their performance as erosion control structures was observed to be worsened by terrace failure. So in cases where terraces were overrun by high-velocity runoff, the erosion rate was found to be severe, probably more severe than it would have been had there been no conservation structures at all. This led to massive loss of soil and destruction of crops. Indeed some of the non-adopters cited the high erosion rates from the failed terraces as one of the reasons that discouraged them from constructing them on their farms. These observations relate to the findings of Bagoora (1997) on the efficacy of terraces in Rukiga. He notes that terraces in Rukiga hardly reduced the gradient of farm plots, except at lower gradients; so the gradient of terraced farm plots remained steep, and as a result, the point of erosion remained high too. He thus observed that the efficacy of terraces in Rukiga increased on lower slopes.

All in all, although the terraces play an important role of trapping runoff and sediments over the upland peasant farms, facilitating accumulation of soil at the upper terrace ends, and therefore reducing sediment pollution in the streams downslope, they have to a large extent been inappropriate to the local farm conditions and are inefficient in controlling on-site point erosion on the upland farms of Bugoye. As such moderate to severe rill and sheet erosion still takes place in between successive terraces, leading to massive loss of soils, destruction of crops, a considerable reduction in soil fertility, reduced land productivity, and thus occasional famine and persistent poverty, a problem that requires agent attention of all conservationists.

4.2 *The Efficiency of Trash Bunds in Controlling Erosion*

Measurements of trash bund dimensions in Table 4 revealed that their average heights range from 0.16 m to 0.41 m, while average spacing ranged from 10.7 m to 19.0 m. The mean height and spacing values for the 16 farm plots with trash bunds

Table 4 Showing trash bund dimensional measurements over five ridges

Slope transect (name of ridge)	Number of farm plots	Mean farm plot slope (%)	Trash bund mean height (m)	Trash bund mean spacing (m)
Nyakajoro	04	25	0.27	13.8
Muramba	03	33	0.21	15.3
Bulindiguru	02	37	0.28	16.9
Kibirizi	01	24.8	0.16	19.5
Kyikubangali	03	29	0.19	11.0
Ndugutu	03	21	0.37	12.2

were 0.27 m and 13.8 m for Nyakajoro, 0.21 and 15.3 for Muramba, 0.28 m and 16.9 m for Bulindiguru, 0.16 m and 19.5 m for Kibirizi, 0.19 m and 11.0 for Kyikubangali, and 0.37 m and 12.2 m for Ndugutu.

As already noted, trash bunds in Bugoye are temporary structures that are destroyed every plowing season and reconstructed in the following season and are common over the steeper middle and upper slopes. It was observed that trash bunds are capable of trapping small amounts of soil and checking the speed of runoff over the upland farms. Their wide spacing and low heights, together with the fact that they are used in isolation, compromise their ability to effectively serve as soil conservation structure over steep slopes where runoff velocities are high. In many cases severe rill and sheet erosion still take place in between the bunds. Over some farm plots, trash bunds were found to be unable to control high-velocity runoff from highly intensive rainfall and had therefore been overrun, submerged, and swept away. Erosion over such plots was found to be even more severe than it could have possibly been had there not been any conservation structures at all (Plate 3). As such, in their present form, trash bunds are largely inefficient in controlling erosion over slopes of Bugoye.

Although trash bunds are meant to trap runoff and reduce erosion rates, old bunds were found to be a liability to soil conservation efforts. Plate 6 shows an old trash bund *(in the fore ground)* that has been submerged and overrun by runoff. As seen in the photograph, water running downslope in well-defined rills submerges and cuts through the already weakened trash bund, causing even more severe erosion.

Plate 3 Severe erosion caused by high-velocity runoff flowing over submerged trash bund over a beans-cassava garden on the Nyakajoro middle slope. (One of the submerged bunds can be seen in the fore found)

This is evidence that trash bunds are unsuitable for controlling erosion over steep slope where water velocities are high. As such where their use is inevitable, trash bunds should not be used in isolation.

4.3 *The Efficiency of the Water Diversion Channels in Controlling Erosion*

Measurements of depths of water diversion channels over the 11 farm plots randomly distributed over 5 transect revealed that the channels over the upland peasant farms of Bugoye were too shallow, ranging from 0.17 m to 0.32 m. A standard diversion channel should be at least 1.5 feet, which is 0.45 m. But as can be seen from the results in Table 5, all the channel depths were below this average and were thus characterized as below standard.

In Bugoye sub-county, the observed diversion channels are meant to dispose runoff from cultivated fields at a nonerosive velocity. Although these structures are capable of controlling erosion in the early stages after their construction when they are still deep enough, their sitting still remains a big problem. In terms of siting, the diversion channels were found to be inappropriately sited. In principle the diversion channels should be dug along the break points of the slope, either where a plateau changes into a steep slope or where along slope changes into a flatter land so as to collect water before it becomes erosive. In Bugoye, however, the channels were found to be sited haphazardly, far below break points in slopes, therefore allowing runoff water to gain velocity and become very erosive, forming rills before it reaches the channel. Even then some of the diversion channels were found to be located at lower sections of the farm plots, thus protecting only a small part of the farm plot from erosive runoff. Observations also revealed that farmers are not always mindful of the need to discharge collected runoff appropriately, and as such runoff is often discharged on to erosive farm plot boundaries and foot paths which later develop into gullies causing severe erosion.

In terms of design, the walls of the channels were found to have been dug in a rectangular form with vertical walls making them susceptible to collapsing under heavy rains, thereby filling up with sediment causing overspills and severe rill erosion. In principle, channel walls are meant to be inclined at an angle to avoid the risk

Table 5 Dimensional measurements of water diversion channels

Slope transect (name of ridge)	Number of farm plots	Farm plot mean slope angle (%)	Mean channel depth (m)	Mean channel width (m)
Nyakojoro	02	12.7	0.32	0.21
Muramba	03	17.4	0.23	0.26
Kibirizi	03	15.5	0.21	0.18
Kyikubangali	01	11.6	0.17	0.23
Ndugutu	02	10.2	0.24	0.22

Plate 4 An over silted water diversion channel initiating severe rill erosion downslope in a beans garden on the Kyikubangali lower slope

of channel walls collapsing. In addition, while diversion channels should be graded in a way that allows them to transport water and sediment off the farm without causing erosion, evidence from the newly constructed channels over the upland peasant farms of Bugoye revealed that most of them were nearly flat, with some depressions that encourage water and sediment accumulation, facilitating channel silting and overspills aggravating the soil erosion problem (Plate 4). The wide spacing between the diversion channels also leaves land between the channels prone to severe erosion. Some channels were also found to have been filled up with sediment and therefore no longer operational as conservation structures.

Water diversion channels in Bugoye were indeed found to betray the very purpose for which they are meant. Plate 4 shows a shallow highly silted diversion channel in a beans garden. As seen in the photograph, runoff from upslope has broken through the diversion channel, thereby creating large and deep rills leading to massive loss of soil and destruction of crops. In fact the crops that were in the areas now occupied by the rills have been washed downslope. Unless the designs of these channels are improved, the structures will remain a liability to the farmers.

Since only well-designed and maintained channels are capable of controlling erosion effectively, the poor maintenance and shallow depths of the diversion channels in Bugoye have limited ability to hold water from intensive storms; and as such, the water diversion channels over the Bugoye upland farms were found to be inappropriate to the local farm conditions and inefficient in controlling erosion.

Table 6 Mean depth values for water collection ditches

Slope transect (name of ridges)	Number of farm plots	Mean farm plot slope angle (%)	Water ditches mean depth (m)	Water ditches mean diameter (m)
Nyakojoro	03	25.4	0.25	0.76
Muramba	02	18.0	0.21	0.71
Bulindiguru	03	15.3	0.28	1.20
Kyikubangali	01	12.2	0.19	1.11
Ndugutu	01	21.9	0.32	0.65

4.4 The Efficiency of Water Collection Ditches in Controlling Erosion

The average depth values of water collection ditches over the ten farm plots were found to be ranging from 0.19 m to 0.32 m, while the diameter values ranged from 0.65 m to 1.2 m. Table 6 shows the mean depth and diameter values for water collection ditches on selected farm plots over the five ridges.

Water collection ditches are meant to collect runoff during a rainstorm, store it, and allow it to infiltrate safely before it becomes erosive. Observations and measurements on existing water collection ditches revealed that most of these were shallow in depth and narrow in diameter. Achan (2001) in her characterization of conservation structures in the banana-based micro-catchments of the Lake Victoria shore region notes that for soak pits to be effective in accommodating water from heavy rain storms, they should be 0.8–1 m in depth and 2–3 m in diameter; but as can be seen from the measurements of the water collection ditches in this study, the depth and diameter values are far below the recommended minimum standards. The observed risk associated with this deficiency is that the pits very often cause overspills during heavy rains, and as the overspilling water flows downslope in a massive way, it often creates large rills in areas of deep soils, aggravating the problem of soil erosion. Generally, although pitting in Bugoye is capable of accommodating water from short and less intense rainstorms, it was observed to be inefficient in controlling erosion during heavy intensive storms, given their shallow depths.

4.5 The Efficiency of Mulching in Controlling Erosion

Measurements of much depth over the six mulched farm plots revealed that average mulch depth values range from 1.5 cm to 2.8 cm, an indication of shallow mulching over all the plots. Most of the mulched plots also had patches of bare ground, and mulches were generally thin and scattered. Table 7 shows the average mulch depth values over the selected farm plots.

Table 7 Measured mulch depth values in relation to slope

Slope transect (name of ridge)	Number of farm plots	Mean farm plot slope (%)	Average mulch depth (cm)
Muramba	02	12.1	1.7
Bulindiguru	01	14.1	2.5
Ndugutu	03	11.7	2.2

Plate 5 Sheet erosion caused by runoff sweeping mulches over a thinly mulched banana farm plot on the Bulindiguru lower slope

Over the farm plots where mulching is practiced, the mulching materials were found to be inefficient in controlling rain splash erosion and checking the speed of runoff to prevent rill and sheet erosion. As seen from Table 7, mulch depth values over the Bugoye farm plots are low, and the mulches were found to be scattered on the farm plots leaving out portions of land unprotected from raindrop effect and offering little resistance to high-velocity runoff which sometimes sweeps the mulches downslope causing sheet erosion (Plate 5). The inefficiency of the mulching practice in controlling erosion is aggravated by the fact that it is in most cases the only conservation practice on the farm plot, yet it is recommended that mulching be used together with physical structures as water diversion channels and terraces to check the speed of runoff.

Although mulching is recommended as suitable for lower slopes, the practice needs a lot of improvement in Bugoye. Plate 8 above shows a thinly mulched banana garden where most of the mulches have been washed downslope by accumulated runoff released from farm plots upslope. The application of mulches in isolation without physical conservation structures exposes mulches to great volumes of runoff, and as such the thin mulches are easily swept away causing severe rill erosion that exposes the banana roots and weakens the banana plants, making them susceptible to falling, a problem that needs attention.

All in all, the performance of the present soil conservation technologies in use over the Bugoye upland farms leaves a lot to be desired; and in their present form, one is right to assert that they are inefficient in controlling erosion over the peasant farm plots. This inefficiency is aggravated by the fact that most of the farmers use one conservation practice in isolation, yet meaningful soil conservation should involve a combination of conservation practices for better results.

5 Conclusion

This study was undertaken in response to the lack of data on the performance of the present soil conservation technologies in the Rwenzori highlands in western Uganda, and as such it generated large amounts of data on the appropriateness and efficiency of existing soil conservation practices. Based on the findings of the study, the following conclusions are made.

The study revealed that farm plots with annual cropping system were the most erodible; this is mainly because most of the annual crops such as beans, cassava, Irish potatoes, and groundnuts are poor soil cover crops, offering very little protection to the soil and thus leaving them prone to severe soil erosion. This poses a serious challenge to soil conservation efforts given that the poor soil cover crops have a cultural bearing making up an important component in the diet of the local population.

As shown by the results, most of the erosion over the slopes of Busongora is attributed to poor land management, particularly lack of adequate permanent cover, poor terrace designs, and improper disposal of runoff. A solution to these problems needs to be sought to retain the soil over these fragile lands so as save the farmers from future starvation.

In the assessment of the efficiency of existing soil conservation technologies, it was noted that terraces, the most dominant conservation practice on the Rwenzori highlands, are inappropriate, that is to say, not constructed according to recommended standards. The fact that most of the terraces were found to be widely spaced, and so brief in height, robes them of the ability to effectively control soil erosion. Indeed the results indicate that despite the existence of these soil conservation structures on peasant farms, soil erosion still remained high. This underscores the fact that consideration of farm plot site-specific factors in the designing of conservation structures is a pre-requisite to their efficiency.

The study indicated that through time, the maintenance of terraces as soil conservation structures presents some problems to the farmers in that the accumulation of soil at the upper terrace end, and the steeping of its lower end by both erosion and continuous digging, results into terrace failure, severe erosion, and destruction of crops.

As such a solution to these problems has to be sought by those designing soil conservation programs.

The study has revealed that though terraces are capable of checking the speed of runoff and impounding soil lost from the upslope arable lands, they do not effectively control erosion. Their wide spacing allows runoff to recollect and gain velocity in between the two successive terraces, thereby becoming highly erosive, creating large rills, and destroying crops. More aggressive conservation measures should be put in place. There is also a need to have adequate vegetation and much cover, particularly during periods of intense rainfall.

The study indicates that besides terraces, other conservation structures like water diversion channels, trash bunds, and water collection ditches only help collect runoff sediment from upslope and do not significantly reduce erosion on portions of the farm plot in between the structures. This leaves point erosion still high, a phenomenon that requires urgent attention.

In essence, the study has revealed that a single conservation measure cannot adequately control erosion. The physical conservation structures applied alone over upland peasant farms have proved inefficient in controlling erosion, and as such a combination of agronomic and soil management measures, together with appropriately designed and well-maintained physical conservation structures such as terraces, cutoff drains, and water collection ditches, is the most viable answer to the erosion problem in the highland areas.

References

Achan LP (2001) Biophysical and socio-economic factors influencing soil and water conservation practices in the banana based micro- catchment of the Lake Victoria Basin. Unpub MA Thesis; Makerere University, Kampala, Uganda

Bagoora FDK (1997) Assessment of runoff and soil loss on peasant upland farms and their implications for environmental conservation in Rukiga Highlands Eastern Kabale District: Unpub phD Dissertation; Makerere University

Beskow S, Mello CR, Norton LD, Curi N, Viola MR, Avanzi JC (2009) Soil erosion prediction in the Grande River Basin, Brazil using distributed modeling. Catena 79:49

Brader L (1994) Sustainable agriculture and environment. In: Achipale E, Bakunda MA (eds) Conservation in Europe. Land Use Policy, vol 27, pp 22–32

Cook RU, Doornkamp JC (1990) Geomorphology in environmental management. Oxford University Press, Oxford

Department of Meterology (2006) Climate data 2001–2006. Department of meterology, Ministry of Water and Environment. Kampala, Uganda

FAO (2000) Guidelines and reference material on integrated soil and nutrient management and conservation for farmer field schools. Land and Nutrient Management Services. Lands and Water Development Department, Rome

Fitz EA (1983) Soils, their formation, classification and distribution. Longman Group, London

GOU (1949) Soil conservation report 1941–1949. Uganda Protectorate

Kuhlman T, Reinhard S, Gaaff A (2010) Estimating the costs and benefits of soil conservation in Europe. Land Use Policy 27(1):22–32. https://doi.org/10.1016/j.landusepol.2008.08.00

Maasha N (1975) The Seismicity of the Rwenzori region in Uganda. In: Nakimera I (ed) The impact of human activities and climate on the vegetation in the Lake Victoria region and on the Rwenzori Mountain and its Neghbourhood. Unpub. PhD Thesis, Faculty of science, Makerere University

Miranda ACR, Silva DP, Mello EL, Pruski FF (2012) Assessment of efficiency and adequacy of retention terraces. Universidade Federal de Viçosa – UFV, Minas Gerais, Brazil

Morgan RPC (1986) Soil erosion and conservation. Longman Group UK Limited, London

Nakileza B (1992) The influence of cropping systems and management practices on soil Erosion on mt Elgon slopes. Unpub Thesis, Makerere University

Nakimera I (2001) The impact of human activities and climate on the Vegetation in the Lake Victoria region and on the Rwenzori Mountain and its Neighbourhood. Unpub. PhD Thesis, Faculty of Science Makerere University

Nsubuga EN (1994) A critical review on the Farmers' Response to soil conservation policy in Uganda. In: Rwakaikara, Silver MC, Bekunda MA, Tenywa JS (eds) Enhancing Farmers' efforts to combat soil degradation: A challenge to soil science technology. Proceedings of the 14th conference of soil science society of East Africa, Mbarara, Uganda, 25–25 November 1994. pp 1–10

O'brien G (1991) Economic aspects of sustainable land management, regional experience Proceedings of the international board of soil research and management IBSRAM NO. 12(2). Education for sustainable land management in the developing world Bangkok: IBSRAM. pp 49–59

Speight JG (1990) Landforms. In: Graciela M, Bernard K, Gernot P (eds) A semi-automated approach for GIS based generation of topographic attributes for landform classification. Global Congresos, Coruna, pp 1–6 (ISBN 0-958-46093-0)

Spomer RG, Shrader WD, Rosenbery PE, Miller E (1973) Level terraces with stabilized back walls on Loessial crop land in the Missouri valley. A cost effective study. J Soil Water Conserv 28:127–13

Tanner PWE (1971) The Stanley Volcanics of the Rwenzori, Uganda. In: Nakimera I (ed) The impact of human activities and climate on the Vegetation in the Lake Victoria region and on the Rwenzori Mountain and its Neghbourhood. Unpub. PhD Thesis, Faculty of science, Makerere University

Temple PH (1972) Sheet wash measurements on erosion plots at Mfumbwe, Eastern Uluguru mountains, Tanzania. In: Morgan RPC, 1986. Soil Erosion and Conservation. Longman Group UK Limited, London

UBOS (2002) National population and housing census results 2002. Uganda Bureau of Statistics. Kampala, Uganda

UBOS (2014) National population and housing census 2014 provisional results. Uganda Bureau of Statistics. Kampala, Uganda

Valentin C, Poesen J, Li Y (2005) Gully erosion: Impacts, factors and control. Catena 63:132–153

Widomski KM (2011) Terracing as a measure of soil erosion control and its effect on improvement of infiltration in eroded environment. Lublin University of Technology, Lublin

Consequences of Land Tenure on Biodiversity in Arabuko Sokoke Forest Reserve in Kenya: Towards Responsible Land Management Outcomes

Tobias Bendzko, Uchendu E. Chigbu, Anna Schopf, and Walter T. de Vries

Abstract This chapter investigates the consequences of land tenure on biodiversity, in the context of identifying responsible land management as an intervention for development. Through a case study of Arabuko Sokoke Forest Reserve in Kenya, it explores how rights to land relate to and affect biodiversity exploitation and argues for employing responsible land management as an intervention to emerging tenure problems. The chapter is theoretically and empirically descriptive (and explorative) in approach. It uses ecological and sociological data collected through primary and secondary sources to infer possible interrelations between social structures and behaviour, ecological constraints and de facto land tenure choices in relation to biodiversity in the study area. The findings of the research point to the modes of land use, land access, land-related cultural practices and exercise of land-based property rights (to mention a few) as factors that influence biodiversity in qualitative and quantitative forms.

Keywords Biodiversity · Land management · Land tenure · Kilifi · Kenya · Social tenure domain model

1 Introduction

How land (or forest) is used can often be contested. On the one hand, some people argue it should provide a sustainable living space for plants, animals and people and a basis for cultivation and food production; on the other hand, people attach value to land for reasons of cultural identity and historical significance and claim rights without any specific economic or ecological benefit. This tension between different values manifests through a number of observable global trends, such as population growth in highly fertile yet urbanised locations, desertification and fast-eroding

T. Bendzko · U. E. Chigbu · A. Schopf · W. T. de Vries (✉)
Chair of Land Management, Technische Universität München, Munich, Germany
e-mail: wt.de-vries@tum.de

Y. Bamutaze et al. (eds.), *Agriculture and Ecosystem Resilience in Sub Saharan Africa*, Climate Change Management,
https://doi.org/10.1007/978-3-030-12974-3_8

landscapes due to climate change and natural disasters and conflicts over land due to increasing pressures from land markets and global land-scale acquisition of land. The result is an increasing scarcity of arable land, increasing pressure on land by growing numbers of stakeholders and overlapping rights and restrictions as a result of attempts to govern and regulate land use. These developments necessitate the adoption of "responsible land management" approaches (de Vries and Chigbu 2017: 69), which must be executed in a highly sensitive political environment (UN-Habitat 2007) for it to produce development outcomes. However, applying responsible land management interventions demands that appropriate land/forest tenure, biodiversity and local cultural practices be followed in ways that improve the living conditions of people and their environments. Biodiversity, as used in this chapter, refers to the general variety and variability of life within a defined land area or ecosystem.

As a way of contributing to the understanding of how land tenure can have consequences on biodiversity, this chapter, through a case study of Arabuko Sokoke Forest Reserve in Kenya, explores how rights to land relate to and affect biodiversity exploitation and argues for employing responsible land management as an intervention to emerging tenure problems. It approaches the subject from both theoretical and empirical perspectives. It uses ecological and sociological data collected through primary and secondary methods to infer possible interrelations between social structures and behaviour, ecological constraints and de facto land tenure choices in relation to biodiversity in the study area. This chapter begins by explaining the concept and practice of responsible land management, identifies critical issues in land tenure in Kenya and provides the institutional and geographic background of the study area. Then it presents and discuses empirical evidence of how land tenure influences land tenure arrangements and activities that in turn affect biodiversity. Finally, it proposes a responsible way forward and concludes.

2 Bringing "Responsible Land Management" into Focus

The need to build sustainable human habitats and maintain natural environments around the world has led to discourses on what is the most practical approach to managing land and natural resources. This has led to the understanding that land management is both a science and a practice. This chapter adopts de Vries and Chigbu (2017: 66) definition of land management as "the science and practice related to the conceptualisation, design, implementation and evaluation of socio-spatial interventions, with the purpose to improve the quality of life and the resilience of livelihoods in a responsible, effective, efficient, consensual and smart manner". Several efforts have been made to (re)define, identify and document best practices of land management in order to conceptualise the best approaches that can apply to specific countries, regions, cities (and villages), communities and neighbourhoods (see Chigbu 2015a; Bendzko et al. 2017; FAO 2017; Chigbu et al. 2018; Speranza and Bikketi 2018). In doing so, several approaches to land management have emerged over the past decades. Most of these approaches to land management have been described using a variety of adjectives, prefixes and catch words including sustainable, transformative, effective, efficient, territorial, pro-poor, gender sensitive, people-centred, minority focused and technologically driven, among many others. This chapter takes a look at the concept of responsible land management in the context of coastal tenure and biodiversity improvement in Kenya.

The term, *responsible*, has gained recognition in land management discourse through publications by Bourgon (2007), Cooper (2012), FAO (2012), Zevenbergen et al. (2016) and Ameyaw et al. (2018), among many others. Introducing a new dimension to the concept of *responsible land management*, de Vries and Chigbu (2017) argued that one of the ultimate characters of any land management intervention should be an outcome that is responsive to the needs and conditions of the people for whom the intervention is being implemented. Land management interventions "are activities put in place to respond to local needs of people in the context of land use, land administration, natural resources use and management" (Ameyaw et al. 2018: 2). According to de Vries and Chigbu (2017: 68),

> Responsible' refers to a basic set of normative notions and goals in terms of structures (of administration, accountability, hierarchy, organisational systems and institutions), processes (of work flows, sequences of steps, inter-organizational dependencies) and outcomes (societal changes, products, infrastructures, services, procedures). The normative goals refer to being responsive to needs of a group of citizens, connected to robust, respected and recognizable institutions, embedded in resilient societal and organisational structures, working using reliable ethics, and relying on reflexive mechanisms and retraceable actions and decisions to seek optimal solutions.

According to Ameyaw et al. (2018: 5), "responsible land management concept highlights the need to recognise socio-economic, customary and environmental factors before the design and adoption of a land management system". This chapter embraces the concept and practice of responsible land management and argues for its tenets to be adopted in biodiversity protection.

3 Identifying Critical Issues Related to Land in Kenya

Kenya's economy is mostly land-based. If not managed responsibly, land (and every other natural resource linked to land) usually fails to be a development factor. This makes land (including land tenure and tenure security) a subject of intense political, economic and social discourse in the country.

Kenya has a land area of about 582,646 square kilometres, which is used for expansive functions including agriculture, mineral exploration, forestry, aquatic activities, extensive livestock production (mainly under nomadic systems) and industrial and human settlement. Land distribution is a very contentious issue in Kenya because of illegal appropriation of public land by the elite during the past decades (Alden 2018). *Kenya's land area of about 582.646 km² comprises of 9.8% of land and 2.2% of water surface.* Of the total land cover, about 2.4% is under indigenous and exotic forests. About 12% of the land has high rainfall and supports the production of cash and food crops – such as tea, coffee, maize, wheat and potatoes. The semiarid area of Kenya covers about 32% of total land and has an average rainfall that supports mixed crop and livestock rearing. Over 50% of the total land cover is arid with extremely low and erratic rainfall and is used for extensive livestock production under nomadic systems.

3.1 The Legal Frameworks Guiding Land

There is a series of legislation that governs land- and natural resource (including agricultural and biodiversity)-related activities in Kenya. Some of them are colonial and postcolonial legislations or statutes that previously governed land that has now been repealed (and replaced). The major ones are two: the Land Act and the Land Registration Act (Government of Kenya 2012). The Land Act (No. 6 of 2012) replaced the Wayleaves Act (Cap 292) and the Land Acquisition Act (Cap 295). The Land Registration Act (No. 3 of 2012) replaced the Indian Transfer of Property Act, the Government Land Act, the Registration of Titles Act, the Land Titles Act and the Registered Land Act. The Land Registration Act (of 2012) applies to the registration of interests in all public land as declared by article 62 of the Constitution; registration of interests in all private land as declared by article 64 of the Constitution; and registration and recording of community interests in land are required by Section 3 of the Land Registration Act of 2012. The Land Registration Act (of 2012) does not prohibit or otherwise affect the system of registration under any law relating to mining or any other rights in respect of public land (see Section 4 of the Land Registration Act of 2012). The Land Act (of 2012) on the other hand applies to public land, private land and community land and any other written law related to community land [Section 3(1) of the Land Act of 2012]. The Land Act (of 2012) adopted the constitutional principles in respect to land management and administration that will be binding on all state organs, state officers, public officers and all people (Section 4 of the Land Act of 2012).

3.2 Renewed Efforts Are Being Made But the Land Challenges Persist

Kenya's constitution (of 2010) provides for conflict resolution. It promotes the devolution of responsibility and authority to communities and encourages the development of homegrown conflict management frameworks about land. Previously, County Land Management Boards conducted land management activities at county levels. The County Land Management Boards were abolished in the amendment of the Land Act (of 2012) in 2015. Community Land Management Committees were established to manage community land. A National Land Commission was established and has been operational since 2013. The National Land Commission was envisioned as a transparent and autonomous institution with responsibility for public land in Kenya.

Kenya's current constitution is still relatively new, and some of the new reforms have been difficult to operationalise. For example, there is ambiguity regarding what constitutes community land. The National Land Policy adopted in 2009 is intended to guide Kenya towards sustainable and equitable land use and to replace the existing patchwork of often incompatible laws. The National Land Policy outlines broad and substantial reforms to the land management and administration system; and it recognises and protects customary rights to land. It also outlines principles of sustainable land use and provides productivity and conservation targets and guidelines, and it calls for reforms of land management institutions, including the establishment of a National Land Commission, district land board and community land boards, as well as legal and institutional frameworks (Government of Kenya 2009). However, the National Land Policy focuses heavily on agrarian land issues at the expense of urban and peri-urban areas. As urbanisation increases, there will be a critical need for policy governing urban land. A comprehensive strategy for effectively implementing the National Land Policy will also have to be adopted. Such a strategy is crucial to tackle the many land resource tenure problems bothering the country. Another noteworthy legislation relevant to this study is the Forest Conservation and Management Act No. 34 of 2016 (Government of Kenya 2017). It joins the many laws on this issue, but which need to be translated into strategy. Of all the challenges faced by Kenyans in the aspect of land tenure security (and sustainable agriculture), the most notable ones include:

- *Gender inequality*: Many scholars have tried to explain the existing gender inequality in Kenya with regard to land ownership and natural resource rights (see Wanyonyi et al. 2010; Chigbu 2015b; Speranza and Bikketi 2018). Inquiries into the many factors that have contributed to gender inequalities in land and natural resource ownership have found that there is a lack of access to land (and the fear of eviction) caused by a lack of equality between men and women in the legal and sociocultural aspects of land tenure practices. Hence, there is a need for a cultural repositioning to cater for women's land resource needs (Chigbu 2015b). Ensuring women's de jure and de facto land and natural resource rights is one of the strategies needed to reduce the historical gender inequality in land resource distribution in Kenya.
- *Increased land and natural resource conflicts*: Insecurity in land and natural resource rights is a problem in Kenya. These conflicts exist due to the ineffective execution of legal frameworks, colonial injustices yet to be resolved or addressed and the embedment of tenure insecurity in the land tenure regimes of Kenya. These have negative consequences on agricultural production, forestry and wildlife management, biodiversity protection and the livelihood activities of the Kenyan citizens.

Understanding the linkages between tenure security, biodiversity and the need for responsible land management in Kenya is important because it may produce the knowledge needed for grasping how to tackle existing and emerging land and natural resource challenges in the country. The case of Arabuko Sokoke Forest Reserve is relevant to knowledge building on this subject because it shows how issues relating to the rights of land owners and users around the area have consequences on biodiversity.

4 Materials and Methods

4.1 Study Area

The study presented in this chapter was done in Arabuko Sokoke Forest Reserve in Kenya, in Kilifi County, which is a coastal area in Kenya. Arabuko Sokoke Forest Reserve lies within the geographical coordinates of 3°20′ south and 39°50′ east. It has an area of 41,600 ha, 5935 ha of which is designated as nature reserve. The area has three distinct vegetation types which are influenced by rainfall patterns, soil types and variations in altitudes (Vieilledent et al. 2016). In terms of biodiversity, it is as an "important Bird Area with endemic bird species numbering about 270 species, including six which are globally threatened and three near-threatened species" (Tarus et al. 2018).

Most of the 104,000 people living in the communities surrounding the forest are small-scale farmers, and more than 62% of their household depend directly on the forest. More than 30% of the households living within 2 km of the forest reserve are engaged in forest-related livelihood activities such as hunting, trapping and collecting non-timber forest products (Tarus et al. 2018). Kilifi, the county where Arabuko Sokoke Forest Reserve is located, covers an area of approximately 12,246 km^2 and has a population of more than one million people (Kilifi County Government 2016.). This explains why the ASFR has been described as a forest reserve suffering high human dependency. Communities around the ASFR depend largely on its resources for their nutritional and livelihood needs. Understanding how all of these affect biodiversity, and suggesting ways of managing the situation responsibly, is the concern of the study presented in this chapter.

4.2 Methods

The study was done during the period of March 2017 to April 2018. During this period, three direct field visits – each friend visit lasted an average of 1 month – were carried out to assess biodiversity. Direct interviews were carried out with land users and farmers to determine the status or nature of their land tenure activities. Five land areas – identified as sampled land areas 1, 2 (a/b), 3, 4 and 5 – were investigated to discern the consequences of land tenure (or tenure activities) on biodiversity. They were distributed randomly around the ASFR.

Table 1 Land areas sampled for the study

Sampled land area	Parcels (*n*)	Total size (*Ha*)	Average size (*Ha*)	Minimum size (*Ha*)	Maximum size (*Ha*)
1	79	156.42 ha	1.98 ha	0.91 ha	15.66 ha
2a	53	105.22 ha	1.99 ha	0.89 ha	2.36 ha
2b	10	15.82 ha	1.58 ha	0.46 ha	5.03 ha
3	102	373.19 ha	3.65 ha	0.04 ha	8.08 ha
4	79	204.61 ha	2.59 ha	0.36 ha	8.09 ha
5	58	175.74 ha	3.03 ha	0.24 ha	6.02 ha
Total	544	1040 ha	1.91 ha	0.04 ha	15.66 ha

A total of 1040 hectares (Table 1) dispersed in five major areas were studied to understand the experience of land right holders/users in these areas in relation to biodiversity changes. Note that land area 2 consists of two patches (a/b) located in separate locations within the same area around the ASFR.

Methods used in data collection included standardised questionnaires ($n = 776$), key informant interviews ($n = 16$) and direct land (and tenure) mapping using "Rubber-Boot Approach (RBA)" (Bendzko et al. 2017). All the respondents (both questionnaire and key informants respondents) were individuals who either used/owned or enjoyed some form of property rights in and around the ASFR. Questionnaire respondents were determined using snowballing methods, while the interview respondents were purposively selected based on their involvement in forest land activities. The questionnaire asked respondents various questions concerning their viewpoints and experiences using/owning land around the vicinity of the ASFR (or depending on it) and the nature of property rights exercised on the land and their encounter with biodiversity changes in the process. The interviews consisted of semi-structured questions related to the same questions used in the questionnaire; however, the interview session provided additional room for direct discussion and engagement with respondents to elicit further details on responses received.

Figure 1 (A, B, C and D) represents a snapshot of the characteristics of the key informant and interview respondents ($n = 792$) in terms of their age distribution, level of education, monthly income earnings and land ownership structures. The characteristics of respondents are important for grasping the socioeconomic status of individuals who are dependent on the land areas (1, 2a/b, 3, 4 and 5) sampled around the ASFR for their livelihood. The RBA is a method used to determine and map the physical and social land tenure of land. It relies on basic mapping methods, which local citizens can apply with relatively limited training in order to describe and locate their own land tenure. This bottom-up process provides for more engagement in the tenure mapping process and ultimately for more tenure security. It uses the approach of the Social Tenure Domain Model (see Lemmen 2013). Documentation of both physical land boundary and documentation of social status and nature of usage (including ownership and rights) of land were used as a means of describing tenure

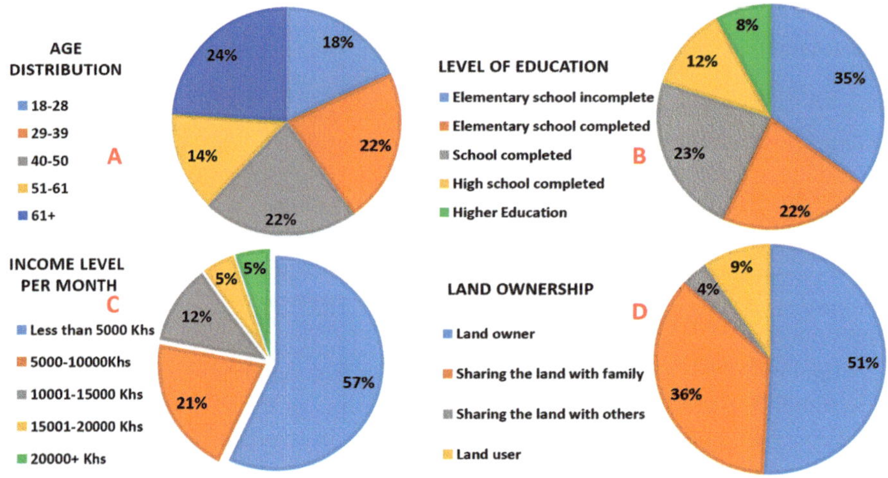

Fig. 1 Characteristics of the respondents

The received questionnaires and interview transcripts were first coded, and their resulting data were subjected to descriptive and interpretative analyses. Land tenure and biodiversity scenarios resulting from the analyses were narrated to some of the local respondents (including key informants) for validation using confirmatory interviews. Emerging information was thematically aligned to land tenure and biodiversity; and the ideas emanating from the literature of responsible land management concept (and practice) were adopted for appropriate recommendations.

5 Consequences of Land Tenure on Biodiversity in ASFR

Although the concern of this chapter is not about tenure security, it is important to note that the respondents stated that the perception of tenure security they have on their property right influences the manner in which they use, hold and manage their land around the ASFR; hence influencing how they exploit the ASFR and contribute to biodiversity changes. In this regard, the study found that 5% of respondents feel insecure, whereas 94% expressed high perceptions of land tenure security. About 1% of the respondents were not sure on whether they felt secure or not. Of utmost importance is that the study produced outcomes that allowed for understanding land tenure arrangements and activities that have consequences on biodiversity in ASFR.

5.1 Tenure Arrangement Consequences on Biodiversity

Land tenure and the qualitative (and quantitative) role it plays in biodiversity changes are complex. The reason is that land tenure, being a bundle of institutional norms and practices, fuels human land use activities which can manifest in ways that make biodiversity either vulnerable or protected. Land tenure is also linked to socioeconomic, political and ecosystem delivery services. Hence, it has the potential to support or hinder biodiversity growth and development in direct and indirect ways.

The first notable issue is that land tenure can appropriately or inappropriately determine land use changes, which then directly influences biodiversity changes. Then there is the issue of abuse of the exercise of property rights embedded in land tenure systems, and this abuse can affect the ecosystem that depends on that land (biodiversity). Land tenure usually serves as a set of rules (and regulations) on land use hold and practices; and where the rule is pro-biodiversity development, its abuse can lead to biodiversity destruction and vice versa. Human activities like bush burning, commonly practised in and around ASFR, which are destructive to biodiversity, are some of the links (or relationships) that biodiversity has with land tenure (see Fig. 2).

Based on interviews with respondents, it was possible to discern the tenure arrangements in ASFR and the biodiversity changes they produce (see Table 2). The key tenure types identified within the ASFR area include use rights, access rights, the rights to livelihood earning through direct or indirect use of resource, exclusion, alienation and management rights.

From Table 1, it is possible to identify the consequences of land tenure, as is dominantly arranged in and around the ASFR in Kenya. These findings add to previous literature on the subject (see Mogoi et al. 2012; Vieilledent et al. 2016; Tarus et al. 2018) which discussed the problems biodiversity faced in ASFR without focus on the contribution of land tenure practices.

One of the commonest ways identified to be altering biodiversity habitat is the infringement of transit routes for animals in the ASFR. The occurrence of deforestation and trafficking of wildlife in the area has life-threatening consequences on animal and plant specifies. Lack of adequate land tenure administration was identified as one of the major reasons deforestation and improper use of land are resulting into wildlife displacement and flora destruction. Hence, the need to incorporate responsible land management strategies is a way forward.

Fig. 2 Images (**a**–**c**) showing bush burning

Table 2 The arrangement of property rights to local community stakeholders

Property rights exercisable	Legal of rights	Prevalent situations	Identified consequences on biodiversity
Use rights	Local communities and the Kenya Forest Service have use rights. However, use rights are limited to dead wood and Non-Timber Forest Products	Squatting, animal poaching and logging	Habitat alterations and habitat pollution
Alienation rights	Kenya Forest Service in consultations with communities. However, they do not make final decision	Courts exercise these rights and call on the communities when needed as witnesses	Reduced access to local knowledge of biodiversity. Leading to poor biodiversity conservation
Access rights	Community members, the government and all stakeholders with known vested interests have unhindered access. Non-locals need permission from Kenya Forest Service	All stakeholders have access to the forest; however, non-locals usually enter the ASFR without permission	Overharvesting and overexploitation of biodiversity
Management rights	Kenya Forest Service, Community Forestry Associations and any stakeholders with approved management plans	Kenya Forest Service has dominant power. Communities have limited powers to monitor but do not exercise it as it has no economic gains	Encourages uncontrolled land use in ASFR, leading to habitat alterations
Rights to earn income (direct or indirect) from the use of forest resource	Local councils, Kenya Forest Service and Community Forest Associations can earn income (directly or indirectly) through approved management plans	Many households depend access income (directly or indirectly) from the forest resources	Over harvesting of Timber and Non-Timber Forest Products. Biodiversity destruction
Exclusion rights	As in management rights	As in management rights	As in management rights

6 Towards "Responsible Land Management" in Biodiversity Protection

Biodiversity protection cannot happen in a void. It can only happen within ecosystems whose sustenance are affected by guided land activities – be it water, forest, air or the physical soil. This makes land tenure conceptions, practices and management a key factor that must be incorporated into biodiversity protection. As a result of its strong land linkage, biodiversity management can benefit from responsible land management interventions as a way to curb all the consequences identified in the case of ASFR – including habitat alterations, habitat pollution, uncontrolled land use, overharvesting of land and natural resources, biodiversity destruction, reduced access to local knowledge of biodiversity and poor biodiversity conservation. Land management has been referred to as intervention in this chapter because (being a science and practice) it has "both the activity of changing a set of physical objects, as well to the process of consultation and deliberation resulting in a legitimate or legal decision" (de Vries and Chigbu 2017: 66). Its application can cause a combination of six types possible changes in biodiversity – including governance, property rights and land use law, social-spatial relations, economic opportunities (and dependencies), human perceptions/beliefs and human and animal behaviours (de Vries and Chigbu 2017: 66). It is obvious that some of these are already being conceived and applied in local and international conservation and nature protection programmes (and projects). However, there is an urgent need to change towards approaches that produce responsible outcomes.

A key aspect of any responsible land management intervention, as stipulated by de Vries and Chigbu (2017: 72), is that any effort put towards improving land-related problems should be done in ways that are respectful, retraceable, recognisable, responsive, reflexive, robust, resilient and reliable. These have been referred to as the eight indicators (8Rs) of responsible management which have been applied in projects in Ghana and Rwanda (de Vries and Chigbu 2017; Ameyaw et al. 2018). If, for instance, the current efforts being made towards the development of sustainable biodiversity in the ASFR are made to adhere to the aforementioned 8Rs, it could help in assessing and designing a land-based management system that is based on the commitment and support of the public land agencies, to achieve a more responsible forest reserve management in Kenya. This is the sustainable way to protecting biodiversity, as it will be guided by a well-coordinated land tenure administration and management system aim at ensuring that people use land and land resources appropriately.

7 Conclusions

This chapter dealt with conceptual issues on responsible land management and used them as a starting point to argue that it is a potential path to curbing the consequences of land tenure on biodiversity. It went on to discuss the general state of land tenure challenges in Kenya (including efforts put in place to improve them) and then the specific dilemma of the ASFR. Using the linkages found between land tenure and biodiversity, it presented further arguments on why "responsible land management" should be incorporated as strategy for biodiversity protection. Despite these efforts, the study presented in the chapter has its share of scholarly limitations. There are critical aspects of land tenure and biodiversity that were not discussed. For instance, tenure security, national maintenance, biosphere conservation, hotspot development and awareness development (among many other subjects) were clearly left untreated. This is because the chapter took a generalist approach to tackling the subject under investigation.

When writing about land tenure or biodiversity, there is always a challenge concerning which aspects should be highlighted and which ones should be left out. The authors chose take a path that contributes ideas that may be helpful in the implementation of the Arabuko Sokoke Strategic Forest Management Plan (2002–2027) which aims at mitigating the challenges faced by ASFR. Unlike other studies which have taken floristic (Wekesa et al. 2017), edaphic (Mwadalu and Gathara 2017) or mammalian (Nyunja et al. 2017) focus, the study presented in this chapter contributes to this array of literature by bringing into focus the responsible land management perspective.

Acknowledgements This research was carried out with the financial support of DAAD. Without their valuable support, the fieldwork would not have been possible.

References

Alden WL (2018) The community land act in Kenya opportunities and challenges for communities. Land 7(1):12

Ameyaw PD, Dachaga W, Chigbu UE, de Vries W, Asante LA (2018) Responsible land management: the basis for evaluating customary land management in Dormaa Ahenkro, in Ghana. World Bank Conference on Land and Poverty, Washington, DC, 19–23 March

Bendzko T, Chigbu UE, Ameyaw PD, Tutu DO (2017) Embracing the rubber-boot approach to securing customary land rights with focus on low-cost land-use inventory. World Bank Conference 2017, Technische Universität München, Munich

Bourgon J (2007) Responsive, responsible and respected government: towards a new public administration theory. Int Rev Adm Sci 73(1):7–26

Chigbu UE (2015a) Ruralisation: a tool for rural transformation. Dev Pract 25(7):1067–1073

Chigbu UE (2015b) Repositioning culture for development: women and development in a Nigerian rural community. Community Work Fam 18(3):336–352

Chigbu UE, Izugbara CO, de Vries WT (2018) Land, culture, culture loss and community: rural insights from sub-Saharan Africa. In: Kenny S, McGrath B, Phillips R (eds) The Routledge handbook of community development: perspectives from around the globe. Routledge, London, pp 98–114

Cooper TL (2012) The responsible administrator: an approach to ethics for the administrative role. Wiley, San Francisco

de Vries W, Chigbu UE (2017) Responsible land management – concept and application in a territorial rural context. FuB – Flächenmanagement und Bodenordnung 79(2):65–73

FAO (2012) Voluntary guidelines on the responsible governance of tenure of land, fisheries and forests in the context of national food security. FAO, Rome

FAO (2017) Land resource planning for sustainable land management: current and emerging needs in land resource planning for food security, sustainable livelihoods, integrated landscape management and restoration. Food and Agriculture Organization of the United Nations, Rome

Government of Kenya (2009) National Land Policy: The Formulation Process – Draft National Land Policy (Ministry of Lands). The Government of Kenya, National Land Secretariat, Nairobi

Government of Kenya (2012) The Land Act of 2012 & The Land Registration Act of 2012. Government of Kenya Printer, Nairobi

Government of Kenya (2017) Forest Conservation and Management Act No. 34 of 2016. Government of Kenya Printer, Nairobi

Kilifi County Government (2016) Kilifi County Government: Demographics. Kilifi County Government, Kenya

Lemmen C (2013) The social tenure domain model - a pro-poor land tool. International Federation of Surveyors (FIG), Copenhagen, p 24

Mogoi J, Obonyo E, Ongugo P, Oeba V, Mwangi E (2012) Communities, property rights and forest decentralisation in Kenya: early lessons from participatory forestry management. Conserv Soc 10(2):182–194

Mwadalu R, Gathara M (2017) Edaphic characterisation of Arabuko Sokoke Forest. In: Ochieng' D, Luvanda A, Wekesa C, Mbuvi MTE, Ndalilo L (eds) Biodiversity status of Arabuko Sokoke Forest, Kenya. Kenya Forestry Research Institute, Nairobi, pp 15–21

Nyunja J, Ogwoka B, Njeri L (2017) Composition and distribution of small and large mammals in Arabuko Sokoke Forest. In: Ochieng' D, Luvanda A, Wekesa C, Mbuvi MTE, Ndalilo L (eds) Biodiversity status of Arabuko Sokoke Forest, Kenya. Kenya Forestry Research Institute, Nairobi, pp 22–33

Speranza CI, Bikketi E (2018) Engaging with gender in water governance and practice in Kenya. In: Fröhlich C, Gioli G, Cremades R, Myrttinen H (eds) Water security across the gender divide. Water security in a new world. Springer, Cham

Tarus GK, Nzioka JM, Githaiga JM, Onwonga R (2018) Projecting the impacts of climate change on tree biomass in Arabuko-Sokoke Forest, Kenya. J Clim Chan Sustain 1:95–107

UN-Habitat (2007) How to develop a pro-poor land policy. Process, guide and lessons. United Nations Human Settlements Programme, Nairobi

Vieilledent G, Gardi O, Grinand C, Burren C, Andriamanjato M, Camara C, Gardner CJ, Glass L, Rasolohery A, Rakoto Ratsimba H, Gond V, Rakotoarijaona JR (2016) Bioclimatic envelope models predict a decrease in tropical forest carbon stocks with climate change in Madagascar. J Ecol 104(3):703–715

Wanyonyi A, Nyadimo E, Kariuki J (2010) Land tenure security in Kenya. In: Kirk M, Antonio D, Mabikke S, Espinoza J, Chigbu UE (eds) Land tenure security in selected countries: global report. UN-Habitat and Global Land Tool Network, Nairobi, pp 27–48

Wekesa C, Otuoma J, Ngugi J, Muturi G (2017) Floristic composition and structure Ofarabuko Sokoke Forest. In: Ochieng' D, Luvanda A, Wekesa C, Mbuvi MTE, Ndalilo L (eds) Biodiversity status of Arabuko Sokoke Forest, Kenya. Kenya Forestry Research Institute, Nairobi, pp 8–14

Zevenbergen J, de Vries WT, Bennett RM (2016) Advances in responsible land administration. CRC Press, London/New York

Abundance and Diversity of Wetland Birds: The Case of Dinder National Park, Sudan

Pasquale Tiberio D. Moilinga and Tahani Ali Hassan

Abstract This study was conducted in Dinder National Park (Sudan). Patterns of bird species diversity, richness, and abundance were documented in four wetland areas inside Dinder National Park, including Ras Amir, Grerisa, Abdel Ghani, and Dabkara *Mayas* (meadows) during the dry seasons starting 2014 through 2016, and, diversity over this period was badly eroded and under serious threats. Timed-species count (TSC) technique was used for counting birds. Rarefaction curves combined with nonparametric estimators of species richness were used to extrapolate species richness beyond the collected data. The Shannon (*H'*) diversity index and the Simpson (*D*) index and the evenness index of Pielou (*J'*) were used to assess alpha diversity and diversity within and between sites, respectively. Finally, chi-square goodness of fit test was used to test the H_0 that bird species frequencies at each site were equally proportional and their mean numbers across the four sites were not significantly different. Sampling efforts of over 90% was attained in each of the four study sites. A total of 203 bird species about 145 (71%) of which were wetland birds, all belonging to 33 families and 11 orders, were identified from a set of 3753 individuals. The mean number of wetland birds was generally low at all the Mayas except at Ras Amir Maya where some species had relatively higher mean number. The orders Anseriformes (especially Anatidae), Charadriiformes (especially Jacanidae and Charadriidae), Ciconiiformes (especially Ardeidae, Threskiornithidae, and Ciconiidae), Coraciiformes (especially Alcedinidae), Passeriformes (especially Motacillidae and Ploceidae), and Pelecaniformes (especially Phalacrocoracidae) were the most important, in terms of both abundance and species richness. Diversity indices ranged, in descending order, from Grerisa Maya, Ras Amir Maya, Abd el Ghani Maya to Dabkara Maya. There was no clear trend in terms of bird numbers and abundance between sites, though water-rich Mayas seemed to support more birds than the drier ones. Likewise, certain species were found only in some Mayas

P. T. D. Moilinga (✉)
Department of Wildlife Science, College of Natural Resources and Environmental Studies, University of Juba, Juba, South Sudan

T. A. Hassan
Department of Wildlife, University of Bahri, Khartoum, Sudan

Y. Bamutaze et al. (eds.), *Agriculture and Ecosystem Resilience in Sub Saharan Africa*, Climate Change Management,
https://doi.org/10.1007/978-3-030-12974-3_9

but not in others such as African jacana (*Actophilornis africanus*), African darter (*Anhinga rufa*), long-tailed cormorant (*Phalacrocorax africanus*), and allies. This study presents baseline data and contributes in important ways to basic knowledge of wetland biodiversity patterns within Dinder National Park which must be protected because they constitute core areas of great conservation importance.

Keywords Maya · Wetland · Diversity indices · Bird abundance · National park

1 Introduction

Wetlands have been defined in many ways, but for the purpose of this study, they are defined as places that hold water for such a time that living organisms (plants and animals) become adapted to aquatic conditions (Chapman and Reich 2007). Numerous systems for the classification of wetlands have also been proposed, usually on a regional or national scale. The system established by the *Office of Biological Services of the Fish and Wildlife Service in the United States* (Cowardin et al. 1977) is of that kind. A worldwide system has also been developed for the implementation of the Ramsar Convention (Mathews 1993).

Wetlands still little understood in the Sudan are used by wildlife for food and water, as breeding habitats, places of refuge, and transition zones especially for migrants, both Afro-tropical and Palaearctic (UNESCO 1994). Furthermore, wetlands act as interface zones between open water bodies and dry uplands and therefore combine attributes for both. This makes them different from each yet highly dependent on both; hence they could be considered the most productive ecosystems on earth (Both et al. 2006).

Wetlands are home to many species of birds, "indeed, around 12 percent of all African bird species are found in and around lakes and swamps. Wetlands have become renowned because of their birdlife and indeed, international action for wetland conservation started with birds and the Convention on Wetlands became one of international importance, especially where waterfowl habitats are concerned" (Mafabi 1999). Dinder National Park was designated as a UNESCO Man and Biosphere Reserve since 1987 (Abdel Hameed et al. 1997). Delineation of the park was completed in 2001, and it was divided into core, transitional and buffer zones following the UNESCO model of Biosphere Reserve (Moilinga 2016).

Sudan boasts some huge assortment of avian fauna over 931 species of birds as documented by Hussein 2015, and 938 species of birds as reported by Dawi 1998, respectively; the bottom line is that there are over 900 species in the country. Birds are an important part of wetland ecosystem since some nest, feed, and roost in wetlands. Yet, others depend wholly or partly on wetlands. They recycle nutrients through the herbivorous food chain. They also return nutrients to the wetland through deposition of guano, which is a rich form of urea (Mafabi 1999; Hadley et al. 2012; Ramchandra 2013; Gunnarsson and Federsel 2014). Apart from providing food, birds that are completely dependent on wetlands are good indicators of wetlands under threat. Species that depend almost entirely on wetlands for their survival are greatly affected by the changes in these wetlands (Mafabi 1999; Brambilla et al. 2011).

It follows therefore that biodiversity inventories need to be designed around the use of effective sampling and estimation procedures, especially for a highly diverse groups of terrestrial organisms (Colwell and Coddington 1994). Studies of ecological communities require ideally that individuals in a sample be properly identified to species level and counted (Hespenheide 1994; Gotelli 2004). However, this is not always possible; hence, individuals are usually identified and separated on the basis of morphotypes treated as equivalent to species in species richness estimations (Magurran 2004; McGill et al. 2007). Species richness is difficult to quantify (Longino et al. 2002) but can be estimated by three different ways: (1) extrapolating species accumulative curves (SACs), (2) using nonparametric estimators to predict the number of "missing" species, or (3) fitting statistical distributions such as species abundance distributions (SADs) (Colwell and Coddington 1994; Chao et al. 2009). SACs allow measurement of within-inventory efficacy and completeness, and estimation of the minimum sampling effort required to reach a satisfactory level of completeness. Nonparametric estimators use the number of rare species in the community to predict the number of missing species (Moreno and Halffer 2000). Based on SACs, effort consists in the total number of individuals collected, or the number of pooled samples, necessary to reach the asymptotic species richness, whereas with species richness estimators, effort is the ratio between species richness observed and theoretical species richness. SADs capture the inequality of species abundances that characterizes ecological communities to highlight their structuring mechanisms (McGill et al. 2007). Ecological diversity is most commonly represented by species diversity, which uses mathematical models broadly known as diversity indices, derived from combining information on richness and evenness (Hamilton 2005; Schowalter 2006).

In this study, diversity of bird communities was assessed within habitat (α diversity) and compared between habitats or landscapes (β diversity) (Magurran 2004; Begon et al. 2006; Stireman 2008). The objective of this paper is restricted to determination of species diversity, spatial distribution, and relative abundance of birds associated with wetlands during the dry season (February/March, 2014, 2015 and 2016), to provide basic ecological knowledge about wetland areas within Dinder National Park.

2 Materials and Methods

2.1 Study Area

Established in 1935, Dinder National Park covers a surface area of about 10,000 km^2 and is located between latitudes 11^0 N and 13^0 N and longitudes 34^0 E and 36^0 E situated along the Sudanese-Ethiopian frontiers. Climate of the area is characterized by wet and dry seasons. The area is described in detail by Holsworth (1968) and Hashim (1984). However, in conformity with the observations made during this study, major habitat types of the park can be broadly classified into two: a mosaic of woodland savannah consisting mainly of *Acacia* species, *Balanites aegyptiaca* and *Combretum* spp. in most parts of the park, and the wetlands represented by the Mayas (shallow oxbow lakes) and semipermanent pools along the Dinder river bed. Grasses are mainly of *Andropogon* spp., *Hyparrhenia* spp., *Panicum*, and *Beckeropsis unisetta*. Forest galleries of discontinuous canopy are found along the rivers of which there are two major ones: Galagu and Dinder (Fig. 1).

2.2 Data Collection

Sampling Period and Localities

Observations were made during the dry seasons from February to March for the years 2014 through 2016 as the dry season was the most suitable period for conducting studies in the park. In the wet season, the park is inaccessible when the black cotton soil swells with water and becomes very sticky making movement almost

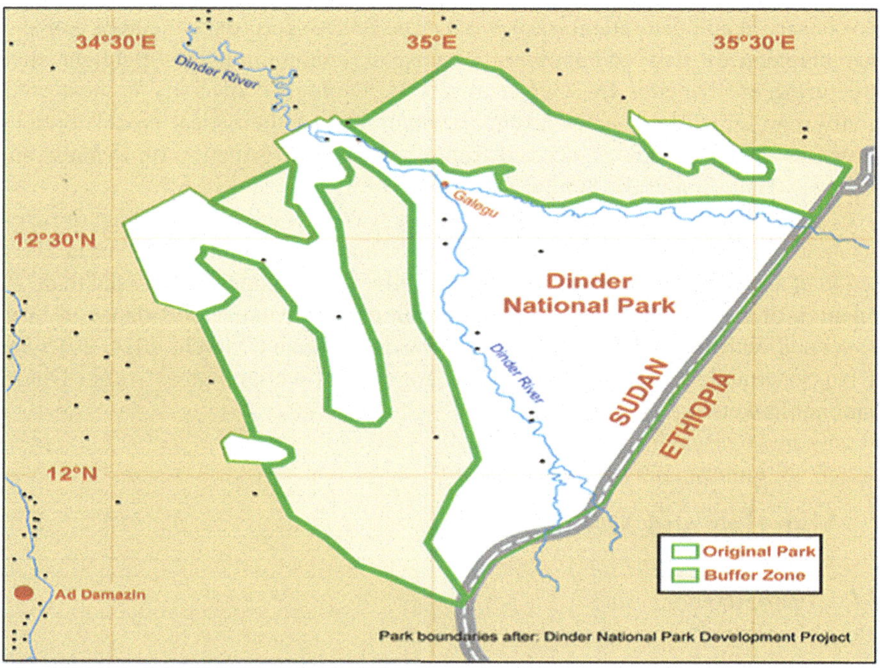

Fig. 1 The Dinder National Park. (Source: https://www.britannica.com/place/Dinder-National-Park appeared on 4/22/2017)

impossible. A choice of sites within the park was made based on the criterion of accessibility and presence of prolific birdlife; as a result four Mayas (meadows) were selected, namely, Abdel Ghani, Dabkara, Grerisa, and Ras Amir; and their features were recorded systematically.

Sampling Methods

The technique used in this study was the timed species count (TSC) (Pomeroy and Tengecho 1986; Dranzoa 1990). The underlying idea here is that first, particular sites are defined where the counts are to be made. The exact size of each area is not important, but around 1 km^2 is convenient (Pomeroy 1992). The sites selected should be homogeneous and typical of the area as a whole and randomly located within the overall area, but the choice in practice is limited by accessibly. Having made the choice of sites of which there were four Mayas, counts were made by moving slowly around the study site, listing any species positively identified in the order seen or heard, anywhere within the Mayas regardless of how far away from the observer(s). This was done within a period of 1 hour per each observation effort. Species flying over were included if they were using the area, i.e., for feeding, displaying, etc., a series of 15 TSCs were made at each study site every year.

It is worth noting that methodological limitation in the use of TSC consists in the difference in detectability between species. Thus, rare but large vocal species may be recorded as more common than common but small and secretive species. In order to minimize the effect of this bias during this study, data were collected from bands with width of less than and greater than 20 m and the results analyzed together combined.

2.3 Data Analysis

Indices of relative abundance from the timed species count (TSC) technique were derived according to Britton (1980) with some modifications. The species were arranged in tables in alphabetic order. A "W" in the first column indicates that the species is a wetland bird and therefore of greater conservation significance. Species were scored 6 if seen in the first 10 minutes of a count (1 hour), 5 when seen between 10 and 20 minutes, and so on to 1 for the last 10 minutes.

Because while counting, some bird species were uncommon, and a few were relatively numerous occurring in flocks thus contributed many birds to the total, or missing them tended to create a bimodal sampling distribution. So, to reduce this type of bias, the count data were transformed into logarithmic values which were obtained after adding one to each original number; since some of the species had zero counts at certain sites and one cannot take logarithm of zero, the extra one is subtracted at the end. All logarithms are to the base 10.

Shannon's (H') diversity index ($H' = -\sum pi \text{ In } pi$) was used to determine alpha diversity. Also Simpson ($\boldsymbol{D} = \sum^{s}_{i=1} ni(ni - 1)/N(N - 1)$), as well as the evenness index of Pielou ($J = H'/\text{In}S$), was used to assess bird diversity within and between sites. In these formulae, *ni* is the number of individuals of each of the *i*th species in the sample, *N* stands for the total number of individuals in the assemblage, *S* is the number of species in the assemblage, and ln is the natural logarithm. All these indices, commonly used in ecological community studies (Magurran 2004), were calculated using PAST 2.12 (PAleontological STatistics) software (Hammer 1999–2011).

Diversity incorporates information about species richness of birds (Magurran 1988; Colwell 2013). Species richness and diversity (alpha and beta) were determined using Estimate *S*, (Magurran 1988; Colwell 2013). Different statistical tools with 100 randomizations of sampling without replacement were used to estimate species richness; and the nonparametric incidence-based estimators of species richness Chao2 was used to estimate the expected number of species (observed or unseen) in the four sites (Gotelli and Colwell 2010). This was because the species richness was influenced by incomplete detection as some species were missed during counting (Gotelli and Colwell 2010). The observed and estimated species richness were then compared to each other to enable the evaluation of the sampling effort in the park. Individual- and sample-based rarefactions were used to generate curves for the estimation of species richness. Jack 1 richness estimator was chosen because its curve was comparatively smooth.

Birds' relative abundance was determined from crude ordinal scales following Bibby et al. (1998) as summarized in Table 1.

Also calculations were made of selection index and proportion of use of each habitat type as follows:

Selection index $w = r/a$
Standardized index as $B = w/\sum w$

If the selection index >1, the habitat is preferred, i.e., usage is greater than availability. Standardizing the indices allows comparison between observations and even studies because they always sum to one. The index approach investigates if resources are used in proportion to their availability (Bookhout 1994). Neu's method is commonly used in this respect. Finally, chi-square goodness of fit tests were used to test

Table 1 Crude ordinal scales

Abundance category (number of individuals per 100 field hours)	Abundance scale	Ordinal scale
<0.1	1	Rare
0.1–2.0	2	Uncommon
2.1–10.0	3	Frequent
10.1–40.0	4	Common
40.0+	5	Abundant

the null hypothesis that bird species frequencies of occurrence at each study site were equally proportional and that their mean numbers across the four sites were not significantly different.

3 Results

3.1 Species Composition and Distribution across the Four Sites

In total 203 bird species were identified at the Mayas; about of 145 (71%) of these were wetland bird species. This number could be more given the short observation period (21 days) each time; at any rate given the seasonality of water availability at the Mayas and their relative sizes, the number could be taken as significant. Ras Amir Maya, approximate area (1500 m × 400 m), had 53 species of which 38 (72%) were wetland birds (Table 2A, 2B, and 2C). Grerisa Maya, approximate area (500 m × 200 m), 68 species were identified of which 51 (75%) were wetland birds (Table 3A, 3B, and 3C). The rest were woodland birds found in the national park.

As for Abdel Ghani Maya, with an approximate diameter 300 m, 48 species were encountered of which 28 (58%) were wetland birds (Table 4A, 4B, and 4C). And finally, 34 species were identified at Dabkara area (approx. 300 m × 100 m), a flat low-lying marshy land richly overgrown with thick carpet of grasses, 28 species (82%) of which were wetland birds (Table 5A and 5B).

At the time of this study, Ras Amir Maya and Grerisa Maya were filled with water from edge to edge for most of the time, except in 2016, where Grerisa Maya was not filled to the edges, whereas Abdel Ghani Maya did not have much open water surface. In fact Abdel Ghani Maya was overgrown with vegetation such that there was no longer open water surface. *Acacia nilotica* trees have invaded the Maya and have formed thick ring in the center of the Maya. Possibly, this fact contributed to the relatively low percentage of its water fowls species. Dabkara was almost dry with no open water surface for most of the study period. It appeared there was high degree of siltation and sedimentation in both Abdel Ghani and Dabkara Maya marshes to the extent that even trees that were normally found at the edges of Mayas on higher grounds were now found deep inside the Mayas such as (*Acacia seyal*), (*Ziziphus* spp.), (*Acacia nilotica*), and (*Crateva adansonii*). The mean number of the water birds was generally low at all the Mayas except at Ras Amir Maya, where some species had relatively higher mean number: cattle egret (*Ardeola ibis*) (Table 2A), grey heron (*Ardea cinerea*) (Table 2B), squacco heron (*Ardeola ralloides*), and spur-winged plover (*Vanellus spinosus*) (Table 2C). Nevertheless, the mean number did not differ significantly across the four study sites ($x^2 = 0.04$, $p = 0.847$, df = 3). But for every Maya, the bird species differed significantly in the frequency with which they occurred (Ras Amir: $x^2 = 202.4$, $p <0.001$, df = 14; Grerisa: $x^2 = 112.7$, $p <0.001$, df = 19; Abdel Ghani: $x^2 = 14.6$, $p <0.001$, df = 9; Dabkara: $x^2 = 68.4$, $p <0.001$, df = 11).

Table 2A Mean number of birds from TSCs at Ras Amir Maya, Dinder National Park, during the dry season

W[a]	Species	Survey scores						Total	Mean	Logarithmic value $(x + 1)$						Mean[b]	Ordinal
		1	2	3	4	5	6	score	score	1	2	3	4	5	6	number	scale
W	*Haliaeetus vocifer*					2		2	2					0.5		0.2	Uncommon
	Coracias abyssinica		10	4	9		3	26	3.25		1	0.7	1		0.6	2.6	Common
	Motacilla aguimp			4	9	2	1	16	6.5			0.7	1	0.5	0.3	1.93	Frequent
W	*Aythya erythrophthalma*	6			3			9	4.5	0.8			1			0.74	Frequent
W	*Anhinga rufa*					2		2	2					0.5		0.2	Uncommon
	Oena capensis	6		4		2		12	4	0.8		0.7		0.5		1.17	Frequent
W	*Actophilornis africanus*						1	1	1						0.3	0.12	Uncommon
W	*Anas platyrhynchos*	6	5		3			14	4.7	0.8	0.8		1			1.34	Frequent
	Laniarius erythrogaster	6				2	1	9	3	0.8				0.5	0.3	1.1	Frequent
W	*Pelecanus onocrotalus*			4		2	1	7	2.3			0.7		0.5	0.3	0.76	Frequent
	Lamprotornis chalybaeus	6		4	3	2	1	16	3.75	0.8		0.7	1	0.5		1.73	Frequent
	Halcyon albiventris	6						6	6	0.1						0.38	Uncommon
	Merops nubicus		10	4	3	4	1	22	4.4		1	0.7	1	0.7	0.3	2.6	Common
	Bubulcus ibis	6	5	12		10		33	8.25	0.8	0.8	1.1		1		3.25	Common
W	*Egretta alba*		5		3	2		10	3.3		0.8		1	0.5		1.04	Frequent
W	*Scopus umbretta*		5			2	1	8	2.7		0.8			0.5	0.3	0.82	Frequent
	Dicrurus adsimilis	6						6	1	0.8						0.14	Uncommon
	Streptopelia lugens		5					5	0.83		0.8					0.35	Common
W	*Anas creca*						1	1	0.17						0.3	0.12	Frequent
W	*Ephippiorhynchus senegalensis*	12						12	2	1.1						0.53	Frequent

W[a]	Species	Survey scores						Total	Mean	Logarithmic value ($x + 1$)						Mean[b]	Ordinal
		1	2	3	4	5	6	score	score	1	2	3	4	5	6	number	scale
W	*Ardea cinerea*	6	10	8	6	6	4	40	6.67	0.84	1	1	0.84	0.84	0.7	6.38	Uncommon
	Tockus nasutus			4		2		6	1			0.7		0.48		0.57	Frequent
W	*Ardea goliath*	6	10	4		2		22	3.67	0.84	1	0.7		0.48		2.23	Uncommon
	Subtotal	**72**	**65**	**52**	**39**	**42**	**15**	**285**									

[a]Wetland bird
[b]Calculated as (anti (mean of 6 log values) – 1)
Table is continued

Table 2B Mean number of birds from TSCs at Ras Amir Maya, Diner National Park, during the dry season

W[a]	Species	Survey scores						Total score	Mean score	Logarithmic value ($x + 1$)						Mean[b]	Ordinal
		1	2	3	4	5	6			1	2	3	4	5	6	number	scale
	Halcyon leucocephala	6						6	1	0.84						0.38	Frequent
W	*Bostrychia hagedash*	6	5			2	1	14	2.33	0.84	0.8			0.48	0.3	1.51	Frequent
	Francolinus icterorhynchus	6						6	1	0.84						0.38	Frequent
W	*Anas acuta*	6	5		6			17	3.5	0.84	0.8		0.84			1.57	Frequent
	Merops pusillus					2		2	0.33					0.48		0.2	Uncommon
W	*Ciconia nigra*			4	3	4	2	13	2.17			0.7	0.6	0.7	0.3	1.42	Common
W	*Leptoptilos crumeniferus*	6	5				2	13	2.17	0.84	0.8				0.5	1.24	Common
	Lbb[c]	18	15	12	9	8	4	66	11	1.28	1.2	1.1	1	0.95	0.7	9.96	Frequent
W	*Phalacrocorax africanus*			4				4	0.67			0.7				0.31	Frequent
W	*Alcedo cristata*			4				4	0.67			0.7				0.31	Uncommon
W	*Nettapus auritus*			4	3		1	8	1.33			0.7	0.6		0.3	0.85	Frequent
W	*Ciconia ciconia*		5					5	0.83		0.8					0.35	Uncommon
W	*Anastomus lamelligerus*	6				2		8	1.33	0.84				0.48		0.66	Frequent
W	*Ceryle rudis*	6	5	4			2	17	2.83	0.84	0.8	0.7			0.5	1.93	Frequent
	Melierax poliopterus	6						6	1	0.84						0.38	Uncommon
W	*Egretta garzetta*		5					5	0.83		0.8					0.35	Uncommon
	Streptopelia capicola	24	25	16	6	6	4	81	13.5	1.4	1.4	1.2	0.84	0.84	0.7	10.75	Abundant
W	*Anas clypeata*			4	6			10	1.67			0.7	0.84			0.8	Frequent
	Lamprotornis purpuropterus	18	10	4			2	34	5.67	1.28	1	0.7			0.5	2.83	Common
W	*Threskiornis aethiopicus*						1	1	0.17						0.3	0.12	Uncommon

W[a]	Species	Survey scores						Total score	Mean score	Logarithmic value $(x + 1)$						Mean[b]	Ordinal
		1	2	3	4	5	6			1	2	3	4	5	6	number	scale
W	*Vanellus spinosus*	6		4	6	2	3	21	3.5	0.84		0.7	0.84	0.48	0.6	2.77	Common
	Bubo africanus					2		2	0.33					0.48		0.2	Uncommon
	Subtotal	**114**	**70**	**60**	**39**	**28**	**22**	**343**									

[a]Wetland bird
[b]Calculated as (anti (mean of 6 log values) – 1)
[c]A group of little birds generally brown not positively identified
Table is continued

Table 2C Mean number of birds from TSCs at Ras Amir Maya, Dinder National Park, during the dry season

		Survey scores						Total	Mean	Logarithmic value $(x + 1)$						Mean[b]	Ordinal
W[a]	Species	1	2	3	4	5	6	score	score	1	2	3	4	5	6	number	scale
	Centropus senegalensis		5					5	0.83		0.8					0.35	Uncommon
	Macrodipteryx longipennis	6	5					11	1.83	0.84	0.7					0.8	Frequent
	Pternistes squamatus	6						6	1	0.84						0.38	Uncommon
W	*Ardeola ralloides*		5	8	3	4	2	22	3.67		0.8	1	0.6	0.7	0.5	2.84	Common
W	*Alopochen aegyptiacus*		5		6			11	1.83		0.8		0.84			0.86	Frequent
W	Species 1		5					5	0.83		0.8					0.35	Uncommon
W	*Tachybaptus ruficollis*					2		2	0.33					0.48		0.2	Uncommon
W	*Sarkidiornis melanotos*			4				4	0.67			0.7				0.35	Uncommon
W	*Anas hottentota*						1	1	0.17						0.3	0.12	Uncommon
	Prinia subflava		10			2		12	2		1.0			0.48		0.79	Frequent
W	*Ciconia abdimii*						1	1	0.17						0.3	0.12	Uncommon
	Acryllium vulturinum		10	4	6	2	2	24	7		1.0	0.7	0.84	0.48	0.5	5.36	Uncommon
W	*Ciconia episcopus*		5					5	1.83		0.8					0.86	Frequent
W	*Ixobrychus minutus*			4	3		1	8	2.33			0.7	0.6		0.3	1.55	Frequent
	Motacilla capensis			4	3			7	1.17			0.7	0.6			0.65	Frequent
W	*Butorides striata*		6					6	1		0.8					0.38	Uncommon
W	*Mycteria ibis*		5					5	0.83		0.8					0.35	Uncommon
	Subtotal	**12**	**61**	**24**	**21**	**10**	**7**	**135**									
	Total	**198**	**196**	**136**	**99**	**80**	**44**	**763**									

[a]Wetland bird
[b]Calculated as (anti (mean of 6 log values) – 1)

Table 3A Mean number of birds from TSCs at Grerisa Maya, Dinder National Park, during the dry season

		Survey scores						Total	Mean	Logarithmic value ($x + 1$)						Mean[b]	Ordinal
W[a]	Species	1	2	3	4	5	6	score	score	1	2	3	4	5	6	number	scale
W	*Anas strepera*						1	1	0.17						0.3	0.05	Uncommon
W	*Ixobrynchus minutus*	12	10	4		2	2	30	5	1.11	1.04	0.7		0.48	0.5	3.31	Common
W	*Nycticorax nycticorax*				3			3	0.5				0.6			0.26	Uncommon
W	*Actophilornis africanus*	30		12	6	6	3	57	9.5	1.49		1.	0.84	0.84	0.6	5.51	Common
W	*Haliaeetus vocifer*		5		3	2		10	1.67		0.78		0.6	0.48		1.04	Frequent
W	*Anhinga rufa*		10					10	1.67		1.04					0.49	Frequent
W	*Gallinago nigripennis*				6			6	1				0.84			0.38	Uncommon
W	*Ephippiorhynchus senegalensis*				3			3	0.5				0.6			0.26	Uncommon
W	*Egretta alba*		5	4	9		1	19	3.17		0.78	0.7	1		0.3	1.91	Frequent
W	Spp. 1		5					5	0.83		0.78					0.35	Uncommon
W	*Himantopus himantopus*					2	1	3	0.5					0.48	0.3	0.52	Frequent
W	*Tachybaptus ruficollis*		5	4	9		1	19	3.17		0.78	0.7	1		0.3	1.91	Uncommon
W	*Sarkidiornis melanotos*	6	10		3			19	3.17	0.84	1.04		0.6			1.59	Frequent
W	*Alopochen aegyptiacus*	18	5	4	9		4	40	6.67	1.28	0.78	0.7	1		0.7	4.54	Frequent
	Bubulcus ibis		25	8	3	8	7	51	8.5		1.41	0.95	0.6	0.95	0.9	5.33	Frequent
	Pycnonotus barbatus		10		3			13	2.17		1.04		0.6			0.88	Common
	Phoeniculus purpureus				3			3	0.5				0.6			0.26	Uncommon
W	*Ardea cinerea*	20	5	4	15	8	5	57	10.2	1.4	0.78	0.7	1.2	0.95	0.8	8.3	Abundant
W	*Ardea goliath*		5		3	2		10	1.67		0.78		0.6	0.48		1.04	Frequent
	Tockus nasutus	6						6	1	0.84						0.38	Uncommon
W	*Plegadis falcinellus*		5	4	6	2	2	19	3.17		0.78	0.7	0.84	0.48	0.5	2.52	Common
	Butastur rufipennis					2		2	0.33					0.48		0.38	Uncommon

(continued)

Table 3A (continued)

		Survey scores						Total	Mean	Logarithmic value ($x + 1$)						Mean[b]	Ordinal
W[a]	Species	1	2	3	4	5	6	score	score	1	2	3	4	5	6	number	scale
W	*Bostrychia hagedash*	6		4	3			13	2.17	0.84		0.7	0.6			1.27	Frequent
W	*Anas acuta*		5	4				9	1.5		0.78	0.7				0.76	Frequent
	Subtotal	**98**	**110**	**52**	**87**	**34**	**27**	**408**									

[a]Wetland bird

[b]Calculated as (anti (mean of 6 log values) – 1)

Table is continued

Table 3B Mean number of birds at Grerisa Maya, Dinder National Park, during the dry season

		Survey scores						Total	Mean	Logarithmic value ($x + 1$)						Mean[b]	Ordinal
W[a]	Species	1	2	3	4	5	6	score	score	1	2	3	4	5	6	number	scale
W	*Anas platyrhynchos*					4	2	6	1					0.7	0.5	0.57	Frequent
W	*Pelecanus onocrotalus*			8		6	2	16	2.67			0.95		0.84	0.5	1.39	Frequent
W	*Nettapus auritus*	19	60	44	18	10	9	160	27.5	1.4	1.78	1.65	1.28	1.04	1	21.82	Abundant
	Lophaetus occipitalis		5					5	0.83		0.78					0.35	Uncommon
W	*Phalacrocorax africanus*		5					5	0.83		0.76					0.34	Uncommon
	Campethera nubica	6		4		2	1	13	2.17	0.84		0.7		0.48	0.3	1.43	Frequent
W	*Ceryle rudis*		15			2		17	2.83		1.2			0.48		0.9	Frequent
W	*Ardea purpurea*	12	3		3	4	1	23	4.17	1.11	0.78		0.6	0.7	0.3	2.82	Common
	Streptopelia semitorquata		5	4		6	1	16	2.67		0.78	0.7		0.8	0.3	1.73	Frequent
W	*Anas clypeata*	36	45	12	18	6	6	123	20.5	1.6	1.67	1.1	1.3	0.8	0.8	15.53	Frequent
W	*Anas capensis*	6	20	24	12	8	4	74	12.33	0.8	1.32	1.4	1.1	1	0.7	10.31	Abundant
	Ploceus luteolus	6	5	4				15	2.5	0.8	0.78	0.7				1.43	Frequent
W	*Anas penelope*		5	4				9	1.5		0.78	0.7				0.76	Frequent
W	*Ardeola ralloides*	36	25	12	18	8	4	103	17.17	1.6	1.41	1.1	1.3	1	0.7	13.79	Abundant
W	Species 1?	6						6	1	0.8						0.38	Uncommon
W	*Plectropterus gambensis*		5		3	2	3	13	2.17		0.78		0.6	0.5	0.6	1.57	Frequent
W	*Vanellus spinosus*		5	4	3			12	2		0.78	0.7	0.6			1.22	Frequent
W	*Anas hottentota*		5					5	0.83		0.78					0.35	Uncommon
W	*Threskiornis aethiopicus*		10		3			13	2.17		1.04		0.6			0.88	Frequent
W	*Alopochen aegyptiacus*			8	3			11	1.83			1	0.6			0.81	Frequent

(continued)

Table 3B (continued)

		Survey scores						Total	Mean	Logarithmic value ($x + 1$)						Mean[b]	Ordinal
W[a]	Species	1	2	3	4	5	6	score	score	1	2	3	4	5	6	number	scale
W	*Ciconia abdimii*		5	4	3	2	2	16	2.67		0.78	0.7	0.6	0.5	0.5	2.21	Common
	Weaver spp. 1?	6					4	10	1.67	0.8						0.38	Uncommon
W	*Ciconia episcopus*		5		6	2	2	15	2.5		0.78		0.8	0.5	0.5	1.69	Frequent
	Subtotal	**133**	**228**	**132**	**90**	**62**	**41**	**686**									

[a]Wetland bird

[b]Calculated as (anti (mean of 6 log values) – 1)

Table is continued

Table 3C Mean number of birds from TSCs at Grerisa Maya, Dinder National Park, during the dry season

		Survey scores						Total	Mean	Logarithmic value $(x + 1)$						Mean[b]	Ordinal
W[a]	Species	1	2	3	4	5	6	score	score	1	2	3	4	5	6	number	scale
W	*Scopus umbretta*		10	4	3		2	19	3.17		1.04	0.7	0.6		0.5	1.95	Frequent
W	*Mycteria ibis*						1	1	0.17						0.3	0.12	Uncommon
W	*Tringa nebularia*	24	10	4	3	8	1	50	8.3	1.4	1	0.7	0.6	1	0.3	5.76	Common
W	*Recurvirostra avosetta*		10	4		4	1	19	3.2		1	0.7		0.7	0.3	1.86	Frequent
W	*Tringa erythropus*		**5**				1	6	1		0.8				0.3	0.51	Frequent
W	*Actitis hypoleucos*	6					1	7	1.2	0.8					0.3	0.55	Frequent
	Milvus migrans						1	1	0.17						0.3	0.12	Uncommon
W	*Calidris minuta*	12					2	14	2.3	1.1					0.5	0.84	Frequent
	Bubulcus ibis				3			3	0.5				0.6			0.26	Uncommon
W	*Calidris ferruginea*		5		3			8	1.3		0.8		0.6			0.7	Frequent
	Batis molitor			4				4	0.67			0.7				0.31	Uncommon
W	*Aythya ferina*		5					5	0.83		0.8					0.35	Uncommon
	Falco alopex					2		2	0.33					0.5		0.2	Uncommon
	Tockus nasutus					2		2	0.33					0.5		0.2	Uncommon
W	*Ardea cinerea*	6		4	9	4	2	25	4.17	0.8		0.7	1	0.7	0.3	2.89	Common
W	*Plegadis falcinellus*				3			3	0.5				0.6			0.26	Uncommon
W	*Butorides striata*						8	8	1.33						1	0.44	Frequent
W	*Egretta garzetta*	6			3		1	10	1.67	0.8			0.6		0.3	0.95	Frequent
W	*Megaceryle maximus*	6	30	8	12	18	2	76	12.7	0.8	1.5	1	1.1	1.3	0.5	9.59	Abundant
	Subtotal	**60**	**75**	**28**	**39**	**38**	**23**	**263**									
	Total	**291**	**413**	**212**	**216**	**134**	**91**	**1357**									

[a]Wetland bird
[b]Calculated as (anti (mean of 6 log values) – 1)

Table 4A Mean number of birds from TSCs at Abdel Ghani Maya, National Park, during the dry season

		Survey scores						Total score	Mean score	Logarithmic value $(x + 1)$						Mean[b]	Ordinal
W[a]	Species	1	2	3	4	5	6			1	2	3	4	5	6	number	scale
	Butastur rufipennis				3			3	0.5				0.6			0.26	Uncommon
W	*Ardea cinerea*	24	5	4	15	8	5	61	10.2	1.4	0.78	0.7	1.2	0.95	0.8	8.3	Abundant
W	*Ardea goliath*		5		3	2		10	1.67		0.78		0.6	0.48		1.04	Frequent
	Tockus nasutus	6						6	1	0.84						0.38	Uncommon
W	*Plegadis falcinellus*		5	4	6	2	2	19	3.17		0.78	0.7	0.84	0.48	0.5	2.52	Common
	Phoeniculus purpureus					2		2	0.33					0.48		0.38	Uncommon
W	*Bostrychia hagedash*	6		4	3			13	2.17	0.84		0.7	0.6			1.27	Frequent
	Falco tinnunculus		5	4				9	1.5		0.78	0.7				0.76	Frequent
	Cypsiurus parvus					4	2	6	1					0.7	0.5	0.57	Frequent
	Apus affinis			8		6	2	16	2.67			0.95		0.84	0.5	1.39	Frequent
	Lbb***	24	60	44	18	10	9	165	27.5	1.4	1.78	1.65	1.28	1.04	1	21.82	Abundant
	Lophaetus occipitalis		5					5	0.83		0.78					0.35	Uncommon
W	*Phalacrocorax africanus*		5					5	0.83		0.76					0.34	Uncommon
	Campethera nubica	6		4		2	1	13	2.17	0.84		0.7		0.48	0.3	1.43	Frequent
W	*Ceryle rudis*		15			2		17	2.83		1.2			0.48		0.9	Frequent
W	*Ardea purpurea*	12	5		3	4	1	25	4.17	1.11	0.78		0.6	0.7	0.3	2.82	Common
	Falco tinnunculus				3			3	0.5				0.6			0.26	Uncommon
	Lbb***	18	20	27	33	14	16	128	21	1	1.3	1.5	1.5	1.2	1.2	20.46	Abundant
W	*Anas strepera*	30	40	20	9	4		103	17.2	1	1.6	1.3	1	0.7	0.3	10.75	Abundant
	Cypsiurus parvus	6			15			21	3.5	1			1.2			3.5	Frequent
	Lophaetus occipitalis				3			3	0.5				0.6			0.5	Uncommon
	Subtotal	**132**	**170**	**119**	**114**	**60**	**38**	**633**									

[a]Wetland bird
[b]Calculated as (anti(mean of 6 log values) – 1)
Table is continued

Table 4B Mean number of birds from TSCs at Abdel Ghani Maya Dinder National Park, during the dry season

W[a]	Species	Survey scores						Total	Mean	Logarithmic value $(x + 1)$						Mean[b]	Ordinal
		1	2	3	4	5	6	score	score	1	2	3	4	5	6	number	scale
	Campethera nubica					2		2	0.33					0.5		0.33	Uncommon
	Nilaus afer				3			3	0.5				0.6			0.5	Uncommon
W	*Anastomus lamelligerus*	6						6	1	1						1	Uncommon
	Struthio camelus		5	4				9	1.5		0.8	0.7				1.5	Frequent
W	*Ardea purpurea*			4		2		6	1			0.7		0.5		1	Frequent
W	*Ciconia nigra*	6						6	1	1						1	Uncommon
W	*Pelecanus onocrotalus*				6	4	1	11	1.83				0.8	0.7	0.3	1.83	Frequent
	Lamprotornis purpuropterus	24	30	20	12	4	2	92	15.3	1	1.5	1.3	1.1	0.7	0.48	15.33	Abundant
	Streptopelia semitorquata					2	1	3	0.5					0.5	0.3	0.5	Uncommon
W	*Ixobrychus minutus*	24	20	16	15	24	7	106	17.7	1	1.3	1.2	1.2	1.4	0.9	17.67	Abundant
W	*Egretta alba*					2	1	3	0.5					0.5	0.3	0.5	Uncommon
W	*Threskiornis aethiopicus*	6	10	8		6		30	5	1	1	1		0.5		5	Common
W	*Dendrocygna viduata*					4		4	0.67					0.7		0.67	Uncommon
	Milvus migrans				3			3	0.5				0.6			0.26	Uncommon
W	*Tringa erythropus*	24	5	4	15	8	5	61	10.2	1.4	0.8	0.7	1.2	0.9	0.8	8.3	Abundant
W	*Ardea goliath*	6	5		3	2		16	1.67	0.8	0.8		0.6	0.5		1.04	Frequent
W	*Scopus umbretta*		5	4	6	2	2	19	3.17		0.8	0.7	0.8	0.5	0.5	2.52	Common
	Bucorvus abyssinicus	6						6	1	0.8						0.38	Uncommon
	Phoeniculus purpureus					2		2	0.33					0.5		0.38	Uncommon
W	*Bostrychia hagedash*	6		4	3			13	2.17	0.8		0.7	0.6			1.27	Frequent
W	*Anas platyrhynchos*		5	4				9	1.5		0.8	0.7				0.76	Frequent

(continued)

Table 4B (continued)

W[a]	Species	Survey scores						Total	Mean	Logarithmic value $(x + 1)$						Mean[b]	Ordinal
		1	2	3	4	5	6	score	score	1	2	3	4	5	6	number	scale
W	*Anas clypeata*					4	2	6	1					0.7	0.5	0.57	Frequent
	Apus affinis			8		6	2	16	2.67			0.9		0.8	0.5	1.39	Frequent
	Subtotal	**108**	**85**	**76**	**67**	**74**	**23**	**425**									

[a]Wetland bird
[b]Calculated as (anti (mean of log values) – 1)
Table is continued

Table 4C Mean number of birds from TSCs at Abdel Ghani Maya, Dinder National Park, during the dry season

W[a]	Species	Survey scores								Logarithmic value ($x + 1$)						Mean[b]	Ordinal
		1	2	3	4	5	6	Total score	Mean score	1	2	3	4	5	6	number	scale
W	*Anas capensis*	24	60	41	18	10	9	162	27.5	1.4	1.8	1.6	1.3	1.0	1	21.82	Abundant
W	*Alopochen aegyptiacus*		5					5	0.83		0.8					0.35	Uncommon
W	*Phalacrocorax africanus*		5					5	0.83		0.8					0.34	Uncommon
W	*Calidris ferruginea*	5		4		2	1	12	2.17	0.8		0.7		0.5	0.3	1.43	Frequent
W	*Actitis hypoleucus*		18			2		20	2.83		1.2			0.5		0.9	Frequent
W	*Sarkidiornis melanotos*	12	5		3	4	1	25	4.17	1.1	0.8		0.6	0.7	0.3	2.82	Common
	Subtotal	**41**	**93**	**45**	**21**	**18**	**11**	**229**									
	Total	**281**	**348**	**240**	**202**	**152**	**72**	**1287**									

[a]Wetland bird
[b]Calculated as (anti (mean of log values) – 1)

Table 5A Mean number of birds at Dabkara Maya, Dinder National Park, during the dry season

W[a]	Species	Survey score						Total	Mean	Logarithmic value $(x + 1)$						Mean[b]	Ordinal
		1	2	3	4	5	6	score	Score	1	2	3	4	5	6	number	scale
W	*Anas platyrhynchos*	24	10	4	3	8	1	50	8.3	1.4	1	0.7	0.6	1	0.3	5.76	Common
W	*Anas clypeata*		10	4		4	1	19	3.2		1	0.7		0.7	0.3	1.86	Frequent
W	*Ciconia abdimii*		**5**				1	6	1		0.8				0.3	0.51	Frequent
W	*Pelecanus onocrotalus*	6					1	7	1.2	0.8					0.3	0.55	Frequent
	Milvus migrans						1	1	0.17						0.3	0.12	Uncommon
W	*Leptoptilos crumeniferus*	12					2	14	2.3	1.1					0.5	0.84	Frequent
	Bubulcus ibis				3			3	0.5				0.6			0.26	Uncommon
W	*Nycticorax nycticorax*		5		3			8	1.3		0.8		0.6			0.7	Frequent
	Batis molitor			4				4	0.67			0.7				0.31	Uncommon
	Pycnonotus barbatus		5					5	0.83		0.8					0.35	Uncommon
	Falco alopex					2		2	0.33					0.5		0.2	Uncommon
	Tockus nasutus					2		2	0.33					0.5		0.2	Uncommon
W	*Ardea cinerea*	6		4	9	4	2	25	4.17	0.8		0.7	1	0.7	0.3	2.89	Common
W	*Plegadis falcinellus*				3			3	0.5				0.6			0.26	Uncommon
W	*Butorides striata*						8	8	1.33						1	0.44	Frequent
W	*Ciconia ciconia*	6			3		1	10	1.67	0.8			0.6		0.3	0.95	Frequent
W	*Alcedo cristata*	6	30	8	12	18	2	76	12.7	0.8	1.5	1	1.1	1.3	0.5	9.59	Abundant
W	*Egretta alba*					2	1	3	0.5					0.5	0.3	0.5	Uncommon
W	*Ixobrychus minutus*	24	20	16	15	24	7	106	17.75	1.4	1.3	1.2	1.2	1.4	0.9	17.67	Abundant
W	*Threskiornis aethiopicus*	6	10	8		6		30	5	1	1	1		0.5		5	Common
W	*Dendrocygna viduata*					4		4	0.67					0.7		0.67	Uncommon
	Subtotal	**90**	**95**	**48**	**51**	**74**	**28**	**386**									

[a]Wetland bird
[b]Calculated as (anti (mean of 6 log values) – 1)
Table is continued

Table 5B Mean number of birds at Dabkara Maya, Dinder National Park, during the dry season

W[a]	Species	Survey score						Total	Mean	Logarithmic value ($x + 1$)						Mean[b]	Ordinal
		1	2	3	4	5	6	score	score	1	2	3	4	5	6	number	scale
W	*Anas penelope*	42	40	28	15	10	4	139	23.2	1.6	1.6	1.5	1.2	1	0.7	17.76	Abundant
	Lophaetus occipitalis			4				4	0.7			0.7				0.31	Uncommon
	Campethera nubica			4				4	0.7			0.7				0.31	Uncommon
	Nilaus afer			4				4	0.7			0.7				0.31	Uncommon
	Cypsiurus parvus	6			3			9	1.5	0.8			0.6			0.26	Uncommon
W	*Ephippiorhynchus senegalensis*	6	10	12	3	4	4	39	6.5	0.8	1	1.1	0.6	0.7	0.7	5.79	Common
W	*Egretta garzetta*	18	55	16	6	18	8	121	20	1.3	1.8	1.2	0.8	1.3	1	15.66	Abundant
W	*Tachybaptus ruficollis*	12	10	4				26	4.3	1.1	1	0.7				1.98	Frequent
W	*Ardeola ralloides*	12	5	8	3		1	29	4.8	1.1	0.8	1	0.6		0.3	3.2	Common
	Centropus senegalensis				3			3	0.5				0.6			0.26	Uncommon
	Colius striatus	6			3			9	1.5	0.8			0.6			0.26	Uncommon
W	*Threskiornis aethiopicus*		5	8	6		2	21	3.5		0.8	1	0.8		0.5	2.22	Common
W	*Anas hottentota*	12	10	4	15	2	3	46	7.7	1.1	1	0.7	1.2	0.5	0.6	5.38	Common
W	*Ciconia episcopus*	6	5		6	2	1	20	3.3	0.8	0.8		0.8	0.5	0.3	2.47	Common
W	*Aythya ferina*		10	4	6	2	1	23	3.8		1	0.7	0.8	0.5	0.3	2.63	Common
W	*Megaceryle maximus*						1	1	0.17						0.3	0.12	Uncommon
W	*Mycteria ibis*		5				1	6	1		0.8				0.3	0.51	Frequent
W	*Tringa nebularia*	24	10	4	3	8	1	50	8.3	1.4	1	0.7	0.6	1	0.3	5.76	Common
W	*Recurvirostra avosetta*		10	4		4	1	19	3.2		1	0.7		0.7	0.3	1.86	Frequent
	Subtotal	**144**	**175**	**104**	**72**	**50**	**28**	**573**									
	Total	**234**	**270**	**152**	**123**	**124**	**56**	**959**									

[a]Wetland bird
[b]Calculated as (anti (mean of 6 log values) – 1)

Table 6 Analysis of TSC diversity, equitability, and species numbers in four Mayas in Dinder National Park

Name of *Mayas*	Diversity	Equability	Mean number of birds	Number of species
Abdel Ghani	8.8	16.77	81.03	48
Dabkara	3.2	0.3286	85.22	34
Grerisa	18.4	0.77	110.34	68
Ras Amir	11.8	0.639	137.81	53

It was also found that Grerisa Maya had the greatest diversity index followed by Ras Amir Maya and then Abdel Ghani Maya and lastly Dabkara Maya (Table 6).

In general no clear trends were apparent between the sites which had plenty of water (Grerisa and Ras Amir) and those with little or almost no open water (Abdel Ghani and Dabkara), with respect to number of species and/or abundance per site. There was a slight tendency of water-rich Maya to support more individuals, but they generally had fewer species than water-deficient Maya. This could be attributed to the fact that toward the edges of the open water surfaces, there are found thick reeds, bushes, and trees which tend to attract many more birds species and woodland as well as water birds.

Certain species were found only on Maya with open water surface such as ducks and geese (Anatidae), especially African pochard (*Aytha erythrophthalma*), spur-winged goose (*Plectopterus gambensis*), African pygmy goose (*Nettapus auritus*), maccoa duck (*Oxyura maccoa*), knob-billed duck (*Sarkidiornis melanotos*), and Egyptian goose (*Alopochen aegyptiaca*); pelicans (Pelecanidae); cormorants (Phalacrocoracidae) especially long-tailed cormorant (*Phalacrocorax africanus*) and African darter (*Anhinga rufa*); kingfishers (Alcedinidae); hamerkops (Scopidae); cranes (Balearicidae); and Jacanas (Jacanidae) especially African jacana (*Actophilornis africanus*) and their allies. Two Palearctic migrant ducks were also observed in the second year of the study at Ras Amir Maya: northern pintail (*Anas acuta*) and northern shoveler (*Anas clypeata*). Yet other species were commonly found on Mayas with no open water surface (i.e., surface overgrown with vegetation) like the shrikes (Laniidae) especially white-crowned shrike (*Eurocephalus ruppelli*) and their allies, weavers, whydah, and sparrows (Ploceidae). In general, a greater number of species occurred on Grerisa Maya and Ras Amir Maya, both of which collectively supported 121 species of birds (i.e., over 73% wetland birds). Fewer total species occurred on both Abdel Ghani and al Dabkara Mayas, 82 species of birds (i.e., of which 68% water birds). There were also certain species of birds which were common to all the sites including bee-eaters (Meropidae) and rollers (Coraciidae), especially carmine bee-eater (*Merops nubicus*), little bee-eater (*Merops pusillus*), and cinnamon-chested bee-eater (*Merops oreobates*); lilac-breasted roller (*Coracias caudata*) and Abyssinian roller (*Coracias abyssinica*); plovers (Charadriidae); wagtails (Motacillidae) and allies; swifts (Apodidae), especially little swift (*Apus affinis*); herons and egrets (Ardeidae); storks (Ciconiidae); ibises and spoonbills (Threskiornithidae); and several dove species (Columbidae).

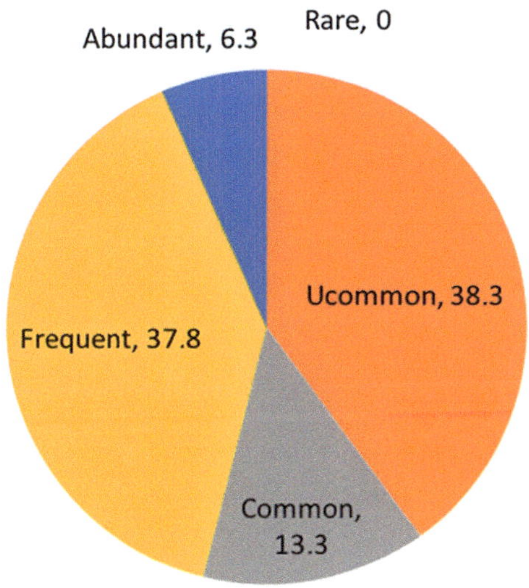

Fig. 2 Ordinal scale of bird abundance in all the four study sites in Dinder National Park

Whereas none of the bird species identified in all the four study sites could be classified as "rare" on the ordinal scale of abundance, the rest constituted different percentages as shown in (Fig. 2) above.

3.2 Species Richness and Diversity (Alpha and Beta Diversity) of Birds in the Four Sites

Species Richness

Results on species richness and diversity of birds indicate that the park has diverse bird species. A total of 203 species of birds were recorded during the sampling at four wetland habitats (sites) which constitute less than 30% of the park area. However, Jack 1 mean estimator of species richness puts the number to be 217 species in the four sites (Fig. 3). Moreover, the same number of species recorded (203 species) and that estimated by Jack 1 mean estimator of species richness (per 1000) were observed for 840 individuals of birds (Fig. 4).

3.3 Bird Species Alpha and Beta Diversity (Species Turnover) between the Four Sites

Overall alpha diversity of birds species expressed as Shannon diversity index was 1.89, and the species turnover across all the four habitats was 0.365. The Grerisa Maya had the highest diversity whereas Dabkara had the lowest (Table 6).

Fig. 3 Sample-based rarefaction curves of expected bird species in wetland habitats, Dinder National Park

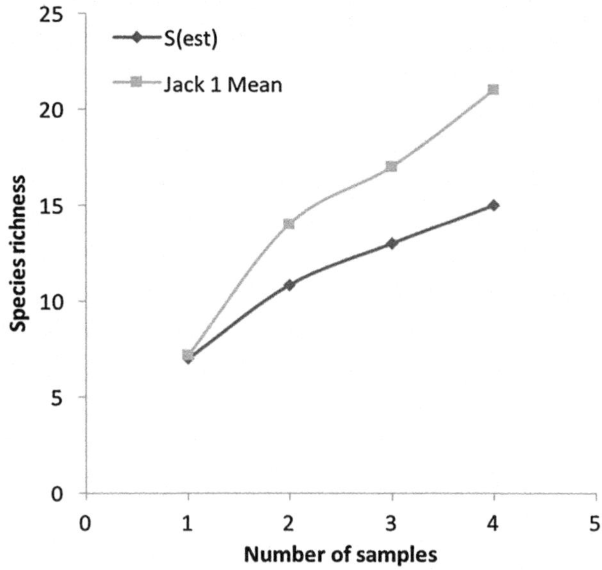

Fig. 4 Individual-based rarefaction curves of expected bird species in wetland habitats, Dinder National Park

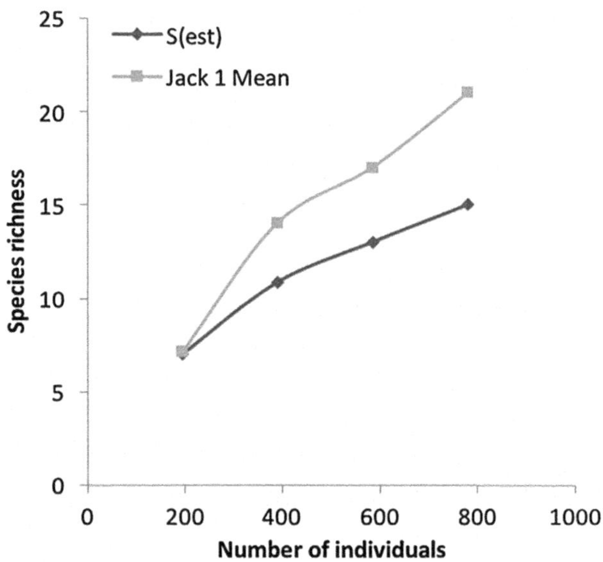

Calculations of selection index and proportion of use of each habitat type by birds showed that, generally, there was no specific habitat preferences i.e. the available resources were used proportionately due to high degree of similarity between the sites; nevertheless there was a slight tendency of birds concentrating in Maya Grerisa and Maya Ras Amir; this could be attributed to availability of much water in both Mayas for most of the observation period (Table 7).

Table 7 Neu's selection index using simulated data from TSC in four Mayas, Dinder National Park

Mayaat	Availability	Usage		Index	
	Proportion a	Records	Proportion r	Selection w	Standardized B
Abdel Ghani	0.255	48	0.236	0.925	0.233
Dabkara	0.208	34	0.167	0.803	0.202
Grerisa	0.315	68	0.335	1.063	0.268
Ras Amir	0.222	53	0.261	1.176	0.296
Total	**1.000**	**203**	**1.000**	**3.967**	**1.000**

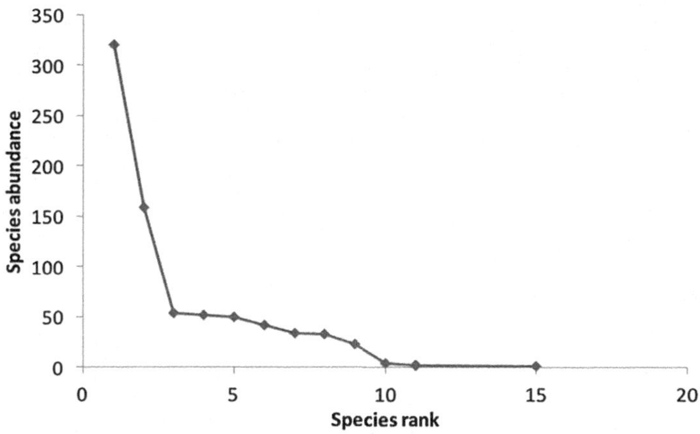

Fig. 5 Species rank-based rarefaction curves of expected bird species in four wetland habitats in Dinder National Park

3.4 Species Rank and Abundance Curves Across the Four Selected Habitats in Dinder National Park

The Overall Rank Abundance Curve Showed that Few Bird Species Were Dominant and the Majority Were Uncommon (Fig. 5)

Comparison of species abundance between four different habitats in the park indicated that few species were common and very few were abundant, while most species were either uncommon or frequent. On the other hand, no species were rare across the four wetland habitats. Abdel Ghani Maya and Dabkara ranked first in having species that were uncommon, whereas Ras Amir and Abdel Ghani Mayas ranked highest in having most of the species that were frequent in abundance then followed by the other two Mayas (Fig. 6).

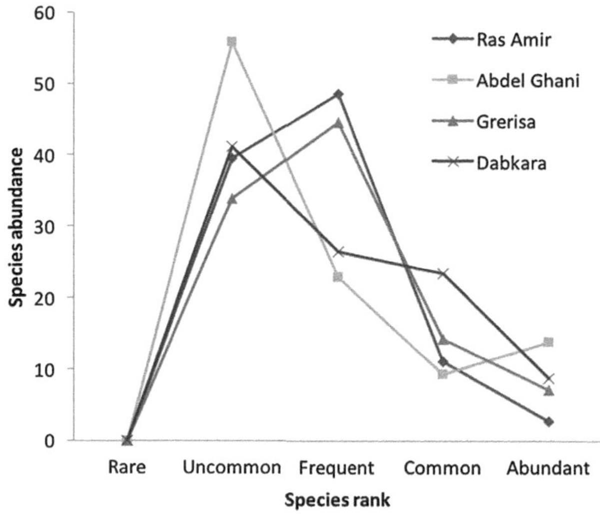

Fig. 6 Bird species rank abundance curves in four wetland habitats in Dinder National Park

4 Discussions

Assessment and evaluation of wetland areas in the context of wetland conservation and management for Dinder National Park, as well as for the whole Sudan, are new topics. Enormous amount of data exists on wetland biodiversity, but most of them are not collected under this subject, for instance (Abdel Hameed et al. 1997; Dawi 1998, Khalid and Salih 2002; Hussein 2015), to mention but a few. Many national parks and forest reserves in the Sudan were established on the basis of their wetland biodiversity (De Jong-Boon 1990). Many national parks and protected areas in Sudan are famous because of their wetland plants and animals including birds: Sanganeb (marine park), Nile Bird Sanctuary, Nimule National Park, and Bandingilo Game Reserve; Zeraf, Shambe, and Mashra Game Reserves (within the Sudd wetlands) are all examples of prime wetland areas characterized by such a diversity, which calls for a closer look and a rethinking with regard to their conservation and management plan. One important goal of sampling is the generation of a complete species list of a locality, along with an unbiased estimate of the abundance of each species (Longino and Colwell 1997). Statistical methods are widely used to estimate the asymptotic number of species present, including species not yet detected (Chao et al. 2009). Species richness estimators assume that the observed species richness is lower than the true richness of the site (Colwell and Coddington 1994). In this study, six nonparametric estimators of species richness were used. The ratio of observed species richness to the average of abundance-based/sample-based species richness estimators revealed a powerful sampling effort higher than 90% in all cases.

Also, the observed number of bird species found in this study in the park was low for a tropical area just a few degrees latitude away from the equator (Parmesan 2006). However, Jack 1 species richness indicated that there were at least 194 bird species in the park. This was further confirmed by both sample- and individual-based rarefaction curves which did not reach an asymptote, suggesting that sampling was incomplete. And usually, rare species may continue to be captured long after the asymptote has been attained (Ramchandra 2013; Schlossberg and King 2008). The overall observed alpha diversity of 1.86 was suggestive of a moderately stable environment of birds where few species dominated the birds' community in Dinder National Park (Buckley and Freckleton 2010; Brambilla et al. 2011).The moderate species (β diversity) among birds is not unique with reference to the observed habitats. This finding was further cross boned by the findings in between sites. Beta diversities which showed less to almost complete overlap between sites suggesting that investigated habitat types might have had both homogenous and heterogeneous sites (Whittingham and Evans 2004). The importance of these wetland areas could be considered in terms of the kind of species found there in; there are species which are restricted to life in wetland fish, frogs, reeds, water lilies, and the numerous invertebrates and species which have a more seasonal dependence on the wetlands (Evans 1994; Brambilla et al. 2011; Hadley et al. 2012). The latter group of species would have a greater use and occupancy of the wetlands in the dry season. So the role of the Mayas for wildlife in Dinder National Park is very crucial. The wetlands of Dinder National Park thus act as refuges for the very much larger faunal species; we can therefore consider them as keystone resources in recognition of this significant role. The findings of this study are similar to those reported by Lee and Rotenberg (2005), Chapman and Reich (2007), Thakare and Zade (2011), and Salah and Idris (2013).

In general, wetlands are still little understood and appreciated in the Sudan; they are used by wildlife for various purposes—for water and food, as breeding habitats, places of refuge, and transition especially for migrants, both Palaearctic and Afrotropical (Bibby et al. 1992). In many instances these waterlogged areas are dominated by vegetation such as reeds, giving the impression that wetlands do not have wildlife species of any significance, especially the large mammals. But the fact is that in Dinder National Park, the Mayas offer the only places where wildlife can get water during dry seasons. Wetlands of Dinder National Park are home to many different kinds of birds. Indeed, this relatively short study shows that around 39.4% of all the birds positively identified are found in and around the Mayas. Over 60% of birds in Dinder National Park can be classified as wetland species because they nest, feed, and roost in wetlands. Some of these species, such as ducks (Anatidae), herons (Ardeidae), gulls (Laridae), and wading birds (Charadriidae), are strongly tied to water therefore depend wholly on aquatic habitats. Others, such as some weavers (Ploceidae), warblers (Sylviidae), and plovers (Charadriidae), are frequently found in aquatic or wet areas but are sometimes seen in dry habitats as well. This observation is in agreement with what was reported by Simon and Okoth (2016), who asserted that even within a single habitat, different species of birds used their surrounding environment differently. Many of the ducks and waders (e.g., pintails) are migrants which breed in Eurasia and are found, together with others, throughout Eastern Africa and Sudan during the northern winter from November to March (Dawi 1998). Other species, such as the squacco heron (*Ardeola ralloides*) and Abdim's stork (*Ciconia abdimii*), are intra-African migrants (Stevenson and Fanshawe 2002). The rest remains here all year round. It is worth mentioning that there were several other bird species identified during this study in other areas away from the *Mayas*, including the crowned cranes (*Balearica regulorum*). Cranes are birds that depend on wetlands for part of their life cycle. Although they are terrestrial birds, cranes breed exclusively in wetlands, especially seasonal swamps (Evans 1994).

5 Conclusions

1. The wetland (Mayas) habitats in Dinder National Park are characterized by high degree of biodiversity; hence, they must be considered as keystone resources that deserve special and intensive management.
2. In general the study showed that the abundance and diversity of the wetland birds was higher in the Mayas with much water than those with relatively less water. The exact reason for the variation of Maya's capacity to store water annually is not known. There is need therefore to study the complicated relationship between the amount of water in the Mayas and the number of birds, as well as the Maya capacity to hold water throughout the dry season, in order to generate knowledge of wetland dynamics in Dinder National Park.
3. Differences in resource availability between the four wetland habitats such as breeding sites, nesting materials, cover, and food and the available amount of water restrict some species of wetland birds to certain habitats while allowing others to be widely distributed.
4. Of the 931 species of the avian fauna documented in Sudan, this study identified 203 species of birds in and around the four Mayas (meadows and oxbow lakes) of Dinder National Park. However, research on birds is very rare in Sudan. As ornithology is not restricted to a theoretical knowledge, the various stakeholders should therefore double their efforts to gain better knowledge of avian fauna in order to establish efficient conservation and management strategies for the birds and their wetland habitats in the Sudan.

Acknowledgment The assistance and encouragement of Professor Ibrahim Mohammed Hashim, of the Wildlife Research Center (Shambat), is highly appreciated. You not only supported the initial idea but offered invaluable advice too. Thanks are also due to the University of Juba administration for providing logistics and financial assistance. Last but not least, we thank the different individuals who facilitated our efforts in one way or the other in carrying out this study, especially, the Wildlife Authority for permitting us to work in the park.

References

Abdel Hameed SM, Awad NM, Nimir MB, Hakim SA, Ahmed EO, Hashim IM (1997) Vegetation succession and its effects on wildlife herbivory in the mayas of Dinder National Park. Sudan, Wildlife Research Center

Begon M, Townsend CR, Harper JL (2006) Ecology: from individual to ecosystems, 4th edn. Blackwell Publishing Ltd., Oxford, UK

Bibby CJ, Collar NJ, Crosby MJ, Heath MF, Imboden C, Johnson TH, Long AJ, Sattersfield AJ, Thirgood SJ (1992) Putting biodiversity on the map: priority areas for global conservation. International Council for Bird Preservation, Cambridge

Bibby C, Jones M, Marsden S (1998) Expedition field techniques BIRD SURVEYS. Expedition Advisory Center, Royal Geographical Society, London

Bookhout TA (ed) (1994) Research and management techniques to wildlife habitat, 5th edn. Wildlife Society, Bethesda

Both C, Bouwhuis S, Lessells CM, Visser ME (2006) Climate change and population declines in long-distance migratory bird. Nature 441:81–83

Brambilla M, Gustin M, Celada C (2011) Defining favourable reference values for bird populations in Italy: setting long-term conservation targets for priority species. Bird Conserv Int 21:107–118

Buckley HL, Freckleton RP (2010) Understanding the role of species dynamics in abundance-occupancy relationships. J Ecol 98:645–658

Britton PL (1980) Birds of East Africa: their habitats, status and distribution. East Africa natural history society. Nairobi, Kenya

Chao ARK, Colwell CWL, Gotelli NJ (2009) Sufficient sampling for asymptotic minimum species richness estimators. Ecology 90:1125–1133

Chapman KA, Reich PB (2007) Land use and habitat gradients determine bird community diversity and abundance in sub urban, rural and reserve landscape of Minnesota, USA. Biol Conserv 135:529–541

Colwell RK (2013) Estimate S: statistical estimation of species richness and shared species from samples. Version 9. Persistent URL purl.oclc.ort/estimates, University of Connecticut, CT

Colwell RK, Coddington JA (1994) Estimating terrestrial biodiversity through extrapolation. Philos Trans R Soc Lond B 345:101–118

Cowardin LM, Carter V, Golet FC, Laroe ET (1977) Classification of wetlands and Deepwater habitats of the United States. United States Fish and Wildlife Service, Washington, DC, + 103 pp

Dawi MH (1998) Bird Fauna of Dinder National Park, Sudan notes and records. 11:187–203

De Jong-Boon C (1990) Environmental poblems in Sudan: a reader (part I), The Hague

Dranzoa C (1990) Survival of forest birds in formerly forested areas. Unpublished MSc thesis, Makerere University, Kampala

Evans MI (1994) Research and management techniques for wildlife and habitats, 5th edn. Wildlife Society, Bethesda

Gotelli NJ (2004) A taxonomic wish-list for community ecology. Philos Trans R Soc Lond B 359:585–597

Gotelli NJ, Colwell RK (2010) Estimating species richness. In: Magurran AE, McGill BJ (eds) Frontiers in measuring biodiversity. Oxford University Press, New York, pp 39–54

Gunnarsson B, Federsel IM (2014) Bumblebees in the city: abundance, species richness and diversity in two urban habitats. J Insect Conserv 18:1185–1191

Hadley SJK, Hadley AS, Betts M (2012) Acoustic classification of multiple simultaneous bird species: a multi-instance multi-label approach. J Acoust Soc Am 131(6):4640–4650

Hammer O (1999-2011) In: Natural History Museum University of Oslo (ed) PAleontological STatistics, Oslo

Hamilton AJ (2005) Species diversity or biodiversity? J Environ Manag 75:89–92

Hashim IM (1984) Meadow use by wild ungulates in Dinder National Park. Ph.D thesis, New Mexico State University, Las Cruces

Hespenheide HA (1994) An overview of formal studies. In: Bawa KS, Hespenheid HA, Hartshorn GS (eds) Lasaha ecology and natural history of neotropical rainforest. The University of Chicago Press, Chicago

Holsworth NW (1968) Dinder National Park. Report to the government of Sudan. No. T4 3457. FAO Rome

Hussein MEAE (2015) Birds of Dinder National Park from 2008–2011. Poult Fish Wildl Sci 3:128. https://doi.org/10.4172/2375-446X.1000128

Khalid MA, Salih, MHE (2002) Preliminary Survey of the Dinder National Park Fish Fauna, diversity and tentative ichthyomass

Lee PY, Rotenberg JT (2005) Relationship between bird species and tree species assemblage in forest habitat of eastern North America. J Biogeogr 32:1139–1150

Longino JT, Coddington J, Colwell RK (2002) The ant fauna of a tropical rain forest: estimating species richness three different ways. Ecology 83:689–702

Longino JT, Colwell RK (1997) Biodiversity assessment using structured inventory: capturing the ant fauna of a tropical rain forest. Ecol Appl 7:1263–1277

Magurran AE (1988) Diversity and species abundance model. In: Magurran AE (ed) Ecological diversity and its measurement. Prince Lion University Print

Magurran AE (2004) Measuring biological diversity. Blackwell, Malden, MA

McGill BJ, Etienne RS, Gray JS, Alonson D, Anderson MJ, Benecha HK, Dornelas M, Enquist BJ, Green JL, Green FH (2007) Species abundance distributions: moving beyond single prediction theories to integration within an ecological framework. Ecol Lett 10:995–1015

Mafabi P (1999) Wetlands and their wildlife. Swara 18(1):15–17

Mathews GVT (1993) Ramsar convention on wetlands: its history and development. Ramsar Convention Bureau, Gland, v + 120 pp

Moilinga P (2016) The olive baboons of Sudan: determinants of structure and survival. LAP Lambert Academic Publishing, Germany. ISBN 978-659-87171-9

Moreno CE, Halffer G (2000) Assessing the completeness of bat biodiversity inventories using species accumulation curves. J Appl Ecol 37:149–158

Parmesan C (2006) Ecological and evolutionary responses to recent climate change. Ann Rev Ecol Evol Syst 37:637–669

Pomeroy DE, Tengecho B (1986) A method of analyzing bird distribution. Afr J Ecol 24:243–253

Pomeroy DE (1992) Counting birds: a guide to assessing numbers, biomass and diversity of afro tropical birds. African wildlife foundation, Nairobi

Ramchandra AM (2013) Diversity and richness of bird species in newly formed habitats of Chandoli National Park in Western Ghats, Maharashtra state, India. Biodivers J 4(1):235–242

Salah O, Idris E (2013) A note on the birds diversity at two sites in Khartoum, Sudan. Egypt Acad J Biol Sci 5(1):1–10

Schlossberg S, King DI (2008) Are shrubland birds edge specialists. Ecol Appl 18(6):1325–1330

Schowalter TD (2006) Community structure. In: Schowalter TD (ed) Insect ecology: an ecosystem approach. Academic Press, San Diego, CA, pp 251–282

Simon GS, Okoth EO (2016) Species richness and abundance of birds in and around Nimule National Park, South Sudan. ScieLett; 2015023 - SL (online first)

Stevenson T, Fanshawe J (2002) Birds of East Africa. T & A D Poyser Ltd, London

Stireman JOI (2008) α and β diversity of a Tachinid parasitoid community over space and time. Ann Entomol Soc Am 101:362–370

Thakare UG, Zade VS (2011) Diversity, abundance and species composition of water beetles (Coleoptera; Dytiscidae, Hydrophilidae and Gyminidae) in Kolkas region of Melghat. Tiger Reserve, Central Indiana. Acad J Entomol 4:64–71

UNESCO (1994) Convention on Wetlands of international importance. Ramsar, 2.2. 1971., as amended by the Protocol of 3.12.1982 and the Amendments of 25.5.1987,--UNESCO. Office of International Standards and Legal Affairs, Paris, 3 pp

Whittingham MJ, Evans KL (2004) The effects of habitat structure on predation risk of birds in agricultural landscapes. Ibis 146(2):210–220

Characterising the Hydrological Regime of a Tropical Papyrus Wetland in the Lake Kyoga Basin, Uganda

Ellen Jessica Kayendeke and Helen K. French

Abstract Papyrus wetlands are predominant in permanently flooded areas of Sub-Saharan Africa, covering approximately 40,000 km^2 in East and Central Africa. Previous studies have quantified and valued ecosystem services of these wetlands, but there is still a need to understand the key processes of the wetlands' hydrology. The study objective was to quantify the seasonal variations in water balance and retention time of a papyrus wetland section. Discharge measurements were carried out to quantify channel flow in upstream and downstream parts of the wetland section. Groundwater fluxes were estimated using borehole triangulation of groundwater levels, while precipitation and other climatic variables for estimating evapotranspiration were monitored using an automatic weather station located 1 km from the study site. The retention time of the wetland section was estimated from tracer experiments and by calculating the volume-discharge ratio. Results show that the water balance of the section is dominated by wetland channel flow, contributing approximately 99.7% of the total inputs, while precipitation and groundwater discharge contribute approximately 0.2% and <0.1%, respectively. Estimated retention time varied between 2 hours and 7 days during periods of high and low flows, respectively. The groundwater gradient showed flow towards the wetland throughout the monitoring period, with average gradients of 0.0074 and 0.0043 on the western and eastern edges of the wetland, respectively. Since wetland channel flow is dominant, the wetland's hydrology is vulnerable to land cover

E. J. Kayendeke (✉)
Faculty of Environmental Sciences and Natural Resource Management, Norwegian University of Life Sciences, Aas, Norway

Department of Environmental Management, Makerere University, Kampala, Uganda
e-mail: ellen.kayendeke@nmbu.no

H. K. French
Faculty of Environmental Sciences and Natural Resource Management, Norwegian University of Life Sciences, Aas, Norway

Y. Bamutaze et al. (eds.), *Agriculture and Ecosystem Resilience in Sub Saharan Africa*, Climate Change Management,
https://doi.org/10.1007/978-3-030-12974-3_10

changes and resultant changes in surface runoff from the upstream catchment. Further research on impacts of land use changes within the upstream catchment on wetland channel flow is recommended.

Keywords Papyrus wetland · Water balance · Retention time

1 Introduction

Papyrus wetlands are predominant in permanently flooded areas of East and Central Africa. Their extent is not known but estimates range from 20,000 to 85,000 km^2. They are dominated by papyrus (*Cyperus papyrus* L.), which is a herbaceous sedge that can grow up to 5 metres (Kipkemboi and van Dam 2018). Papyrus can grow firmly rooted at water edges or in waterlogged areas but is physically adapted to flooded areas because its culms and rhizomes have large quantities of air-filled plant tissue (Kipkemboi and van Dam 2018; Mburu et al. 2015), which enables it to form floating mats in permanently inundated areas (Fig. 1). This ability gives papyrus a competitive advantage over other aquatic plants that are less adapted to permanent flooding (Kipkemboi and van Dam 2018). The floating mats are made up of interconnected roots, rhizomes, and organic matter (Azza et al. 2000). The plants spread

Fig. 1 Floating papyrus wetland in Eastern Uganda, with open water at inlet culvert. The depth of the water column beneath the papyrus mat varies between 1 and 2 metres

through both rhizome propagation and seed germination (Gaudet 1977; Terer et al. 2014) and thrive in low energy environments where wind velocity and water level fluctuations are moderate (Mburu et al. 2015; Morrison et al. 2013; Terer et al. 2012).

Papyrus wetlands provide several goods and services like food (fish and crops at wetland edges), fuel wood, raw materials for crafts and building, and water for domestic and irrigation use. In addition, they have various regulatory functions including carbon storage, flood mitigation, and water purification (Donaldson et al. 2016; Mburu et al. 2015; Saunders et al. 2013; Terer et al. 2014; van Dam et al. 2007). The potential of papyrus wetlands in removal of faecal coliforms and nutrients from water has been documented (Kansiime and Nalubega 1999; Kyambadde et al. 2004; Mburu et al. 2015; Okurut 2000). For example, measured values of nitrogen uptake for papyrus wetlands fringing Lake Victoria vary between 0.047 g N m^{-2} d^{-1} and 0.21 g N m^{-2} d^{-1} (Gaudet 1977; Kansiime and Nalubega 1999; van Dam et al. 2007), while estimated carbon uptake varies between 0.48 kg C m^{-2} y^{-1} and 1.6 kg C m^{-2} y^{-1} (Jones and Humphries 2002; Saunders et al. 2007). The nutrients and carbon are removed permanently from the wetland system when the papyrus plant is harvested (Kansiime et al. 2007).

Despite their provisioning and regulatory functions, papyrus wetlands in East Africa are not protected due to poor implementation of existing laws and are managed by neighbouring communities according to their livelihood needs. Activities in the wetlands are often unregulated, which has led to over-exploitation and permanent conversion of the wetlands for agricultural activities. Estimates of papyrus loss within the East African region range from 0.5% to 5% per annum (Kipkemboi and van Dam 2018; van Dam et al. 2014). Uganda alone had a 30% loss in total wetland area between 1994 and 2009 (Turyahabwe et al. 2013; WMD 2009).

Research on papyrus wetlands until now has mainly dealt with water purification potential, economic value of ecosystem services, papyrus biomass and reproduction, carbon storage, and fish production among others (Emerton et al. 1999; Gaudet 1977; Jones and Humphries 2002; Kansiime and Nalubega 1999; Opio et al. 2014; Saunders et al. 2013; Ssanyu et al. 2014; Terer et al. 2014). Nonetheless there is still a need to identify the key processes of the wetlands' hydrology since this would aid with designing suitable management options (Rasmussen et al. 2018; van Dam et al. 2014).

The overall study objective was to quantify the seasonal variations in the water balance of a papyrus wetland section, and the specific objectives were (i) to identify the main components of the wetland's water budget and (ii) to estimate the wetland's retention time in the dry and wet seasons. The hypothesis was that the hydrological regime including retention time is driven by flow in the main channel, with discharge fluctuations mirroring seasonal rainfall variations and highest retention time occurring in periods of low flows.

2 Materials and Methods

2.1 Description of the Study Site

This study was carried out in a section of the Naigombwa wetland, located within the Mpologoma River basin in Eastern Uganda (Fig. 2). The Mpologoma basin is approximately 8900 km^2 and is part of the Nile basin. Mpologoma River has four tributaries including Namatala, Manafwa, Malaba, and Naigombwa rivers. The rivers have vast expanses of papyrus vegetation, which forms floating mats in permanently flooded areas but is rooted at river edges. These rivers are referred to as wetlands from here on, following the hydrogeomorphic classification scheme for wetlands (Semeniuk and Semeniuk 1995; Semeniuk and Semeniuk 1997). The upstream areas of these wetlands have been extensively drained for agricultural expansion, and the natural wetland vegetation has been replaced with rice farms (Namaalwa et al. 2013). In addition, the rivers are diverted through channels that provide irrigation water for rice paddies (NEMA 2008). The mean annual rainfall and temperature from historical records (1961–1990) of stations in the Mpologoma basin is 1300 mm distributed between two rainy seasons (March to May and September to November) and 21 °C, respectively (Kigobe et al. 2014). We define the onset of the rainy season as the first of two or more consecutive days of at least 1 mm of rainfall each, whose accumulated total over the period is greater than 20 mm and less than 5 consecutive dry days in subsequent days (MacLeod 2018). The main land use in the catchment is subsistence agriculture, with crops like maize, beans, and cassava. The Manafwa and Malaba wetlands have large-scale rice irrigation schemes, but small-scale rice farms were also started immediately downstream of the irrigation schemes. Upstream areas of Namatala and Naigombwa wetlands have also been converted for small-scale rice farming.

The Naigombwa wetland is located in Iganga District and drains northwards into the Mpologoma River (Fig. 2). The Naigombwa sub-catchment is relatively flat, with elevation varying from 1056 to 1348 metres above sea level, and the soils are mainly ferralitic with reddish-brown sandy loams. The main land use is subsistence agriculture, and common crops include maize, beans, cassava, and sweat potatoes among others.

Our study section was chosen because it is demarcated by a highway and railway line which improves accessibility to the wetland (Fig. 2). In addition, major inflows and outflows in culverts of the two causeways can be quantified more easily. The wetland section has an area of 0.18 km^2 and drains an area of 734 km^2. The groundwater divide for this wetland section is assumed to coincide with the topography and is at approximately 250 m and 170 m from the western and eastern edges of the wetland, respectively. Therefore, the groundwater contributing area is less than 0.1 km^2 on both sides of the wetland (<0.02% of total catchment area). The section has one large culvert and six small ones along the highway in the southeast, which are the main flows into the wetland. A railway line crosses the downstream end in the North West and has two large culverts for outflows. The dominant vegetation in

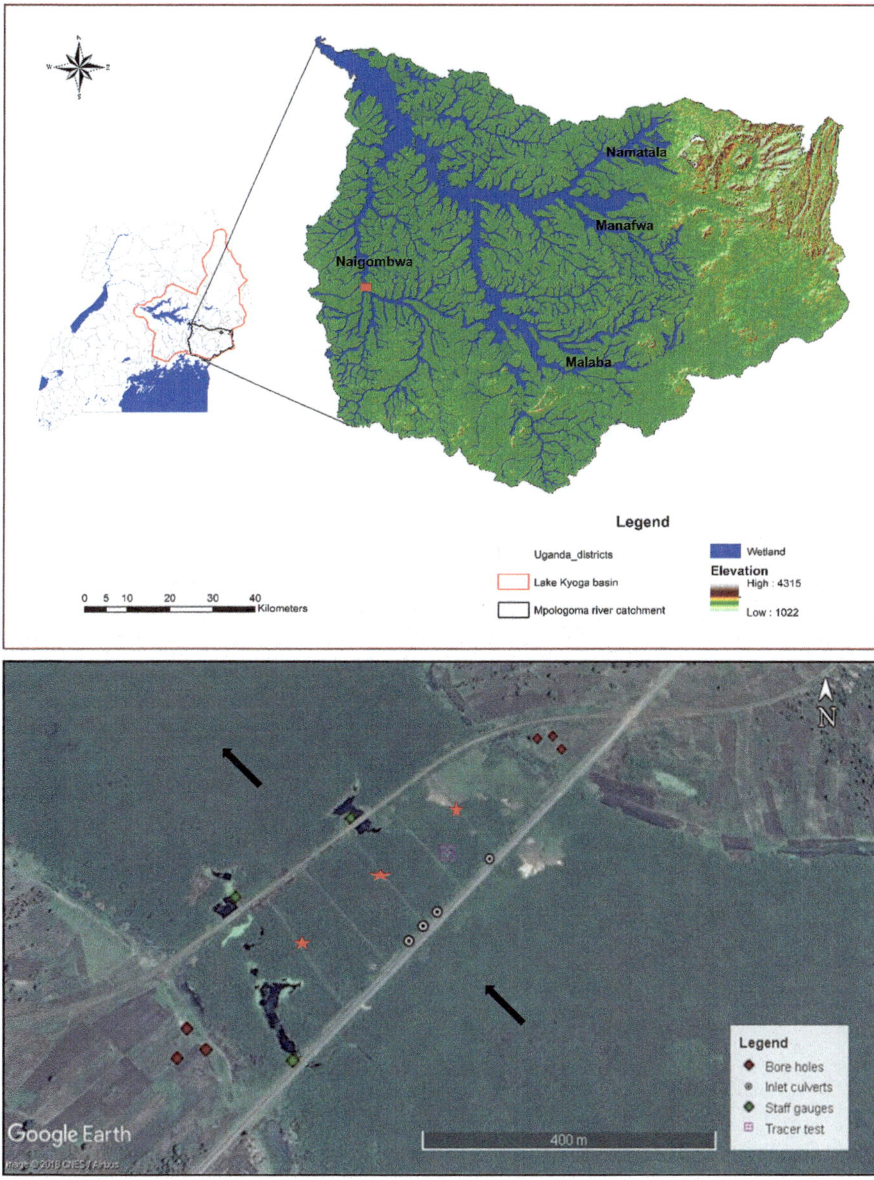

Fig. 2 (**a**) The Mpologoma River basin of Eastern Uganda. The red square shows location of the study site along the Naigombwa wetland. (**b**) The instrumented section of the wetland along Naigombwa River; bright red stars represent locations of automatic water divers

the wetland section is papyrus although palm trees and a variety of grasses are present at the edges of the wetland. Less than 1% of the total wetland section is open water, which occurs near inlets and outlets (Fig. 2).

2.2 Description of Instrumentation at Wetland Site

We installed six shallow groundwater monitoring wells, three on either side of the wetland section (Fig. 2) to monitor groundwater responses to rainfall and wetland level changes. The wells were also used to calculate the direction of groundwater flow and its gradient relative to the wetland water level. Soil characteristics at the borehole locations are illustrated in Fig. 3.

The borehole depths ranged from 1.21 to 1.81 metres below ground level (Table 1). The screen length and diameter of all wells were 1 m and 0.051 m, respectively. We estimated hydraulic conductivity by conducting both falling head (July 2015) and rising head (January 2016) slug tests. We used the Bouwer and Rice (1976) equation to calculate saturated hydraulic conductivity:

$$K_s = \frac{r_c^2 \ln\left(R_e / r_w\right)}{2L} \frac{1}{t} \ln \frac{y_0}{y_t} \tag{1}$$

where K_s = saturated hydraulic conductivity, r_c = radius of the casing, R_e = effective radius, r_w = horizontal distance from well centre to undisturbed aquifer (the width

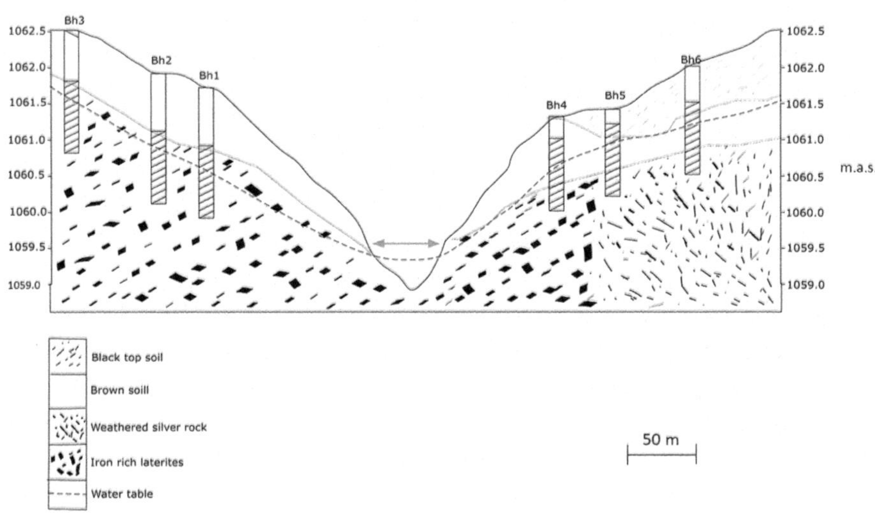

Fig. 3 Simplified vertical section through the wetland and its edges, showing soil characteristics at wetland edges where boreholes were installed. Note that the width of wetland channel (blue double arrow) is not to scale; the actual width is about 700 m

Table 1 Borehole logs and saturated hydraulic conductivity (K_S)

Location	Site	Elevation (masl[a])	Depth (mbgl[b])	Soil unit	K (m/s)	K (m/d)
Western side of wetland edge	Bh1	1061.7	1.81	Clay loam, laterite	3.4E-7	2.9E-2
	Bh2	1061.9	1.69	Clay loam, laterite	5.0E-8	4.0E-3
	Bh3	1062.5	1.73	Clay loam, laterite	1.3E-7	1.1E-2
Eastern side of wetland edge	Bh4	1061.3	1.29	Clay loam, laterite	1.2E-7	1.0E-2
	Bh5	1061.4	1.21	Black topsoil, clay loam, weathered silver rock	3.2E-6	2.8E-1
	Bh6	1062	1.53	Black topsoil, clay loam, weathered silver rock	2.3E-7	2.0E-2

[a]Ground surface elevation, metres above sea level
[b]Metres below ground level

of the backfill material was measured using a tape measure), L = height of well screen, t = time, y_0 = water level at initial time, and y_t = water level at time t.

The units of the calculated hydraulic conductivity were converted to metres per day (md^{-1}) to correspond to units of other water balance components (Table 1).

The calculated conductivities are lower than estimates by Owor et al. (2011) for boreholes at the southern shores of Lake Kyoga (0.4–15 md^{-1}). However, the low values can be attributed to the presence of material that was deposited during construction of the railway line. Similar values (<0.02 md^{-1}) were estimated for a site that is close to a railway pier on the northern shore of Lake Victoria (Owor et al. 2011). Staff gauges were installed near the main inlet culvert of the highway and outlet culverts at the railway to monitor wetland levels during the study period (Fig. 2). Water velocity measurements were carried out by staff of the Directorate of Water Resources Management (DWRM) five times over a 10-month period, spread out during periods of low and high flows. Since this method requires open water surface, measurements were done at all inlet and outlet culverts where there is a free water surface (Fig. 2).

The measurements were done using an OTT current metre except the measurements on July 2015, which were done using an acoustic Doppler current profiler (ADCP). The wetland stage was recorded on all measurement dates. In addition the width and depth of water in the culverts were measured to calculate the effective cross-sectional area. The rating equation, which provides a relationship between wetland stage and discharge for this particular section of the wetland, was derived by DWRM staff using Hydata software (Institute of Hydrology 1992), based on discharge measurements and corresponding gauge readings.

We obtained data on precipitation, air temperature, wind speed, relative humidity, and dew point temperature from a weather station (model: Davis Vantage Vue 6250) located approximately 1 km from the wetland section. The climate data was logged at 30-minute intervals. We also monitored variations in water depth at the wetland centre (Fig. 2) at 30-minute intervals using automatic divers (supplier: Geonor AS, model Micro-Diver DI6). The elevation and shape of the wetland bottom was surveyed in January 2016. Table 2 shows the duration of monitoring for the different datasets.

Table 2 Monitoring datasets at the wetland site

	Time period																			
	2015								2016											
Datasets	5	6	7	8	9	10	11	12	1	2	3	4	5	6	7	8	9	10	11	12
Climate			x	x	x	x	x	x	x	x	x	x	x	x	X	x	x	x	x	x
Wetland stage	x	x	x	x	x	x	x	x	x	x	x	x	x	x	X	x	x	x	x	x
Groundwater l.		x	x	x	x	x	x	x	x	x	x	x	x	x	X	x	x	x	x	x
Discharge[a]	x		x		x			x		x										
Survey of wetland bathymetry									x											
Water depth									x	x	x	x	x	x	X	x	x	x	x	x

[a]No discharge measurements done after February 2016 because of construction works on culverts

2.3 *Estimation of Water Balance Components*

Conceptual Model of Water Balance for the Wetland Section

We calculated a monthly water balance for the wetland section (0.18 km^2) by considering major flows in and out of the section (Eq. 2). The conceptual water balance for the wetland section is illustrated in Fig. 4.

$$ds / dt = Q_{in} + P + Qg_{in} + Q_f - Q_{out} - ET - e\left(\text{all terms in m}^3 / \text{day}\right) \qquad (2)$$

where ds/dt = change in storage, Q_{in} = channel inflow, P = precipitation, Qg_{in} = groundwater discharge, Q_f = overland flow from edges of wetland section, Q_{out} = channel outflow, ET = evapotranspiration, and e = residual error.

Daily precipitation was recorded at the weather station. We assumed that overland flow from edges of wetland section (Q_f) was negligible due to the small size of the wetland section; therefore it was excluded from the calculations. Surface flows from the upstream catchment area are included in channel inflow calculations (Q_{in}). The water balance was calculated for a period of 8 months, from July 2015 to February 2016 because there were no discharge measurements beyond February 2016 (Table 2).

Groundwater Gradient and Direction

Groundwater discharge Qg_{in} [m^{3}d^{-1}] from the wetland edges was calculated using Darcy's law as shown in Eq. 3:

$$Qg_{in} = -K_s A \frac{dh}{dl} \qquad (3)$$

where Qg_{in} = groundwater discharge [m^{3}d^{-1}], K_s = saturated hydraulic conductivity [md^{-1}], A = cross-sectional area [m^2], and dh/dl = hydraulic gradient.

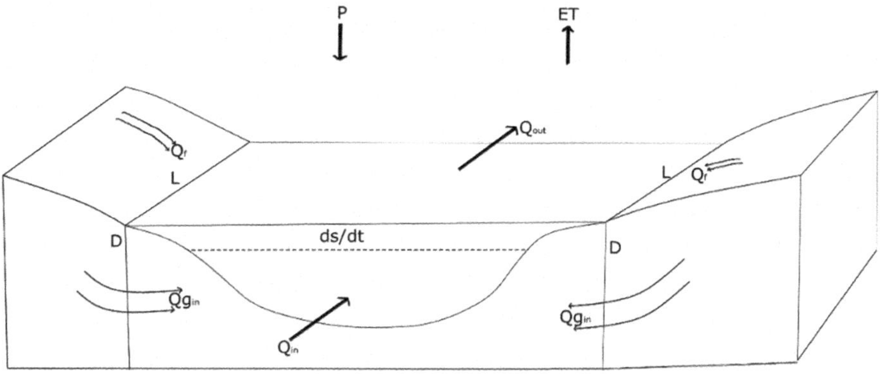

Fig. 4 Vertical section through the wetland section perpendicular to the main flow direction, illustrating the water balance components of the wetland section

The saturated hydraulic conductivities (K_s) for individual boreholes (Table 1) were averaged on each side of the wetland, which gave an average value of 0.015md^{-1} and 0.104md^{-1} for the west and east side of the wetland, respectively. The cross-sectional area (A [m^2]) perpendicular to the direction of flow was calculated by multiplying the length of wetland edge (L = 258 m and 174 m on the west and east sides, respectively) with an estimated depth (D = 9 m) of the unconfined aquifer (Fig. 4). The depth of the unconfined aquifer was estimated from local borehole logs, which show that the average depth to bedrock is 9 m below the ground surface. We calculated daily values of the magnitude and direction of hydraulic gradient [dh/dl] from the groundwater time series data using the 3-point estimation tool (3PE) as described by Beljin et al. (2014).

Estimation of Flow in the Main Channel

Wetland channel inflow (Q_{in}) and outflow (Q_{out}) through the culverts were estimated using the rating equations derived from discharge measurements and corresponding gauge measurements (eqs. 4 and 5, respectively).

$$Q_{in} = 6.03\left(H_1 + 0.096\right)^{2.8} \tag{4}$$

$$Q_{out} = 8.75\left(H_2 - 0.648\right)^{2.8} \tag{5}$$

where Q_{in} = channel inflow (m^3/d), Q_{out} = channel outflow (m^3/d), H_1 = stage at inlet (m), and H_2 = stage at outlet (m) (average of two outlets).

We estimated flow in the causeways (highway and railway) by using Darcy's law and assuming saturated hydraulic conductivity and hydraulic gradients in the same order of magnitude as those observed at wetland edges. The contributing area was calculated from the length of the causeway multiplied by the height of the water level near the causeways. Based on these assumptions, this flow was much smaller than groundwater flow from the wetland edges (<0.1%). We therefore assumed, for the water balance calculations, that all channel flow into and out of the wetland section was directed through the culverts.

Estimation of Evapotranspiration

We estimated potential evapotranspiration over the wetland section using a modification of the FAO 56 Penman-Monteith equation, which calculates a reference evapotranspiration for alfalfa (Zotarelli et al. 2010):

$$ET_{sz} = \frac{0.408\Delta\left(R_n - G\right) + \gamma \frac{C_n}{T+273} u_2 \left(e_s - e_a\right)}{\Delta + \gamma\left(1 + C_d u_2\right)} \tag{6}$$

where ET_{sz} = potential evapotranspiration rate (mm d^{-1}), Δ = slope of the saturated vapour pressure curve [δes/ δT, where es = saturated vapour pressure (kPa) and Tmean = daily mean temperature (°C)], R_n = net radiation flux (MJ m^{-2} d^{-1}), G = sensible heat flux into the soil (MJ $m^{-2}d^{-1}$), γ = psychrometric constant (kPa °C^{-1}), C_n = the numerator constant for the reference crop type (alfalfa) at 24-hour time step (1600), C_d = denominator constant for the reference crop type (alfalfa) at 24-hour time step (0.38), T = mean daily air temperature (°C), u_2 = wind speed (m s^{-1}), e_s = saturation vapour pressure (kPa), e_a = actual vapour pressure (kPa), and the constant 0.4808 is a conversion factor for net radiation to mm/day.

The soil heat flux (G) is considered negligible in some heat balance studies because it is the smallest component of all the terms and also because the most incoming energy into the soil is eventually lost to the atmosphere in the night as long-wave radiation (Anadranistakis et al. 1997; Sauer and Horton 2005). In our study section, water is above ground level at all times, and its depth varies between 1 and 2 metres; therefore, we used a value of zero for the soil heat flux because we assumed soil heating mechanisms to have minimal influence on the evapotranspiration rate.

Independent Estimates of Wetland Storage

We surveyed the wetland's bathymetry to obtain a digital elevation model (DEM) and used the acquired DEM to obtain a depth-volume relationship using ArcMap 3D analyst tools. The details of the methods used are described in Kayendeke et al. (2018). The goal of obtaining the depth-volume relationship was that wetland water volume could be estimated from the wetland depth data recorded at the wetland site. However, we did not have depth data from July to December 2015, since depth monitoring did not start until January 2016 (Table 2). Since wetland stage data spans the entire monitoring period, we estimated water depth for July–December 2015 using the relationship between water depth and wetland stage. We then calculated daily nominal residence times for the wetland using a ratio of volume to channel discharge (Kadlec and Knight 1996; Zahraeifard and Deng 2011), to evaluate the magnitude and variation of retention time over the 8-month period.

2.4 *Wetland Flow Dynamics and Velocity*

We conducted salt tracer experiments to get information on micro-scale flow dynamics within the study section using a portable electrical conductivity (EC) metre (Orion Star A329). The experiments were done in the dry season on the 17th and 18th of January 2017, at the main inlet culvert and in an undisturbed papyrus area (Fig. 2b), respectively. At the inlet culvert, we recorded baseline electrical conductivity (EC) on both sides of the highway (260 µs/cm). The tracer was prepared by adding 500 g of salt (NaCl) to 20 litres of water taken from the same wetland and stirred until all the salt was completely dissolved. The EC of the prepared water was 1050 µs/cm, and this water was then released on the upstream end of the highway culvert. The EC measurements were recorded at the point of release as well as at the downstream side of the highway culvert at 1-minute intervals (Fig. 5).

For the papyrus plot, we made 12 holes through the papyrus root mat. The holes were 0.75 m apart except the diagonals which were 1.5 m apart (Fig. 5). We choose to have minimal distance between the holes since we had observed very low velocity at the inlet culvert the previous day. An additional objective of the tracer experiment was to understand vertical flow dynamics between the papyrus root mat and the water column beneath it. In addition, we were interested in velocity differences between the mat and water column. We recorded the baseline EC values at all points. We then prepared the tracer by adding 100 g of salt to 10 litres of water and released it at the centre point (0) where the released tracer solution entered into the root mat as well as the free water column beneath it. We proceeded to measure EC values at all points in a clockwise spiral shape (from 0 to 12) until EC values returned to the recorded baseline values. Individual EC measurements were taken both in the water column beneath the papyrus root mat and within the root mat.

3 Results and Discussion

3.1 *Precipitation Patterns*

The rainfall dataset begins just before the beginning of the second rain season (September to December) of 2015. The frequency of successive rain days increased in mid-August to the end of December, and the total rainfall amount for that season

Fig. 5 Top: EC measurements at the inlet culvert. Bottom: set up of experimental plot in the papyrus zone

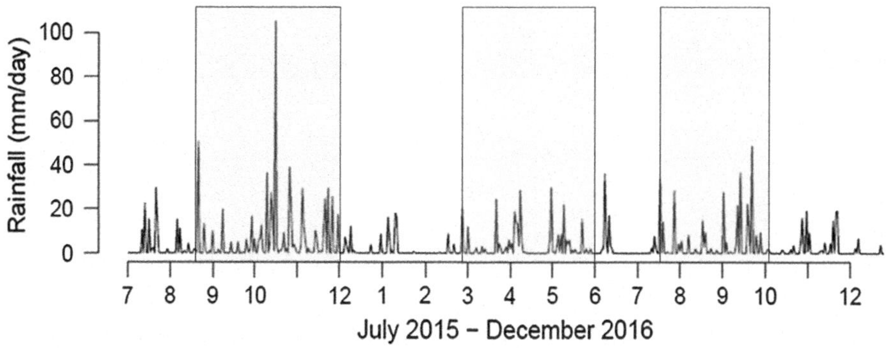

Fig. 6 Observed precipitation during the monitoring period (18 months) with rain seasons in grey

was 555 mm (Fig. 6). The rainfall seasons in 2016 had lower frequency and quantity of rainfall, with the first (March to June) and second (August to October) seasons having a cumulative rainfall of 297 mm and 290 mm, respectively.

3.2 *Groundwater Gradient and Direction*

We observed a negative groundwater gradient towards the wetland on either side of the wetland edge throughout the monitoring period, indicating groundwater discharge into the wetland. The average hydraulic gradients were −0.0074 and −0.0043 on the western and eastern sides of the wetland, respectively (Fig. 7). The gradient on the western side of the wetland was in south-eastern direction towards borehole 1, whereas on the eastern side, the gradient was in the south-western direction towards borehole 5. Despite the angular direction of flow of groundwater towards the wetland, the discharge is calculated assuming a groundwater flow perpendicular to both the wetland edge and to the flow direction in the main wetland channel. The groundwater discharge pattern towards the wetland is most likely controlled by heterogeneities in the hydrogeological properties of the surrounding slopes, but it was neither the scope nor the resources of this study to map this spatial variability in groundwater flow pattern. It is however interesting to note that the direction is fairly constant in time on both sides of the wetland.

The groundwater gradient had a significant correlation with both local rainfall and wetland stage although there was a higher correlation to wetland stage (0.67, $p = 000$) than to local rainfall (0.22, $p = 000$) (Fig. 8). High rainfall amounts led to an immediate response in the hydraulic gradient. In addition, the gradient increased in periods of cumulative rainfall but declined in periods with longer dry spells (Fig. 8). On the other hand, there was a 2-week lag between hydraulic gradient and wetland water level (Fig. 8).

The methods used give an idea of the groundwater gradient at the wetland site. However, because of the small number of boreholes used, the groundwater flow

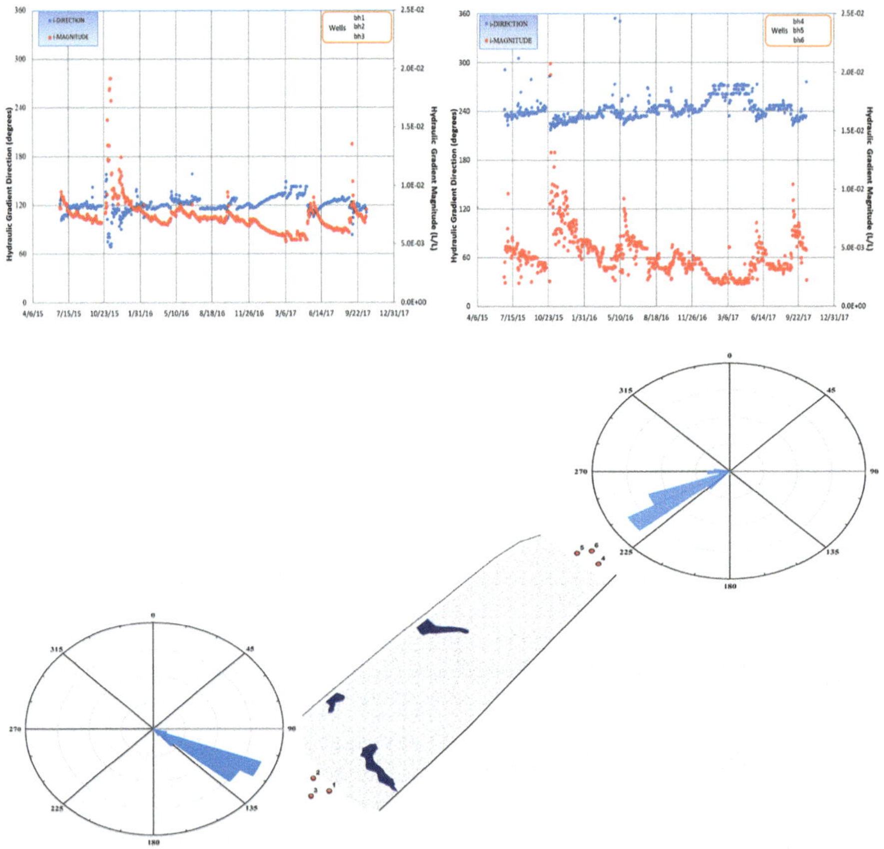

Fig. 7 Variation in direction and magnitude of hydraulic gradient on the west and eastern side of wetland edge (top); dominant direction of hydraulic gradient during the monitoring period (bottom). Note that direction is measured in degrees from north clockwise

patterns at a larger scale are still undefined. Since surface and groundwater interaction are complex, further insight into the dynamics of the system requires an extensive monitoring network (Krasnostein and Oldham 2004; Trask et al. 2017). Additional monitoring networks (e.g. piezometer nests) within the study area should give a better understanding of vertical groundwater gradients at the wetland edge as well as subsurface hydrogeological properties.

3.3 *Wetland Water Balance*

The calculated water balance (Table 3) shows that the wetland is dominated by wetland channel flow contributing approximately 99.7% of the total inputs into the wetland, while direct precipitation and groundwater discharge contributed

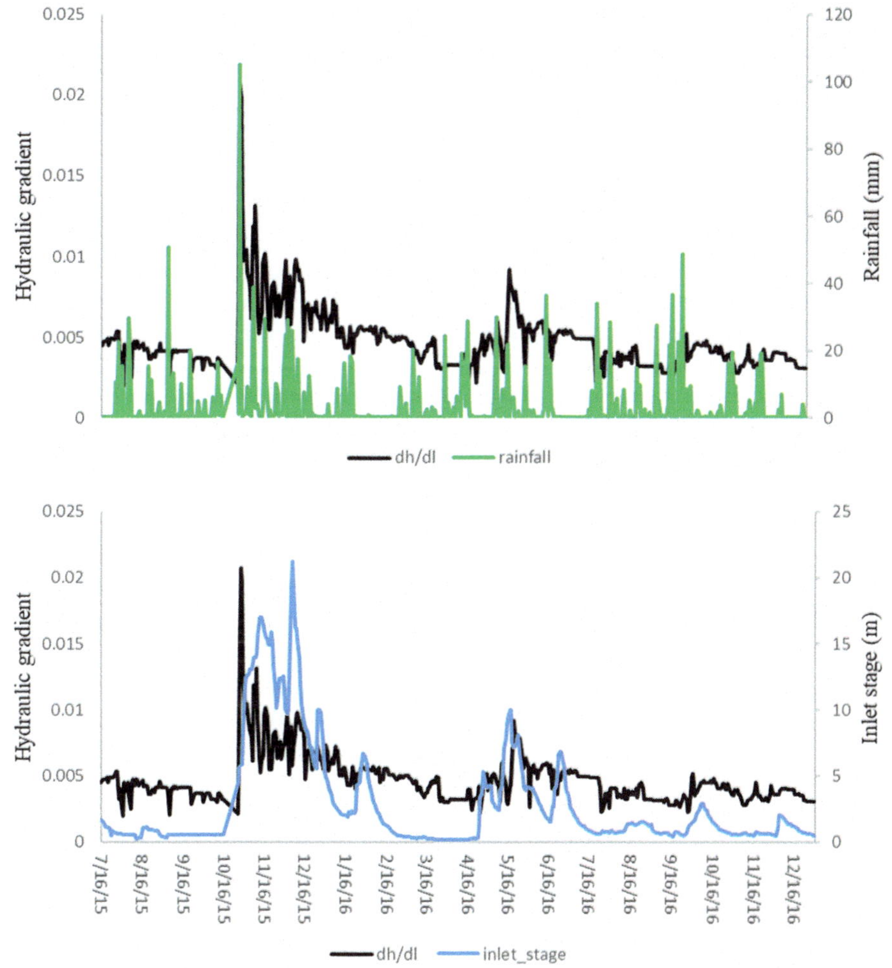

Fig. 8 Hydraulic gradient (black lines) on eastern edge of wetland, plotted together with rainfall (top) and wetland stage (bottom)

Table 3 Monthly water balance of Naigombwa wetland section (m^3)

	Q_{in}	R	Q_G	Q_{out}	ET	dS/dt
Jul-15	169,529	554	1.0	170,674	781	−1371
Aug-15	53,597	481	0.9	62,032	855	−8808
Sep-15	39,960	635	0.9	42,364	893	−2661
Oct-15	127,290	1503	1.5	131,313	923	−3442
Nov-15	1,195,913	950	1.5	1,240,389	886	−44,410
Dec-15	917,534	804	1.5	825,320	865	92,155
Jan-16	293,435	423	1.2	254,652	877	38,329
Feb-16	135,422	65	1.1	140,176	940	−5627
8-month average	372,824	689	1.2	364,360	883	2766

approximately 0.2% and <0.1%, respectively. Estimates of total potential evapotranspiration were 30% higher than total precipitation during the study period. Between July and November 2015, outflows exceeded inflows, and there was a net loss in storage during this time. The months of December 2015 and January 2016 had higher inflows than outflows, with a net increase in storage; but the flow reversed again in February 2016. The water balance average over the 8-month period shows that there was a net increase of storage of approximately 2700 m^3.

We observed an increase in precipitation from September to December 2015. However, channel flows did not start increasing until October 2015, which implies a 1-month lag between rainfall and channel flows. The lag could be due to storage effects within the catchment, where infiltrated water is discharged into the river after rainfall has ceased (Tomasella et al. 2008). We hypothesised that channel flows through the wetland mirror rainfall variations, this is partially true, but results also show a 1-month lag between rainfall and channel flows.

Since channel flow dominates the water balance of the wetland section, this illustrates that the hydrological regime is vulnerable to land use changes in the upstream catchment because these influence the magnitude and timing of flows into the wetland (Krasnostein and Oldham 2004). Land use changes can play an important role because they influence infiltration rates, wetland water levels, and flow dynamics within the catchment. For example, the increasing extent of rice farms in the Mpologoma River catchment reduces infiltration rate especially during the growing season. This is because the fields are maintained with standing water, yet areas with saturated soils have little additional storage capacity (Acreman and Holden 2013). Such areas are therefore expected to generate saturation excess overland flow more quickly compared to soils that are not saturated. Therefore, having a large expanse of rice paddies could increase the risk of flash floods in rainy seasons.

The water balance estimates give an indication of the key processes of the wetland's hydrological regime. Since the monitored water dynamics cover a short time period, continued monitoring is important to get further insight into annual variations in hydrology. Although water balance estimates were affected by the small number of discharge measurements, the stage-discharge relationship can be updated as more data is collected. Another issue of concern is that the method we applied to estimate discharge was developed for use with open water channels (Gore and Banning 2017) and could only be used at the inlet and outlet culverts. Therefore, the effect of water spreading into the remaining wetland area and the drag on water flow from the floating vegetation may not be correctly reflected in the calculations.

3.4 Independent Estimates of Wetland Storage

There was a good correlation between wetland stage measured at the culvert and water depth at the centre of the wetland in the dry season ($r = 0.97$, $R^2 = 0.93$), but the correlation in the wet season was lower (0.61, $R^2 = 0.37$) as shown in Fig. 9. Therefore, the overall correlation and coefficient of determination for the two

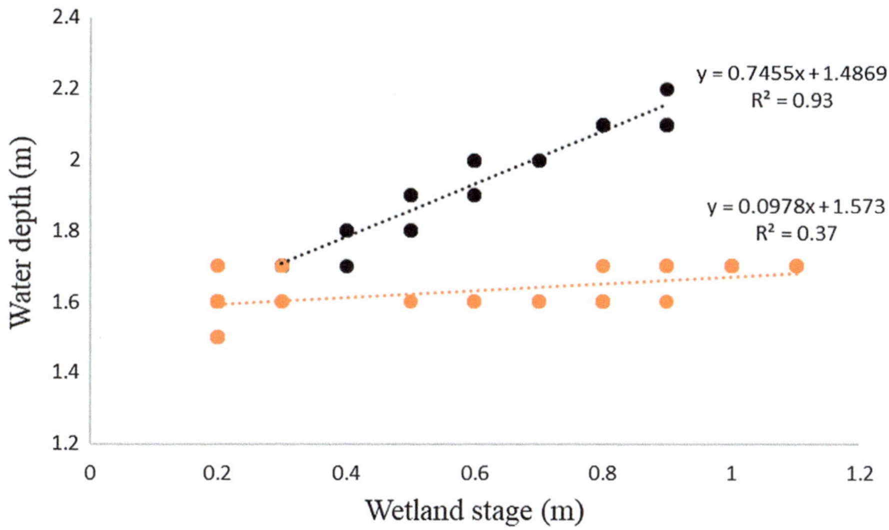

Fig. 9 Relationship between wetland stage and water depth at the wetland centre. The upper and lower limbs of the data represent correlations in the dry and wet seasons, respectively

variables were low (r = 0.49, R^2 = 0.24). The poor relationship between the two variables in the wet season could be due to delay in water level rise between the culvert and the wetland centre that is dominated by floating papyrus. This is because water velocity at the inlet and outlet culverts increases in the wet season, which creates fast flow between the inlet and outlet culvert. On the other hand, velocity at the wetland centre is slowed down due to the presence of floating vegetation, which reduces the correlation between the variables. Because of the low correlation in the wet season, the accuracy of corresponding volume estimates was very low. Consequently, change in storage for the water balance was estimated as part of the residual term as shown in Table 3.

Despite the high uncertainty in the water depth estimates, we utilised them to get crude daily estimates of water volume and calculated retention time using the volume-discharge ratio. Based on these estimates, the shortest and longest retention times were 2 hours and 7 days, observed between November to December 2015 and February to April 2016, respectively.

Since the water balance estimates at monthly time steps may smooth out some of the hydrodynamic interactions between precipitation and discharge, we analysed the cumulative precipitation, discharge, and retention time on a daily basis (Fig. 10), to explore their relationship in more detail. In Fig. 10 we see that the peaks in discharge correspond to periods with a high accumulation of precipitation. During the same period, the retention time (the bottom part of Fig. 10) is only a few hours, while it increases to around 1 week in periods of low flow. Retention time is highest between February and April 2016 during low flows.

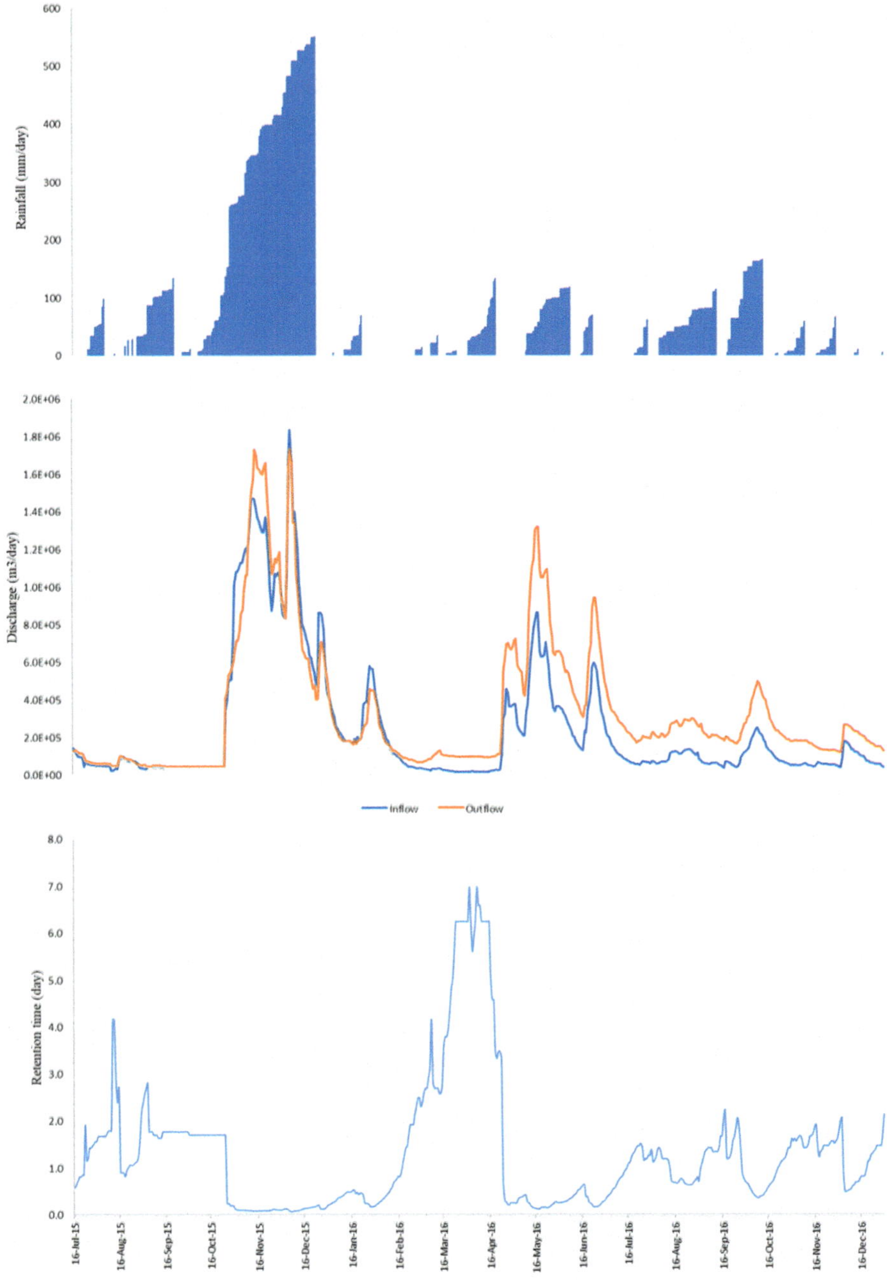

Fig. 10 Top, accumulated precipitation separated by dry spells (defined as days with less than 1 mm/day and more than 5 successive days of no rain); middle, discharge in and out of wetland section; bottom, estimated retention time for wetland section. Note that the time period extends beyond the water balance period in Table 3

The observed retention times indicate that the flow to volume ratio is very high, which is because of the small size of the wetland section relative to its catchment area. It's important to note the flows are modified by the position of the culverts. This is another indication of preferential flow and higher velocities from the inlet culvert at the highway to the downstream culverts at the railway, while other parts of the wetland with floating papyrus vegetation have lower velocities. The estimated retention time is therefore not representative of the wetland as a whole.

3.5 *Water Flow Patterns*

There was a low velocity at the inlet culvert during the tracer experiment; therefore EC at the downstream side of the highway increased only slightly above baseline values (Fig. 11). We used the time (6 hours, 36 minutes) at which the highest EC value was recorded at the downstream end to estimate water velocity through the culvert underneath the highway. The width (travel distance) of the road is 13 m, which gives a velocity of 5.47×10^{-4} m/s.

If the same velocity is assumed throughout the wetland section, and a travel distance of 290 m between the inlet and outlet culvert, this gives an estimated retention time of 6 days. This estimate is comparable to the highest observed retention time of 7 days that was estimated using the volume-discharge ratio. However, as mentioned in the previous section, this retention time represents water flow between the inlet and outlet culvert and not the flow in other areas of the wetland. We therefore estimated flows in other wetland areas using tracer measurements in the papyrus plot.

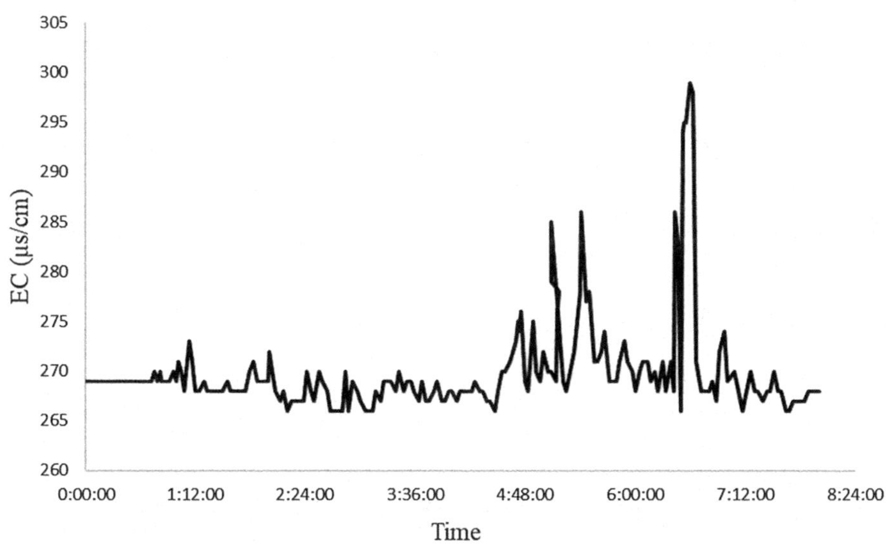

Fig. 11 EC variation with time at outlet of highway culvert

For the papyrus plot, we used contour plots to visualise changes in EC around the centre point (where the tracer was released) at specific time intervals (Fig. 12). There was some increase in EC at points 10 and 11 that are west and north of the centre point, respectively. The increases were observed both within the root mat (Fig. 12) and in the water column beneath it (Fig. 12b). However, EC did not increase much beyond these two points which indicates that there was little movement of water. Since EC was monitored manually by moving from one point to another on top of the floating papyrus mat, the effect of walking on the root mat might have caused the observed dispersion pattern of the tracer from the centre point. We propose that a larger water velocity would have showed a more skewed pattern of the tracer movement. We therefore conclude that there is low velocity in this part of the wetland.

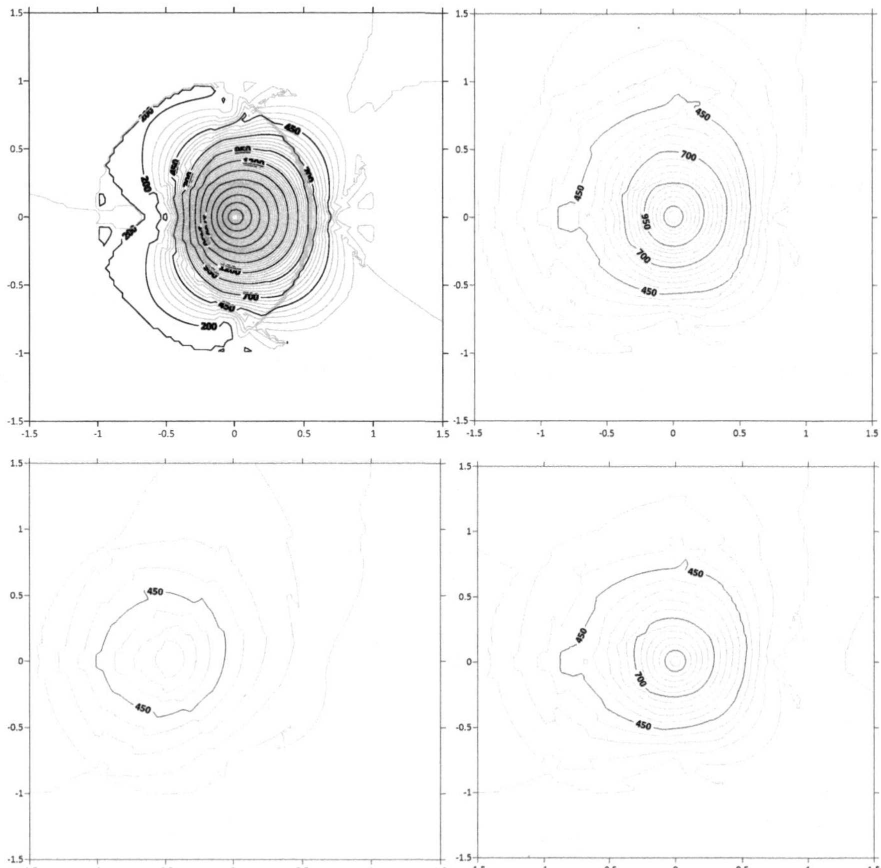

Fig. 12 (**a**) EC (μs/cm) distribution within the papyrus root mat. The contour plots from top left in clockwise direction represent time intervals at 0–20 minutes, 48–1:04 hours, 1:36–2:02 hours, and 2:28–2:40 hours, respectively. (**b**) EC (μs/cm) distribution below the root mat. The contour plots from top left in clockwise direction represent time intervals at 0–20 minutes, 48–1:04 hours, 1:36–2:02 hours, and 2:28–2:40 hours, respectively

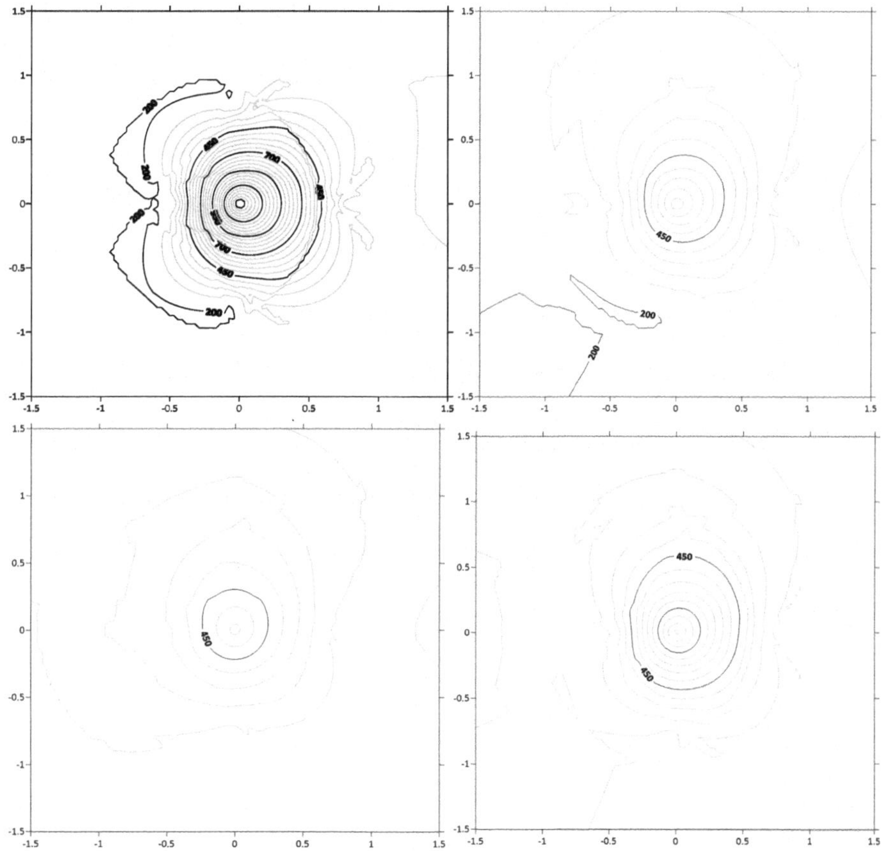

Fig. 12 (continued)

4 Conclusion

In this study, we characterised the water balance components of a section of a papyrus wetland and identified flow in the main channel as the dominating factor, contributing 99.7% of total inputs. The wetland's retention time between the inlet and outlet culverts varied between 2 hours and 7 days during periods of high and low flows, respectively. Further, hydraulic gradients on either side of the wetland indicate groundwater flow towards the wetland throughout the monitoring period. The observed dynamics of the groundwater flow towards the wetland implies that the wetland plays a major role as a boundary condition for the local groundwater system and that draining the wetland would impact groundwater levels in the areas along the wetland edges in the long run. Although the water balance approach has its limitations, it gives an initial understanding of the key processes in the hydrological regime of a small wetland section.

For future research, we propose that more comprehensive water flow measurements through areas of floating papyrus vegetation can be obtained by using automatic sensors to reduce impact of movement during manual measurements on the experimental results. In addition, installation of piezometer nests (for better vertical characterisation) and monitoring wells further away from the wetland edges will lead to an improved characterisation of hydrogeological properties and groundwater flow dynamics. We also recommend further research using catchment-based models to assess possible impacts of land use changes within the upstream catchment on channel flows.

Acknowledgements This work was funded by the 'Norwegian Programme for Capacity Development in Higher Education and Research for Development (NORHED)'; Grant number: 3303010034. The funding is through the 'Regional Capacity Building for Sustainable Natural Resource Management and Agricultural Improvement under Climate Change (CAPSNAC)' Project housed at Makerere University, Uganda. We appreciate Dr. Michael Owor and Dr. Philip Njenje for advice and assistance in installation of monitoring boreholes. We also thank Mr. Bright Twesigye and Mr. Kukutu Solomon Collin for their assistance in the data collection exercise in and around the Naigombwa wetland. We thank Mr. Frank Kigozi for help with wetland gauge installation and for carrying out the discharge measurements. Finally, Mr. Kisitu Francis is greatly appreciated for providing information about drinking water wells in the areas surrounding the wetland section.

References

Acreman M, Holden J (2013) How wetlands affect floods. Wetlands 33:773–786. https://doi.org/10.1007/s13157-013-0473-2

Anadranistakis M, Liakatas A, Alexandris S, Aggelides S, Kerkides P, Rizos S, Poulovassilis A (1997) Soil Heat Flux in the Penman-Monteith Evapotranspiration Equation. Paper presented at the ISHS Acta Horticulturae: II International Symposium on Irrigation of Horticultural Crops, Chania, Crete, Greece

Azza NGT, Kansiime F, Nalubega M, Denny P (2000) Differential Permeability of Papyrus and Miscanthidium Root Mats in Nakivubo Swamp, Uganda. Aquat Bot 67:169–178. https://doi.org/10.1016/S0304-3770(00)00093-0

Beljin M, Ross RR, Acree SD (2014) 3pe: a tool for estimating groundwater flow vectors. (EPA/600/R-14/273, 2014). U.S. Environmental Protection Agency, Washington, DC

Bouwer H, Rice RC (1976) A slug test for determining hydraulic conductivity of unconfined aquifers with completely or partially penetrating wells. Water Resour Res 12(3):423–428

Donaldson L, Woodhead AJ, Wilson RJ, Maclean IM (2016) Subsistence use of papyrus is compatible with wetland bird conservation. Biol Conserv 201:414–422. https://doi.org/10.1016/j.biocon.2016.07.036

Emerton L, Iyango L, Luwum P, Malinga A (1999) The Present Economic Value of Nakivubo Urban Wetland, Uganda. IUCN, Nairobi

Gaudet JJ (1977) Natural Drawdown on Lake Naivasha, Kenya and the Formation of Papyrus Swamps. Aquat Bot 3:1–47. https://doi.org/10.1016/0304-3770(77)90002-X

Gore JA, Banning J (2017) Chapter 3 - Discharge measurements and streamflow analysis. In: Hauer FR, Lamberti GA (eds) Methods in stream ecology, vol 1, 3rd edn. Academic Press, Boston, pp 49–70. https://doi.org/10.1016/B978-0-12-416558-8.00003-2

Institute of Hydrology (1992) Hydata operation manual. Hydrological data processing system, version 3.1. Institute of Hydrology, Wallingford

Jones M, Humphries S (2002) Impacts of the C4 Sedge *Cyperus papyrus* L. On Carbon and Water Fluxes in an African Wetland. Hydrobiologia 488:107–113. https://doi.org/10.1007/978-94-017-2031-1_10

Kadlec RH, Knight R (1996) Treatment Wetlands. CRC Baca, Raton, FL

Kansiime F, Nalubega M (1999) Wastewater treatment by a natural wetland: the Nakivubo Swamp, Uganda. Processes and Implications. (PhD Thesis), Wageningen Agricultural University, Netherlands

Kansiime F, Sanders MJ, Loisella SA (2007) Functioning and dynamics of wetland vegetation of Lake Victoria: an overview. Wetl Ecol Manag 15:443–451. https://doi.org/10.1007/s11273-007-9043-9

Kayendeke EJ, Kansiime F, French HK, Bamutaze Y (2018) Spatial and temporal variation of papyrus root mat thickness and water storage in a tropical wetland system. Sci Total Environ 642:925–936. https://doi.org/10.1016/j.scitotenv.2018.06.087

Kigobe M, Wheater H, McIntyre N (2014) Statistical Downscaling of Precipitation in the Upper Nile: use of Generalized Linear Models (Glms) for the Kyoga Basin. In: Melesse AM, Abtew W, Setegn SG (eds) Nile River Basin: ecohydrological challenges, climate change and hydropolitics. Springer, Cham, pp 421–449. https://doi.org/10.1007/978-3-319-02720-3_22

Kipkemboi J, van Dam AA (2018) Papyrus wetlands. In: Finlayson CM, Milton GR, Prentice RC, Davidson NC (eds) The wetland book: II: distribution, description, and conservation. Springer Netherlands, Dordrecht, pp 183–197. https://doi.org/10.1007/978-94-007-4001-3_218

Krasnostein A, Oldham C (2004) Predicting wetland water storage. Water Resour Res 40:W10203

Kyambadde J, Kansiime F, Gumaelius L, Dalhammar G (2004) A comparative study of *cyperus papyrus* and *miscanthidium violaceum*-based constructed wetlands for wastewater treatment in a tropical climate. Water Res 38:475–485. https://doi.org/10.1016/j.watres.2003.10.008

MacLeod D (2018) Seasonal predictability of onset and cessation of the East African Rains. Weather Clim Extrem 21:27. https://doi.org/10.1016/j.wace.2018.05.003

Mburu N, Rousseau DPL, van Bruggen JJA, Lens PNL (2015) Use of the macrophyte *cyperus papyrus* in wastewater treatment. In: Vymazal J (ed) The role of natural and constructed wetlands in nutrient cycling and retention on the landscape. Springer, Cham, pp 293–314. https://doi.org/10.1007/978-3-319-08177-9_20

Morrison EH, Upton C, Pacini N, Odhiambo-K'oyooh K, Harper DM (2013) Public perceptions of papyrus: community appraisal of wetland ecosystem services at Lake Naivasha, Kenya. Ecohydrol Hydrobiol 13:135–147. https://doi.org/10.1016/j.ecohyd.2013.03.008

Namaalwa S, Van dam AA, Funk A, Ajie GS, Kaggwa RC (2013) A characterization of the drivers, pressures, ecosystem functions and services of Namatala Wetland, Uganda. Environ Sci Pol 34:44–57. https://doi.org/10.1016/j.envsci.2013.01.002

NEMA (2008) Pilot integrated environment assessment of the Lake Kyoga Catchment Area. National Environment Management Authority (NEMA), Kampala

Okurut TO (2000) A pilot study on municipal wastewater treatment using a constructed wetland in Uganda. PhD Thesis, Wageningen University Netherlands

Opio A, Jones M, Kansiime F, Otiti T (2014) Growth and development of *cyperus papyrus* in a tropical wetland. Open J Ecol 4:113–123. https://doi.org/10.4236/oje.2014.43013

Owor M, Taylor R, Mukwaya C, Tindimugaya C (2011) Groundwater/surface-water interactions on deeply weathered surfaces of low relief: evidence from Lakes Victoria and Kyoga, Uganda. Hydrogeol J 19:1403–1420. https://doi.org/10.1007/s10040-011-0779-1

Rasmussen TC, Deemy JB, Long SL (2018) Wetland hydrology. In: Finlayson CM, Everard M, Irvine K, McInnes RJ, Middleton BA, van Dam AA, Davidson NC (eds) The wetland book: I: structure and function, management, and methods. Springer Netherlands, Dordrecht, pp 201–216. https://doi.org/10.1007/978-90-481-9659-3_71

Sauer TJ, Horton R (2005) Soil heat flux. In: Hatfield JL, Baker JM (eds) Micrometeorology in agricultural systems. Asa monograph, vol 47. American Society of Agronomy, Crop Science Society of America, and Soil Science Society of America, Madison, WI, pp 131–154. https://doi.org/10.2134/agronmonogr47.c7

Saunders MJ, Jones MB, Kansiime F (2007) Carbon and water cycles in tropical papyrus wetlands. Wetl Ecol Manag 15:489–498. https://doi.org/10.1007/s11273-007-9051-9

Saunders MJ, Kansiime F, Jones MB (2013) Reviewing the carbon cycle dynamics and carbon sequestration potential of *cyperus papyrus* L. wetlands in Tropical Africa. Wetl Ecol Manag 21:143. https://doi.org/10.1007/s11273-013-9314-6

Semeniuk CA, Semeniuk V (1995) A geomorphic approach to global classification for inland wetlands. Vegetatio 118:103–124. https://doi.org/10.1007/bf00045193

Semeniuk V, Semeniuk CA (1997) A geomorphic approach to global classification for natural inland wetlands and rationalization of the system used by the Ramsar Convention – A discussion. Wetl Ecol Manag 5:145–158. https://doi.org/10.1023/a:1008207726826

Ssanyu GA, Kipkemboi J, Mathooko JM, Balirwa J (2014) Land-use impacts on small-scale mpologoma wetland fishery, Eastern Uganda: a socio-economic perspective. Lakes Reserv Res Manag 19:280–292. https://doi.org/10.1111/lre.12073

Terer T, Muasya AM, Higgins S, Gaudet JJ, Triest L (2014) Importance of seedling recruitment for regeneration and maintaining genetic diversity of *cyperus papyrus* during drawdown in Lake Naivasha, Kenya. Aquat Bot 116:93–102. https://doi.org/10.1016/j.aquabot.2014.02.008

Terer T, Triest L, Muasya AM (2012) Effects of harvesting *cyperus papyrus* in undisturbed Wetland, Lake Naivasha, Kenya. Hydrobiologia 680:135–148. https://doi.org/10.1007/s10750-011-0910-2

Tomasella J, Hodnett MG, Cuartas LA, Nobre AD, Waterloo MJ, Oliveira SM (2008) The water balance of an amazonian micro-catchment: the effect of interannual variability of rainfall on hydrological behaviour. Hydrol Process 22:2133–2147. https://doi.org/10.1002/hyp.6813

Trask JC, Fogg GE, Puente CE (2017) Resolving hydrologic water balances through a novel error analysis approach, with application to the Tahoe Basin. J Hydrol 546:326–340. https://doi.org/10.1016/j.jhydrol.2016.12.029

Turyahabwe N, Tumusiime DM, Kakuru W, Barasa B (2013) Wetland use/cover changes and local perceptions in Uganda. Sustain Agric Res 2:95. https://doi.org/10.5539/sar.v2n4p95

van Dam AA, Dardona A, Kelderman P, Kansiime F (2007) A simulation model for nitrogen retention in a papyrus wetland near Lake Victoria, Uganda (East Africa). Wetl Ecol Manag 15:469–480. https://doi.org/10.1007/s11273-007-9047-5

van Dam AA, Kipkemboi J, Mazvimavi D, Irvine K (2014) A synthesis of past, current and future research for protection and management of papyrus (*cyperus papyrus* L.) wetlands in Africa. Wetl Ecol Manag 22:99–114. https://doi.org/10.1007/s11273-013-9335-1

WMD (2009) Mapping a better future: how spatial analysis can benefit wetlands and reduce poverty in Uganda. Wetlands Management Department, Ministry of Water and Environment

Zahraeifard V, Deng Z (2011) Hydraulic residence time computation for constructed wetland design. Ecol Eng 37:2087–2091. https://doi.org/10.1016/j.ecoleng.2011.08.011

Zotarelli L, Dukes MD, Romero CC, Migliaccio KW, Morgan KT (2010). Step by step calculation of the Penman-Monteith evapotranspiration (Fao-56 method). Institute of Food and Agricultural Sciences. University of Florida

Ecological Sustainability: Miombo Woodland Conservation with Livestock Production in Sub-Saharan Africa

Peter Rogers Ruvuga, Ismail Said Selemani, and Anthony Zozimus Sangeda

Abstract Miombo woodlands cover approximately 3.6 million km^2 in sub-Saharan Africa and have been identified as one of the global priority areas for conservation. Nonetheless, it is debatable whether the presence of the pastoralists and livestock in Miombo woodlands is linked to the ecological degradation and resources use conflicts. Besides, the impact of Miombo woodlands utilization in relation to biodiversity and ecosystem service provision is not well understood. The current review describes the various concurrent issues underlying livestock production in Miombo woodlands and ecosystem resilience. Analytical review established that livestock in Miombo woodlands have a crucial role in enhancing food security and animal source protein requirement in the future. Ecologically, pastoralists' presence in the Miombo serves biodiversity conservation, enhances nutrient recycling, and regulates wildfires. On the other hand, pastoralists are faced with several challenges including pests and diseases and persistent drought that leads to scarcity of water and pasture resources. Other setbacks include high enteric methane emission due to poor livestock productivity, unsustainable rangeland management practices, land tenure, and poor water management. Observations of wildlife interactions with the natural ecosystem elsewhere provide crucial evidence for the potential of livestock-woodland beneficial symbiotic interaction. It is concluded that, since there is high ecological compatibility of livestock production in Miombo woodlands, there is need to reform management policy to promote livestock interaction on the Miombo woodlands.

P. R. Ruvuga (✉)
Swedish University of Agricultural Sciences, Department of Animal Nutrition and Management, Uppsala, Sweden

University of Dar es Salaam, Institute of Resource Assessment, Dar es Salaam, Tanzania
e-mail: peter.ruvuga@slu.se

I. S. Selemani · A. Z. Sangeda
Sokoine University of Agriculture, Department of Animal, Aquaculture and Range Sciences, Morogoro, Tanzania
e-mail: sangeda@sua.ac.tz

Y. Bamutaze et al. (eds.), *Agriculture and Ecosystem Resilience in Sub Saharan Africa*, Climate Change Management,
https://doi.org/10.1007/978-3-030-12974-3_11

Keywords Livestock-woodland interaction · Miombo woodlands · Livestock production · Ngorongoro Conservation Area · Food security · Animal genetic conservation · Multiple land use

1 Introduction

Miombo woodlands have been identified as one of the unique areas for conservation considering their distribution, resilience, high endemicity, and shelter provision for rare species (Jew et al. 2016). This phyto-region covers approximately 3.6 million km^2 in sub-Saharan Africa; it consists mostly of trees (*Brachystegia* species) and understory grasses (Abdallah and Monela 2007; Njana et al. 2013; Sangeda and Maleko 2018a). They are located in the semiarid rangelands of Eastern, Central, and Southern Africa (Malmer 2007). Based on amount and distribution of rainfall, Miombo ecosystems are categorized into dry and wet. The dry Miombo woodlands receive <1000 mm rainfall/year, while wet Miombo woodlands are located in the regions receiving >1000 mm/year (Abdallah and Monela 2007). Some important ecological services offered by Miombo include provision of forage for wild and domestic animals, water catchments, carbon sequestration, and biodiversity conservation (Lawton 1982; Williams et al. 2008: Munishi et al. 2010). Additionally, Miombo woodlands have cultural values to local communities such as provision of thatching grasses for house roofing, edible wild fruits (e.g., *Tamarindus indica*, *Adansonia digitata*, *Vitex* spp., and *Parinari curatellifolia*), honey, and wild vegetables. Another remarkable Miombo woodlands role is its richness in medicinal plants. Various plant species in Miombo (including *Cassia abbreviata* and *Combretum zeyheri)* are used in curing of different anomalies for humans and livestock particularly in the areas where health services are inadequate (Njana et al. 2013; Olekao and Sangeda 2018).

Therefore, the immediate rural communities surrounding Miombo woodlands are direct beneficiaries and derive their daily livelihood through these woodlands. Others communities also benefit indirectly through intrinsic forest services in the global systems and consumption of forestry products (Monela and Abdallah 2007). Nevertheless, in rural communities, agriculture is still a common socioeconomic activity, and it combines both crop cultivation and livestock husbandry. Crop cultivation is mostly under subsistence farmers in form of either small-scale sedentary or shifting cultivation (Nduwamungu et al. 2008). Livestock on the other hand are reared in extensive traditional system managed by pastoralists and agropastoralists. These traditional herders move around with their herd on the communal rangelands as dictated by seasonal and spatial varied forages and water resources (Nindi et al. 2014; Sangeda and Maleko 2018b). Moreover, upon forage scarcity, pastoralists tend to trespass and graze into conserved areas. For that matter, pastoralists and livestock in Miombo woodlands are blamed for causing ecological degradation (Nduwamungu et al. 2008; Benjaminsen et al. 2009; Nindi et al. 2014; Mtimbanjayo and Sangeda 2018).

There are sufficient resources in Miombo to support the livestock population within ecological carrying capacity (Fynn and O'Connor 2000; Aubault et al. 2015; Mtimbanjayo and Sangeda 2018). The balance between livestock production and conservation in Miombo could be sustained through good rangeland management practices (Alkemade et al. 2013). Proper grazing management within Miombo could be of ecological value as it favors tree regrowth and masks fire effects (Gambiza et al. 2000). On the other hand, uncontrolled grazing intensity could lead to change in land cover and could also result into disputes among different land users (Benjaminsen et al. 2009; Alkemade et al. 2013; Njana et al. 2013; Nindi et al. 2014). This chapter addresses the possibilities of combining forestry activities and livestock production while sustaining the ecological capacities of the land in the sub-Saharan Africa. At the time of this research, many of the forested lands, specifically Miombo woodlands, are being converted into protected areas by either central or local governments. The chapter reviews the key issues underlying livestock production in Miombo woodlands and seeks to propose a balance between the livestock productivity and the ecosystem resilience.

2 Opportunities of Livestock Production in Miombo Woodlands

2.1 *Food Security*

Human population has been increasing exponentially; at the current population growth rate, global human population is estimated to reach 9.15 billion by 2050 from the current 7.2 billion (Thornton 2010; Pimentel and Burgess 2015). This tremendous increase in human population, requires a 50% increase in the current food production to meet the demand of the growing population in the upcoming years (Philipsson et al. 2011). Unfortunately there has been a diminishing return in terms of yield of the major cereal crops globally, affecting mostly poor developing countries in sub-Saharan Africa and Asia (Ehrlich and Harte 2015). Despite livestock greenhouse gas emission, animal husbandry might offer the alternative and ensure global food security in the near future. Luckily, sub-Saharan Africa is endowed with a large livestock population. In 2014 it was estimated that the region comprised about 100 million ruminants (44.8 million cattle, 43.4 million goats, and 11.7 million sheep) collectively reared in the countries with vast Miombo woodlands (Fig. 1), with Tanzania at the top of the list (FAOSTATS 2017a). This large livestock herd could provide animal source of protein that is needed to balance diet and combating malnourishment, especially in the poor rural areas (Ehrlich and Harte 2015).

The average per capita consumption of animal products is approximately 11 kg of meat and 27.2 kg of milk in sub-Saharan Africa (Otte and Chilonda 2002). Growing human population has kept pace with growing demand of food of animal origin, while per capita production is only marginally increasing. It is estimated that

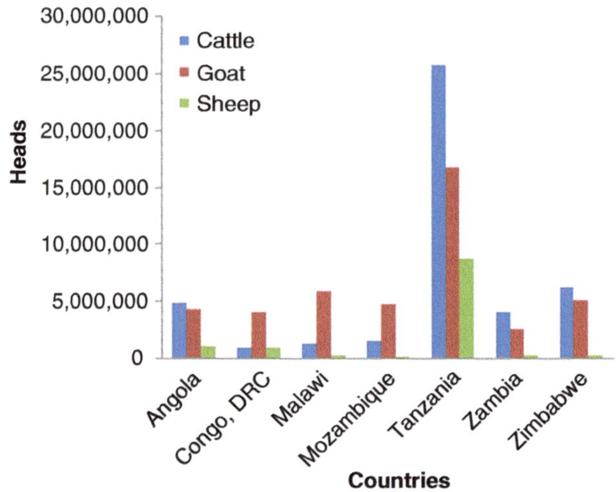

Fig. 1 Livestock herds distribution in some countries with Miombo woodlands in 2014. (Adopted from FAOSTAT (2017a))

the demand for animal source protein will increase by 63% in the future, because of the increased population size and purchasing power (Philipsson et al. 2011). Expansion in livestock production can significantly contribute to food security through improving supply of meat and milk (Philipsson et al. 2011; Peter 2017).

2.2 Pastoralism, Animal Genetic Resources, and Biodiversity Conservation

There is a difference in genetic makeup of farm animals among and within the breeds; this is attributed to natural selection and human influence (Wollny 2003). In sub-Saharan Africa, traditional herders have been, for a long time, selecting animals which are adapted to their environment and meet their demands. Disease resistance and drought tolerance are some of the selection criteria (Zander et al. 2009; Philipsson et al. 2011; Chenyambuga and Lekule 2014). Nonetheless, 30% of the farm animal breeds are at risk of becoming extinct worldwide, mostly because they are considered unproductive and culled in favor of exotic productive breeds, which are less adaptive to tropical environment (Drucker et al. 2001; Wollny 2003). Conservation of appropriate farm animal genetic resources is very crucial due to changes in market preferences and environmental requirement for livestock diversity (Philipsson et al. 2011). Temperature rise is expected to range between 0.85 and 1.06 °C courtesy of global warming (Shaffril et al. 2017). This might lead to changes in livestock diseases etiology and decline in forage quantity and quality, and hence it is likely that animals with prior exposure and adapted to these environments will survive compared to their counterparts (Philipsson et al. 2011).

Sustainability of livestock production in the Miombo woodlands is favored by ecological adaptation and community-based conservation (CBC) of the indigenous livestock breeds (Zander et al. 2009). Although indigenous breeds in sub-Saharan Africa such as Nguni and Zebu are considered to have poor growth performance and relatively low tender meat, yet they have high disease resistance and drought tolerance (Muchenje et al. 2009). In addition, keeping of these indigenous livestock breeds in extensive farming system is advantageous in several ways. It is a relatively cheap and reliable means of farm animals' genetic resource conservation, unlike in vivo conservation that is subjected to policy shift and often operates on government subsidies, which is difficult to conduct in developing countries (Philipsson et al. 2011; Paiva et al. 2016). The Convention on Biological Diversity (2010) acknowledges traditional herders for their contribution in sustainable use and conservation of biodiversity. They noted that rearing of indigenous livestock breeds under extensive traditional system and foraging on the mixed fodder simulates their corresponding wild herbivores grazing patterns and interactions with the environment.

2.3 Forage Diversity and Composition

Miombo woodlands are very diverse in terms of botanical composition amidst trees, browsing plants, and grasses which could be used to support different class of livestock (Mtimbanjayo and Sangeda 2018). The dominant woody species in Miombo such as *Brachystegia* spp. and *Harrisonia abyssinica* are potential browsed plants, and dominant grasses (Table 1) such as *Setaria*, *Digitaria*, *Dactyloctenium*, *Brachiaria*, and *Hyparrhenia* are potential feed source for grazers (Nyathi and Campbell 1994; Mtambanengwe and Kirchmann 1995; Baloyi et al. 1997; Chidumayo and Kwibisa 2003; Mbwambo et al. 2008; Nampanzira et al. 2016). For example, due to high protein content, *Brachystegia spiciformis* (12.13 g/kg DM) can be utilized as source of the livestock feed (Nyathi and Campbell 1994; Baloyi et al. 1997).

Table 1 Forages with livestock consumption potential in Miombo

Trees	Browsing plants	Grasses
Brachystegia spiciformis	*Harrisonia abyssinica*	*Dactyloctenium aegyptium*
Acacia spp.	*Flueggea virosa*	*Setaria* spp.
		Digitaria spp.
		Andropogon spp.
		Hyparrhenia spp.
		Brachiaria spp.

Adapted and modified from Mtambanengwe and Kirchmann 1995; Baloyi et al. 1997; Chidumayo and Kwibisa 2003; Mbwambo et al. 2008; Nampanzira et al. 2017

Additionally, inclusion of *Harrisonia abyssinica* has shown to boost crude protein content in goat ration, which in return improves feed intake and carcass quality (Nampanzira et al. 2017). The detailed information on the nutritive values of the Miombo woodlands forage species are shown in Table 2.

On the other hand, aboveground biomass in wet Miombo woodlands is ranging between 66.9 g/m^2 and 110.7 g/m^2 (Chidumayo and Kwibisa 2003). There is growing evidence that these grasses contribute significantly to the topsoil organic carbon through litter decomposition (Mtambanengwe and Kirchmann 1995).

2.4 Grazing Potential

In sub-Saharan Africa, Miombo woodlands have potential for livestock grazing and wildlife habitat. The clear symbiotic relationship exists between Miombo woodland and grazing herbivores. While Miombo woodlands provide cheap forage resources for animals, the controlled grazing in Miombo would contribute to the control of bush encroachment and reduce frequent and intensive burning of the grazing lands through reduction of fuel load (García et al. 2012; Tsegaye et al. 2013; Sangeda and Maleko 2018a). Seed dispersal and germination in Miombo woodland is one of the ecosystem services provided by grazing herbivores. Cattle are considered as the most dispensers of seeds and also facilitate seed germination through endozoochory process where seeds germinate after passing through animals' guts (Mouissie et al. 2005). Moderate grazing pressure is associated with high biodiversity level, while a high grazing pressure momentarily could be a good tool in weed control (García et al. 2012). Proper grazing management is the subject of harmony between utilization and ecological conservation of the rangelands within Miombo woodland (Gambiza et al. 2000; Alkemade et al. 2013). It is widely agreed that keeping livestock within the ecological carrying capacity not only improves productivity but also conserves environment (Fynn and O'Connor 2000; Aubault et al. 2015; Sangeda and Maleko 2018b). Holechek et al. (1998) attributed high financial returns from

Table 2 Nutrition composition of selected forage species in Miombo woodlands

Forage	DM (%)	Ash (g/kgDM)	CP (g/kgDM)	NDF (g/kgDM)	ADF (g/kgDM)	Digestibility (%)
Brachystegia spiciformis	93.8	32.0	10.1	441	333	42.6
Harrisonia abyssinica	39.1	63.0	16.9	296	245	74.3
Acacia angustissima	92.1	57.8	22.9	354	171	34.3
Brachiaria brizantha	24.0	36.7	18	620	350	62.0
Setaria anceps	13.5	67.0	14.9	682	445	46.5
Digitaria decumbens	22.6	107.4	7.2	790	442	62.8

Adapted and modified from Kabuga and Darko 1993; Nyathi and Campbell 1994; Baloyi et al. 1997; Merkel et al. 1999; Archiméde et al. 2000; Hove et al. 2001; Rubanza et al. 2007; Tikam et al. 2010; Fukushima et al. 2015; Nampanzira et al. 2016; Chibinga and Nambeye 2016; Nampanzira et al. 2017.

livestock keeping, trend in ecological condition, forage productivity, and soil stability with appropriate grazing intensity. For the Miombo woodland to offer sustainable grazing, it is recommended to use conservative stocking rate.

Conservative stocking rate is defined as the level of grazing in which forage utilization does not exceed 35% (Holechek et al. 1999). However, for a while the question among range ecologist has been the accurate ways of estimating livestock carrying capacity (Gillson and Hoffman 2007). Some scholars propose the use of livestock population against vegetation composition a model known as equilibrium, while others want to focus on the abiotic factors like rainfall hence non-equilibrium model (Vetter 2005; Gillson and Hoffman 2007). It is still debatable however as to which model, between the equilibrium and non-equilibrium, is appropriate for sustainable grazing management; Vetter (2005) proposed integration of both models in highly variable arid and semiarid ecosystems. Derry and Boone (2010) on their spatial models demonstrated a decrease in number of animals in rangelands, with the increased variation in annual rainfall, hence support for the role of abiotic factor as stipulated in non-equilibrium models. Furthermore, it is argued (Sasaki 2010) that integration of equilibrium and non-equilibrium models is appropriate rangeland ecosystems approach that provides desirable outcome. Nevertheless, relying only on the one model in Miombo woodlands with high variable and unpredictable rainfall is not recommended. It is common knowledge that in the forested areas, livestock density is highly influenced by rainfall amount (Boone and Wang 2007). Hence in dry Miombo woodlands, livestock production can only be productive within the balanced ecosystem and well-planned management operation. Generally, the impact of livestock grazing on ecological thresholds needs to be investigated regularly (Sasaki 2010).

From the above, it is clear that livestock production in forested lands like Miombo has several benefits, ranging from improving human livelihoods and animal welfare to enhancement of environmental conservation. These multiple benefits are often realized when proper rangeland management principles are followed. The benefits aside, livestock production in Miombo is attended by some challenges, some of which are the subject of the next section.

3 Challenges of Livestock Production in Miombo Woodlands

3.1 Tsetse Fly and African Trypanosomiasis

Miombo ecosystem consists of various forms of life such as pests and vectors for animal diseases. Tsetse (*Glossina morsitans*), the annoying biting fly, is one of the biological vectors of zoonotic disease trypanosomiasis affecting both human and animals (Lawton 1982; Simukoko et al. 2011). *Trypanosoma congolense* is the common causative agent of African animal trypanosomiasis in Miombo; its prevalence ranges from 2.36% to 33.5%, and cattle are the most affected among the farm animals (Simukoko et al. 2007; Malulu et al. 2017). It is urged that disease incidences are determined by several predisposing factors. These include a proportion of meals by tsetse on the host species of interest, tsetse fly density, and trypanosomal infections prevalence within the tsetse fly (Simukoko et al. 2011). Several efforts to eradicate tsetse fly have been made, including the huge forestry clearance program during colonial era in the 1920s (Barrow and Shah 2011). While the tsetse control program was reported to be successfully achieving the long-term fly suppression, less has been documented on its effect on the Miombo ecosystem (Campbell 1996). Although, on the continental scale, wildlife has been reported to be affected by anti-tsetse operation program (Matthiessen and Douthwaite 1985), there is no clear projection of the future impact on Miombo ecosystem as tsetse suppression methods are changing over time. Tsetse flies and trypanosomiasis control efforts in Miombo woodlands face various challenges, including observed changes in feeding behavior of the tsetse fly and manifestation of drug resistance in affected human and animal hosts (Delespaux and de Koning 2007; Torr and Vale 2015). The major challenge facing the range ecologists and animal health experts if the Miombo woodlands are to be utilized for animal husbandry today is how to balance between production and control of tsetse flies without altering the entire ecosystem.

3.2 Greenhouse Gas Emission

Livestock production contributes 18% of the global anthropogenic greenhouse gas emission (GHG). These GHG include methane as the by-product of the ruminal fermentation (Table 3), nitrous oxide from manure, and carbon dioxide from respiration (Steinfeld et al. 2006; Eckard et al. 2010; Morgavi et al. 2010). However it is methane gas that caught attention of most range nutrition scientist as it is strongly influenced by the livestock diet (Knapp et al. 2014). Despite its lower emission levels in relation to carbon dioxide, methane has almost 28 times global warming potential compared to the latter (Du et al. 2017).

Table 3 Enteric methane emission in countries with Miombo woodlands as per 2014

Country	Methane (Gg)	CO_2– eq. (Gg)
Angola	167.8301	3524.43
Congo, DRC	48.2779	1013.836
Malawi	65.3455	1372.256
Mozambique	61.9952	1301.9
Tanzania	676.6869	14,210.42
Zambia	131.5575	2762.71
Zimbabwe	146.6	3078.60

Adapted and modified from FAOSTAT (2017b)

CO_2 – eq. (Carbon dioxide equivalent of methane emission)

The enteric methane production is associated with energy loss in the feed consumed; there could be up to 12% loss of gross energy intake, which could affect both feed utilization efficiency and animal productivity (Stergiadis et al. 2016). There are several proposed methodologies of reducing enteric methane emission, though the practical one in sub-Saharan Africa would be to increase the production efficiency (Grainger and Beauchemin 2011; Knapp et al. 2014; Mushi et al. 2015). Poor productivity is associated with increased methane emission (up to 450 g methane/cow/day) in per unit of produce in the developing countries (Philippe and Nicks 2015; Chibinga et al. 2012; Mushi et al. 2015). Unfortunately, most pastoralists rearing livestock in Miombo woodlands are considered inefficient in terms of resources use (Monela and Abdallah 2007; Nduwamungu et al. 2008). They keep large number of livestock and add the burden to already scarce resources in these dry woodlands and hence reduce production potential (Monela and Abdallah 2007). For instance, the average mature live weight of 20 kg is attained within 2 years by the goats raised in the extensive traditional system (Mushi 2004; Shija et al. 2013). This is owing to the seasonal fluctuation of the forage in form of quantity and quality (Adjorlolo et al. 2014). However, underlying cultural and ecological trade-offs among pastoralists need to be addressed before adoption of the intensive livestock production system in Miombo woodlands that will reduce per unit enteric methane emission (Wollny 2003; Zander et al. 2009).

3.3 Unsustainable Rangeland Management Practices

Pastoralists in sub-Saharan Africa are characterized by large number of livestock exceeding the grazing limit of the rangeland (Holechek et al. 2017). Miombo woodlands in sub-Saharan Africa are experiencing heavy grazing pressure associated with overstocking. Despite all of these, little is done by pastoralists to alleviate the pressure on the depreciating resources. While some pastoralists normally move away during dry season to allow recovery on the grazed land, mismanagement of communally grazed rangelands leads to increased bush encroachment risks and soil compaction, with severe effects on soil water (Solomon et al. 2007; Treydte et al. 2017). Other than migration, pastoralists employ bushfire as the tool in order to improve rangelands and control of encroached bushes. However, uncontrolled fire in Miombo woodlands has negative effects, regardless of availability of fire-resistant tree species (Gambiza et al. 2000; Malmer 2007; Solomon et al. 2007). The young growing trees, dry litter, topsoil humus, and dry herbaceous layer are vulnerable to wildfire (Sangeda and Maleko 2018a). On the contrary, if the fire is controlled, the carbon and nitrogen losses are minimized compared to effects of wildfire (Malmer and Nyberg 2008). However, other scholars argue that low to moderate intensity fire causes increase in soil water repellency in the semiarid grazing lands and also affects water and air quality in the surrounding areas (Stavi et al. 2017). Generally, uncontrolled fire is responsible for ecological and economical damage of approximately 350 million hectares of global land yearly and renders them unfit for other uses (García et al. 2012). Nevertheless, it should be known that stocking density has indirect influence on bush fire. Very low stocking rate may result into huge accumulation of fuel load, which is a risk factor for wildfire. Although prescribed fire is very useful for controlling bush encroachment, application of controlled fire may not be relevant in the arid and semiarid grazing land, provided that grazing intensity is still high (Lohmann et al. 2014; Sangeda and Maleko 2018a).

3.4 Land Tenure and Water Resources Management

In sub-Saharan Africa, Miombo woodlands are categorized into two forms of ownership: general land (open access) and conserved areas. The general is a source of continuous concern due to lack of proper management and is susceptible to anthropogenic interventions leading to forest degradation (Abdallah and Monela 2007; Monela and Abdallah 2007; Murwira et al. 2010). Management of Miombo woodland requires strong ownership and regulated access to the land. Continuous modification of Miombo woodland through anthropogenic activities like clearance of trees for commercial cultivation, charcoal production, fuel wood collection, and rampant fire is associated with poor land tenure institutions. Miombo woodlands suffer from high illegal wood utilization, pressure stemming from the need to fulfill alternative energy demands created by persistent electricity shortages, and an unstable economic environment in sub-Saharan African countries. Lack of land tenure and poor institutional arrangement contributes to unsustainable Miombo woodlands utilization. This results in decline of the fertile soil for various activities and increases more pressure on the already scarce resources in Miombo including water (Benjaminsen et al. 2009; Nindi et al. 2014).

With increased grazing, the soil will be compacted and impede water from infiltrating into the soil, where it would be accessible for growing plants and regeneration of ground water. Dry Miombo woodlands are located in the semiarid areas with lower rainfall and long dry seasons (Malmer 2007). It is widely known that groundwater recharge and dry season stream flow are highly influenced by the silvicultural practices such as succession of some young indigenous species after deforestation (Malmer 2007; Malmer and Nyberg 2008). Ultimately in addressing this in dry woodlands, it is important to design the ecological model that will consider water resources, especially now when the global water demand is high (Ran et al. 2016).

It might be difficult to conclude whether the analysis of benefits accrued from livestock production in Miombo outweighs the challenges or vice versa due to the subjectivity and interests of stakeholders in question. However, to reduce this anomaly, Tanzania (a country highly populated with livestock among Miombo woodlands states) is used as a case study to draw experiences and lessons from multiple land uses in Ngorongoro Conservation Area, to explain the livestock-woodland interaction that has existed for more than seven decades.

4 Lessons from Multiple Land Uses in Tanzania and Ngorongoro Conservation Area

Tanzania is one of the world's biological richest countries, harboring millions of endemic flora and fauna species. It is approximated that around 11,000 species of plants, many which are endemic, are found in Tanzania's forestry and woodlands particularly within Miombo woodland (IUCN 2017). It is approximated that forestry and woodlands cover 48.1 million hectares of Tanzania land (NAFORMA 2015). Tanzania is committed to conservation of biological diversity through establishment of protected areas under different conservation categories. A total land area of 731,806.24 km^2 has been set aside as protected areas of different categories equivalent to 33.5% of terrestrial land of the country (IUCN 2017). Expansion of these protected areas, coupled with increasing human population, has created serious land use conflicts in the country, particularly between pastoralists and conservation authorities (Nindi et al. 2014). There is growing debate among ecologists, scientists, and politicians on whether integration of livestock within protected areas under multiple land use system is ecologically compatible and socially acceptable (Mtimbanjayo and Sangeda 2018). It is hypothesized that, coexistence of grazing livestock within Miombo woodland will reduce land use conflicts and enhance ecosystem services. To discuss these debates, this section has drawn on the experiences of multiple land uses in Ngorongoro Conservation Area (NCA) in Tanzania and recommends policy reforms that could promote integration of livestock in Miombo woodland in a compatible manner.

Ngorongoro Conservation Area (NCA) is the only world heritage site and biosphere reserve in Tanzania with multiple land use, where both wildlife conservation and selected human activities such as livestock grazing are allowed (Masao et al. 2015). The NCA (Fig. 2) is located in the northern part of Tanzania, largely occupied by Maasai pastoralists who are also the main participants of livestock grazing in Miombo. The NCA has total area of 8300 km^2 lying between the rift valley and Serengeti plains. The heart of this area is Ngorongoro crater, a large volcanic caldera within the area.

The NCA contains renowned wildlife population and increasing human population with diversified economic activities (Boone et al. 2006). The area has been managed under multiple land uses (wildlife conservation, ecotourism, and livestock grazing) since 1959 when it was separated from Serengeti National Park (Byers 1994). Despite the growing population of Maasai, the livestock population has remained stable, wildlife conservation has generally been successful, and ecotourism has expanded dramatically (Byers 1994). The unique scenic, conservation, and cultural and ecological value of NCA have led to international recognition, and thus in 1979 it was accepted as a UNESCO World Heritage, and in 1982 it was approved as biosphere reserve (Boshe 1989). All these happened when wildlife, humans, forest, and livestock resources are in a common area.

Since the World Heritage Convention by UNESCO 1972 encourages identification, protection, and preservation of cultural and natural heritage around the world,

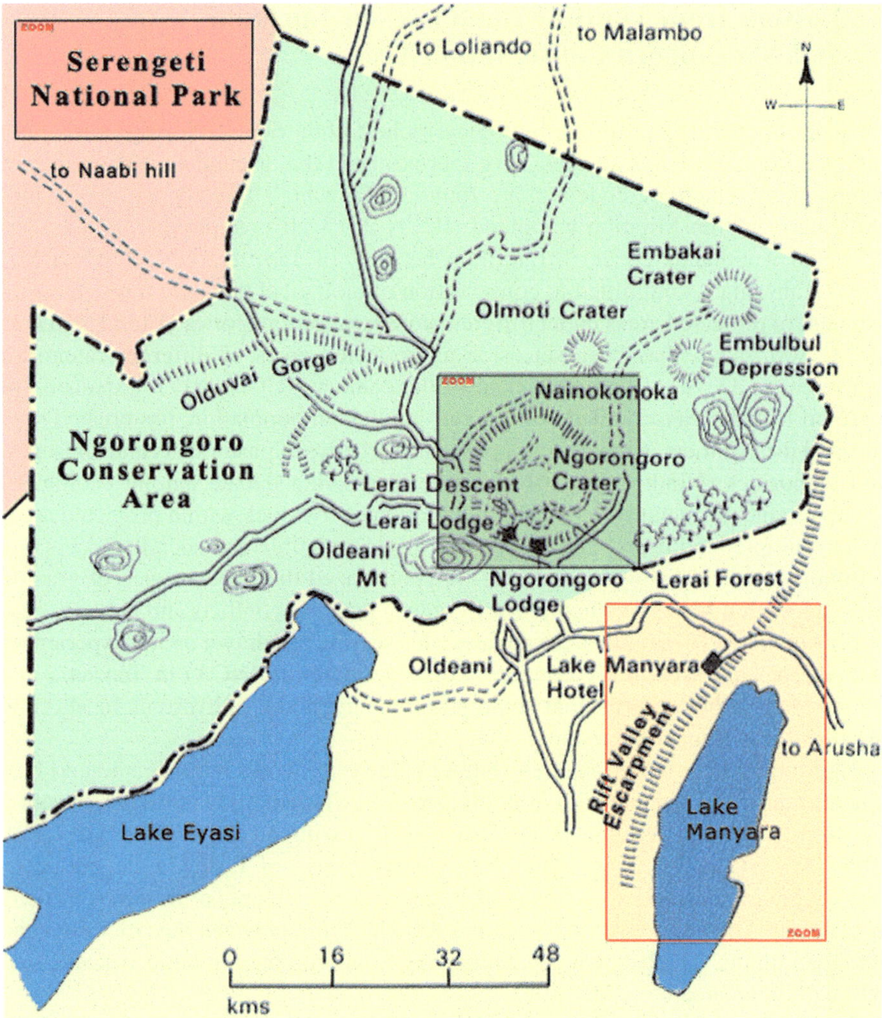

Fig. 2 Map of Ngorongoro Conservation Area in Northern Tanzania. (Source: ESA 2017)

the dynamic relationship between Maasai people and nature in NCA is of fundamental need (Masao et al. 2015). Despite the attractive advantages of integrating conservation with human development, various ecological, economic, social, and political factors need to be harmonized. For example, the sustainability of multiple land uses in NCA is in the hands of community-based conservation, a concept that integrates biodiversity conservation by engaging people in the conservation process. People, who daily live with natural ecosystem like Maasai, are inseparable from nature, and thus it is important that conservation efforts should integrate and promote their livelihood needs (Olekao and Sangeda 2018). Maasai are typically pastoralists and have managed to survive in natural ecosystems with minimal retrogressive impacts on ecosystems. Nelson (2012) pointed out that there is a historical influence of pastoralists on savanna ecosystems, and therefore, there is a general compatibility between pastoralists and wildness. Pastoralists in NCA provide economically viable ecological services by conserving wildlife on their land, which in turn helps to sustain the natural assets that tourism industry depends on.

The key lesson from NCA is the high degree of overlap and coexistence between nature conservation and human development. Pastoralists adapted to highly variable and unpredictable natural ecosystem through various coping strategies such as mobility and economic diversification (Olekao and Sangeda 2018). Nelson (2012) highlighted that pastoralists could still respond to non-equilibrium condition by moving between different local pastures, depending on rainfall distribution and pattern. NCA experience can be a useful conservation strategy in similar natural ecosystems including Miombo woodlands. Miombo ecosystems are vital natural resources with more or less similar economic role as those of wildlife conservation in the rest of the country; and therefore, integrating conservation efforts of Miombo and human development is imperative. One of the possible and compatible options for Miombo-human interaction is through sustainable grazing. Livestock grazing and biodiversity conservation of Miombo woodland can be complementary activities if well managed.

5 Future Prospects and Possible Solutions

Livestock should play a central role in the food security and in the economic growth of sub-Saharan Africa (Philipsson et al. 2011). Diminishing grazing land due to other land uses will put increased pressure on livestock productivity. Therefore, the continued presence of the pastoralists in natural woodlands, particularly Miombo, is necessary. However, the vital question is how to address environmental impact of their practices. Land use plan initiatives that involve local community in sustainable forest utilization may provide possibilities for livestock-Miombo woodland integration (Gmünder et al. 2014). Currently, lands assigned for grazing in these areas do not cope well with the large livestock herd, since most of these communal semiarid rangelands are less productive (Vetter 2005; Nindi et al. 2014). Thus the wide adoption of the land use program and institutional intervention for technical support or direct management of the general managed land could be beneficial as it has been demonstrated elsewhere (Basupi et al. 2017). The lesson from multiple land use in the NCA Tanzania provides empirical evidence for possible integration of sustainable conservation with human development. The well-managed pastoralists' herds within acceptable carrying capacity can enhance the livelihoods of pastoralists while maintaining the stable woodland ecosystems.

Furthermore, traditional breeds selection and breeding procedures offer sustainable solutions to the tsetse fly (*Glossina morsitans*) control problem, instead of previously adopted forest clearing method (Mapedza 2007; Murwira et al. 2010; Munang'andu et al. 2012). Trypano-tolerant livestock breeds, such as Djallonke sheep as well as N'dama and Muturu cattle, could remain productive in the infested areas without trypanocidal drugs. They suffer low casualty and most importantly contribute to efforts toward vaccine development through deciphering of the parasites' complex antigen behavior (Yaro et al. 2016). Further understanding of the vector behavior among veterinary entomologist is expected to add efforts in turning the disease trends. It is already known that *G. morsitans* is more active during the beginning of the rain season in Miombo woodlands (Simukoko et al. 2011). All of these tools individually could be combined in several fronts to eradicate animal trypanosomiasis in Miombo woodlands and thus foster both livestock production and Miombo woodland conservation.

The remaining question is one of sustainability and the place of pastoralism in a changing world, with anticipated more mouth to feed, scarce resources, and constantly degraded rangeland (Vetter 2005; Pimentel and Burgess 2015). On top of this, there is the critical situation in the form of climate stress through greenhouse gas emission (Mushi et al. 2015; Benjaminsen et al. 2009). The solution in the future would be to improve the production system through sustainable grazing management within the ecological threshold comprising of both biotic and abiotic components (Sasaki 2010; Nindi et al. 2014). Efficient utilization of available resources is crucial including adoption of enclosures (like *Olalili* and *Ngitiri* among Maasai and Sukuma pastoralists in Tanzania) and use of forage conservation techniques when they are abundant so they could be used during critical dry season (Maleko and Koipapi 2015; Treydte et al. 2017). Important point of consideration among policy makers and scholars in case of intervention in the future is to capitalize on pastoralists' traditional ecological knowledge (Olekao and Sangeda 2018). As it is manifested in many areas of Africa, some pastoralists are coming up with adaptation strategies for their challenges including change in their herd composition by increasing browsing animals like goats and camels (Coppock et al. 2018; Sangeda and Malole 2014). This is done in response to changes in the rangeland vegetation. Woody plants encroachment has increased and understory herbaceous species for grazers declined (Liao et al. 2016). These strategies are important for sustainable livestock-woodland conservation.

6 Conclusions and Recommendations

Inappropriate rangeland management practices such as overgrazing and wildfire could result into deterioration of Miombo woodlands. However, there are many positive ecological roles of livestock and pastoralists presence in Miombo, such as farm animal gene conservation, nutrient recycling, and reduction of wildfire intensity. Therefore, livestock production in Miombo woodlands could be perceived as both competitive and complementary. Nevertheless, the future is very uncertain with food insecurity challenges and increased grazing land degradation, all of which call for efficient utilization of the available Miombo woodland resources. This could be obtained through integrated resource management in form of sustainable land use plan, accurate estimation of allocated grazed land carrying capacity involving both biotic and abiotic factors, and forage and water conservation techniques. Based on the facts that there is ecological compatibility between livestock production in Miombo woodland, there is a need for policy reform for the sub-Saharan Africa region on the conventional Miombo woodland management that promotes livestock interaction.

Acknowledgment This work is part of the SIDA-funded Food Security Program in collaboration with the University of Dar es Salaam, Tanzania. Authors are grateful for their financial assistance that led to successful completion of this work. Authors appreciate the internal technical review provided by Gert Nyberg that benefited the smooth accomplishment of this chapter.

References

Abdallah JM, Monela GG (2007) Overview of Miombo woodlands in Tanzania. Work Pap Finnish Forest Res Inst 50:9–23

Adjorlolo LK, Adogla-Bessa T, Amaning-Kwarteng K, Ahunu BK (2014) Seasonal effect on rumen function in sheep on range in the Accra Plains of Ghana. Trop Anim Health Prod 46:1223–1228

Alkemade R, Reid RS, van den Berg M, de Leeuw J, Jeuken M (2013) Assessing the impact of livestock production on biodiversity in rangeland ecosystems. Proc Natl Acad Sci U S A 110(52):20900–20905

Archiméde H, Boval M, Alexandre G, Xanddé A, Aumont G, Poncet C (2000) Effect of regrowth age on intake and digestion of *Digitaria decumbens* consumed by Black-belly sheep. Anim Feed Sci Technol 87(3–4):153–162

Aubault H, Webb NP, Strong CL, McTainsh GH, Leys JF, Scanlan JC (2015) Grazing impacts on the susceptibility of rangelands to wind erosion: the effects of stocking rate, stocking strategy and land condition. Aeolian Res 17:89–99

Baloyi JJ, Ngongoni NT, Topps JH, Ndlovu P (1997) Chemical composition and degradability of *Brachystegia spiciformis* (musasa) leaves and stems harvested over 4 months from three sites in Zimbabwe. Anim Feed Sci Technol 69:179–186

Barrow E, Shah A (2011) TEEBcase: traditional forest restoration in Tanzania. Available at: TEEBweb.org

Basupi LV, Quinn CH, Dougill AJ (2017) Using participatory mapping and a participatory geographic information system in pastoral land use investigation: impacts of rangeland policy in Botswana. Land Use Policy 64:363–373

Benjaminsen TA, Maganga FP, Abdallah JM (2009) The Kilosa killings: political ecology of farmer-herder conflict in Tanzania. Dev Chang 40(3):423–445

Boone RB, Wang G (2007) Cattle dynamics in African grazing systems under variable climates. J Arid Environ 70(3):495–513

Boone RB, Galvin KA, Thornton PK, Swift DM, Coughenour MB (2006) Cultivation and conservation in Ngorongoro conservation area, Tanzania. Hum Ecol 34:809–828

Boshe J (1989) Wildlife conservation in Ngorongoro conservation area in Tanzania: social and ecological implication of increasing pastoralists and declining per capita livestock population. In: Transaction of XIXIUGB Congress, International Union of Game Biologists, September, 1989, Trondheim, Norway, pp 84–100

Byers BA (1994) Population growth, environmental degradation, and conflict in Ngorongoro conservation area, Tanzania. Paper prepared for workshop on the state of environment, conflict in the North East Africa

Campbell B (1996) The Miombo in transition: woodlands and welfare in Africa. Center for International Forestry Research. Malaysia

Chenyambuga SW, Lekule FP (2014) Breeding preferences and breeding practices for goats in agro-pastoral communities of semi-arid and sub-humid areas in Tanzania. Livest Res Rural Dev 26(6):117

Chibinga OC, Nambeye E (2016) Nutritive value of some selected indigenous livestock browse species in the drylands of Southern Province, Zambia. Int J Appl Pure Sci Agric 2(1):61–66

Chibinga OC, Musimba NRK, Nyangito MM, Simbaya J, Daura MT (2012) Climate variability; enhancing adaptive utilization of browse trees for improved livestock production among agro-pastoralists communities in Southern Zambia. Afr J Environ Sci Technol 6(7):267–274

Chidumayo EN, Kwibisa L (2003) Effects of deforestation on grass biomass and soil nutrients status in Miombo woodlands, Zambia. Agric Ecosyst Environ 96(1–3):97–105

Coppock DL, Bailey D, Ibrahim M, Tezera S (2018) Diversified investments of wealthy Ethiopian pastoralists include livestock and urban assets that better manage risk. Rangel Ecol Manag 78(1): 138–148

Delespaux V, de Koning HP (2007) Drugs and drug resistance in African trypanosomiasis. Drug Resist Updat 10(1–2):30–50

Derry JF, Boone RB (2010) Grazing systems are a result of equilibrium and non-equilibrium dynamics. J Arid Environ 74(2):307–309

Drucker AG, Gomez V, Anderson S (2001) The economic valuation of farm animal genetic resources: a survey of available methods. Ecol Econ 36(1):1–18

Du M, Peng C, Wang X, Chen H, Wang M, Zhu Q (2017) Quantification of methane emissions from municipal solid waste landfills in China during past decade. Renew Sust Energ Rev 78:272–279

Eckard RJ, Grainger C, de Klein CAM (2010) Options for the abatement of methane and nitrous oxide from ruminant production: a review. Livest Sci 130(1–3):47–56

Ehrlich PR, Harte H (2015) Opinion: to feed the world in 2050 will require a global revolution. Proc Natl Acad Sci U S A 112(48):14743–14744

European Space Agency (ESA) (2017) Ngorongoro Conservation Area. https://www.esa.int/SPECIALS/Eduspace_Global_EN/SEMX9X05VQF_0.html

FAOSTAT (2017a) Statistical database of the Food and Agriculture Organisation of the United Nations on agriculture, nutrition, fisheries, forestry, food aid, land use and population. http://www.fao.org/faostat/en/#data/QA. Last accessed 3rd July, 2017

FAOSTAT (2017b) Statistical database of the Food and Agriculture Organisation of the United Nations on agriculture, nutrition, fisheries, forestry, food aid, land use and population. http://www.fao.org/faostat/en/#data/GE. Last accessed 27th July, 2017

Fukushima RS, Bacha CB, Fuzeto AP, Port ACR, Herling VR, Velásquez AV (2015) Utilisation of equations to predict carbohydrate fractions in some tropical grasses. Anim Feed Sci Technol 208:12–22

Fynn RWS, O'Connor TG (2000) Effect of stocking rate and rainfall on rangeland dynamics and cattle performance in a semi-arid savanna, South Africa. J Appl Ecol 37:491–507

Gambiza J, Bond W, Frost PGH, Higgins S (2000) A simulation model of Miombo woodland dynamics under different management regimes. Ecol Econ 33:353–368

García RR, Celaya R, García U, Osoro K (2012) Goat grazing, its interactions with other herbivores and biodiversity conservation issues. Small Rumin Res 107(2–3):49–64

Gillson L, Hoffman MT (2007) Rangeland ecology in a changing world. Science 315:53–54

Gmünder S, Zah R, Rainhard J, Charron-Doucet F (2014) Transforming Tanzania's charcoal sector: life cycle assessment component. *Quantis*, Dübendorf, Switzerland

Grainger C, Beauchemin KA (2011) Can enteric methane emissions from ruminants be lowered without lowering their production? Anim Feed Sci Technol 166–167:308–320

Holechek JL, Gomes H, Molinar F, Gait D (1998) Grazing intensity: critique and approach. Rangelands 20(5):15–18

Holechek JL, Gomes H, Molinar F, Gait D (1999) Grazing studies: what we learned? Rangelands 21(2):12–16

Holechek JL, Cibils AF, Bengaly K, Kinyamario JI (2017) Human population growth, African pastoralism and rangelands: a perspective. Rangel Ecol Manag 70(3):273–280
Hove L, Topps JH, Sibanda S, Ndlovu LR (2001) Nutrient intake and utilization by goats fed dried leaves of the shrub legumes *Acacia angustissima*, *Calliandra calothyrsus* and *Leucaena leucocephala* as supplements to native pasture hay. Anim Feed Sci Technol 91(1–2):95–106
IUCN (2017) Governance of protected and conserved areas in Tanzania, Phase 1 workshop as part of an IUCN-assisted process of assessment and action to enhance governance for conservation and sustainable livelihoods, Dar es Salaam, Tanzania 21–22 March 2017
Jew EKK, Dougill AJ, Sallu SM, O'Connell J, Benton TG (2016) Miombo woodland under threat: Consequences for tree diversity and carbon storage. Forest Ecology and Management 361:144–153
Kabuga JD, Darko CA (1993) In Sacco degradation of dry matter and nitrogen in oven dried and fresh tropical grasses and some relationships to in vitro dry matter digestibility. Anim Feed Sci Technol 40:191–205
Knapp JR, Laur GL, Vadas PA, Weiss WP, Tricarico JM (2014) Invited review: enteric methane in dairy cattle production: quantifying the opportunities and impact of reducing emissions. J Dairy Sci 97(6):3231–3261
Lawton RM (1982) Natural resources of Miombo woodland and recent changes in agricultural and land-use practices. For Ecol Manag 4:287–297
Liao C, Ruelle ML, Kassam KS (2016) Indigenous ecological knowledge as the basis for adaptive environmental management: evidence from pastoralist communities in the Horn of Africa. J Environ Manag 182:70–79
Lohmann D, Tietjen B, Blaum N, Joubert DF, Jeltsch F (2014) Prescribed fire as a tool for managing shrub encroachment in semi-arid savanna rangelands. J Arid Environ 107:49–56
Maleko DD, Koipapi ML (2015) Opportunities and constraints for overcoming dry season livestock feed shortages in communal semi-arid rangelands of Northern Tanzania: a case of Longido district. Livest Res Rural Dev 27(4):70
Malmer A (2007) General ecological features of Miombo woodlands and considerations for utilisation and management. Work Pap Finnish Forest Res Inst 50:34–42
Malmer A, Nyberg G (2008) Forest and water relations in Miombo woodlands: need for understanding of complex stand management. Work Pap Finnish Res Inst 98:70–86
Malulu DJ, Kimbita E, Tuntufye H, Kinugh'i S, Nyingilili H, Mbilu T, Kaboya J, Lyaruu E, Malele I (2017) An investigation on *Glossina* species and the prevalence of trypanosomiasis in cattle in Meatu district, Tanzania. J Parasitol Vector Biol 9(2):13–18
Mapedza E (2007) Forestry policy in colonial and postcolonial Zimbabwe: continuity and change. J Hist Geogr 33(4):833–851
Masao CA, Makoba R, Sosovele H (2015) Will Ngorongoro conservation area remain a world heritage site amidst increasing human footprint? Int J Biodivers Conserv 7(9):394–407
Matthiessen P, Douthwaite B (1985) The impact of tsetse fly control campaigns on African wildlife. Oryx 19(4):202–209
Mbwambo L, Valkonen S, Kuutti V (2008) Structure and dynamics of Miombo woodland stand at Kitulangalo Forest Reserve, Tanzania. Work Pap Finnish Forest Res Inst 98:10–19
Merkel RC, Pond KR, Burns JC, Fisher DS (1999) Intake, digestibility and nitrogen utilization of three tropical tree legumes: I. As sole feeds compared to *Asystasia intrusa* and *Brachiaria brizantha*. Anim Feed Sci Technol 82(1–2):91–106
Monela GC, Abdallah JM (2007) Principle socio-economic issues in utilisation of Miombo woodlands in Tanzania. Work Pap Finnish Forest Res Inst 50:115–122
Morgavi DP, Forano E, Martin C, Newbold CJ (2010) Microbial ecosystem and methanogenesis in ruminants. Animal 4(7):1024–1036
Mouissie AM, Vos P, Verhagen HMC, Bakker JP (2005) Endozoochory by free ranging, large herbivores: ecological correlates and perspectives for restoration. Basic Appl Ecol 6:547–558
Mtambanengwe F, Kirchmann H (1995) Litter from savanna woodlands (Miombo): chemical composition and C and N mineralisation. Soil Biol Biochem 27(12):1639–1651

Mtimbanjayo JR, Sangeda AZ (2018) Ecological effects of cattle grazing on Miombo tree species regeneration and diversity in Central-Eastern Tanzania. J Environ Res 2(1):3

Muchenje V, Dzama K, Chimonyo M, Strydom PE, Hugo A, Raats JG (2009) Some biochemical aspects pertaining to beef eating quality and consumer health: A review. Food Chemistry 112 (2):279–289

Munang'andu HM, Siamudaala V, Munyeme M, Nalubamba KS (2012) A review of ecological factors associated with the epidemiology of wildlife trypanosomiasis in the Luangwa and Zambezi valley ecosystems of Zambia. Interdiscip Perspect Infect Dis 2012:372523

Munishi PKT, Mringi S, Shirima DD, Linda SK (2010) The role of the Miombo woodlands of the Southern highlands of Tanzania as carbon sinks. J Ecol Nat Environ 2(12):261–269

Murwira A, Skidmore AK, Huizing HJG, Prins HHT (2010) Remote sensing of the link between arable field and elephant (*Loxodanta africana*) distribution change along a tsetse eradication gradient in the Zambezi valley, Zimbabwe. Int J Appl Earth Obs Geoinf 12(1):123–130

Mushi DE (2004) Studies on marketing of meat goats and carcass composition – a case study of Gairo auction markets. MSc. Dissertation, Sokoine University of Agriculture, Morogoro, Tanzania

Mushi DE, Eik LO, Bernués A, Ripoll-Bosch R, Sundstol F, Mo M (2015) Reducing GHG emissions from traditional livestock system to mitigate changing climate and biodiversity. In: Lal R (ed) Sustainable intensification to advance food security and enhance climate resilience in Africa. Springer International Publishing, Switzerland

NAFORMA (2015) National Forest Resources monitoring and assessment of Tanzania Mainland. Tanzania Forestry Service Agency

Nampanzira DK, Kabasa JD, Katongole CB, Okello S, Tabuti JRS (2016) Natural gastro properties of *Ficus natalensis*, *Rhuss natalensis* and *Harrisonia abyssinica* in native East African goats. Livest Sci 187:31–34

Nampanzira DK, Kabasa JD, Katongole CB, Okello S, Tabuti JRS (2017) Ameliorative effects of *Ficus* and *Harrisonia* on Small East African goat meat yield. Small Rumin Res 153:189–193

Nduwamungu J, Bloesch U, Munishi PKT, Hagedorn F, Lulu K (2008) Recent land cover and use changes in Miombo woodlands of Eastern Tanzania. Tanzania J Forestry Nat Conserv 78(1):50–59

Nelson F (2012) Natural conservationists? Evaluating the impact of pastoralist land use practices on Tanzania's wildlife economy. Res Policy Pract 2(15):1–19

Nindi SJ, Maliti H, Bakari S, Kija H, Machoke M (2014) Conflicts over land and water resources in the Kilombero Valley Floodplain, Tanzania. African Study Monographs Suppl. 50:173–190

Njana MA, Kajembe GC, Malimbwi RE (2013) Are Miombo woodlands vital to livelihoods of rural households? Evidence from Urumwa and surrounding communities, Tabora, Tanzania. Forests, Trees and Livelihoods 22(2):124–140

Nyathi P, Campbell P (1994) Leaf quality of *Sesbania sesban*, *Leucaena leucocephala* and *Brachystegia spiciformis*: potential agroforestry species. For Ecol Manag 64:259–264

Olekao SK, Sangeda AZ (2018) Traditional ecological knowledge in management of dryland ecosystems among the Maasai pastoralists in Kiteto district, Tanzania. J Environ Res 2(1):2

Otte MJ, Chilonda P (2002) Cattle and small ruminant production systems in sub-Saharan Africa : a systematic review. Rome: FAO

Paiva SR, McManus SM, Blackburn H (2016) Conservation of animal genetic resources - a new tact. Livest Sci 193:32–38

Peter AS (2017) Cyber resilience preparedness of Africa's top 12 emerging economies. Int J Crit Infrastruct Prot 17:49–59

Philippe F-X, Nicks B (2015) Review on greenhouse gas emissions from pig houses: production of carbon dioxide, methane and nitrous oxide by animals and manure. Agric Ecosyst Environ 199:10–25

Philipsson J, Zonabend E, Bett RC, Okeyo AM (2011) Global perspectives on animal genetic resources for sustainable agriculture and food production in the tropics. In: Ojango JM, Malmfors B, Okeyo AM (eds) Animal genetics training resource, version 3, 2011. International

Livestock Research Institute, Nairobi, Kenya, and Swedish University of Agricultural Sciences, Uppsala, Sweden
Pimentel D, Burgess M (2015) World human population problems. Reference Module in Earth Systems and Environmental Sciences, https://doi.org/10.1016/B978-0-12-409548-9.09303-9
Ran Y, Lannerstad M, Herrero M, Van Middelaar CE, De Boer IJM (2016) Assessing water resource use in livestock production: a review of methods. Livest Sci 187:68–79
Rubanza CDK, Shem MN, Bakengesa SS, Ichinohe T, Fujihara T (2007) The content of protein, fibre and minerals of leaves of selected Acacia species indigenous to North-Western Tanzania. Arch Anim Nutr 61(2):151–156
Sangeda AZ, Maleko DD (2018a) Regeneration effectiveness post tree harvesting in natural Miombo woodlands, Tanzania. J Plant Sci Agric Res 2(1):10
Sangeda AZ, Maleko DD (2018b) Rangeland condition and livestock carrying capacity under the traditional rotational grazing system in northern Tanzania. Livest Res Rural Dev 30:Article #79
Sangeda AZ, Malole JL (2014) Tanzanian rangelands in a changing climate: impacts, adaptations and mitigation. Net J Agric Sci 2(1):1–10
Sasaki T (2010) Paradigm integration between equilibrium and non-equilibrium concepts for evaluating vegetation dynamics in rangeland ecosystems. Global Environ Res 14:17–22
Secretariat of the Convention on Biological Diversity (2010) Pastoralism, nature conservation and development: a good practice guide. Montreal, 40 + iii pages
Shaffril HAM, Samah AA, D'Silva JL (2017) Adapting towards climate change impacts: strategies for small-scale fishermen in Malaysia. Mar Policy 81:196–201
Shija DS, Mtenga LA, Kimambo AE, Laswai GH, Mushi DE, Mgheni DM, Mwilawa AJ, Shirima EJM, Safari JG (2013) Chemical composition and meat quality attributes of indigenous sheep and goats from traditional production system in Tanzania. Asian Australas J Anim Sci 26(2):295–302
Simukoko H, Marcotty T, Phiri I, Geysen D, Vercruycce J, Van den Bossche P (2007) The comparative role of cattle, goats and pigs in the epidemiology of livestock trypanosomiasis on the plateau of eastern Zambia. Vet Parasitol 147(3–4):231–238
Simukoko H, Marcotty T, Vercruycce J, Van den Bossche P (2011) Bovine trypanosomiasis risk in an endemic area on the eastern plateau of Zambia. Res Vet Sci 90(1):51–54
Solomon TB, Snyman HA, Smit GN (2007) Cattle-rangeland management practices and perceptions of pastoralists towards rangeland degradation in the Borana zone of Southern Ethiopia. J Environ Manag 82(4):481–494
Stavi I, Barkai D, Knoll YM, Glion HA, Katra I, Brook A, Zaady E (2017) Fire impact on soil-water repellency and functioning of semi-arid croplands and rangelands: implications for prescribed burnings and wildfires. Geomorphology 280:67–75
Steinfeld H, Gerber P, Wassenaar T, Castel V, Rosales M, de Haan C (2006) Livestock's long shadow environmental issues and options. Food and Agriculture Organisation, Rome, Italy
Stergiadis S, Zou C, Chen X, Allen M, Wills D, Yan T (2016) Equations to predict methane emissions from cow fed at maintenance energy level in pasture-based systems. Agric Ecosyst Environ 220:8–20
Thornton PK (2010) Livestock production: recent trends, future prospects. Philosophical Transactions of the Royal Society B: Biological Sciences 365 (1554):2853–2867
Tikam T, Mikled C, Vearaslip T, Phatsara C, Sűdekum K (2010) Digestibility of nutrients and evaluation of energy of Pangola Grass in sheep compared with Napier Grass. Conference of international research on food security, natural resources management and rural development ETH Zurich, September 14–16, 2010
Torr SJ, Vale GA (2015) Know your foe: lessons from the analysis of tsetse fly behaviour. Trends Parasitol 31(3):95–99
Treydte AC, Schmiedgen A, Berhane G, Tarekegn K (2017) Rangeland forage availability and management in times of drought – a case study of pastoralists in Afar, Ethiopia. J Arid Environ 139:67–75

Tsegaye D, Vedeld P, Moe SR (2013) Pastoralists and livelihoods: a case study from northern Afar, Ethiopia. J Arid Environ 91:138–146
Vetter S (2005) Rangelands at equilibrium and non-equilibrium: recent developments in the debate. J Arid Environ 62:321–341
Williams M, Ryan CM, Rees CM, Sambane E, Fernando J, Grace J (2008) Carbon sequestration and biodiversity of re-growing Miombo woodlands in Mozambique. For Ecol Manag 254(2):145–155
Wollny CBA (2003) The need to conserve farm animal genetic resources in Africa: should policy makers be concerned? Ecol Econ 45(3):341–351
Yaro M, Munyard KA, Stear MJ, Groth DM (2016) Combatting African Animal Trypanosomiasis (AAT) in livestock. The potential role of trypanotolerance. Vet Parasitol 225:43–52
Zander KK, Drucker AG, Holm-Müller K (2009) Costing the conservation of animal genetic resources: the case of Borana cattle in Ethiopia and Kenya. J Arid Environ 73(4–5):550–556

Impacts of Dams on Downstream Riparian Ecosystems' Health and Community Livelihoods: A Case of the Lesotho Highlands Water Project

Patrick Gwimbi and Thabo E. Rakuoane

Abstract Environmental and social impact assessments and reviews for the Lesotho Highlands Water Project (LHWP) dams were completed and approved, thus providing rationale for the construction of the dams. However, little is known about the downstream effects of the dams on river flow regulations and riparian ecosystems' health and subsequent impact on river-dependent communities' livelihoods. This study assessed the impacts of Phase 1 LHWP on downstream river flow regulation and riparian ecosystems' health and their subsequent impacts on river-dependent community livelihoods. The methodology included a review of secondary data and primary data derived from a questionnaire administered to a case study community. Data sets included in-stream flow requirements (IFR), riparian vegetation and responses of river-dependent communities on the impact of dams on their livelihoods. Construction of LHWP dams resulted in modification of the flow pattern of downstream rivers and overall condition of the riparian ecosystems. Excessive livestock grazing also caused extensive riparian zone degradation and soil erosion, thereby threatening people's livelihoods. The riparian zone was characterized by loss of bushes and grass and invasion of the channel sloughs by woody vegetation. Community members downstream of LHWP dams reported experiencing reduction in fish populations in rivers, grass for thatching and crafts making, wild fruits, livestock grazing pastures, food sources, medicinal plants and timber. Livestock overgrazing was, however, also identified as significantly contributing to riparian ecosystems' degradation long before the construction of the LHWP. The study concluded that reduced flows downstream of dams represent an escalating problem on riparian ecosystems' health, threatening livelihood sources of river-dependent communities. This review is relevant in understanding of the linkages between dams, downstream river flow and its effects on riparian ecosystems' health and subsequent impact on river-dependent community livelihoods. The paper concludes with recommendations for improving the integration of environmental flow considerations into riparian ecosystems' management to sustain livelihoods of local communities.

P. Gwimbi (✉) · T. E. Rakuoane
Department of Environmental Health, National University of Lesotho, Roma, Lesotho

Y. Bamutaze et al. (eds.), *Agriculture and Ecosystem Resilience in Sub Saharan Africa*, Climate Change Management,
https://doi.org/10.1007/978-3-030-12974-3_12

Keywords Lesotho Highlands Water Project (LHWP) · IFR policy · Downstream · Riparian ecosystems' health · Livelihoods

1 Introduction

There is growing concern in the science of environmental flows that downstream riparian ecosystems' health is at risk from dam-induced alterations of river flows (Richter et al. 2010; Lu et al. 2015; Marcinkowski and Grygoruk 2017). In recent years, rivers have been increasingly segmented by dams, altering the natural distribution and timing of stream flows downstream, resulting in reduction of flood peaks and frequency and extent and duration of floodplain inundation (Duvail and Hamerlynck 2003), posing threats to riparian ecosystems' health (Alldredge and Moore 2012; Marcinkowski and Grygoruk 2017). These changes pose serious threats to river-dependent communities whose livelihood sources depend on the healthy functioning of riparian ecosystems (Richter et al. 2010).

Healthy riparian ecosystems are among the most diverse and productive systems supporting flora and fauna that perform a variety of biophysical functions capable of providing important flood attenuation, nutrient cycling, carbon dioxide sequestration, sediment deposition, timber production, recreation and wildlife habitat functions that humans and the natural environment depend on (Lin 2011; Alldredge and Moore 2012; Webb 2017). Riparian areas are shaped and maintained through seasonal flooding and soil fertility deposits that are supported by natural river flows. However, in recent years, dam-induced changes in river flows downstream have been identified as the largest anthropogenic threat to riparian ecosystems' health through flow reductions and flooding disruptions (O'Connor 2001; Schneider et al. 2017), blocking of water and nutrient exchange between river and floodplain (Lejon et al. 2009; Talukdar and Pal 2018), lowering of the water table and base flows (Bouwes et al. 2016) and hampering the renewal of soil fertility (Poff et al. 2012; Talukdar and Pal 2018). These changes pose many challenges to more than 470 million worldwide river-dependent people downstream (Marcinkowski and Grygoruk 2017; Timpe and Kaplan 2017).

Worldwide, it is estimated that about 470 million people are negatively affected by downstream riparian ecosystems' health deterioration due to dam constructions (Timpe and Kaplan 2017). In Africa, millions of people who are subsistence users of riparian ecosystems for food, medicines, nutritional supplements, firewood, construction materials, thatching grass and livestock grazing are the worst affected downstream of dams (Duvail and Hamerlynck 2003; Kafumbata et al. 2014).

Lesotho, a small land-locked country completely surrounded by South Africa, is home to the Lesotho Highlands Water Project (LHWP), a binational interbasin transfer scheme, involving the construction of five dams and transferring the water to Gauteng province of South Africa while generating hydropower in Lesotho as a spin-off. The LHWP was planned to be implemented in four phases. Phase I of the project was divided into two subphases: Phase 1A was composed of the construction of the Katse Dam, while Phase 1B consisted of the construction of the Mohale Dam (Devitt and Hitchcock 2010). The impact of these dams on downstream river flows and subsequent riparian ecosystems as well as people dependent on them is the focus of this study. The affected communities incurred loss of arable and grazing land, houses, graves, forests and fruit trees, medicinal and other indigenous plants, cultural roots, functions and values and control of their natural resources to the project (Ramaili 2006).

Katse Dam, the highest dam in Africa at 180 m and on the Malibamatšo River, with a storage capacity of 1950 million m^3 was completed in 1998 (Moleko et al. 2011). Mohale Dam, built on the Senqunyane River and the largest rock-filled embankment dam on the African continent at 144 meters in height and storage capacity of 946 million m^3, was completed in 2004 (Hitchcock 2015). According to Mashinini (2010), the two dams jointly occupied 5000 hectares of grazing land and 17 hectares of garden land. In addition, other valuable resources used in sustaining livelihoods by the affected people including trees, medicinal plants, wild vegetables, thatching grass, river sand and fish were lost due to the adverse environmental impacts of reduced river flows (Hoover 2001). Some of the displaced people settled downstream of the dams as they opted to remain in the area with their livestock production (Devitt and Hitchcock 2010). However, no comprehensive data-supported research on the downstream effects of the LHWP has been conducted so far. The main objective of this study was to assess the impacts of Phase 1 LHWP on downstream river flow regulation and riparian ecosystems' health and their subsequent impacts on river-dependent community livelihoods. A general conceptual framework showing the relationship between river flow and riparian ecosystems' health and subsequent impact on local communities is provided.

2 Conceptual Framework Linking Riparian Ecosystems' Health to Human Well-Being Downstream of Dams

The conceptual framework used in this study is situated within the context that an inextricable link exists between dams, downstream river flows and riparian ecosystems' health and is essential for the sustainable livelihoods' well-being of river-dependent people. It is well acknowledged in literature that dams may impact the health of downstream riparian vegetation communities through flow modifications such as decreased flood frequency and duration (Overton et al. 2014). Understanding the linkages between dams, river flow and health of riparian ecosystems is thus necessary for developing cause-and-effect relationships between these components. This conceptual framework first delves into the concepts of riparian ecosystem and human health before discussing the link between the two and the determinant factors.

The term riparian zone is largely defined by the narrow strip of land located along the banks of rivers, streams and water networks (Fu et al. 2017). Such areas range from small isolated patches rarely more than 300 m wide (Varty 1990) to fragmented patches of mosaic forests within corridors defined by a hydrologic regime, floods and meandering pattern of the stream (Medley 1992). Riparian zones provide many ecosystem services, such as flood attenuation, nutrient cycling, carbon dioxide sequestration, sediment deposition, timber production, recreation and wildlife habitat that are all critical for human well-being (Reinecke 2013). In these zones, hydrology plays the role of the 'master variable' of vegetation communities, so flood regime changes are often associated with the decline of riparian vegetation. Riparian vegetation refers to the riverine plant community that is adjacent to the active channel and is sustained by generally moist conditions along river margins.

River flow is considered the master variable responsible for the health of the riparian ecosystem as it directs river channel structure, moisture regime and the life history of plants that grow there (Reinecke 2013). It is essential for the survival of the riparian areas while also being the driving force that shapes them (Ford et al. 2015). Understanding the links between dams, river flows and riparian ecosystem health aids the appreciation of how riparian vegetation communities change in response to altered river flow regimes (Reinecke 2013; James et al. 2016).

In Lesotho, riparian ecosystems are classified into three categories of wetlands, namely, palustrine, lacustrine and riverine, distinguishable by a variety of hydrologic, geomorphologic, chemical and biological characteristics (Olaleye 2012). Upland palustrine is the dominant type and includes mires (bogs and fens), most of which are found at high altitude. Bogs are fresh water systems that have highly organic soils and exist on high water table. They form sponges of river systems as they accumulate rainwater and seepage coming down the screes, which are released as a regular flow (Van Zinderen Bakker and Werger 1974). The bogs are dominated by sedges and grasses. The bogs vary in size from about 1 ha to several square kilometres. It has been acknowledged by various authors that these wetlands of Lesotho are unique (Van Zinderen Bakker and Werger 1974; Grobbelaar and Stegmann 1987; Grundling et al. 2015).

Lacustrine wetlands are defined as swamps associated with the shallow part of a lake where aquatic vegetation or water-tolerant plants are growing amidst water (Jay et al. 2013).The presence of stagnant water is the distinguishing characteristic of lacustrine wetlands. Lacustrine wetlands are found in artificial lakes and dams (Olaleye 2012). Both the aquatic and terrestrial sections of lacustrine wetlands are replenished by the lake water (Jay et al. 2013).

Riverine wetlands are found along the river valleys and streams (Olaleye 2012). They are provided by rivers and landscapes that are hydrologically connected to rivers. In such cases river flow variables drive riparian ecosystem processes important for plant growth, such as groundwater recharge, nutrient deposition and sediment transport and substrate deposition (Schachtschneider 2010). Riverine ecosystems are subject to the influence of the water level, shrinking at a higher water level and vice versa (Jay et al. 2013). According to Renofalt et al. (2010), river flow components control physical and biological processes that are important for maintaining the integrity of riparian ecosystems.

Construction of dams is deemed to reduce the magnitude and intensity of river flow to the point that riparian ecosystems' health deteriorates significantly.

To address the challenge of river flow downstream of dams, in-stream flow requirement (IFR) has been used to maintain the flow releases, thus ensuring the health of riparian ecosystem (Brown and King 2012). IFR is defined as the water that is left in a river system, or released into it, for the specific purpose of managing the ecological condition of that river (Brown and King 2012). The IFR is based on the premise that flow is a key determinant of the ecological characteristics of rivers and their riparian zones.

3 Methodology

3.1 Study Area

The study focused on areas downstream Katse and Mohale Dams in the Lesotho Highlands (Fig. 1). The Lesotho Highlands contain some of the highest mountains in southern Africa and play an important role in the surface water resources of the region (Sene et al. 1998). The hydrology of the Lesotho Highlands is characterized by a dendritic drainage pattern with high yields due to rapid runoff from steep slopes and a highly variable flow regime. Runoff contributes almost half of the flow into the rivers, and in the pre-dam era, the rivers were subject to seasonal floods which

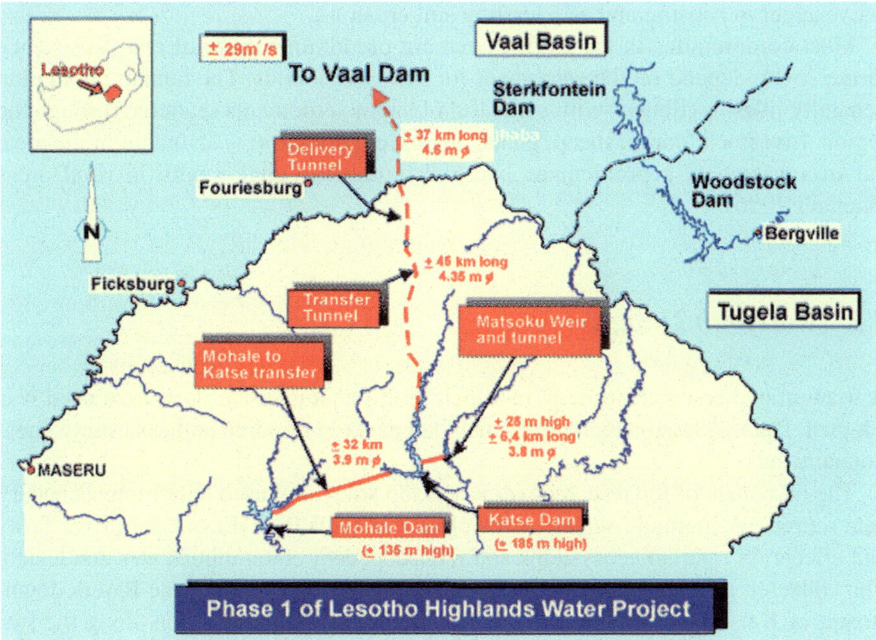

Fig. 1 Phase 1 of Lesotho Highlands Water Project. (Source: Letsebe 2012)

were considered the most important regular phenomenon shaping the riparian wetlands. The highlands occupy roughly two-thirds of the Lesotho and cover 30,588 km^2 (Sene et al. 1998). The altitudes range from around 2800 m up to 3482 m (Boelhouwers and Sumner 2003).

The climate of the Lesotho Highlands is tropical mountain, with summer rainfall season from October to March and a pronounced dry season giving minimum rainfall in June and July (Sene et al. 1998). The average annual rainfall is 1600 mm (Van Zinderen Bakker and Werger 1974). Snowfall is common in winter, occurring on average five to ten times during the months of June and July (Sene et al. 1998). The maximum temperature does not exceed 16 °C, while the nightly minimum temperatures at soil level are below freezing point throughout the year (Van Zinderen Bakker and Werger 1974).

The vegetation characterizing Katse and Mohale catchment areas is described by Moleko et al. (2011) as largely alpine with shrubs, grasslands and some woodland in varying degrees of prevalence. Mires and wetlands form the headwaters of the rivers feeding the two dams with water. These mires and wetlands have been acknowledged by various authors as unique (Van Zinderen Bakker and Werger 1974; Grundling et al. 2015).

The grass species in the Lesotho Highlands are an important contributor to local animal husbandry. Overgrazing has however been identified as one of the causes for wetland degradation (Grundling et al. 2015). Due to the steepness of their slopes, the soils are generally shallow and poorly suited for crop cultivation (Moleko et al. 2011). Small mammals such as mice also dig burrows in the ground and are an active agent in causing and accelerating soil erosion.

Most communities in the LHWP area are predominantly rural and subsistence farmers who depend on the catchment for their livelihoods. The human population generally lives in villages, with a handful of larger settlements scattered through the region. Livestock (cattle, sheep, goats, horses and donkeys) is abundant in the area, but overgrazing is a problem as livestock numbers equal wealth in rural areas (Ramaili 2006).

3.2 Research Design

A triangulated research strategy in which multiple sources of data were used was adopted. Data collection techniques included desktop research and case study questionnaire survey.

The first stage of the research was a desktop study and used data from secondary data sources. A literature search of the effects of LHWP on downstream river flows and effects on riparian ecosystems and river-depedent communities was conducted. The collected data included IFR from Malibamatšo and Senqunyane Rivers downstream of Katse and Mohale Dams, respectively, riparian vegetation along the two rivers and impact of the LHWP on downstream river-dependent communities' livelihoods. A qualitative analysis of the downstream effects of the dams on river flow, riparian habitat conditions and river-dependent community livelihoods was done using established methods, enabling comparisons across studies. The vast majority of studies were conducted in the Lesotho Highlands and covered the dams under study. In the second stage, primary data were collected from purposively sampled residents of Bokong, adjacent to Katse Dam, using a closed and open-ended questionnaire.

3.3 Data Collection

Data was collected using a mixed-research approach and a case study design. The study commenced with a desktop review of data on three thematic areas: IFR downstream of Katse Dam, downstream riparian vegetation responses to Katse and Mohale Dams' constructions and impact of riparian vegetation conditions on river-dependent communities' livelihoods. Primary data was collected through a structured questionnaire, with residents of Bokong villages adjacent the Katse Dam.

In-Stream Flow Requirement (IFR) of Rivers Downstream LHWP Dams

In-stream flow requirement (IFR) data were obtained from the Lesotho Highlands Development Authority website (LHDA 2002a, 2003, 2006, 2007). The IFR were implemented with the aim of minimizing the impacts of reduced flow downstream of dams by striking a balance between the water resources development goals, ecosystem conservation and protection goals as well as social needs of communities (LHDA 2016). The IFR policy sets appropriate river condition targets and provided for water release programme to realize the predetermined conditions for riverine ecosystem downstream of the impoundments. These data were used to study the flow patterns in rivers downstream Katse and Mohale Dams, in an effort to relate them to the riparian vegetation structure (Fig. 2). The rivers of concern in this study were Malibamatšo and Senqunyane. The actual flow volumes in million cubic metres (MCM) were compared with the required IFR policy.

The relationship between actual flow releases downstream Katse and Mohale Dams and required IFR policy amounts were later used to examine the riparian vegetation conditions downstream the dams.

Riparian Vegetation Response to Flow Releases Downstream LHWP Dams

The river condition classification comprises a set of qualitative descriptions of the state of the riverine ecosystem. A time series analysis of riparian vegetation conditions in the pre- and post-dams' construction was done. Maps showing areas under inundation in the pre- and post-dam construction were also analysed to examine any changes. Key riparian vegetation parameters evaluated included species richness, tree regeneration, diversity and density, dominance and frequency among other variables.

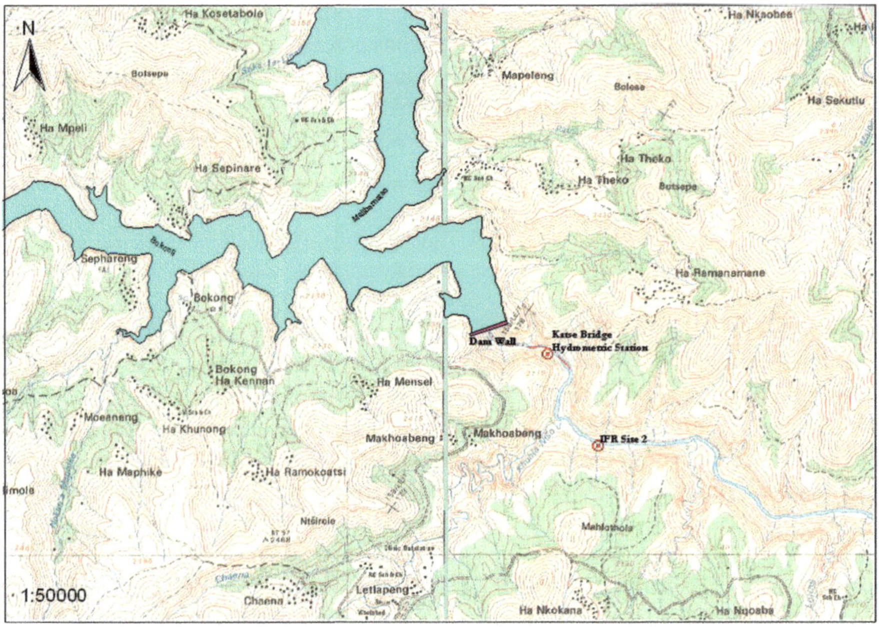

Fig. 2 Katse Dam and the Katse Bridge hydrometric station where flow release is recorded. (Source: LHDA 2016)

Impacts of LHWP Dams on Downstream Community Livelihoods

The method for this portion of the study entailed reviewing literature on the impacts experienced by local people downstream of the two dams as a result of their construction. The desk research involved a review of relevant literature on dam impacts on livelihoods.

Case Study: Impact of Katse Dam on Bokong Community Livelihoods

Purposive sampling was employed to select 50 respondent residents in Bokong, based on their knowledge of the area regarding riparian areas and context in which they were affected by Katse Dam. A questionnaire and observations were used to collect data.

4 Data Processing

The data were analysed qualitatively. Content analysis was used to analyse the qualitative data. Patterns of downstream river flow releases, riparian ecosystem responses to river flow patterns and quantification of their impact on livelihoods were captured and described.

5 Results

5.1 Flow Releases Downstream of the Lesotho Highlands Water Project Dams

The LHWP IFR policy that became operational in December 2002 indicates that the average release from Katse Dam should be 1.3 m^3/s. The results showed that the actual flows were generally below this IFR policy. The actual average flow release downstream of Katse Dam was generally 0.75 $m^3/s \pm 10\%$. The flow releases however varied from high flows to low flows over different time scales. Between 2003 and 2004, the flow releases at 47.67 MCM was 29.4% lower than the IFR release requirement of 67.55 MCM. An IFR audit carried out in 2007 found that, on average, actual downstream flow releases were 60% compliant with the IFR policy (Brown 2008).

Distinguished high flows were noted during the month of February to April at all cases (Fig. 3). The river receives lowest flows between September and December.

In the 2015/2016 hydrological year, the total IFR release at Katse Dam site was 28.78MCM against a target of 36.12MCM. This was 80% of the target. In the Mohale Dam catchment, a total IFR release of 26.07MCM was made at dam site against a target of 29.97MCM. This was 98.23% of the target (Fig. 4).

The dams may have altered the downstream flux of water and sediment that could have modified the biogeochemical cycles as well as the structure and dynamics of aquatic and riparian habitats.

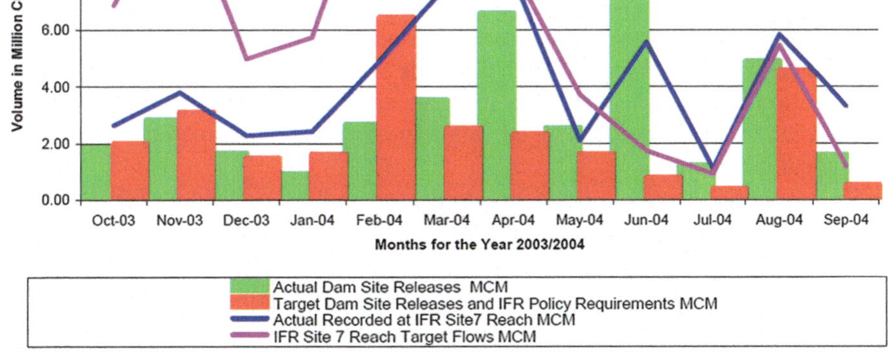

Fig. 3 Flow releases downstream Mohale Dam (2003/2004). (Source LHDA 2006)

Fig. 4 Katse Dam downstream dry season flow rate

5.2 *Riparian Vegetation Response to Flow Releases Downstream LHWP*

Prior to the construction of Katse and Mohale Dams, most commissioned studies predicted declines in downstream riparian vegetation (Metsi Consultants 2002; LHDA 2002b). Metsi Consultants (2002) predicted that modification of flows of the Malibamatšo and Senqunyane Rivers would result in reduction in bush vegetation due to reduced quantity and quality of water in the rivers. The LHDA (2002b) projected declines in thatching grass, crafts riparian grass, medicinal and wild vegetables eaten or sold in urban areas by households due to decreased river flow. Reviews of vegetation photographs downstream Katse and Mohale Dams however showed that riparian bushes had not changed, while in other cases the riparian woody vegetation had increased.

The response of riparian vegetation to upstream dams showed a shift in woody vegetation species composition and density on the channel sloughs (Brown and King 2012). A spatial analysis of the response by Brown and King (2012) showed that trees increased in number on the basal area of the Malibamatšo River (Fig. 5). Terrestrial species such as *A. afra* and *R. rubiginosa* tended to colonize riparian zones following the river impoundment, while species such as *G. virgatum* and *S. mucronata* declined in abundance. The decreased flow downstream of Katse Dam

Fig. 5 Tree colonization of downstream river channels. (Source: Letsebe 2012)

resulted in invasion of the channel by woody vegetation because of the lack of spawning floods. These patterns explain apparent contradictions in conclusions to other studies and provide a framework for predicting some of the effects of dams on woody vegetation.

5.3 *Livestock Overgrazing Threats to Riparian Vegetation*

Destruction of mires in Lesotho has however been reported in literature during the pre-LHWP, with livestock overgrazing the explanation (Van Zinderen Bakker and Werger 1974; Backeus and Grab 1995; Du Preez and Brown 2010). Van Zinderen Bakker and Werger (1974) mention wetland degradation from livestock overgrazing way back in 1969. Backeus and Grab (1995) noted that prior to the construction of the LHWP, mires were already heavily grazed by livestock. Du Preez and Brown (2010), while acknowledging the LHWP contribution to the degradation of mires, write that grazing pressure has always been a major threat to their existence (Fig. 6).

Rampai (2017) attributes erosion of wetlands in the Lesotho Highlands to overgrazing and the presence of burrowing ice-rats.

Grazing areas in Lesotho are communal and controlled primarily by local chiefs (Letšela et al. 2003). The open-access nature of wetlands often leads to failure in

Fig. 6 Overgrazing observed downstream in one of the LHWP dams during field visit

appreciation of the important values of wetlands in decisions relating to their use and conservation (Moqekela 2016).

Some of the displaced people also opted to settle among communities downstream the dams, displacing grazing areas and putting pressure on riparian vegetation. The dam constructions also affected traditional farming practices, such as transhumance, preventing livestock from rotating grazing lands during winter months.

5.4 Households' Perceptions of the Impacts of LHWP on Their Livelihoods

Analyses of previous studies revealed that household livelihood sources such as grazing areas, grasses, wild fruits and arable land were adversely impacted by the construction of LHWP (Letsebe 2012; Lewis et al. 2015). Katse Dam catchment area, for example, experienced a lot of overstocking of cattle, sheep and goats, leading to overgrazing and reduction in vegetation cover, causing siltation of surface water resources due to increased soil erosion (Lewis et al. 2015).

Communities that depended on riverine areas for food, medicine and timber for fuel have also been recorded indicating that they experienced reduction in vegetative resources in riparian zones when the flows of dammed rivers were altered (Letsebe 2012; Manwa 2014).

Box 1 Case Study

A Review of Riverine Ecosystem Along Bokong River and Its Impact on Local Livelihoods

Introduction

The pre-Katse Dam construction Bokong riparian zone was home to a rich array of biodiversity that played an important role in supporting local peoples' livelihoods (Lewis et al. 2015). Wetlands and riparian areas provide harvestable resources and play significant socio-cultural significances in local communities. This case study highlights how the Katse Dam affected downstream river-dependent communities whose livelihoods revolved around the seasonal flow regimes of the Bokong River. Bokong River feeds water into Katse Dam (Fig. 7).

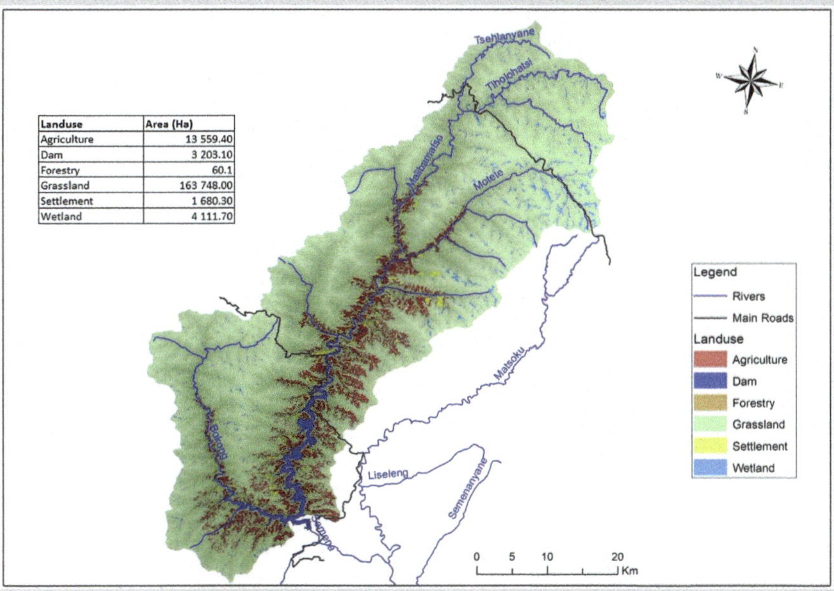

Landuse	Area (Ha)
Agriculture	13 559.40
Dam	3 203.10
Forestry	60.1
Grassland	163 748.00
Settlement	1 680.30
Wetland	4 111.70

Fig. 7 Land use activities within the Katse Dam catchment area. (Source: Lewis et al. 2015)

This affected communities that experienced the impact of Katse Dam as well as its implications on their livelihoods.

Condition of the Bokong Riverine Vegetation

Overgrazing and trampling by domestic livestock in Bokong riverine areas have been the major threats to the riparian ecosystems (Lewis et al. 2015; Mathebula 2015). Common indicators of riparian vegetation stress included reduced biodiversity, reduced grass productivity and increased erosion as reflected by the increasing number of gullies.

(continued)

Box 1 Case Study (continued)
Impact of Katse Dam on Bokong Riparian Ecosystem and Community Livelihoods

The inundation of arable land by Katse Dam was raised as having worsened the condition of the riparian ecosystem. Katse Dam flooded fertile valleys, displacing areas onto communal alpine areas (Grundling et al. 2015).

Thirty five of the fifty (70%) respondents indicated that riparian had been degrading due to inadequate land for grazing animals. Pre-dam conditions provided several livelihood opportunities to downstream communities, including fishing, farming craftwork and medicinal plants. Post-dam construction resulted in dwindling of the resource base, contributing to household impoverishment.

The forms of riparian degradation expressed by respondents are shown in Fig. 8. When asked to describe the quality of the riparian vegetation before the project started, the highlighted characteristic features of the riparian vegetation were grass species and the wetlands. Ten respondents did not have knowledge as they mentioned that they were too young at that time. Twenty eight of the forty (70%) respondents who had knowledge cited that the riparian was dominated by dense *Festuca caprina* grass species and *Merxmuellera disticha* grass species. Respondents were able to indicate that these grass species were evergreen grasses.

Concerning the quality of the riparian areas after Katse Dam construction, thirty five of the fifty (70%) respondents indicated that due to the inundation of land by dam water, riparian areas had been degrading due to inadequate land for grazing animals. Respondents emphasized that there had been increased pressures on pastures.

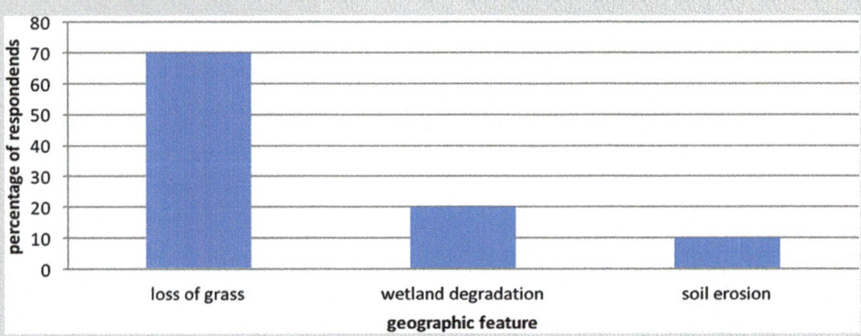

Fig. 8 Forms of riparian vegetation degradation reported by respondents

The main source of livelihood is subsistence agriculture with more than 50% of households in both areas citing it as an important source of livelihood. The communities of Bokong harvest natural resources for a variety of needs, mainly firewood, handcrafts, medicine, food, construction and socio-cultural amenities (Letšela et al. 2003).

(continued)

Box 1 Case Study (continued)

Handcrafts grass are exported as raw materials to South Africa and Australia (Letšela et al. 2003) and are locally used to make products, such as hats and ornaments, that are largely targeted at the tourist market. Other traditional handcraft items, such as mats (meseme), beer strainers (metlhotlo), food baskets (Liroto) and grain baskets (lisiu), have been reported in literature (Letšela et al. 2003).

Field observations revealed declining thatching grass along the Bokong river channels due to severe overgrazing (Fig. 9).

Fig. 9 Overgrazing has resulted in thatching grass decline

Traditional healers use the riparian areas as sources of medicinal plants for their healing. Species such as *Aloe polyphylla* Schonl. ex Pillans, *Bulbine narcissifolia* Salm-Dyck and *Alepidia amatymbica* Eckl. and Zeyh. have been decreasing and continue to decrease in the area compared to the past (Letšela et al. 2003).

Conclusion

The outcomes of this study showed that the local communities in Bokong had knowledge of their environment before and after the construction of Katse Dam. Katse Dam reduced the grazing areas of the Bokong communities thus resulting in overgrazing in the remaining areas. Overgrazing has negatively affected the species composition and especially biomass production leading to a decrease in resilience of the ecosystem.

6 Discussion

Results from this study showed that dams can induce river flow changes downstream, potentially altering the riparian ecosystems' health and adversely impacting the well-being of river-dependent communities. The riverine ecosystem downstream LHWP was deemed healthy and able to sustainably support river-dependent communities depending on them for their livelihoods. The construction of the LHWP dams invariably resulted in the inundation of land and destruction of riparian ecosystems. These findings are in tandem with previous studies elsewhere (Wang et al. 2016; Divine et al. 2017; Mul et al. 2017). According to Wang et al. (2016) alteration of flow threatens the biodiversity and ecosystem functions of rivers. Divine et al. (2017) note that dams have resulted in species diversity being endangered downstream of dams. Mul et al. (2017) similarly noted that ecosystem services downstream are under pressure due to river impoundments. However, notwithstanding the aforementioned threats from dams on downstream riparian vegetation, increased grazing pressure on riparian areas led to soil being eroded causing a tremendous loss in biological diversity and ecosystem goods and services (Letšela et al. 2003; Mathebula 2015; Rampai 2017). Overstocking has not been taken as a serious threat to downstream riparian ecosystems in Lesotho Highlands, yet studies show otherwise. According to Chakela (2014) riparians in Lesotho are threatened by, among other factors, overgrazing and fuel collection which affect the pasture productivity, woody biomass and biodiversity. Palmer (2014) also reports that many districts in Lesotho are heavily overstocked. Rampai (2017) argues that livestock overgrazing has caused extensive land degradation and soil erosion, thereby threatening people's livelihoods in the Lesotho Highlands.

Socioeconomically, livelihood sources of local communities downstream LHWP decreased due to downstream riparian ecosystems' degradation. The communities downstream of LHWP, for a long time dependent on grazing areas, grasses, wild fruits and medicinal plants in riparian zones for their livelihoods, are adversely affected by LHWP. To address the interacting processes linking dams, ecosystem and human health, new ways of working across policies and strategies are urgently required (Ford et al. 2015). Previous researchers have argued that sustainable grazing management systems should enhance ecosystem functions and ecosystem goods and services for present and future generations (Chakela 2014; Rampai 2017). The current scenario, however, according to Rampai (2017) provides a typical example of the 'tragedy of the commons' (Hardin 1968) as increased livestock numbers serve only to maximize individual benefits from the use of rangeland at the expense of rangeland health, a cost that is shared by all rangeland users.

The challenges identified in this study regarding riparian ecosystems' degradation call for implementation of policies and governance mechanisms for the protection of riparian ecosystems downstream of dams. The key to improving riparian ecosystems' health lies in sustainable riparian areas' management practices that allow the areas time to recover after grazing and ensure the functioning of main ecosystem processes (Ford et al. 2015; Rampai 2017). In view of the study findings, addressing the need for restoration of degraded riparian vegetation to improve grazing pastures could invariably enhance livelihoods of communities downstream of LHWP. Livestock management strategies can be applied to limit stress and provide sufficient opportunity for plant growth and regrowth downstream LHWP areas. Strategies such as introducing highly nutritional grasses and maintaining the livestock carrying capacity by limiting livestock grazing in riparian areas have been used elsewhere and could lead to recovery of endangered riparian areas (Richter et al. 2010; Swanson et al. 2015). It is critical to ensure that local people are involved in decision-making and planning of these strategies. Public participation is therefore crucial to the success of livelihood projects downstream of LHWP.

7 Conclusion

This study was based on secondary and primary data which identified that the LHWP altered the downstream river flow discharges negatively impacting on riparian ecosystems' health and livelihoods of river-dependent communities. Livestock overgrazing also negatively impacted on riparian ecosystems and communities' livelihood source. Riparian ecosystem management focusing on improving the resilience of riparian ecosystems in the face of reduced downstream flow releases and overgrazing to improve their health conditions could sustainably support livelihoods downstream of dams. Policies and strategies focusing on long-term riparian vegetation restoration and monitoring as well as maintaining livestock carrying capacity should be prioritized by the dam authorities to deal with the challenges identified in this study. The inclusion of local communities is critical for their sustainability. Public involvement and participation is crucial to the success of livelihood projects downstream of LHWP.

References

Alldredge B, Moore G (2012) Assessment of riparian vegetation sensitivity to river hydrology downstream of a major Texas dam. River Res Applic 30:230–244

Backeus I, Grab S (1995) Mires in Lesotho. Gunneria 70:243–250

Boelhouwers JC, Sumner PD (2003) The palaeo environmental significance of southern African block fields and blockstreams. In: Phillips M, Springman SM, Arenson LU (eds) Permafrost. Swets & Zeitlinger, Lisse, pp 73–78. ISBN 90 5809 582 7

Bouwes N, Weber N, Jordan CE, Saunders WC, Tattam IA, Volk C, Wheaton JM, Pollock MM (2016) Ecosystem experiment reveals benefits of natural and simulated beaver dams to a threatened population of steelhead (*Oncorhynchus mykiss*). Sci Rep 6:28581

Brown C (2008) Let it flow: reducing the environmental legacy of dams. Int Rivers 23(2):6

Brown C, King J (2012) Modifying dam operating rules to deliver environmental flows: experiences from southern Africa. Int J River Basin Manag 10:13. https://doi.org/10.1080/15715124.2011.639304

Chakela QK (ed) (2014) The Lesotho national conference on sustainable land management. Book of Proceedings, Ministry of Forestry and Land Reclamation, Maseru

Devitt P, Hitchcock RK (2010) Who drives resettlement? The case of Lesotho's Mohale Dam. Afr Stud Monogr 31(2):57–106

Divine OA, Michael S, Bridgette F, Henrietta A (2017) Environmental and socioeconomic perturbations of a dam project on catchment communities, Ghana. Glob Environ Health Saf 1(2):13

Du Preez PJ, Brown LR (2010) Impact of domestic animals on ecosystem integrity of Lesotho high altitude peatlands. In: Grillo O (ed) Ecosystems biodiversity. Janeza Trdine 9, 51000 Rijeka, Croatia. ISBN: 978-953-307-417-7

Duvail S, Hamerlynck O (2003) Mitigation of negative ecological and socio-economic impacts of the Diama dam on the Senegal River Delta wetland (Mauritania), using a model based decision support system. Hydrol Earth Syst Sci 7(1):133–146

Ford AES, Graham H, White PCL (2015) Integrating human and ecosystem health through ecosystem services frameworks. EcoHealth 12:660–671. https://doi.org/10.1007/s10393-015-1041-4

Fu B, Li Y, Wang Y, Campbell A, Zhang B, Yin S, Zhu H, Xing Z, Jin X (2017) Evaluation of riparian condition of Songhua River by integration of remote sensing and field measurements. Sci Rep 7, 2565. https://doi.org/10.1038/s41598-017-02772-3

Grobbelaar JU, Stegmann P (1987) Limnological characteristics, water quality and conservation measures of a high altitude bog and rivers in the Maluti Mountains, Lesotho. Water SA, Vol. 13(3)

Grundling P, Linström A, Fokkema W, Grootjans AP (2015) Mires in the Maluti Mountains of Lesotho. Mires Peat 15:1–11

Hardin G (1968) Tragedy of the commons. Science 162:1243–1248

Hitchcock RK (2015) The Lesotho Highlands Water Project: dams, development, and the world bank. Soc Anthropol 3(10):526–538. https://doi.org/10.13189/sa.2015.031004

Hoover R (2001) Pipe dreams: the world bank's failed efforts to restore lives and livelihoods of dam-affected people in Lesotho. International Rivers Network, Berkeley

James CS, Mackay SJ, Arthington AH, Capon SJ, Barnes A, Pearso B (2016) Does stream flow structure woody riparian vegetation in subtropical catchments? Ecol Evol 6(16):5950–5963

Jay G, Xi-lai L, Brierley G, Cheung A, Yang Y-w (2013) Geomorphic-centered classification of wetlands on the Qinghai-Tibet Plateau, Western China. J Mt Sci 10(4):632–642. https://doi.org/10.1007/s11629-013-2561-4

Kafumbata D, Jamu D, Chiotha S (2014) Riparian ecosystem resilience and livelihood strategies under test: lessons from Lake Chilwa in Malawi and other lakes in Africa. Phil Trans R Soc B 369:20130052. https://doi.org/10.1098/rstb.2013.0052

Lejon AGC, Renöfält BM, Nilsson C (2009) Conflicts associated with dam removal in Sweden. Ecol Soc 14:4

Letsebe PH (2012) A study of the impact of Lesotho Highlands Water Project on residents of Khohlo-Ntso: is it too late for equitable benefit sharing? Unpublished Thesis, Faculty of Arts, University of Wits

Letšela T, Witkowski ETF, Balkwill K (2003) Plant resources used for subsistence in Tsehlanyane and Bokong in Lesotho. Econ Bot 57(4):619–639

Lewis F, Browne M, Oosthuizen S, Peerbay K (2015). Mapping climate change vulnerability and potential economic impacts in Lesotho. A case study of the Katse Dam catchment. Institute of Natural Resources NPC, Pietermaritzburg

LHDA (2002a) Analysis of the minimum degradation, treaty, design limitation and fourth scenarios for phase 1 development. Summary Report compiled by Metsi Consultants. Report No. LHDA 678-F-002. LHDA, Maseru, Lesotho

LHDA (2002b) Analysis of the minimum degradation, treaty, design limitation and fourth scenarios for phase 1 development, section 3: social consequences. Report compiled by H. Sabet and P. Martel. Report No. LHDA 678-F-002. LHDA, Maseru, Lesotho

LHDA (2003) Lesotho Highlands Water Project: phase 1: policy for instream flow requirements (2nd Edition Incorporating Corrigenda). July 30. Maseru

LHDA (2006) Annual flow releases for in-stream flow requirement (IFR) implementation. Report N0.5. LHDA, Maseru, Lesotho
LHDA (2007) Annual flow releases for in-stream flow requirement (IFR) implementation. Report N0.7. LHDA, Maseru, Lesotho
LHDA (2016) IFR monitoring report for 2015 to 2016. LHDA, Maseru, Lesotho
Lin Q (2011) Influence of dams on river ecosystem and its countermeasures. J Water Resour Prot 3:60–66
Lu YR, Wang Y, Zhang H, Su P, Wang A, Jenkins RC, Ferrier M, Bailey SG (2015) Ecosystem health towards sustainability. Ecosyst Health Sustain 1(1):2. https://doi.org/10.1890/EHS14-0013.1
Manwa H (2014) Impacts of Lesotho Highlands Water Project on sustainable livelihoods. Mediterr J Soc Sci 5(15). https://doi.org/10.5901/mjss.2014.v5n15p640
Marcinkowski P, Grygoruk M (2017) Long-term downstream effects of a dam on a lowland river flow regime: case study of the Upper Narew. Water 9:783
Mashinini V (2010) The Lesotho Highlands Water Project and sustainable livelihoods: policy implications for SADC. (AISA Policy Brief no. 22).
Mathebula B (2015) Assessment of the surface water quality of the main rivers feeding the Katse Dam, Lesotho. MSc Thesis. University of Pretoria, Pretoria
Medley KE (1992) patterns of forest diversity along the Tana River, Kenya. J Trop Ecol 8:353–371
Metsi Consultants (2002) Summary of main findings for Phase 1 development. Report 678-F-001 Metsi Consultants 2002. Additional scenarios and production of a new final report (vol II). Report 678-F-002
Moleko L, Thokoa M, Dlamini Z (2011) Challenges of managing communal compensation downstream of the LHWP dams as part of the Instream Flow Requirements (IFR) Policy and Procedures. WIT Trans Ecol Environ 153:387–395
Moqekela MC (2016) A valuation based approach for sustainable utilization of Khubelu Wetlands in Lesotho. Master of (Integrated Water Resources Management) Dissertation University of Dar es Salaam
Mul M, Pettinotti L Amonoo NA, Bekoe-Obeng E, Obuobie E (2017) Dependence of riparian communities on ecosystem services in northern Ghana. Colombo, Sri Lanka: International Water Management Institute (IWMI). 43p. (IWMI Working Paper 179). doi: https://doi.org/10.5337/2018.201
O'connor TG (2001) Effect of small catchment dams on downstream vegetation of a seasonal river in semi-arid African savanna. J Appl Ecol 38:1314–1325
Olaleye AO (2012) Mountain watershed in Lesotho: water quality, anthropogenic impacts and challenges. In: Křeček J, Haigh MJ, Hoffer T, Kubin E (eds) Management of mountain watersheds. Springer, Netherlands, pp 139–152
Overton IC, Smith DM, Dalton J, Barchiesi S, Acreman MC, Stromberg JC, Kirby JM (2014) Implementing environmental flows in integrated water resources management and the ecosystem approach. Hydrol Sci J 59(3–4):860–877. https://doi.org/10.1080/02626667.2014.897408
Palmer A (2014) Towards assessing rangeland condition and production for The Kingdom of Lesotho. In: Chakela QK (ed) Sustainable land management in Sub-Saharan Africa: increasing land productivity. Book of Proceedings. Janeza Trdine 9, 51000 Rijeka, Croatia
Poff B, Koestner KA, Neary DG, Merritt D (2012) Threats to western United States riparian ecosystems: a bibliography. Gen. Tech. Rep. RMRS-GTR-269. Fort Collins, CO: U.S. Department of Agriculture, Forest Service, Rocky Mountain Research Station: 78
Ramaili LG (2006) Impact of the Lesotho Highlands Water Project in poverty alleviation in Lesotho. MA Thesis. University of Stellenbosch, South Africa
Rampai RI (2017) Effects of different grazing systems on ecosystem functions in Lesotho and Iceland. United Nations University Land Restoration Training Programme [final project]. http://www.unulrt.is/static/fellows/document/rampai2017.pdf
Reinecke MK (2013) Links between lateral riparian vegetation zones and flow. PhD Thesis, Faculty of Agri Sciences, Stellenbosch University, South Africa

Renofalt BM, Jansson R, Nilsson C (2010) Effects of hydropower generation and opportunities for environmental flow management in Swedish riverine ecosystems. Freshw Biol 55:49–67

Richter BD, Postel S, Revengan C, Scudder T, Lehner B, Churchill A, Chow M (2010) Lost in development's shadow: the downstream human consequences of dams. Water Alternatives 3(2):14–42

Schachtschneider K (2010) Water sourcing by riparian trees along ephemeral riverbeds. PhD Thesis, Department of Botany, University of Cape Town

Schneider C, Flörke M, De Stefano L, Petersen-Perlman JD (2017) Hydrological threats to riparian wetlands of international importance – a global quantitative and qualitative analysis. Hydrol Earth Syst Sci 21:2799–2815

Sene KJ, Jones DA, Meigh JR, Farquharson FAK (1998) Rainfall and flow variations in the Lesotho Highlands. Int J Climatol 18:329–345

Swanson S, Wyman S, Evans C (2015) Practical grazing management to maintain or restore riparian functions and values on rangelands. J Rangeland Appl 2:1–28

Talukdar S, Pal S (2018) Application of frequency ratio and logistic regression models for assessing physical wetland vulnerability in Punarbhaba river basin of Indo-Bangladesh. Hum Ecol Risk Assess 24(5):1291–1311. https://doi.org/10.1080/10807039.2017.1411781

Timpe K, Kaplan D (2017) The changing hydrology of a dammed Amazon. Sci Adv 3:1–13

Van Zinderen Bakker EM, Werger MJA (1974) Environment, vegetation and phytogeography of the high altitude bogs of Lesotho. Vegetation 29:37–49

Varty N (1990) Ecology of the small mammals in the riverine forests of the Jubba Valley, southern Somalia. J Trop Ecol 6:179–189

Wang Y, Rhoads BL, Wang D (2016) Assessment of the flow regime alterations in the middle reach of the Yangtze River associated with dam construction: potential ecological implications. Hydrol Process 30:3949–3966

Webb AD (2017) Fire effects and management in riparian ecosystems of the Southwestern United States and Mexico. MSc Thesis, University of Arizona. http://hdl.handle.net/10150/626146

Part II
Landscape Processes, Human Security and Governance

The Fragility of Agricultural Landscapes and Resilience of Communities to Landslide Occurrence in the Tropical Humid Environments of Kigezi Highlands in South Western Uganda

Denis Nseka, Yazidhi Bamutaze, Frank Mugagga, and Bob Nakileza

Abstract This chapter examines the influence of agricultural land uses on the occurrence of landslides in the humid tropical environments of Kigezi highlands in South Western Uganda. Analysis of the agricultural land use practices is a prerequisite to understanding landscape fragility and community resilience to landslide hazards. An analysis of agricultural land use patterns was undertaken through interpretation of Sentinel 2A images for 2016. The imagery data was acquired from the European Space Agency (ESA) Sentinels Scientific Data Hub. Field surveys and investigations were also carried out to establish and map the spatial distribution of landslide hazards. Six agricultural land use categories were identified, namely, annuals, perennials, grazing, fallow, wood lots and agroforestry. Considering the agricultural land use patterns, annual crops are the dominant agricultural land use type spanning 69% of the total area followed; by perennials (13%). Grazing land and fallows covered 11% and 5%, respectively. The study revealed that annual crop land is the most affected agricultural land use category. Out of the 65 landslide scars mapped, 31% occurred on annual crop areas. Wood lot areas experienced the least landslide occurrence (4%). A close spatial distribution of agricultural land use practices and landslide occurrence is discernible. It is recommended that a comprehensive tree planting campaign be carried out in areas prone to landslides since it was inferred that wood lots suffered least from landslides. It is also recommended that farmers be encouraged to restore and manage terrace bunds which could help to check on water velocity and consequently increase on slope stability.

Keywords Agricultural land uses · Landslides · Tropical environments

---

D. Nseka (✉) · F. Mugagga · B. Nakileza
Department of Geography, Geo-Informatics and Climatic Sciences, School of Forestry, Environmental and Geographical Sciences, Makerere University, Kampala, Uganda

Y. Bamutaze
Department of Geography, Geo-Informatics, Makerere University, Kampala, Uganda

Y. Bamutaze et al. (eds.), *Agriculture and Ecosystem Resilience in Sub Saharan Africa*, Climate Change Management,
https://doi.org/10.1007/978-3-030-12974-3_13

1 Introduction

Landslides are typical hydro-geotechnical processes dominant in highland terrains of tropical and subtropical regions (Broothaerts et al. 2012; Kirschbaum and Zhou 2015). Steep slopes, high annual rainfall, increasing population pressure and deforestation make most areas in the humid tropical world very sensitive to landslides (Glade and Crozier 2005). Landslides and the associated processes are some of the most costly and damaging natural hazards of the tropical world. The East African highlands are considered to be prone to landslides due to their humid tropical climate, steep topography and high population densities (Knapen et al. 2006; Kitutu et al. 2009; Mugagga et al. 2012). Most landslide studies (e.g. Ohlmacher and Davis 2003; Ayalew and Yamagishi 2005) have focused on quasi-static factors such as geology, topography and soil strength properties, which do not change in the considered time frame. Less attention has been focused on the effect of triggering variable factors including agricultural land use practices and their implications for hillslope hydrology and soil properties (Ali et al. 2014). The conversion of forests and natural grasslands to agricultural land in highland environments is on the increase in developing countries (Breuer et al. 2009), including Kigezi highlands in Uganda. Human actions can have a significant impact on ecosystems due to the interdependence of human and ecological systems (Alexander 2013). Human activities lead to environmental degradation which increases landslide risk and contributes to vulnerability (Bahadur et al. 2013). Degradation of hillsides leads to rapid water runoff which can cause flash floods and landslides (Begueria 2006; Karlsi et al. 2009; NEMA 2012).

The majority of people in Uganda depend on agriculture for livelihood. Loss of soils and farmlands therefore renders them highly impoverished (NEMA 2014). Agricultural systems are increasingly vulnerable to landslide occurrence (Kitutu et al. 2011), yet strategies to reduce risk and vulnerability have not been greatly explored (Liesbet et al. 2015). Agriculture is the main driver of habitat destruction and may increase susceptibility to landslide damage (e.g. Begueria 2006). Landslides have degraded farmlands and the environments in Uganda's highland areas (Knapen et al. 2006; NEMA 2010; Kato and Mutonyi 2011; Kitutu et al. 2011). Landslides are a serious hazard in Uganda's highland areas, with great socio-economic and environmental impacts (Kitutu et al. 2009). Several people in these highlands have been forced to settle elsewhere (NEMA 2012).

The Kigezi highland region of South Western Uganda is an agricultural landscape prone to several degradation issues (Carswell 1997; NEMA 2010). Increased demand for agricultural and grazing land has led to destruction of the natural land cover (Farley 1996; Nkonya 2001; NEMA 2014). Increased land cover degradation is one of the major conditions for landslide occurrence in these highlands (Bagoora 1993; Lindblade and Carswell 1998). Landslide occurrence in these highlands is expected to increase in the near future given the current land pressure (NEMA 2012). Most landslides in the Kigezi highlands are occurring where there is preexisting human interference, especially agricultural practices. There is a need to investigate the prevailing agricultural land uses on the slopes and their influence on landslide occurrence. Understanding the factors that control agricultural land use patterns in a region susceptible to landslide occurrence is therefore essential. The main objective of this study was to investigate the fragility of agricultural landscapes to landslide occurrence and analyse the resilience of agricultural communities to these hazards. The results could help to draw attention of government as well as volunteer organisations in the affected areas, which is essential for future land management and landslide risk reduction of the area.

2 Resilience Framework

Resilience is the ability of a system or community to resist or to change in order that it may obtain an acceptable level in functioning and structure (UNDP 2013; USAID 2013). According to Maxwell et al. (2015), resilience emphasises the multiple ways a system can respond to hazard occurrence. Hazard resilience covers pre-event measures that seek to prevent hazards (Bahadur et al. 2015). It has been indicated by Brooks et al. (2014a) that resilience assessment helps land use planners to understand which sectors, regions and communities are the least resilient and how they can improve the adaptive capacity to respond to hazards and disturbances while sustaining a functional state of resilience. The capacities of the systems to respond to change and to create lasting well-being for people and place are some features that closely define resilience (Brooks et al. 2014b). Resilience can therefore be understood as an identifiable condition, embodying a core set of principles.

There are a number of frameworks for assessing resilience of ecosystems and societies to environmental hazards. These frameworks include DFID's disaster resilience framework (Cabell and Oelofse 2012), the UN Disaster Risk Management Framework (UNDP 2013), Oxfam's people-centred resilience approach and dimensions of resilience (OXFAM 2013; Jeans et al. 2015), USAID's conceptual framework for resilience (USAID 2013) and the Food and Agriculture Organization framework for maximising the nutritional impact of resilience programmes (FAO 2014). These frameworks have arisen due to differing organisational programming, policy priorities and purposes, as well as the fact that there is no agreed definition of resilience among the international community (FAO 2008). This study adopted the UN Disaster Risk Management Framework that demonstrates the continuum between risks, disasters and longer-term development. This framework considers disaster risk management as a continuum – an ongoing process of interrelated actions, which are initiated before, during and after disaster situations (UNDP 2013). The disaster risk management actions are aimed at strengthening the capacities and resilience of households and communities that are vulnerable to recurrent hazards to protect their livelihoods through measures to prevent or mitigate effects of hazards (Fig. 1). This framework promotes a holistic approach to disaster risk management and demonstrates the relationships between hazard risks/disasters and development (FAO 2016).

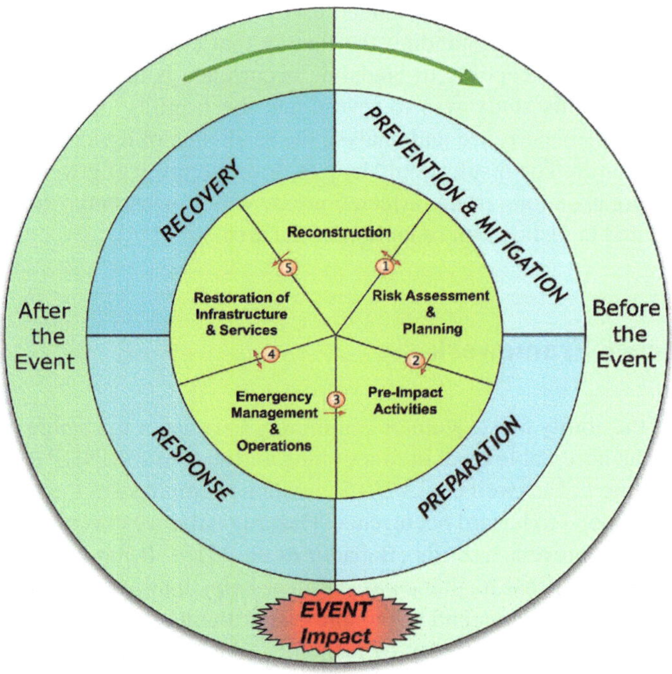

Fig. 1 UN Disaster Risk Management Framework. (Source: FAO 2008)

The UN disaster risk management framework emphasises that in order to enhance community resilience, it is necessary to carry out risk and vulnerability assessment before a major hazard in order to identify the recovery initiatives. This would help in hazard prevention, mitigation and community preparedness (Fig. 1). The framework focuses on reconstruction, which includes land use planning that can be influenced by risk assessment and vice versa. It emphasises the pre-impact scenario analysis which is often conducted based on previously constructed emergency planning scenarios. Lessons learnt from emergency experience feed back into pre-impact planning, as well as exploiting existing communication strategies. Emergency management can prioritise the restoration of services during the response. The concept of community resilience is becoming a factor framework for enhancing community-level disaster preparedness, response and recovery in the short-term (Arbon et al. 2012) and environmental change adaptation in the longer-term (Cabell and Oelofse 2012). Resilience enhances the ability of a community to prepare and plan for, absorb, recover from and more successfully adapt to actual or potential adverse events in a timely and efficient manner (Bene et al. 2012; Jeans et al. 2015). Resilience also includes the restoration and improvement of basic functions and structures (Choptiany et al. 2015). A system that is effective in managing risk through a range of options is likely to become more resilient to shocks and stresses. Managing risk in this context means reducing, transferring and sharing risk, preparing for recurrent, infrequent or unexpected events and responding or recovering efficiently (Maxwell et al. 2015).

3 Data and Methods

3.1 Study Area

The study was conducted in the humid tropical environments of Kigezi highlands in South Western Uganda situated between 01° 21′ 25″ and 0° 58′ 08″ South and 29° 43′ 30″ and 30° 05′ 51″ East. Rukiga catchment (Fig. 2) was delineated for detailed study of agricultural land use practices and landslide occurrence. In order to decide on the study area, an inventory was carried out with the guidance of local people to identify visible landslide scars in relation to agricultural land use practices. The topography of Kigezi highlands comprises mainly extensive flat-topped ridges and hills (Fig. 3), broken by short, steep-sided dip valleys and numerous subsidiary strike valleys (Ollier 1969). The highlands are characterised by distinctive advanced erosion surfaces represented particularly by the main steeply rising slopes (NEMA 2010). The geology of these highlands is composed of sedimentary rock system of the Precambrian age (Bagoora 1989). They are collectively grouped into Karagwe-Ankolean system, underlain by metamorphic gneiss and granite intrusions, as well as shale and phyllite which have given rise to clay deposits (Bagoora 1993). The study area comprises numerous highland streams which drain valleys incised in the ridges and hills (NEMA 2012). The dense network of streams serves as an index of

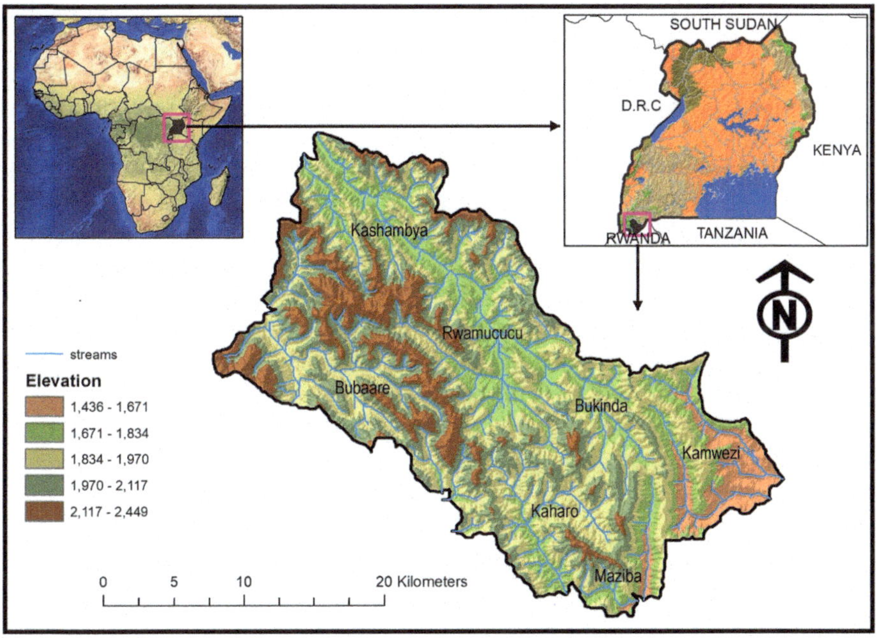

Fig. 2 Location of the study area in Uganda

Fig. 3 The landscape of the study area, 2017. (Photo credit: Nseka)

topography and morphology of the area and efficient drainage which is a characteristic of highland environments (NEMA 2010).

The climate of Kigezi highlands is warm to cool humid characterised by a bimodal rainfall pattern with annual rainfall of 994 mm (Kabale Meteorological Station, weather data, 2015: WMO No. 63726, National No. 91290000), which can be classified as moderate. Rainfall however, increases to 1250–1540 mm or more in higher areas (NEMA 2012).

The main rainfall season is from mid-February to May, with a peak in March–April, and the second season is from September to December, with a peak in October/November (NEMA 2010). The vegetation cover of these highlands was until about 500 years ago characterised by montane forests comprising mainly Mahogany and other hard woods, including species such as *Juniperus proce, Podocarpus milanjianus and Hagenia abyssinica* (Bagoora 1993). Centuries of human interference has led to serious vegetation cover degradation (Farley 1996; Carswell 1998; NEMA 2014). The conducive ecological conditions of the highlands have attracted a sedentary farming population (Nkonya 2001; NEMA 2010). The population density (409.8/km^2) of the region is one of the highest in Uganda (UBOS 2017) and has put tremendous pressure on natural resources, leading to hillslope degradation (NEMA 2014). This may be responsible for the occurrence of the observed landslide processes.

3.2 Data Sets and Sources

The agricultural land uses for Rukiga catchment were analysed using a Sentinel image for 2016. Sentinel 2A image of S2MSI1C type for 01/05/2016 was acquired from the European Space Agency (ESA) website (https://scihub.copernicus.eu/), through the Sentinels Scientific Data Hub archive. One cloud-free Sentinel 2 MSI Level 1C image covering Rukiga catchment in Kigezi highlands of South Western Uganda was selected and downloaded for agricultural land use estimation. Six agricultural land use classes were generated from the Sentinel image, including annuals, perennials, grazing, fallow, woodlots and agroforestry (Table 1). The European Space Agency's (ESA) satellite constellations Sentinel 2 programme was used to combine both high spatial and temporal resolution. Sentinel 2 sensor acquires images using 10 spectral bands, four bands at 10 m spatial resolution, featuring blue (0.49 μm), green (0.56 μm), red (0.665 μm) and near-infrared (0.842 μm). Sentinel 2 spectral range offers cirrus (0.443 μm), water vapour (0.945 μm) and aerosol (1.38 μm) bands, at 60 m spatial resolution, which have been dedicated to atmospheric monitoring. For this study, ten bands were used with the exception of cirrus, water vapour and aerosol bands. All bands at 20 m were resampled to 10 m spatial resolution using nearest neighbour resampling in Sentinels Application Platform (SNAP) environment. The image was subjected to atmospheric correction effects using the Sentinel 2 correction prototype processing tool in SNAP. The already orthorectified Sentinel 2 image was geometrically corrected in Universal Transverse Mercator projection and World Geodetic System (WGS) 84.

3.3 *Mapping the Spatial Distribution of Landslide Hazards*

Detailed field investigations were carried out to map the spatial distribution of landslide hazards. Landslide dimensions were measured during the, field surveys (Fig. 4). Field surveying and in situ measurement of landslide sites was done to collect data on landscape positions and geographic location of landslide scars. Landslide morphology was established by way of measuring key features including average width, depth and overall length from head to toe using a tape measure. Geographic location of landslide features was taken using a handheld GPS receivers.

Coordinates for the mapped landslide features were imported to ArcGIS 10.1 software to produce landslide distribution map for the study area. Through an overlay in a GIS environment, the landslide distribution and agricultural land use maps

Table 1 Description of applied agricultural land uses

Agricultural land use classes	Class description
Annuals	Maize, sorghum, beans, potatoes, vegetables,
Perennials	Bananas, coffee, tea, sugarcane,
Grazing	Grasslands, shrubs and thickets
Fallow	Abandoned land left to grow into bushes
Woodlots	Eucalyptus and pine trees
Agroforestry	Fruit trees, e.g. avocados, mangoes, apples and oranges intermixed with crops

Fig. 4 Field surveys to map agricultural land uses and landslide distribution, 2015. (Photo credit: Nseka)

were overlain to derive the requisite relationships. A geospatial analysis was conducted to determine the agricultural practices contributing to landslide occurrence in the landscape.

4 Results and Discussion

4.1 Patterns of Agricultural Practices

The agricultural land uses for the catchment are presented in Figs. 5, 6, and 7 and Appendix 1. There are six agricultural land uses observed in the catchment.

Considering the agricultural land use patterns, annual crops are the dominant agricultural land use type spanning 69% of the total area followed by perennials (13%). Grazing land and fallows covered 11% and 5%, respectively (Fig. 6).

The hillslopes appear to be intensively cultivated with gardens dominated by annual and perennial crops running from the valley bottoms to the ridge tops (Fig. 7). Increases in agricultural production in the last decades have been achieved through intensifying agricultural practices, such as increasing the frequency of cultivation at the expense of natural fallows and through expanding the cultivated areas,

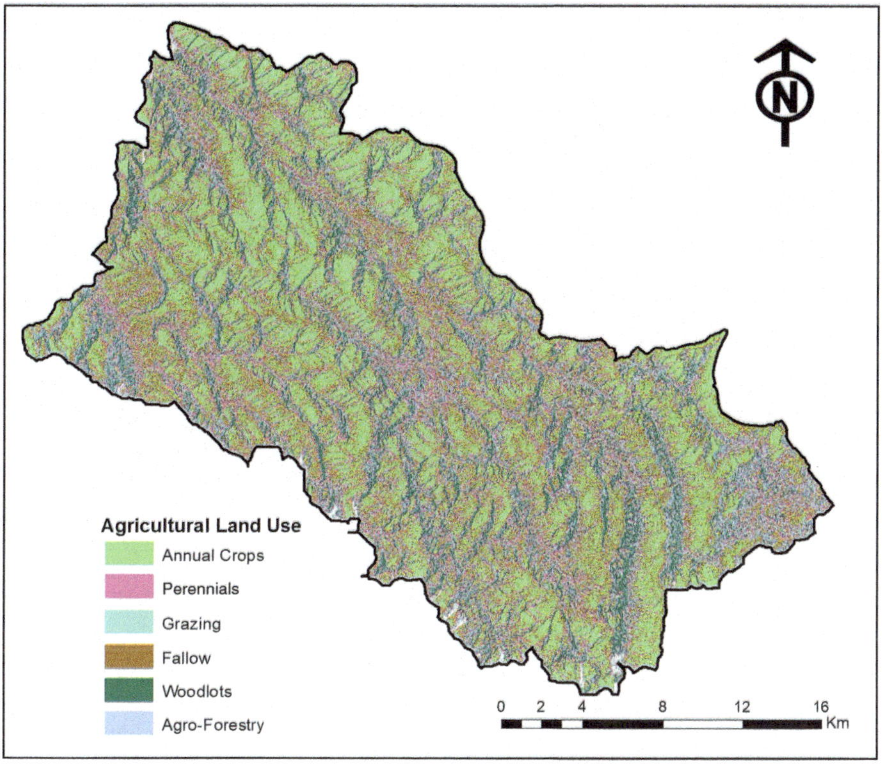

Fig. 5 Agricultural land use practices in Rukiga catchment

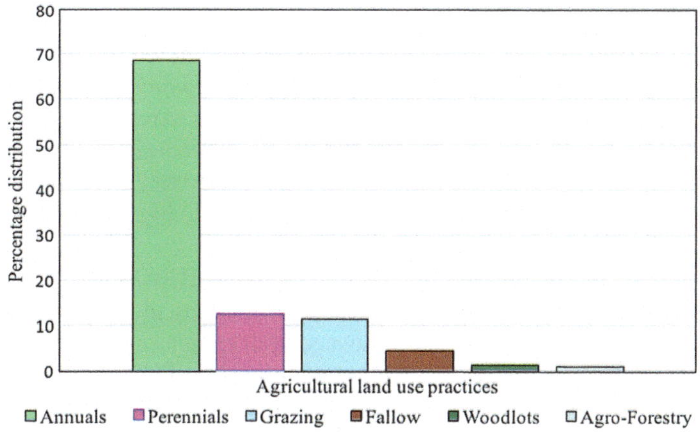

Fig. 6 Distribution of agricultural land use practices

Fig. 7 Intensively cultivated slopes with high vulnerability to landslide occurrence, 2017. (Photo credit: Nseka)

especially into fragile environments such as steep hillslopes. The study established that most areas in the catchment that had been under pasture or wetlands have been converted to cultivation (Fig. 7), and most fields are being managed with only short fallows. In the study area, most farms are small ranging between 0.5 and 2.2 hectares (UBOS 2016) and intensely cultivated, often on steep slopes.

This study established that arable farming is dominant along the topographic hollows from the valley bottoms to the upper slopes, with mainly annual crops grown including sorghum (*Sorghum bicolor*), Irish potatoes (*Solanum tuberosum*), sweet potatoes (*Ipomoea batatas*), maize (*Zea mays*), beans (*Phaseolus vulgaris*), peas (*Pisum sativum*), finger millet (*Eleusine coracana*) and vegetables. On the other hand, perennial crops such as banana (*Musa paradisiaca*), coffee (*Coffea* spp.), tea (*Eragrostis abyssinica*), sugarcane (*Saccharum officinarum*) and mangoes (*Mangifera indica*) are grown for commercial purposes. Areas abandoned due to severe land degradation are normally left under woodlots consisting of mainly eucalyptus (*Eucalyptus globulus*) and pine (*Pinus leiophylla*) trees. The spur slopes and ridge tops are dominated by cattle grazing, due to the dominance of shallow soils which do not favour crop production.

The Kigezi region population of 1,393,981 persons (UBOS 2014) threatens to wreck the fragile environment through unsustainable agricultural practices. Every available space has been intensively cultivated, and gardens appear like a continuous carpet for kilometres in the valleys and on hillslopes (Fig. 8). Almost up to the ridge top, the slopes have been invaded and stripped of natural vegetation. Increased cultivation on steep upper slopes is one of the major factors influencing landslide occurrence in the study area. Cultivation has been noted to affect the soil structure by weakening the internal cohesive forces (Selby 1993; Begueria 2006; Karlsi et al.

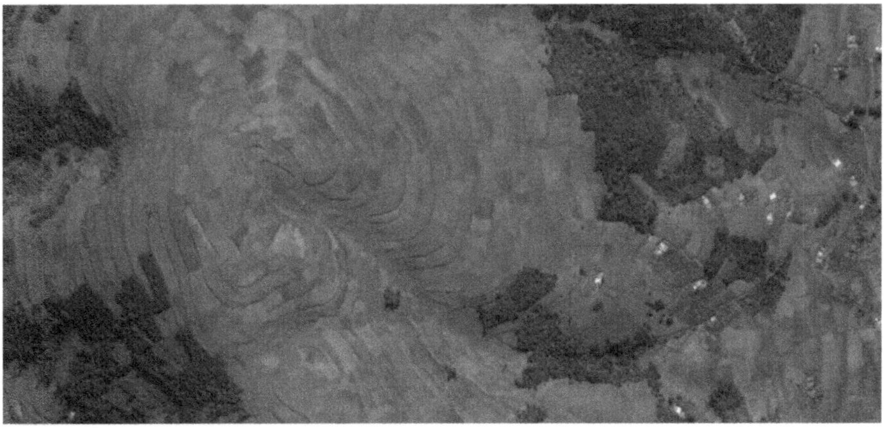

Fig. 8 Agricultural land use distribution in North Eastern part of the catchment based on Google Earth image (October 2016)

2009; Promper and Glade 2012). There is vivid evidence of land mismanagement as a result of poor methods of cultivation that could bring catastrophe to this fragile highland region as has already been experienced in the recent years (NEMA 2014). There are already signs of many hills becoming rocky after losing the top soils, forcing the peasants to look elsewhere for cultivation especially the sensitive upper steep slope elements.

In the study area, farmers have induced deep changes in the landscape to obtain terrain suitable to farming. In steep areas, the lack of suitable terrain for farming has induced farmers to modify the slopes using terraces (NEMA 2012). Terraces may be regarded as a human interference with the geomorphic system, which drives the evolution of the terrestrial surface (Sidle and Ochiai 2006). This modification has been done by varying the profile of the terrain to gain subhorizontal surfaces and managing the flow of the rainfall runoff (Bagoora 1993). Crop production is done on terraced benches developed between 1930 and 1940 (Carswell 1997). Terracing has been used for centuries to maximise agricultural productivity on slopes and can reduce erosion compared to slopes without terraces (Lindblade and Carswell 1998; Nkonya 2001). Terrace bunds help to check on the speed of runoff and hold the soils on the intensively cultivated steep slopes (NEMA 2010). However, terraces are also often associated with landslides when they concentrate water or are not properly made or maintained (Bagoora 1993). The farmers are, however, increasingly destroying contour bunds on terraces in order to get more land for cultivation (NEMA 2014). Destruction of the terrace bunds increases water velocity and susceptibility to landslide occurrence in the highlands. In many areas of Kigezi highlands, farmers have also abandoned the practice of terracing that helped to hold soils on the intensively cultivated steep slopes (NEMA 2012). The result has been the continued denudation of the hillslopes, leading to landslide occurrence.

4.2 Agricultural Practices and Fragility of the Landscapes

The agricultural land use practices involve manipulations of natural ecosystems through species removals or additions (Sidle et al. 2006). Agricultural practices usually involve changes in species composition and density, which can have both positive and negative effects on slope stability, depending on the relative change in root and canopy characteristics or flammability (Meusburger and Alewell 2008). Most of the agricultural practices in the study area, especially cultivation, involves removal of trees and other vegetations during land opening. This removal can destabilise slopes by the loss of a protective cover to intercept rain, increased soil water from reduced evapotranspiration and damage to stabilising root systems (Sidle and Ochiai, 2006). Conversion of forest to crop land is likely to decrease overall root strength (Lin et al. 2007). In the study area, most forests have been converted into farm lands dominated by annual crops. Such areas covered by annual crops have decreased soil strength and are therefore vulnerable to landslides. Walker and Shiels (2013) indicate that slope instability increases in areas where annuals replace perennials. Sidle and Ochiai (2006) observe that the conversion of forests to crops such as potatoes (India), tea (India and Japan) and corn (Honduras) has been associated with increased sediment loss. Grazing just like intensive cultivation, also leads to slope instability due to soil compaction. Soil compaction reduces water infiltration, and sustained grazing can have cumulative and destabilising effects on slope stability (Karlsi et al. 2009). Overgrazing further exacerbates erosion and can increase landslide frequency and severity (Glade 2003). According to Bagoora (1993), the increasing problem of surface runoff, erosion and landslides in Kigezi highlands is due to intense pressure on natural resources resulting from agricultural practices. This has been further aggravated by recent development of infrastructures such as road networks in unstable highland terrain without proper engineering and safety measures (NEMA 2014).

4.3 Agricultural Practices and Landslide Hazards

The spatial occurrence of landslides on agricultural land uses is illustrated in Figs. 9, 10, and 11 and Table 2.

A close spatial relationship between the agricultural land use practices and landslide occurrence is discernible (Figs. 9, 10, and 11).

As illustrated in Figs. 10 and 11, annual crop land is the most affected agricultural land use category, while wood lot areas experience the least landslide occurrence. Out of the 65 landslide scars analysed, 31% occurred on annual crop areas, while woodlots and agroforestry areas experienced only 4% and 5%, respectively (Fig. 10). In the study area, more than 60% of the landslides in the highlands occur on hillslope zones with slope gradients between 20° and 35° (Table 2). Such slope segments are characterised by active or recently cultivated agricultural fields

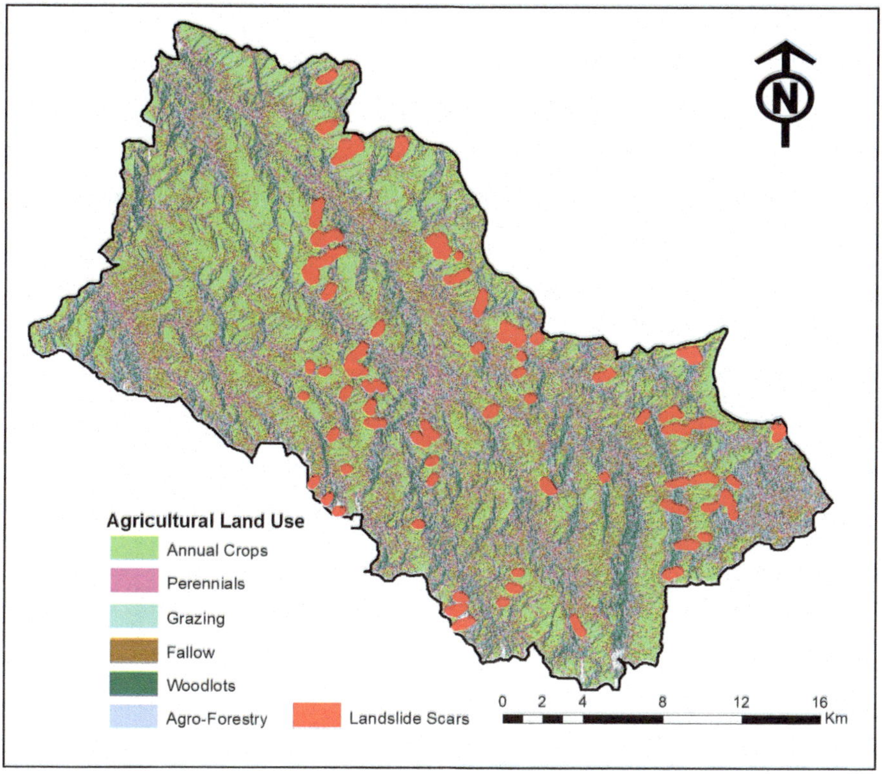

Fig. 9 A spatial distribution of agricultural land use practices and landslide occurrence

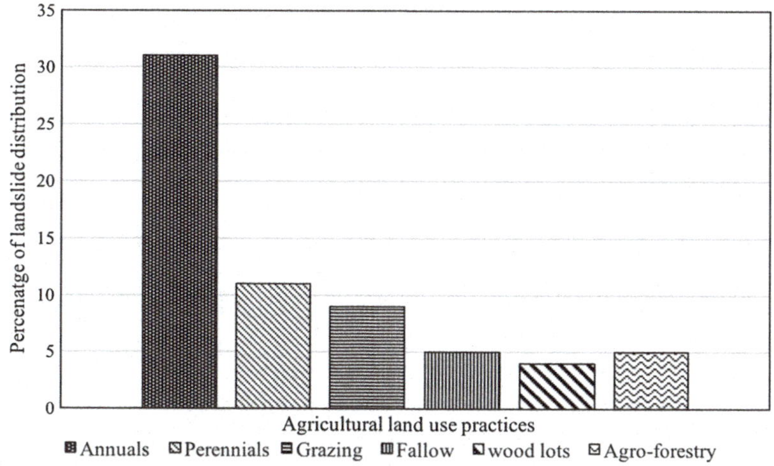

Fig. 10 Distribution of landslides with agricultural land use practices

Fig. 11 Shallow landslides on degraded and intensively cultivated slopes, 2016. (Photo credit: Nseka)

dominated by annual crops. Landslide occurrence is less pronounced in sections with gradients lower than 20°. This study established that steeper zones covered with woodlots and other forest covers were less susceptible to landslides compared to those covered by annual crops. Slope elements with high tree cover and better agroecological practices resist and recover from the effects of landslides (e.g. Selby 1993; Sidle and Ochiai 2006).

There has been a drastic decimation of the natural land cover, due to increased farming on the hillslopes. Agricultural land use practices in the study area have accentuated degradation of the hillslopes. The most obvious effect of farming on slope stability is the destruction of stabilising vegetation by the removal or reduction of canopies, ground cover and roots (Promper and Glade 2012). Thus, the landscape of the area is now dominated by various landslide scars and bare hills with few traces of terraces (Figs. 11 and 12).

Land cover conversion from forest and grassland to cultivated fields permanently reduces slope stability (Selby 1993). Agricultural practices in the study area have affected slope stability and resulted into reduced soil cohesion. Soil cohesion is lost once tree roots, which help to bind the soil materials together, are removed due to destruction of vegetation cover. According to Ferreira et al. (2015), cultivation and burning on hillslopes are the most important contributing factors for landslide occurrence. Several studies point out that the effects of hydrology and mechanism of slope failure are influenced by land use practices especially agricultural patterns (Bagoora 1993; Begueria 2006; Petley et al. 2005). Agricultural practices can cause

Table 2 Landslide scar characteristics and agricultural land uses

Slide scar no.	Scar dimensions in metres: Average width	Average depth	Total length	Landslide scar area in m^2	Volume of the scar in m^3	Average gradient (in degrees) at slide failure zone	Agricultural land use classes	Agricultural land uses
1	3.7	1.7	402	1487.4	2461.7	27	Annuals	Sorghum, beans and potatoes
2	9.66	2.3	463.5	4477.4	6064.3	30	Annuals	Maize and potatoes
3	17.5	0.74	350	6125	4426.5	34	Fallow land	Fallows with bushes
4	2.1	1.2	602	1264.2	3772.72	27	Grazing	Grasslands and shrubs
5	8.5	2.245	14.1	119.9	257.98	33	Annuals	Sorghum, potatoes, beans
6	10	5	600	6000	30,000	35	Woodlots	Eucalyptus
7	10	5	400	4000	20,000	34	Grazing	Grasslands and shrubs
8	10	5.3	498	4980	26,394	33	Grazing	Grasslands and shrubs
9	16.6	4.3	315	5229	22,600	35	Annuals	Maize, beans, sorghum
10	10	0.5	12.5	125	62.5	25	Woodlots	Eucalyptus and pines
11	5.6	0.85	525.1	2940.6	1676.2	45	Annuals	Potatoes and beans
12	5.8	2.3	530	3074	10508.3	42	Fallow land	Fallows with bushes
13	2.7	1.8	885	2389.5	3669	36	Perennials	Bananas
14	3.14	1.52	786	2468	4340.54	33	Annuals	Potatoes, beans and sorghum
15	2.95	1.6	784	2312.8	4517.94	37	Annuals	Sorghum, maize and beans
16	6.2	2.8	835	5177	14495.6	37	Grazing	Grasslands, shrubs
17	4.33	2.8	752	3256.2	9412.5	32	Annuals	Sorghum and beans
18	5	2.5	600	3000	7500	35	Perennials	Coffee, bananas

(continued)

Table 2 (continued)

Slide scar no.	Scar dimensions in metres			Landslide scar area in m²	Volume of the scar in m³	Average gradient (in degrees) at slide failure zone	Agricultural land use classes	Agricultural land uses
	Average width	Average depth	Total length					
19	2.5	2.5	653	1632.5	4081.3	37	Annuals	Sorghum and potatoes
20	1.7	1.9	268	455.6	865.64	31	Perennials	Bananas and coffee
21	1.2	1.4	198	237.6	332.64	27	Annuals	Maize and sorghum
22	0.9	2.1	213	191.7	402.57	33	Woodlots	Eucalyptus and pines
23	2.4	2	201	482.4	964.8	30	Fallow land	Fallows with bushes
24	1.7	1.4	341	579.7	811.58	19	Perennials	Bananas
25	2.3	1.9	189	434.7	825.93	23	Annuals	Potatoes, beans, sorghum
26	2.8	1.7	244	683.2	1161.44	29	Grazing	Grasslands, shrubs
27	1.9	1.2	196	372.4	446.88	34	Fallow land	Fallows with bushes
28	2.4	1.5	204	489.6	734.4	18	Annuals	Beans and sorghum
29	2.8	1.9	302	845.6	1606.64	22	Annuals	Maize and beans
30	1.9	1.6	194	368.6	589.76	32	Annuals	Potatoes, sorghum and beans
31	2.1	2	219	459.9	919.8	29	Fallow land	Fallows with bushes
32	1.3	1.8	142	184.6	332.28	26	Annuals	Maize, beans and sorghum
33	2.7	2.2	408	1101.6	2423.52	31	Annuals	Potatoes and beans
34	2.5	2	386	965	1930	37	Grazing	Grasslands, thickets
35	1.6	0.9	125	200	180	29	Grazing	Grasslands and shrubs
36	1.7	2.1	184	312.8	656.88	23	Perennials	Bananas and tea
37	2.4	2.2	296	710.4	1562.88	26	Annuals	Potatoes and sorghum

(continued)

Table 2 (continued)

Slide scar no.	Scar dimensions in metres: Average width	Average depth	Total length	Landslide scar area in m^2	Volume of the scar in m^3	Average gradient (in degrees) at slide failure zone	Agricultural land use classes	Agricultural land uses
38	1.8	1.2	202	363.6	436.32	31	Perennials	Bananas and coffee
39	1.4	1.7	182	254.8	433.16	34	Fallow land	Fallows and bushes
40	2.7	2.2	501	1352.7	2975.94	28	Grazing	Grasslands and shrubs
41	2.1	1.7	234	491.4	835.38	21	Annuals	Potatoes, sorghum and beans
42	1.5	1.8	267	400.5	720.9	32	Annuals	Maize, beans and sorghum
43	2.2	2.2	58	127.6	281	27	Perennials	Bananas, coffee and tea
44	1.9	2.7	135	256.5	693	25	Annuals	Sorghum, beans, potatoes
45	2.8	1.2	196	548.8	658.6	33	Fallow land	Fallows with bushes
46	2.3	1.9	243	558.9	1062	31	Fallow land	Fallows with bushes
47	4.2	0.9	55	231	207.9	30	Grazing	Grasslands and shrubs
48	2.9	1.1	129	374.1	411.5	35	Woodlots	Eucalyptus
49	3.2	0.7	231	739.2	517.4	32	Agroforestry	Fruit trees with crops
50	3.1	0.8	89	275.9	220.7	22	Annuals	Sorghum, beans, potatoes
51	2.8	2.1	197	551.6	1158.4	21	Perennials	Bananas, coffee and tea
52	3.4	1.7	238	809.2	1375.6	19	Annuals	Potatoes, beans
53	3.2	1.2	345	1104	1325	34	Fallow land	Fallows with bushes
54	2.8	0.8	118	330.4	264.3	18	Grazing	Grasslands and thickets
55	1.8	1.1	102	183.6	202	20	Agroforestry	Fruit trees with crops

(continued)

Table 2 (continued)

Slide scar no.	Scar dimensions in metres: Average width	Average depth	Total length	Landslide scar area in m^2	Volume of the scar in m^3	Average gradient (in degrees) at slide failure zone	Agricultural land use classes	Agricultural land uses
56	3.6	2.8	189	680.4	1905.1	32	Grazing	Grasslands
57	3.9	3.1	213	830.7	2575.2	28	Agroforestry	Fruit trees with crops
58	3.2	2.7	96	307.2	829.4	29	Fallow land	Fallows
59	1.8	1.2	47	84.6	101.5	25	Grazing	Grasslands
60	1.2	0.8	66	79.2	63.4	22	Annuals	Maize, beans, potatoes
61	5.9	2.1	138	814.2	1709.8	29	Perennials	Banana, tea
62	3.6	1.9	123	442.8	841.3	34	Annuals	Potatoes, beans, maize
63	6.2	3.2	84	520.8	1666.6	32	Perennials	Banana, coffee, beans
64	7.0	1.7	73	511	868.7	25	Woodlots	Eucalyptus
65	4.2	1.3	144	604.8	786.2	30	Agroforestry	Fruit trees and crops

Fig. 12 Landslide reactivation due to cultivation around the landslide scars, 2016. (Photo credit: Nseka)

large alterations in the hydro-morphological functioning of hillslopes (Meusburger and Alewell 2008). This is through its effects on rainfall partitioning, infiltration characteristics and runoff production (Sidle and Ochiai 2006).

The study area has experienced increasing population leading to high demand for agricultural land (UBOS 2014). The increasing population has resulted into acute shortage of land for cultivation, especially in the lowlands which are favourable for settlements, forcing people to encroach on the sensitive upper slope elements (NEMA 2014). During field investigations, it was observed that intensive cultivation is evident on the upper sensitive slope zones with gradients greater than 45°. Due to the acute shortage of land in the study area, farmers have continued to cultivate within and around the landslide scars (Fig. 12). Some slopes in the study area have already suffered landslide recurrence due to continuous cultivation around the landslide scars (Fig. 12). This is rendering most of the already affected slopes susceptible to more failures. This study established that most failures occur on slope elements that are heavily cultivated and grazed. The observed landslides were as a result of intensive cultivation on both the hilltop and steeper hillside slopes, vegetation clearance along the terraces living the soil bear and prone to sliding. Therefore, agricultural practices have an influence on landslide occurrence in the Kigezi highlands. Similar observations were made by Mugagga et al. (2012) on the slopes of mount Elgon in Eastern Uganda; they also inferred on the influence of cultivation of steep concave section on landslide occurrence.

Landslide hazards are likely to increase in this highland region given the current land pressure, which is increasing the cultivation of steep uplands. Agricultural practices also affect the water-repellent properties in some soils (Wasowski et al. 2010), which may either increase landslide frequency by promoting overland flow or decrease frequency by reducing infiltration (Sidle and Ochiai 2006). According to Karlsi et al. (2009), continuous cultivation can also promote down slope movement of individual soil and rock particles, because of the loss of cohesion with organic matter.

As noted already, increased cultivation of steep concave slopes can affect soil material shear strength (e.g. Glade 2003; Meusburger and Alewell 2008; Mugagga et al. 2012) leading to increased landslide occurrence. Wasowski et al. (2010) also observed that higher frequency and susceptibility to landslides in South Eastern Italy was a consequence of new ploughing on steep slopes for EU-sponsored wheat cultivation.

4.4 Landslide Impacts on Agricultural Land Uses

Landslides are among the most widespread geological hazards that threaten lives and cause destruction of property globally (Broothaerts et al. 2012; Kirschbaum and Zhou 2015). They continuously result in human suffering, environmental degradation, property damage and destruction of infrastructure. The majority of people in Kigezi highlands like the rest of Uganda depend on agriculture for livelihood

(UBOS 2014). Agricultural systems in these highlands are, however, increasingly becoming vulnerable to landslide occurrence (NEMA 2014). The main environmental impacts of landslides in the Kigezi highlands are damage to agricultural lands and destruction of socio-economic infrastructures. Landslides in the Kigezi highlands destroy farms and burry crops. They also affect the land quality due to intense soil loss while land managers struggle to conserve the soil (Bagoora 1993). The loss of soils and farmlands leaves these landscapes highly impoverished. Landslide occurrence in the study area also leads to flooding of the lowlands (Fig. 13a). The floods result from blockage of stream and other drainage channels by the debris flow. The floods destroy farmlands and damage crops, especially the root tubers. They also destroy other socio-economic infrastructures including roads, thus affecting the livelihoods of the agricultural communities in this fragile tropical environment. In addition, they lead to demise of the limited prime agricultural lands due to the debris deposition (Fig. 13b). Landslides also remove the top productive soils and render the post landslide areas less productive. The landslides in Kigezi highlands just like any other tropical highland environments alter the landscape morphology and render agricultural modernisation difficult (Bagoora 1993). As a result, they undermine poverty reduction strategies among rural communities, which depend on agriculture in these highlands (NEMA 2012). The need therefore to build community resilience to landslide hazards in these tropical highland environments cannot be overemphasised.

4.5 Resilience of Communities to Landslide Hazards

Resilience is the ability of a system and its component parts to anticipate, absorb, accommodate or recover from the effects of a hazardous event in a timely and efficient manner, including ensuring the preservation, restoration or improvement of its

Fig. 13 Landscape demise by (**a**) floods and (**b**) landslide debris, 2017

essential basic structures and functions (Arbon et al. 2012). Resilience is also associated with the socioecological system (Bahadur et al. 2013) and implies "spring back from" disturbances (Brooks et al. 2014b). After a shock occurs in a system, the processes of response and recovery can happen subsequently or can be overlapped (Choptiany et al. 2015). In order to reduce the impacts of landslides on society, it is necessary to quantitatively estimate the risk from potential landslides (Kirschbaum et al. 2016). Landslides will occur in the future under the same circumstances that caused them in the past (Giuseppe et al. 2016). The agricultural communities of Kigezi highlands are especially predisposed to suffer more harm from landslides due to poor accessibility and high dependence on natural resources for their livelihoods. According to Chao-Yuan et al. (2017), the first key steps toward developing an effective landslide risk management programme is understanding vulnerability factors and landslide coping mechanisms of the population. This is because the impacts suffered by the community depend upon its inherent vulnerability and the characteristics of the perturbation to which it is exposed (Kirschbaum and Zhou 2015).

This study found that the agricultural communities of Kigezi highlands are increasingly becoming resilience to landslide hazards. The communities are adopting a number of strategies to respond to the increasing occurrence of landslides in this fragile humid environment. The study established that there is increasing farmer's awareness of the landslide problems. This is evidenced by the farmer's knowledge on the seasonality and return periods of landslides. The farmers very well know which seasons of the year are most vulnerable to landslides. Such seasons correspond with the rainfall patterns of the area. The agricultural communities are also aware of which rainfall events can lead to landslides as well as their return periods. Such knowledge on landslide seasonality is increasing the resilience of these farmers to the landslide hazards. The farmers are able to prepare for the likely landslide impacts and consequently reduce on the damage. This study also found that the local communities clearly know where landslides are likely to be experienced within the landscape. Using indigenous knowledge and experience, the local communities can clearly identify the landslide path within the landscape. They have consequently responded by avoiding such areas which are susceptible to landslide hazards. This study established that slope segments likely to experience landslides in this landscape have been left to fallow into bushes and in some cases colonised by shrubs and thickets. Such adaptation strategies are also helping to build community resilience to landslide occurrence in these highlands.

This study also found that farmers have started using afforestation and reforestation strategies to cope with the increasing occurrence of landslides. The woodlots consisting of mainly eucalyptus (*Eucalyptus globulus*) and pine (*Pinus leiophylla*) trees help to hold soil materials against the shear stress forces leading to their stability. The woodlots covering the hillsides also help in reducing runoff and its associated effects (Bagoora 1993). This study established that slope segments covered by woodlots are less susceptible to landslides. This coping strategy to reduce landslide impacts has also been reported by Kato and Mutonyi (2011) on the slopes of Mount Elgon in Eastern Uganda. Tree planting strategy has therefore increased the resilience of agricultural communities to landslide hazards in this tropical highland environment.

The study further established that farmers in the Kigezi highlands have responded to the increasing problem of landslides by adopting better and sustainable farming practices. There is a high-level agricultural intensification and widespread use of soil conservation practices in many areas of the highlands. Such agricultural practices include mixed cropping, crop rotations using legumes and a good vegetation cover on plots. They are also practicing growing of cover crops, perennial crops, agroforestry and bush fallowing. These agricultural practices are important for proper land management of these highlands. Such agricultural practices are helping to create healthy ecosystems, which are capable of withstanding the landslide impacts. It has been noted by Cabell and Oelofse (2012) that community resilience to landslides can be improved through practicing agroforestry where crops are grown alongside trees to increase slope stability.

It was noted during the field surveys that farmers were aware of the landslide risk and runoff in the highlands. This explains why they dug and maintain storm storage trenches along the hillslopes. These storm storage trenches help to carry away water and materials from the landslide, hence protecting the farms and homesteads. In some areas of the highlands, farmers are also digging horizontal trenches in the "drag-down hoeing" practice (locally known as "dig down") across the fields to control the velocity of running water that generates landslides. All these adaptation options are helping to build resilience of these agricultural communities to landslide hazards. This study therefore recommends a comprehensive programme to develop such trenches within the entire landscape.

Agricultural communities in Kigezi highlands have reduced landslide impacts through avoidance of landslide hazard areas. Individuals in these highlands have also reduced their exposure to landslide hazards by educating themselves on the past hazard history of a site and therefore avoiding such areas. Susana et al. (2017) note that local governments can reduce landslide effects through proper land use policies and regulations that promote healthy ecosystems. Healthy ecosystems can withstand the shocks of environmental change and lead to increased community resilience to natural hazards (Alexander 2013) including landslides. Regulations on hilly and mountainous areas also need to be better enforced to ensure good management practices that benefit both the communities and the environment in general. According to Cabell and Oelofse (2012), proper land use planning and management can lead to healthy ecosystems which provide protection against hazards. Productive ecosystems support sustainable livelihoods and can be important assets for communities in the aftermath of disasters (Walker and Shiels 2013). Individuals in these degraded hillslopes can manage and replenish their environments through measures such as reforestation, environmentally sustainable farming practices and terracing to prevent runoff and landslides (Bene et al. 2012; Barrett and Constas 2014).

5 Conclusion

The dominant agricultural land use category in the catchment is annual crops, while woodlots is the least. Increased cultivation on steep upper slopes is one of the major factors increasing the fragility of the landscape to landslide occurrence in the study area. Annual crop areas are the most sensitive and vulnerable agricultural category to landslide occurrence. Areas covered by woodlots and agroforestry are less susceptible to landslides. Slopes with well-managed terrace bunds are relatively stable than those without. A close spatial distribution of agricultural land use practices and landslide occurrence is discernible in these tropical humid highlands.

6 Recommendations

There is a need to carry out comprehensive tree planting campaigns in areas prone to landslides within this fragile tropical environment. This study inferred that woodlots suffered least from landslides. The tree roots have a binding effect and help to hold soil materials together, hence increasing on the stability (Selby 1993; Schmaltz et al. 2017). The woodlots would improve the root density in the soil, which will help to hold soil materials together (Schmaltz et al. 2017). Root reinforcement could increase soil strength, thereby reducing slope failures (Begueria 2006; Yalcin 2011; Schmaltz et al. 2017). In order to increase community resilience to landslide hazards, it is also recommended that when planting trees on the hillslope, there must be proper space to allow under growth of other plants. It is this undergrowth which helps to hold materials together and thus reduces on slope failure (Bagoora 1993). When the trees are close to each other, they limit undergrowth, especially if it is eucalyptus which are known to consume a lot of water (Lin et al. 2007). They limit moisture for other plants, which negatively affects undergrowth leading to landslides. Therefore trees like eucalyptus should be well spaced as recommended by the forest department. The undergrowth will help to further increase on the stability of the slope materials.

It was noted in the present study that landslide occurrence and damage was low in sections dominated by perennial crops. Mixed perennials crops including bananas, tea, sugarcane and coffee help to reduce on water velocity and can therefore deter landslide occurrence. It is therefore recommended that farmers in the highlands should be encouraged to use mixed perennial crops, especially in the landslide-prone areas.

Acknowledgment The authors gratefully appreciate the research grants from Makerere University- Sweden international development agency (Sida) Phase IV (2015/2020 Agreement)-Building Resilient Ecosystems and livelihoods to Climate Change and Disaster Risk (BREAD) project 33.

Appendix 1: Agricultural Land Use Distribution

Agricultural practices	Area in hectares	Percentage
Annuals	42282.8	68.6
Perennials	7695.8	12.5
Grazing	7119.6	11.6
Fallow	2843.2	4.6
Woodlots	937.3	1.5
Agroforestry	723.4	1.2
Total	61602.1	100

References

Alexander D (2013) Resilience and disaster risk reduction: an etymological journal. Nat Hazards Earth Syst Sci 13:2707–2716

Ali G, Birkel C, Tetzlaff D, Soulsby C, Mcdonnell JJ, Tarolli P (2014) A comparison of wetness indices for the prediction of observed connected saturated areas under contrasting conditions. Earth Surf Process Landf 39(3):399–413. https://doi.or/10.1002/espe.3506

Arbon P, Gebble K, Cusack L (2012) Developing a model and tool to measure community disaster resilience. Torrens Resilience Institute, Adelaide

Ayalew L, Yamagishi H (2005) The application of GIS-based logistic regression for landslide susceptibility mapping in the Kakuda- Yahiko Mountains, Central Japan. Geomorphology 65:15–31

Bagoora FDK (1989) A preliminary investigation into the consequences of inadequate conservation policies on steep slopes of the Rukiga highlands, South Western Uganda. In: Thomas DB, Biamah EK, Kilewe AM (eds) Soil conservation in Kenya. Dept of Agri. Er, Univ. of Nairobi, Kenya

Bagoora FDK (1993) An assessment of some causes and effects of soil erosion hazard in Kabale Highland, South Western Uganda, and peoples attitude towards conservation. In: Abdellatif (ed) Resource use and conservation. Faculty of Social Sciences, Mohammed V. University, Rabat Morroco. Mountain Research and Development, vol 8

Bahadur A, Ibrahim M, Tanner T (2013) Characterising resilience: unpacking the concept for tackling climate change and development. Clim Dev 5(1):55–65

Bahadur A, Peters K, Wilkinson E, Pichon F, Gray K, Tanner T (2015) The 3As: tracking resilience across BRACED. Working paper. ODI, London

Barrett C, Constas M (2014) Toward a theory of resilience for international development applications. PNAS 111(40):14625–14630

Begueria S (2006) Changes in land cover and shallow landslide activity: a case study in the Spanish Pyrennes. Geomorphology 36:25–32

Bene C, Wood R, Newsham A, Davies M (2012) Resilience: new utopia or new tyranny? Reflection about the potentials and limits of the concept of resilience in relation to vulnerability reduction programmes. Institute of Development Studies, Brighton

Breuer L, Huisman JA, Willems P, Bormann H, Bronstert A, Croke BFW, Frede HG, Graff T, Hubrechts L, Jakeman AJ, Kite G, Lanini J, Leavesley G, Lettenmaier DP, Lindstrom G, Seibert J, Sivapalan M, Viney NR (2009) Assessing the impact of land use change on hydrology by ensemble modelling (LUCHEM). I: Model inter-comparison with current land use. Adv Water Resour 32:129–146

Brooks N, Aure E, Whiteside M (2014a) Measuring resilience: progress to date and proposed methodology: presentation assessing the impact of ICF programmes on household and communities resilience to climate variability and climate change. http://dx.doi.org/10.12774/eod_tg.March2014.sturgess2

Brooks N Aure E Whiteside M (2014b) Final report: assessing the impact of ICF programmes on household and community resilience to climate variability and climate change. Evidence on Demand. https://doi.org/10.12774/eod_cr.june2014.brooksetal

Broothaerts N, Kissi E, Poesen J, Van Rompaey A, Getahun K, Van Ranst E, Diels J (2012) Spatial patterns, causes and consequences of landslides in the Gilgel Gibe catchment, SW Ethiopia. Catena 97:127–136. https://doi.org/10.1016/j.catena.2012.05.011

Cabell J, Oelofse M (2012) An indicator framework for assessing agroecosytem resilience. Ecol Soc 17(1):18

Carswell G (1997) African farmers in colonial Kigezi, Uganda, 1930–1962: opportunity, constraint and sustainability. PhD thesis, University of London

Carswell G (1998) Soil conservation policies in colonial Kigezi, Uganda. Successful implementation and an absence of resistance. In: Beinart W, McGregor J (eds) Social history and African environment. Heinemann and James Currey (in press)

Chao-Yuan L, Cheng-Yu L, Chuphan C (2017) Risk-based models for potential large-scale landslide monitoring and management in Taiwan, geomatics. Nat Haz Risk 8(2):1505–1523. https://doi.org/10.1080/19475705.2017.1345797

Choptiany J, Garub B, Phillips S (2015) Self-evaluation and holistic assessment of climate resilience of farmers and pastoralists (SHARP). FAO, Rome

FAO (2008) Disaster risk management systems analysis: a guide book. UN FAO, Rome. Available at: http://www.fao.org/3/a-i0304e

FAO (2014) Strengthening the links between resilience and nutrition in food and agriculture: a discussion paper. United Nations FAO, Rome. Available online at: http://www.fao.org/3/a-i3777e.pdf

FAO (2016) RIMA-II: moving forward the developing of the resilience index measurement and analysis model. Available at fao.org/3/a-i5298e.pdf

Farley C (1996) Smallholder knowledge, soil resource management and land use change in the highlands of southwest Uganda. PhD Dissertation, University of Florida

Ferreira CSS, Walsh RPD, Steenhuis TS, Shakesby RA, Nunes JPN, Coelho COA, Ferreira AJD (2015) Spatiotemporal variability of hydrologic soil properties and the implications for overland flow and land management in a peri-urban Mediterranean catchment. J Hydrol 525:249–263. https://doi.org/10.1016/j.jhydrol.2015.03.039

Giuseppe F, Giovanna C, Pasquale V (2016) Evaluating performance of simplified physically based models for shallow landslide susceptibility. Hydrol Earth Syst Sci 20:4585–4603

Glade T (2003) Landslide occurrence as a response to land use change: a review of evidence from New Zealand. Catena 51:297–314

Glade T, Crozier MJ (2005) The nature of landslide hazard impact. In: Glade T, Anderson M, Crozier M (eds) Landslide hazard and risk. Wiley, Chichester, pp 43–74

Jeans H, Thomas S, Castillo G (2015) Oxfam framework and guidance for resilient development. Oxfam

Karlsi F, Atasoy M, Yalcin A, Reis S, Demir O, Gokceoglu C (2009) Effects of land use changes on landslides in a landslide-prone area (Andean, Rize, NE Turkey). Environ Monit Assess 156:241–255

Kato S, Mutonyi R (2011) The challenges of managing increasing landslides vulnerability on mount Elgon ecosystem, Uganda: a case of human interactions with its environment on the verge of collapsing

Kirschbaum DB, Zhou Y (2015) Spatial and temporal analysis of a global landslide catalog. Geomorphology. https://doi.org/10.1016/j.geomorph.2015.03.016

Kirschbaum DB, Stanley T, Yatheendradas S (2016) Modelling landslide susceptibility over large regions with fuzzy overlay. Landslides 13:485–496. https://doi.org/10.1007/s10346-015-0577-2

Kitutu MG, Muwanga A, Poesen J, Deckers JA (2009) Influence of soil properties on landslide occurrence in Bududa district, Eastern Uganda. Afr J Agric Res 4(7):611–620

Kitutu MG, Poesen JM, Deckers J (2011) Farmer's perception on landslide occurrences in Bududa District, Eastern Uganda. Afr J Agric Res 6:7–18

Knapen A, Kitutu MG, Poesen J, Breugelmans W, Deckers J, Muwanga A (2006) Landslides in a densely populated county at the footsteps of Mount Elgon (Uganda): characteristics and causal factors. Geomorphology 73:149–165

Liesbet J, Olivier D, Jean P, Damien D, Wim T, Matthieu K (2015) The Rwenzori Mountains, a landslide prone region. J Int Consortium Landslides:1612–510X. https://doi.org/10.1007/s10346-015-0582-5

Lin SH, Lai CF, Chen YJ (2007) A study on the root system of Makino's bamboo forest and its effect on the subsequent landslide. J Soil Water Conserv 39:173–187

Lindblade K, Carswell G (1998) Mitigating the relationship between population growth and land degradation. Land-use change and farm management in South Western Uganda. Ambio 27:564–571

Maxwell D, Constas M, Frankenberger T, Klaus D, Mock M (2015) Qualitative data and subjective indicators for resilience measurement. Resilience measurement technical working group. Technical series no. 4. Food Security Information Network, Rome

Meusburger K, Alewell C (2008) Impacts of anthropogenic and environmental factors on the occurrence of shallow landslides in an alpine catchment (Urseren Valley, Switzerland). Nat Hazards Earth Syst Sci 8:509–520. http://www.nat-hazards-earth-syst-sci.net/8/509/2008/

Mugagga F, Kakembo V, Buyinza M (2012) Land use changes on the slopes of Mount Elgon and the implications for the occurrence of landslides. Catena 90:39–46

NEMA (2010) National State of environment report for Uganda for 2009/2010. National Environment Management Authority, Kampala, Uganda

NEMA (2012) National State of environment report for Uganda for 2011/12. National Environment Management Authority, Kampala, Uganda

NEMA (2014) National State of environment report for Uganda for 2013/14. National Environment Management Authority, Kampala, Uganda

Nkonya E (2001) Soil conservation practices and non-agricultural land use in the South Western highlands of Uganda. SCRIP phase II report submitted to the USAID Uganda Mission. International Food Policy Research Institute, Washington, DC

Ohlmacher GC, Davis JC (2003) Using multiple logistic regression and GIS technology to predict landslide hazard in northeast Kansas, USA. Eng Geol 69:331–343

Ollier CD (1969) Terrain classification and Data storage, Uganda land System, MEXE report no: 959:0xford; University of Oxford Press

OXFAM (2013) A multidimensional approach for measuring resilience. OXFAM GB Working Paper. http://policypractice.oxfam.org.uk/publications/a-multidimensional-approach-tomeasuring-resilience-302641

Petley DN, Dunning SA, Rosser NJ (2005) The analysis of global landslide risk through the creation of a database of worldwide landslide fatalities. In: Proceedings of the international conference on landslide hazard and risk. Taylor & Francis Ltd., Vancouver, pp 350–358

Promper C, Glade T (2012) Land cover changes for landslide risk evolution first results from Lower Austria. In: Eberhardt E, Froese C, Turner AK, Leroueil S (eds) Proceedings of the 11th International Symposium on landslides and 2nd North American Symposium on landslides and Engineered slopes protecting society through improved understanding, Banff, Canada, pp 409e413

Schmaltz E, Mergili M, Steger S, Glade T (2017) Computational slope stability experiments implementing root system morphology in a generic hillslope environment. EGU General Assembly Conference Abstracts, 2017

Selby MG (1993) Hillslope materials and processes. Oxford University Press, New York

Sidle RC, Ziegler AD, Negishi JN, Nik AR, Siew R, Turkelboom F (2006) Erosion processes in steep terrain – truths, myths, and uncertainties related to forest management in Southeast Asia. For Ecol Manag 224:199–225

Sidle RC, Ochiai H (2006) Landslides: processes, prediction, and land use. American Geophysical Union, Water Resources Monograph 18, Washington, DC

Susana A, Elizabeth AH, Francesca P, Thorsten W (2017) Dealing with deep uncertainties in landslide modelling for disaster risk reduction under climate change. J Natl Hazards Earth Syst Sci 17:225–241

UBOS (2014) Uganda national household survey 2014. Ministry of Finance, Planning and Economic Development, Uganda. http://www.ubos.org.

UBOS (2016) Statistical abstracts 2016. Ministry of Finance, Planning and Economic Development, Uganda. http://www.ubos.org

UBOS (2017) Statistical abstracts 2017. Ministry of Finance, Planning and Economic Development, Uganda. http://www.ubos.org

UNDP (2013) Community Based Resilience Assessment (CoBRA) conceptual framework and methodology. United Nations Development Programme (UNDP) Drylands Development Centre. http://www.disasterriskreduction.net/File admin/userupload/drought/docs

USAID (2013) The resilience agenda: measuring resilience in USAID. https://www.usaid.gov/sites/default/files/Documents/1866/TechnicalNote

Walker LR, Shiels AB (2013) Living with landslides for landslide. In: Landslide ecology. Cambridge University Press, USDA National Wildlife Research Center – Staff Publications, New York, p 1637. https://digitalcommons.unl.edu/icwdm_usdanwrc/1637

Wasowski J, Lamanna C, Casarano D (2010) Land-use and climate change impacts on landslides. Q J Eng Geol Hydrogeol 43:387–401

Yalcin A (2011) A geotechnical study on the landslides in the Trabzon province, NE, Turkey. Appl Clay Sci 42:11–19

The Loss of Agricultural and Ecological Resilience in Abandoned Lands of the Eastern Cape Province, South Africa

Vincent Kakembo

Abstract Land abandonment is a widespread phenomenon in the communal areas of the Eastern Cape Province, South Africa, despite the exponential population increase. Non-recovery of abandoned lands is explained by a unique suite of soil properties derived predominantly from mudstone and shales, which render soils highly erodible. Soil properties associated with mudstone and shales include high proportions of silt and fine sand; pockets of duplex, sodic, and illitic soils susceptible to hard-setting; soil dispersion; crusting; and high runoff. Pockets of duplex soils have greater proportions of clay in the B horizon of up to 45% than the fine sand and silt surface samples. Soil organic carbon of many areas of the Province is as low as 0.28%. Inherent to the emergent patchy vegetation patterns on abandoned lands within the drier 200–500 mm isohyetic zone are soil crusting, infiltration inhibition, runoff generation and connectivity and erosion intensification. Such feedback mechanisms and the consequent ecohydrological impacts have engendered the total loss of agricultural and ecological resilience of abandoned lands in communal areas of the Province.

Enabling rural communities to take ownership of land rehabilitation processes should be the starting point of rehabilitating eroded abandoned lands. A dedicated programme has to be developed within the land degradation neutrality framework to rehabilitate and restore abandoned lands, as degradation hotspots. Promoting disconnectivity on invaded hillslopes and applying standard gully control measures will initiate and enhance rehabilitation from a degraded state.

Keywords Abandoned lands · Soil properties · Ecohydrological impacts · Emergent vegetation patterns · Disconnectivity

---

V. Kakembo (✉)
Department of Geosciences, Nelson Mandela University, Port Elizabeth, South Africa
e-mail: Vincent.Kakembo@mandela.ac.za

Y. Bamutaze et al. (eds.), *Agriculture and Ecosystem Resilience in Sub Saharan Africa*, Climate Change Management,
https://doi.org/10.1007/978-3-030-12974-3_14

1 Introduction

Agricultural abandonment has been reported by several studies (see Andrew 1992; Andrew and Fox 2004; Kakembo and Rowntree 2003; Mzobe 2013; van der Waal and Rowntree 2018) as a widespread phenomenon in the communal lands of the Eastern Cape Province of South Africa. In all the instances, historical trend studies and evidence from aerial photography indicate that abandonment intensified mainly in the early 1950s. According to Kakembo (2014), the intensification of this trend in the face of exponential population increase signifies the decreasing dependency on land as a source of livelihood in communal areas of the Province.

The land use change trend is not unique to the Eastern Cape Province. It has been highlighted in South Africa's KwaZulu-Natal Province by Watson (1996) and several studies of the Mediterranean Europe (Cammeraat and Imeson 1999; López-Bermúdez et al. 1998; Kosmas et al. 2000; Grove and Rackham 2001; Kosmas et al. 2002; Obando 2002; Dunjó et al. 2003; Koulouri and Giourga 2007; Arnaez et al. 2010). In southern Bolivia, Preston et al. (1997) explain the trend as a response to population decrease in rural areas due to migration and the decreasing importance of farming.

However, a distinct contrast is drawn between the Eastern Cape and other areas in terms of the implications of land abandonment for agricultural and ecological resilience. In KwaZulu-Natal, Watson (1996) noted that eroded previously cultivated land in the Mfolozi catchment contracted during the 1970s wet spell. Notwithstanding the cases of terrace failure and consequent moderate to severe erosion on abandoned land (Lesschen et al. 2008; Cammeraat et al. 2005), recovery of pedological properties, increasing vegetation cover, decreasing erosion and an increase in organic matter on abandoned land were identified by the respective studies in Mediterranean Europe pointed out earlier. A decrease in soil erosion and recovery of vegetation were also identified in southern Bolivia by Preston et al. (1997). Agricultural fields were reported to recover after abandonment. Therefore, the main objective of this chapter is to unravel the key underpinnings of the loss of agricultural and ecological resilience of abandoned lands in the Eastern Cape Province (Fig. 1).

The contrast of the province to other areas where land abandonment has occurred could certainly be explained by a multifactorial setting. That notwithstanding, this chapter will focus on the pedo-ecological underpinnings and vegetation-erosion interactions, which are considered key in the total loss of agricultural and ecological resilience of most abandoned lands. In particular, shifts in ecohydrological conditions of these lands are reviewed. As a starting point, the chapter explores the reasons for land abandonment in the Eastern Cape Province. An understanding of these aspects and related land-based processes should provide the basis for a discussion of land restoration options.

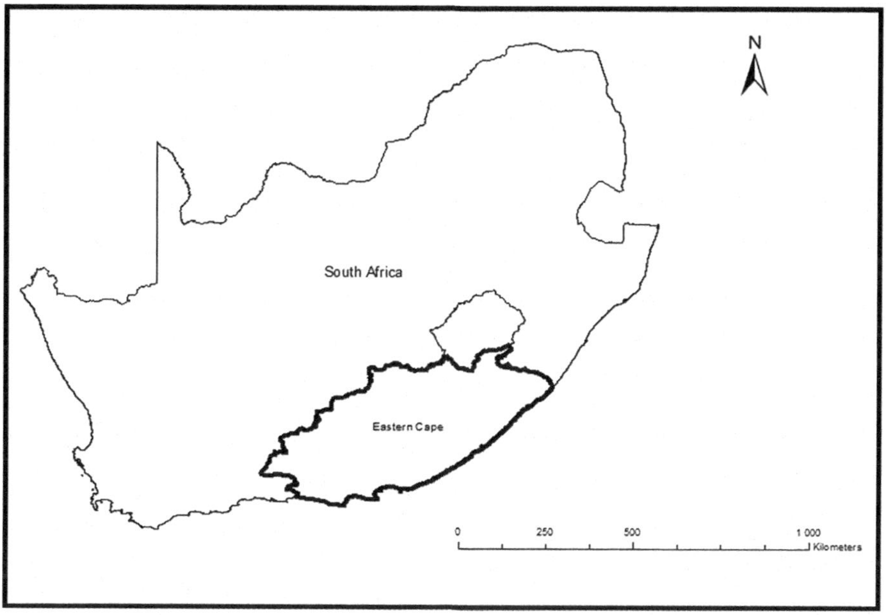

Fig. 1 The Eastern Cape Province of South Africa

2 Reasons for Land Abandonment

The major drivers of land abandonment are categorised under ecological, socio-economic and injudicious agricultural systems and mismanagement (Benayas et al. 2007; Haddaway et al. 2013; Cerdá et al. 2018). The first category includes constraining aspects such as deterioration of soil fertility, physiography, soil erosion and climate change. The migrant labour phenomenon, land tenure security and farm characteristics constitute the second category (Benayas et al. 2007). However, as observed by Boardman et al. (2012), the reasons for land abandonment in the Eastern Cape Province are unclear. Paradoxically, the intensification of abandonment in the Province coincided with an exponential increase in population (Kakembo and Rowntree 2003).

A decline in the importance of peasant production is traced by Bundy (1988) back to the period between the 1900s and 1930s, which he attributes to economic changes in the South African economy and state policies. Government subsidies that favoured white farmers and effectively excluded black peasant farmers from agricultural markets explain the origins of the decline. In their case study of the Transkeian coastal belt, Andrew and Fox (2004) pointed out the two most cited reasons for field abandonment as (1) fields getting too old, hence losing productivity, and (2) livestock damage to crops. However, they caution that farmers in this region have not abandoned fields completely. Instead, it is a shift in strategy from less productive distant fields to food gardens, whose yields per unit area are much higher than fields.

In the Keiskamma and Great Fish River catchments of the Province, divergent explanations for land abandonment have been obtained. Elderly local residents interviewed during field surveys by Kakembo (1997) ascribe this phenomenon to the long and recurrent droughts that struck the area between the early 1940s and 1970. Besides other adverse drought effects, ploughing was done using ox-drawn ploughs, in which oxen were lost during acute drought spells. This hindered the resumption of cultivation even during the short rainy spells that punctuated drought periods.

Socio-economic considerations, for instance, the pension system in the early 1960s, are alternative reasons given for cultivation abandonment. Pension payments to the rural elderly were increased in the early 1960s to match those of their urban counterparts. This provided ample source of livelihood to them, hence cultivation abandonment. The migrant labour phenomenon also targeted the young and energetic, leaving the elderly to work the land.

A survey was also conducted by Kakembo (2014) to obtain first hand information from community members in some of the Province's catchments, where severely eroded abandoned lands are widespread. The survey targeted an exclusive group of elderly residents ($n = 60$), aged between 60 and above 80 years. It is such ageing population group that was actively engaged in cultivation before its abandonment in the late 1950s and early 1960s. Asked to provide reasons why cultivated land was abandoned, the respondents singled out drought as the main reason, alongside other related problems (Fig. 2). It is discernible that drought, bad soil and loss of oxen, as the most cited reasons, are not mutually exclusive.

3 Ecohydrological Impacts of Land Abandonment

Shifts in hydrological processes and ecosystem functioning engendered by land abandonment are manifest in the modification of soil surface conditions, redistribution of water on and in the soil and consequently vegetation on the land surface. The impacts

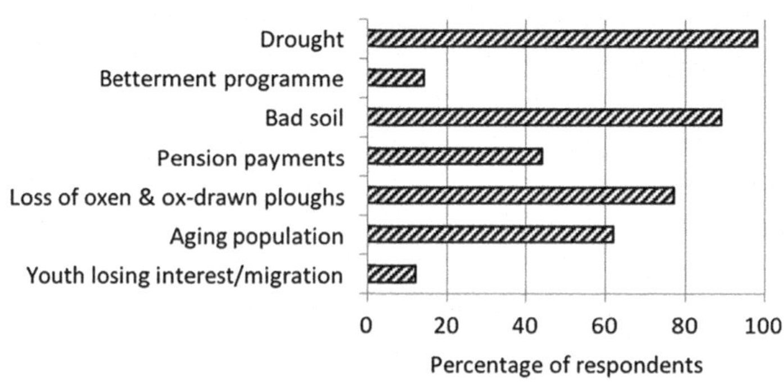

Fig. 2 Reasons for land abandonment

could either be positive or negative, depending on land use history and presence or absence of regular management systems. Both positive and negative shifts reported in literature range from impacts on biodiversity and soil properties to vegetation development. On the one hand, threats of land abandonment on biodiversity include both habitat loss and patchiness decrease, invasion of non-native plants and competitive exclusion (Plieninger et al. 2014). In a broader sense, Benayas et al. (2007) enumerate increase of fire frequency, soil erosion and desertification, reduction of water provision and landscape diversity, as well as loss of aesthetic values as the main negative impacts.

On the other hand, processes that enhance ecosystem recovery, for example, secondary succession, soil recovery, water retention and habitat diversity increase, are listed in literature (Benayas et al. 2007; Bowen et al. 2007; Haddaway et al. 2013; Plieninger et al. 2014) as benefits related to abandonment. It is noteworthy that such positive impacts may not overcome the constraints of low primary productivity, particularly in semiarid environments. Non-recovery explained by such constraints is the trajectory that characterised most abandoned lands of the Eastern Cape Province. The main constraints to recovery identified on abandoned lands of the Province are its inherent edaphic attributes.

However, it is important to distinguish between cause and effect in the endeavour to understand constraints to recovery. When asked whether land was eroded first before it was abandoned, all respondents in the survey by Kakembo (2014), alluded to earlier, indicated that land was abandoned first. They also confirmed that land was eroded because it was abandoned. By implication, impairment of the inherent edaphic attributes subsequent to land abandonment is an effect rather than a cause. The interaction of these attributes with emergent vegetation patterns, particularly in the more arid parts of the Province, is unravelled in this chapter.

3.1 Soil Properties

The inherent properties of soils on most abandoned lands of the Eastern Cape rendered them particularly vulnerable to the total loss of agricultural resilience (Fig. 3). The soils of most severely eroded abandoned lands are reported to be highly erodible, developed from Beaufort mudstones and shales (Kakembo 1997; Kakembo and Rowntree 2003; Laker 2004; Kakembo et al. 2007; Mzobe 2013; de Jager 2015; van der Waal and Rowntree 2015; Manjoro et al. 2017; van der Waal and Rowntree 2018). The shales and mudstones of the Beaufort series give rise to fine textured dispersive soils (Garland et al. 2000). The specific soil properties of the Province and their implications are explored in the subsequent subsections.

Soil Texture

Particle-size analyses of soil samples derived from shales in several catchments of the Province have exhibited high silt and very fine sand proportions, with clay content as low as 3% on average (Kakembo et al. 2012). According to Morgan (2005)

Fig. 3 Gullied abandoned lands on highly dispersive soils derived from mudstones

and Laker (2004), soils with high silt and fine sand proportions, with low clay content, are the most erodible. The particle binding role of clay is non-existent, rendering the soils easily detachable. The shallow Mispah soil form (Leptosols) derived from mudstones and shales in many parts of the Province has a high silt content, with a high erodibility factor.

Site-specific variations in soil texture, characterised by abundance of the clay fraction, particularly at gullied sites have, however, also been identified. Subsurface soil samples collected by de Jager (2015) from several catchments at gullied abandoned land sites displayed greater proportions of clay of up to 45% than the fine sand and silt surface samples. This confirmed the duplex nature of the soils at the sites, whose implications for erosion are explained below. It is also a well-known fact that a clay fraction well above 10% constitutes a problem soil; the clay proportion is adequate to supply colloids to sustain dispersive piping (Mhangara 2011).

Duplex soils that are highly erodible with widespread gully erosion are also common in the Province (Le Roux et al. 2010). Owing to an increase in clay in the B horizon, compared to the overlying horizons, such soils characteristically impede water movement and root growth and are highly susceptible to erosion (Fey 2010). The soils are also associated hard-setting when dry, owing to a weak A horizon with a dominance of silt and fine sand. As observed by De Santis et al. (2010), dispersive soils have severe problems with crusting and reduced porosity. Consequently, soil surface conditions characterised by crusting (Fig. 4) are widespread on many abandoned and overgrazed lands of the Eastern Cape.

Fig. 4 Soil surface crusting on abandoned land

Crusting triggers ecohydrological feedback mechanisms that entail infiltration impediment, enhancement of runoff generation and connectivity, particularly on patchy vegetation, as explained in the section on emergent vegetation patterns. The biophysical attributes of soil on abandoned lands must have become impaired with time, especially during the drought periods. Thus exposed, soil crusting and age-long hardening will have considerably impeded infiltrability and promoted runoff at the onset of the wet period.

Clay Mineralogy

Illite has been identified by several studies (Kakembo et al. 2007; Mhangara 2011; De Jager 2015) as the dominant clay mineral in soils on eroded, usually severely gullied abandoned lands of the Province. The mineral, which tends to predominate in soils derived from shales and mudstones, has a gross structure and 2:1 lattice similar to that of montmorillonite. High runoff, erosion and sensitivity to surface crusting are associated with illitic soils (Laker 2004). This in part explains the widespread nature of crusted soil surfaces in most Eastern Cape soils.

Subsurface soil sample analyses by de Jager (2015) identified the presence of kaolinite (Fig. 5), as well as smectite in the clay mineral assemblage of the deepest samples collected from some abandoned land gullies. The phenomenon of 'kaolinite contamination by smectites' (Laker 2004) could therefore be a possible scenario.

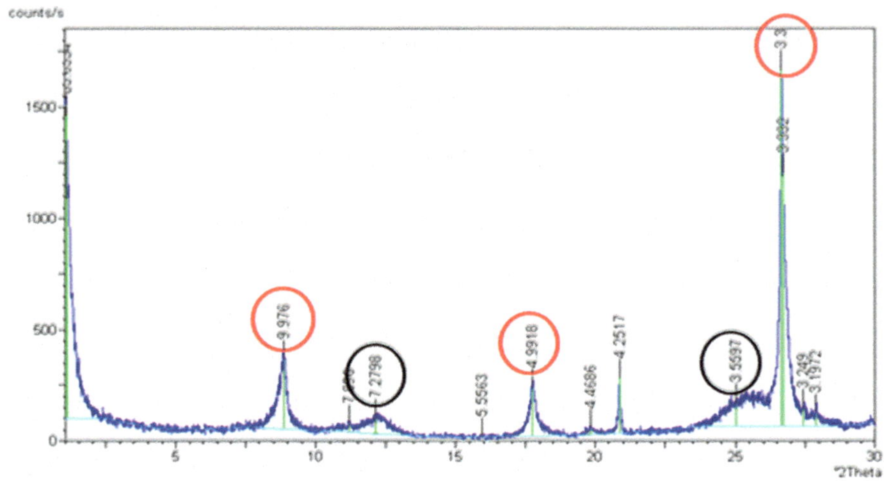

Fig. 5 X-ray diffractogram of the dominant clay minerals present in three catchments of the Eastern Cape Province. Illite and kaolinite peaks are encircled in red and black, respectively. (After de Jager 2015)

Sodicity

Soils on many abandoned lands of the Province have been identified as sodic, hence dispersive, owing to the Exchangeable Sodium Percentage (ESP) exceeding 5% in most of the soil samples. Advanced piping and gullying in the former Transkei homeland was identified by Beckedahl (1996) in soils with ESP values of up to 19%. Similarly, soil analyses by Mhangara (2011) in the former Ciskei revealed ESP values of most soil samples in the A and B horizons ranging from 5% to 15%. Elevated sodicity levels of up to 35% were distinct in the B horizon, particularly in areas dominated by soil piping (Fig. 6).

According to Laker (2004), soil dispersion is linked to sodium as the most dispersive cation in soils. Sodicity enhances swelling and dispersion of clay particles and hence the deterioration of the physical conditions of the soil (Sumner 1993). It also inhibits infiltration, promotes soil piping and reduces plant-available water capacity (Peverill et al. 1999).

However, Laker (2004) cautions that the effect of sodium on dispersion and soil erodibility must not be viewed simplistically, as no single threshold ESP value may define erodibility. On the one hand, he singles out a study of highly weathered sesquioxic South African soils where ESP exceeding 40% did not induce dispersion. On the other hand, ESP values of just 2.5% were associated with severe erosion in soils derived from Beaufort mudstones and shales. Conversely, highly erodible melanic soils with low organic matter were noted as highly erodible with ESP values as low as 0.12. Therefore, sodium may be surpassed by other factors as a determinant of dispersion. By implication, parent material, for instance, the Beaufort mudstones and shales, determines the influence of sodium on erodibility of the derived soils.

Fig. 6 Piping and gully erosion associated with highly sodic soils derived from mudstones and shales in the Keiskamma catchment, Eastern Cape Province

Other factors that are not necessarily unique to the Eastern Cape have also contributed significantly to the loss of agricultural resilience of abandoned lands. It is a well-known fact that organic matter binds soil particles together, providing protection against erosion. Almost 60% of South Africa's soils have less than 0.5% organic carbon, and only 4% contain more than 2% organic carbon (du Preez et al. 2011). Organic carbon of soil samples from several catchments of the Eastern Cape, analysed by Kakembo et al. (2007), Mhangara (2011) and de Jager (2015), ranged between 0.28% and 1.65%. In the absence of judicious land stewardship in most communal areas of the Eastern Cape, a combination of the soil properties reviewed above will have precluded the recovery of abandoned lands.

3.2 Emergent Vegetation Patterns

The emergence of *Stipa tenacissima* vegetation patterns on abandoned fields in Mediterranean semiarid SE Spain was investigated by Cammeraat and Imeson (1999) under conditions of drought and grazing pressure. Similarly, the invasion of abandoned and overgrazed lands by *Pteronia incana* patchy karroid shrub has been examined by Kakembo (2004), Kakembo et al. (2006, 2007) and Kakembo (2009). In both the Mediterranean SE Spain and Eastern Cape Province, South Africa, the major characteristics inherent to the emergent patchy patterns were identified as bare inter-shrub areas (Fig. 7), soil surface crusting and patchiness loss.

Fig. 7 *P. incana* patchy invader vegetation on abandoned lands of the Eastern Cape Province. Small pockets of grass remnants can be seen middle foreground microtopographic hollows

The ecohydrological impacts of such patchy spatial vegetation patterns in terms of soil surface conditions and redistribution of water on and in the soil have been the focus of different studies (see Bedford and Small 2008; Seyfried and Wilcox 2006; Ludwig et al. 2005; Holm et al. 2002; Bryan and Brun 1999; Cammeraat and Imeson 1999; Macdonald et al. 1999; Valentin et al. 1999; Bergkamp et al. 1996; Tongway and Hindley 1995).

The interaction of soil attributes described in the foregoing sections with *P. incana* patchy vegetation patterns in the semiarid 200–550 mm isohyetic zone of the Eastern Cape Province (see map, Fig. 8) is explored in this section. The implications for soil surface conditions, hydrologic constraints, runoff generation and erosion are discussed.

Soil Surface Conditions

Soil crusting is inherent to vegetation-bare patch mosaics. The phenomenon typifies *P. incana*-bare inter-patch mosaics. The predisposition to crusting has been the focus of many patchy vegetation studies (Kinast et al. 2014; Kakembo 2009; Bryan and Brun 1999; Cammeraat and Imeson 1999; Valentin and d'Herbès 1999; Bromley

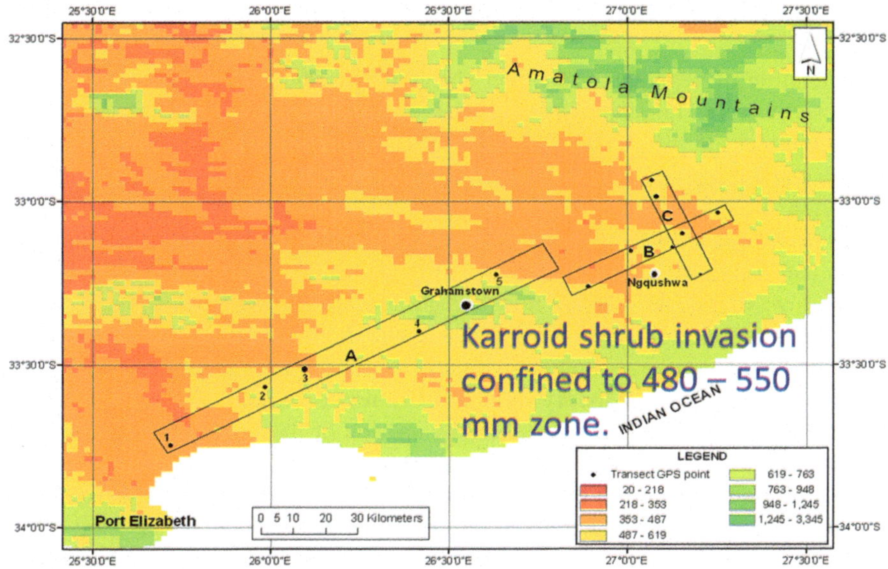

Fig. 8 Transects along which *P. incana* shrub invasion, confined to 20–550 mm isohyetic zone, were surveyed. (After Odindi 2009)

et al. 1997; Bergkamp et al. 1996; Dunkerley and Brown 1999). The susceptibility of soils of the Eastern Cape Province to crusting was alluded to earlier. Soil crusts of up to 1 mm thick, observed as erosional crusts, according to the description by Valentin and Bresson (1998), are a widespread phenomenon on abandoned lands of the Province, particularly in the *P. incana* inter-patch bare areas.

Before vegetation-bare patterns emerge, land disturbances and the attendant physico-chemical dispersion precede age-long hardening. This results from earlier disturbance of particle-to-particle bonds, as one of the effects of drought on bare soil surfaces. It is noteworthy that abandoned lands were identified on sequential aerial photography by Kakembo and Rowntree (2003), as distinctly bare and, in most instances, eroded surfaces. Crusting precedes the development vegetation-bare zone mosaics in prior-to-disturbed environments. It is also true that vegetation-bare zone mosaics do promote soil crust formation, particularly on overgrazed lands. Kakembo (2009) confirmed that soil surface conditions in terms of crusting under *P. incana* are significantly greater than under grass.

The significance of soil crusts in the development of patchy vegetation lies in reduced infiltrability, inhibiting seedling emergence of particularly grasses and run-off generation (Valentin and Bresson 1998). The differences in soil moisture dependencies between *P. incana* and grasses give the former a competitive advantage under crusted soil surface conditions. A spatial relationship between soil crusts and soil moisture was confirmed by Kakembo (2009) and Valentin and d'Herbès (1999). The ecohydrological constraints associated with *P. incana* patchy vegetation are unravelled in the subsequent subsection.

Soil Moisture Constraints

The alteration of vegetation cover on abandoned lands to *P. incana*-bare surface mosaics constrains hillslope hydrology, as demonstrated by the consistently less moisture under the invader tussocks than grasses, particularly after rainfall events (Odindi and Kakembo 2011; Weiss (2010); Kakembo 2009). This signifies reduced infiltration and increased runoff under the invader vegetation. Similarly, Weiss (2010) noted wetting fronts under grass were 3.2 and 4 times deeper than under *P. incana* and inter-shrub bare patches. By implication, the main water pathways of throughflow, saturation overland flow and groundwater flow are constrained on the one hand, while Hortonian overland flow is enhanced on the other. As observed by Kakembo (2013), this constitutes a disruption of field water cycle.

The two-layer moisture theory in the context of grasses and *P. incana* was articulated by Kakembo (2009). The removal by erosion of the topsoil layer, which the grass species are largely dependent on, and the formation of erosion crusts whose implications were discussed earlier, restrains the soil moisture budget. Under such conditions, the two-layer soil moisture theory is largely inhibited. Surface soil moisture is limited, and it is only the deep-rooting shrubs, in this case *P. incana*, which are able to capture some of the surface runoff generated on crusted soil surfaces that will survive. This explains the competitive advantage and subsequent blanket invasion *P. incana* gains over grass species on abandoned lands.

A distinct divide between soil moisture dependencies for *P. incana* and grass species is illustrated in Fig. 9. Soil moisture readings were taken during the dry season from grass patches and *P. incana* clusters. A calibrated Delta-T ML2 metre, which provides soil moisture values for surface soil up to 7 cm, was used. Normalised Difference Vegetation Index (NDVI) values were generated from high-resolution (1x1m) infrared imagery, corresponding to the respective grass patches and *P. incana* clusters.

Grass patch NDVI values are distinctly positive and coincide with high soil moisture as opposed to the negative values for *P. incana.* As can be noted from the figure, grass species are, under the given conditions, confined to areas where surface soil moisture volume is on average 10% and above.

Runoff Generation and Erosion

Bare inter-patch areas and often crusted zones between vegetated patches are characteristically runoff connectivity and generation zones. As pointed out by Janeau et al. (1999), a strong feedback relationship exists between bare surface crusts and vegetated zones, which in turn determines the local runoff pattern. Bare crusted zones enhance runoff flowpaths, which could coalesce into larger pathways. The phenomenon of patchiness loss and its implications for runoff enhancement and erosion has been investigated by several studies (Kakembo 2009; Holm 2000; Cammeraat and Imeson 1999).

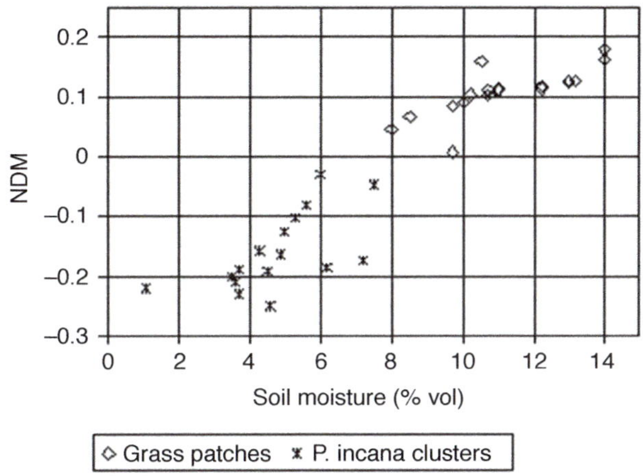

Fig. 9 Divide between soil moisture dependencies for *P. incana* and grass

Fig. 10 An increase in soil erosion in worst category, as shrub invasion intensifies on abandoned land

Owing to patchiness loss, an analysis of satellite imagery by Manjoro et al. (2012) noted an increase in soil erosion in the worst category, as invader shrub invasion intensified on abandoned lands (Fig. 10).

Bare surfaces identified in 1998 deteriorated into severely gullied areas in 2008. Enhanced connectivity, patchiness loss, erosion intensification and conversion of hillslopes to dysfunctional landscapes were also discerned by Kakembo et al. (2012) from sequential 1x1 m high-resolution imagery between 2001 and 2009. Therefore, loss of agricultural and ecological resilience of abandoned lands can be explained as a manifestation of an ecohydrological positive feedback mechanism that entails a chain of processes elucidated above.

4 Land Restoration Options

In the quest to achieve sustainable development goal (SDG) 15, the United Nations Convention to Combat Desertification (UNCCD) adopted the concept of land degradation neutrality (LDN), which strives to achieve a land degradation neutral world by 2030. The UNCCD set among other targets the reduction of degraded land and the impact of invader species. Mainstreaming of LDN in land use planning, restoration and rehabilitation of degraded land are recommended.

South Africa in particular has set specific targets within the LDN framework to ensure the restoration of degraded ecosystems by 2030, increasing land cover and productivity, while contributing to climate change mitigation and adaptation (LDN National Working Group Report 2015). The working group also commits to developing a communication and awareness-raising strategy and implementation plan for LDN by 2020. Whereas sectoral programmes to support LDN, such as 'Landcare Programme', 'Working for Water', 'Working for Wetlands' and 'Working for Ecosystems' are in place, the report identifies a major weakness as the lack of clearly defined targets and processes. For example, no dedicated programme has been developed to rehabilitate and restore abandoned lands, as degradation hotspots. There is a need to rehabilitate eroded abandoned lands to functional landscapes and ultimately restore them productive farmland.

Restoration seeks to re-establish ecosystem structure and function to its state before disturbance (Bradshaw 2002), as illustrated in Fig. 11. It is essentially a long-term process. What can be achieved in the period up to 2030 is rehabilitation, which is a shorter-term ideal that comprises actions of improvement from a degraded state and seeks to reinstate ecosystem functionality (Bradshaw 1987).

Whereas rehabilitation and restoration of abandoned lands are feasible within the LDN framework, it is the former that is achievable within the set timeframes. Follow-up efforts would then take degraded ecosystems to restoration. As conceded by the National Working Group Report, it is indeed unrealistic to expect South Africa to have restored all degraded ecosystems by 2030.

Community participation is pointed out by the report as one of the mechanisms that can influence the achievement of LDN. However, LDN is described by Oettle (2015), as a top-down concept whose meaning and implications rural South Africans are unware of. No governance initiatives exist that would enhance community understanding of degradation processes and serve to improve the success of rehabilitation efforts (Palmer and Bennett 2013). In the case of the Eastern Cape where land is communally owned, rehabilitation and restoration are achievable by linking

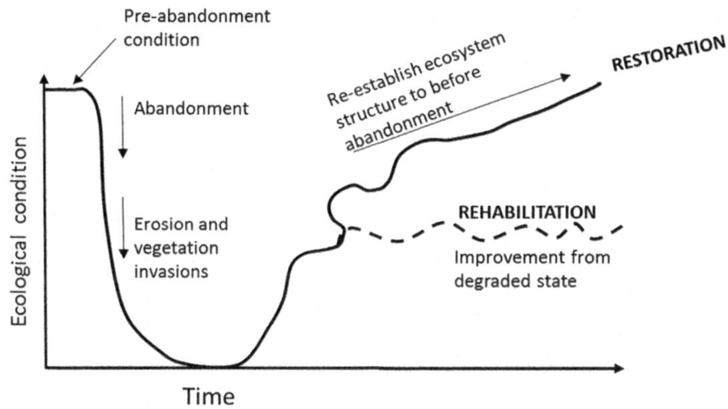

Fig. 11 A schematic illustration of rehabilitation and restoration. (Adapted from Bradshaw (1987)

communities to landscapes, which entails among other things sensitising and encouraging rural communities to take ownership of the rehabilitation process by way of active participation, and should be adopted as a strategy in the rehabilitation process. As stated by Kakembo (2014), communal farmers should be sensitised and encouraged to take ownership of the land restoration process.

Against the background of the moisture dependencies of shrubs and grasses, efforts to restore surface soil moisture conditions were noted by Kakembo (2013) as conducive to grass regeneration. Enhanced connectivity on hillslopes, owing to patchiness loss was highlighted earlier. Laying brush piles and mulching using cleared invader shrubs on hillslope sections with wide and narrow inter-patch areas, respectively, are recommended. The measures would effectively promote disconnectivity and trap sediment between the various components of the affected hillslopes (see Kakembo 2013). Identified as run-on zones, microtopographic concavities on hillslopes would be the focal points of the mulching effort.

Gullied areas on most abandoned lands are usually complex gully fans dissecting colluvium accumulations on lower slope elements (Kakembo 2009). Promotion of disconnectivity upslope of the gully sites would control runoff and headward extension of gullies. Coupled to standard gully control measures that include brush fills, loose stone and gabion check dams on the lower sections of gully fans would stabilise gullies and control excessive sedimentation of adjacent streams. Community ownership of rehabilitation processes would restore the functionality and ecological resilience of abandoned lands, most of which are currently in a state of dysfunctional landscapes of the Eastern Cape Province.

5 Conclusions

This chapter has presented the implications of land abandonment for agricultural and ecosystem resilience in the communal lands of the Eastern Cape Province. The phenomenon is largely attributed to the recurrent droughts that hit the province between 1940 and 1970. Contrary to the positive impacts of abandonment identified by some studies, non-recovery explained by constraints of low primary productivity characterised abandoned lands of the Eastern Cape. A predominant derivation of properties from mudstone and shales, characterised by high silt and fine sand proportions, renders soils of the Province highly erodible. Duplex soils, characterised by abundance of clay in the B horizon and hard-setting when dry, are widespread in the Province. Illitic soils, also associated with the predominant shales and mudstone, are susceptible to crusting and high runoff. Soil dispersion, strongly linked to sodicity, is also a derivation of shales and mudstones.

Patchy vegetation species, particularly *P. incana* emergent vegetation patterns, typify abandoned lands in the 200–550 mm isohyetic zone of the Eastern Cape Province. Ecohydrological constraints associated with the patch vegetation patterns comprise reduced infiltration, hence a soil moisture deficit, inhibition of throughflow, enhancement of Hortonian overland flow and consequent disruption of the field water cycle. Patchiness loss enhances runoff connectivity, erosion intensification and eventual conversion of abandoned fields to dysfunctional landscapes, signifying total loss of agricultural and ecological resilience.

Restoration of degraded land by 2030 within the LDN framework is unrealistic. There is a need to develop a dedicated programme to rehabilitate and restore eroded abandoned lands of the Province. Linking communities to landscapes should be adopted as a strategy to ensure community participation in the rehabilitation process. Specifically, laying brush piles and mulching, particularly within bare interpatch zones and microtopographic concavities, would enhance run-on zones on degraded and invaded abandoned hillslopes. Besides standard gully control measures, promotion of disconnectivity upslope of gully sites would serve to control headward extension of gullies.

References

Andrew M (1992) A geographical study of agricultural change since the 1930s in Shixini location, Gatyana District, Transkei. MA thesis. Rhodes University Grahamstown

Andrew M, Fox RC (2004) Undercultivation' and intensification in the Transkei: a case study of historical changes in the use of arable land in Nompa, Shixini. Dev South Afr 21:687–706. https://doi.org/10.1080/0376835042000288851

Arnaez J, Lasanta T, Errea MP, Ortigosa L (2010) Land abandonment, landscape evolution, and soil erosion in a Spanish Mediterranean mountain region: the case of Camero Viejo. Land Degrad Dev 22:537–550

Beckedahl HR (1996) Subsurface soil erosion phenomena in Transkei and southern KwaZulu-Natal, South Africa. Ph.D thesis. University of Natal

Bedford DR, Small EE (2008) Spatial patterns of ecohydrologic properties on a hillslope-alluvial fan transect, central New Mexico. Catena 73:34–48

Benayas JMR, Martins A, Nicolau JM, Schulz JJ (2007) Abandonment of agricultural land: an overview of drivers and consequences. Pers Agr Vet Sci Nut and Nat Res 2:1–14

Bergkamp G, Cammeraat LH, Martinez-Fernandez J (1996) Water movement and vegetation patterns on shrubland and an abandoned field in two desertification-threatened areas in Spain. Earth Surf Process Landf 21:1073–1090

Boardman J, Hoffman MT, Holmes PJ, Wiggs GFS (2012) Soil erosion and land degradation. In: Holmes P, Meadows M (eds) Southern African geomorphology: recent trends and new directions. African Sun Media. pp. 307–328. ISBN: 978-1-920382-02-5

Bowen ME, McAlpine CA, House APN, Smith GC (2007) Regrowth forests on abandoned agricultural land: a review of their habitat values for recovering forest fauna. Biol Conserv 140:273–296

Bradshaw AD (1987) Restoration: an acid test for ecology. In: Jordan WR, Gilpin ME, Aber HJD (eds) Restoration ecology: a synthetic approach to ecological research. Cambridge University Press, Cambridge, UK, pp 23–29

Bradshaw AD (2002) Introduction and philosophy. In: Perrow MR, Davy AJ (eds) Handbook of ecological restoration, Principles of restoration. Cambridge University Press, Cambridge, pp 3–9

Bromley J, Brouwer J, Barker T, Gaze S, Valentin C (1997) The role of surface water redistribution in an area of patterned vegetation in south west Niger. J Hydrol 198:1–29

Bryan RB, Brun SE (1999) Laboratory experiments on sequential scour/ deposition and their application to the development of banded vegetation. In Valentin C, Poesen J, (eds) Special issue: the significance of soil, water and landscape processes in banded vegetation patterning. Catena 37:147–163

Bundy C (1988) The rise and fall of the South African peasantry, 2nd edn. David Phillip, Cape Town

Cammeraat LH, Imeson AC (1999) The evolution and significance of soil-vegetation patterns following land abandonment and fire in Spain. In Valentin C, Poesen J (eds) Special issue: the significance of soil, water and landscape processes in banded vegetation patterning. Catena 37:107–127

Cammeraat LH, van Beek LPH, Kooijman A (2005) Vegetation succession and its consequences for slope stability in SE Spain. Plant Soil 278:135–147

Cerdá A, Rodrigo-Comino J, Novara A, Brevik EC, Vaezi AR, Pulido M, Giménez-Morera A, Keesstra SD (2018) Long-term impact of rainfed agricultural land abandonment on soil erosion in the Western Mediterranean basin. Prog Geogr 42:1–18

De Jager M (2015) An analysis of soil properties associated with badland and gully erosion in rural catchments of the Ngqushwa District, Eastern Cape Province. MSc thesis, Nelson Mandela Metropolitan University

de Santis F, Giannossi ML, Summa V, Tateo F (2010) Impact of physico-chemical soil properties on erosion features in the Aliano area (Southern Italy). Catena 81:172–181

du Preez CC, van Huyssteen CW, Mnkeni PNS (2011) Land use and soil organic matter in South Africa 1: a review on spatial variability and the influence of rangeland stock production. S Afr J Sci 107(5/6):8. https://doi.org/10.4102/sajs.v107i5/6.354

Dunjó G, Pardini G, Gispert M (2003) Land use change effects on abandoned terraced soils in a Mediterranean catchment, NE Spain. Catena 52:23–37

Dunkerley DL, Brown KJ (1999) Banded vegetation near Broken Hill, Australia: significance of surface roughness and soil physical properties. In Valentin C, Poesen J (eds) Special issue: the significance of soil, water and landscape processes in banded vegetation patterning. Catena 37:75–88

Fey M (2010) Soils of South Africa. Cambridge University Press. ISBN: 9780511782183

Garland GG, Hoffman MT, Todd S (2000) Soil degradation. In: Hoffman MT, Todd S, Ntshona Z, Turner S (eds) A national review of land degradation in South Africa. South African National Biodiversity Institute, Pretoria, South Africa, pp 69–107

Government of South Africa Report (2015). Land Degradation Neutrality (LDN) legal and institutional environment. LDN National Working Group South Africa Report

Grove AT, Rackham O (2001) The nature of mediterranean Europe, an ecological history. Yale University Press, New Haven

Haddaway NR, Styles D, Pullin AS (2013) Environmental impacts of farm land abandonment in high altitude/mountain regions: a systematic map of the evidence. Environ Evid 2:18

Holm AM (2000) A study of degradation processes at patch to landscape scale within the arid shrubland of Western Australia. PhD thesis, University of Western Australia, Perth

Holm AM, Bennett LT, Loneragan WA, Adams MA (2002) Relationships between empirical and nominal indices of landscape function in the arid shrubland of western Australia. J Arid Environ 50:1–21

Janeau JL, Mauchamp A, Tarin G (1999) The soil surface characteristics of vegetation stripes in Northern Mexico and their influences on the system hydrodynamics. An experimental approach. In Valentin C, Poesen J (eds) Special issue: the significance of soil, water and landscape processes in banded vegetation patterning. Catena 37:165–173

Kakembo V (1997) A reconstruction of the history of land degradation in relation to land use change and land tenure in Peddie District, former Ciskei. MSc thesis, Rhodes University, Grahamstown

Kakembo V (2004) Factors affecting the invasion of Pteronia Incana (Blue bush) onto hillslopes in Ngqushwa (formerly Peddie) District, Eastern Cape. Ph.D thesis. Rhodes University

Kakembo V (2009) Vegetation patchiness and implications for landscape function: the case of *Pteronia incana* invader species in Ngqushwa Rural Municipality, Eastern Cape, South Africa. Catena 77:180–186

Kakembo V (2013) Land disturbance, vegetation invasions and implications for land degradation. In: Daniels AJ (ed) Advances in environmental research, vol 29. Nova Science Publishers Inc, New York, pp 183–200

Kakembo V (2014). Less dependency on land, more erosion! The twenty-first century in the Eastern Cape Province, South Africa. Professorial inaugural lecture NMMU Inaugural Lecture Series, ISBN No. 978–1–920508-36-4

Kakembo V, Rowntree KM (2003) The relationship between land use and soil erosion in the communal lands near Peddie Town, Eastern Cape, South Africa. Land Degrad Dev 14:39–49

Kakembo V, Palmer AR, Rowntree KM (2006) The use of high resolution digital camera imagery to characterise the distribution of Pteronia Incana Invader Species in Ngqushwa (Formerly Peddie) District, Eastern Cape, South Africa. Int J Remote Sens 27:2735–2752

Kakembo V, Rowntree KM, Palmer AR (2007) Topographic controls on the invasion of *Pteronia incana* (Blue bush) onto hillslopes in Ngqushwa (formerly Peddie) district, Eastern Cape. Catena 70:185–199

Kakembo V, Ndlela S, Cammeraat E (2012) Trends in vegetation patchiness loss and implications for landscape function: the case of *Pteronia incana* in the Eastern Cape. Land Degrad Dev 23:548–556

Kinast S, Ashkenazy Y, Meron E (2014) A coupled vegetation–crust model for patchy landscapes. Pure Appl Geophys 173:983–993

Kosmas C, Danalatos N, Gerontidis S (2000) The effect of land parameters on vegetation performance and degree of erosion under Mediterranean conditions. Catena 40:3–17

Kosmas C, Danalatos NG, López-Bermúdez F, Romero Diaz MA (2002) The effect of land use on soil erosion and land degradation under Mediterranean conditions. In: Geeson NA, Brandt CJ, Thornes JB (eds) Mediterranean desertification: a mosaic of processes and responses. Wiley, Chichester, pp 57–70

Koulouri M, Giourga C (2007) Land abandonment and slope gradient as key factors of soil erosion in Mediterranean terraced lands. Catena 69:274–281

Laker MC (2004) Advances in soil erosion, soil conservation, land suitability evaluation and land use planning research in South Africa, 1978–2003. S Afr J Plant Soil 21:345–368

Le Roux JJ, Nkambule VT, Mararakanye N, Pretorius DJ (2010) Provincial mapping of gully erosion at the field scale using high resolution satellite imagery (SPOT 5). ISCW Report No. GW A/2009/04. ARC-Institute for Soil, Climate and Water: Pretoria, South Africa

Lesschen JP, Cammeraat LH, Nieman T (2008) Erosion and terrace failure due to agricultural land abandonment in a semi-arid environment. Earth Surf Process Landf 33:1574–1584

López-Bermúdez F, Romero-Díaz A, Martínez-Fernandez J, Martínez-Fernandez J (1998) Vegetation and soil erosion under a semi-arid Mediterranean climate: a case study from Murcia, Spain. Geomorphology 24:51–58

Ludwig JA, Wilcox BP, Breshears DD, Tongway DJ, Imeson AC (2005) Vegetation patches and runoff-erosion as interacting ecohydrological processes in semiarid landscapes. Ecology 86:288–297
Macdonald BCT, Melville MD, White I (1999) The distribution of soluble cations within chenopod-patterned ground, arid western New South Wales, Australia. In Valentin C, Poesen J, (eds) Special issue: the significance of soil, water and landscape processes in banded vegetation patterning. Catena 37:89–05
Manjoro M, Kakembo V, Rowntree KM (2012) Trends in soil erosion and woody shrub encroachment in Ngqushwa district, Eastern Cape Province, South Africa. Environ Manag 49:570–579
Manjoro M, Rowntree KM, Kakembo V, Foster I, Collins AL (2017) Use of sediment source fingerprinting to assess the role of subsurface erosion in the supply of fine sediment in a degraded catchment in the Eastern Cape, South Africa. J Environ Manag 194:27–41
Mhangara P (2011) Land use/cover change modeling and land degradation assessment in the Keiskamma catchment using remote sensing and GIS. PhD thesis, Nelson Mandela Metropolitan University
Morgan, RPC (2005) Soil erosion and conservation, 3rd edn. Blackwell Publishing. ISBN: 1–4051–1781-8
Mzobe PN (2013) Sediment linkages in a small catchment in the Mount Fletcher southern Drakensberg region, Eastern Cape, South Africa. MSc thesis, Rhodes University
Obando J (2002) The impact of land abandonment on regeneration of semi-natural vegetation: a case study from the Guadalentin. In: Geeson NA, Brandt CJ, Thornes JB (eds) Mediterranean desertification: a mosaic of processes and responses. Wiley, Chichester, pp 269–276
Odindi OJ (2009) The invasion of *Pteronia incana* (Blue bush) along a range of gradients in the Eastern Cape Province: Its spectral characteristics and implications for soil moisture. PhD thesis Nelson Mandela Metropolitan University
Odindi OJ, Kakembo V (2011) The hydrological response of *Pteronia incana* invaded areas in the Eastern Cape Province, South Africa. Ecohydrology 4:832–840
Oettle N (2015) Land degradation neutrality and a new fund to advance the concept. The adaptation network. Available: http://www.adaptationnetwork.org.za/wp-content/uploads/2013/07/AN-Newsletter-September2015.pdf. Accessed 25 April 2018
Palmer AR, Bennett JE (2013) Degradation of communal rangelands in South Africa: towards an improved understanding to inform policy. Afr J Range Forage Sci 30:57–63
Peverill KI, Sparrow LA, Reuter DJ (1999) Soil analysis: an interpretation manual. CSIRO Publishing, Collingwood
Plieninger T, Hui C, Gaertner M, Huntsinger L (2014) The impact of land abandonment on species richness and abundance in the Mediterranean Basin: a meta-analysis. PLoS One 9:1–12
Preston D, Macklin M, Warburton J (1997) Fewer people, less erosion: the twentieth century in southern Bolivia. Geogr J, 163:198–205
Seyfried MS, Wilcox BP (2006) Soil water storage and rooting depth: key factors controlling recharge on rangelands. Hydrol Process 20:3261–3276
Sumner ME (1993) Sodic soils: new perspectives. Aust J Soil Res 31:683–750
Tongway D, Hindley N (1995) Manual for soil condition assessment of tropical grasslands. CSIRO, Canberra
Valentin C, Bresson LM (1998) Soil crusting. In: Lal R, Blum WH, Valentin C, Stewart BA (eds) Methods for assessment of soil degradation. CRC Press, New York, pp 89–107
Valentin C, d'Herbès JM (1999) Niger tiger bush as a natural water harvesting system. In Valentin C, Poesen J (eds) Special issue: the significance of soil, water and landscape processes in banded vegetation patterning. Catena 37:231–256
Valentin C, d'Herbès JM, Poesen J (1999) Soil and water components of banded vegetation. In Valentin C, Poesen J (eds) Special issue: the significance of soil, water and landscape processes in banded vegetation patterning. Catena 37:1–24
van der Waal B, Rowntree KM (2015). Assessing sediment connectivity at the hillslope, channel and catchment scale. Report to the Water Research Commission. WRC Report No 2260/1/15

ISBN 978–1–4312-0715-2. Available: http://www.wrc.org.za/Knowledge%20Hub%20 Documents/Research%20Reports/2260-1-15.pdf. Accessed 22 April 2018

van der Waal B, Rowntree KM (2018) Landscape connectivity in the upper mzimvubu river catchment: an assessment of anthropogenic influences on sediment connectivity. Land Degrad Dev 29:713–723

Watson HK (1996) Short and long term influence on erosion of settlement by peasant farmers in KwaZulu-Natal. S Afr Geogr J 78:1–6

Weiss N (2010) The effect of Bluebush invasion on soil hydrologic characteristics in the Ngqushwa (formerly Peddie) district, Eastern Cape, South Africa. MSc thesis, University of Amsterdam

Catching Rain: Sand Dams and Other Strategies for Developing Locally Resilient Water Supplies in Semiarid Areas of Kenya

Wayne S. Teel

Abstract Water is one of the key limiting factors in arid, semiarid, and subhumid lands throughout Africa. Reducing the labor demand involved in the daily efforts to find and haul water releases available time to women to pursue productive agricultural activities and care for their families. This, in turn, improves resilience at the household and community level. Sand dams have proven an excellent way to do this in the subhumid and semiarid areas of Kitui, Machakos, and Makueni Counties in Kenya and are spreading further inside and outside the country. The key to building a sand dam is community involvement from the earliest stage in the process. Locally controlled and organized NGOs are the primary promoters of this activity. The community development groups, called self-help groups, site the dam, prepare the supporting infrastructure, build the dam, and coordinate local control of the new resource. This then drops the water collection time from more than 3 hours per day to as little as 15 minutes. The work then continues with promotion of drought-resistant crops, better fodder supply, and income security.

Most of this article is based on my work with two nongovernment organizations, SASOL and the Utooni Development Organization (UDO) in 2009, 2011, and 2013, spending time building dams, visiting even more sites, and conversing with their leaders, field workers, and area farmers. Field work was done with field agents of UDO and five students from the Kenya Field School, a study abroad program of James Madison University. All the pictures are mine.

W. S. Teel (✉)
James Madison University, Department of Integrated Science and Technology,
Harrisonburg, VA, USA
e-mail: teelws@jmu.edu

Y. Bamutaze et al. (eds.), *Agriculture and Ecosystem Resilience in Sub Saharan Africa*, Climate Change Management,
https://doi.org/10.1007/978-3-030-12974-3_15

1 Defining the Water Problem in Machakos, Makueni, and Kitui

Semiarid land is defined as region with low and highly variable rainfall with correspondingly high-risk agriculture. In many cases, the historic land use was pastoralism, but as population grew, agriculture spread to marginal landscapes capable of producing crops only in above-average rainfall years. The potential for failure is always high and risks are great. Resilience in these situations comes if the supply of landscape stored water resources is readily available. There are two critical types of stored water: soil water captured near the surface and readily available for crops and deeper groundwater or vadose water that moves underground to springs, which supplies water in the dry seasons. Both are dependent on capturing rainfall at the surface and allowing it to percolate into the soil and deeper strata.

The problem across subhumid and semiarid lands in both Kenya and many regions of Africa is that population growth, combined with political instability and poor agricultural practices, has contributed to a denuding of landscapes (Teel 1994). This often means that rain, especially early season rains, falls on nearly bare soils capped with dry clays. Instead of penetrating, this water runs off the surface to the nearest stream, carrying some of that soil and organic matter with it. The loss of organic material is critical as it has the primary responsibility for storing soil water (Lal 2016). This lowers landscape resilience especially in lower rainfall seasons, making it even more difficult for farmers to get a crop. At the same time, the increased flush of runoff contributes to a slow and steady lowering of the water table, resulting in dry springs and greater labor requirements to fetch water (Tiffen et al. 1994; Mutiso, personal communication).

Resilience in any system depends on the ability to address the major limiting factors in a system. In semiarid lands water is often the most important limiting factor, but the reasons for that shortfall are not always immediately apparent. A spring no longer supplying water year-round may have shrunk because of actions far higher in the watershed. Conversion of a higher elevation forest to agricultural land is a common cause, and this could be exacerbated by grazing pressure from cattle, sheep, and goats and subsequent loss of ground cover. Ironically the ability to address these problems is compromised by a lack of labor. Women are normally tasked with fetching water in rural areas. If the local spring is dry, this entails carrying water much further, decreasing the labor available for agricultural or household tasks. In the semiarid agricultural lands of Kenya, it is not uncommon for this task to take more than 3 hours per day (Utooni Development Organization, personal communication). Any effort to improve resilience of an area has to start with reducing this labor demand, or they will founder.

Expansion of the water resource has happened locally in unexpected and highly creative ways in the land of the Kamba-speaking people in the counties of Machakos, Makueni, and Kitui in Kenya. Here the landscape is hilly, with the highest points reaching nearly 2000 meters. This region lies between 0 and 3 degrees south latitude and 37 and 39 degrees east longitude, and rainfall patterns are complex and often insufficient to produce crops. The intertropical convergence zone, the equatorial phenomenon that produces weather in the tropics, passes over the region twice a year, giving it two rainy seasons. The long rains occur in March, April, and into May. The short rains last from October to early December. However, these two 60-day periods of rain are inconsistent, occasionally fail completely, and sometimes do not last long enough to allow a crop to complete its growth cycle. As a result, the region has a chronic food deficit, broken only in the rare good season when rains permit growth of maize.

In order to paint an accurate picture of this region's rain profile, it is important to take into account the rainfall variation across the region's geography. The topography slopes from high points in Machakos and northwestern Makueni, lowers toward Kitui to the north, and gradually drops in elevation and gets correspondingly drier as you move south and east toward Kenya's Indian Ocean Coast (see Fig. 1). The upper regions receive 800 to 1200 mm of rainfall per year on average, while the lowest areas of Kitui and Makueni receive less than 500 mm annually. Generally, maize requires 600 mm of rainfall in one season to guarantee a crop. Only in the hilltops of the upper region in normal or above-average seasons are rainfalls sufficient. Yet every year, the Kamba people persist in planting maize in the hope that they will catch a good rainy season.

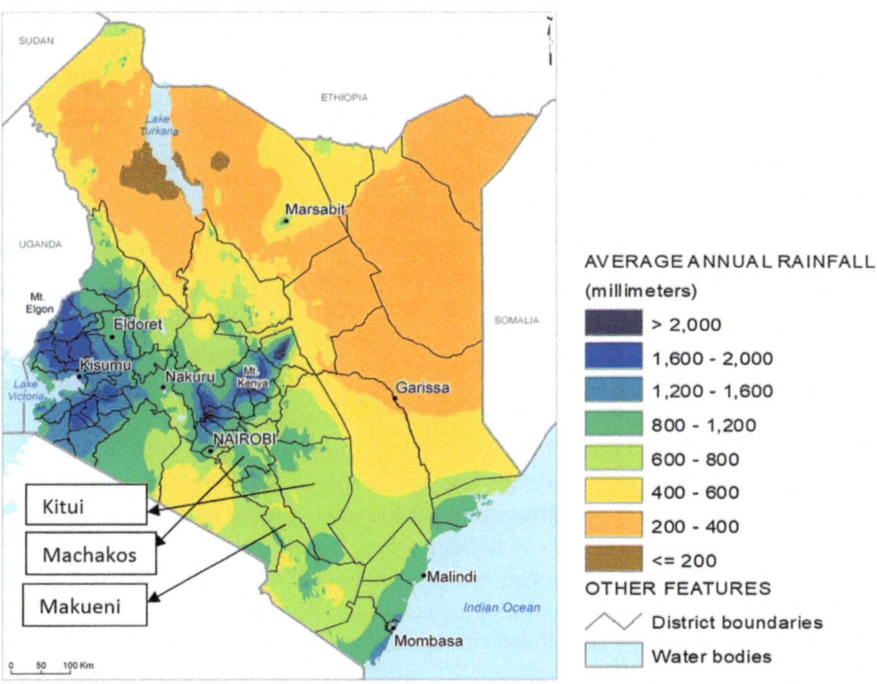

Fig. 1 Rainfall map of Kenya. (World Resources Institute (n.d.))

2 Harnessing Water

Like the rest of Kenya, population growth among the Kamba exploded in the period just before and following independence in 1964, reaching a maximum of 4% in the early 1980s. This put a lot of pressure on the landscape. Even prior to the end of British colonialism, erosion was named as one of the most severe problems. Yet because labor-intensive terracing was mandated by the colonial government, many Kamba abandoned the practice following independence. The problem persisted, and many resumed terracing and pursued tree planting by the 1980s, a fact noted in some detail in Mary Tiffen's book, *More People, Less Erosion* (Tiffen et al. 1994). Yet this effort did not address what was recognized by some as a deeper and more pressing problem, the availability of water. Water is the key limiting factor throughout this region. Resilience, defined as the ability to recover after periods of high environmental stress often caused by drought, depends on a steady supply of locally available water for a variety of reasons.

In the 1970s, Joshua Mukusya, working with the National Christian Council of Kenya (NCCK), was very aware of the dual problems of erratic rainfall and poor supply systems. Mukusya grew up near Kola, a town not far from the present border between Machakos and Makueni Counties. As a young boy, he was often sent down the hill to fetch water in whatever container he could find and then carry it back uphill to his mother. If he spilled a drop on the way, he received a sharp verbal or physical reminder of why he should be careful. As he grew, he wondered why it is that rain falls on his house and runs off downhill, and then later he has to run downhill to fetch it. Why not catch it before it runs away?[1]

In 1978, Joshua and his wife, Rhoda, joined with five other families and founded a self-help group (SHG) they called the Utooni SHG. They wanted to work on issues related to water, agriculture, firewood, trees, and local economic development, including housing. The fundamental importance of water, primarily pushed by the women as they did most of the fetching, kept them focused, as did the lessons from development failures that Joshua witnessed during his time with NCCK. The self-help group is still functioning today and has grown into the Utooni Development Organization (UDO), which works throughout Machakos and Makueni Counties and has even expanded into Kajiado County among the Maasai. They primarily build sand dams, but they never lost focus on Joshua's original vision of not having to chase water downhill. One look at their compound in Kola tells the story (Fig. 2).

The UDO compound consists of four large buildings made of rock and cement and a covered parking area. It stands on a slope that is now terraced gardens with lots of trees, which is somewhat unusual for the area. All the buildings have metal roofs, and every roof has gutters to collect water. Hardly a drop of rain landing on the roof ever hits the ground; instead, water is channeled to a number of water tanks. Five of these tanks are the large black plastic variety: three holding 10,000 liters and two of 5000 liters each. These, however, are dwarfed by the 220,000 liter capacity

[1] Joshua Mukusya, personal communication July 2011.

Fig. 2 The UDO compound's main water storage tank, which holds 220,000 liters. The square tank on the left holds 60,000 liters. The main office building is on the right. (Photo by author 2011)

large cement tank that holds nearly all the runoff from the office building, the covered parking area, and two ends of the long, narrow buildings where guests stay (Fig. 2). A smaller cement tank next to it holds an additional 60,000 liters. Joshua calculated that the 320,000 liters of water storage would only be completely filled during a heavy rainy season, but he wanted to catch all possible water, and he does. UDO seldom has to buy water, nor do they need to expend extra energy to fetch it. UDO models its own philosophy through example: resilience means making the best use of the resources close to home.

UDO has taken this philosophy to the field. A self-help group near the town of Mtito Andei in Makueni County built a catchment system on a large rock outcropping. These rocky outcroppings are primarily hunks of granitic rock very resistant to erosion that were remaining after the surrounding plain eroded. It is impossible to know if the barren nature of these slopes is natural or a result of past overgrazing, but to UDO they became a resource. The Miamba Mitamboni SHG and UDO designed a catchment on the rock face, outlined by a low rock-and-cement wall ranging between 15 and 30 cm high, depending on the slope and location (see Fig. 3 below). The wall channels rainfall to an outlet pipe at the lowest section of the approximately 3500m^2 catchment. This water flows to two 150,000 liter storage tanks. Because there is no forage on the catchment rock, animals stay away; and the captured water is quite clean and much lower in salt compared with the water found

Fig. 3 Catchment area of the Miamba Mitamboni SHG in southern Makueni County. The low cement and rock walls converge at a point behind the shrub in the lower left and feed into a pipe leading to the storage tanks. (Photo by author 2013)

in nearby streams or wells.[2] In fact, it is the preferred water supply for all the families within donkey cart distance of the storage tanks (Fig. 4).

The members of Miamba Mitamboni SHG sell the water for 3 Kenya shillings per 20 liter jerry can. This money goes into a group fund to finance group projects like their community garden, which is irrigated by catchment water, as well as a revolving microloan fund available to SHG members.

3 The Role of Sand Dams

Although catchment systems are impressive, the rock required for their creation is relatively rare; and so most of UDO's energy is focused on building sand dams. The idea of the sand dam arose from work that Joshua Mukusya did with a man named Ndunda who had built dams with the British in the 1950s. Joshua visited some of these dams with Ndunda and saw that many were filled with sediment, primarily sand. When he dug in the sand behind the dam, there was water even in the driest of seasons. The thought came to him, "why not build dams intentionally to capture the

[2] We tested some sand dams near this rock catchment on a large, ephemeral stream. These dams are managed by the Kithito kya Atumia na Iveti SHG and the Ilikoni Beekeepers SHG and averaged just under 3000 ppm salinity. The full data is found in Table 1 toward the end of this chapter.

Fig. 4 Miamba Mitamboni water storage tanks with the native acacia woodland in the background, southern Makueni County. (Photo by author 2013)

sand?" During his work with NCCK, he had seen literature about subsurface dams that slowed the flow of underground water and allowed pastoralists to water their animals without the need to dig deeply and lift the water to the animal. Instead, the subsurface dams enabled the animals to drink the water directly without additional human labor. Mukusya knew that a subsurface dam like this one would not work very well in Machakos, since the river had too steep a slope and the amount of sand trapped would be minimal. However, by constructing a dam between 1 and 3 meters high, they could capture more sand and therefore more water in the pore space between the sand grains. Joshua and the Utooni Development Self Help Group built their first trial dam in 1978, and it worked.

Sand dams are of necessity bulky structures like the one seen in Fig. 5. They stand across a stream and hold a considerable volume of water and sand that varies depending on the height of the dam and the degree of slope in the stream. In order to hold fast, they must anchor to the bedrock of the stream. Across the entire width of the stream, loose bedding material, including sand, is cleared all the way to the bedrock. This involves digging through a lot of soil to anchor the wings of the dam. Most dams measure between 1 and 4 meters in height above bedrock and are highly variable in width. Fig. 5 shows a typical dam, this one built by the Kitandi Fruit Tree Growers SHG. In general the dams are just less than 2 meters thick at their base and slope inward on their downstream side until they narrow to just 1 meter of thickness at the top. The upstream side is vertical and straight. The dam in Fig. 6 contains approximately 85 cubic meters of cement, sand, water, and rock. These materials

Fig. 5 This sand dam was built in 2009 by the Kitandi Fruit Tree Growers SHG on the Kaiti River in northern Makueni County. It was extended 1 year later to reach its present height. It has a 17.9 meter spillway and holds nearly 11,000 cubic meters of sand and 4000 cubic meters of water when saturated. Note the strong stand of napier grass on the right stream bank. It is also planted as cuttings on the left bank. (Photo by the author, July 2013)

Fig. 6 A newly constructed sand dam by the Kitito Self Help Group, Machakos County, completed in June 2013. The dam has a 17.5 meter spillway. The 1.4 meter high left wing adds an additional 7.7 meters and the right wing another 6.0 meters. It stands 2.8 meters above bedrock and is 1.8 meters thick at the base, 1.0 meter at the top of the spillway, and 0.68 meters at the top of each wing. The entire structure has a volume of approximately 85 cubic meters. The dam was built by hand with more than 100 community members involved in the process. (Photo by author 2013)

are anchored to bedrock using 18 mm rebar that extends from the rock to just below the top of the dam. Forms for the cement are made after the rebar is secured and the process of building the dam begins.

The sand in the dam, seen in Fig. 5, builds up naturally from the bed of the stream. Natural erosion moves sand as water flows every rainy season. The dam slows the movement of water, and the sand particles, larger than silt or clay and unable to float in slow-moving water, drop out rather quickly. Each time it rains, more sand moves downstream, and the dam fills up in consecutive layers. According to UDO field officers, this takes between 1 and 3 years. Early in the process, the sand dam will have some clay and silt, most of which concentrates on the surface of the sand, but this is flushed over the dam in the next rainy season. When fully mature, the dam fill is primarily sand. The dam shown in Fig. 5 was measured in July of 2013 and found to contain nearly 11,000 cubic meters of sand. Since sand has a porosity of between 35% and 42%, this means it contains 3800 cubic meters of water at full capacity. Moreover, this does not include the elevated water levels in the natural water table on both sides of stored sand. The total amount of water available for extraction from this dam could potentially be greater than ten million liters (Hut et al. 2007).

Perhaps the most impressive aspect of the Kitito SHG dam in Fig. 6 was the community involvement in building it. UDO stresses the importance of community action in any dam they help build. First any community group must be officially recognized as a SHG by the government, and this requires the raising and paying of a fee that is often too taxing to a poor community. Once this is done, UDO will come to survey the area for appropriate locations for a sand dam and select a site. At this point the community is required to do two more things. First they must build two terraces on each side of the potential sand dam's sand catchment area, which in Kitito SHG's case meant 4, 250 meter long terraces. These require about 1 person-day of labor for every 2 meters of terrace. Second they must collect all locally required materials for the dam, primarily sand and rock. These make up over 90% of the volume of the dam, so between 75 and 80 cubic meters of material, all collected by hand. These two activities do a great deal to build community relationships in a common task, contributing overall to community resilience. At this point they can build the dam, which in many ways is a celebratory act, the culmination of community building activities.

The water newly available from these dams has transformative effects, even if they are not immediately visible. UDO has documented a number of these changes, as has Sahelian Solutions (SASOL)—another NGO working on sand dams in the neighboring county of Kitui. The first impact is felt by women, since they are normally the ones who fetch water for their households. It is not unusual for a woman to cut the time used to fetch water by 3 hours a day, though the average used by UDO is between 1 and 2 hours per day. The shortened distance to water can also allow women to transfer the job of fetching water to other family members, increasing time available to work on the farm, cook, care for children, and participate in community activities. Most water is collected from holes dug in the sand and then scooped into containers. These holes are often protected using thorn branches so animals cannot contaminate the water. Additionally, since the water is cleaner, instances of diarrheal disease drop. SASOL has also documented an increase in school attendance after sand dams are fully functional (Mutiso 2002) primarily because illness drops and nutrition improves. These immediate benefits lead to longer-term gains. The decrease in labor required to manage water means the household is able to more easily handle stresses that arise in other areas such as job loss.

When water is available, opportunities change for the entire household. The general pattern in rural Kenya is for women to manage household activities and for men to seek work for cash. Income from farming in semiarid areas is minimal and highly inconsistent across seasons, so many men seek work in cities. Having water changes local employment prospects. This is especially true for households that have land a short distance from a sand dam. When walking and measuring sand dams in July of 2013, a number of farmers were seen in close proximity to the stream: men and women were preparing and planting vegetable gardens in small terraced plots. Their primary crops were sukuma wiki (a vegetable related to collards and kale), tomatoes, onions, and Swiss chard (which they call "spinach" when translating from Kamba). These vegetables are all highly marketable and are often sold in the early morning and taken to local cities like Wote and Machakos. In these smaller gardens, the farmers carry 20-liter jerry cans of water from the sand dam to their plots to water the vegetables, which they grow in shallow depressions. Though I did not see this personally, the UDO field officers said that a number of households, as far as half a kilometer uphill from the dam, plant similar gardens around their houses. They often use a donkey to carry up to four jerry cans of water per load from the sand dam with children between 8 and 14 years of age leading the donkey.

Also observed, in July of 2013, were farmers who had installed pumps, although these were less common since level land near the streams is rare. For instance, Justin, chair of the Mkuta Mwea SHG, has a diesel-powered pump at the base of a dam in an enhanced natural depression in the river. He pumps to a garden where he raises tomatoes, French beans, Swiss chard, green peppers, sukuma wiki, maize, and green gram (an Indian legume sometimes called mung beans). Though he uses some to feed his family, most of his product is sold. However, he does not have to take the crop to the market himself. His reputation as a quality farmer is high, and those who sell his goods in Wote come to him to get their produce. His pump was paid off completely in the first year of his operation.

Another farmer, chair of the Kitandi Fruit Tree Growers SHG, has a large farm adjacent to the stream that is connected with a pump. True to the name, he has planted a lot of fruit trees. Asked if other members also plant fruit trees, he said yes, but most grow them near their homes and bring water from the dams to get them started. All of these other farmers were using the large nursery on his property to grow the seedling trees for their farms. The species grown include avocado, citrus, mango, guava, papaya, banana, and even macadamia nut. Most of the trees are young and have not reached their primary production years; but despite this, the avocado production was impressive. Also impressive was the quantity of vegetables growing on the farm. The vegetables are watered by gravity from two 1000-liter tanks that are filled by the pump. These serve as a reservoir for the water and help conserve fuel. The chair had hired a woman who had no land of her own to water and weed the vegetables. Most of his product was sold in nearby towns.

While the number of vegetables and fruit trees generally appearing around sand dams is significant, the most common plant grown near the dams is Napier grass (Fig. 5 on the right of the dam), or *Pennisetum purpureum*.[3] This is a very deeply rooted fodder grass with an ability to spread slowly via rhizomes and stolons but mostly planted through root cuttings. When established it can stabilize the steep banks of terraces, and it does very well on stream banks as well, but its primary purpose is as a fodder grass. It is not uncommon to bring the animals to the sand dam to drink and then load them with fodder cut along the edge of the sand to carry back to their stalls. Keeping the animals in stalls enables farmers to collect and store manure for use in their rainy season agriculture. It also avoids the overgrazing problem common in the region. The chair of Kitandi Fruit Tree Growers explained that the grass provides cut-and-carry fodder for lactating cows and goats, extending the milking periods and improving the health of calves, kids, and people, especially children. This otherwise unavailable protein supply is one of the most important benefits of the sand dams. When combined with its erosion control qualities, Napier grass is arguably the most important sand dam crop.

Another important long-term benefit of the sand dams is trees. No one has done more than Patrick Musyimi of the Makuta Mwea SHG, who has planted over 10,000 trees and is still working on his forest. Other than fruit trees, the most commonly planted species in the higher elevation areas of Machakos and Makueni are *Eucalyptus camaldulensis*, *Croton megalocarpus*, *Grevillea robusta*, *Warburgia ugandensis*, and *Prunus africana*.[4] The last two are native Kenyan plants with medicinal properties. In lower, drier areas, *Melia volkensii* and *Azadirachta indica* are more important. Most trees need little or no supplemental water after they are established. The trick is getting them started, and the sand dam provides the water to enable them to thrive. Through time this promotes reforestation of the region, a trend that is already visible in satellite data (Ryan 2012).

As UDO helps communities to build sand dams, other communities both inside and outside the county have expressed interest, including the Maasai in neighboring Kajiado. The Maasai primarily want to improve water availability for their animals, and sand dams have distinct advantages over boreholes. First, sand dams provide water over a long area, not just at one distinct point, so the damage to the landscape is reduced. One Maasai made a very simple comment to Arnold, the field officer for Kajiado, which illustrates this point: "At our sand dam, there is no need to queue." Second, they are far cheaper than a borehole and are built using primarily local resources. An average sand dam costs around $15,000, excluding labor, sand, and rock, all of which are provided by the community SHG doing the building. A borehole,

[3] *FAO Pennisetum purpureum*: http://www.fao.org/ag/agp/AGPC/doc/gbase/data/Pf000301.HTM

[4] *Eucalyptus* and *Grevillea* are both Australian imports used for firewood. *Grevillea* does well when close to crops as it does not compete for nutrients and its shade is light. *Croton megalocarpus* is a favorite native shade tree whose leaves make good mulch.

on the other hand, costs an average of $75,000.[5] Most local communities can afford neither, so outside assistance is required, but since communities participate in building a sand dam, the sense of local ownership is much higher, and therefore the sense of responsibility for the outcome is higher as well. Third, sand dams have no running costs and rarely need repairs. Since starting construction of sand dams in 1978, UDO has built nearly 1500, of which only four have failed. Perhaps the greatest advantage for the Maasai is the impact that the higher water table near seasonally dry streams has on nearby vegetation. Over time, this causes riparian vegetation to improve, and this becomes an important source of dry season fodder. Since the Maasai are almost entirely dependent on milk for sustenance, any improvement in cattle, goat, and sheep production results in significant increase in community resilience.

3.1 How Much Water Is Captured by Sand Dams?

The use of water in the sand dams for irrigation, in addition to the other uses, gives rise to the question of volume. How much irrigation could a sand dam support without compromising the need for household water and water for animals? The question is not an easy one to answer. While it is relatively easy to measure the surface area of sand behind a dam, the volume is more difficult. Stream beds are not flat with even widths at the top and the bottom. Unless detailed measurements are done before a dam is built and fills with sand, any measure of sand volume is going to be an estimate. Accurate measurements require sophisticated equipment, which is not readily available. In our work in July 2013, we measured a number of dams to estimate volume.

We used a simple formula. We first estimated that the depth of sand equaled the dam height times a fraction related to its increased distance from the dam, growing shallower as you get further away. Surface width at specific distances from the dam was measured, usually in 30 to 50 meter increments. This surface width was multiplied by 0.7 to get a value for width of the pre-dam stream bed. This value was based on observations and measurements of values in undammed portions of streams and is actually conservative. While stream shapes are variable, the intermittent flow and intensity of the runoff have produced stream beds that are wide and flat with steep banks, especially in larger streams. The variability comes from intruding rock. The width and height measurement produces a cross-sectional trapezoid which when multiplied by the distance between measurements provides an approximate volume. Most dams were measured every 30 meters moving upstream from the damsite, so a dam backing up 300 meters of sand would have ten trapezoids whose size is inversely proportional to the distance from the dam. These are added together to get a final measurement.

Table 1 below shows the final data of total sand volume for each of the dams measured. These are divided according to the self-help group (SHG) that built the dam.

[5] Both of these figures come from materials in the UDO offices in Kola and are based on their costs for building a sand dam and local average costs for construction of a borehole with a diesel-powered pump and an aboveground storage tank.

Table 1 Sand Dam Variables: volume, water storage, pH, and salinity

Dam #	Year built	Dam height meters	Spillway width meters	Distance meters	Sand volume m3	Water storage m3	Average pH	Average salinity ppm
Kitandi Fruit Tree Growers Self Help Group								
548	Jan-09	4	9.15	210	None	None	8.03	350
544		3.2	12.2	467	7954	2658	7.38	484
545	Feb-09	1.72	14.8	150	1621	567	7.69	499
546	Mar-09	3.6	21.7	250	8530	2986	7.37	358
528	Oct-08	1.5	14.3	240	4306	1507	7.72	334
566	Aug-09	2.8	17.9	300	10,887	3810	7.43	313
573	Dec-09	1.4	11.6	225	2515	880	7.45	406
583	Oct-09	2	11.5	777	4318	1511	7.67	339
587	Nov-09	2.45	14.2	216	6357	2223	7.62	556
624		2.2	16.8	240	5785	2025	7.54	279
Uvaani Self Help Group								
581	Oct-09	3.7	20	722	41,704	14,421	7.83	303
Mkuta Mwea SHG								
532	Jan-09	3	22.5	239	7708	2698	7.33	1459
64	Jun-06	1.35	12.6	404	2649	927	7.67	892
63		3.5	10.8	253	9307	2357	7.41	914
Matondoni SHG								
1		3.1	15.15	163	2506	877	7.59	1290
2	Sep-08	3	10.4	181	4211	1474	7.28	374
3		1.4	14	227	2859	1001	7.18	369
Nzaayamuisyo SHG								
70	Jun-05	4.5	22.9	240	18,377	6432	7.08	397
67	May-05	2	21	230	5062	1772	7.34	492
Kithito kya Atumia na Iveti SHG								
	Jul-09	2.1	22.3	297	9064	3173	7.95	2990
	Jan-08	1.5	27.7	660	16,600	5810	7.87	2624
Ilikoni Beekeepers SHG								
	Jul-06	2.9	37.9	600	21,120	7392	7.81	3460
	Apr-07	1	15.1	180	3299	1155	7.97	2345

The information in this table was collected in July 2013 by the author, five students from James Madison University and two staff from UDO

Kitandi Fruit Tree Growers SHG has the largest number of dams, while we only measured one from the Uvaani SHG, but it is a large dam, and the group was actively involved in building a second dam during our visit. Sand volume is given in the adjacent column. This volume is based on soil science literature showing that pure sand has porosity between 30% and 40% depending on sand granularity. We chose to use the medium value of 35% to make this calculation. The water volume given represents a completely filled pore space after a rainfall event or the immediate end of the rainy season. For most dams the actual volume observed was something less than this. Usually the holes observed, or that we dug, had water between 3 and 60 cm below the surface. In two cases water was flowing slowly on the surface, so the dams could be considered saturated. It was rare that we did not find water in a hole dug to 60 cm. Dam 583 of the Kitandi SHG proved an exception; no water was found in the sand closest to the dam, but water was found between 350 and 700 meters from the dam. No explanation for this was identified.

Table 1 also shows an average of measurements taken for salinity and pH behind each dam. These measurements were taken either in holes we dug on the width transect line measured or in existing holes dug by the community within 5 meters of the transect line. A Hach multimeter for pH and salinity was used to make the measurements. In most cases salinity values were below 500 ppm (or mg/liter), which is the level at which humans begin to taste salt. A few measurements rise above the level in the higher elevation dams measured, most notably Mkuta Mwea dam number 532. This dam was not used for drinking water and in fact was little used by the community at all. We found very few holes already dug in the sand, and UDO's local field officer stated that he had not seen people using the dam during his visits to the site.

The other dams where salinity is a definite issue are located in the southeastern portion of Makueni District not far from the town of Mtito Andei. This area has the lowest annual average rainfall of any area we visited, around 400 mm/year, being located in the semiarid agroclimatic zone 6, meaning that its potential evapotranspiration rate exceeds annual rainfall by at least a factor of 4. The two self-help groups, Illikoni Beekeepers and Kithito kya Atumia na Iveti (meaning in Kamba, Efforts of Women and Men), are on the same wide river, and all salinity measurements exceeded one part per thousand, reaching as high as five parts per thousand in one place. People prefer not drinking the water from the river if they have a choice, but do use the river regularly for watering their animals. No attempt was made to use the water for irrigation even though there were numerous failed attempts at dryland agriculture in the local area.

Acidity proved a very consistent variable. We saw no pH reading below 7.0, and the measurements higher than 8.0 were rare except in the Mtito Andei area dams. Acidity is not a major concern in the Machakos and Makueni area.

3.2 Thinking Big by Thinking Small

A question almost always asked by people new to sand dams and the idea of catching water at high points in a watershed is: what about the people downstream? Don't they have less water available because the people upstream are keeping their water? This is a good question and deserves a solid answer. SASOL and UDO have started sorting this out. The starting point is the ecosystem; what does the watershed, the entire area that drains water to the dam or series of dams in a stream, really look like? How does water move across this landscape? What did the natural system look like before humans altered the landscape? How did the flow patterns of water change in the altered landscape? Historical information is difficult to find in this region of Kenya, and what we do know is contained in the memories of the oldest community members, some of which is documented (Tiffen et al. 1994). Dr. Gideon Mutiso, a co-founder of SASOL and longtime development consultant and academic who grew up in Kitui County, interviewed many of the old farmers in Machakos and Kitui and captured pieces of this lost information (personal communication—unpublished).

The Kitui, Machakos, and Makueni region has always been dry, specifically subhumid to semiarid. What has changed is the number of trees. Previously, it was a near-continuous deciduous tropical forest in its higher regions and savanna woodland in the lower regions. These lower reaches were the home of large mammals like elephant, black-and-white rhinoceros, and numerous antelopes from eland to dik-dik, Cape buffalo, giraffe, lions, and leopards. Stream flow throughout the region was low, but continuous. Rainfall did not rush off the landscape when it fell; instead it hit a vegetated surface, slowed, and flowed through the soil and in the water table. Springs were common and continued to exist in much of the region up to the end of the 1940s. That is when the dramatic changes really hit. The local population expanded rapidly. The big game was almost entirely wiped out and confined to newly created parks by the colonial administration and the new Kenya government. The highlands were deforested and converted to agriculture. The lowlands changed with the loss of browsing game like black rhinos, and acacia/commiphora scrub took over from savanna grasses. Rainfall more commonly hit bare soil in the highlands and ran off quickly, carrying soil with it, and failed to reach the water table at all. The water table lowered, springs dried up or became ephemeral, and streams widened, with an increased bed load of sand, silt, and clay.

The renewed landscape promoted by UDO, SASOL, and others seeks to restore a modified original ecology while maintaining a productive agriculture for people. They recognize the need to supply water as a first step, thus the sand dams. They also see the loss of perennial vegetation as a key aspect of degradation, so they promote tree planting and the growing of grasses. They also know that people have to grow crops in a way that reduces or eliminates further damage to the soil. Thus they promote terraces, advocate no-till and no burn agriculture, and push for drought-tolerant crops. These changes capture more rainfall and put it back in the ground. In fact, a dam captures only a tiny fraction of the water flowing off the landscape: researchers from Vrije University, Amsterdam, studied this in a watershed south of Kitui town and calculated a capture of 1–3% of seasonal flow by the dams (Borst et al. 2006; Hut et al. 2007). However, this water is not lost; it is merely slowed and stored for more controlled release. As the sand dams rise and groundwater begins to swell, the dams raise the stream level. This rising water level combines with the terraces and trees, which slow surface runoff and allow more water to enter the soil, and eventually the water reaches the spring outlet points. The total amount of water flowing down the river does not change; what really changes is the timing of the flow. Instead of rushing in quick, energetic, and damaging bursts, water flows more gently over a longer period both on the surface and underground. When the original underground water profile is restored, the amount of water available downstream at any given time goes up, not down. The most significant change is that the size of the flood flow drops, while the base flow is restored. Whenever soil water supply rises, overall ecological resilience improves as well.

4 Conclusion

Sand dams are a locally conceived, low maintenance method of capturing runoff on a landscape. The captured water improves local ecosystem functions and at the same time enables local residents to collect water more conveniently throughout the year. In some cases the dams also provide water for dry season gardening and the growing of fruit trees and other multipurpose tree species. The fact that the sand dams and other forms of water catchment were locally developed and are locally owned improved the rate of adoption and made this a success story that has high potential to expand to similar regions in East Africa and beyond.

References

Borst L, de Haas SA (2006) Hydrology of sand storage dams: a case study in the Kiindu catchment, Kitui District, Kenya. 1 February 2006. Master Thesis Hydrogeology (Code 450122, 27 ECTS), Faculty of Earth and Life Sciences, Vrije University, Amsterdam

Hut R, Ertsen M, Joeman N, Vergeer N, Winsemius H, van de Giesen N (2007) Effects of sand storage dams on groundwater levels with examples from Kenya. Phys Chem Earth 33(2008):56–66

Lal R (2016) Soil health and carbon management. Food Energy Sec 5:212. https://onlinelibrary.wiley.com/doi/full/10.1002/fes3.96

Mutiso GCK (2002) Kitui sand dams: social and economic impacts. Muticon, Nairobi

Ryan C (2012) The potential for sand dams to increase the adaptive capacity of drylands to climate change. Thesis, Masters of Science in Climate Change Management, Birkbeck College, University of London

Teel WS (1994) Catching the rain: agroforestry and soil conservation in Nampula, Mozambique. Dissertation, Cornell University, Ithaca, NY

Tiffen M, Mortimere M, Gichiki F (1994) More people, less Erosion: environmental recovery in Kenya. Wiley, Chichester

Utooni Development Organization (n.d..) http://www.utoonidevelopment.org/

World Resources Institute (n.d.) Chapter 1: ecosystems and ecosystem services (slide show – slide #4). http://www.slideshare.net/WorldResources/naturesbenefits-kenya-01

Are Livestock Keepers in and Around Forests Key Stakeholders in Forest Management? Experiences from Mabira Central Forest Reserve, Uganda

Dorothy K. Nampanzira, Constantine B. Katongole, Vincent B. Muwanka, and John R. S. Tabuti

Abstract Globally, forests play an important role in supporting livelihoods of local communities that surround them. However, livestock rearing is hardly considered an important livelihood activity supported by forests. Forests can be a source of pastures which are key feed resources for livestock especially ruminants. There is little information on how a forest reserve affects livestock production especially in sedentary systems. In the current study, the status of livestock production in and around Mabira forest reserve, Uganda, was studied, to characterize the livestock production systems and determine the level of reliance on the forest for forages. A cross-sectional survey was conducted using a structured questionnaire, and a total of 80 households were interviewed. Results revealed that over 70% of the respondent farms had more than one livestock type. Cattle (71%), pigs (49%), chickens (47%) and goats (40%) were the most frequently kept livestock types. Most respondents fed cattle (54%) and pigs (81%) under the stall-feeding system, while 68% of the farms tethered goats. Chickens are mainly fed under free ranging feeding system (66%). Firewood, water, poles for construction and forages were the four forest products of significance importance to households rearing livestock around Mabira forest. Among the key determinants of level of reliance on forages from Mabira forest was negatively and significam, household's distance to Mabira forest (P <0.01), household size (P <0.05) and landholding size owned by the household (P <0.05) were the variables found to be statistically significant. In conclusion, livestock farmers in and around Mabira forest rear a diversity of livestock types. Forest forages contribute substantially to the feed resource base of a significant proportion of

D. K. Nampanzira (✉)
Department of Livestock and Industrial Resources, Makerere University, Kampala, Uganda

C. B. Katongole
Department of Agricultural Production, Makerere University, Kampala, Uganda

V. B. Muwanka · J. R. S. Tabuti
Department of Environmental Management, Makerere University, Kampala, Uganda

Y. Bamutaze et al. (eds.), *Agriculture and Ecosystem Resilience in Sub Saharan Africa*, Climate Change Management,
https://doi.org/10.1007/978-3-030-12974-3_16

households rearing livestock in and around Mabira forest. Therefore, livestock keepers in and around this Forest reserve are important stakeholders in the forest estate since they depend substantially on it for livestock forage.

Keywords Livestock production · Forage resource · Forest communities · Co-benefits

1 Introduction

Forests in the tropics and the developing world are threatened and are declining in extent of area cover. Between 2000 and 2010, there was a net forest loss of 7 million hectares per year and a net gain in agricultural land of 6 million hectares per year (FAO 2016). The main causes of this trend are population growth and agriculture. In Uganda, forested areas are declining rapidly. The country's annual forest loss over 15 years (1990–2005) was approximately 90,000 ha/year (NFA 2009). This has resulted in a reduction of forest cover from an estimated 4.9 to 3.6 million ha in the country. The loss of forests has very dire consequences on human welfare and the global environment. For example, this status presents a risk to food security, energy security and income generation and has significant negative effect on climate change. For this reason, the global world has undertaken many interventions and developed a number of policies and strategies to conserve forests. Among which is the REDD+ multilateral policy. REDD+ is aimed at mitigating climate change and also supporting livelihoods and converting biodiversity (Brown et al. 2008). The implementation of REDD+ is expected to have costs for the people that live in and around forests. However, such policies must tread carefully not to restrict access to forest products for local communities (Brown et al. 2008). It is important to manage and reduce these costs if people are going to participate and support mechanisms like REDD+. To be able to do this requires identification of key stakeholders who depend on forest resources for consultation. Forest stakeholders in and around forests are many. They include livestock keeping small-scale farmers, hunters and loggers (García-Nieto et al. 2015). These depend on forests for the gathering of many different products. In many parts of the world, livestock ranching is known as a key driver for deforestation (Hosonuma et al. 2012). In Africa, fuelwood is the most important driver (Ahrends et al. 2010), but the contribution of livestock farmers is not well known (Rudel 2013). Such information would be necessary to guide the designing of appropriate and efficient strategies for decreased dependence on forests and emissions reductions. This study sought to answer two research questions: (1) What livestock production systems exists in and around Mabira Forests Reserve? (2) To what extent does the livestock community depend on the forest for forage?

2 Materials and Methods

2.1 Description of Study Area

Mabira forest reserve (Fig. 1) is located in Buikwe District, 54 km from Kampala City and 26 km from Jinja town (0°24′ and 0°35 N and 32°52′ and 33°07′E), at an altitude of between 1070 and 1340 m above sea level. It covers an area of 306 sq. km (31,293 ha). The study was conducted in two sub-counties of Buikwe District (Najjembe and Kawolo). The two sub-counties were purposively selected based on the existence of livestock production activities. According to the local government administrative system in Uganda, the lowest administrative unit is a village; several villages constitute a parish, and several parishes constitute a sub-county. The reserve is predominantly occupied by tropical high-forest communities classified as type D1 *Celtis–Chrysophyllum* medium altitude, moist, semi-deciduous forest, while the rest is classified as *Piptadeniastrum–Albizia–Celtis* medium altitude, moist, evergreen forest (Langdale-Brown et al. 1964). The Mabira forest reserve is divided into compartments according to the forest working cycles, namely, the strict nature reserve, recreation/buffer zone, production (low impact) and production (encroachment area). According to Baranga 2007, signs of former human activities such as cultivation are still obvious within the forest and especially in the recreation/buffer zone. Abandoned banana plantations (*Musa sapientum*) and jackfruit (*Artocarpus heterophyllus*) form an integral part of the forest. Timber is logged mainly from

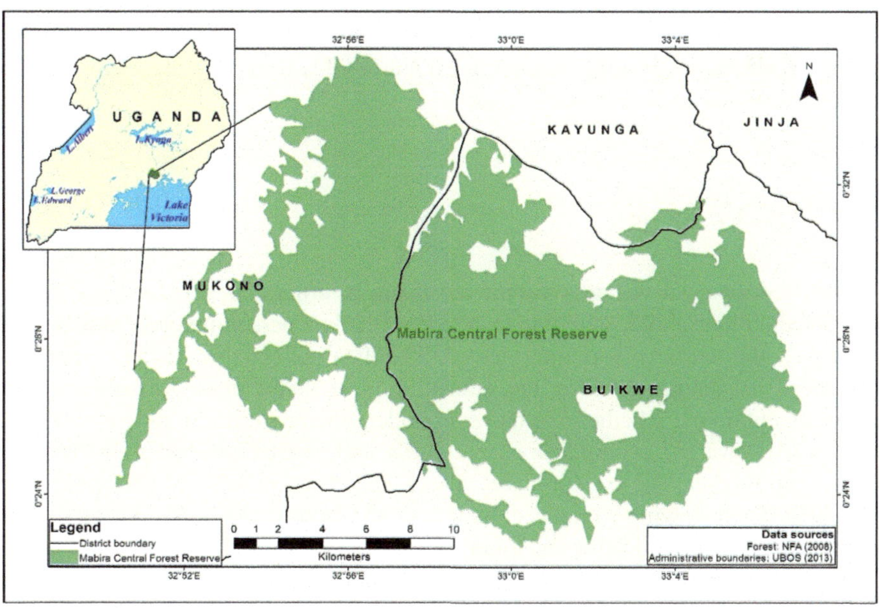

Fig. 1 Mabira forest reserve

M. eminii and *Cordia africana*, while *F. elastica* is harvested to provide building poles and firewood. *Pseudospondias macrocarpa*, *Bridelia micrantha* and *Albizia zygia* are tree species cut for charcoal burning. Wood is also extracted for domestic firewood and commercial purposes, poles for building and construction of various household items and shrubs for fencing and boundary marking. Residents in the village enclaves extract various medicinal plants from the forest. Nontimber products include materials for basket weaving, forest food items (mushrooms, honey, etc.) as well as food crops harvested from many abandoned cultivations within the forest.

2.2 Selection of Respondents

The study population consisted of households rearing livestock. According to the local government administrative system in Uganda, the lowest administrative unit is a village; several villages constitute a parish; and several parishes constitute a sub-county.

With the help of the district extension workers, two sub-counties one within Mabira forest (Najjembe) and another around Mabira forest (Kawolo) were identified as the study areas basing on their livestock rearing activities. Three parishes were purposively selected from each sub-county. This resulted in a total of six parishes. For sample selection purposes, extension workers were asked to provide lists of all households rearing cattle, goats/sheep, pigs and chickens (irrespective of herd size) from the selected parishes in their respective sub-counties of jurisdiction. Based on these lists, a sample of households to be interviewed was randomly selected for each sub-county. A total of 80 households from 23 villages distributed in the 2 sub-counties were interviewed.

2.3 Data Collection

A cross-sectional survey was conducted using a structured questionnaire. From each selected household, only one respondent, the key person involved in the daily feeding of the animals, was interviewed. The data collected included respondent characteristics, such as gender, age, education level, status in the household, years at the current home, whether one had ever undergone any training in livestock production and awareness about any forest laws and regulations for protecting and conserving forests. The questionnaire also asked about household characteristics, including the following main occupation of the household head, distance from Mabira forest, household size and main source of income for the household. Finally, the respondents were asked to provide the following farm characteristics: farmland ownership, total landholding size, herd/flock size and composition, feeding system, main reason for rearing livestock and manure handling system. Questions were also asked to obtain information on access/extraction of forest resources (including forages for feeding livestock). The contribution that forest resources make to the household requirements was estimated using a rating of "negligible contribution", "little contribution", "moderate contribution" and "substantial contribution".

2.4 Statistical Analysis

The data were analysed using the SPSS 18.0 statistical package (2010). The analysis included Pearson chi-square tests of association. The key determinants of level of reliance on forages from Mabira forest were explored using the regression equation: LevelReliance = b_0 + b_1*Education(secondary)* + b_2*HhOff-farmEmp* + b_3*HDistance* + b_4*HSize* + b_5*Sex of respondent* + *bRuminantOwn* + b_6*HMainIncome* + b_7*FarmSize* + b_8*CattleOwn* + b_9*GoatOwn* + b_{10}*PigOwn* + b_{11}*ChickenOwn* + *e*. The variables used in the regression equation are defined in Table 1.

Table 1 Definition of variables included in the regression model

Variable	Definition
Dependent	
LevelReliance	Level of reliance on forages from Mabira forest (little, moderate or much = 1; otherwise = 0)
Independent	
Sex	Sex of respondent
Respondent characteristics	
Education(secondary)	Respondent education level (completed secondary level = 1; otherwise 0)
AwareFLaws	Respondent awareness about forest laws and regulations (aware = 1; otherwise 0)
Household characteristics	
HhOff-farmEmp	Household head off-farm employment (off-farm employment = 1; otherwise = 0)
HDistance	Household's distance to Mabira forest
Hsize	Number of household members
HMainIncome	Main contributor to total household income (livestock = 1; otherwise = 0)
Farm characteristics	
RuminantOwn	Household owned cows or goats (yes = 1; otherwise = 0)
FarmSize	Total landholding size owned by the household
CattleOwn	Household owned cattle (yes = 1; otherwise = 0)
GoatOwn	Household owned goats (yes = 1; otherwise = 0)
PigOwn	Household owned pigs (yes = 1; otherwise = 0)
ChickenOwn	Household owned chickens (yes = 1; otherwise = 0)

3 Results

3.1 Respondent and Household Characteristics

Respondents were composed of 64% men and 36% women. The average respondent age was 44.7 ± 13 years. Over 90% of the respondents had been residing at their current homes for 5 and above years. All the respondents, except one, had some kind of formal education, see Table 2. Almost half of the respondents (48%) had ever attended at least one training in livestock production. The majority of households (55%) were located within 1 km from Mabira forest. On average each household consisted of 7.5 ± 4 members, with over 60% of the households having 5–10 members. The main contributors to the total household income were livestock, crops, salaried employment, business/trading, casual labour, sale of firewood from Mabira forest and hand-outs from relatives as shown in Table 2.

4 Farm Characteristics

4.1 Land and Livestock Ownership

Close to 76% of the respondent households owned their farmlands under the "Kibanja" (squatter) tenure category. The other households either leased (17%) or hired on a seasonal basis their farmlands (7%). Nearly one half of the respondent farms (48%) operated on more than 1 acre of land; only 19% operated on less than 0.5 acres. Over 70% of the respondent farms had more than one livestock type. Cattle (71%), pigs (49%), chickens (47%) and goats (40%) were the most frequently

Table 2 Respondents and household characteristics

Respondent characteristic	Percentage	Household characteristic	Percentage
Relation to household head		Distance of household from forest	
Household head	71	Within forest	15
Spouse	23	Within 1 km from forest	55
Son/daughter	4	Beyond 1 km	30
Other relative	1		
Education		Main source of household income	
No formal education	1	Livestock	47
Lower primary	8	Crops	22
Upper primary	32	Salaried employment	11
Lower secondary	41	Business/trading	10
Upper secondary	8	Casual labour	5
College	8	Firewood from Mabira forest	4
University	8	Handouts from relatives	1

kept livestock types. The herd sizes ranged between 1 and 10 (for cattle), 1 and 13 (for pigs), 3 and 1000 (for chickens) and 1 and 10 (for goats). The main reason for rearing the different livestock types (cattle, goats, pigs and chickens) was cash income generation either through the sale of live animals or their products (particularly milk and eggs). Slaughtering for family consumption was more common among respondent farms with chickens (32%). Nearly all respondent households (95%) were also involved in crop farming.

4.2 *Livestock Breeds and Feeding Systems*

The major cattle breed categories were undefined crossbreeds (75%), indigenous breeds (21%) and pure exotic breeds (12%), in that descending order of predominance (Table 2). The goat herds and chicken flocks were predominantly composed of indigenous breeds (86 and 88%, respectively). Only 18% of the respondent farms had exotic chicken breeds. About 45% of the respondent farms had indigenous pig breeds, and 56% had undefined crossbreeds. Only 12% of the farms had exotic pig breeds.

Stall feeding and tethering were the major feeding systems for cattle, pigs and goats, while free ranging and total confinement were the major feeding systems for chickens. Majority of respondent farms fed cattle (54%) and pigs (81%) under the stall-feeding system, while 68% of the farms tethered goats. Only 6.0, 3.1 and 7.1% of the respondent farms allowed cattle, pigs and goats (respectively) to scavenge (free range). Chickens were kept under the free ranging feeding system (66%), total confinement (24%) or under the semi-confinement feeding system (10%).

Various feed resources were cited by the respondents, of which elephant grass, naturally growing plants (grasses, legumes, herbs and shrubs), sweet potato vines and banana pseudo-stems were the most commonly used feed resources for cattle (in that order of importance). For goats, naturally growing plants, elephant grass, banana peels and paper mulberry (*Broussonetia papyrifera*) foliages from Mabira forest were the most commonly used feed resources. The naturally growing plants were mainly harvested from roadsides (58%), farmers' own lands (21%), Mabira forest (11%) or neighbours' lands (11%). Pig farmers cited maize bran, cocoyams (*Colocasia* and *Xanthosoma* species), sweet potato vines and brewer's waste, whereas chicken farmers cited scavengeable feed resources, compounded feeds and maize bran as their most commonly used feed resources.

4.3 *Housing and Manure Management*

In the majority of the farms, night-time housing structures were specifically constructed for goats (50%), cattle (56%), chickens (65%) and pigs (80%). The kitchen was used in 22.6% and 17.9% of the farms for chickens and goats, respectively.

In some farms, the cattle (19%), goats (18%) and pigs (15%) would be tethered under trees for night shelter. Only 3.6% and 3.2% of the farms accommodated goats and chickens (respectively) within their family houses. Different practices were reported in the handling of livestock manures. The vast majority of respondent farms spread/incorporated the fresh manure from cattle (65%), goats (80%), pigs (61%) and chickens (73%) directly into the soils in their gardens. A few farms composted the manure (less than 15%) or used it for biogas production (only 2%) (Table 3).

Table 3 Livestock ownership characteristics

	% of responses			
	Cattle	Goats	Pigs	Chickens
Breed				
Indigenous breeds	21	86	44	88
Undefined crossbreeds	75	24	56	12
Exotic breeds	12	3	12	18
Feeding system				
Stall feeding/zero-grazing (total confinement)	54	25	81	24
Free ranging (scavenging)	6	7	3	66
Tethering	34	68	16	–
Communal herding[a]	4	–	–	–
Semi-stall feeding (semi-confinement)	2	–	–	10
Reason for keeping livestock				
Milk/eggs for cash income	63	–	–	14
Milk/eggs for family consumption	26	–	–	5
Live animals for cash income	11	86	100	50
Slaughter for family consumption	–	7	–	32
Slaughter for cultural ceremonies	–	4	–	–
Livestock manure	–	4	–	–
Housing for livestock				
Under a tree	19	18	14	–
Animal structure	56	50	80	65
Backyard kraal	25	4	6	–
Within family house	–	4	–	3
Kitchen	–	18	–	23
Balcony	–	4	–	–
Store on family house	–	–	–	10
Livestock manure handling system				
Composting (aerobic degradation)	6	12	9	14
Biogas production (anaerobic degradation)	2	–	–	–
Spreading/incorporation of fresh manure in soil	67	80	61	73
Stacking before spreading/incorporation in soil	18	4	30	–
Sell to fellow farmers	4	–	–	14
Simply ignore	2	4	–	–

[a]A herder is paid (either on a daily or monthly basis) to graze several herds on open access lands (roadsides, undeveloped plots, etc.)

4.4 *Reliance on Forest Forages*

Besides firewood, water and poles for construction, forages were the fourth forest product of significant importance to households rearing livestock in and around Mabira forest (Table 4). The level of reliance on forages from Mabira forest was indicated as negligible 49.2%, little 21.3%, moderate (9.8%) and much (19.7%). Naturally growing plants, foliages of paper mulberry (*Broussonetia papyrifera*) and Ficus tree species were the most frequently cited forages (in that order of importance) extracted from Mabira forest. These forages were used in the cut-and-carry system (forages are cut and carried to the animals but not fed from within the forest).

Results of the regression equation developed to explore the key determinants of level of reliance on forages from Mabira forest are presented in Table 5. Household's distance to Mabira forest ($P < 0.05$), household size ($P < 0.01$) and landholding size owned by the household ($P < 0.05$) were the variables found to be statistically significant. The regression coefficient of awareness about household size, sex of respondents and maximum level of education was positive, while the coefficients of distance to the forest, ownership of ruminant animal and landholding size were negative.

5 Discussion

5.1 *Farm Characteristics*

Most of the respondents owned more than one livestock type, the majorly consisting of cattle, pig, chicken and goats. This is a common practice in mixed farming communities in Africa. Keeping more than one livestock type is regarded as a financial security and risk response, especially in cases of disease outbreak. This result is in accordance with earlier studies (Malla et al. 2003) who indicated that livestock is a

Table 4 Respondents' perceived level of reliance on resources from Mabira forest

	% of responses			
	Zero or negligible	Little	Moderate	Much
Firewood	39.4	1.5	0.0	59.1
Water	53.7	1.9	7.4	37.0
Poles for construction	59.3	7.4	9.3	24.1
Forages	49.2	21.3	9.8	19.7
Wild foods	70.0	12.0	6.0	12.0
Timber	86.3	2.0	7.8	3.9
Medicinal plants for animal	85.2	1.9	9.3	3.7
Medicinal plants for humans	71.4	7.1	17.9	3.6
Hunting	96.1	2.0	2.0	0.0

Table 5 Key determinants of level of reliance on forages from Mabira forest

Variables	Coefficient	*t*-value	*P*-value
Education(secondary)	0.143	0.885	0.318
HhOff-farmEmp	0.063	0.509	0.613
HDistance	−0.275**	−2.821	0.007
HSize	−0.035*	−2.130	0.038
HMainIncome	0.035	0.283	0.778
FarmSize	−0.183*	−2.044	0.047
CattleOwn	0.166	1.103	0.276
GoatOwn	0.065	0.507	0.615
PigOwn	0.173	1.346	0.185
ChickenOwn	0.146	1.346	0.185
RuminantOwn	−0.028	−0.232	0.817
Sex	0.145	−1.117	0.270
Constant	1.037*	2.420	0.019
Model statistics			
R^2	0.392		
Adjusted R^2	0.249		
F-value	2.750		
P-value	0.008		

*, **Significance at 5% and 1%, respectively.

major capital asset among forest communities. In addition to livestock production, almost all respondents were involved in crop production. This is an indication that the community has alternative activities they rely on for food and income as opposed to depending on the forest products solely. Alternative activities are known to reduce pressure on the forest and thereby promote the restoration and conservation of natural resources (Fisher 2004). This could also mean that the forest products available to the community may not necessarily be those products which are most needed by the people or may not be equitably distributed in the case were they are needed (Timsina 2002; Neupane 2003).

5.2 Livestock Breeds and Feeding Systems

Stall feeding and tethering were the major feeding systems for cattle, pigs and goats in this community. This finding is similar to that reported by Prabhu et al. (2003) in Zimbabwe. This result implies that these animals are not freely grazed in the forest; rather, forage is cut and carried to the livestock, or the livestock are tied around a particular tree so they can graze on the nearby vegetation in the radius of the rope. This reduces on the effect of trampling on the forest, overgrazing, uncontrolled grazing and the likelihood of cutting down trees to create grazing land. This finding is contrary to Delcurto et al. (2005), who indicated that livestock exert a lot of pressure on the forest through their grazing behaviour. This study has demonstrated that livestock production can be practiced in forest communities without threatening the degradation of the forest if pasture harvesting is practiced sustainably. This should be benchmarked by other forest communities that have been grazing directly in the forest. The study further proves that livestock production is an alternative economic activity that can be adopted by forested communities' in order to reduce pressure on the forests. This component should further be included in the training programmes on forest sustainability use, restoration and co-existence with humans.

5.3 Housing and Manure Management

Manure from the livestock was incorporated into the soils of the gardens while fresh by the majority of the respondents. This practice is detrimental to the environment due to its potential to emit carbon gases to the atmosphere. According to Hao et al. (2005), manure disposal practices are one of the avenues through which livestock contribute to the increase in carbon gases. These carbon gases (N_2O and CH_4) are produced when anaerobic (without oxygen) decomposition of manure takes place. When manure is handled as a solid or deposited naturally on grassland, it decomposes aerobically (with oxygen) and creates little methane emissions. These greenhouse gases can be directly or indirectly produced and emitted at each stage of the manure management process, including manure stores, manure pits, manure treatment and manure spreading/incorporation to land (Chadwick et al. 2011).

Therefore farmers should be trained in manure handling practices that are friendly to the environment such as creation of energy through biogas, frequent and complete manure removal (Johnson et al. 2007) and recycle to form other animal feeds. This will contribute to the clean climate mechanisms while improving incomes, food security and ultimately better livelihoods for the communities.

5.4 Reliance on Forest Forages

Besides firewood, water and poles for construction, forages were the fourth forest resource of significant importance to households rearing livestock in and around Mabira forest. Extraction of forest forages has also been reported by previous studies on forest dependency by local households (Adhikari et al. 2004; Mamo et al. 2007; McElwee 2010). Results of this study indicated that the contribution that forages from Mabira forest make to the livestock feed resource base decreased with household's distance to Mabira forest ($P < 0.01$), number of household members ($P < 0.05$) and landholding size owned by the household ($P < 0.05$). As expected, distance to the forest was negatively and significantly associated with the contribution that forages from Mabira forest make to the livestock feed resource base ($P < 0.01$). This implies that the closer a livestock farmer is from Mabira forest, the more likelihood to rely on forest forages. The negative and significant relationship observed for distance to Mabira forest is in agreement with previous studies (Mamo et al. 2007; Illukpitiya and Yanagida 2008; Kamanga et al. 2009) on extraction of forest resources or income from forests. Generally, increasing distance to the forest reduces extraction of forest resources.

Farm size was found to be negatively and significantly associated with the contribution that forages from Mabira forest make to the livestock feed resource base ($P < 0.05$), implying that households with better access to farmland are better able to circumvent the feed scarcity constraint, compared to households with smaller land acreages. However, the negative association observed contradicts previous forest dependency studies in Ethiopia (Babulo et al. 2008) and Nepal (Adhikari et al. 2004), whereby livestock rearing households with more access to grazing land had the likelihood of greater dependence on forest fodder. Unlike intensive farms (where livestock are under total confinement), extensive farms (where livestock are let out to the grazing areas) tend to be highly dependent on freely available environmental resources. Thus, in extensive systems (as in many parts of Ethiopia and Nepal) more access to grazing land tends to be related to bigger livestock herds, which in turn implies greater use of environmental resources such as forest forages. In this study, the vast majority of the respondent farms had small herd sizes and were particularly based on zero-grazing and tethering systems. Thus, in intensive systems, larger acreage households are more likely to access forages from their own farms as opposed to lower acreage households and hence the likelihood of lesser dependence on forest forages.

6 Conclusion

Livestock farmers in and around Mabira Central Forest Reserve rear a diversity of livestock types. Forest forages contribute substantially to the feed resource base of a significant proportion of households (29.5%) rearing livestock in and around Mabira forest. Therefore livestock keepers in and around this reserve are important stakeholders in the forest estate since they depend substantially on it for livestock forage.

Acknowledgements The authors acknowledge the financial support from NORAD (UGA-13/0019).

References

Adhikari B, Di Falco S, Lovett JC (2004) Household characteristics and forest dependency: evidence from common property forest management in Nepal. Ecol Econ 48(2):245–257

Ahrends A, Burgess ND, Milledge SAH et al. (2010) Predictable waves of sequential forest degradation and biodiversity loss spreading from an African city. Proc Natl Acad Sci USA 107:14556–14561

Babulo B, Muys B, Nega F, Tollens E, Nyssen J, Deckers J, Mathijs E (2008) Household livelihood strategies and forest dependence in the highlands of Tigray, Northern Ethiopia. Agric Syst 98(2):147–155

Brown D, Seymour F, Peskett L (2008) How do we achieve REDD co-benefits and avoid doing harm? In: Angelsen A (ed) Moving ahead with REDD: Issues, options and implications. Center for International Forestry Research (CIFOR), Bogor. http://www.cifor.org/publications/pdf_files/Books/BAngelsen080111.pdf

Chadwick D, Sommer S, Thorman R, Fangueiro D, Cardenas L, Amon B, Misselbrook T (2011) Manure management: implications for greenhouse gas emissions. Anim Feed Sci Technol 166:514–531

DelCurto T, Porath M, Parsons CT, Morrison JA (2005) Management strategies for sustainable beef cattle grazing on forested rangelands in the Pacific northwest. Rangel Ecol Manag 58(2):119–127

Food and Agriculture Organization of the United Nations (2016) State of the world's forests: Forests and agriculture, Land-use challenges and opportunities, FAO.

Fisher M (2004) Household welfare and forest dependence in Southern Malawi. Environ Dev Econ 9(2):135–154

García-Nieto AP, Quintas-Soriano C, García-Llorente M, Palomo I, Montes C, Martín-López B (2015) Collaborative mapping of ecosystem services: the role of stakeholders′ profiles. Ecosyst Serv 13:141–152

Hao X, Larney FJ, Chang C, Travis GR, Nichol CK, Bremer E (2005) The effect of phosphogypsum on greenhouse gas emissions during cattle manure composting. J Environ Qual 34(3):774–781

Hosonuma N, Herold M, De Sy V, De Fries RS, Brockhaus M, Verchot L, Romijn E (2012) An assessment of deforestation and forest degradation drivers in developing countries. Environ Res Lett 7(4):044009

Illukpitiya P, Yanagida JF (2008) Role of income diversification in protecting natural forests: evidence from rural households in forest margins of Sri Lanka. Agrofor Syst 74(1):51–62

Johnson JMF, Franzluebbers AJ, Weyers SL, Reicosky DC (2007) Agricultural opportunities to mitigate greenhouse gas emissions. Environ Pollut 150(1):107–124

Kamanga P, Vedeld P, Sjaastad E (2009) Forest incomes and rural livelihoods in Chiradzulu District, Malawi. Ecol Econ 68:613–624

Langdale-Brown I, Osmaston HA, Wilson JG (1964) The vegetation of Uganda and its Bearing on Land Use. Government of Uganda Printer, Uganda

Malla YB, Neupane HR, Branney PJ (2003) Why aren't poor people benefiting more from community forestry. J For Livelihood 3(1):78–92

Mamo G, Sjaastad E, Vedeld P (2007) Economic dependence on forest resources: a case from Dendi District, Ethiopia. Forest Policy Econ 9(8):916–927

McElwee PD (2010) Resource use among rural agricultural households near protected areas in Vietnam: the social costs of conservation and implications for enforcement. Environ Manag 45(1):113–131

Neupane HR (2003) Contested impact of community forestry on equity: some evidence from Nepal. For Livelihood 2(2):55–61

NFA (2009) Draft land Cover technical report of Uganda 2005. National Forestry Authority

Prabhu R, Haggith M, Mudavanhu H, Muetzelfeldt R, Standa-Gunda W, Vanclay JK (2003) ZimFlores: a model to advise co-management of the Mafungautsi Forest in Zimbabwe. Small-scale. Forestry 2(2):185–210

Rudel TK (2013) The national determinants of deforestation in sub-Saharan Africa. Phil Trans R Soc B 368(1625):20120405

Timsina N (2002) Empowerment or marginalization: a debate in community forestry in Nepal. For Livelihood 2(1):27–33

A Review of Studies Related to Charcoal Production, Consumption, and Greenhouse Gas Emissions in Tanzania

Greyson Zabron Nyamoga and Birger Solberg

Abstract Production and consumption of charcoal play a significant role in enhancing the livelihoods of people in Tanzania but may also lead to adverse environmental impacts. In this chapter, a review is presented of studies of charcoal production and consumption in Tanzania, and promising new research tasks are identified. Many interesting and valuable studies have been done, and it is clearly seen how important charcoal consumption and production are in a social, ecological, and economic perspectives. However, many of the reviewed studies lack clear hypotheses and specifications of behavior theories to be used for developing realistic and testable hypotheses. More research is needed on factors effecting charcoal demand – like changes in prices, income, and policies – and for that, using national household surveys is recommended. More research is needed also about tree regeneration (time and volumes) in miombo woodlands; how various forms of land ownerships influence miombo woodlands management; the possibilities and preferability in Tanzania of establishing forest plantations for producing charcoal; total and distributional impacts of policies; GHG impacts of charcoal production and consumption; and the development of bio-economic models which make possible consistent analyses of ex ante defined interesting changes from the present economic and policy situation.

Keywords Bioenergy · Climate impacts · Agent behavior · Urbanization · Deforestation · East Africa

---

G. Z. Nyamoga (✉)
Norwegian University of Life Sciences, Faculty of Environmental Science and Nature Management (MINA), Department of Ecology and Natural Resources Management (INA), Ås, Norway

Sokoine University of Agriculture – SUA, College of Forestry, Wildlife and Tourism, Department of Forest and Environmental Economics, Morogoro, Tanzania
e-mail: gnyamoga@sua.ac.tz; greyson.nyamoga@nmbu.no

B. Solberg
Norwegian University of Life Sciences, Faculty of Environmental Science and Nature Management (MINA), Department of Ecology and Natural Resources Management (INA), Ås, Norway

Y. Bamutaze et al. (eds.), *Agriculture and Ecosystem Resilience in Sub Saharan Africa*, Climate Change Management,
https://doi.org/10.1007/978-3-030-12974-3_17

1 Introduction

1.1 Overview

In Tanzania and most sub-Saharan countries, biomass accounts for more than 70% of the consumed energy (Mwampamba 2007; Felix and Gheewala 2011; Schure et al. 2013; Lusambo 2016) with firewood and charcoal being the most common (Mshandete and Parawira 2009; Dasappa 2011; Al-Mulali and Sab 2012). High population growth and inefficient stoves escalate the demand of charcoal and firewood; and together with agricultural expansion, overgrazing, illegal logging, and improper land tenure systems lead to deforestation and forest degradation in rural poor communities (Hosier et al. 1990, 1993; Chidumayo 1993; Hofstad 1997; Chidumayo and Gumbo 2013).

In urban areas in Tanzania, charcoal is used by more than 80% of the population as the dominant source of energy for cooking (Zahabu 2010; Chidumayo and Gumbo 2013). The high demand of charcoal there is caused by demographic factors as well as its reliability and affordability compared to other energy sources (Martin et al. 2009; Felix and Gheewala 2011). The urban population of Tanzania constitutes at present about 32% of the country's total population and is growing at a rate of 7% p.a. (URT 2017a, b; NBS 2017 and WB 2018), whereas the total population of Tanzania is increasing at the rate of 2.7% p.a. The high charcoal consumption in urban areas, high population growth, and high urbanization rate are important factors for the future demand of charcoal in Tanzania and consequently for the country's wood extraction for charcoal production. At present about 55% of mainland Tanzania is covered by forests (Tomppo et al. 2010, 2014; Malimbwi and Zahabu 2014), and the annual deforestation is estimated at 3728 km^2, equivalent to 1.1% of the country's total forest area (Bahamondez et al. 2010; Malimbwi and Zahabu 2014).

Electricity, gas, or kerosene substitute for using charcoal in cooking in some cases. In 2007 less than 15% of Tanzania's population were connected to electricity (Mwampamba 2007). By 2016, according to REA (2015) and URT (2017a, b), about 33% of all households in mainland Tanzania were electrified, covering about 64% of the households in urban areas and 17% in rural areas. However, in Tanzania, electricity, kerosene, and gas are very expensive compared to charcoal, and their use is rather limited.

Charcoal production and consumption play a significant role in deforestation, land degradation, and economic livelihood in Tanzania. Many studies have been done to obtain increased knowledge about this role, but these studies differ with respect to factors like objectives, geographical coverage, sample size, methodology, and results obtained. It is thus of high interest to compare these studies and identify information mostly needed for improved land-use planning and policy making in Tanzania.

This study was undertaken against this background, and its main objectives were to (i) give an overview of previous studies of charcoal production and consumption in Tanzania with particular reference to documenting behavior theories, sample size and statistical methods applied, geographical area covered, and main results obtained and (ii) identify where improved data and research are mostly needed.

2 Methodology

2.1 Selection of Publication and Data

We applied Google Scholar as search engine for finding publications to be included in our study. The main selection criteria were (i) having charcoal in Tanzania as main element and covering at least one of the above stated sub-objectives; (ii) published in peer-reviewed research journals or in governmental or consultancy reports which are publicly available and are of sufficiently high scientific quality; and (iii) published after 1990. Regarding criterion (i) we made exceptions (in particular regarding studies on GHG emission) for a few studies which cover countries in East Africa or sub-Saharan Africa but are relevant also for Tanzania.

For each of the selected studies we focused on presenting methodology and main results. Regarding methodology, we emphasized on behavior theories assumed, geographical area covered, sample size, main variables studied and statistical method applied. Including behaviour theory in this overview was done because any statistical study of consumption or production to be realistic ought to be based – implicitly or explicitly – on factors which reflect on the behavior of the producers or consumers studied. Our search resulted in 16 articles published in peer-reviewed research journals and 5 governmental or consultancy reports, as shown in Table 1. Only very few studies relevant for Tanzania were found covering GHG emissions from charcoal production or consumption.

3 Results and Discussion

In Table 1 an overview of the main findings is presented for each of the reviewed studies. In the first column, the title and year of publishing are shown. In the second column, we show geographical coverage, sample size, whether statistical analyses have been done and if so which, if any behavior theory is used for justifying hypotheses, and the chosen explanatory variables applied in the statistical analyses. In the third column, the main results of the studies are presented, placing emphasis on quantified results but also including qualitative results which are found to be of particular interest. The information presented in Table 1 is self-explanatory in many respects. Because of space limitations, we therefore concentrate our discussion on charcoal consumption, charcoal production, the emission of Greenhouse Gases (GHGs), and, finally, a more general discussion.

Table 1 Summary of the main findings in selected publications regarding production, GHG emissions, and consumption of charcoal in Tanzania

No	Authors and article title	Methodology	Main results
1.	Hosier (1993) *Charcoal production and environmental degradation: Environmental history, selective harvesting and post-harvest management*	• No behavioral theory stated • Interviewed 180 charcoal producers and visited 19 production sites in Dar es Salaam, Mbeya, and Shinyanga region • Environmental degradation and wood recovery assessed based on *soil type, visible soil erosion, land use, cultivation, vegetation cover, land management practices, tree harvesting height, kiln damage and distribution of resprouting trees* • No statistical analysis done	• Selective harvesting, species mix, and growth are very important for natural forest management and regeneration • Post-harvesting management is an important policy measure for sustainable forest management and enhancing regeneration in the miombo woodlands • The woodland regenerations and time depend on the harvesting intensity and the disturbances • In most sites, nothing grew back in the site areas where the charcoal kiln was built • Fire management plays a critical role in determining woodland regeneration • Multiple burning and land exhaustion (extended agriculture and overgrazing) can rather much retard regeneration in the woodland by affecting soil fertility • Numerous agricultural clearance mixed with heavy grazing pressure and long-lasting erosion problem reduces the ability of the woodland resources to recover • Increased efforts are needed focusing on improving post-harvest management and efficiency of charcoal production in the areas located within the effective harvesting and transport distances • No regeneration times were quantified, but the author makes the observation that the miombo woodland has strong regenerative capacity and that some of the visited sites were reported by the village guides to have been harvested for charcoal production "three times in an individual's lifetime"

(continued)

Table 1 (continued)

No	Authors and article title	Methodology	Main results
2.	Hosier and Kipondya (1993) *Urban household energy use in Tanzania: prices, substitutes and property*	• Behavioral theory stated: *Economic theory of utility maximization* • Interviewed 1600, 620, and 450 households in Dar es Salaam, Mbeya, and Shinyanga, respectively, using structured questionnaire • Statistical analysis done • Variables used: (i) Dependent variable: *Energy consumed* (ii) Independent variables: *Income, household size, market price, effective price*	• Household energy consumption constituted about 80% of Tanzania's energy use • Electricity usage in Dar es Salaam, Mbeya, and Shinyanga were exclusively for lightning • The household consumption of charcoal per capita per year was 176 kg, 195 kg, and 245 kg in Dar es Salaam, Mbeya, and Shinyanga in the same order • Regarding firewood the annual per capita household consumption was 452 kg, 817 kg, and 784 kg for the same cities, respectively • During the 5-year period 1985–1990, 4.5% of the respondents reported that they had shifted from charcoal to another energy carrier (1.2% to electricity, 1.8% to kerosene, and 1.3% to firewood). But also 4% reported they had moved to charcoal use from another energy source (2% from firewood, 1% from kerosene, 0.6% from LPG), making the total consumption of charcoal unchanged during this 5-year period • Energy consumption did not differ by income, and electricity and LPG behaved as a normal good, while kerosene was an inferior good (consumption decreased with income) • Woodfuel behaved as a normal good in low-income groups and as an inferior good in a high-income category • Charcoal was the only main reliable source of energy in all the three selected cities • Electricity was found to be the cheapest energy source when calculated as price per gross energy unit delivered, and even when compared as cost per effective unit of energy, it was still the cheapest

(continued)

Table 1 (continued)

No	Authors and article title	Methodology	Main results
			• Charcoal was found the cheapest energy source if improved cooking stove (*named Jiko*) was used, subsidies and duties are not included, and foreign exchange is accounted for, and firewood was the next cheapest • The author state that "Heavy reliance on modern fuels is the manifestation of misallocation of resources resulting from deviation of financial prices from the economic costs" • The energy ladder or energy transition theory is based loosely on the economic theory of household behavior and the assumptions that modern fuels are normal goods, while traditional energy are inferior goods • It is difficult for the energy ladder theory to work in Tanzania mainly because of large geographical differences regarding energy supply and the seasonal unreliability of electricity • Energy-poverty linkage do exist, but in Tanzania energy scarcity rarely causes poverty • Lifeline subsidy is recommended as the best policy option for poor households to afford modern energy types and hence reduce pressure on the forests. These households would rather go for fuelwood if no subsidy is available • Subsidy for kerosene seems more likely to help both the urban and rural poor to switch away from woodfuels • Household fuel mix seems the cheapest and best policy option for supplying energy at national perspective

(continued)

Table 1 (continued)

No	Authors and article title	Methodology	Main results
3.	Boberg (1993) *Competition in Tanzanian woodfuel markets*	• No behavioral theory stated • No statistical analyses done • Backward linkage approach was used (*the path of the fuel was followed back from end user to the producers*) • Series of structured questionnaire surveys in 1990 interviewing 1600, 620, and 450 households in Dar es Salaam, Mbeya, and Shinyanga, respectively • In Dar es Salaam and Mbeya, a subsample of 10% of the respondents were randomly drawn for more detailed analyses, while in Shinyanga 7% were drawn • Field visit (excursion) was made to the selected production sites	• 75%, 79%, and 85% of the households in, respectively, Dar es Salaam, Mbeya, and Shinyanga used charcoal. For firewood the corresponding figures were, respectively, 17%, 59%, and 14% • Dar es Salaam residents consume more charcoal per capita per year (279 kg) than in Mbeya (215 kg) and Shinyanga (196 kg) • Regarding firewood the annual per capita consumption was 395 kg, 400 kg, and 104 kg for the same cities, respectively • Secondary traders have a great influence on the charcoal trade and are prevalent in all the three urban centers • Average charcoal and firewood transport distance was highest in Shinyanga (173 km and 105 km, respectively), followed by Mbeya (116 km and 20 km, respectively) and Dar es Salaam (102 km and 69 km, respectively) • The distance from producer site to nearest road was in all areas on average 5% of the total transport distance for charcoal and 15% of the total transport distance for firewood • Charcoal is transported exclusively by truck and lorries • The market for woodfuel in Dar es Salaam is competitive with many traders, wholesalers, and producers because there are many alternatives for transports than in other areas, while Mbeya and Shinyanga are dominated by a few large integrated wholesalers, transporters, and retailers • The demand for charcoal is relatively inelastic; hence traders in Mbeya and Shinyanga are flexible for raising charcoal prices

(continued)

Table 1 (continued)

No	Authors and article title	Methodology	Main results
			• The margins for charcoal producers and traders varied between the three regions because they sell in different units and different market segments • In all the three regions, transport was the largest cost component (23–43% of retail price), surpassing the producer price • Taxes and fees summed to about 7–8% of the per sack retail price for charcoal, but a larger percent of these taxes were found not collected • The prices of the alternative energy sources are controlled, and the supply is often limited or inadequate leading to shortage and aftermarket sales at prices much higher than the official rates • Woodfuel supply systems in Tanzania are not well integrated, and a better integrated supply may increase efficiencies in coordination of transport and facilities for storing
4.	Monela et al. (1993) *Socio-economic aspects of charcoal consumption and environmental consequences along the Dar es Salaam-Morogoro highway, Tanzania*	• No behavioral theory stated (but it is mentioned that the price of charcoal drives its production and that deforestation near cities is caused by the profitable charcoal business at the expense of environmental protection) • No statistical analyses done	• Consumers believe that the greater the distance from the highway into the woodlands the better the charcoal • High-quality charcoal is believed to be *from tree species of the genera Terminalia*, *Combretum*, *Brachystegia*, and *Dalbergia* hence extensive deforestation in those areas with these species

(continued)

Table 1 (continued)

No	Authors and article title	Methodology	Main results
		• Data collection: – Unstructured interviews of 750 charcoal traders using Ubungo charcoal checking point for Dar es Salaam and 250 charcoal producers randomly selected – Unstructured interviews of some electricity using households in Dar es Salaam to map electricity cooking costs	• During rainy season, less charcoal is produced since manpower is shifted to agricultural activities and also some charcoal kilns are not accessible at this time • The regeneration time of miombo forest to reach a harvestable size after selection felling is estimated to be about 35 years or more • Assuming 62% of the wood coming from high-stocked miombo woodland with an average growing stock of 45m^3 per ha and 38% of the wood coming from low-stocked miombo woodland with an average growing stock of 10 m^3 per ha and that it takes 7 m^3 of wood to produce 1 ton of charcoal, one gets an average use of forest land (deforestation rate) of 0.2208 ha per ton charcoal produced • Each of the 1177 households in the surveyed area who produced charcoal made an average of 36 kilns per year at an average production rate of 10 bags per kiln • With an average producer price of Tshs 95 per bag of charcoal this gave an annual income of Tshs 32,260. The average operational time per kiln was 10 days • A household of 5 people consumed on average about 21 bags of charcoal annually equivalent to about 0.6 tons • Total urban household annual expenditure for using charcoal was calculated to Tshs 12,160, while that of electricity was Tshs 12,193 • But charcoal was still preferred mainly because of the high investment costs of electricity stove and higher reliability of using charcoal

(continued)

Table 1 (continued)

No	Authors and article title	Methodology	Main results
5.	Hofstad (1997) *Woodland deforestation by charcoal supply to Dar es Salaam*	• Economic behavioral assumed: 1. *Charcoal production is the result of profit maximization and is consumed in urban areas only* 2. *Charcoal consumption is a function of charcoal price, number of urban families, and average family income* • A theoretical demand/ supply model study • Variables included in the model: distances, production costs, transport costs, prices, biomass quantity harvested, and size of areas deforested • No statistical analysis done	• Model results showed that charcoal price increased from Tshs 1800 to Tshs 1958 per bag in 10 years, while the supply area increased from 3416 to 6886 km^2, and harvest increased from 2.05 to 2.66 million m^3 per year • Degraded wedge will increase together with charcoal price until the steady state is reached when increment within the area is equal to consumption • The steady state is reached at a price of Tshs 3371 per bag and a degraded area of 91,518 km^2 • Land area used for charcoal production increases over time but at different rates depending on the population growth rates • The volume is not reduced to zero at any location because cost of wood collection then becomes prohibitive • As long as the cost of household energy through charcoal stays below that of kerosene, the price elasticity of charcoal demand is the most important factor on the demand side in affecting the rate of deforestation • Reduced demand for charcoal and shift to other forms of energy are the factors controlling deforestation but is only possible at high prices of charcoal causing consumers to shift to other forms of energy • An increase in the real price of charcoal is likely as a consequence of rapidly increasing urban population • It is argued that the income elasticity of demand for charcoal is likely less than one at present income levels; at higher-income levels, it may even be negative; and it is realistic to assume that the future charcoal demand will increase less than estimated in this study if urban household income grows in the future

(continued)

Table 1 (continued)

No	Authors and article title	Methodology	Main results
			• Subsidizing the supply of substitutes (kerosene, electricity, or plantation-grown wood) may be an interesting policy measure
6.	Luoga et al. (2000a) *Subsistence use of wood products and shifting cultivation within a miombo woodland of Eastern Tanzania, with some notes on commercial uses*	• No behavioral theory stated, but the authors write that "The interactions between local communities, natural resource base, markets and the socio-political environment contribute to deforestation" • Data collected by structured interviews of 80 rural households in two villages in Morogoro Region • Systematic sampling of different wealth groups, age classes, and gender • Focused group discussion/ interviews with key informants • No statistical analyses done • Data were analyzed using descriptive statistics and content analysis	• 96% of the respondents used firewood for domestic fuel and 4% used charcoal • On average one household used about 162 ± 11 (SE) headloads of firewood per year, each weighing 29.2 ± 1.4 (SE) kg and having a volume of 0.048 ± 0.002 m^3, corresponding to 4730 kg and 7.8 m^3 of wood per year • The annual per capita firewood consumption was 1.5 ± 0.17 m^3 per year, implying an average of about 7.8m^3 for the household size of 5.2 persons • The durability of poles and hence the longevity of houses ranged from 3 to 15 years depending on the natural resistance of the poles to termites and other biodegraders • The woodland is important in subsistence farming where the cultivation of food crops goes along with collection of other food materials of fruits, edible tubers, and leaves from the woodlands • Shifting cultivation is common in the miombo woodland in Morogoro region – practiced by 68% of the respondents (*and probably this is similar to many parts of Tanzania*) • Timber for furniture and construction purposes and charcoal are the commercialized resources in the area • Charcoal was produced mainly for the markets in the urban areas of Dar es Salaam and Morogoro and was the most reliable cash-generating activity in the area

(continued)

Table 1 (continued)

No	Authors and article title	Methodology	Main results
			• About 54% of the households were involved in charcoal production, but the participants move in and out of the business depending on the conditions • With five persons per household, four houses per household, and average house life span of 8 years, the per capita consumption of construction wood was 0.138 ± 0.01 m^3 per year • The volume used for subsistence purposes of fuels and housing then became 1.64 m3 per year per capita • The high price for the high-quality timber causes a shift to lesser known timber species for household items • A sack of charcoal weighing 35 kg was at USD 2.00 at kiln site, USD 2.50 at highway site, and about USD 5.00 to urban end consumers
7.	Luoga et al. (2000b) *Economics of charcoal production in miombo woodlands of Eastern Tanzania: Some hidden costs associated with commercialization of the resources*	• No behavioral theory stated explicitly, but the study is linked to economic theory of externalization • Structured and unstructured interviews of key informants in two rural villages 50 km East of Morogoro – 8 charcoal producers – 3 charcoal wholesalers – 3 village headmen – 3 forest guards – 1 regional forestry officer • 10 unburnt charcoal kilns were sampled for volume estimation • Focused group interviews	• High charcoal production in the area studied by using traditional earth mound kilns and household male labor • Labor is the major production input – all other costs at kiln site are negligible • On average it took 100 person days per kiln for making charcoal (including felling, log piling, kiln plastering, roofing, unloading of kiln, and loading sacks) • The standing wood volume of 16.7 m^3 cleared for charcoal production can produce 61 bags of charcoal ha^{-1}

(continued)

Table 1 (continued)

No	Authors and article title	Methodology	Main results
		• No statistical analyses done	• On average a household of about 5 people constructed 5 kilns per year with each kiln requiring 10.2 ± 2.02 (SE) m^3 of wet wood and having a mean production of 1.2 ± 0.26 tons of charcoal equivalent to 44.2 ± 8.67 bags of charcoal (≈27.1 kg per bag). This means 8.16 m^3 of wood per ton of charcoal • 42 different tree species were used in the charcoal production • More than 56% of the harvested trees in the study area (ranging between 2.4 and 68.6 cm trunk diameter at breast height) were felled for charcoal burning • Sensitivity analysis indicated that charcoal business would still be profitable if more tax was paid by the charcoal producers • The profit realization and employment creation associated with charcoal production were high but accomplished at the expense of other potential uses of the woodland
8.	Sem (2004) *Supply/demand chain analysis of charcoal/firewood in Dar es Salaam and Coast Regions and differentiation of target groups*	• No behavioral theory stated explicitly • Questionnaire interviews of about 170 respondents • Direct field observation (field survey and visits) • Literature review of documented reports, information, and studies relevant to the study • Personal communications • Consists of two main parts – the first describing present woodfuel consumption, production costs, and main constraints in the supply/demand value chain and the second part providing information of efficiency, costs, and constraints of various types of stoves	• About 90% of the population in Dar es Salaam depends on charcoal as first choice of energy for cooking • The average daily consumption of charcoal in Dar es Salaam was estimated to be 2.8 kg per household or 24,000 bags per day, but only 10–20% of this amount passed through legal checkpoints • Reported average charcoal prices (in 2004) along the supply chain were – At production site 1000–1500 Tshs per bag – At nearest main road 1500–2500 " – At wholesalers in Dar es Salaam 4500 " – Retail price Dar es Salaam 5500 "

(continued)

Table 1 (continued)

No	Authors and article title	Methodology	Main results
		• No statistical analysis done	• The main types of stoves used by urban dwellers are charcoal stoves and ovens, while rural dwellers use mainly firewood stoves dominated by inefficient traditional three-stone fireplace • Low-income communities located in rural and urban areas form a potential user group of charcoal and woodstoves • None of those low-income groups who are earning less than Tshs 45,000 per month was using electricity as main type of energy • Affordable stoves are those with prices ranging between Tshs 1350 and Tshs 5000 • By using improved charcoal stoves, the survey has recorded the savings among the users of up to 50 percent • The adoption level of improved stoves is higher in the urban households as compared to rural households • The simple traditional kilns are capable of making charcoal at a conversion rate ranging from 2 to 5.2 bags of charcoal from $1m^3$ of fuelwood (*one study reported 2–3 bags and another one reported 2.84–5.20*) • Charcoal dealing is a purely male-dominated activity as no women dealers were found during the research period • A large potential is reported for improved institutional stoves to reduce fuel consumption in community centers and thereby reduce deforestation as well as health hazards • Improved institutional woodstoves designed at the University of Dar es Salaam and installed at some schools in Tanzania have indicated reduced fuelwood consumption, with fuel saving between 60% and 80%

(continued)

Table 1 (continued)

No	Authors and article title	Methodology	Main results
			• Durability of the stoves is reported as the main aspect of concern among the interviewed persons and should be emphasized when planning future modifications
9.	Malimbwi et al. (2005) *Charcoal potential of miombo woodlands at Kitulangalo, Tanzania*	• No behavioral theory stated • Inventory Data at Kitulangalo Training Forest using 46 plots in the forest reserves and 30 plots in adjacent public land • All the plots were chosen by stratified random sampling	• Twelve species were found in the forest reserve, while eight species only were found in the public lands • The average volumes and basal areas per ha were 46.2 m^3 ha^{-1} and 7 m^2 ha^{-1} in public lands and 78.8 m^3 ha^{-1} and 10 m^2 ha^{-1} in forest reserve, respectively • The known suitable tree species for charcoal making in miombo woodlands, i.e., *Julbernardia globiflora* and *Brachystegia boehmii*, appeared to be abundant in the forest reserve and less available in public lands • The per ha volume and basal area increased with distance from the highway, while stem numbers per ha showed a reverse trend meaning that the woodland along the roadside had been depleted mostly for charcoal extraction due to easy accessibility compared with woodlands away from the highway • Average standing wood volume was 24.5 m^3 per ha and 56.5 m^3 per ha in, respectively, public land and forest reserve land • The mean annual increment (MAI) for the period of 3 years (1996–1999) was 2.35 m^3 ha^{-1} $year^{-1}$ • By using a conversion factor of 0.85 for fresh wood volume to wood biomass and kiln efficiency of 23%, the weight of charcoal that can be extracted from the woodland at the roadside was 0.29 m^3 ha^{-1} (fresh wood) × 0.85 × 0.23 = 56 kg of charcoal, equivalent to only one bag of charcoal

(continued)

Table 1 (continued)

No	Authors and article title	Methodology	Main results
			• About 54 and 125 bags may be extracted at 5 km distance and from beyond 10 km from the highway, respectively • In the forest reserve, it will take about 23 years, 16 years, and 8 years for the woodland at, respectively, the roadside, 5 km and 10 km away from the highway to attain the forest conditions of 53.4 m^3 ha^{-1} of preferred tree species for charcoal making • In public lands, the recommended years to attain 35m^3 per ha standing volume is 15 years and 8 years, respectively, 5 km and 10 km from the roadside
10.	Monela et al. (2007) *Socio-economics of charcoal extraction in Tanzania: A case of Eastern part of Tanzania*	• No behavioral theory explicitly stated, but the study mention that *poverty, unemployment, urbanization, low prices and high demand are the main drivers of charcoal production and consumption* • Data collection: – Structured questionnaire used in interviewing 113 charcoal makers – Focused group discussion using a set of checklists – Participant observation • Statistical analyses done • Variable: (i) Dependent variable: *The amount (bags) of charcoal produced per month for sale* (ii) Independent variables: *Age, gender, education level, ethnic group, and number of wives*	• High rate of migration to charcoal production areas (along the Dar es Salaam and Arusha highways) • Twenty different tribes originating from different parts of the country were found in the study sites where charcoal is produced • Main economic activities are agriculture and charcoal production • Household members are the main source of labor for charcoal production • Main species favorable for charcoal production are *Julbernardia globiflora, Brachystegia boehmii, Tamarindus indica, Acacia nigrescens, Acacia gerrardii, Combretum adenogonium, Combretum molle, Combretum zeyheri, Combretum collinum, Diospyros kirkii, Xeroderris stuhrmanii, Mimusops kummel, Albizia harvey, Acacia goetzei* subsp. *goetzei, and Lonchocarpus capassa*

(continued)

Table 1 (continued)

No	Authors and article title	Methodology	Main results
			• On average the producers got 3 bags of charcoal per tree felled, varying from 1.7 to 11 bags per tree • The average amount of charcoal produced for sale was about 354 bags (each about 28 kg) per household per year • Age, sex, and number of wives had statistical significant coefficients, hence impacts on charcoal extraction • Most charcoal producers used rectangular traditional charcoal kiln yielding about 29 bags of charcoal per kiln • Most of the charcoal produced were sold at the production site and in the production village where dealers from Dar es Salaam and other urban centers come to collect charcoal bags for their business • Price of charcoal was Tshs 1500/=, 1400/=, and 1000/= at roadside, village center, and kiln site, respectively • On average it used 40.6 days per kiln for wood cutting, kiln preparation, carbonization period, and unloading • For own consumption each household per year used on average about 100.3 headloads and 3.3 charcoal bags • About 67% of respondents indicated that charcoal is more scarce today than 10 years ago; at the same time, the tree cover was also found to be less today than 10 years back • Charcoal extraction in the woodlands is the most important economic activity providing employment and income to many households in both rural and urban centers • The high number of species preferred for charcoal extraction found in the study area is a clear indication of the available high potential for charcoal extraction

(continued)

Table 1 (continued)

No	Authors and article title	Methodology	Main results
			• There is strong link between charcoal extraction and ecological balance of the woodland resource • Poverty is a compelling factor for the decision to engage in charcoal extraction • In the longer term, a costly and painstaking process of adopting improved technologies which require capital investment is unavoidable
11.	Van Beukering et al. (2007) *Optimization of the charcoal chain in Tanzania*	• The stated behavioral theories: 1. *Improved charcoal production can increase its sustainability* 2. *Profit maximization is the main motive for charcoal production* • 360 observations by interviewing different stakeholders • Data collected using semi-structured questionnaires, interviews, surveys, GIS, and value chain analysis • No statistical analyses done	• About 28% of the households are involved in charcoal production, earning between 71% and 81% of the household cash income • Charcoal producers make the least profit than other stakeholders along the charcoal value chain • Very low efficiency of the traditional kiln in the study area ranging between 10% and 20% • The total income from charcoal is estimated to be about 17.6 billion (USD 17.6 million) in 2005 • The commercial sector (small eating places, restaurants, small-scale industries, agro-processing industries {tobacco curing, tea drying, beeswax processing, etc.}) is estimated to use 31% of the total charcoal consumption in Tanzania, and the remaining 69% is for household consumption • While charcoal can be produced from a variety of different tree species, most of the trees used for charcoal are from the natural Miombo woodlands • Land use change around Kazimzumbwi area shows that plantation forest and cultivation with tree crops has increased considerably at the expense of natural forest and bush lands

(continued)

Table 1 (continued)

No	Authors and article title	Methodology	Main results
			• Increased agriculture at the expense of forested areas in other parts of the catchments could be a major contribution to the effects seen in the area around Kazimzumbwi forest • Interviewed people confirmed that production of charcoal had done most damage to the forest • Due to relatively low efficiencies, a large percentage of fuelwood is diverted to GHGs like CO, CO_2, NO_x, and SO_2 • The ban from the government had little effect on charcoal production. Producers continued to manufacture charcoal despite the ban, and with traders loath to buy, stocks of charcoal increased in the production areas. After the ban, the increased demand from the consumers and little stock in the cities caused the producers to double the prices from the pre-ban level • The following main conclusions are drawn: – Because of its vast magnitude changes in the charcoal sector can only be realized gradually, sudden interventions such as a ban on charcoal production and trade are counter-effective – Despite high environmental awareness among the charcoal producers, their poverty leaves no alternative but continuing the profession of charcoal making – Projects improving the extremely low kiln efficiency would be beneficial both for local communities and the environment

(continued)

Table 1 (continued)

No	Authors and article title	Methodology	Main results
			– Payments for environmental services (PES) could be considered to reduce externalities – Current policies directed at the charcoal chain are inefficient in many ways. The command and control policies dominating the current government policies need to be supplemented by market-based approaches
12.	Mwampamba (2007) *Has the woodfuel crisis returned? Urban charcoal consumption in Tanzania and its implications to present and future forest availability*	• No behavioral theory stated directly, but the author links "*per capita charcoal consumptions with* per capita *income"* *and also "a high population growth and high reliance on charcoal as a major cause of environmental degradation*" • Households survey in six selected regions • 244 observations • Scenario analysis is done using population and consumption to project future charcoal demand to year 2100	• Per capita household consumption ranges between 3.12 and 6.01 bags per person per year (equivalent to 93.6–180.3 kg per person per year) • Low- and middle-income groups did not have significant effect on the amount of charcoal consumed • Increase in household size caused lower per capita consumption per household • High charcoal consumption in almost all selected regions (*Dar es Salaam*, *Mwanza*, *Morogoro*, *Mbeya*, and *Arusha*) with price variations between them • Mtwara and Zanzibar had the lowest per capita consumption • Lindi had the lowest price, while other towns in mainland Tanzania had almost the same price (Tshs 4683/bag) • The highest price was in Zanzibar where no or little production is taking place (Tshs 5280 per bag) • In the scenario analyses, 80% of the urban population are assumed to use charcoal for cooking • Important to reduce charcoal consumption by improving stove efficiency or kiln efficiency in the production • Promote alternative energy sources for cooking in Tanzania to reduce the negative impacts of charcoal production and consumption

(continued)

Table 1 (continued)

No	Authors and article title	Methodology	Main results
			• More efficient cooking stoves are considered important for reducing charcoal consumption in the country as well as reducing the price of the alternative energy sources mainly gas, kerosene, and electricity • A coherent policy framework governing charcoal production, trade, and use in Tanzania does not exist, and reliable statistic on the sector rarely exists • Various policies are presented based on Tanzanian conditions and experiences in other countries having a significant charcoal sector
15.	Msuya et al. (2011) *Environmental Burden of Charcoal Production and Use in Dar es Salaam, Tanzania*	• No behavioral theories stated • Used a simulation model (STELLA) doing mass balance accounting ecological modelling tool • Estimated charcoal consumption and forest loss using population growth as basis for the estimation	• Charcoal consumption in Dar es Salaam ranges from 1600 to 2200 tons per day causing a significant loss of forest due to charcoal production • Projected charcoal demand in 2009 in Dar es Salaam was 1904 tons/day totaling 694,960 tons per year • About 105,303 ha of forest are lost each year due to charcoal production for Dar es Salaam only totaling to about 150,433 ha nationwide • The annual charcoal consumption up to 2030 is projected to emit about 49.7 million tons of CO_2 • Emission of other GHGs will also be very high if no measure will be in place to minimize the emission • By 2030 more than 2.8 million ha of forest will have to be cut due to charcoal production for Dar es Salaam only
16.	Felix and Gheewala (2011) *A review of biomass energy dependency in Tanzania*	• No behavioral theory stated • A study based on literature review	• Charcoal and firewood are the main biomass energy sources for most households in both rural and urban areas and constitute 90% of all energy use in Tanzania

(continued)

Table 1 (continued)

No	Authors and article title	Methodology	Main results
			• The total charcoal consumption in Tanzania was 750,000 tons annually in the year 2000 • Local wood consumption for charcoal in 2000 is in the study reported to be 222.37 million m^3, for a population of about 33 million people and average household size of 5 to 7 people, while firewood consumption is reported to be 55.5 million m^3 • The average charcoal consumption for each household was 30.05 m^3 and for firewood 7.5 m^3 • Charcoal is consumed by 94% of the households either alone or mixed with other fuels, and only about 6% of the households are estimated to not use charcoal • The use of energy-efficient charcoal stoves in Tanzania is minimal due to high initial installation cost that cannot be afforded by households with low income • Energy efficiency stoves for burning firewood and charcoal are not easily available in Tanzania due to the lack of government support and poor biomass energy policies • Charcoal has a higher calorific value per unit weight than firewood, which is about 31.8 MJ per kg of completely carbonized charcoal with about 5% moisture content as compared to about 16 MJ per kg of firewood with about 15% moisture content on dry matter basis • The overreliance on charcoal and the excessive use of firewood are the major causes of deforestation and land degradation of about 91,000 hectares of land annually

(continued)

Table 1 (continued)

No	Authors and article title	Methodology	Main results
			• One way of reducing wood fuel consumption is to improve charcoal production techniques as well as charcoal cooking stoves in the households • More emphasis should be directed toward the use of energy-efficient charcoal and firewood stoves for cooking and the use of mixed fuel like liquefied petroleum gas (LPG) and biogas to reduce the burden on the forests
17.	Schaafsma et al. (2012) *Towards transferable functions for extraction of Non-timber forest products: A case study on charcoal production in Tanzania*	• Behavioral theory: Economic valuation of non-timber forest products • 1176 household observations from 4 different surveys done in earlier studies along the Eastern Arc Mountain (EAM) in Morogoro • These observations were put in a geographic information system (GIS) transfer modelling frame	• 80% of the interviewed households rely on agriculture as a main source of income • 94% of the sampled households uses firewood as a main cooking fuel, while 73% of the houses in the study were made of poles which mainly originate from the nearby forests • Households whose main source of income is from timber and NTFPs are more likely to produce charcoal • The available survey data suggest that 60% of the households producing charcoal use wood from protected forests and woodlands like protected areas and forest reserves, 20% from open access forests and woodland and 45% from farmland • The estimated total annual household production of charcoal from the EAM was about 2.9 million 30 kg bags (1.45 million 60 kg bags) equivalent to approximately 11% of the combined annual charcoal consumption in Dar es Salaam and the cities of Morogoro and Tanga, the main markets for charcoal from the EAM blocks • Prices vary from Tshs 4000 to Tshs 45,000 per 60 kg bag across the study area, with a mean price of Tshs 30,088 (USD 21) per bag in Dar es Salaam and Tshs 16,584 (USD 12) elsewhere

(continued)

Table 1 (continued)

No	Authors and article title	Methodology	Main results
			• Several factors contribute to the variation of the market prices including distance from the market to Dar es Salaam (transportation costs), dummy for prices (to cover taxes, levies, and bribes and the year of data collection and inflation) • The total value of the annual extraction of charcoal from the EAM was approximately Tshs 21 billion per year in 2010 prices or USD 14 million, including charcoal sold as well as any charcoal consumed domestically
18.	Sander et al. (2013) *Enabling reforms: Analyzing the political economy of the charcoal sector in Tanzania*	• No explicit stated behavioral theories • Used political economy (PE) analysis and a so-called Net-Map tool to identify key actors and power networks prevailing in the charcoal sector in Dar es Salaam • Assessed the interactions between actors in the charcoal value chain in Tanzania • Used focus group discussions and key informant interviews • Interviewed 200 different stakeholders	• Tanzania charcoal sector is characterized by weak governance, limited low enforcement capability, and other regulatory capacity constraints • Comprehensive policies, strategies, and legal frameworks addressing charcoal sector are absent or missing • Overlapping responsibilities between different central government agencies are common in the country which in most cases are unnecessary duplicates performing similar or related duties • Charcoal production and trading is characterized by overlapping responsibilities between different central government agencies (Ministry of Natural Resources, Ministry of Energy, and the Vice President's Office under the Division of Environment, Prime Minister's Office, District Authorities, Village Authorities, TRA, etc.) • Most of the collected revenues (81%) goes to the National Authorities, and very little 17% and 2% are retained by the District and Village Authorities, respectively

(continued)

Table 1 (continued)

No	Authors and article title	Methodology	Main results
			• Tanzania's charcoal sector does not function as envisaged in government strategies and policies due to the complex, formal governance framework of the charcoal sector in Tanzania and the associated incentives and disincentives • Well-known tax avoidance strategies are used to bypass formal sector regulations and to integrate government officials or institutions in an informal benefit sharing mechanism • The fiscal disempowerment of village and district governments creates substantial disincentives for formalization of the business and sustainable management of the charcoal sector • Lack of effective benefits sharing mechanism, unclear ownerships of forest assets, and low capacity of low enforcement are among the reasons for the loss of charcoal revenues at different levels • There is also lack of government control over charcoal business, and the only government-level authority linking and interacting directly with charcoal producers and traders is the village-level government • In many instances, one person tends to play many roles in the charcoal value chain, hence difficult to control them • There is a strong divergence between the de jure and de facto power relations between the government and other charcoal stakeholders • It is therefore vital to strengthen vertical accountability and exchange of information, engage charcoal dealers (producers, transporters, wholesalers) network, empower responsible institutions, and enhance regulatory transparency along the entire value chain

(continued)

Table 1 (continued)

No	Authors and article title	Methodology	Main results
			• Reforming the charcoal sector in Tanzania requires a strong political economy consideration and willingness to change the illegal communication to formal and legal channels of information sharing
19.	Zulu and Richardson (2013) *Charcoal, livelihoods, and poverty reduction: Evidence from sub-Saharan Africa*	• No behavioral theory stated • Analyzes the linkage between charcoal production and poverty alleviation and the negative narrative that poverty is causing forest loss and environmental degradation • Explains the different dimensions of poverty *(material deprivation, voiceless and powerlessness, vulnerability and exposure to risks, poor education and health)* • The study is based purely on literature review • No statistical analyses	• Growing demand for charcoal has increased opportunities for income generation, rural livelihood support (from production and trading), and poverty alleviation • Charcoal production and trading pose different challenges including unsustainable production, environmental degradation, and negative health impacts for material-deprived households • The overexploitation of forest resources for charcoal production is mainly due to weak, misguided, neglected, underdeveloped, disjointed, overly prohibitive, contradictory, or nonexistent woodfuel policies and laws, combined with poor enforcement and regulatory capacity • The market for charcoal is described as dispersed, poorly developed, and weakly regulated • Charcoal economy is extensive and links to numerous enterprises and supports livelihoods in urban and rural areas • The primary actors in the charcoal value chain are *producers, wholesalers, retailers, and transporters* and *end users (consumers)* • Charcoal production and market plays a significant role in generating seasonal and full-time employment in regional value chains

(continued)

Table 1 (continued)

No	Authors and article title	Methodology	Main results
			• Lower-income households generally consume more charcoal per capita; wealthier households also use charcoal • Lower-income households often pay a higher price per kilogram for charcoal because they buy it in smaller packages; wealthier households will typically purchase larger quantities for a lower price per kilogram • Rapid population growth, urbanization, and improved incomes are generally associated with decreases in firewood use and increases in charcoal consumption • For low-income households, both firewood and charcoal are assumed to be normal goods but are considered inferior goods for high-income households • Charcoal production increases vulnerability and exposure to risks by contributing to environmental degradation through deforestation, soil erosion, and increases in greenhouse gas emissions as negative impacts of poverty reduction • Charcoal production and trading undermining agricultural productivity by diverting male labor into charcoal production hence overburdening women on food crop production • The relationships between charcoal and poverty are very complex, and although there may exist positive relations, there are limits to charcoal-based poverty alleviation, and it therefore requires multifaceted and integrated approaches both on the production and demand sides

(continued)

Table 1 (continued)

No	Authors and article title	Methodology	Main results
20.	D'Agostino et al. (2015) *Socio-economic determinants of charcoal expenditures in Tanzania: Evidence from panel data*	• No clearly stated behavioral theories, but the hypotheses are indirectly based on economic theory • Used two panel data sets collected in 2008/2009 and 2010/2011, with altogether 6367 household observations • Statistical analyses done • Variables used (i) Dependent variable: Household charcoal expenditure (ii) Independent variables: *Income, household size, location (rural and urban), and various dummy variables as so-called control variables* • Charcoal price is however not included as it is argued that it will be too uncertain to include	• Household income has a strong positive association with charcoal consumption, while household size does not • Rural households consume less charcoal than those located in urban areas • In the total sample with all control variable included, a 10% increase in total expenditure is estimated to be associated with 3.9% increase in charcoal expenditure, meaning an income elasticity of 0.424. This suggests that charcoal is a basic good • Panel data from Dar es Salaam indicates that household size is significant, and income elasticity there was estimated to be 1.11 • The estimated income elasticities varied according to which sample is used, being 1.11 in the sample which included only Dar es Salaam households (2178 observations), 0.69 in the sample having only urban households (82,178 observations), 0.39 in the total sample (6367 observations), and 0.32 in the sample where the households in Dar es Salaam were excluded (5322 observations) • For the betterment of the forests and the forest sector in general, policy makers should focus on finding sustainable alternatives and substitutes to replace charcoal • Targeting the urban centers is important because of the large effects income has on charcoal consumption in those areas

(continued)

Table 1 (continued)

No	Authors and article title	Methodology	Main results
21.	Nyamoga et al. (2019) *Econometric analysis of urban household charcoal consumption in Tanzania: The case of Morogoro, Dodoma and Mtwara*	• Based on economic theory • Data collected by structured questionnaire of 360 households randomly chosen in Morogoro, Dodoma, and Mtwara • Statistical analyses done for three income classes and the whole sample • Variables included (i) Dependent variable: *Charcoal consumption* per capita (ii) Independent variables: *Household income, charcoal price, price of other energy sources, household size, and geographical location*	• In the low-income group, statistical significant elasticities for annual charcoal per capita demand were found to be −0.40 and −0.60 for, respectively, charcoal price and household size • In the middle-income group, only household size was found significant with −0.83 as estimated elasticity • In the high-income group, elasticities of 0.20 for per capita income and −0.42 for household size were found significant • Statistically significant differences between regions were found in the low-income group and the whole sample. Price of electricity or gas did not become statistically significant variables. • The findings indicate that price of charcoal, household size, per capita income, and regional differences are key factors influencing urban charcoal consumption in Tanzania • The per capita charcoal consumption was significantly positive in the high-income group which contradicts with the so-called energy ladder theory • The study recommends further surveys (preferably a national survey) and research activities in other regions to investigate the long-run implication of charcoal production and consumption in the country. In this regards, masters and PhD students need to be involved intensively

3.1 Charcoal Consumption

Methodology

Sixteen of the 21 reviewed studies deal with charcoal consumption. Five of these studies include some kind of statistical analysis, but only three have stated some behavior theory to give hypotheses and guide for which variables to include in the statistical analyses.

Quantity

The estimates of charcoal quantities consumed per capita vary quite a lot. Boberg (1993) estimates the annual household per capita consumption in Dar es Salaam, Mbeya, and Shinyanga to be 280 kg, 220 kg, and 200 kg, respectively. Nyamoga et al. (2019) estimate the annual per capita consumption in Dodoma, Morogoro, and Mtwara to be 165 kg, 140 kg, and 145 kg, respectively, while the total average was 142 kg. Monela et al. (1993) estimate this consumption to be 117 kg for Dar es Salaam, whereas Malimbwi et al. (2007) estimate it to be 164 kg, and Hofstad (1997) assumes it to be 174 kg for Dar es Salaam. There are obviously many possible reasons for this variation in consumption estimates, like differences regarding sample selection, sample size, location, and questionnaire used, as well as different time and prices reflecting different supply and demand situations.

Two studies have estimated the use of charcoal in Dar es Salaam in the commercial and public sectors (hotels, bars, agro-industrial enterprises, schools, hospitals, army, etc.): Van Beukering et al. (2007) estimate that these sectors account for 31% of the total end use of charcoal in Dar es Salaam, whereas Malimbwi et al. (2007) estimate this share to be 22% (corresponding to a daily use of 4200 bags of charcoal from hotels, bars, and vendors, 2000 bags from schools, and 33 bags from hospitals and the army).

Prices

Van Beukering et al. (2007) observe that charcoal demand among commercial enterprises in Dar es Salaam is rather inelastic with respect to price. In a survey among these enterprises, most respondents claimed that they would continue to buy charcoal even if the price doubled. Van Beukering op cit. argue that this inelasticity of charcoal price can be directly related to the prices of alternative fuels and presents annual average fuel prices (measured as Tshs per Kcal energy produced) which show that in all years during the period 1995–2006, the price for charcoal in Dar es Salaam has been much lower than for kerosene, LPG, and electricity.

3.2 *Charcoal Production*

Methodology

Seven of the 21 reviewed studies deal with charcoal production. Of these, only Monela et al. (2007) include statistical analysis, with profit maximization referred to as behavior theory. Nearly all of the reviewed studies have a regional or local area focus. Different stakeholders are interviewed and the studies have from 8 to about 400 respondents. No national survey is found.

Quantities Produced

Several of the studies include interviews of kiln producers and analyses of their charcoal productions. Almost all of the charcoal production seems to be done with earth mound kilns having an energy recovery of 10–19%, meaning that 81–90% is wasted energy. The estimates of wood input requirement vary from $7m^3$ to $8.5m^3$ wood input per ton of charcoal produced.

None of the studies give estimates of charcoal export to or import from Tanzania, and presumably both might be small and/or counterbalance each other. Malimbwi et al. (2007) estimate that 70% of the charcoal consumption in Zanzibar comes from Mainland Tanzania, corresponding to 10,500 bags of charcoal per day, each bag weighing 53 kg, and that about 7500 bags are transported illegally.

3.3 *Emission of GHG*

Fuelwood consumption (firewood and charcoal) is among the critical environmental problems in many sub-Saharan countries including Tanzania (Sulaiman et al. 2017). It has been estimated globally that emission of greenhouse gases (GHGs) from biomass burning can exceed those emitted from fossil fuel-based GHG in many less developed countries (Bailis et al. 2003). To fulfil this fuelwood demand, tree cutting is necessary leading to significant deforestation and land degradation. The process of charcoal production can cause emission of different greenhouse gases including carbon dioxide (CO_2), carbon monoxide (CO), nitrogen oxides (NO_2), SO_2, and methane (CH_4) (Bailis et al. 2003; Bailis 2009; Msuya et al. 2011; Chidumayo and Gumbo 2013). The emitted greenhouse gases, especially carbon monoxide and carbon dioxide, can be very poisonous and detrimental to the nervous and the brain system, resulting into severe illness and possibly death. The emitted nitrogen oxides, CO, formaldehyde, and carcinogens react with sunlight in the atmosphere leading into air pollution (ibid). It has been reported that the smokes produced from charcoal burning tend to augment those from diesel engines and industrial chimneys (NorConsult 2002). Msuya et al. (2011) report that charcoal production and consumption could lead to the production of about 50 million tons of CO_2 and more than 20 million tons of CO by the year 2030 and that about 9.8, 1.1, and 12.5 million tons of NO_2, SO_2, and CH_4, respectively, will be emitted by 2030 in Dar es Salaam alone, if the country continues using charcoal in the same way as today.

By using appropriate technologies, it is possible to reduce emissions from charcoal at both production and consumption points in the charcoal life cycle, through provision of high-efficiency and low-emissions charcoal stoves with improved combustion and heat transfer efficiency (Bailis et al. 2003). Except for Msuya et al. (2011), no studies from Tanzania were found of GHG emission related to charcoal consumption and production. We therefore have used FAO (2017) for getting more information here, and our best estimates for Tanzania are as shown below, assuming earth kiln in the production and Kenyan average charcoal stove in the consumption of charcoal. The uncertainty in these estimates is of course high.

1. Regarding emission from charcoal production, we assumed Kenyan earth mound kiln 2 with yield efficiency of 21.6% as stated in FAO (2017:151) and originally documented by Pennise et al. (2001), getting the following emission factors:

CH_4	35.2 kg GHG per ton of charcoal produced
CO_2	1992 kg GHG per ton of charcoal produced
CO	207 kg GHG per ton of charcoal produced
TNMHC or TNMOC[a]	90.3 kg GHG per ton of charcoal produced
N_2O	0.20 kg GHG per ton of charcoal produced
NOx	0.12 kg GHG per ton of charcoal produced
TSP	41.2 kg GHG per ton of charcoal produced

[a] *TNMHC* total non-methane hydrocarbons, *TNMOC* total non-methane organic carbon (Pennise et al. 2001)

2. Regarding emission from cook stoves, we assumed the average Kenyan charcoal stove as stated in FAO (2017:156) and originally documented by Bailis et al. (2003), getting the following emission factors:

CH_4	18 kg per ton of charcoal produced
CO_2	2280 kg per ton of charcoal produced
CO	260 kg GHG per ton of charcoal produced
TNMOC	3.2 kg GHG per ton of charcoal produced

In addition some GHG emissions from, e.g., transport and net GHG emissions from forest soils and biomass growth changes are also possible, but are not reported because the information here has not been found (not certain here, but this is what seems logical).

3.4 More General Discussions

Income Generation and Overexploitation

This review has revealed that charcoal production and consumption contribute significantly to the incomes of people in Tanzania and the rate of production along the miombo woodlands is increasing (Luoga et al. 2000a, b; Malimbwi et al. 2007) in spite of scarce literatures on its value chain. The increasing number of charcoal producers may be attributed to the free access to the miombo woodland and the low capital required for starting charcoal production among other factors. Although the investment in tree plantations for charcoal production is increasing in tropical regions including Tanzania, most biomass for charcoal production still comes from natural forests which tend to regenerate naturally (Chidumayo and Gumbo 2013). As the area of woodland decreases, the marginal value of each tree increases because the demand exceeds supply (Hofstad 1997; Luoga et al. 2000a, b). Most miombo woodlands are regarded as common pool resources, freely available to everyone. Therefore, the profit-maximizing individuals or households will continue producing charcoal as long as profit is maximized.

Despite the importance of charcoal for income generation in the households, inefficient production technologies such as the traditional earth kiln with very low efficiency of about 11–30% are used, leading to higher wood biomass consumption (Luoga et al. 2000a, b). With the low kiln efficiencies, about 70–80% of the wood are lost along the production process, signifying an increase in deforestation. Any innovation for improving kiln and cooking stove efficiencies may have a significant contribution in reducing the deforestation rates in Tanzania (Zein-Elabdin 1997; Sanga and Jannuzzi 2005).

Statistics indicate that Tanzania is among the top 10 global charcoal producers, producing about 3% of the world's total charcoal (FAO 2010). Rearranging and addressing holistically the entire value chain of charcoal can make charcoal production and consumption contribute to sustainable development and poverty alleviation in Tanzania (Neufeldt et al. 2015; Luvanda 2016). Provision of proper information on charcoal production and value chain would most likely help to identify possible opportunities for more efficient ways of organizing the charcoal markets and institutions arrangement for enhancing better returns to all stakeholders in the value chain (Shively et al. 2010). Empirical evidence from the field and previous studies indicate that charcoal production and consumption will dominate the energy sector for many years in sub-Saharan Africa (Hosier et al. 1990, 1993; Hosier 1993; Hosier and Kipondya 1993).

Charcoal production and the entire business in general have been perceived negatively because of the historical unsustainable production technologies employed. Due to unsustainable production techniques in Tanzania, charcoal production tends to be linked to deforestation and forest degradation, which in turn affects the livelihoods of people due to reduced land and ecosystem productivity (Jones 2002; Kissinger et al. 2012; Cerutti et al. 2015). According to Zulu and Richardson (2013), the overexploitation of forest resources for charcoal production is mainly due to weak, misguided, neglected, underdeveloped, disjointed, overly prohibitive, contradictory woodfuel policies and laws combined with poor enforcement and regulatory capacity. This is mainly due to the fact that most of the miombo woodlands in Tanzania are managed under general land with open and free access and subjected to many management challenges. The management challenges emerge because of the undefined land ownership that allows free access to the forest land and overexploitation (Schlager and Ostrom 1992). New regulations have been put in place for supporting sustainable charcoal production (Delahunty-Pike 2012). Charcoal production increases vulnerability and exposure to risk by contributing to environmental degradation through deforestation, soil erosion, and increase in greenhouse gas emissions (Kaimowitz 2003). This is an indicator that charcoal production and consumption can have both positive and negative impacts, like poverty reduction, deforestation, and land degradation. It may also undermine agricultural productivity by diverting male labor into charcoal production, hence overburdening women on food crops production. One of the reasons for charcoal production's contribution to deforestation is the use of traditional kilns with very low efficiency, which may require as much as 10 kg of wood for producing 1 kg of charcoal (Adam 2009). Besides demanding large amount of raw materials, these traditional kilns also release large amounts of greenhouse gases during the carbonization process. The efficiency problem also exists on the consumption side where the burning of charcoal in traditional cooking stoves is inefficient, resulting into increased wasteful consumption of charcoal as well as gas emissions inside houses, directly influencing peoples' health. Efficiency improvement is therefore an important aspect of both the production and consumption stages in order to reduce the negative effects (Sanga and Jannuzzi 2005; Adam 2009).

In Tanzania, cooking energy is dominated by biomass-based fuels and is primarily firewood and charcoal accounting for more than 90% of primary energy supply which is estimated at $1m^3$ of round wood per capita per year (Felix and Gheewala 2011; Lusambo 2016). The high preference for charcoal is due to its high calorific value per unit weight, which is about 31.8 MJ per kg of completely carbonized charcoal with about 5% moisture content compared to about 16 MJ per kg of firewood with about 15% moisture content on dry basis (Felix and Gheewala 2011). It is also preferred because of convenience in transporting, storing, and non-susceptibility to infections by fungi. Studies indicate that a household of about 5 people can consume 21 bags of charcoal annually which is equivalent to about 0.6 tons (Monela et al. 1993). Previous models developed by Faustmann and Hartman have suggested that harvesting of even-aged stand tends to be influenced by, among other factors, the opportunity costs for delaying the harvest and whether land owners pay management fee and property tax to government (Koskela and Ollikainen 2001; Chang and Gadow 2010; Deegen et al. 2011). The implementation of these models depends on factors like proper land ownership, forest characteristics, and vegetation types (Amacher et al. 2009; Kant and Alavalapati 2014) which are scarce in most miombo woodlands in Tanzania. The miombo woodlands are characterized by uneven-aged wood stands although charcoal and timber producers harvest the trees to maximize utility of the forest resources (Frost 1996; Luoga et al. 2002). Furthermore, unlike in developed countries, the amenity values for forest stands in Tanzania are less recognized by the local communities or does not exist at all; it is hence difficult to implement policy measures suggested by the models like those developed by Faustmann and Hartman.

The high demand for charcoal in Tanzania is mainly due to high human population growth, high rate of urbanization, and the high economic growth which in turn increases the population of the middle-income group that consumes charcoal in the urban centers (Malimbwi et al. 2007; Mwampamba 2007; Msuya et al. 2011). About 90% of the urban population depend on charcoal as the main source of energy for cooking (Hosier and Kipondya 1993; Hosier et al. 1993; Mwampamba 2007; Lusambo 2016). Most of the charcoal (>70%) consumed in Zanzibar comes from mainland Tanzania, and more than half of the charcoal is traded illegally (Mwampamba 2007). About 3 million tons (3.6 million m^3) of wood are needed to produce the 28,759 bags (each weighing 56 kgs) consumed daily in Dar es Salaam alone (Malimbwi et al. 2005, 2007). The World Bank consultancy report of 2009 indicates that Tanzania consumed about 2650 metric tons a day, totaling to about 1 million tons of charcoal per year (WB 2009). Although charcoal is considered as a transitional energy to the more clean energy, research results show that it is still the main source of cooking energy consumed by the majority of people in towns, including the wealthier families. This is because of its reliability, availability, and easiness to trade and transport (Nyamoga et al. 2016).

The positive impacts of the growing demand for charcoal are the increased opportunities for income generation for both rural and urban population, through production and trading, thereby contributing to poverty alleviation (Zulu and Richardson 2013). In urban areas where woodfuel is scarce and alternative energy sources expensive, charcoal has enhanced the expansion of domestic markets and provided opportunities for households' savings. However, it also poses some challenges for sustainable production, environmental degradation, and negative health impacts on resource-poor households. Generally, charcoal economy and business encompasses numerous enterprises that support livelihoods in urban and rural areas (Boberg 1993; Van Beukering et al. 2007). Although lower-income households consume more charcoal per capita at relatively higher price per unit quantity (because of buying in small quantities), the wealthier households also consume substantial amounts of charcoal (Zulu and Richardson 2013). The high demand and consumption of charcoal makes it a potential income-generating activity along the value chain in both rural and urban areas.

Market and Marketing Systems in Tanzania

Several actors and institutions are involved in the charcoal value chain (Van Beukering et al. 2007); and in Tanzania about 2 million people are estimated to be involved in the entire value chain of charcoal (WB 2009). Most of the traded charcoal in Tanzania is produced locally and traded illegally through informal channels (Milledge and Elibariki 2005; Milledge et al. 2007). The informal charcoal trading channels tend to benefit producers, traders, and transporters but cause significant loss of government revenues through tax evasion (Milledge and Kaale 2005; Milledge et al. 2007; Sander et al. 2013). Formalizing charcoal trade is therefore essential to ensure that the necessary government revenues are collected efficiently (Schure et al. 2013). The formalization of charcoal trading is also necessary and advantageous for the government to have a record of all main agents in the charcoal value chain. Under the current system, charcoal is traded under informal conditions, characterized by dispersed, poorly developed, and weakly regulated markets (Zulu and Richardson 2013). Under informal systems, the key primary actors in the charcoal value chain such as producers, wholesalers, retailers, and transporters may be poorly coordinated and inefficient (Van Beukering et al. 2007). The World Bank (2009) reported that the value of the charcoal business in Tanzania was about USD 650 million, which was higher than the combined incomes from cotton, coffee, and tea in that year (WB 2009).

Charcoal Transport and Market Prices

Both commercial and noncommercial traders are involved in charcoal transportation using trucks, motorcycles, and bicycles and sometimes carrying on heads and shoulders (Malimbwi et al. 2007). Although the sector employs about 2 million people along the value chain (WB 2009), large profits are concentrated in the hands of wholesalers and transport agents, leaving the other agents more deprived (Van Beukering et al. 2007). Historical data show that the price of charcoal has been rising each year since 2000, and in 2004–2007 the market price of charcoal increased by about 160%, while the retail price increased by about 400% from Tanzania shillings 5000–25,000/= (per bag?) between 2003 and 2008 (WB 2009). The price increases are significant and provide incentives for producers and other stakeholders to engage in charcoal production and trading. As distance from the production center to the market increases, the prices per bag of charcoal at those production centers decreases (Hofstad and Sankhayan 1999). According to Hofstad and Sankhayan (1999), transport cost is an important component for deciding the price of charcoal in the urban markets. Since charcoal trading is mainly under the informal sector, there is minimal or no control by the government; hence no guidelines exist on wholesale and retail prices. Wholesalers and retailers tend to set prices based on the purchasing prices, transport, and other indirect costs incurred. The charcoal traders interviewed explained that the price of charcoal in the market sometimes depends on the tree species used and where the charcoal was produced, both of which are linked to the quality of charcoal. Currently motorcycles and bicycles are the major modes of transport for charcoal, especially in peri-urban centers thereby making it hard for the government officials to control.

Illegal logging and tree cutting for charcoal production are common in Tanzania, although very few published studies exist which have quantified these activities. However, reports from the Traffic revealed an illegal timber harvesting of about

30–70% in the Miombo woodlands (Milledge and Elibariki 2005; Milledge and Kaale 2005; Milledge et al. 2007). This amount is rather high and if excluded from regional or national analysis may mislead the conclusions and policy recommendations.

4 Conclusions and Future Research

The main objectives of this study were to give an overview of previous studies of charcoal production and consumption in Tanzania and to identify where improved information and research are mostly needed.

Tanzania's high population, urbanization, and economic growth influence strongly the country's consumption of charcoal. Since alternative energy sources are rather expensive, most urban households are expected to continue using charcoal as the main source of energy for cooking. Likewise, lack of affordable alternatives will force rural households to continue using firewood for cooking and heating. Producers will continue to engage in charcoal business as long as it is profitable, irrespective of the rates of deforestation.

This review shows that many interesting and valuable studies have been done about charcoal production and consumption in Tanzania. It is clearly seen how important these two topics are in social, ecological, and economic perspectives. However, many of the reviewed studies lack clearly specified hypotheses and specifications of behavior theories to be used for developing realistic and testable hypotheses. This should be improved in future research, because agent behavior assumptions are important for getting useful results in studies of charcoal consumers and producers. More studies are needed on factors effecting charcoal demand, like changes in prices, income, and policies. Using national household surveys like in D'Agostino et al. (2015) seems promising, but price and quantities of charcoal should be included as variables.

Regarding supply, more information is needed about regeneration (time and volumes) in miombo woodland for various tree species and locations. How various forms of land ownerships influence miombo woodlands management is another area where additional and improved information is needed. We also know very little about the possibilities and preferability in Tanzania of establishing forest plantations for producing charcoal – which species are suitable, which land is available, and what the costs would be.

Both charcoal demand and supply depend on policies – they make the frame within which supply and demand agents operate. Many policies are proposed in the reviewed studies, but little is known about their total and distributional impacts. More knowledge on this matter would be very interesting for policy makers.

Very few studies have been done in Tanzania about GHG impacts of charcoal production and consumption. These impacts might be high and rather decisive for the future land use in Tanzania, and more research on these matters should be implemented. Another interesting research area is the health impacts of using charcoal in urban households.

Charcoal production and consumption in Tanzania are linked to many urgent challenges – like land use, employment, income, global climate change, human health, energy security, and water availability just to mention a few. Many charcoal supply and demand factors interact, and it would be very useful to develop bio-economic models which include at least parts of these interactions and make possible consistent analyses of ex ante defined interesting changes from the present situation. This is a challenging but very interesting research task.

References

Adam J (2009) Improved and more environmentally friendly charcoal production system using a low-cost retort–kiln (Eco-charcoal). Renew Energy 34(8):1923–1925

Al-Mulali U, Sab CNBC (2012) The impact of energy consumption and CO 2 emission on the economic growth and financial development in the Sub Saharan African countries. Energy 39(1):180–186

Amacher GS et al (2009) Deforestation and land use under insecure property rights. Environ Dev Econ 14(3):281–303

Bahamondez C et al (2010) Global forest resources assessment 2010: main report. Food and Agriculture Organization of the United Nations (FAO). http://www.fao.org/forestry/fra/fra2010

Bailis R (2009) Modeling climate change mitigation from alternative methods of charcoal production in Kenya. Biomass Bioenergy 33(11):1491–1502

Bailis R et al (2003) Greenhouse gas implications of household energy technology in Kenya. ACS Publications, Easton

Boberg J (1993) Competition in Tanzanian woodfuel markets. Energy Policy 21(5):474–490

Cerutti PO et al (2015) The socioeconomic and environmental impacts of wood energy value chains in Sub-Saharan Africa: a systematic map protocol. Environ Evid 4(1):12

Chang SJ, Gadow KV (2010) Application of the generalized Faustmann model to uneven-aged forest management. J For Econ 16(4):313–325

Chidumayo EN (1993) Zambian charcoal production. Energy Policy 21(5):586–597

Chidumayo EN, Gumbo DJ (2013) The environmental impacts of charcoal production in tropical ecosystems of the world: a synthesis. Energy Sustain Dev 17(2):86–94

D'Agostino AL et al (2015) Socio-economic determinants of charcoal expenditures in Tanzania: evidence from panel data. Energy Econ 49:472–481

Dasappa S (2011) Potential of biomass energy for electricity generation in sub-Saharan Africa. Energy Sustain Dev 15(3):203–213

Deegen P et al (2011) The Faustmann model as a model for a forestry of prices. Eur J For Res 130(3):353–368

Delahunty-Pike A (2012) Gender equity, charcoal and the value chain, Western Kenya

FAO (2010) Global forest resources assessment 2010. The main report. Food and Agriculture Organization of the United Nations, Rome

FAO (2017) The charcoal transition: greening the charcoal value chain to mitigate climate change and improve local livelihoods, by J. van Dam. Food and Agriculture Organization of the United Nations, Rome

Felix M, Gheewala SH (2011) A review of biomass energy dependency in Tanzania. Energy Procedia 9:338–343

Frost P (1996) The ecology of miombo woodlands. In: Campbell BM (ed) The Miombo in transition: woodlands and welfare in Africa. CIFOR, Bogor, pp 11–57

Hofstad O (1997) Woodland deforestation by charcoal supply to Dar es Salaam. J Environ Econ Manag 33(1):17–32

Hofstad O, Sankhayan PL (1999) Prices of charcoal at various distances from Kampala and Dar es Salaam 1994–1999. South Afr For J 186(1):15–18
Hosier RH (1993) Charcoal production and environmental degradation: environmental history, selective harvesting, and post-harvest management. Energy Policy 21(5):491–509
Hosier RH, Kipondya W (1993) Urban household energy use in Tanzania: prices, substitutes and poverty. Energy Policy 21(5):454–473
Hosier RH et al (1990) Energy planning and wood balances. Nat Res Forum 14:143–154, Wiley Online Library
Hosier RH et al (1993) Future energy development in Tanzania: the energy costs of urbanization. Energy Policy 21(5):524–542
Jones S (2002) A framework for understanding on-farm environmental degradation and constraints to the adoption of soil conservation measures: case studies from Highland Tanzania and Thailand. World Dev 30(9):1607–1620
Kaimowitz D (2003) Not by bread alone... forests and rural livelihoods in sub-Saharan Africa. Forests in poverty reduction strategies: capturing the potential. EFI Proc 47:45–63
Kant S, Alavalapati J (2014) Handbook of forest resource economics. Routledge, London
Kissinger G et al (2012) Drivers of deforestation and forest degradation: a synthesis report for REDD+ policymakers. Lexeme Consulting, Vancouver
Koskela E, Ollikainen M (2001) Forest taxation and rotation age under private amenity valuation: new results. J Environ Econ Manag 42(3):374–384
Luoga E et al (2000a) Economics of charcoal production in miombo woodlands of eastern Tanzania: some hidden costs associated with commercialization of the resources. Ecol Econ 35(2):243–257
Luoga E et al (2000b) Subsistence use of wood products and shifting cultivation within a miombo woodland of eastern Tanzania, with some notes on commercial uses. S Afr J Bot 66(1):72–85
Luoga EJ et al (2002) Harvested and standing wood stocks in protected and communal miombo woodlands of eastern Tanzania. For Ecol Manag 164(1):15–30
Lusambo L (2016) Household energy consumption patterns in Tanzania. J Ecosys Ecograph 5:2
Luvanda A (2016) Impact and assessment of charcoal marketing system through community associations in Kitui County, Kenya. Octa J Environ Res 4(2):86–91
Malimbwi R, Zahabu E (2014) NAFORMA process and biophysical results. Proceedings of the international conference on reducing climate change challenges through forestry and other land use practices. http://hdl.handle.net/123456789/450. Pg 11–29
Malimbwi R et al (2005) Charcoal potential of miombo woodlands at Kitulangalo, Tanzania. J Trop For Sci 17:197–210
Malimbwi R et al (2007) Situation analysis of charcoal sector in Dar es Salaam; charcoal supply and consumption. Department of Forest Mensuration and Management, Faculty of Forestry and Nature Conservation, Sokoine University of Agriculture, Morogoro
Martin M et al (2009) Biofuel development initiatives in Tanzania: development activities, scales of production and conditions for implementation and utilization. J Clean Prod 17:S69–S76
Milledge S, Elibariki R (2005) The status of logging in Southern Tanzania. TRAFFIC East/Southern Africa. Technical report submitted to WWF East Africa Coastal Forests Ecoregion Programme, 69 pp
Milledge S, Kaale B (2005) Bridging the gap: linking timber trade with infrastructural development in Southern Tanzania-Baseline data before completion of the Mkapa Bridge. TRAFFIC East/Southern Africa, Dar es Salaam, Tanzania. ISBN 958402590
Milledge SA et al (2007) Forestry, governance and national development: lessons learned from a logging boom in southern Tanzania, Traffic. Traffic East/Southern Africa, Dar es Salaam
Monela GC et al (1993) Socio-economic aspects of charcoal consumption and environmental consequences along the Dar es Salaam-Morogoro highway, Tanzania. For Ecol Manag 58(3–4):249–258
Monela G et al (2007) Socio-economics of charcoal extraction in Tanzania: a case of eastern part of Tanzania

Mshandete AM, Parawira W (2009) Biogas technology research in selected sub-Saharan African countries-a review. Afr J Biotechnol 8(2):116–125

Msuya N et al (2011) Environmental burden of charcoal production and use in Dar es Salaam, Tanzania. J Environ Prot 2(10):1364

Mwampamba TH (2007) Has the woodfuel crisis returned? Urban charcoal consumption in Tanzania and its implications to present and future forest availability. Energy Policy 35(8):4221–4234

National Bureau of Statistics (2017) Tanzania total population by regions and district, 2016_2017. http://www.nbs.go.tz/nbs/takwimu/census2012/Tanzania_Total_Population_by_District-Regions-2016_2017r.pdf. (https://nbs.go.tz/nbs)

Neufeldt H et al (2015) From transition fuel to viable energy source: improving sustainability in the sub-Saharan charcoal sector. ICRAF Working Paper-World Agroforestry Centre(196)

Norconsult (2002) The true cost of charcoal: a rapid appraisal of the benefits of substituting LPG for charcoal as an urban fuel in Tanzania. http://coastalforests.tfcg.org/pubs/NORConsult_charcoal_vs_LPG.pdf

Nyamoga GZ et al (2016) Potentials for rehabilitating degraded land in Tanzania. In: Lal R, Kraybill D, Hansen DO et al (eds) Climate change and multi-dimensional sustainability in African agriculture: climate change and sustainability in agriculture. Springer, Cham, pp 149–170

Nyamoga GZ, Sjølie HK, Latta G, Ngaga YM, Malimbwi R, Solberg B (2019) Econometric analysis of urban household charcoal consumption in Tanzania: the case of Dodoma, Morogoro and Mtwara

Pennise DM et al (2001) Emissions of greenhouse gases and other airborne pollutants from charcoal making in Kenya and Brazil. J Geophys Res Atmos 106(D20):24143–24155

Rural Energy Agency (2015) Annual Report for the financial year 2014/15. http://www.rea.go.tz/. Visited on 05/08/2018

Sander K et al (2013) Enabling reforms: analyzing the political economy of the charcoal sector in Tanzania. Energy Sustain Dev 17(2):116–126

Sanga GA, Jannuzzi GD (2005) Impacts of efficient stoves and cooking fuel substitution in family expenditures of urban households in Dar es Salaam, Tanzania. Energy Discussion Paper(2.59): 1

Schaafsma M et al (2012) Towards transferable functions for extraction of non-timber forest products: a case study on charcoal production in Tanzania. Ecol Econ 80(Supplement C):48–62

Schlager E, Ostrom E (1992) Property-rights regimes and natural resources: a conceptual analysis. Land Econ 68:249–262

Schure J et al (2013) Formalisation of charcoal value chains and livelihood outcomes in Central- and West Africa. Energy Sustain Dev 17(2):95–105

Sem N (2004) Supply demand chain analysis of charcoal. Firewood in Dar Es Salaam and coast region and differentiation of target groups. Tatedo, Dar es Salaam

Shively G et al (2010) Profits and margins along Uganda's charcoal value chain. Int For Rev 12(3):270–283

Sulaiman C et al (2017) Wood fuel consumption, institutional quality, and forest degradation in Sub-Saharan Africa: evidence from a dynamic panel framework. Ecol Indic 74:414–419

Tomppo E et al (2010) Report to the food and agriculture organization of the United Nations (FAO) in support of sampling study for National Forestry Resources Monitoring and Assessment (NAFORMA) in Tanzania. FAO, Rome

Tomppo E et al (2014) A sampling design for a large area forest inventory: case Tanzania. Can J For Res 44(8):931–948

United Republic of Tanzania (2017a) Energy access situation report, 2016 Tanzania mainland. http://www.nbs.go.tz/nbstz/index.php/english/other-statistics/849-energy-access-situation-report-2016-tanzania-mainland. visited on 01/08/2018

United Republic of Tanzania (2017b) Tanzania's forest reference emission level submission to the United Nations framework convention on climate change-UNFCCC/TAR/2017/TZA. https://www.ncmc.sua.ac.tz/?p=493

Van Beukering P et al (2007) Optimization of the charcoal chain in Tanzania. Institute for Environmental Studies, Amsterdam, pp 2007–2003

World Bank (2009) Environmental crisis or sustainable development opportunity? Transforming the charcoal sector in Tanzania: a policy note. Washington, DC. © World Bank. https://openknowledge.worldbank.org/handle/10986/18603 License: CC BY 3.0 IGO

World Bank (2018) https://data.worldbank.org/country/tanzania?view=chart. Visited on 05/08/2018

Zahabu E (2010) Rogers Malimbwi, Emmanuel Chidumayo, Eliakim Zahabu, Stephano Kingazi, Salome Misana, Emmanuel Luoga and Jean Nduwamungu. The Dry Forests and Woodlands of Africa: 155

Zein-Elabdin EO (1997) Improved stoves in Sub-Saharan Africa: the case of the Sudan. Energy Econ 19(4):465–475

Zulu LC, Richardson RB (2013) Charcoal, livelihoods, and poverty reduction: evidence from Sub-Saharan Africa. Energy Sustain Dev 17(2):127–137

Part III
Climate Risk Management in Tropical Africa

Aboveground Species Diversity and Carbon Stocks in Smallholder Coffee Agroforestry in the Highlands of Uganda

Namaalwa Justine, Balaba Susan Tumwebaze, Kigonya Ritah, and Gorettie Nabanoga

Abstract Types of agroforestry systems and their capacity to sequester carbon vary globally, and the extent of carbon sequestered greatly depends on environmental conditions and system management. This study aimed at investigating the species composition and determining the aboveground carbon stocks of coffee agroforests at low (1240–1320 m a.s.l.) and medium (1321–1504 m a.s.l.) elevations of Manafwa District in Uganda. For each elevation, the agroforest structures were described and the aboveground carbon (AGC) stocks estimated using allometric models for all measured shade and coffee trees. Two coffee varieties were cultivated with SL-14 extending up to 40 years, while LWIL-11, a more recently introduced variety, extended up to 7 years only. Therefore, the estimated AGC stocks were significantly greater for the SL-14 (0.250–2.317 tons ha^{-1}) than LWIL-11 (1.044–2.099 tons ha^{-1}) and were significantly higher at the medium versus the low elevation. The analysis for shade trees indicated no significant differences in the species diversity for the elevation sites, but with significant variations in mean DBH and thus AGC stocks. Farms at low elevation were characterized by smaller (2.037 ± 0.131 tCO2e ha^{-1}) and significantly high (2.037 ± 0.131 tCO2e ha^{-1}) mean AGC stocks per unit area for coffee and shade trees, respectively, as compared to the medium elevation farms. While the variation in the coffee trees within the elevation sites could be attributed to the uneven distribution within the age groups, the AGC stocks in the shade trees were attributed to the generally large sizes of the trees that dominated. Irrespective of the differences in elevation attributes, coffee agroforests can potentially provide carbon sinks and thus contribute to climate change mitigation.

Keywords Carbon sequestration · Coffee agroforest · Coffee-banana · Uganda

N. Justine (✉) · B. S. Tumwebaze · K. Ritah · G. Nabanoga
School of Forestry, Environmental and Geographical Sciences, Makerere University, Kampala, Uganda
e-mail: namaalwa@caes.mak.ac.ug; nabanoga@caes.mak.ac.ug

Y. Bamutaze et al. (eds.), *Agriculture and Ecosystem Resilience in Sub Saharan Africa*, Climate Change Management,
https://doi.org/10.1007/978-3-030-12974-3_18

1 Introduction

Climate change has been on top of the agenda for environment and development. Through efforts to enhance mitigation and adaptation to its negative effects, the role of forests is increasingly being recognized (Locatelli et al. 2010). This has been emphasized in several decisions by the United Nations Framework Convention on Climate Change (UNFCCC). Forests are a large sink of carbon (C), and their role in carbon cycles is well recognized (e.g., Dixon et al. 1994; Lorenz and Lal 2010). In order to enhance climate change mitigation, Uganda has in the past put much emphasis on natural and plantation forests. However, evidence is emerging that agroforestry systems are promising management practices to mitigate greenhouse gas emissions through increased aboveground and soil carbon stocks (Mutuo et al. 2005).

According to Pandey (2002), agroforestry systems are a better climate change mitigation option than oceanic and other terrestrial options. Agroforestry systems (with majorly tree-crop interactions) within the tropical latitudes have carbon storage potential in the multiple plant species and soil (e.g., Dixon 1995; Montagnini and Nair 2004). Dixon (1995) estimated the carbon storage potential at 12–228 Mg C ha^{-1}. In addition, agroforestry systems can have an indirect effect on carbon sequestration in reducing the pressure on natural forests (Soto-pinto et al. 2010), as Dixon (1995) estimated that 1 ha of sustainable agroforestry (with practices that minimize soil and plant disturbance) within the tropical latitudes could potentially offset 5–10 ha of deforestation. Agroforestry systems also have secondary environmental benefits such as helping to attain food security and secure land tenure in developing countries, increasing farm income, restoring and maintaining aboveground and belowground biodiversity, and providing corridors between protected forests. The systems have also been recognized to be of special importance because of their applicability in agricultural lands as well as in reforestation programs (Ruark et al. 2003).

Among the prominent agroforestry systems that have been recognized for sequestering carbon and contributing to climate change mitigation is the coffee agroforestry system. Soto-Pinto et al. (2010) indicated that these systems have emerged as promising land use systems for reducing or offsetting deforestation. As a result, shade-grown coffee systems have continued to be recognized as viable afforestation and reforestation strategies under the Clean Development Mechanism (CDM) of the Kyoto Protocol (Watson et al. 2000). However, to date, there have been few empirical, peer-reviewed investigations of carbon dynamics within coffee-forest landscapes (Schmitt-Harsh et al. 2012), and hence the potential for agroforestry as a strategy for carbon sequestration has not yet been fully recognized (Montagnini and Nair 2004). This is clear as several studies have indicated that more rigorous research results are required for agroforestry systems to be used in global agendas of carbon sequestration (Oelbermann et al. 2004; Ramachandran Nair et al. 2010). Albrecht and Kandji (2003) also emphasized that the significance of agroforestry with regard to carbon sequestration and other CO2 mitigating effects is being widely recognized, but there is paucity of quantitative data on specific systems which may include crops integrated with remnant trees or crops integrated in planted trees.

Coffee agroforests are a common system in Uganda where coffee is the major cash crop for the country. Coffee is traditionally grown under the shade of trees, forming coffee agroforests. Although this system has been reported to have the potential of contributing to climate change mitigation through carbon sequestration (e.g., Mutuo et al. 2005; Ramachandran Nair et al. 2010; Soto-Pinto et al. 2010; Schmitt-Harsh et al. 2012), there are limited studies about the subject in Uganda. Available studies have focused on (1) carbon stocks in agroecosystems (Mukadasi et al. 2007; Gwali et al. 2015; Kyarikunda et al. 2017; Kabiru et al. 2018), (2) carbon payments and their role in tree diversity and carbon stocks (Nakakaawa et al. 2010), and (3) soil organic carbon stocks under coffee agroforestry systems (Tumwebaze and Byakagaba 2016). ICRAF (International Centre for Research in Agroforestry) and its collaborators concluded that the greatest potential for carbon sequestration in the humid tropics is aboveground, not in the soil (Montagnini and Nair 2004). This chapter evaluates the aboveground carbon (AGC) stocks for the coffee agroforestry system in Manafwa District, Uganda. The main objective of the study was to determine the AGC storage levels for coffee agroforests at two different elevations.

The results are expected to inform project proponents interested in promoting shade trees in the region (and thus link coffee farmers to voluntary carbon markets) about the desirable tree species for carbon sequestration. As Luedeling et al. (2011) put it, studies on the C sequestration potential of agroforestry are often conducted with a view to creating opportunities for smallholder farmers to benefit from international carbon payment schemes. The findings may also inform the development and implementation of the National REDD+ strategy in which agroforestry is expected to contribute to the enhancement of forest carbon stocks. Evaluating the carbon storage capacity of coffee agroforestry systems with different shade tree species will further contribute to a better understanding of the role that these ecosystems can play in REDD+ programs.

2 Study Area

Arabica coffee is grown in the highland areas of the country including the slopes of Mount Elgon in the Eastern part of Uganda. The investigation was undertaken in Manafwa District located on the slopes of Mount Elgon. It is located at coordinates $0^{\circ}53$ North and $34^{\circ}20$ East, bordered by Bududa District in the North, Mbale District in the West, Tororo District in the South, and Kenya in the East. The district occupies a total surface area of 602.1 sq. km and is divided into three topographic regions including the lowland with 1300–1500 m a.s.l., the upland with 1500–1600 m a.s.l., and the mountainous landscapes above 1600 m a.s.l. (Mugagga 2017). The district lies within the mixed montane forest zonal belt, below an elevation of 2500 m a.s.l.

(Scott 1994). The district experiences a bimodal rainfall pattern, with the heaviest rains being in the first season of March–June and the second season in September–November. The area receives a mean annual rainfall of 1500 mm, with mean maximum and minimum temperatures of 23° and 15°, respectively (Mugagga et al. 2012). The volcanic soils are fertile and rich in calcium, sodium, and potassium (Mugagga et al. 2010).

The district is dominated by small-scale farmers, and it is among the five major districts growing Arabica coffee on the slopes of Mt. Elgon, the others being Bududa, Bulambuli, Mbale, and Sironko (MWE 2015). Arabica coffee varieties grown in the area include SL-14, SL-28, and Nyasaland (Liebig et al. 2018), as well as LWIL-11 which was reportedly introduced in 2010. The coffee is predominantly cultivated under the coffee-banana system with integration of other crops such as maize, sorghum, potatoes, and beans (Kimaiyo et al. 2017). Planting additional crops and adding shade in the coffee systems are adaptation strategies used by the local farmers to mitigate the effects of climate variabilities such as droughts and pest and disease infestation in the coffee farms, as well as to obtain short-term benefits like food and income (Mugagga 2017).

3 Data

3.1 Data Collection Methods

A total of 16 coffee agroforest sites were investigated and categorized based on the major coffee variety grown and elevation. The structure for each site was described in terms of the variety and age of coffee and the tree sizes and composition of the shade trees. The two coffee varieties, that is, LWIL-11 and SL-14, have been introduced in the area at different times, with LWIL-11 as the most recent variety. Within each variety, coffee plants at different (currently available) ages were considered. For each of the 16 farms, a random point was established, and an area ranging from 1 to 2.5 ha was mapped off using a GPS. In this area (total inventory area of 24.5 ha), all the shade trees (≥10cm) and saplings (3–9.9 cm) were enumerated for diameter at breast height (DBH), total height and crown diameter, and the species identified. Within the inventoried area for each site, 600 m^2 (20*30 m) rectangular plots were randomly established targeting five plots per hectare. In each plot, stump diameter for each coffee plant was measured at 5–10 cm from the ground (depending on the forking point) and the number of forks (for the forked crop) counted. Fork diameter was measured at 10 cm after forking point for utmost three forks per plant. The crown diameter and crop height were measured for each of the coffee stand using a measuring rod calibrated in meters. A total of 15 plots were established for each of the LWIL-II and SL-14 varieties.

3.2 Data Analysis

The analysis was done to identify differences in the plant species composition and thus carbon sequestration levels for coffee agroforests at 1240–1320 m a.s.l. (low elevation) and 1321–1504 m a.s.l. (medium elevation). For each cluster, the agroforest structures (tree sizes and composition) were described, and the Above Ground Biomass (AGB) and carbon stocks were estimated. The differences were then sought between the clusters. The agroforest structure was defined through descriptive statistics of the key parameters including DBH, height, dominant species, and species diversity for the shade trees as well as the variety and age of coffee plants. For all the measured shade trees and coffee trees, allometric models were used to estimate the AGB and carbon stocks. For all trees, saplings, and coffee trees, the estimated biomass (tons/ha) was used to derive the carbon sequestration potential using the IPCC defaults where carbon is estimated to be a constant proportion (0.47) of live and standing biomass (IPCC 2006).

For the coffee plants, within each variety, the stands were clustered in age groups with a 2-year interval, the minimum being 1–3 years and maximum >30 years. But this took consideration of the fact that the highest class for LWIL-11 was 4–6 years, while SL-14 has up to 13–15 years, with only one farm in the category of 40 years. The mean stump diameters for the coffee trees of different varieties but the same age group and/or elevation were compared.

The AGB for coffee plants was derived using an allometric biomass model developed by Tumwebaze and Byakagaba (2016). The model is as specified below:

$$\mathrm{Ln}(\mathrm{AGB}) = -1.553 + 0.86\mathrm{Ln}(\mathrm{SD}) + 0.20\mathrm{Ln}(\mathrm{HT}) + 0.47\mathrm{Ln}(\mathrm{CD}) + 0.21\mathrm{Ln}(\mathrm{BA})$$

where:

AGB = aboveground biomass in kg
SD = stump diameter in cm at 15 cm from the ground
HT = shrub height in m
CD = crown width in m
BA = basal area in m^2

The estimated biomass was corrected using the Baskerville (1972) correction factor (1.082) to remove the biasness of antilog of biomass.

For the shade trees, descriptive statistics of the key parameters including DBH and height were generated. Further, species diversity was assessed and the Shannon-Wiener index of diversity was used, designated as:

$$H' = \sum p_i \log p_i$$

where:

H' = Shannon index of diversity
$\sum$ = summation of values for s species

p_i = the proportion of individuals or the abundance of species I in the sample
$\log_a$ = base of any logarithm

The differences in the mean DBH values for tree species occurring at both low and mean elevation were investigated using a t-test.

In order to estimate AGB and thus carbo stocks, the revised nondestructive allometric equation described by Chave et al. (2014) was used to estimate the aboveground live biomass (AGB, kg) contained within each tree, given as a function of DBH (cm), height (H, m), and wood-specific gravity (ρ, g cm^{-3}):

$$\text{AGB} = 0.0673^{*}\left(D^2 H \rho\right)^{0.976}$$

Wood density values were obtained from the National Biomass Survey (NBS) data base and the study by Buyinza et al. (2014). The estimated AGC values for the coffee and tree components of the agroforests in low and medium elevation were compared using descriptive statistics.

4 Results

The coffee agroforests in Uganda are mainly established through integrating coffee and other crops mainly bananas in a landscape with remnant trees. Low elevation was characterized by a mean value of 1291 ± 9.68 m a.s.l, while medium elevation was characterized by a mean value of 1357 ± 21.12 m a.s.l. In terms of composition, the farms at the low elevation were mainly coffee-banana systems with remnant trees (mainly the large trees with expanse crowns), while at the medium elevation, most of the farms had coffee integrated with remnant trees and less of crops. The estimates of the tree components of the system are presented first separately duly considering the effect of elevation. The estimates are then combined to reflect the biomass and carbon stocks for the low/medium elevation agroforest systems.

4.1 *The AGB and Carbon Stocks for Coffee*

The enumerated coffee plants ranged between 1 and 5 years for LWIL-11 and 1 and 40 years for SL-14. The management of the agroforestry farms, and specifically the coffee plants, was not uniform; and as a result, there were variations in the stump diameters of plants of the same variety and age. Within each variety, the stands were further clustered into age groups of a 2-year interval, and the stump diameters varied between 2.79 ± 0.91 and 2.93 ± 0.86 cm for LWIL-11 and 1.73 ± 0.42 and 6.065 ± 2.08 cm for SL-14. The descriptive statistics for the carbon sequestered by the coffee plants of different varieties and age and grown at different elevations is presented in Table 1.

Table 1 Descriptive statistics for carbon sequestered by coffee plants (tons/ha)

Coffee variety	Elevation	Age range	Mean	Standard error	Minimum	Maximum
LWIL-11	Low	4–6	1.044	0.169	0.711	1.262
	Medium	1–3	1.315	0.251	0.604	2.774
		4–6	2.099	0.104	1.944	2.297
SL-14	Low	1–3	0.250			
		7–9	2.198	0.816	1.296	3.827
		10–12	0.729			
		13–15	0.930	0.001	0.929	0.930
	Medium	4–6	2.301	0.206	1.868	2.821
		7–9	2.317	0.667	1.650	2.984
		>30	1.462			

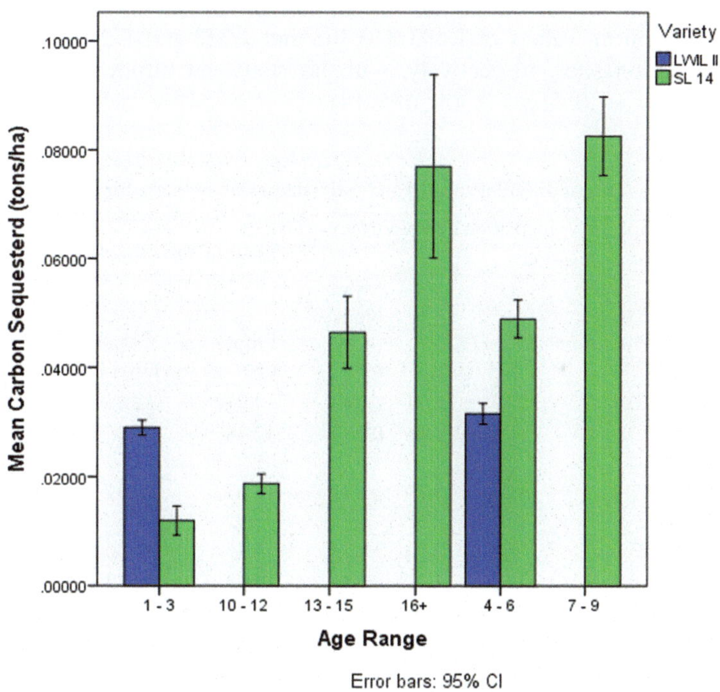

Fig. 1 Mean carbon content of coffee plants across different age classes

Despite the uneven representation of coffee plants in the different age groups for the different varieties, it was observed that SL-14 had generally higher C stocks in the comparable age group of 4–6 years under medium elevation. The analysis of variance for carbon stocks revealed a significant interaction ($p = <0.001$) between age range and variety, implying that the amount of carbon sequestered for each variety of coffee depends on the age of the plant (Fig. 1). There were, however, a nonsignificant ($P = 0.776$) difference among slopes and nonsignificant ($P = 0.112$) interaction between slope and variety.

4.2 The Shade Trees

Twenty-seven shade tree species were identified in the landscape, including *Cordia millenii*, *Ficus natalensis*, *Albizia coriaria*, *Ficus sur*, *Terminalia glaucescens*, *Combretum collinum*, and *Rhus vulgaris*. While there were some similar species in both the low and medium elevation sites, there were characteristic species that were found exclusively in the medium slope sites, such as *Combretum collinum, Acacia* spp., and *Lannea barteri* (Table 2). These were indicative of a wooded savannah vegetation that was selectively cleared to establish coffee. The species diversity index revealed mean values of 1.493 ± 0.368 and 1.722 ± 0.472 for the low and medium elevation sites, respectively, with no statistical difference between the mean values ($t = -1.055$; $p = 0.309$; df = 14).

Table 2 Proportion (%) of different species and estimated mean biomass for the shade trees

	Species composition and biomass estimates					
	Low elevation			Medium elevation		
Species	Proportion (%)	Mean (±SD) DBH (cm)	Mean biomass (tons/tree)	Proportion (%)	Mean (±SD) DBH (cm)	Mean biomass (tons/tree)
Albizia spp.	20.4	29.91 ± 20.71	0.745	4.91	22.65 ± 10.05	0.211
Ficus natalensis	19.4	24.67 ± 11.66	0.259	12.38	20.55 ± 9.09	0.172
Cordia millenii	17.0	33.41 ± 13.39	0.831	16.67	18.39 ± 7.70	0.190
Ficus spp.	13.5	43.77 ± 22.38	1.124	5.43	30.56 ± 29.62	0.736
Grevillea robusta	9.7	26.20 ± 11.90	0.439	4.13	16.82 ± 4.57	0.118
Maesopsis eminii	6.8	23.02 ± 8.02	0.246	1.57	28.65 ± 11.08	0.315
Persea americana	4.7	22.79 ± 6.38	0.181	1.04	12.08 ± 3.23	0.030
Terminalia glaucescens				13	15.37 ± 5.20	0.084
Combretum spp.				10	16.08 ± 4.88	0.087
Acacia spp.				7	24.13 ± 8.71	0.273
Croton megalocarpus				5	11.93 ± 1.97	0.043
Lannea barteri				2	13.95 ± 7.43	0.053

The mean DBH for the trees in the low elevation sites was 32.29 ± 2.07 as compared to 19.22 ± 1.323 for the trees in the medium elevation sites. A pairwise comparison test was conducted, and it revealed a mean difference of 11.838 ± 1.490 ($p = 0.000$). For both sites, the high diameter classes were dominated by *Cordia millenii* and *Ficus* spp. which were also the predominant species in this landscape. The DBH for *Ficus* spp. varied between 13.5 and 99.5 cm, with a mean of 44.6 cm, while for *Cordia millenii* varied between 10 and 58.8 cm, with a mean of 25.9 cm. Species dominance within the landscape and biomass estimates is presented in Table 2.

For the species that occurred predominantly at both elevations, the mean DBH values were generally higher for the trees found at low elevation. A pairwise comparison test revealed statistical differences between the mean values ($t = 2.956$; $p = 0.012$; df = 6). The high occurrence of the large diameter species in the low elevation sites as compared to the medium elevation sites explains significant differences in mean biomass stocks. The AGB for shade trees varied between 0.0034 tons/tree (*Lannea barteri*) and 4.205 tons/tree (*Ficus sur*).

4.3 Aggregated Attributes for the Coffee Agroforestry System at Different Elevations

The attributes of the system components indicated that the biomass for coffee plants was approximately two orders of magnitude smaller than that for shade trees with mean values varying between 1.044 and 2.099 tons ha^{-1} for LWIL-II and 0.250 and 2.317 tons ha^{-1} for SL14 depending on the age class and elevation. The amount of carbon sequestered varied widely across the sites within the low and medium elevation (Table 3).

Farms at low elevation were characterized by smaller AGC stocks per unit area of coffee, but with significantly high AGC stocks per unit area of shade trees. While the variation in the coffee trees within a given subsystem (low/medium elevation) could be attributed to the uneven distribution within the age groups, the AGC stocks in the shade trees were attributed to the characteristics (species and sizes) of the trees that dominated the subsystem. The mean aboveground carbon stored at the time the measurements were taken was 42.01 tons ha^{-1} and 12.48 tons ha^{-1} at the low and medium elevation farms, respectively.

5 Discussion

Coffee agroforests have a great potential to sequester carbon and thus contribute to climate change mitigation. However, the amount of carbon sequestered varies with the site- and system-specific characteristics, mainly elevation and tree characteristics, especially size. The farms at the low elevation were mainly the coffee-banana

Table 3 Summary statistics for coffee agroforestry systems in the low and medium elevation

	Sites	
	Low elevation	Medium elevation
Coffee plants per ha		
LWIL-11	1091	970
SL-14	494	570
Shade trees		
Species count	19	29
Mean tree count/ha	38 ± 23.9	35 ± 13.98
Species diversity	1.493 ± 0.368	1.722 ± 0.472
Mean Dbh (cm)	31.34 ± 8.66	17.76 ± 5.80
Mean height (m)	12.93 ± 1.56	8.35 ± 2.60
Mean CO2e (tons ha^{-1})		
Trees	39.98 ± 13.98	10.03 ± 1.59
Coffee trees		
LWIL-11	1.898 ± 0.117	2.233 ± 0.047
SL-14	2.177 ± 0.144	2.679 ± 0.162
Overall	2.037 ± 0.1305	2.456 ± 0.104

systems with large diameter remnant trees and relatively lower species diversity as compared to the sites at the medium elevation. Farms located at the medium elevation were mainly composed of coffee and trees, with less of banana and other crops integrated.

The coffee species grown in the area were not distinct by elevation but rather by preferences and access to planting material, and thus no defined distribution pattern across the two elevations. The variety LWIL-11 is more recently introduced in the area, dating back to the year 2010. Therefore, the analysis for carbon stocks in coffee plants was more interesting in relation to age groups across a given variety as indicated in Table 2. Given the variations in the management practices, there was no clear pattern on the rate of carbon accumulation in the coffee plants with increasing age.

While in some instances the species for the shade trees were similar for both low and medium elevation sites, the maximum attained diameter and height varied and thus the carbon stocks. While tree species such as *Cordia millenii*, *Ficus* species, and *Albizia* spp. were identified at both elevations, savannah species such as *Combretum*, *Acacia*, *Rhus vulgaris*, and *Lannea barteri* were identified in the medium elevation farms. The test for species diversity confirmed significant differences, which pronounces the inherent differences in carbon accumulation levels over time. While the coffee trees have greatest C stocks at the medium elevation, shade trees in the low elevation have more C stocks as compared to the medium elevation trees. Some previous studies have highlighted the significance of the large diameter trees in agroforestry systems. For example, Ehrenbergerová et al. (2016) reported AGB of shade trees varying between 9.9 kg (*Erythrina edulis*) and 18,400 kg (*Carinian adecandra*) per tree (DBH ≥ 10 cm). Manjunatha et al. (2017) clearly showed the contribution of dominant tree species of coffee-based agroforestry systems with regard to carbon stocks.

Considering the system as a whole rather than the individual tree components, mean AGC stocks of 42.01 tons ha^{-1} and 12.48 tons ha^{-1} for the agroforests in the low and medium elevation, respectively, were estimated. Schroeder (1994) estimated average carbon of 21 and 50 Mg C ha^{-1} for agroforestry systems in subhumid and humid regions. This clearly indicates the variations with respect to elevation, implying that elevation has an effect on the productivity of vegetation and thus biomass stocks. Girardin et al. (2010) stated that AGC in tropical forests typically decreases with elevation, due to temperature and productivity gradients. Therefore, site-specific characteristics are important to consider when estimating the existing or likely contribution of forestry systems to climate change mitigation. Further, while this study focused on AGC stocks only, Tumwebaze and Byakagaba (2016) reported that soil carbon accumulation in coffee stands is far much higher than aboveground carbon. This affirms the role coffee agroforests can play in storing carbon, most especially when all carbon pools are considered.

6 Conclusions

Coffee agroforests have a great potential to sequester carbon and thus contribute to climate change mitigation. However, the amount of carbon sequestered varies with the site- and system-specific characteristics, mainly elevation and tree characteristics. Despite the identified differences in the shade tree characteristics at the low and medium elevation sites, the potential of the agroforest systems at the varying elevations to sequester carbon needs to be recognized. This would be as important as is often reported for either the low and highly stocked tropical forests or the natural forests and woodland forests. While it is evident that different coffee varieties in the Mt. Elgon are either grown as a monocrop or with shade from tree species or bananas, it is important to appreciate the contribution of shade trees in carbon sequestration and thus mitigation of climate change. And thus, as Uganda continues to prepare and implement its National REDD+ strategy, agroforestry systems need to be considered as important carbon sinks, especially in cases where remnant trees exist or/and integration of trees on farmlands has been/can be implemented. However, the capacity of different tree species to sequester carbon needs to be considered against the benefits the different tree species yield to the system, such as provision of shade and litter.

References

Albrecht A, Kandji ST (2003) Carbon sequestration in tropical agroforestry systems. Agric Ecosyst Environ 99(1–3):15–27. https://doi.org/10.1016/S0167-8809(03)00138-5

Baskerville GL (1972) Use of logarithmic regression in the estimation of plant biomass. Can J For Res 2:49–53

Buyinza J, Tumwebaze, SB, Namaalwa J, Byakagaba P (2014) Above-ground biomass and carbon stocks of different land cover types in Mt. Elgon, Eastern Uganda. International Journal Of Research On Land-Use Sustainability, 1(2):51–61

Chave J, Réjou-Méchain M, Búrquez A, Chidumayo E, Colgan MS, Delitti WBC, Duque A, et al. (2014). Improved allometric models to estimate the aboveground biomass of tropical trees. Glob Chang Biol 20(10):3177–3190

Dixon R (1995) Agroforestry systems: sources or sinks of greenhouse gases? Agrofor Syst 31:99–116. https://doi.org/10.1007/BF00711719

Dixon RK, Solomon AM, Brown S, Houghton RA, Trexier MC, Wisniewski J (1994) Carbon pools and flux of global forest ecosystems. Science 263(5144):185–190. https://doi.org/10.1126/science.263.5144.185

Ehrenbergerová L, Cienciala E, Kučera A, Guy L, Habrová H (2016) Carbon stock in agroforestry coffee plantations with different shade trees in Villa Rica, Peru. Agrofor Syst 90(3):433–445. https://doi.org/10.1007/s10457-015-9865-z

Girardin CAJ, Malhi Y, AragãO LEOC, Mamani M, Huaraca Huasco W, Durand L, Feeley KJ, Rapp J, Silva-Espejo JE, Silman M, Salinas N, Whittaker RJ (2010) Net primary productivity allocation and cycling of carbon along a tropical forest elevational transect in the Peruvian Andes: net primary productivity from Andes to Amazon. Glob Change Biol 16(12):3176–3192. https://doi.org/10.1111/j.1365-2486.2010.02235.x

Gwali S, Agaba H, Balitta P, Hafashimana D, Nkandu J, Kuria A, Pinard F, Sinclair F (2015) Tree species diversity and abundance in coffee farms adjacent to areas of different disturbance histories in Mabira forest system, central Uganda. Int J Biodivers Sci Ecosyst Serv Manag 11(4):309–317. https://doi.org/10.1080/21513732.2015.1050607

IPCC (2006) Guidelines for national greenhouse gas inventories: generic methodologies applicable to multiple land-use categories

Kabiru S, Hassan S, Umar U, Musab I (2018) An inventory of agroforestry practices in Butta Sub-County, Manafwa District, Uganda. Asian J Environ Ecol 5(4):1–7. https://doi.org/10.9734/AJEE/2017/39518

Kimaiyo J, Kiptot E, Tanui J, Oduol J, Kegode H, Isubukalu P, Buyinza J, Chemangei A, Nyangas S, Okia C (2017) Livelihood analysis of households in Manafwa and Kapchorwa. World Agroforestry Centre, Nairobi

Kyarikunda M, Nyamukuru A, Mulindwa D, Tabuti JRS (2017) Agroforestry and management of trees in Bunya County, Mayuge District, Uganda. Int J For Res 2017:1–9. https://doi.org/10.1155/2017/3046924

Liebig T, Babin R, Ribeyre F, Läderach P, van Asten P, Poehling H-M, Jassogne L, Cilas C, Avelino J (2018) Local and regional drivers of the African coffee white stem borer (*Monochamus leuconotus*) in Uganda: Local and regional drivers of *M. leuconotus*. Agric For Entomol 20:514. https://doi.org/10.1111/afe.12284

Locatelli B, Brockhaus M, Buck A, Thompson I (2010) Forests and adaptation to climate change: challenges and opportunities. IUFRO

Lorenz K, Lal R (2010) The importance of carbon sequestration in forest ecosystems. In: Lorenz K, Lal R (eds) Carbon sequestration in forest ecosystems. Springer Netherlands, Dordrecht, pp 241–270

Luedeling E, Sileshi G, Beedy T, Dietz J (2011) Carbon sequestration potential of agroforestry systems in Africa. In: Kumar BM, Nair PKR (eds) Carbon sequestration potential of agroforestry systems. Springer Netherlands, Dordrecht, pp 61–83

Manjunatha M, Devakumar A, Nivedita M, Kushalappa C (2017) Carbon stock present in dominant tree species of coffee based agroforestry system. Int J Agrofor Silvic 5(5):311–314

Montagnini F, Nair PKR (2004) Carbon sequestration: An underexploited environmental benefit of agroforestry systems. Agrofor Syst 61:281–295

Mugagga F (2017) Perceptions and response actions of smallholder coffee farmers to climate variability in montane ecosystems. Environ Ecol Res 5(5):357–366. https://doi.org/10.13189/eer.2017.050505

Mugagga F, Buyinza M, Kakembo V (2010) Livelihood diversification strategies and soil erosion on Mount Elgon, Eastern Uganda. A socio-economic perspective. Environ Res J 4(4):272–280
Mugagga F, Kakembo V, Buyinza M (2012) Land use changes on the slopes of Mount Elgon and the implications for the occurrence of landslides. Catena 90:39–46. https://doi.org/10.1016/j.catena.2011.11.004
Mukadasi B, Kaboggoza JR, Nabalegwa M (2007) Agroforestry practices in the buffer zone area of Mt Elgon National Park, eastern Uganda. Afr J Ecol 45(s3):48–53. https://doi.org/10.1111/j.1365-2028.2007.00857.x
Mutuo PK, Cadisch G, Albrecht A, Palm CA, Verchot L (2005) Potential of agroforestry for carbon sequestration and mitigation of greenhouse gas emissions from soils in the tropics. Nutr Cycl Agroecosystems 71(1):43–54. https://doi.org/10.1007/s10705-004-5285-6
MWE (2015) Economic assessment of the impacts of climate change in Uganda; Arabica coffee production in the Mount Elgon region (Bududa District). Ministry of Water and Environment, Kampala
Nakakaawa C, Aune J, Vedeld P (2010) Changes in carbon stocks and tree diversity in agro-ecosystems in south western Uganda: what role for carbon sequestration payments? New For 40(1):19–44. https://doi.org/10.1007/s11056-009-9180-5
Oelbermann M, Paul Voroney R, Gordon AM (2004) Carbon sequestration in tropical and temperate agroforestry systems: a review with examples from Costa Rica and southern Canada. Agric Ecosyst Environ 104(3):359–377. https://doi.org/10.1016/j.agee.2004.04.001
Pandey DN (2002) Carbon sequestration in agroforestry systems. Clim Pol 2(4):367–377
Ramachandran Nair PK, Nair VD, Mohan Kumar B, Showalter JM (2010) Carbon sequestration in agroforestry systems. Adv Agron (108):237–307
Ruark GA, Schoeneberger MM, Nair PKR (2003) Agroforestry–helping to achieve sustainable forest management. UNFF (United Nations Forum for Forests) intersessional experts meeting on the role of planted forests in sustainable forest management, 24–30 March 2003, New Zealand
Schmitt-Harsh M, Evans TP, Castellanos E, Randolph JC (2012) Carbon stocks in coffee agroforests and mixed dry tropical forests in the western highlands of Guatemala. Agrofor Syst 86(2):141–157. https://doi.org/10.1007/s10457-012-9549-x
Schroeder P (1994) Carbon storage benefits of agroforestry systems. Agrofor Syst 27(1):89–97. https://doi.org/10.1007/BF00704837
Scott P (1994) An assessment of natural resource use by communities from Mt. Elgon National Park. UNDP
Soto-Pinto L, Anzueto M, Mendoza J, Ferrer GJ, de Jong B (2010) Carbon sequestration through agroforestry in indigenous communities of Chiapas, Mexico. Agrofor Syst 78(1):39–51. https://doi.org/10.1007/s10457-009-9247-5
Tumwebaze SB, Byakagaba P (2016) Soil organic carbon stocks under coffee agroforestry systems and coffee monoculture in Uganda. Agric Ecosyst Environ 216:188–193. https://doi.org/10.1016/j.agee.2015.09.037
Watson RT, Noble IR, Bolin B, Ravindranath N, Verardo DJ, Dokken DJ (2000) Land use land-use change, and forestry. International Panel on Climate Change

Contribution of Smallholder Tree Growers to Increasing Tree Cover in Kaliro District

Derick Kisegu, David Kagaha, Cory Whitney, and John R. S. Tabuti

Abstract Many stakeholders including the Ugandan government are growing trees and contributing to increasing tree cover in Uganda. Whereas the contribution of large-scale tree growers to increasing tree cover has been documented, that of smallholder tree growers (STGs) is not known. But because STGs make up the majority of tree growers, it is possible that their contribution to tree cover in the country is significant. In this study we have addressed this gap for Kaliro District. We have also explored the factors that influence tree planting by STGs. Data was collected in the months of January–June 2017 using interviews with 206 tree growers. Included in the survey were all smallholder farms with 20 or more trees. The tree growers planted 39 species, in the period 1997–2016, the most popular of which were *Pinus* spp., *Eucalyptus* spp., *Grevillea robusta* and *Maesopsis eminii.* Few STGs planted trees in that time (n = 206) and those who did tended to plant few trees (median 175 trees). Despite the average low contribution by STGs, tree cover is increasing due to a few STGs who planted very many trees. Growers who planted the most trees were male or engaged in small-scale business, possessed a university degree and received support from the government. New approaches may be necessary that provide better access to a greater diversity of smallholder farmers such as women and those without low income and education access. STGs affiliated to tree-growing associations appear to have planted fewer trees than those outside tree-growing associations. Tree growers associations should be redesigned to provide maximum benefit to STGs and local ecology. Income generation was cited as the STGs planters' main motivation for planting trees. However, the key immediate benefit was firewood. These may be important points to consider when developing interventions that target tree planting in the region and throughout Uganda. We conclude that the contribution of STGs to tree growing in Kaliro District is low and that

D. Kisegu · D. Kagaha · J. R. S. Tabuti (✉)
Makerere University, Kampala, Uganda
e-mail: jtabuti@caes.mak.ac.ug

C. Whitney
University of Bonn, Department of Horticultural Sciences (INRES), Bonn, Germany

Y. Bamutaze et al. (eds.), *Agriculture and Ecosystem Resilience in Sub Saharan Africa*, Climate Change Management,
https://doi.org/10.1007/978-3-030-12974-3_19

tree planting campaigns should be inclusive and focus on STGs perceived benefits in order to achieve successful reforestation.

Keywords Reforestation · Smallholder tree growers · Afforestation · Tree planting · Incentives

1 Introduction

Native forests and other forested areas are declining rapidly in Uganda. According to the National Forestry Authority (2009), Uganda's annual forest loss over a 15-year period 1990–2005 was nearly 90,000 ha/year resulting in a reduction of forest cover from an estimated 4.9 in 1990 to 3.6 million ha in 2005 in the country. This loss was fastest in land outside protected areas where forest area declined from 3.46 million ha in 1990 to 2.3 million ha in 2005. The underlying causes of forest area loss in Uganda is the rapid population growth (3.2% per annum) and low levels of economic development and high levels of poverty. These factors lead to clearing of forests for agricultural land and/or degradation of forests through fuelwood production and timber harvesting. Fuelwood is used by over 90% of the Ugandan population. The demand for timber is high, and it is estimated that Uganda will require 150,000 ha of well-managed timber plantations by 2020 (Jacovelli 2009).

To counter deforestation and bridge the demand gap for timber and other resources, the Uganda Government and bilateral partners have supported tree growing by private tree planters, government organisations and civil organisations. National institutions actively involved in tree planting are the Uganda Wildlife Authority (UWA), the Forest Sector Support Department (FSSD) and the National Forestry Authority (NFA). Two important government interventions for tree planting are the Sawlog Production Grant Scheme (SPGS) and the Farm Income Enhancement and Forest Conservation (FIEFOC). For example, the SPGS founded in 2003 provided grants to people establishing tree plantations primarily for timber production, and by 2009, over 10,000 hectares had been established through its support (Jacovelli 2009).

Because there are very many smallholder tree growers (STGs) in Uganda, it is possible that their contribution to increasing tree cover through tree growing is significant in the country. However, this contribution has not been documented and is missing from the national records. In order to understand their contribution, it is important to determine how many of these tree growers are in operation and determine which species they are interested in, how many trees they have planted and what motivates them to plant trees. We undertook this study with the main objective of documenting the contribution of STGs to tree planting in Kaliro District; the secondary objective was to characterize the tree growers. We selected Kaliro District because it is one of the districts with the fastest loss of tree cover in Uganda and also because it is a rural area and agriculture is culturally important.

2 Study Area and Methods

Kaliro District is located in the Eastern Region of Uganda occupying an area of about 1000 km^2 (N 00° 54.694′ and E 033° 28.043′). The region is generally flat and at an altitude of 1045–1075 m a.s.l., with scattered rocky outcrops that are not suitable for crop agriculture but serve as important habitats for plants. The vegetation is classified as moist *Combretum* wooded grassland and dry *Combretum* wooded grassland (van Breugel et al. 2015). The district has two Central Forest Reserves (CFRs), the Kaliro CFR and Namalemba CFR, and one Local Forest Reserve (LFR) Namukooge. The agroecological zone of Kaliro is classified as the Banana/Millet/Cotton system. The climate is hot and dry, and rainfall averages 1430 mm per annum (International Food Policy Research Institute (IFPRI) and Datawheel 2017). There are two wet seasons, March to June and August to October. The region's soils are considered to be of low productivity are dominated by the Mazimasa complex of catenas derived from ancient lake deposits. This soil type is a shallow grey or brown sandy loam on laterite base rock (Ollier and Harrop 1959). The other types are mineral hydromorphic soils influenced by permanent or seasonal waterlogging and organic hydromorphic soils (Department of Lands and Survey 1962). Kaliro District has four major land use/land cover categories: (i) non-uniform small-scale farmland (67%), (ii) wetlands (16%) dominated by *Cyperus papyrus*, (iii) woodlands (4%) and (iv) bushlands (1%). Other land uses include settlements.

Kaliro is an agricultural community with 76% of the people practicing subsistence agriculture as their main source of livelihood (Kaliro District Local Government 2012). Agricultural practices consist of fallow cultivation and permanent cultivation farming systems. Average landholdings in the region are approximately 2 ha per homestead.

The district has about 42,000 homesteads and an estimated 236,199 people. The population density is high 303 people per km^2. The population growth rate is 3.5% per annum. The majority of the people (64%) are below 20 years of age (UBOS 2016); many (40%) are illiterate (UBOS 2016). People of eastern Uganda, including Kaliro District, are among the poorest in Uganda (UNDP 2007), and according to the Kaliro District statistical abstract of 2011/2012, 42% of the population is below the poverty line. The Uganda National Household Survey from the Uganda Bureau of Statistic (UBOS 2016) indicates that 27% of Uganda's population (ten million people) live in poverty. This is severe in eastern Uganda where poverty has increased by 27% since 2013.[1]

The residents and institutions of Kaliro depend heavily on wood products, e.g. for fuelwood, construction poles, etc. (Tabuti 2007). Firewood is the principal fuel used for cooking in 97% of the households and organisations such as prisons, police stations and schools (Tabuti 2007). Kaliro District is urbanizing, and this is creating

[1] Oketch, M.L. 2017. 3.4 million more Ugandans slip into poverty. http://www.monitor.co.ug/News/National/34-million-Ugandans-poverty-income-prices/688334-4115106-mulfd7z/index.html

new and acute pressures for production of tree products. The rural community of Kaliro District is experiencing rapid woody vegetation loss arising mainly out of land use conversion and over-exploitation. The underlying factors for tree loss are the rapidly growing population and the accompanying high levels of poverty. Trees have been rampantly destroyed mostly because of increasing crop agriculture and charcoal production. In the period between 1990 and 2005, Kaliro District lost more than 86% of its tree stock (from 29 tons per ha to just above 4 tons per hectare) (National Forestry Authority 2009). Creation of space for commercial sugarcane production is also a new risk to trees in Kaliro. A new sugar factory was recently established in the district, and sugar cane is now intensively grown and is expected to affect tree cover negatively in Kaliro District.

As is the case throughout Uganda, Kaliro does not have sufficient extension staff to oversee tree growing. Kaliro has an estimated 20 agricultural extension workers (AfranaaKwapong and Nkonya 2015), which leaves an urgent gap for active interventions to promote tree growing.

2.1 Data Collection

According to Mercer (2004), there are many factors that influence attitudes concerning tree growing, which include household attributes, resource endowments and institutional frameworks. Household attributes that influence tree planting include age, education, gender, off-farm or paid labour and anticipated benefits, while resource endowments or the assets available to farmers for investing in tree growing include land, labour available for tree growing and savings. Institutional frameworks on their part include arrangements to access seedlings, technical support and information.

We developed hypotheses following a human ecology theoretical approach (cf. Whitney et al. 2018a, b) and based on the above literature and our experience in the field. Our hypotheses were that household attributes (age, gender, education, source of livelihood/labour) and institutional frameworks (access to seedlings, technical support and information) influence the number of trees planted by a tree planter.

Selection of respondents was done purposively. With the help of the local chairperson and regional officers, we identified and interviewed all smallholder tree growers who had planted 20 or more trees and had managed them for at least 1 year. Interviews took place throughout Kaliro District (Fig. 1) between the months of January–June 2017, with a total of 206 tree growers including 16 institutions, 25 women and 164 men. We restricted our survey to a 20-year time period, i.e. from 1997 to 2016.

During the interviews, we documented socio-economic information including gender, age, employment, level of education, household size, land tenure and membership in any tree growers associations. We also documented silvicultural practices, including the number and species that were planted, the year of planting, the number that survived and sources of planting materials. We also sought to learn the

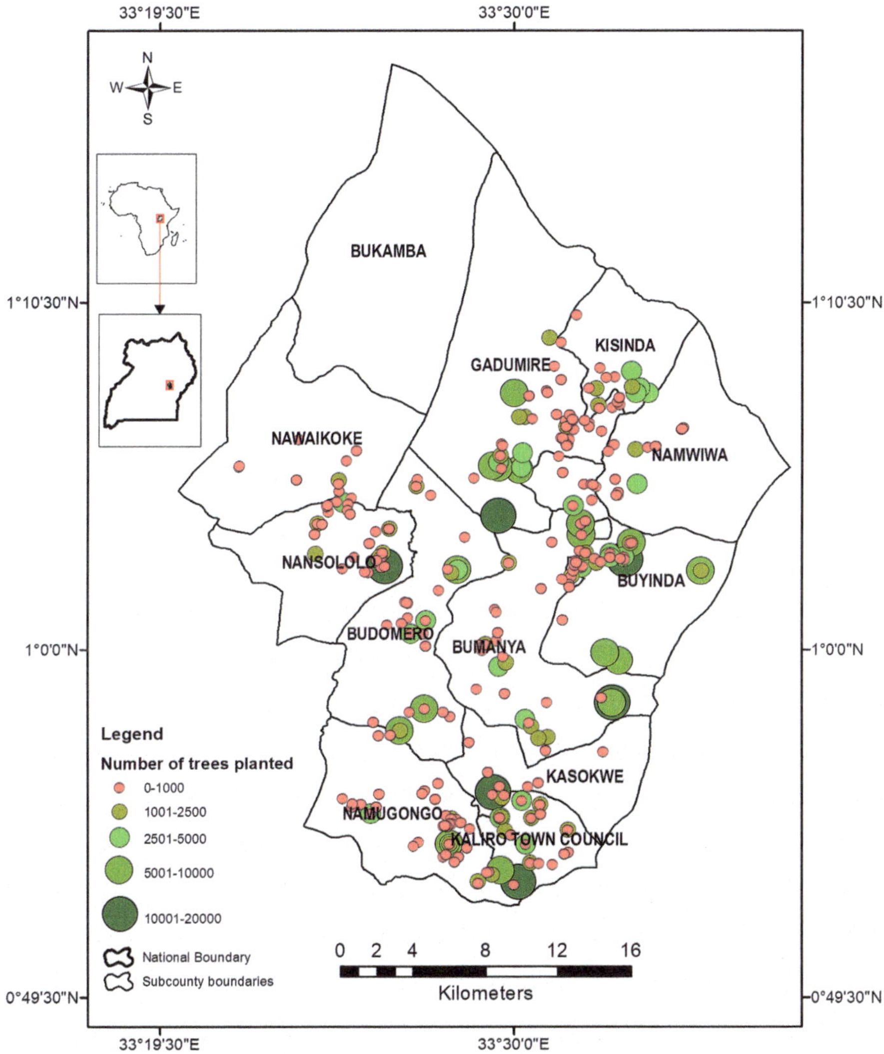

Fig. 1 Map of the Kaliro District generated using data from this survey. The map shows the locations of the tree gardens/plantations and the number of trees planted for the smallholder tree growers. Insets of Africa and Uganda are shown

tree growers' motivations for planting trees, as well as perceived benefits from trees, sources and forms of technical support concerning tree planting, management plans and the risks and challenges for tree planting. The survey included observations on the general condition of the trees and the tree plantation including the quality of the maintenance.

The average number of people in the homestead was eight (ranging from one to 29 people). Almost all tree growers, whether individuals or institutions, owned the

Table 1 Social economic attributes of the tree growers

Attribute	Number
Gender	
Female	25
Male	164
Education	
None	4
Primary level	33
Lower secondary level	81
Upper secondary level	6
Tertiary/vocational certificate	22
University degree	40
Occupation	
Farmer	107
Civil servant	68
Business man	34
Artisan	13
Politician	9
Other (priest 2, crime preventer 1, student 1)	4
Affiliated to tree-growing association	
No	176
Yes	25
Affiliated to institution	
Individual	189
Institution	16

land on which they grew trees, and only two rented the land (NFA reserves) they used for planting trees. Only 25 tree planters were affiliated to a tree-growing association (Table 1). The respondents had an average age of 49 years (range 21–85) and were employed as farmers (107), civil servants (68), business men (34) and artisans (13). One respondent was still a student. The majority had attained up to lower secondary level education (i.e. 11 years of formal education; 40 of the tree planters had a university degree).

2.2 *Data Analysis*

We used the Wilcoxon rank sum test with continuity correction and Kruskal-Wallis sum test to test for differences between the independent variables in relation to the number of trees grown by individual tree growers. Specifically, the Wilcoxon test was used to test for differences in tree growing observed between gender and affiliation to a tree growers association. We also used the same to test whether being an individual or an institution influenced the number of trees grown.

The Kruskal-Wallis sum test was used to test for differences between livelihood activity, education level, source of seedlings and institutional support. We also tested for associations using the generalized linear model (GLM) and chi-squared contingency table tests. All analyses were performed in the R programming language (R Core Team 2017). Statistical level of significance was set at 95%.

3 Results

Altogether 206 tree growers took part in the study, including 189 individuals and 16 institutions across Kaliro District (Fig. 1). Trees were planted in configurations of small woodlots and sometimes in mixtures with crops. The interviewed tree growers reported that they had planted a total of 457,763 trees since 1997, and that of these 314,388 were still surviving on their farms (a survival rate of 61%). The average (median) number of trees planted per tree grower was 175 (range 1–15,500), and the median number of trees surviving was 74.5 per grower. There were wide variations in the reported numbers of trees planted by tree growers. A few growers planted very many trees, up to 15,500 in some instances (Fig. 2).

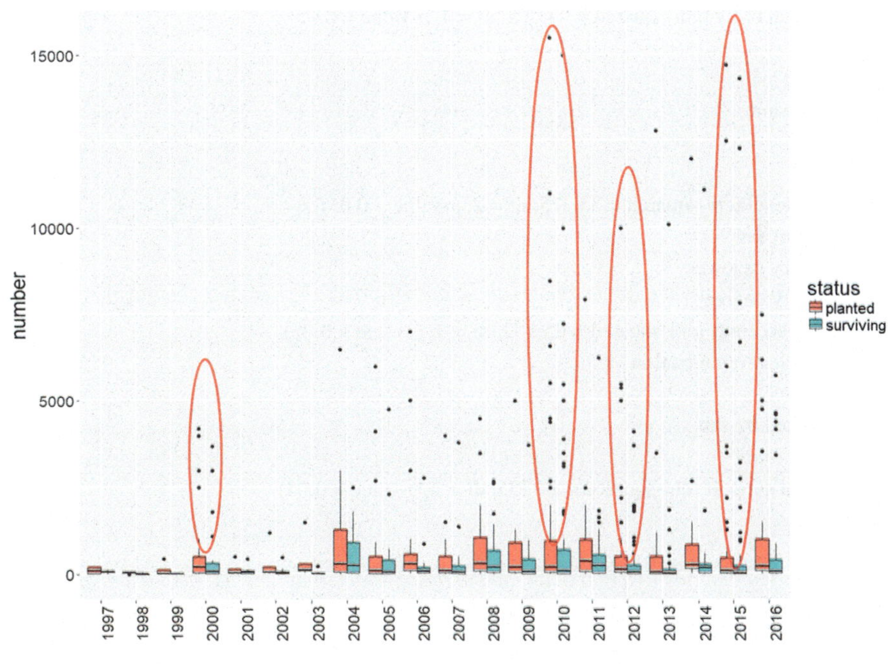

Fig. 2 Tree planting by smallholder tree growers in Kaliro District between 1997 and 2016. Shown are number of trees planted (red boxplots) and those surviving (blue boxplots) by year. Peaks (outliers) of intensive tree-planting farmers in tree growing in the years 2000, 2010, 2012 and 2015 highlighted by the red ellipses corresponding to the time when Farm Income Enhancement and Forest Conservation and a local NGO SUPD were active

The absolute number of trees planted by tree growers in Kaliro District was found to have increased over the selected time period, from a low of 427 trees planted in 1997 to 44,142 in 2016. It peaked in 2010 (92,656 trees), then in 2012 (53,094 trees) and 2015 (62,958 trees). (Fig. 2). In the years 2010 and 2012, the government provided free seedlings for tree planting under the Farm Income Enhancement and Forest Conservation (FIEFOC); again between 2014 and 2017, a local NGO Sustainable Use of Plant Diversity (SUPD) facilitated tree planting by providing free seedlings, tree propagation materials and training to tree growers to propagate and grow trees. These interventions resulted in a weak but significant relationship between increased availability of seedlings to tree growers and number of trees planted ($R^2 = 13.7$, $p < 0.05$). (Fig. 2).

The type of livelihood/activity, education level, source of seedlings, institutional support, gender and affiliation to a tree-growing association was found to influence the number of seedlings planted by tree growers (Table 2). Those planters whose

Table 2 Influence of intrinsic and extrinsic household factors on the success of tree growing. Influence of livelihood/activity, education level, source of seedlings and institutional support analysed using Kruskal-Wallis, gender and affiliation to a tree growers association analysed using Wilcoxon rank sum test with continuity correction

Variable	Median
Livelihood/activity (chi-squared = 74.708, df = 4, *p*-value = 0.05)	
Artisan	300
Business man	1000
Civil servant	350
Farmer	100
Student	183
Education level (chi-squared = 35.725, df = 2, *p*-value = 0.05)	
Skilled artisan	100
Secondary or lower	123
University degree	500
Source of seedlings (chi-squared = 25.882, df = 3, *p*-value = 0.05)	
Civil society organization	60
District	200
Propagated by self	245
Other	100
Institutional support (chi-squared = 14.355, df = 2, *p*-value = 0.05)	
District	500
Other	183
Self	100
Gender (W = 8552.5, p-value = 0.05)	
Female	100
Male	200
Affiliation to a tree growers association (W = 22,367, *p*-value = 0.05)	
No	220
Yes	60

Table 3 Numbers of different tree species planted by smallholder tree growers in Kaliro District, Uganda. *Tectona grandis* is not listed here since it was planted by few growers

Species	Total number of trees	Median number by tree grower	No. (%) of tree growers
Pinus caribaea Morelet	208,374	755	113 (53%)
Eucalyptus grandis W. Hill	90,968	500	62 (29%)
Eucalyptus unnamed hybrid	61,604	3626	14 (7%)
Eucalyptus camaldulensis Dehnh.	26,287	200	35 (17%)
Grevillea robusta A. Cunn. ex R. Br.	26,101	100	85 (40%)
Maesopsis eminii Engl.	15,349	100	70 (33%)

main occupation was small-scale business owner, or who possessed university education, or received support from the district (government) planted a significantly larger number of seedlings than for other levels of these factors (Table 2). Gender was found to be an important factor, as men tended to plant more trees than women. People affiliated to tree-growing associations tended to plant fewer seedlings than those outside tree-growing associations; however they planted more species (median = 5) than non-affiliates (median = 3); $W = 12{,}004$, p-value = 0.0003.

Tree growers in the survey planted 39 different tree species (Appendix 1). The most popular of these, by proportion of tree growers planting the species and by number of trees planted, are *Pinus caribaea* Morelet (53%), three *Eucalyptus* spp. (53%), *Grevillea robusta* A. Cunn. ex R. Br. (40%) and *Maesopsis eminii* Engl. (33%) (Table 3). The most popular species were fast-growing timber trees (Fig. 3).

Tree growers reported that tree seedlings were acquired through purchases using personal funds (40%) or were provided free of charge by the district (34%) or by NGOs (8%). Most plantations were found to be poorly managed but still had trees with good boles. Most tree growers (89%) did not have management plans for their plantations, and management activities were limited to thinning, slashing/weeding and pruning. Tree growers stated that they acquired awareness to promote tree planting and technical information about tree growing and management from the District Forestry Office (DFO), local non-governmental organizations and civil society organizations.

Tree growers cited several sources of motivation that inspired them to plant and manage their trees. The most important motivations included future and current sources of income (42% and 25%, respectively), land tenure (i.e. secure land rights, 21%), environmental protection and restoration (20%), firewood (11%), timber (10%), windbreak (9%) and shade (7%). There were also additional incentives including firewood and timber (Fig. 4).

There was a slight difference between the most important incentive and most important benefits. Unlike the incentives that motivate tree growers, where income was the most important value, the most frequently cited actual benefits enjoyed were firewood mentioned by 61% of the tree growers, shade for people and animals

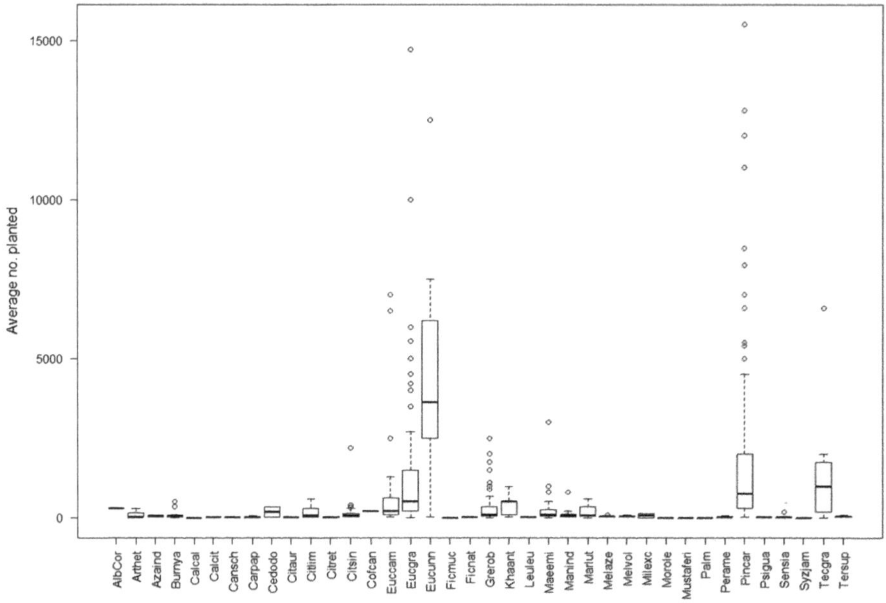

Fig. 3 Box plot of average numbers of individual trees of tree species grown by tree growers in Kaliro District, Uganda. *Eucalyptus* spp. (Euccam, Eucunn, Eucgra), *Pinus caribaea* (Pincar) and *Tectona grandis* (Tecgra). Albcor (*Albizia coriaria* Welw. ex Oliv.), Arthet (*Artocarpus heterophyllus* Lam.), Azaind (*Azadirachta indica* A. Juss.), Burnya (*Burtdavya nyassica* Hoyle), Calcal (*Calliandra calothyrsus* Meisn.), Calcit (*Callistemon citrinus* [Curtis] Skeels), Cansch (*Canarium schweinfurthii* Engl.), Carpap (*Carica papaya* L.), Cedodo (*Cedrela odorata* L.), Citlim (*Citrus limon* [L.] Osbeck), Citaur (*Citrus aurantiifolia* [Christm.] Swingle), Citret (*Citrus reticulata* Blanco), Citsin (*Citrus sinensis* [L.] Osbeck), Cofcan (*Coffea canephora* Pierre ex A. Froehner), Euccam (*Eucalyptus camaldulensis* Dehnh.), Eucgra (*Eucalyptus grandis* W. Hill), Eucunn (*Eucalyptus* unnamed hybrid), Ficmuc (*Ficus mucuso* Welw. ex Ficalho), Ficnat (*Ficus natalensis* Hochst.), Grerob (*Grevillea robusta* A. Cunn. ex R. Br.), Khaant (*Khaya anthotheca* (Welw.) C. DC.), Leuleu (*Leucaena leucocephala* (Lam.) de Wit), Maeemi (*Maesopsis eminii* Engl.), Manind (*Mangifera indica* L.), Marlut (*Markhamia lutea* (Benth.) K. Schum.), Melaze (*Melia azedarach* L.), Melvol (*Melia volkensii* Gürke), Milexc (*Milicia excelsa* [Welw.] C.C. Berg), Morole (*Moringa oleifera* Lam.), Mustaferi, Palm (Palm not identified), Perame (*Persea americana* Mill.), Pincar (*Pinus caribaea* Morelet), Psigua (*Psidium guajava* L.), Sensia (*Senna siamea* [Lam.] H.S. Irwin & Barneby), Syzjam (*Syzygium jambos* [L.] Alston), Tecgra (*Tectona grandis* L. f.), Tersup (*Terminalia superba* Engl. & Diels)

(30%), followed by income generation and windbreak (27% each) (Fig. 5). Other key benefits were windbreak, mentioned by 27% of the respondents, environmental restoration and secure land tenure (18% each). Taking the incentives and realised benefits together, the most important tree values in Kaliro appear to be income, firewood, security of land tenure, windbreaks, shade and to support a good environment.

Tree growers reported several important challenges that impact tree growing. The most important of which was the loss of seedlings and trees to pests (75%) and

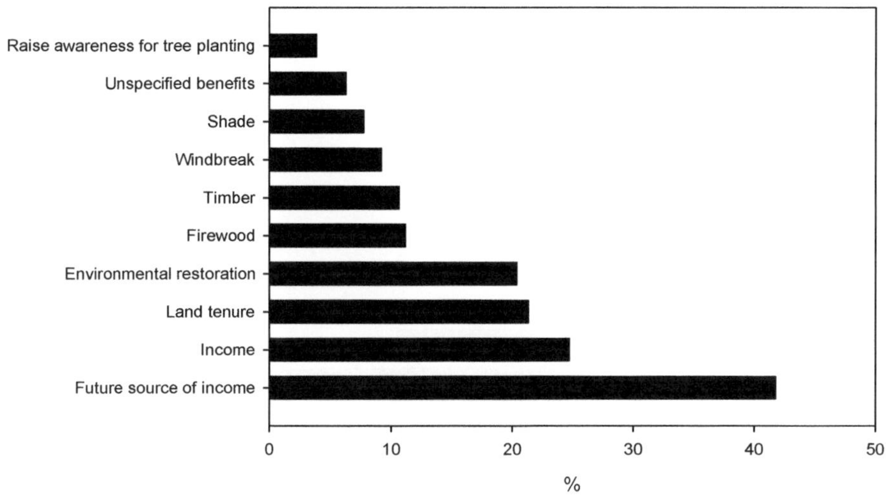

Fig. 4 Incentives for tree planting cited by tree growers in Kaliro District, Uganda

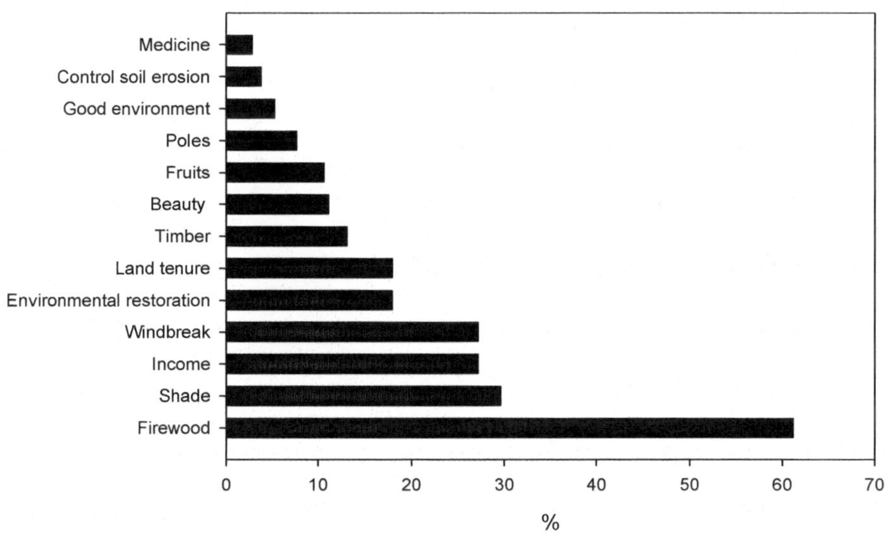

Fig. 5 Benefits from trees cited by tree growers in Kaliro District, Uganda

livestock damage (51%). The most commonly known pests are termites mentioned by 90% of the respondents that mentioned pests as a problem. Drought was also a major issue, mentioned by 49% of the planters. Other problems reported were vandalism from other community members (33%) and theft of seedlings and tree products (namely, firewood, edible fruits, bark for repelling mosquitoes, etc.) (Fig. 6).

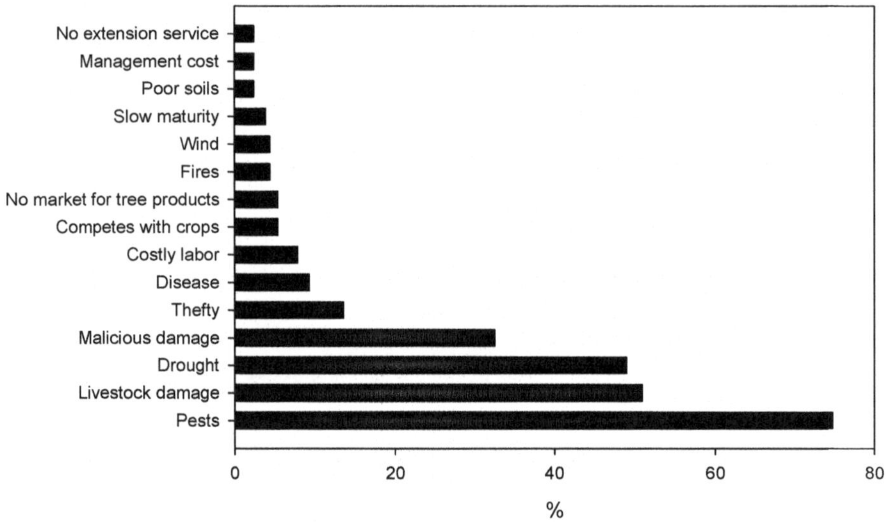

Fig. 6 Challenges, cited by tree planters, that hamper tree planting in Kaliro District. The main pests that were mentioned were termites

4 Discussion

The tree planting effort in Kaliro District appears to be low with only 189 participating tree growers (or homesteads), and 16 institutions, out of the estimated 42,000 homesteads in Kaliro District, and with tree growers planting an average of 175 trees per tree grower. If we assume that a hectare can be planted with 900 trees, then every tree grower has planted on average 0.2 hectares. This notwithstanding, tree cover appears to be increasing because of a few tree growers who plant very many trees. Some of these tree growers have planted in excess of 15,000 seedlings.

Planters with the most trees have the attributes of being male and having attained university level education. Education enables higher incomes with which planters can afford to access the necessary resources including land, equipment, labour and materials such as pesticides that are required to manage the plantation. People with sufficient financial resources are not constrained by the long periods required for the trees to mature before they can be sold for cash income (Kallio et al. 2011). On the other hand, poor tree growers who depend on cash incomes from the farm often rely on short-maturing crops that can be harvested in a single season and sold for cash (Kallio et al. 2011).

Planters involved in business as a livelihood occupation also planted more trees than other tree growers, who were involved in other livelihood activities such as farming. It is not clear why businessmen were more active in tree planning than others. We can speculate, however, that the current wood supply gap in Uganda, due to the increased demand of tree products (FAO and DFID 2016), is driving business-minded persons to invest in tree planting.

Contrary to what is known from the literature that affiliation to tree planting associations promotes tree growing (e.g. Kallio et al. 2011), affiliation to tree growers associations in our case had a negative effect on tree growing. Planters in tree-growing associations planted fewer trees than those outside of these associations. They on the other hand planted more species than those outside of tree-growing associations. That tree growers affiliated to tree-growing associations planted fewer number of trees than those not affiliated to associations is surprising, as it is widely believed that affiliation to groups should improve the success of an activity as it improves access to resources and technical skills among other benefits. This aspect needs further exploration to determine the reasons why tree growers in associations planted fewer trees.

Tree planting has changed with time and a number of historical factors may have influenced these shifts. For example, in the years 2010 and 2012, government provided free tree seedlings under the Farm Income Enhancement and Forest Conservation (FIEFOC) project. Then, between 2014 and 2017, the local NGO Sustainable Use of Plant Diversity (SUPD) facilitated tree planting by providing free seedlings and tree propagation materials and trained tree growers to propagate and grow trees. This support by government and CSOs resulted in a small but statistically significant increase in the number of trees grown. It appears, therefore, that institutional support in the form of easing access to seedlings (whether sold or otherwise) is important for successful tree growing and suggests that institutions are important for tree growing. Furthermore, institutions can raise awareness about the need for and benefits of tree planting. However, it is important to note that planters were active players who to a very large extent (40%) invested their own funds to buy seedlings.

Results of this study seem to agree with the Induced Innovation Theory and the Livelihood Strategy Theory described in Scherr (1995). Briefly, the Livelihood Strategy Theory proposes that farmers' or tree growers' tree-growing strategies are determined by their overall household livelihood strategies and resource base, e.g. tree growers may grow timber trees as a form of saving if they have no superior strategy for saving. On the other hand, the Induced Innovation Theory asserts that farmers tree growing can be induced by historical changes in socio-economic conditions that include (1) declining access to tree products, (2) increasing demand for tree products, (3) declining farm sizes that create a need for planting fences or boundary markers and (4) declining land quality that causes the planting of trees to restore soil fertility, for example. We see this in the nature of species selected by tree growers for planting, the motivations for planting and benefits realised from the planted trees. For instance, the most frequently grown and thus the most preferred species in Kaliro District (Table 3) are fast-maturing pole and timber species, e.g. *Eucalyptus* spp. that are much in demand and have a ready market. With four out of the six priority species being species that produce poles and timber for sale, the implication is that the primary incentive for tree planting is income generation. Indeed, when asked why they planted trees, most tree growers reported income generation more frequently than other factors.

There appears to be a difference between the incentives that motivate tree planting and the benefits that are actually realised from trees. Whereas the key incentive was income, it turns out that the most frequently mentioned benefit was firewood and income generation coming second. We can speculate that even though income is the key motivating factor, but that during the period it takes for trees to mature to a stage where they can provide commercial products, people harvest and utilize firewood from them. Firewood is becoming increasingly scarce and a critical resource in this rural and poor community that people are now forced to buy it (SUPD unpublished report). So firewood has now moved beyond a subsistence product in this community to a marketable product.

The main challenges cited by the planters were biophysical in nature. These included pests, drought and livestock damage. There were also a number of socio-economic problems such as vandalism and theft. These are common challenges to tree planting that have been widely cited in the literature.

It is important to note that lack of market was not one of the challenges mentioned by tree growers. We can speculate that the market for timber and poles is well established and the demand is high. However, the market for horticultural crop products (e.g. mangoes and oranges) is not that well developed and that people are not aware of its potential. This may explain why few fruit trees were planted by tree growers in this study. There is a need therefore to create awareness about the market potential of horticultural crops.

5 Conclusion and Recommendations

This study suggests that the contribution of SHF smallholder tree growers to tree planting is very low at an average of 175 trees per tree grower and with only 206 tree growers (individuals and institutions). There is a need, therefore, to promote more tree growing. This notwithstanding, tree cover has increased in Kaliro District since 1997 because of a few farmers who planted more intensively. Characteristics that differentiate tree growers are that they are men, who are mostly engaged in business, have a university degree and are not affiliated to a tree growers association. For interventions aimed at promoting tree growing to succeed in Kaliro District, institutional support is needed that promotes access to seedlings and increases awareness and skills for tree planting. New approaches may be necessary that provide better access to a greater diversity of smallholder farmers such as women and those with low levels of access to other sources of income and to education. The structure and functions of tree growers associations may need to be revisited to ensure that it provides the maximum benefits to members and local ecology.

The cited difference between realized or anticipated benefits and motivating incentives is an important insight gained through this study. Whereas the key incentive for tree planting is primarily cash incomes, the key benefits were firewood. It is important that these benefits are considered when interventions for tree planting are being designed. Interventions should make available materials necessary for planting, e.g. seedlings, pesticides, etc. and also avail skills for propagating seedlings and also raise awareness about the value of tree management. Additional recommendations from this study are that tree growers should be made aware about the potential market for horticultural crops, because this will not only stimulate tree growing but also alleviate poverty.

Acknowledgments This work was supported by NORAD through the NORHED project (UGA-13/0019). Permission to conduct the project was provided by the Uganda National Council for Science and Technology (NS – 511). We acknowledge with thanks all tree growers of Kaliro District who provided information for this project.

Appendix 1

Tree species planted by smallholder tree growers in Kaliro District, Uganda from 1997–2017 sorted by number of farmers that planted the species

Species	Total trees	Median	# of famers
Pinus caribaea Morelet	208,374	755	113
Eucalyptus grandis W. Hill	90,968	500	62
Eucalyptus unnamed hybrid	61,604	3626	14
Eucalyptus camaldulensis Dehnh.	26,287	200	35
Grevillea robusta A. Cunn. ex R. Br.	26,101	100	85
Maesopsis eminii Engl.	15,349	100	70
Tectona grandis L. f.	12,469	1000	8
Citrus sinensis (L.) Osbeck	5147	60	28
Khaya anthotheca (Welw.) C. DC.	2700	500	7
Mangifera indica L.	2160	55	19
Burtdavya nyassica Hoyle	1411	50	13
Citrus limon (L.) Osbeck	1080	70	6
Markhamia lutea (Benth.) K. Schum.	674	70	3
Artocarpus heterophyllus lam.	517	21	6
Terminalia superba Engl. & Diels	384	37	10
Cedrela odorata L.	363	181.5	2
Melia azedarach L.	308	50	7
Albizia coriaria Welw. ex Oliv.	300	300	2
Senna siamea (Lam.) H.S. Irwin & Barneby	262	20	6
Melia volkensii Gürke	240	40	5
Persea americana Mill.	219	20	7
Coffea canephora Pierre ex A. Froehner	200	200	1
Milicia excelsa (Welw.) C.C. Berg	135	67.5	2
Azadirachta indica A. Juss.	125	62.5	2
Carica papaya L.	83	10.5	4
Citrus reticulata Blanco	77	13	4

Species	Total trees	Median	# of famers
Callistemon citrinus (Curtis) Skeels	61	30.5	2
Ficus natalensis Hochst.	42	21	2
Psidium guajava L.	30	30	1
Leucaena leucocephala (Lam.) de Wit	21	21	1
Citrus aurantiifolia (Christm.) Swingle	20	20	1
Canarium schweinfurthii Engl.	15	15	1
Syzygium jambos (L.) Alston	13	6.5	2
Moringa oleifera Lam.	6	6	1
Calliandra calothyrsus Meisn.	5	5	1
Ficus mucuso Welw. ex Ficalho	5	5	1
Mustaferi	5	5	1
Palm not identified	3	3	1

References

AfranaaKwapong N, Nkonya E (2015) Agricultural extension reforms and development in Uganda. J Agric Ext Rural Dev 7:122–134

Department of Lands and Survey (1962) Atlas of Uganda. Department of Lands & Survey, Uganda

FAO and DFID (2016) State of Uganda's Forestry. Ministry of Water and Environment, Uganda Government, Kampala, Uganda

International Food Policy Research Institute (IFPRI) and Datawheel, 2017. https://DataAfrica.io. Accessed 15 March 2018

Jacovelli PA (2009) Uganda's Sawlog production Grant scheme: a success story from Africa. Int For Rev 11:119–125

Kaliro District Local Government (2012) Kaliro Local Government: Statistical Abstract. The Republic Of Uganda

Kallio MH, Kanninen M, Rohadi D (2011) Farmers' tree planting activity in Indonesia—case studies in the Provinces of Central Java, Riau, and South Kalimantan. Forests Trees Livelihoods 20:191–209

Mercer DE (2004) Adoption of agroforestry innovations in the tropics: a review. Agrofor Syst 61–62:311–328

National Forestry Authority (2009) National Biomass Study: Technical Report 2005. National Forestry Authority

Ollier CD, Harrop F (1959) The Soils of the Eastern Province of Uganda: a reconnaissance Survey. Department of Agriculture. Memoirs of the Research Division. Series I: Soils, Number 2

R Core Team (2017) R: A language and environment for statistical computing. R Foundation for Statistical Computing, Vienna, Austria

Scherr SJ (1995) Meeting household needs: farmer tree-growing strategies in western Kenya. In: Arnold JEM, Dewees PA (eds) Tree management in farmers strategies: responses to agricultural intensification. Oxford University Press, Oxford, pp 141–173

Tabuti JRS (2007) The uses, local perceptions and ecological status of 16 woody species of Gadumire sub-county, Uganda. Biodivers Conserv 16:1901–1915

UBOS (2016) National population and housing census 2014: subcounty report – eastern region. Uganda Bureau of Statistics, Kampala

UNDP (2007) Uganda human development report 2007: rediscovering agriculture for human development. United Nations Development Programme (UNDP), Uganda

van Breugel P, Kindt R, Lillesø J, Bingham M, Demissew S, Dudley C, Friis I, Gachathi F, Kalema J, Mbago F, Moshi H, Mulumba J, Namaganda M, Ndangalasi H, Ruffo CK, Védaste M, Jamnadass R, Graudal L (2015) Potential Natural Vegetation Map of Eastern Africa (Burundi, Ethiopia, Kenya, Malawi, Rwanda, Tanzania, Uganda and Zambia). Version 2.0. http://vegetationmap4africa.org

Whitney CW, Luedeling E, Hensel O, Tabuti JRS, Krawinkel M, Gebauer J, Kehlenbeck K (2018a) The role of homegardens for food and nutrition security in Uganda. Hum Ecol 46(4):497–514. https://doi.org/10.1007/s10745-018-0008-9

Whitney CW, Luedeling E, Tabuti JRS, Nyamukuru A, Hensel O, Gebauer J, Kehlenbeck K (2018b) Crop diversity in homegardens of southwest Uganda and its importance for rural livelihoods. Agric Hum Values 35(2):399–424. https://doi.org/10.1007/s10460-017-9835-3

Climate Change Adaptation Through Aquaculture: Ecological Considerations and Regulatory Requirements for Tropical Africa

Oyediran Olusegun Oyebola and Olufunmike Martha Olatunde

Abstract Aquaculture has been identified to have potentials for adapting climate impacts, and it is being leveraged upon in some developing tropical African countries for food production, employment, food security and poverty eradication. However, it is also vulnerable to climate impacts, linked with certain ecological challenges and competition from anthropogenic factors. This chapter utilized secondary and primary data from various sources to address review questions bordering on the vulnerability of tropical Africa to impact of climate change, potentials of utilizing aquaculture for enhanced climate impact adaptation and food security, aquaculture-related ecological issues and the required management, policy and regulatory actions for its sustainable utilization in this regard. The study revealed that tropical Africa is vulnerable to climate change; aquaculture has viable elements for climate impact adaptation and food security but could contribute to environmental challenges. Aquaculture however has capacity to adjust to the environmental claims, achievable through adequate monitoring, control and surveillance for adherence to ecological considerations. There is the need to specifically formulate policies and provide a strong institutional framework to cater for the nuances of aquaculture, climate impact adaptation and food security. And to ensure for sustenance, it is necessary to strengthen policies and management frameworks with human and financial capacity, within the implementing agencies at local, national and regional levels in tropical Africa.

Keywords Sustainable ecosystem · Food security · Fish farming · Climate adaptation

---

O. O. Oyebola (✉) · O. M. Olatunde
Department of Aquaculture and Fisheries Management, University of Ibadan, Ibadan, Nigeria

Y. Bamutaze et al. (eds.), *Agriculture and Ecosystem Resilience in Sub Saharan Africa*, Climate Change Management,
https://doi.org/10.1007/978-3-030-12974-3_20

1 Introduction

Africa is located between latitudes 37° N and 34° S. The larger part of Africa lies within the intertropical zone, and it is confined to subtropical and Mediterranean areas at its most northern and southern tips. Generally, Africa tends to have a hot, warm climate (Andy 2014). Africa's climate and vegetation cut across equatorial rainforests, tropical deserts, savanna grassland and the Mediterranean. The continent straddles the equator, incorporating both the Tropic of Cancer and Capricorn; and it is surrounded by the Mediterranean Sea, Red Sea, Indian Ocean and Atlantic Ocean. Africa is the second-largest and second most populous continent on Earth, currently hosting about 16.87% of the world population (Worldometers 2018). Africa covers 6% of Earth's total surface, and it is home to 54 recognized sovereign states and countries, 9 territories and 2 de facto independent states, and the total population estimate for the continent is about a billion people (World Population Review 2018). However, Africa is the world's poorest and least developed continent, and it characterized by disease and malnutrition. The income per capita or gross domestic product per capita of many African countries falls to the bottom of the list of the nations of the world (Adeyeye et al. 2017). This situation is further complicated by the recurrent conflicts and emergencies, wide inequalities, depleted resources, ill health and premature mortality in Africa (Atinmo et al. 2009). Moreover, the African continent is one of the most vulnerable to the impacts of climate change (New 2016), as well as food and nutrition insecurity (Olaifa 2015).

There is a global demand for actions geared towards food security, especially in this era of global climate change impacts. This call is more important in tropical African countries where there is high level of underdevelopment, with implications on low resilience to climate impact. The shift in the Earth's climatic condition is already taking place (Freeman 2017; Oladokun et al. 2015; Ipinjolu et al. 2014; Agbarevo 2013; Idowu et al. 2011). Climatic changes manifest in increase in temperature, change in snow and rainfall patterns and more drastic uncontrolled events. Climate-related challenges often include heavy rainstorms, record-breaking temperatures, drought and flood. Accelerated sea level rise and the increased frequency of extreme events such as coastal flooding, cyclones and storm surges have also been highlighted (IPCC 2014). Most of the changes in climatic conditions are connected to the increase in carbon dioxide and greenhouse gases from burning of fossil fuels and other anthropogenic activities (US EPA 2016). The circumstances that result in climate change have been highlighted to include increasing burning of fossil fuel and changes in land use that continuously emit quantities of greenhouse gases such as carbon dioxide (CO2), methane (CH4) and nitrogen dioxide (N2O) into the atmosphere (UNFCCC 2007).

Accumulation of greenhouse gases into the atmosphere and water has been linked to a number of phenomena, including gradual changes in water temperature, acidification of waters, ocean current changes and rising of sea levels. These physical variations affect the frequency, intensity and location of dangerous weather events and ecological functions within the aquatic environment (Cochrane et al. 2009).

Meanwhile, health and adequate functioning of aquatic ecosystems is conjoined to productivity of its fishery (MAB 2009). The ecosystem of a water body is more at risk to fluctuation in the primary production and the way such production is transferred through the aquatic food web. The aquatic system is also vulnerable to fluctuations in the physical and chemical entities, including temperature, salinity, acidity and water current and flow. The highlighted climatic parameters play essential roles in productivity and survival of biological organisms, including fish and other invertebrates.

Temperature changes are an important indicator of climate change, and its fluctuation could induce migration in marine and aquatic invertebrates in an attempt to regain preferred internal temperature (Roessig et al. 2004). Ideally, primary productivity of the marine environment depends on availability of nutrients in such an environment, and this relies on freshwater run-off and the ocean mixing with lights and temperature catalysing the activities. Climate change has been predicted to decline at lower latitude (FAO 2008a), while consistent extreme weather happenings will affect fishermen and harm their homes, services and population structure, most especially, in coastal areas (IPCC 2007). Aquatic ecosystems are highly vulnerable to climate impacts (Friggens and Woodlief 2015; Glen 2010; Boon and Raven 2012; Capon et al. 2013). Ecological responses have been associated with changes in patterns of host-parasite interactions (Marcogliese 2001; Paull et al. 2012), food webs (Ledger et al. 2013) and structure of body size (Yvon-Durocher et al. 2011). Wilby et al. (2010) on the other hand reported that ecological responses to changes are highly uncertain and gross generalization and prediction of the impacts is not ideal. Climate change is expected to exacerbate the current threats to freshwater ecosystems, yet multifaceted studies on the potential impacts of climate change on freshwater biodiversity at scales that inform management planning are lacking (Markovic et al. 2017).

Climate change is impacting on several of the oceans, coasts and freshwater environments, with side effects on fisheries resources (Cochrane et al. 2009). Studies have shown that climate change impacts on marine fisheries and coastal aquaculture (Handisyde et al. 2006; De Silva and Soto 2009; Brander 2010; Ahmed and Diana 2015; Shameem et al. 2015). It is obvious that artisanal fishers, fish farmers and seaside dwellers would be more impacted through unstable livelihood, changes in quantity and availability of fish food and drastic elevation in risks to their health and home safety. The vulnerability of fishery-based livelihoods to climate change has increased over time (Islam et al. 2014). Meanwhile, the communities that are dependent on fisheries are usually vulnerable to risky conditions as a result of their low income level and lack of socio-economic infrastructure. The fragility of these communities is further weakened by overexploited fisheries resources and destroyed ecosystems. The challenges of climate change would be more devastating to livelihood of this vulnerable societal group. The number of people employed in fisheries and aquaculture has increased faster than global population increase in general (FAO 2014). Meanwhile fish capture from marine and inland waters are declining, while aquaculture is advancing. Aquaculture is the most rapidly growing food-producing sector in the world, with an average annual growth rate of 8.6%

since the late 1980s (FAO 2014); it provides employment and offers reliable income for often underprivileged communities (Pant et al. 2014).

Aquaculture focuses on aquatic ecosystems, and the sector is geographically favoured to thrive successfully in tropical and subtropical Africa. The sector is being leveraged upon in some developing tropical African countries for food production, employment, food security and poverty eradication. Despite these potentials, the sector is also vulnerable to climate impacts (Oyebola and Fada 2017; De Silva and Soto 2009). Climate risk to aquaculture can be direct, impacting on natural resources such as water, land, seed, feed and energy required for farming practice (Easterling et al. 2007). On the other hand, changes in rainfall will cause a ripple effect on the quantity of water; extending from droughts and famines to flooding. This phenomenon would reduce water quality, causing increased salinity of groundwater supplies due to dehydration, while flushing of saline water from upstream would reduce available freshwater resources for fish farming (IPCC 2007). Unpredictable heavy rainfall culminating from climate change increases run-off and brings nutrients from fertilizer, sewage and agricultural farm wastes to the adjoining aquatic environment. This activity leads to algal blooms with devastating reduction in dissolved oxygen which results in fish kills (Diersing 2009). Meanwhile, this activity and its effect are heightened by anthropogenic factors. Apart from the influences of climate change, aquaculture has also been linked with certain ecological challenges which have potential of limiting its use for climate adaptation and food security.

Policy and legal framework is an essential tool in resource management as sustainable management of natural resources needs to be supported by coherent policies (Jamart and Rodeghier 2009). Efforts to share the rights and responsibilities of managing natural resources seldom work in the long run in a national context that is not shaped by enabling legislation and policies. Open Development Cambodia (2015) observed that policy instrumentation has capacity for boosting natural resources production, and there is the need to integrate principles of sustainable development into the country's policies and programmes in order to reverse loss of environmental resources. Policy instruments could create enabling environments favourable to co-management of natural resources (Jamart and Rodeghier 2009). Meanwhile, legislation on policy statements makes room for litigation, which serves as control for potential erring individuals or organization in resource management.

The ability of a state to regulate, control and allocate its resources and ensure efficient and proper use for sustenance is usually challenged without appropriate legal framework (Salman and Daniel 2006). However, policy and legal actions on natural resources management have to be dynamic because production, utilization and management of natural resources are not static. Therefore, Mitike et al. (2016) aver the need for periodic review of policy, regulatory and organizational frameworks on environment and health issues. It is ideal for states to provide the legal and policy frameworks that are necessary for successful and sustainable natural resource management. Availability of policy and regulatory tools will inform the readiness of a state to circumvent environmental/ecological challenges in order to ensure sustainability of its natural resources especially in the fisheries and aquaculture systems.

Fish farming, especially caged fish farming, is sporadically increasing in tropical African countries, and it promises to be a good resource for climate adaptation and food security. However, there is the need to create awareness on the dimensions of the highlighted issues, for sustainable use of fish farming in adapting to climate change impact in tropical Africa. Therefore this review focuses on issues affecting sustainable utilization of aquaculture for climate impact adaptation in tropical Africa. It focuses on vulnerability of tropical Africa to climate change impacts, potentials of utilizing aquaculture for enhanced climate impact adaptation and food security, aquaculture-related ecological issues and the required policy and regulatory actions for its sustainable utilization in this regard.

Recommendation: content editor should take another look at this introduction. It is too longwinded and repetitive to make a good introduction. My mandate was language editing, and this in itself was problematic.

2 Vulnerability to Climate Change in Tropical Africa

Changes in temperature, rainfall pattern, humidity, sea level, greenhouse gases (GHG), continental drifts, deviation in the Earth's orbit and activities of man are some of the compounded evidence of climate change (Yazdi and Shakouri 2010). Climate change affects agricultural production systems with reflections on food insecurity. There is empirical evidence that there will be changes in the supply and demand of food commodities as a result of low yields, resulting mainly from drought and flooding events (Adeleke and Omoboyeje 2016). Hence, it is important to address climate impact on agriculture and water resources for the purpose of sustainable food security and human survival. Climate change is presently a reality in Africa; there are extended and intensified droughts in East Africa, unexpected flooding in West Africa, deteriorations in the rain forests in equatorial Africa and a rise in the salinity and acidity in oceans around the southern coast. About 14 countries, namely, Burkina Faso, Chad, Ethiopia, Ghana, Kenya, Liberia, Mali, Niger, Nigeria, Senegal, Sudan, Togo, Uganda and Rwanda, are the worst hit by flood in the African continent (BBC News 2007).

According to IPCC (2014), the recent anthropogenic emissions of greenhouse gases are the highest in history; many of the observed changes in climate parameters are unprecedented as the atmosphere and ocean have warmed, amounts of snow and ice have diminished, and sea level has risen. Relative to 1850–1900, global surface temperature change for the end of the twenty-first century (2081–2100) is projected to exceed 1.5 °C for RCP4.5, RCP6.0, and RCP8. *Warming is likely to exceed 2 °C for RCP6.0 and RCP8.5, more likely than not to exceed 2 °C for RCP4.5 but unlikely to exceed 2 °C for RCP2.6.* Content editor – Check to ensure that it makes the intended sense! Rainfall variability is estimated to rise, resulting in recurrent flooding and drought (Hulme et al. 2001; Gladys 2017). Increase in the ocean temperatures and acidity dictates that the capability of the ocean carbon sink will become more vulnerable, giving the increase in global concerns (UNEP 2010; Harrould-Kolieb and Dorothée Herr 2011; Secretariat of the Convention on Biological Diversity 2014). Agricultural production and food security, energy security, water and health are immensely altered by weather patterns and extreme climatic conditions, which in turn challenge the ability for Africa to sustainably grow and develop as the IPCC (2007) report, average temperature in African countries is predicted to increase by 1.5–3 °C in the year 2050. The increase in temperature in Africa would probably be higher than other continents. This indicates that Africa is highly vulnerable to climate change impact. This corroborates the report by New (2016) which implies that African countries are among the most vulnerable to climate change, and this is having a negative impact on agriculture, the food production sector.

According to Martin (2015) Pope Francis was quoted as saying: 'Many poor people live in areas particularly affected by phenomena related to heating, and their livelihoods strongly depend on natural reserves and so-called ecosystem services, such as agriculture, fisheries, and forestry. Many of these people also live in the countries near the Equator; it is inhabitants of the tropics who will feel the effects the soonest, and who will suffer the most'. (Which page numbers are we talking about here?) Martin (2015) reported annual climate change vulnerability index of the UK-based risk analysis firm; Maplecroft lists the top 32 countries at 'extreme risk' from climate change of which the top 10 are all tropical countries: Bangladesh, Sierra Leone, South Sudan, Nigeria, Chad, Haiti, Ethiopia, the Philippines, the Central African Republic and Eritrea. Pages?

Exposure implies degree to which a system is exposed to a hazard, perturbation or stress caused by the changing climatic conditions; sensitivity highlights the degree to which a system is affected by, or responsive to, climate change stimuli (Smit et al. 2000, pages?); while adaptive capacity or adaptability is the potential or capability of a system to adapt to climatic stimuli. Vulnerability of aquaculture and fisheries to climate change could take ecological, socio-economic and technical dimensions, and these factors could combine. The reviewed FAO-commissioned case studies on the vulnerability of fisheries and aquaculture to climate change complemented with additional information were utilized to evolve a framework on the link between ecological and socio-economic vulnerability in the sector. The result of this action presented in Fig. 1 indicates a refined and precise cause of vulnerability, while focusing on both the fish production ecosystem's resilience and the adaptive capacity of its linked human/institutional system.

In the ecological domain, exposure and sensitivity create impact potential. Impact potential and recovery potential (ecological resilience) together form ecological vulnerability. This vulnerability together with socio-economic exposure and sensitivity results in socio-economic impact potential which combines with human and institutional adaptive capacity to result in social ecological vulnerability which gives feedback to the ecological system. However, Signgh and Purohit (2014) reported that dependence on agriculture, and low level of financial, technical and institutional ability to adapt, makes impact of climate change to be more evident among developing countries. This highlights the fact that although the table identified ecological and socio-economic vulnerability, climate impact vulnerability

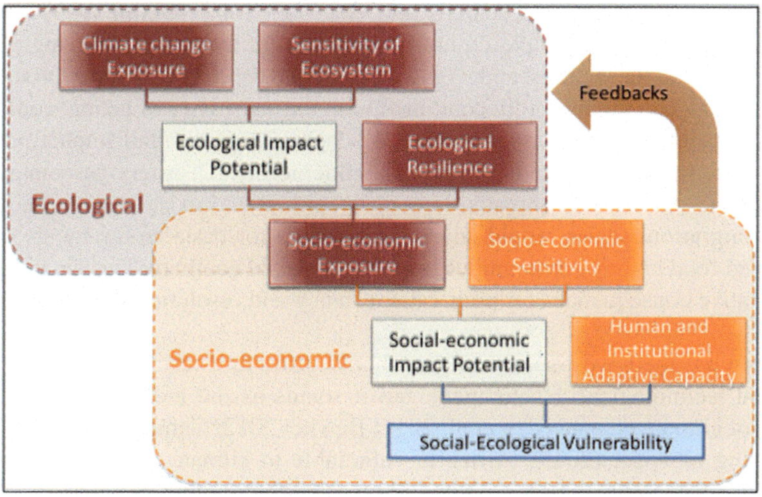

Fig. 1 Framework linking ecological and socio-economic vulnerability in aquaculture and fisheries production. (Source: Cinner et al. (2013), p. 4 – Adapted from Marshall et al. (2010))

could take diverse dimensions which could include agriculture production system and technology, socio-economic and institutional structure as well as ecological/environmental system.

Climate change affects ecosystems and their productivity through the changing patterns in temperature and precipitation, droughts, floods, heavy winds and other extreme events, representing both new threats for some regions and opportunities for others. It impacts generally on agriculture (Pereira 2017). Meanwhile, fish farming system can be exposed to climate change impact such as flooding through closeness of fish farm to river environment, use of swampy environment for fish farming, location of fish farm at valley adjoining slopy and poorly drained land, exposure of farm area to increased unpredictable rainfall and technical faults such as poor pond construction and poor fish farm location and management. The system could be sensitive through levels of awareness on diverse manifestations of climate change impacts of flood and drought occurrences in fish farming systems. Meanwhile, adaptive capability of fish farming could be in socio-economic including average annual income, livelihood diversifying into non-farming activities, membership of social groups, educational qualification and training, availability of compensation for loss and insurance and availability of communication technology. Climate-smart techniques are important in African agriculture (Zougmoré et al. 2016). Technical innovations on climate-smart aquaculture including use of climate variation-tolerant fish species for stocking would be of immense advantage to climate impact adaptation through aquaculture.

Climate change represents a new major security threat for the world, particularly for Africa (Brown et al. 2007). Ecology-related disasters threaten human security, as populace would now have to compete for limited basic human needs, and this bring in the theory of environment, scarcity and violence (Homer-Dixon 1999) into play. The marginalized poor people do not have the buffer to handle recurring climatic disasters even at the smallest and would lack enough time for recovery (Olsson et al. 2014). Ecological vulnerability combined with the sensitivity of people could form the impact potential for a society. Social adaptive capacity and impact potential would together create social ecological vulnerability. Disasters associated with ecology and climate that threaten human livelihood and existence can encourage forced migrations which would create competition for basic requirements such as water and food among communities resulting in social conflicts. This creates potential negative consequences for political constancy and resolution of conflicts (CIGI 2009).

Africa has the lowest source of greenhouse gases as a result of receding industrial and technological development, but it stands as the most vulnerable to the effects of climate change (Beg et al. 2011; Bewket 2012; Gemeda and Sima 2015). According to Vogel (2015), Africa is vulnerable to climate change because it is exposed to damaging climate risks including extreme droughts, flooding and storms; the continent also has low adaptive capacity making it particularly vulnerable and exposed because of high rates of poverty, financial and technological constraints as well as a heavy reliance on rain-fed agriculture. It is commonplace in the discourse on global climate change to say that Africa will be the most vulnerable location worldwide, because of poverty, environmental degradation and conflict with highest concentration of climate impact vulnerability located along the West African coastal states such as Nigeria, Liberia, Sierra Leone and Guinea-Bissau (Joiner et al. 2012). It would be ideal to indicate here that the current socio-economic condition of the populace, the existing institutional structure and the technological capability of the vulnerable African countries seem not strong enough to cope with the challenging climate variability impacts on their food security and livelihood.

2.1 Case Scenario on Climate Change Vulnerability in Selected Tropical African Countries

Vulnerability of fisheries and aquaculture to climate change has been carried out across continents. However, case scenarios of two of the leading vulnerable countries in tropical Africa were presented by Connolly-Boutin and Smit (2016). This is highlighted here.

Nigeria

Vulnerability calculations of composite vulnerability show Nigeria to be highly vulnerable to climate change in two primary locations: the northern Nigeria and the Niger Delta. The Niger Delta is a low-elevation coastal zone that has historically experienced violence from ethnic diversity, oil wealth and land degradation issues. At the household vulnerability level, subnational calculations showed that socio-economic vulnerability expressed in high concentration of poverty, illiteracy and poor access to healthcare contributed to the stark contrast and inequality between southern and northern Nigeria. The highly vulnerable Niger Delta and northern Nigeria, Jos and Lagos are population centres being challenged by unique vulnerability. Religious divisions are responsible for violence and crisis at locations such as Jos, while Lagos, a mega city with large poor population, is located in a low-elevation coastal zone. Meanwhile, oil and mineral dependence and ethno-religious diversity seem to have magnified Nigeria's vulnerability to climate-related hazards and household-level stressors.

Nigeria's physical exposure to climate-related hazards concentrates in the Niger Delta and the region around the city of Lagos, due to their high potential for flooding. The Niger Delta is vulnerable to flooding due to lowlands and presence of many waterways. Flooding could spread waters contaminated with oil into new areas of the Delta, affecting the population's access to water and the fragile ecosystem that currently supports fishing and agriculture. This also impairs fish farming as ideal water quality for fish culture is changed, invasive fish are transferred to culture environment and farmed fish are often carried away. Sea level rise has potential economic effect on the Nigeria population. Sea level rise would increase erosion, change soil composition and decrease agricultural yields. The intrusion of seawater into the freshwater ponds would change water chemistry and ecosystem structure which consequentially impairs growth and development of farmed fish in such area.

Northern Nigeria has greater household vulnerability than the rest of Nigeria. This variation could be traced to socio-economic divergences. Literacy level was below the national average; majority of the nation's university students originated in the south, and most of the technocracy within the national bureaucracy are southerners. The average incomes in southern states nearly double the average incomes in the northern states, and southern populations have greater access to mobile phones, television and the Internet. These scenarios indicate potential divergence in vulnerability across the northern and southern Nigeria. The disparities manifest in response to health challenges reflecting in infant mortality rates and incidence of underweight children. Drought ravages the north; flood threatens the south. These are climate risks that majorly challenge fish farming in Nigeria. It is important to stress that the spreading desert affects roughly 35 million people in northern Nigeria and the region's population faces many household-level challenges to adapt to their changing circumstances. The overall inequality of development makes northern Nigeria fish farmers to be less able to adapt to drought and desertification, while the south farmers struggle with menace of flood in fish farming.

Estimations of climate-related hazard exposure, population density, household vulnerability and governance revealed Jos is facing high vulnerability to climate change, the Niger Delta is a hotspot because of its vulnerability to climate-related hazards, the city of Warri faces high vulnerability to climate change, and Lagos also faces high vulnerability to climate change. The city's low elevation threatens to make Lagos the site of widespread flooding and displacement due to sea level rise – circumstances the population will be ill-equipped to face without substantial government assistance. It was concluded that Nigeria contains some of West Africa's most vulnerable locations. These locations would require intensive government initiative to adapt to climate change. Flooding and sea level rise along the southern coast are natural occurrences that may become worse as time passes. The lack of development in northern Nigeria makes this region particularly vulnerable to climate-related hazards since the populations in the region do not have many resources to prepare for a possible increase in droughts. Meanwhile, the religious and ethnic dimensions of the population complicate Nigeria's political economy which could impair objective policymaking in resource allocation for climate impact adaptation, if not well managed.

Guinea-Bissau

Guinea-Bissau consistently ranked among the most vulnerable countries in West Africa. The country is characterized by low adaptive capacity for climate change impact at the national and community level. The country's vulnerability is primarily driven by physical exposure such as low-elevation coastal zones, dependence on agriculture and fishing, ethnic polarization and poor governance. Increased flooding and saltwater intrusion due to global sea level rise could potentially affect this coastal community, and the population would quickly feel the losses because they rely on mangrove rice cultivation for livelihood. The country faces rainy seasons and long dry seasons with abrupt transitions, less stable West African monsoon and long droughts. Northern Guinea-Bissau borders the Sahel region having possibility of increased long droughts.

Livelihood of the population is vulnerable to climate change. Although little industry exist indicating possible lower GHG emission, the lack of industrial development limits the level of employment leaving most of the population to subsistence farming. Meanwhile, climate change has already begun to affect coastal farmers through saltwater encroachment. Some ethnic groups, such as the Diola in the north, the Fula and the Mandinka, say that the weather variation is abnormal and blame anthropogenic climate change. Livelihood diversification is important in climate adaptation. However, the disruption of traditional livelihood sources resulted in substantive potential problem. For instance, farmers who can no longer grow rice due to saltwater encroachment have shifted into cashew production. This can lead to a crippling dependence on a single crop for entire communities with potentials of putting their livelihoods and even their ability to feed themselves at the mercy of harvest from a single crop, such as cashew and of international cashew prices. Although substituting cash crops for subsistence crops may boost incomes, this could increase food insecurity and make populations more vulnerable to price fluctuations.

Climate change has adverse effects on the fisheries and fishing in Guinea-Bissau. Rising sea temperatures and changes in the oceans' dynamics such as acidification and loss of nursery areas are predicted to reduce fish populations. The destruction of coral reefs and mangroves through anthropogenic activities destroys fish spawning grounds, decreasing the availability of fish, limiting the livelihoods of fisherfolks and leading to precarious food security. Declining fish populations will reduce government revenue through fishing licences, and it has pushed some fisherfolks into human and drug trafficking which has begun to grow in the poorly regulated space of Guinea-Bissau. Governance and adaptive capacity with respect to climate impact adaptation were rated low for Guinea-Bissau due to ethnic conflict and struggle for power which often occurs over ethnic lines. Rivalries between civilian leaders and the military, fuelled by ethnic tensions, seem to be a dominant feature of Guinea-Bissau politics. Ethnic, personal and factional rivalries have led to several coups or coup attempts, and these have not allowed the country to build strong institutions especially for climate impact adaptation.

It was concluded that climate change could cause a humanitarian disaster in Guinea-Bissau with flood displacing people and damaging infrastructure while the government lacks resources to respond effectively. Over a longer period, the climate-related degradation of Guinea-Bissau's agricultural economic base would heighten the country's aid dependence and drive more of the populace to illicit occupations due to lack of opportunity in the legitimate economy. Falling revenues from agriculture and fishing would further diminish government adaptive capacity, while the concentration of production in a few crops at the expense of subsistence methods will diminish household adaptive capacity. Meanwhile, ethnic infighting and decreasing government revenues will leave the government unable to address the needs of its population. Climate change would impact on the country's food security, humanitarian disasters, regional instability and greater contribution to transnational threats.

3 Potentials of Aquaculture in Climate Change Adaptation and Food Security in Tropical Africa

3.1 Need for Aquaculture in Climate Impact Adaptation and Food Security in Africa

Fish is generally acceptable as a food to all regardless of region, religion, race, gender and age (Gurung 2016). Fish including shellfish gives the essential nutrition for over 3 billion of people; and it provides 50% of animal protein and minerals to half a million people in the poorest countries (Tubiello 2012). According to de Graaf and Garibaldi (2014), the value added by the fisheries sector in Africa as a whole was estimated at more than US$24 billion in 2011; this accounts for 1.26% of the GDP of all African countries; aquaculture is still developing in Africa, but it already produces an estimated value of almost US$3 billion per year; the fisheries sector as a whole employs 12.3 million people as full-time fishers or full-time and part-time processors in which 7.5% work in aquaculture, while about 27.3% of the people engaged in fisheries and aquaculture are women; FAO (2014) reported 1.49 million tonnes production from African aquaculture despite huge potentials.

There is strong relatedness of fish demand and supply outlook with food security (HLPE 2014; World Bank 2013; OECD-FAO 2013; Hall et al. 2011). Specifically, the critical role of fish in food and nutrition security was highlighted by HLPE, High Level Panel of Experts on Food and Nutrition Security (2014). According to the report, bioavailability of fish protein is approximately 5–15% higher than that from plant sources. Fish contains several amino acids essential for human health, especially the usually limiting lysine and methionine, and contains unique long-chain, polyunsaturated fatty acids (LC-PUFAs) with many potential beneficial effects for adult health and child development, and fish is an important source of essential micronutrients – vitamins D, A and B and minerals (calcium, phosphorus, iodine, zinc, iron and selenium) which are especially found in many small fish species that are consumed whole. Nutritional benefits of fish consumption have a positive link to increased food security and decreased poverty rates in developing states (Lehane 2013).

Fish is a product for local, national and international trade, consumed for household nutrition security and human health benefits. Fish is also utilized as feed ingredient for production of other livestock. However, these benefits are provided through capture fisheries and aquaculture. Gurung (2016) observed that fisheries production is important for global food security and nutrition. Capacities of capture fisheries and aquaculture to deliver the highlighted functions are usually limited when the environment, production ecosystems and/or the resources bases (fish stocks) are degraded or overexploited (Agardy and Alder 2005; FAO/NACA 2012). Meanwhile, aquaculture seems to have relatively better potential compared to capture fisheries in this regard. It would therefore be a better option as instrument for food security in these days of climate change impact in tropical Africa. This is especially important going by the recent trend of capture fisheries production in which global capture fisheries are currently witnessing stagnation in the face of increasing global appetite for fish (Msangi and Batka 2015).

Capture fisheries and aquaculture systems are exposed to climate change impacts (Williams and Rota 2012; WorldFish Center 2009; Medugu 2009; Halls 2009; IPCC 2007). Natural climate cycle has adversely affected agriculture sector including aquaculture in Africa (Ziervogel et al. 2006). Climate change has been confirmed to impact greatly on capture fisheries production in Africa (FAO 2016; Rhodes et al. 2014). Climate change impacts on distribution, species composition, seasonality and production of natural/wild fish stocks (Ficke et al. 2007; Daw et al. 2008; Pörtner and Peck 2010). WorldFish Center (2009) reported that fisheries as well as the quality and availability of habitats that support it are sensitive to climate change effects, and many fishery-dependent communities in Africa region are highly exposed. The long-term mismanagement of the world's fisheries has led to the overexploitation of capture fisheries resources, with consequences of reduced available fish stocks for production; important coastal habitats, which support numerous fisheries, are being affected through sea level rises and extreme weather events such as flood in Africa (Allison et al. 2005),

Climate change would normally alter capture fisheries distribution and productivity which would impact negatively on livelihood of the people that depend on aquatic environments. Climate change impacts are usually fiercer on systems that depend more on natural phenomena such as the capture fisheries. Hence, capture fisheries would be a relatively weak instrument for improving fish production for food security in these days of challenging climate change and food insecurity in tropical Africa, when compared to aquaculture.

3.2 Aquaculture: Advancing in the Face of Climate Change Impacts

Aquaculture plays an important role as economic activity and livelihood component of rural communities (FAO 2007a, b; Mbugua 2008). Aquaculture provides protein and fish consumption needs (Na-Nakorn and Brummett 2009), provides viable source of income for rural communities and can be easily integrated into existing farming systems (World Bank and FAO 2010). About 48 countries and 5 island nations are practising some form of aquaculture, often at a very low level in Africa, but the potentials for expansion are considerable (Machena and Moehl 2001). Growth in the aquaculture sector over the last two decades has been tremendous, making it the fastest-growing food sector in the primary production (FAO 2007b). It is of note that while capture fishery is declining (Omitoyin 2007; Faturoti 1999), aquaculture is advancing (Olaifa 2015; Belton and Thilsted 2014; Gabriel et al. 2007). Aquaculture is the world's fastest-growing food production system, growing at 7% annually. Although WorldFish Center (2010) warns that aquaculture is also vulnerable to climate change impact, Coulibaly et al. (2007) opined that aquaculture could be utilized in adapting climate change impacts, while Rothuis et al. (2014) reported that aquaculture has the potential to make significant contribution to food security and income generation in tropical Africa. The advancement of aquaculture despite the established climate change impacts in the region corroborates the reports of Coulibaly et al. (2007) and Rothuis et al. (2014) that the aquaculture subsector has potentials for climate impact mitigation and food security especially in tropical Africa. Adedeji and Okocha (2011) reported that the growth in fish production is due to increased activities of aquaculture, while the need for aquaculture arose from the decrease in supply from ocean fisheries as a result of overfishing, habitat destruction and pollutions. It would however be ideal to highlight the trend of growth of the aquaculture subsector in the face of climate change impacts and to highlight its features of relative relevance in this regard.

The relative growth of aquaculture compared to capture fisheries could be hinged on its relative resilience against several manifestations of climate change and other anthropogenic factors. It is understandable that climate change would impact on aquaculture due to its direct or indirect impact on natural resources such as land, water, seed, feed, etc. required for farming operations. Pond fish culture could be affected by solar radiation, wind velocity, air temperature and water turbidity causing higher vaporization and probable reduction in solar radiation getting to ponds which can lead to dinoflagellate algal blooms and red tides on water surface with resultant effect of fish kills. Climate change would cause natural flood or drought disaster, thereby altering availability or quality of water for aquaculture; but aquaculture could utilize, retain, conserve and recycle limited water, thus been less prone to the factors to which the capture fishery is highly vulnerable. Fish stocking and harvesting can be reprogrammed to eschew flood/drought periods/months, and farming facilities could be manipulated to better cope with the climate risks in aquaculture.

Water resources in capture fisheries are usually limited in quality and quantity due to climate change impacts. This is contrary to the scenarios of aquaculture where some level of control and management of water quality is relatively more feasible. Aquaculture could deliberately intervene in improving fish yield in natural environment through culturing and restocking of depleted wild fish species, which would ultimately result in increased capture fisheries production. Aquaculture maximizes fish yield by manipulating growth, reproduction, recruitments and natural mortality rate. Capture fisheries production depends on inland river and marine/estuary water sources. These water sources are usually threatened by human activities such as pollution which is difficult to control due to multiple use nature of the water resources. Meat production through aquaculture seems to be relatively favourable in climate scenario. However, it would be ideal to highlight the basis for growth of the subsector vis-a-vis other meat production sector of agriculture in the face of climate change impacts and to highlight some of the features of aquaculture that permit this development in the subsequent sections.

3.3 Growth of Aquaculture in the Face of Climate Change in Tropical Africa

Growth of aquaculture began at a time when global sustainability in the use of natural resources and the accompanying environmental degradation issues was becoming a priority. Aquaculture's positive influence on issues such as climate change has gone unheeded, and society at large needs to consider that all food production has environmental costs (Bartley et al. 2007; Cao 2012; van Cleef et al. 2010) which have to be compared in a fair way with environmental sustainability (Cooke et al. 2016). In line with this, an attempt is made below to outline the positive contributions of aquaculture towards the global problem of climate change. Carbon emissions/greenhouse gases from animal husbandry and aquaculture in one form or the other, driven by anthropogenic activities, could contribute to climate change (Brook et al. 1996; Flannery 2005; Friedlingstein and Solomon 2005; IPCC 2007). Hence, all mitigating measures revolve around reducing the carbon emissions. Considering the degree of carbon emissions of the various animal food production sectors with a view to gauging the degree to which aquaculture contributes to this primary cause is necessary. It is conceded that accurate and/or even approximate estimations of total emissions from each of the sectors are difficult, if not impossible, to compute. However, any approximation will bring to light the indirect role that aquaculture plays in this regard.

Steinfeld et al. (2006) observed that 18% of global GHG emissions could be attributed to animal products alone. For the EU, a figure of 29% of all consumption-derived GHG emissions are food related (EIPRO 2006). Methane is an important contributor to GHG emission. Meanwhile, the US Environmental Protection Agency, US EPA (2016), recognized major sources responsible for methane emissions in the USA and ranked enteric fermentation and manure management from animal husbandry as the third and fifth highest emitters, respectively. The emissions from these two animal food production sources were 117.9 and 114.8 and 31.2 and 39.8 Tg CO_2 equivalents for years 1990 and 2002, respectively. Domesticated livestock, the ruminant animals (cattle, buffalo, sheep, goats, etc.), produce significant amounts of methane in the rumen in the normal course of food digestion, through microbial fermentation (enteric fermentation) that is discharged into the atmosphere. The solid waste produced (manure) needs to be managed, and this process results in the emission of significant amounts of methane. It is to be noted that the atmospheric methane level has increased from 715 ppb in the pre-industrial revolution period to 1775 ppb at present with comparable trends being recorded from ice cores from Greenland (Brook et al. 1996). Continual increase in this emission rate is not acceptable for sustenance of life in the planet Earth.

It has been reported that the world's livestock accounts for 18% of greenhouse gases emitted, more than all transport modes put together, and most of this is contributed by 1.5 billion cattle (Lean 2006). Rearing of cattle produces more greenhouse gases than driving of cars (UN Report 2006); greenhouse gas emissions from dairy open lot and manure stockpile in northern China (Ding et al. 2016). Although efforts towards reducing this emission is ongoing (Basarab et al. 2012), by convention, the global warming potential of methane is estimated to be 23 times that of carbon dioxide. Hence, on the overall, the livestock sector is estimated to account for 37% of all human-induced methane emissions. The processing of other livestock sources also seems to consume more energy compared to fish. The farmed aquatic organisms do not emit methane and therefore are not direct contributors to these causative problems. However, Allsopp et al. (2008) observed that this unique attributes of fish production has not been taken into account, particularly by those who tend to advocate against aquaculture.

It is estimated that in the developing world, the per capita meat consumption rose from 15 kg in 1982 to 28 kg in 2002 and is expected to reach 37 kg by 2030 (FAO 2003). There is a growing appetite for fish, and the World Bank (2014) had clamoured for raising more fish to meet rising demand. According to Voegele (2014), there is a major opportunity for developing countries that are prepared to invest in better fisheries management and environmentally sustainable aquaculture. The increasing demand for animal food products in developing countries would normally contribute to accelerated rate of production. Any analysis of GHG emission in food production units has to revolve around human food needs and the proportionate contribution of each food-producing sector to greenhouse gas emissions. It is therefore understandable that there would be relative growth in aquaculture production as the world is requiring more animal food products, fuelled by rising incomes and urbanization, particularly in the developing world.

However, sustainable growth in aquaculture is being challenged in some quarters. Among the challenges of aquaculture is the use of fishmeal and fish oil, obtained through reduction processes of raw material supposedly suitable for direct human consumption (Naylor et al. 1998; Aldhous 2004). Need for coastal aquaculture has resulted in mangrove clearing for shrimp farming (Primavera 1998, 2005). It has been estimated that less than 5% of mangrove areas have been lost due to shrimp farming, most losses occurring due to population pressures and clearing for agriculture, urban development, logging and fuel (GPA 2008). A counter argument is that positive contributions from aquaculture may not have been totally quantified alongside these challenges. The relative benefits of aquaculture compared to the alternative animal protein sources may have been underestimated.

3.4 *Features for Adapting to Climate Change Through Aquaculture*

Climate change impacts through anomalies in rainfall and evaporation with resultant impact on availability, distribution and quality of water. Therefore, availability of technologies which gives room for efficient water use would be useful for adaptation.

Fish Farming Systems

Fish farming is traditionally categorized under extensive, semi-intensive and intensive systems based on level of investment and dependence on natural resources. The extensive system is less expensive but grossly depends on natural conditions of the culture system; the intensive system depends more on provision of basic resources outside nature and is thus more a climate-smart approach, while the semi-intensive is a system that is in between these two extremes. Adaptation to climate change in these farming systems is usually enhanced by the type of culture facility, culture period, cultured species and adoption of value addition strategies. The integrated fish farming and polyculture systems are of importance in climate adaptation using aquaculture.

Integrated Fish Farming/Aquaculture System

Integrated multi-trophic aquaculture is traditionally found in West Africa, India, Indonesia and Vietnam, as a climate impact mitigation tactic. This technique results in better irrigation water efficiency and quality. Decreasing methane emissions and waste through the integration of rice and fish culture systems is realistic, and this system can also provide new sources of income while improving performance of cultivated agroecosystems, and it enhances human well-being. It is a form of livelihood diversification that ensures that wastes are transformed to wealth through aquaculture integration with other agriculture products. Animal wastes from feed and other livestock's are utilized in fish farming. The decreasing nitrogen releases from feed and animal wastes can lead to better groundwater quality and reduced loss of biodiversity, improving agroecosystem function and resilience to climate extremes by enriching soil fertility and soil water retention (Schimel et al. 2007; Schimel and Schaeffer 2012). More information on integration will be obtained under the section on polyculture.

Polyculture

Polyculture is the rearing of multiple fish species together in the same rearing facility. This is synonymous to mix cropping in agriculture. The stocking of multiple fish species with probably differential resilience to climate change sound advantageous. This approach also helps in fish product diversification. Combination of the African catfish, *Clarias gariepinus*, with Nile tilapia, *Oreochromis niloticus*, is a popular polyculture option in some tropical countries such as Nigeria. The catfish is a very hardy species of great potential to be resilient to climate change impact, while the Nile tilapia has high reproductive viability. Having seen the advantages of polyculture and integrated fish farming. There is a growing trend of combining both systems for enhanced productivity. This approach is also relevant in fish farming in the face of climate change. Limbu et al. (2017) investigated the yield and economic benefits of such combination in a polyculture of African sharptooth catfish, *Clarias gariepinus*, and Nile tilapia, *Oreochromis niloticus*, reared in earthen ponds for 270 days integrated with Chinese cabbage, *Brassica rapa chinensis*, farmed for 45 days to small-scale farmers in Tanzania. The integrated aquaculture-agriculture (IAA) system involving fishes and vegetables had 3 and 2.5 times higher net yield than the culture of fishes alone and farming of non-integrated vegetables, respectively. This system has also been utilized in Bangladesh (Jahan et al. 2013). Periphyton has been combined with polyculture of carps and small indigenous species with good results (Jha et al. 2018). Integrating fisheries and agriculture aims at maximizing the synergistic and minimizing the antagonistic interactions between the two sectors. The synergy is mainly derived from the recycling of nutrients arising in the course of agriculture-livestock-fish production processes, from integrated pest management and from the optimal use of water resources. The antagonistic interactions from application of pesticides and herbicides harm aquatic living organisms. Nutrient run-off from agriculture and soil erosion causes eutrophication of inland and near-shore coastal waters. Meanwhile, the increased sediment load causes alterations in hydrological regimes, drainage of wetland and swamps, and obstruction of fish migration routes.

Fish Culture Facilities

Culture facilities are also important in building resilience to climate change and food security through aquaculture. There are many culture facilities with varied degree of vulnerability to climate impact. Important facilities in this regard are discussed.

Earthen Pond

Citing of earthen fish ponds at slopy/valley areas has been most common in rural poor settings of tropical Africa, but this system has unfortunately been the most vulnerable to climate-induced flooding with great impact on their fish production, food security and livelihood. A case study on Nigeria and Uganda (*pers. comm*) revealed that majority of respondent flood-prone farmers in these two selected tropical African countries was fish farming practitioners that utilize earthen pond system. Citing of earthen pond-based fish farms at far away location from such terrain is ideal for management of climate flood scenarios; however, these poor peasant local fish farmers who had their contribution to food security are often limited due to land distribution system and lack of fund to acquire such ideal but competitive farm land. Fish must be produced despite climate-induced flood and poor terrain. The flexibility of fish farming system had enabled this set of farmers to develop some innovations. The encountered technical innovations among these fish farmers in southwestern Nigeria were observed to be based on single/combined diverse techniques of ponds' fencing, pond embankment modifications, netting and caging, pond slope manipulations and river bank modification. These adaptive responses to climate impact among flood-prone fish farmers were similar to those encountered indigenous techniques reported by Bordoloi and Muzaddadi (2014); according to this author, the technical adaptation to fish farming in flooded areas of Dhemeji district of Assam, India, includes manipulation of pond dyke to be tall and wide for protection from flood; fencing of ponds with fine-meshed nets to prevent fish from escaping during flood; turfing and plantation above the dyke; and the use of clay soil from neighbouring areas for dyke construction to protect dykes from erosion. There is the need to compile more of these adaptive techniques and to spread awareness of it across flood-prone areas in tropical African countries.

Tank and Land Enclosures

The plasticity of fish farming in the face of hazards is also reflected in the evolution of culture of fish in tanks and different types of enclosures. These fish production techniques are more resilient to flood, are relatively easily manipulated and managed in times of hazards but are relatively more expensive. The tank systems are diverse and are becoming more popular in metropolitan communities of the tropical African countries where fish culture is popular. The tanks are usually placed above ground, and it conserves quantity and quality of water in use. This enables fish culture in situations of limited water emanating from climate change impact. It can also be placed indoor where climatic variables can be controlled. Different forms of this system are in operation; however, balanced diets are provided in this system since cultured fish are eschewed from access to natural nutrient sources which the earthen pond system utilizes.

Tanks are usually placed above ground and can be of plastic materials, reinforced block/concrete, etc. The tanks can utilize water from borehole, wells and streams. Honfoga et al. (2017) highlighted the challenges of modern fish farming in West Africa and found that the use of aboveground tanks (AGT) was not profitable enough to reward production and recover capital costs, should the investments be fully bank loan-funded. This is because feed prices would not allow a cost-effective meet of market demand; meanwhile lake water-fed pond (LWP) fish farming of *Clarias gariepinus* was tenfold more profitable (profit rate of 57.7%) than AGTs. The author thus recommended that, although its profit rate is still far below the potential performance level in the subregion, it should be promoted among lake village cooperatives or young rural entrepreneurs and among urban farmers to meet the growing fish demand; financial support should be made available to face the high costs of initial investments. There is a need to install quality feed production enterprises to promote both fish production systems towards sustainability, food security and economic development.

Fish Culture in Installed Cages

Fish culture in cages is generally known as 'cage culture' and 'pen culture' as both terms are often used interchangeably. These are generally referred as 'enclosure culture' which involves holding organisms captive within an enclosed space while maintaining a free exchange of water (Miller 1979). Water availability may be limited by climate impacts, but this system ensures rearing of fish in enclosure placed in a body of water thus ensuring maximal use of the available water. Fish culture in cages is advancing in tropical African countries in these days of climate change impact. Although the advancement of the culture system may not be traceable to direct adaptation to climate impact, its operation is quite beneficial to climate mitigation and food security. Cages made of variety of materials ranging from local to imported materials are usually placed on water bodies and stocked with indigenous or exotic fish species, fed and grown to maturity and subsequently harvested. This culture system uses simple technology; it can be used as a method for producing high-quality protein at cheap rate, to clean up eutrophicated waters through the culture and harvesting of caged planktivorous species and to improve conditions in acid lakes. Cage culture is growing in sea environments and inland lakes, and the growth in this sector is rapid.

The culture system is in operation in Asia (Das et al. 2009; Beveridge and Muir 1997), Indonesia (Syandri et al. 2015), West Africa (Oluwatobi et al. 2017; Olapoju and Edokpaye 2014; Ikotun and Omoloyin 1979) and East Africa (Aura et al. 2018;

Nwebaza-Ndawula et al. 2017; Kifuko 2015); cages in freshwaters are used for food fish culture and for fry to fingerling rearing (Gupta et al. 2012; Thordarson 2007). However, there are growing concerns on the environmental challenges posed by this method. Cage fish farming creates challenges in terms of mass escapes, introduction of genetically modified fish, spread of infectious diseases, parasite infestation, reliance upon toxic chemicals, contamination and bioaccumulation of organochlorine pesticides such as dioxins and PCBs (Oyebola and Fada 2017; Staniford 2002; Milewski 2001). These challenges call for responsible aquaculture, and this has been highlighted by Tacon et al. (2010). A 71-participant FAO technical committee on cage culture in Africa unanimously agreed that cage culture is an important development opportunity for many African countries but will require effective policy frameworks to ensure that structural constraints to development are overcome and that development is equitable and sustainable (FAO 2006). Policy needs are generally discussed under a separate section.

4 Ecological Considerations in Climate Change Adaptation Through Aquaculture

Aquaculture play important roles in food security and livelihoods, but its sustainability would depend on stable and healthy aquatic ecosystems. This reality is often ignored and undervalued, and lessons can be drawn from dynamic aquaculture producers in other regions to guide fish farmers in Africa (Msangi and Batka 2015). Aquaculture industry has developed and incorporated technological advances, moving from extensive to intensive systems. This development is accompanied by increase in the potential threat to the ecological equilibrium in streams, reservoirs and oceans.

There is the need to take cognizance of concerns over sustainability, environmental degradation and food safety in food production sources (Capper 2013) and especially in aquaculture (McAllister et al. 1997; The World Bank 2007). This is especially important in these days of global climate change (Anthony 2012). Nutrient-loaded wastes from inland aquaculture can cause algal blooms in riparian areas resulting in immense damages to the ecosystem and the caged fisheries, resulting in great economic losses, and caged fisheries may suffer from unprecedented pollution. Recycled aquaculture system (RAS) could be energy-consuming, support GHG emission and expensive, while integrated aquaculture could act as vector for disease pathogen. Caged fish culture is a potential threat to indigenous wild fisheries, but it is increasing in the tropical African countries. Although this system is greatly beneficial; it could constitute ecological threat if not well managed. Suffice to highlight the ecological threats of this valuable climate mitigation aquaculture strategy. The extensive arguments on the issues are presented in Staniford (2002). Improper management of cage fish culture has been associated with mass farmed fish escapes, genetically modified (GM) fish, the spread of infectious diseases, parasite infestation, reliance on toxic chemicals, contamination of the seabed and the bioaccumulation of organochlorine pesticides such as dioxins and PCBs (Milewski 2001; Staniford 2002).

Issues around sustainable aquaculture have been on waste management and eutrophication (Berry 1996; Davies 2000; Staniford 2002; OSPAR 2001; HELCOM 2001); management of fish escapes from aquaculture (McGinnity et al. 1997, 2002; Clifford et al. 1998; Fleming et al. 2000; McGinnity et al. 2002; Oyebola and Fada 2017); transmission of diseases and parasites through intensive culture (Meikle 2002); use of chemicals in fish production (Intrafish 1998; Barnett 2000; Cameron 2002a, b; Jensen 2001); uncontrolled use of aquafeeds; and feed contaminants (Edwards 2002; Vliet and Katan 1990; George and Bhopal 1995; Paone 2000). Most of the studies addressing these issues are from the developed world where intensive aquacultures have been mostly practiced. The increasing aquaculture intensification in tropical Africa would have to be backed up with adequate data gathering and researching on the dimensions of the listed threat to sustainable aquaculture in the region. Despite all these, outright rejection of animal source foods to save the planet may not be possible (Capper 2013), looking especially on the other benefits of fish as a food substance and as an important element for food security and climate mitigation. It is important to note that there is no climate mitigation approach that does not have its shortcoming, and trade-offs would have to be managed. Climate impact mitigation using aquaculture would require adherence to aquaculture best management practices backed up with policy directions and measures in controlling the listed challenges.

Globally, issues around mitigation and adaptation always inform need for trade-offs, holistic management and policy direction. Studies and policy instruments for ensuring this would help in sustenance of the sector, and this is advancing. Investigating dynamics of poverty in Senegal, Dang et al. (2014) observed that there are repercussions on any mitigation and adaptation strategies, and this can decide survival. In areas where climate change will trigger significant shifts in the hydrological regime, but where hydropower potentials are still available, there would be increase in competition for water, especially if climate change adaptation efforts in various sectors are implemented. There would be competition for surface water resources between the use of water for irrigation in order to cope with climate change impacts in agriculture, increased demand for drinking water, increased demand for cooling water for the power sector and the use of the water for cage fish culture. This confirms the importance of integrated land and water management strategies for river basins, to ensure the optimal allocation of scarce natural resources, that is, land and water. This scenario has resulted in various communal crises between different resource users bringing the theory of resource scarcity and conflict into play.

Studies confirm potential clashes between water supply, flood control, hydropower and minimum stream flow required for ecological and water quality purposes under changing climatic and hydrological conditions (Christensen et al. 2004; Van Rheenen et al. 2004). There is need to develop stream flow projections in climate change mitigation and adaptation (Miller et al. 2011). In water scarcity scenarios there will be need for increasing reuse of wastewater and the associated treatment, deep-well pumping and especially large scale desalination, which would increase energy use in the water sector (Boutkan and Stikker 2004). Meanwhile, changes in energy use pattern would generate GHG emissions, calling for other climate smart measures (Lomborg 2013). Climate impact mitigation and adaptation strategy such as use of aquaculture has to be evaluated with biasness to explicit trade-offs in order to optimize economic investments while fostering sustainable development. In this regard, there is the need to highlight the following concepts of which there is scarcity of information on their dimensions in the use of aquaculture in climate impact mitigation. The concepts include:

- *Reversibility*: A subsequent change in management can reverse the gains made in carbon sequestration over a similar period of time. However, many agricultural mitigation options such as reduction in N2O and CH4 emissions, avoided emissions as a result of agricultural energy efficiency gains or substitution of fossil fuels by bioenergy are reversible.
- *Reference*: The greenhouse gas net emission reductions need to be assessed relative to a reference baseline. The selection of an appropriate baseline to measure management-induced soil carbon changes is still an obstacle in some mitigation projects.
- *Uncertainty*: Uncertainty about the complex biological and ecological processes in agricultural systems makes investors more wary of land-based mitigation options compared to more clearly cut industrial mitigation activities. This barrier can be reduced by investment in research. In addition, high variability at the farm level can be reduced by increasing the geographical extent and duration of the project.
- *Unclear leakage*: Adopting certain agricultural mitigation practices may reduce production within implementing regions, leading to increased production and emissions outside the project region.
- *Transaction costs*: Under an incentive-based system such as a carbon market, the amount of money farmers receive is not the market price, but the market price less brokerage cost. This may be substantial and a serious entry barrier for smallholders. Pooling many activities together can serve to lower transaction costs of participating farmers.
- *Measurement and monitoring costs*: Measurement costs per carbon credit sold decrease as the quantity of carbon sequestered and area sampled increase. Methodological advances in measuring soil carbon may reduce costs and increase the sensitivity of change detection. Development of remote sensing may offer opportunities to reduce costs.
- *Property rights*: Property rights, landholdings and the lack of clear single-party land ownership in certain areas may inhibit implementation of management changes.
- *Other barriers*: Other barriers include availability of capital, rate of capital stock turnover, rate of technological development, risk attitudes, need for research and

outreach, consistency with traditional practices, pressure for competing uses of agricultural land and water, demand for agricultural products and high costs for certain enabling technologies.

5 Required Actions for Sustainable Climate Adaptation and Food Security Through Aquaculture

Food security crisis would multiply in tropical Africa under climate change scenarios. Precisely climate change may accelerate the food import dependence of most countries. Stakeholders involved in aquaculture in such countries may not be able to adapt to climate change impacts as a result of reduced adaptive capacity and higher environmental vulnerability. Water scarcity will make production unsustainable, leading to vicious migrations from poor rural communities that will no longer sustain them. However, these risks are not yet properly or adequately represented in Africa. Studies on climate change and aquaculture nexus need to be increased in Africa (Tubiello 2012). Climate change will affect both challenges and opportunities for aquaculture, and this would have to be extensively assessed in tropical African countries.

Sustainable utilization of aquaculture in climate impact adaptation would require arrangements towards climate impact mitigation through actions on reduction of pressure on water quality and quantity. Rise in seawater volume causes an interference of saltwater to freshwater, limiting the quantity of freshwater available for freshwater practices (Brown et al. 2009). Adapting climate change impact through aquaculture in the coastal areas could be threatened by the continuous increase in the temperature of the Earth causing a rise in sea levels and attendant fluctuating water chemistry. This would impair coastal aquaculture productivity. Climate change causes reduced water availability in main rivers and streams in developing countries of Asia and Africa (IPCC 2007) causing major effects in fish spawning, migration and seed availability for cage famers. Reduced water availability could lead to increased drought periods and less water retention time in non-perennial water bodies (Goswami et al. 2006; IPCC 2007). Appropriate mitigation and adaptation strategies in this dimension would assist sustainable aquaculture development in African countries.

The interaction between climate change mitigation measures and water is a give-and-take one. Climate impact adaptation/mitigation processes can impact on water resources and their management, and it is important to know and accept this when developing and evaluating mitigation options. Water resources management can also influence greenhouse gas emissions and their respective mitigation practices. However, the shift to mitigation in aquaculture is needed at all levels (individual, business, community, national and regional) and time scales. Climate impact mitigation system to support food security and livelihood with cognizance of the requirements for adaptation is needed in tropical Africa.

Climate-smart aquaculture would be needed. However, this strategy requires sustainable use of natural resources to produce fish and aquatic goods using climate resilient approaches; preserving the attributes of the aquatic body and the communities that depend on them to enable the system to continue to function sustainably and understanding ways to reduce the vulnerability of those most prone to these impacts are needful. These approaches could include ensuring increased efficiency for fish production through better integrated options; improved quality feed and feeding practices; and reduced feed loss to environment and reduced fish mortality through disease outbreak. There is the need to ensure decrease of post-harvest and production losses; and rationale for sustainable regional trade must be developed. All invested parties from both private and public sectors will require being in the development of context-specific options to ensure that the aquaculture practices is climate-smart. Markets and trade may benefit a cushioning effect from the impact of variations in production that affect food security, food prices and demand-supply gaps. However, the effects of climate change impacts and climate change policies on the entire fish supply and value chain need to be better understood, while suitable policy measures on climate-smart aquaculture need to be defined and implemented.

In the meantime, any climate mitigation/adaptation measures in fisheries and aquaculture is expected to improve management and ensure integrity of aquatic ecosystems, respond to the ecological threats and opportunities to food and livelihood security and help the sector to reduce greenhouse (Magawata and Ipinjolu 2014). Climate adaptation/mitigation through aquaculture would require adoption of procedures for aquaculture best management practices, avoiding or displacement of emission, reducing postharvest losses, use of fishing practices that adhere to principles of the code of conduct for responsible fisheries and replanting mangroves in coastal aquaculture areas. There is the need to cushion the effects of climate change by reducing pressure on both the social and economic developments, slowing down population growth of developing tropical African countries, adopting forest conservation measures including the prohibition of forestland for non-forest purposes and encouraging afforestation and reforestation. There is the need to include climatic studies in school curriculum in order to create awareness from grassroots to national levels. This action is needed across educational institutions that offer aquaculture as a course for study. Student's research in aquaculture study programmes should focus on energy reduction strategies applicable to aquaculture sector. Economic planning and budget should include tangible provisions for promoting policies and programmers of climate change as it relate to aquaculture. Direct reduction of anthropogenic emissions or enhancement of carbon sinks that are necessary for limiting long-term climate damage would also be necessary. Policy instrument is very important in any climate adaptation/mitigation approach, and there is the need to develop significant policy underlying adaptation and mitigation actions.

Climate change adaptation strategies should aim at maintaining or even increasing food production in key exporting developed and developing regions or in regions key to regional food security. Sustainability of relevance of aquaculture to food security and nutrition in the face of climate change would require domestication of a large number of aquaculture species, and reducing the dependency on wild caught seed, there is the need for development of sophisticated, efficient farming systems and husbandry practices besides the traditional integrated crop-fish-livestock production systems. Feed accounts for about 60% production cost in fish farming; hence, improved feed management through development of cost-effective and nutritionally complete feeds with reduced fishmeal use, along with efficient feed management systems, is needed for improved fish production and food security with resultant impact on livelihood. Production of improved breeds through application of genetics and selective breeding programmers for a number of species and application of biotechnology and better fish health management practices would be helpful. According to Costa-Pierce (2007) supported by Byron and Costa-Pierce (2013), aquaculture is on the road of sustainability as experts are already working on strategies to overcome the listed challenges. However there is the need to strengthen these actions through manpower development and translation of appropriate climate adaptation technologies. Appropriate policy dimensions in this regard would also be necessary.

6 Policy Needs and Regulatory Frameworks for Sustainable Climate Change Adaptation Through Aquaculture in Tropical Africa

Adaptation and mitigation under climate change would require dynamic policies to cope with high level of uncertainty in the timing and magnitude of potential climate changes and the rapidly evolving knowledge and techniques. This would have to be implemented with specific reference to the aquaculture sector in tropical Africa. It is critical that climate change adaptation policies on aquaculture would have to interact with sustainable development in order to protect environment and ensure wild fish health and genetic structure and ensure properly coordinated fisheries governance and political rights. Supporting policies and regulatory frameworks on appropriate interactions of these factors is necessary for the use of aquaculture in broad resilience to climate risk.

Cage aquaculture and integrated fish farming systems are promising fish culture systems in the face of climate change impact in Africa. These systems are advancing, but there is critical need for appropriate policy and regulatory framework for their sustenance. It is important for policy support in ensuring zero waste in aquaculture through general adoption of integrated aquaculture across African region. Some dimensions of frameworks for policy issues on cage culture in tropical Africa have been reported (Aura et al. 2018; Blow and Leonard 2007; FAO 2006; Salman and Daniel 2006). However, these reports seem not to capture the policy dimension in line with the information on the nuances of climate change, food security and the use of aquaculture for adaptation. There is an urgent need to evolve policies that will support adoption of climate-smart aquaculture as a means of ensuring food security across nations of Africa.

Information and awareness creation on climate change impact adaptation is required to guide livelihood and food security in tropical Africa. The suitability of aquaculture in climate impact adaptation with the attendant potential ecological threats indicates the need for monitoring, surveillance and control. There is the need to evolve climate-related policies as efficient and sustainable ways to mitigate and adapt to the overall impacts of climate change in general and, especially, in the use of aquaculture in this regard. Policies supporting establishment of functional national and regional regulatory bodies for co-management of the aquaculture and environmental systems will be required for proper monitoring and control of ecological risks of aquaculture especially cage culture in the face of climate change in tropical Africa. There is the need to ensure coordinated national, regional and international actions needed to ensure that issues of food security are part of climate impact adaptation and to ensure that discussions on the topic address issues of importance to rural poor people and provide effective integration with sensible adaptation and or mitigation policies.

There is the need to guide actions of investors in aquaculture to comply with existing policies and regulations on environmental safety and health for sustenance of the indigenous fisheries which are the store room for sustainable and viable aquaculture development. This could be in form of strong institutions coordinating and regulating their compliance. Such institution would be charged with responsibilities to guide investors in aquaculture to embark on long-term investments in fish breeding programmes for climate risk adaptive fish species of the cultured and the underutilized culturable species.

Effective framework would include capacity building in climate-smart aquaculture and quarantine systems. The watersheds adjoining water bodies would have to be well monitored to ensure compliance with the water catchment protection rules, while establishing forest plantations for watershed management would strengthen the action. Policies and management frameworks should support flood adaptation, water storage and recycling systems. Policies for maximizing societal welfare under future climate risk will also be needed.

Climate-smart approaches would afford opportunity for enhanced food security in tropical Africa, and this is expected to be strengthened by appropriate policies and management frameworks in African countries. Such endeavours would have to be well structured to address, especially, the rural poor agrarian communities, being the major contributor to agriculture production in these countries, and they are highly vulnerable to risks due to the fragility of their socio-economic status.

Knowledge on climate impact and food insecurity vulnerability indices of specific rural fish farming communities would provide opportunity to establish effective policies and management frameworks, but there is dearth of such data in most of the tropical African countries. There is the need for establishment of adequate institutional framework for data gathering, research activities and establishment of technical/operational requirements of farmers across Africa. African fish farmers are usually more of smallholders and were subsistence. Subsistent farmers would suffer locally specific impacts of climate change that will be hard to predict. The complexity of impacts and adaptation solutions would be heightened by strains of cultured fish produced and the importance of non-market relations, while small farm sizes, technological deficiency, low capital and diverse non-climate stressors would increase vulnerability. These issues would have to be captured in formulating effective policies and management frameworks.

Although the rural poor farmers could be vulnerable, they could also be resilient in some instances, such as family labour, diversity of fish species, patterns of diversification away from agriculture and possession of indigenous knowledge that may be useful in designing less vulnerable production systems. Policies that favour these would be useful. In rural poor communities, farms are generally often held under traditional or informal tenure and are in marginal or risk-prone environments. Soil-related constraints to productivity are widespread, severe and increasing. These issues need to be put into consideration for effective policy and management framework.

Production systems, exploited species, types of integration, production objectives and management institutional arrangement in rural communities are complex and diverse, and risks may be differentially felt by individual households or entire communities. There is the need to generate data for better understanding of the impacts of climate change on smallholder and subsistence aquaculture and related livelihoods in tropical Africa. Frameworks will require harnessing the growing understanding of the biological processes involved, as climate change impacts on fish production and food security. This will have to be done in relation to the specific features of the livelihoods of the community.

An ideal framework would recognize the complexity and high location specificity of the fish production systems. It should incorporate knowledge on non-climate stressors in rural livelihoods and their contribution to vulnerability and recognize that climate change impact on livelihoods of rural poor communities is manifold and scale dependent with focus on biological processes affecting fish production at the levels of individual organisms or farms; environmental and physical processes affecting production, watershed or community levels; and impacts of climate change on human health and non-fishery livelihoods.

Challenges to food production from aquatic environment could be multispatial (Oyebola and Fashae 2017). It is important to consider both spatial and temporal climate stressors and reliability of the information with the scale and nature of the decision-making on management.

Development of management framework would need to utilize bottom-up assessments of the potential technical effectiveness of farm-level adaptations in the face of long-term climate impact projections.

6.1 Required Critical Policy Instruments for Ensuring Aquaculture Thrives in the Era of Climate Change

African aquaculture requires several enabling factors that include a positive perception of aquaculture, sound policies at the national level, strong public institutions, availability of nutrient inputs, conducive investment policies to attract increased private-sector participation and access to credit for commercial-scale enterprises (Machena and Moehl 2001). Sustainable aquaculture would require proper policy guidance on aquaculture best management practices (BMP) in production, utilization, marketing and distribution with focus on sustenance of production and socio-economic benefits with minimal GHG and other environmental nuisance. Such policies would have to be holistic with actions cutting across individuals, national, subregional and regional spheres in Africa. The following could be critical in this regard.

(a) There is the need for policy guidelines facilitating increased advocacy for the use of aquaculture as means of adapting climate change impact and food insecurity in Africa.
(b) There is the critical need for policy statements necessitating climate-smart aquaculture across African countries.
(c) The existing environmental laws should be supported by appropriate sections that could address trending environmental challenges of aquaculture especially cage aquaculture.
(d) Trans-country laws establishing strong institutions for monitoring, control and surveillance for compliance to environmental rules for fish culture are critically required.
(e) Regulations on low energy use options in commercial aquaculture would further strengthen the sectors sustainable use in climate impact adaptation and mitigation.
(f) There is need for law ensuring low tariffs on imported fish farm inputs especially fish feed since feed normally accounts for over 60% of production cost in aquaculture.
(g) Establishment of laws facilitating development of the alternative indigenous input supply network is critical to sustainable development of aquaculture in climate adaptation.
(h) There is the need to legislate on means of increasing private sector participation in aquaculture across Africa. This could be in the form of establishment of government institutions who would be creating link between end user, practitioners and researcher in aquaculture for evolving demand-driven aquaculture products in commercial scale.
(i) Most of aquaculture practitioners have limited access to credit facility, and most farms are usually not insured. There is critical need for policies supporting low cost credit and insurance for fish farming sector.
(j) All African countries should legislate on adoption of aquaculture best management practices (BMP) in production, utilization, marketing and distribution of aquaculture products with focus on sustenance of production and socio-economic benefits with minimal GHG and other environmental nuisance.
(k) There is immediate need for comprehensive policies in support of provisions for adequate requisite human and financial capacity within the implementing agencies for fisheries, aquaculture and environment in African countries.

7 Conclusions and Recommendations

Climate change greatly threatens tropical African countries with extended negative impact on current and future food security. Aquaculture has viable elements for climate impact adaptation and food security in Africa, but its sustainable use in this regard require adequate monitoring, control and surveillance for adherence to environmental/ecological considerations. The critical specifically policy and management needs to cater for this in Africa were highlighted in this review. Sustenance of the use of aquaculture for climate impact adaptation and food security in Africa is important, but aquaculture and environment-related policy implementation institutions would need to be strengthened with adequate policy and management instruments. Such instruments will have to aim at supporting the utilizing of aquaculture for enhanced production with less environmental hazards in the face of threatening climate impact and food insecurity in Africa. However, development of comprehensive policy and management frameworks for African aquaculture would need to be strengthened with human and financial capacity within implementing agencies at local, national and regional levels in tropical Africa.

References

Adedeji OB, Okocha RC (2011) Constraint to aquaculture development in Nigeria and way forward. Veterinary Public Health and Preventive Medicine, University of Ibadan, Nigeria

Adeleke ML, Omoboyeje VO (2016) Effects of climate change on aquaculture production and management in Akure Metropolis, Ondo State, Nigeria. Acad J Agric Res 4(6):319–325

Adeyeye SAO, Adebayo-Oyetoro AO, Tiamiyu HK (2017) Poverty and malnutrition in Africa: a conceptual analysis. Nutr Food Sci 47(6):754–764. https://doi.org/10.1108/NFS-02-2017-0027

Agardy T, Alder J (2005) Coastal systems. In: Hassan R, Scholes R, Ash N (eds) Ecosystems and human well-being: current state and trends, Volume 1. Findings of the condition and trends Working Group of the Millennium Ecosystem Assessment. Millennium Ecosystem Assessment Series 1, Island Press: Washington. ISBN 1-55963-227-5. pp 513–549

Agbarevo MNB (2013) Climate change in Nigeria: an evaluation of farmers' adoption of climate change adaptation technologies –a case study of Obudu Agricultural zone of cross-river state, Nigeria. Int J Sci Technol Res 2(12):270–273

Ahmed N, Diana JS (2015) Threatening "white gold": impacts of climate change on shrimp farming in coastal Bangladesh. Ocean Coast Manag 114:42–52

Aldhous P (2004) Fish farms still ravage the sea. Sustainable aquaculture takes one step forward, two steps back. Nature. online, 17: February 2004. https://doi.org/10.1038/news040216–10

Allison EH, Adger WN, Badjeck MC et al (2005) Effects of climate change on the sustainability of capture and enhancement fisheries important to the poor: analysis of the vulnerability and adaptability of fisherfolk living in poverty. Fisheries Management Science Programme Department for International Development, DFID, 167p

Allsopp M, Johnston P, Santillo D (2008) Challenging the aquaculture industry on sustainability; technical overview. Greenpeace Research Laboratories Technical Note 01/2008, 59p

Andy G (2014) List of tropical countries–tropical. A travel community, https://www.hobotraveler.com/. https://www.scribd.com/document/202335264/List-of-Tropical-Countries. Accessed 7 April 2018

Anthony R (2012) Taming the unruly side of ethics: overcoming challenges of a bottom-up approach to ethics in the areas of climate change and food security. J Agric Environ Ethics 25(6):813–841

Atinmo T, Mirmiran P, Oyewole OE et al (2009) Breaking the poverty/malnutrition cycle in Africa and the Middle East. Nutrition Reviews 67(Suppl 1):S40–S46. https://doi.org/10.1111/j.1753-4887.2009.00158.x

Aura CM, Musa S, Yongo E et al (2018) Integration of mapping and socio-economic status of cage culture: Towards balancing lake-use and culture fisheries in Lake Victoria, Kenya. Aquac Res 49(1):532–545

Barnett A (2000) Illegal poison' used on salmon – chemical treatment at fish farms is hazard to health and marine life, claims ex-employee. The Observer, 30th April. http://www.guardianunlimited.co.uk/Archive/Article/0,4273,4013345,00.html

Bartley DM, Brugère C, Soto D et al (2007) Comparative assessment of the environment costs of aquaculture and other food production sectors: methods of meaningful comparisons. Rome. FAO Fisheries Proceedings, No 10, 240 pp.

Basarab J, Baron V, López-Campos O et al (2012) Greenhouse gas emissions from calf- and yearling-fed beef production systems, with and without the use of growth Promotants. Animals (Basel) 2(2):195–220

BBC News (2007) Rains threaten flood-hit Africa. http://news.bbc.co.uk/2/hi/africa/6994995.stm

Beg N, Morlot JC, Davidson O et al (2011) Linkages between climate change and sustainable development. Clim Policy 2(2–3):129–144

Belton B, Thilsted SH (2014) Fisheries in transition: food and nutrition security implications for the global South. Glob Food Sec 3(1):59–66

Berry AW (1996) Aquaculture and sea loch nutrient ratios: a hypothesis. In: Black KD (ed) Aquaculture and sea lochs. Scottish Association of Marine Science, Oban. http://www.mar.dfompo.gc.ca/science/mesd/he/lists/phycotoxins-1/msg00360.html

Beveridge MCM, Muir JF (1997) Environmental impacts and sustainability of cage culture in Southeast Asian lakes and reservoirs. In: van Densen W, Saidin T, Verdegem M (eds) Ecological aspects of fish production in SE-AsianLakes and Reservoirs. University of Wageningen, The Netherlands

Bewket W (2012) Climate change perceptions and adaptive responses of smallholder farmers in Central Highlands of Ethiopia. Int J Environ Stud 69(3):507–523

Blow P, Leonard S (2007) A review of cage aquaculture: sub-Saharan Africa. In: Halwart M, Soto D, Arthur JR (eds) Cage aquaculture – regional reviews and global overview. pp 188–207. FAO Fisheries Technical Paper. No. 498. Rome, FAO. 2007, 241pp

Boon PJ, Raven PJ (2012) River conservation and management. Wiley-Blackwell, London

Bordoloi R, Muzaddadi AU (2014) Indigenous technical knowledge associated with disaster management and fisheries related activities in the highest flood affected district (Dhemaji) of Assam, India. Indian J Tradit Knowl 14(3):407–415

Boutkan E, Stikker A (2004) Enhanced water resource base for sustainable integrated water resource management. Nat Res Forum 28(2):150–154. https://doi.org/10.1111/j.1477-8947.2004.00082.x

Brander K (2010) Impacts of climate change on fisheries. J Mar Syst 79:389–402

Brook EJ, Sowers T, Orchardo J (1996) Rapid variation in atmospheric methane concentration during past 110 000 years. Science 273:1087–1990

Brown O, Hammill A, McLeman R (2007) Climate change as the "new" security threat: implications for Africa. Int Aff 83(6):1141–1154

Byron CJ, Costa-Pierce B (2013) Carrying capacity tools for use in the implementation of an ecosystems approach to aquaculture. In: Ross LG, Telfer TC, Falconer L, Soto D, Aguilar-Manjarrez J (eds) Site selection and carrying capacity for inland and coastal aquaculture, pp. 87101. FAO/Institute of Aquaculture, University of Stirling, Expert Workshop, 6–8 December 2010. Stirling, the United Kingdom of Great Britain and Northern Ireland. FAO Fisheries and Aquaculture Proceedings No. 21. Rome, FAO. 282

Brown R, Keath N, Wong T (2009) Urban water management in cities: historical, current and future regimes. Water Sci Technol 59(5):847–855

Cameron F (2002a) Shetland firm which used illegal chemical bows out of quality scheme. Intrafish, 30th January. http://www.intrafish.com/article.php?articleID=20246

Cameron F (2002b) UK consumers call for more medicine residue tests. Intrafish, 3rd April http://www.intrafish.com/article.php?articleID=21928

Cao L (2012) Farming shrimp for the future: a sustainability analysis of shrimp farming in China. PhD dissertation, Natural Resource and Environment, University of Michigan

Capon SJ, Chambers LE, MacNally R et al (2013) Riparian ecosystems in the 21st century: hotspots for global change adaptation. Ecosystems 16:359–381. https://doi.org/10.1007/s10021-013-9656-1

Capper JL (2013) Should we reject animal source foods to save the planet? A review of the sustainability of global livestock production. S Afr J Anim Sci 43(3):233–246

Christensen NS, Wood AW, Voisin N et al (2004) The effects of climate change on the hydrology and water resources of the Colorado River basin. Clim Chang 62:337–363. https://doi.org/10.1023/B:CLIM.0000013684.13621.1f

CIGI (Centre for International Governance Innovation) (2009) Climate change in Africa: adaptation, mitigation and governance challenges. Edited by Hany Besada and Nelson Sewankambo. Waterloo, Canada

Cinner J, McClanahan T, Wamukota A et al (2013) Social ecological vulnerability of coral reef fisheries to climatic shocks. FAO Fisheries and Aquaculture Circular No. 1082. Rome, FAO. 63:4 Also available at www.fao.org/docrep/018/ap972e/ap972e00.htm

Clifford S et al (1998) Genetic changes in Atlantic salmon (*Salmo salar*) populations of Northwest Irish river resulting from escapes of adult farmed salmon. Can J Fish Aquat Sci 55(2):358–363. http://www.watershedwatch.org/ww/publications/sf/Genetic_Change_Irish_rivers.pdf

Cochrane K, De Young C, Soto D et al (eds) (2009) Climate change implications for fisheries and aquaculture: overview of current scientific knowledge. FAO Fisheries and Aquaculture Technical Paper No. 530. Rome, FAO. 212 pp. Available at www.fao.org/docrep/012/i0994e/i0994e.pdf

Connolly-Boutin L, Smit B (2016) Climate change, food security, and livelihoods in sub-Saharan Africa. Reg Environ Chang 16:385–399. https://doi.org/10.1007/s10113-015-0761-x

Cooke SJ, Allison EH, Jr DB et al (2016) 2016 On the sustainability of inland fisheries: finding a future for the forgotten. Ambio 45(7):753–764

Coulibaly A, Ouattara IN, Koné T et al (2007) First results of floating cage culture of the African catfish *Heterobranchus longifilis* Valenciennes, 1840: effect of stocking density on survival and growth rates. Aquaculture 263:61–67

Costa-Pierce B (2007) Ecology as the paradigm for the future of aquaculture. Ecological aquaculture: The evolution of the blue revolution, pp. 337–372. https://doi.org/10.1002/9780470995051.ch12

de Graaf G, Garibaldi L (2014) The value of African fisheries. FAO Fisheries and Aquaculture Circular. No. 1093. Rome, FAO, 76 pp.

Dang HH, Lanjouw PF, Swinkels R (2014) Who remained in poverty, who moved up, and who fell down? An investigation of poverty dynamics in Senegal in the late 2000s. Policy Research Working Paper 7141. JEL: C15, I32, O15

Das AK, Vass KK, Shrivastava NP et al (2009) Cage culture in reservoirs in India. (A Handbook) WorldFish Center Technical Manual No. 1948. The WorldFish Center, Penang, Malaysia, 24 pp.

Davies IM (2000) Waste production by farmed Atlantic salmon (*Salmo salar*) in Scotland. ICESCM O:01, 1–6

Daw T, Adger WN, Badjeck M (2008) Climate change and capture fisheries-impacts, adaptation, mitigation, and the way forward. Background paper to the FAO Workshop on Climate Change Implications for Fisheries and Aquaculture. Rome: Overseas Development Group, UEA, Norwich

De Silva SS, Soto D (2009) Climate change and aquaculture: potential impacts, adaptation and mitigation. In: Cochrane K, De Young C, Soto D (eds) Climate change implications for fisheries and aquaculture: overview of current scientific knowledge. FAO Fisheries and Aquaculture Technical Paper No. 530, pp 151–212

Diersing N (2009) Phytoplankton blooms: the basics. Florida Keys National Marine Sanctuary, Key West, Florida, USA, 2. Available at: http://floridakeys.noaa.gov/pdfs/wqpb.pdf

Ding L, Cao W, Shi Z et al (2016) Carbon dioxide and methane emissions from the scale model of open dairy lots. J Air Waste Manage Assoc 66(7):715–725

Easterling WE, Aggarwal PK, Batima P et al (2007) Food, fibre and forest products. Climate change 2007: impacts, adaptation and vulnerability. In: Parry ML, Canziani OF, Palutikof JP, van der Linden PJ, Hanson CE (eds) Contribution of working group II to the fourth assessment report of the intergovernmental panel on climate change. Cambridge University Press, Cambridge, UK, pp 273–313

Edwards R (2002) Study proves cancer-link chemicals in farm salmon: call for consumer boycott as groundbreaking study finds evidence of major pollutants in the food chain. The Sunday Herald, 7th July http://www.sundayherald.com/26081

EIPRO (2006) Environmental Impact of Products (EIPRO), European Commission Joint Research Centre. Available at http://ec.europa.eu/environment/ipp/pdf/eipro_report.pdf

FAO (2003) World agriculture: towards 2015/2030: a FAO perspective. Earthscan. Rome, FAO, 432 pp

FAO (2006) FAO regional technical expert workshop on cage culture in Africa, Entebbe Uganda, 20–23 October, 2004. FAO fisheries proceedings No. 6, Rome, FAO 2006, 113p

FAO (2007a) Role of aquaculture to food security. http://www.fao.org. Date accessed 28th May 2010

FAO (2007b) The state of food and agriculture 2007; Paying farmers for environmental services. Food and Agriculture Organization of the United Nations, Rome

FAO (2008a) The state of the world fisheries and aquaculture-Sofia 2008. Fisheries and Aquaculture Dept. FAO, Rome. http://www.fao.org/docrep/fao. Accessed January 2018

FAO (2008b) Climate change implications for fisheries and aquaculture. In: The State of Fisheries and Aquaculture 2008. FAO, Rome, Italy, 2008, 87–91. 13

FAO (2014) The state of the world fisheries and aquaculture, opportunities and challenges. Food and Agriculture Organization of the United Nations, Rome, p 2014

FAO (2016) Case studies on climate change and African coastal fisheries: a vulnerability analysis and recommendations for adaptation options, edited by Jim Anderson and Timothy Andrew. FAO Fisheries and Aquaculture Circular No. 1113, Rome, Italy

FAO/NACA (2012) Farming the waters for people and food. In: Subasinghe RP, Arthur JR, Bartley DM, De Silva SS, Halwart M, Hishamunda N, Mohan CV, Sorgeloos P (eds) Farming the waters for people and food, pp 137–166. Proceedings of the Global Conference on Aquaculture 2010. Phuket, Thailand. 22–25 September 2010. Rome, FAO, and Bangkok, NACA, 896 p

Faturoti EO (1999) Fisheries potentials and investment opportunities in Nigeria. Paper presented at the Fisheries Society of Nigeria (FISON) Investors Workshop held at the Nigerian Institute for Oceanography and Marine Research (NIOMR) Lagos, April 22, 1999

Ficke AD, Christopher AM, Lara JH (2007) Potential impacts of global climate change on freshwater fisheries. Rev Fish Biol Fish 17(4):581–613

Flannery T (2005) The weather makers. Text Publishing, Melbourne

Fleming IA et al (2000) Lifetime success and interactions of farm salmon invading a native population. Proc R Soc Lond B 267(1452):1517–1523

Freeman OE (2017) Impact of climate change on aquaculture and fisheries in Nigeria; a review. Int J Multidiscip Res Develop 4(1):53–59

Friedlingstein P, Solomon S (2005) Contributions of past and present human generations to committed warming caused by carbon dioxide. PNAS 102(31):10832–10836. Also available at www.pnas.org/cgi/doi/10/1073/pmas0504755102

Friggens MM, Woodlief CK (2015) Final report: synthesis of aquatic climate change vulnerability assessments for the interior west. March 31st, 2015, 30pp

Gabriel UU, Akinrotimi OA, Belibele DO et al (2007) Locally produced fish food: potentials for aquaculture development in Sub-Saharan Africa. Afr J Agric Res 2(7):287–295

Gemeda DO, Sima AD (2015) The impacts of climate change on African continent and the way forward. J Ecol Nat Environ 7(10):256–262. https://doi.org/10.5897/JENE2015.0533

George R, Bhopal R (1995) Fat composition of free living and farmed sea species: implications for human diet and sea-farming techniques. Br Food J 97:19–22

Gladys KV (2017) Rainfall and temperature variability and its effect on food security in Kitui county, Kenya. Int J Dev Sustain 6(8):924–939

Glen G (2010) The impact of global change on European lakes. Springer, New York

Goswami BN, Venugopal V, Sengupta D, Madhusoodanan MS, Xavier PK (2006) Increasing trend of extreme rain events in a warming environment. Science 314:1442–1445

GPA (2008) Physical alterations and destruction of habitats. Restoration of mangrove areas. Global Programme for Action for the Protection of the Marine Environment from Land-based Activities

Gupta N, Haque MM, Khan MJ (2012) Growth performance of tilapia fingerling in cage in ponds managed by Adivasi households: an assessment through length-weight relationship. Bangladesh Agril Univ 10(1):149–155

Gurung TB (2016) Role of inland fishery and aquaculture for food and nutrition security in Nepal. Agri Food Secur 5:18. https://doi.org/10.1186/s40066-016-0063-7

Hall SJ, Delaporte A, Phillips MJ et al (2011) Blue frontiers: managing the environmental costs of aquaculture. The WorldFish Center, Penang, Malaysia

Halls AS (2009) Addressing fisheries in the climate change and adaptation initiative. Fish Res Dev Melkong Region 15(1):12–16

Handisyde NT, Ross LG, Badjeck MC et al (2006) The effects of climate change on world aquaculture: a global perspective. Final Technical Report, DFID Aquaculture and Fish Genetics Research Programme, Stirling Institute of Aquaculture, Stirling, UK, 151 pp. Available at www.aqua.stir.ac.uk/GISAP/pdfs/Climate_full.pdf

Harrould-Kolieb ER, Dorothée Herr D (2011) Ocean acidification and climate change: synergies and challenges of addressing both under the UNFCCC. Clim Pol 12(3):378–389. https://doi.org/10.1080/14693062.2012.620788

HELCOM (2001) Measures aimed at the reduction of discharges from marine fish farms. Helsinki Commission, Finland

HLPE (2014) Sustainable fisheries and aquaculture for food security and nutrition. A report by the High Level Panel of Experts on Food Security and Nutrition of the Committee on World Food Security, Rome

Homer-Dixon TF (1999) In: Dessler D (ed) The environment, scarcity, and violence. Princeton University Press, Princeton, 253 pp.

Honfoga BG, Tognon IA, Chikou A (2017) Profitability and sustainability of modern fish farming in Benin: an on-farm experimental appraisal of two production systems of *Clarias gariepinus*. J Dev Agric Econ 9(9):243–249

Hulme M, Doherty RM, Ngara T (2001) African climate change: 1900–2100. Clim Res 17(2):145–168

Idowu AA, Ayoola SO, Opele AI (2011) Impact of climate change in Nigeria. Iran J Energy Environ 2(2):145–152

Ikotun SJ, Omoloyin O (1979) Aquaculture and prospects of cage and pen culture in Nigeria. In Proceedings of the International Workshop on Pen Cage Culture of Fish, 11–12 February 1979, Tigbauan, Iloilo, Philippines (pp. 97–98). Tigbauan, Iloilo, Philippines: Aquaculture Department, Southeast Asian Fisheries Development Center; International Development Research Centre

Intrafish (1998) Using banned chemical – one of Shetland's biggest salmon firms was fined a thousand pounds yesterday for using the banned fish farm chemical Cypermethrin. Intrafish, 30th April http://www.intrafish.com/article.php?articleID=809

IPCC (2007) Summary for policymakers. In: Parry ML, Canziani OF, Palutikof JP, van der Linden PJ, Hanson CE (eds) Climate change 2007: impacts, adaptation and vulnerability: contribution

of working group II to the fourth assessment report of the intergovernmental panel on climate change. Cambridge University Press, Cambridge, UK. ISBN-13: 9780521880107, pp 7–22
IPCC (2014) Climate change 2014: synthesis report. In: Core Writing Team, Pachauri RK, Meyer LA (eds) Contribution of working groups I, II and III to the fifth assessment report of the intergovernmental panel on climate change. IPCC, Geneva, Switzerland, 151 pp
Ipinjolu JK, Magawata I, Shinkafi BA (2014) Potential impact of climate change on fisheries and aquaculture in Nigeria. J Fish Aquat Sci 9:338–344
Islam MM, Sallu S, Hubacek K et al (2014) Vulnerability of fishery-based livelihoods to the impacts of climate variability and change: insights from coastal Bangladesh. Reg Environ Change 14:281–294. https://doi.org/10.1007/s10113-013-0487-6
Jahan KM, Crissman C, Antle J (2013) Economic and social impacts of integrated aquaculture-agriculture technologies in Bangladesh. CGIAR Research Program on Aquatic Agricultural Systems. Penang, Malaysia. Working Paper: AAS-2013-02
Jamart C, Rodeghier M (2009) Sharing power: the sustainable management of natural resources is a political choice, What policy instruments create environments favorable to co-management? Online resources base Natural Resources Governance around the world. http://www.agter.org/bdf/en/corpus_chemin/fiche-chemin-29.html. Retrieved 8 September 2018
Jensen BA (2001) Illegal medicine used in Norwegian marine farming. Intrafish, 3rd August. http://www.intrafish.com/articlea.php?articleID=14738.http://www.biomar.no/niv2/produkter/sertiikat/Malachite_eng.html
Jha S, Rai S, Shrestha M, Diana JS et al (2018) Production of periphyton to enhance yield in polyculture ponds with carps and small indigenous species. Aquaculture Reports 9:74–89
Joiner E, Kennedo D, Sampson J (2012) Climate Change and African Political Stability (CCAPS) vulnerability to climate change in West Africa: adaptive capacity in the regional context. Student working paper No. 4; 31pp
Kifuko R (2015) The state of cage fish farming in Uganda: actors, enabling environment, challenges and way forward. Int J Educ Res 3(3):483–488
Lean G (2006) Cow 'emissions' more damaging to planet than CO2 from cars. INDEPENDENT, Environment Editor Sunday 10 December 2006 00:00 GMT. https://www.independent.co.uk/environment/climate-change/cow-emissions-more-damaging-to-planet-than-co2-from-cars-427843.html
Ledger ME, Brown LE, Edwards FK et al (2013) Drought alters the structure and functioning of complex food webs. Nat Clim Chang 3:223–227. https://doi.org/10.1038/nclimate1684
Lehane S (2013) Fish for the future: aquaculture and food security. FDI Research Analyst strategic analysis paper, 26 August, 2013. http://futuredirections.org.au/wp-content/uploads/2013/08/Fish_for_the_Future_Aquaculture_and_Food_Security_2708.pdf. Accessed 7 April 2018
Limbu SM, Shoko AP, Lamtane HA et al (2017) Fish polyculture system integrated with vegetable farming improves yield and economic benefits of small-scale farmers. Aquac Res 48(7):3631–3644
Lomborg B (2013) Smart solutions to climate change: comparing costs and benefits. Cambridge University Press, Amazon.com. ebook available at https://books.google.com.ng/books?id=CZET42QPOQEC&pg=PA232&lpg=PA232&dq=energy+use+in+the+water+sector+(Boutkan+and+Stikker,+2004)&source=bl&ots=PtRMyt68KX&sig=1BjHYis9Wcz5CiB0io7gHrJvyS8&hl=en&sa=X&ved=0ahUKEwjO1i4_8XaAhWqDcAKHVUoCZIQ6AEINjAE#v=oepage&q=energy%20use%20in%20the%20water%20sector%20(Boutkan%20and%20Stikker%2C%202004)&f=false
MAB (Multi-Agency Brief) (2009) Fisheries and aquaculture in a changing climate. Available at: ftp://ftp.fao.org/FI/brochure/climate_change/policy_brief.pdf
Machena C, Moehl J (2001) Sub-Saharan African aquaculture: regional summary. In: Subasinghe RP, Bueno P, Phillips MJ, Hough C, McGladdery SE, Arthur JR (eds) Aquaculture in the third millennium. Technical proceedings of the conference on aquaculture in the third millennium, Bangkok, Thailand, 20–25 February 2000, pp. 341–355. NACA, Bangkok and FAO, Rome
Magawata I, Ipinjolu JK (2014) Climate change: mitigation and adaptation strategies in fisheries and aquaculture in Nigeria. J Fish Aquat Sci 9:257–261. https://doi.org/10.3923/jfas.2014.257.261

Marcogliese DJ (2001) Implications of global change for parasitism of animals in the aquatic environment. Can J Zool 79:1331–1352. https://doi.org/10.1139/z01-067
Markovic D, Carrizo SF, Kärcher O et al (2017) Vulnerability of European freshwater catchments to climate change. Glob Chang Biol 23(9):3567–3580
Marshall et al (2010) In Cinner J, McClanahan T, Wamukota A, et al (2013) Social ecological vulnerability of coral reef fisheries to climatic shocks. FAO Fisheries and Aquaculture Circular No. 1082. Rome, FAO, 63 pp. Also available at www.fao.org/docrep/018/ap972e/ap972e00.htm
Martin R (2015) Sustainable energy climate change: why the tropical poor will suffer most. MIT Technology Review June 17, 2015
Mbugua MH (2008) Aquaculture in Kenya: opportunities and challenges. http://www.fisheries.go.ke/index.php?option=com_docman&task=cat_view&gid=27&Itemid=94. Date accessed 4th October2012
McAllister DE, Hamilton AL, Harvey B (1997) Global freshwater biodiversity: striving for the integrity of freshwater ecosystems. Seawind 11:1–139
McGinnity P et al (1997) Genetic impact of escaped farmed Atlantic salmon (*Salmo salar*) on native populations: use of DNA profiling to assess freshwater performance of wild, farmed and hybrid progeny in a natural river environment. ICES J Mar Sci 54:998–1008
McGinnity P et al (2002) A comparison of the fitness of farmed/wild "hybrid" salmon with native fish. Paper presented at the 6th International Atlantic Salmon Symposium in Edinburgh, 16–18th July
Medugu NI (2009) Climate change: a threat to Nigeria's development. Daily Trust Newspaper, Nigeria, July 14, 2009
Meikle J (2002) Search for BSE type disease turns to fish farms. The Guardian, 15th March. http://www.guardian.co.uk/Archive/Article/0,4273,4374667,00.html
Milewski I (2001) Impacts of salmon aquaculture on the coastal environment: a review. Paper presented at a conference in New Brunswick, Canada. http://eastern.penbay.org/downloads/mmilewski.pdf
Miller D (1979) Recent developments in cage and enclosure aquaculture in Norway. In: Pillay TVR, Dill WA (eds) Advances in aquaculture. Fishing News Books Ltd, United Kingdom. Halls, pp 447–452
Miller WP, Piechota TC, Gangopadhyay S, Pruitt T (2011) Development of streamflow projections under changing climate conditions over Colorado River basin headwaters. Hydrol Earth Syst Sci 15:2145–2164. www.hydrol-earth-syst-sci.net/15/2145/2011/doi:10.5194/hess-15-2145-2011
Mitike G, Motbainor A, Kumie A et al (2016) Review of policy, regulatory, and organizational frameworks of environment and health in Ethiopia. Ethiop J Health Dev 30(1 Spec Iss):42–49
Msangi S, Batka M (2015) The rise of aquaculture: the role of fish in global food security. In 2014–2015 Global food policy report. Chapter 8, pp 61–71. Washington, DC: International Food Policy Research Institute (IFPRI). http://ebrary.ifpri.org/cdm/ref/collection/p15738coll2/id/129078
Na-Nakorn U, Brummett RE (2009) Use and exchange of aquatic genetic resources for food and agriculture: Clarias. FAO/NACA Expert Meeting on the Use and Exchange of Aquatic Genetic Resources Relevant for Food and Agriculture. Chonburi, Thailand. 31st March to 2nd April
Naylor R, Goldburg RJ, Mooney H et al (1998) Nature's subsidies to shrimp and salmon farming. Science 282:883–884
New M (2016) Understanding, reducing and managing African climate risk. www.axa.com
Nwebaza-Ndawula L, Kiggunda V, Megazi G, Ocaya H (2017) Effects of cage fish culture on water quality and selected biological communities in northern Lake Victoria, Uganda. Uganda J Agric Sci 14(2):61–75
OECD-FAO (2013) Agriculture outlook 2013–2022 – highlights. Paris, OECD, and Rome, FAO, 119
Oladokun OJ, Adedara TM, Adedadamola JO (2015) Tourism and climate change: combating climate change effects on tourism participation in Nigeria. J Tour Hosp Sports 5:1–6

Olaifa FE (2015). Food security: the perspective of aquaculture and fisheries. 2014/2015 Faculty lecture, Faculty of Agriculture & Forestry, University of Ibadan, delivered on 20 May, 2015. Ibadan University Press, 58pp

Olapoju OA, Edokpaye CA (2014) Ecological study on the impact of mariculture fish cage site on the physico chemical characteristics of the Lagos lagoon, Southwestern Nigeria. J Ecol Nat Environ 6(12):398–405

Olsson LM, Opondo P, Tschakert A et al (2014) Livelihoods and poverty. In: Climate change 2014: impact, adaptation and vulnerability, Part A: contribution of working group II to the 4th assessment report of the IPCC

Oluwatobi AA, Mutalib HA, Adeniyi TK et al (2017) Possible aquaculture development in Nigeria: evidence for commercial prospects. J Agric Sci Technol B 7:194–205. https://doi.org/10.17265/2161-6264/2017.03.007

Omitoyin BO (2007) Introduction to Fish Farming in Nigeria (ed). Ibadan University Press Publishing House, University of Ibadan, Nigeria, 90pp

Open Development Cambodia (2015) Overview of policy and legal framework. 7 September 2015. https://opendevelopmentcambodia.net/topics/overview-of-policy-and-legal-framework/. Retrieved 7 September, 2018

OSPAR (2001) Nutrient discharges from fish farming in the OSPAR Convention area. Oslo and Paris Commission, Copenhagen. http://www.ospar.org/v_publications

Oyebola OO, Fada PO (2017) Climate change: a potential risk to cage-culture and genetic structure of endemic fisheries in fish cage installed inland lake. African Great Lakes Conference 2017, Entebbe, Uganda. 2–5 May, 2017. www.africangreatlake.org

Oyebola OO, Fashae OA (2017) Remote sensing and spatial analysis revealed status of multidimensional threats to fisheries in a hydrodynamic African lake. African Great Lakes Conference 2017, Entebbe, Uganda. 2–5 May, 2017. www.africangreatlake.org

Pant J, Barman BK, Murshed-E-Jahan MK et al (2014) Can aquaculture benefit the extreme poor? A case study of landless and socially marginalized Adivasi (ethnic) communities in Bangladesh. Aquaculture 418–419:1–10

Paone S (2000) Farmed and dangerous: human health risks associated with salmon farming. Friends of Clayoquot Sound, Tofino. http://www.ancientrainforest.org/reports/salmon_farming_health_risks.pdf

Paull SH, LaFonte BE, Johnson PTJ (2012) Temperature-driven shifts in a host-parasite interaction drive nonlinear changes in disease risk. Glob Chang Biol 18:3558–3567. https://doi.org/10.1111/gcb.12018

Pereira L (2017) Climate change impacts on agriculture across Africa. Framing concepts in environmental science, environmental issues and problems, Agriculture and the Environment Online Publication, Mar 2017. https://doi.org/10.1093/acrefore/9780199389414.013.292

Pörtner HO, Peck MA (2010) Climate change effects on fishes and fisheries: towards a cause-and – effects understanding. J Fish Biol. https://doi.org/10.1111/j.1095-8649.2010.02783.x

Primavera JH (1998) Tropical shrimp farming and its sustainability. In: De Silva SS (ed) Tropical mariculture. Academic Press, London, pp 257–289

Primavera JH (2005) Mangroves, fishponds, and the quest for sustainability. Science 310:57–60

Rhodes E, Axsen J, Jaccard M (2014) Does effective climate policy require well-informed citizen support? Glob Environ Chang 29(2014):92–104

Roessig JM, Woodley CM, Cech JJ et al (2004) Effects of global climate change on marine and estuarine fishes and fisheries. Rev Fish Biol Fish 14:251–275

Rothuis A, Turenhout M, van Duijn A et al (2014) Aquaculture in East Africa; a regional approach. Wageningen, LEI Wageningen UR (University & Research centre), LEI Report IMARES C153/14|LEI 14-120.54 pp.

Salman MA, Daniel B (2006) Regulatory frameworks for water resources management: a comparative study. Law, justice, and development. Washington, DC: World Bank. © World Bank. https://openknowledge.worldbank.org/handle/10986/7054

Schimel JP, Schaeffer SM (2012) Microbial control over carbon cycling in soil. Front Microbiol 3. https://doi.org/10.3389/fmicb.2012.00348

Schimel JT, Balser C, Wallenstein M (2007) Microbial stress-response physiology and its implications for ecosystem function. Ecology 88:1386–1394

Secretariat of the Convention on Biological Diversity (2014) In: Hennige S, Roberts JM, Williamson P (eds) An updated synthesis of the impacts of ocean acidification on marine biodiversity. Montreal, Technical Series No. 75, 99 pages

Shameem MIM, Momtaz S, Kiem AS (2015) Climatic change local perceptions of and adaptation to climate variability and change: the case of shrimp farming communities in the coastal region of Bangladesh. Climate Change 133(2):253–266. https://doi.org/10.1007/s10584-015-1470-7

Signgh A, Purohit B (2014) Public health impacts of global warming and climate change. Peace Rev J Soc Justice 26(1):112–120

Smit B, Burton I, Klein RJT, Wandel J (2000) An anatomy of adaptation to climate change and variability. Clim Chang 45:223–251

Staniford D (2002) Sea cage fish farming: an evaluation of environmental and public health aspects (the five fundamental flaws of sea cage fish farming). Excerpts from a paper presented by Don Staniford at the European Parliament's Committee on Fisheries public hearing on 'Aquaculture in the European Union! Present Situation and Future Prospects' https://studylib.net/doc/8487456/sea-cage-fish-farming--an-evaluation-of-environmental-and...

Steinfeld H, Gerber P, Wassenaar T et al (2006) Livestock's long shadow. In: Environmental issues and options. FAO, Rome

Syandri H, Elfiondri J, Azrita (2015) Social status of the fish-farmers of floating-net-cages in Lake Maninjau, Indonesia. J Aquac Res Develop 6:391. https://doi.org/10.4172/2155-9546.1000391

Tacon A, Turchini GM, Metian M, De Silva SS (2010) Responsible aquaculture and trophic level implications to global fish supply. Rev Fish Sci 18(1):94–105. https://doi.org/10.1080/10641260903325680

The World Bank (2007) Changing the face of the waters agriculture and rural development: The Promise and Challenge of Sustainable Aquaculture. The International Bank for Reconstruction and Development. Document 41694. The World Bank 1818 H Street NW, Washington, DC20433. http://documents.worldbank.org/curated/en/322181468157766884/pdf/416940PAPER0Fa18082137015501PUBLIC1.pdf

The World Bank (2014) Raising more fish to meet rising demand, The World bank IBRD-IDA, February 6, 2014. http://www.worldbank.org/en/news/feature/2014/02/05/raising-more-fish-to-meet-rising-demand

Thordarson J (2007) Economical basis for fingerling production in cage culture in Sri Lanka. Final Project Munugoda Hewage Soma Ariyaratne National Aquatic Resources Research & Development Agency Crow Island, Mattakkuliya, Colombo-15 Sri Lanka

Tubiello F (2012) Climate change adaptation and mitigation: challenges and opportunities in the food sector. Natural Resources Management and Environment Department, FAO, Rome. Prepared for the High-level conference on world food security: the challenges of climate change and bioenergy, Rome, 3–5 June 2008

U.S. Environmental Protection Agency (USEPA) (2016) Climate change indicators in the United States, 2016. Fourth edition. EPA 430-R-16-004. www.epa.gov/climate-indicators

UN Report (2006) Rearing cattle produces more greenhouse gases than driving cars, UN report warns. UN News https://news.un.org/en/story/2006/11/201222-rearing-cattle-produces-more-greenhouse-gases-driving-cars-un-report-warns

UNEP (2010) UNEP emerging issues: environmental consequences of ocean acidification: a threat to food security

UNFCCC (United Nations Framework Convention on Climate Change) (2007) Climate change: impacts, vulnerabilities and adaptation in developing countries. https://unfccc.int/resource/docs/publications/impacts.pdf

van Cleef BA, Verkade EJ, Wulf MW et al (2010) Prevalence of livestock-associated MRSA in communities with high pig-densities in the Netherlands. PLoS One 5:e9385

Van Rheenen NT, Wood AW, Palmer RN, Lettenmaier DP (2004) Potential implications of PCM climate change scenarios for Sacramento-San Joaquin River Basin Hydrology and Water Resources. Clim Chang 62:257–281. https://doi.org/10.1023/B:CLIM.0000013686.97342.55

Vliet TV, Katan MB (1990) Lower ratio of n-3 to n-6 fatty acids in cultured than in wild fish. Am J Clin Nutr 51:1–2

Voegele J (2014) There is a major opportunity for developing countries that are prepared to invest in better fisheries management and environmentally sustainable aquaculture. The World Bank (2014), Raising More Fish to Meet Rising Demand, The World bank IBRD-IDA, February 6, 2014. http://www.worldbank.org/en/news/feature/2014/02/05/raising-more-fish-to-meet-rising-deman

VogelC(2015)WhyAfricaisparticularlyvulnerabletoclimatechange.TheConservation,May20,2015. http://theconversation.com/why-africa-is-particularly-vulnerable-to-climate-change41775

Wilby RL, Orr H, Watts G et al (2010) Evidence needed to manage freshwater ecosystems in a changing climate: Turning adaptation principles into practice. Sci Total Environ 408:4150–4164. https://doi.org/10.1016/j.scitotenv.2010.05.014

Williams L, Rota A (2012) Impact of climate change on fisheries and aquaculture in the developing world and opportunities for adaptation. IFAD Fisheries Thematic Paper: Tool for Project Design. http://www.ifad.org/lrkm/pub/fisheries.pdf

World Bank (2013) Fish to 2030: prospects for fisheries and aquaculture. World Bank Report No. 83177 GLB. Washington, DC, 102 p

World Bank & FAO (2010) The Hidden Harvests, the global contribution of capture fisheries, p. 111, Agriculture and Rural Development Department Sustainable Development Network. Food and Agricultural Organization of the United Nations. World Bank

World Population Review (2018) Africa Population 2018. http://worldpopulationreview.com/continents/africa-population/. Retrieved 7 September 2018

WorldFish Center (2009) Climate change: research to meet the challenges facing fisheries and aquaculture. Issues Brief No. 1915. WorldFish Center, Penang, Malaysia, 6 pp. Available at: http://www.worldfishcenter.org/resource_centre/WF_2167.pdf. Date accessed: 6 December 2010

WorldFish Center (2010) Climate change challenges facing fisheries and aquaculture. WorldFish Center, Pennang, Malaysia, pp 1–12

Worldometers 2018 Population of Africa (2018 and historical). www.Worldometers.info. Retrieved 7 September, 2018

Yazdi SK, Shakouri B (2010) The effects of climate change on aquaculture. Int J Environ Sci Develop 1(5):378–382

Yvon-Durocher G, Montoya JM, Trimmer M et al (2011) Warming alters the size spectrum and shifts the distribution of biomass in freshwater ecosystems. Glob Chang Biol 17:1681–1694. https://doi.org/10.1111/j.1365-2486.2010.02321.x

Ziervogel G, Bharwani S, Downing TE (2006) Adapting to climate variability: pumpkins, people and policy. Nat Res Forum 30:294–305

Zougmoré R, Partey S, Ouédraogo M et al (2016) Toward climate-smart agriculture in West Africa: a review of climate change impacts, adaptation strategies and policy developments for the livestock, fishery and crop production sectors. Agric Food Secur 5:26. https://doi.org/10.1186/s40066-016-0075-3

Livelihood Resilience, Climate Risk Management and Agriculture in the Mid-Zambezi Valley, Zimbabwe

Admire Nyamwanza

Abstract This chapter is a research article which utilizes the livelihood resilience and climate risk management constructs to explore climatic sensitivities and response strategies to these sensitivities within African agricultural systems using a case study of the mid-Zambezi valley area in northern Zimbabwe. Discussions in this chapter are based on a 6-year longitudinal qualitative primary fieldwork research focused on livelihood resilience, climate change adaptation and climate information needs in the case study area. The chapter positions research results within a broader conversation of how to adequately deal with barriers to and effectively stimulate the enablers of livelihood resilience and climate risk management within rural agricultural and livelihood systems in Africa.

Keywords Livelihood resilience · Climate sensitivities · Climate risk management · Agriculture · Zimbabwe

1 Introduction

Building smallholder farmer resilience in sub-Saharan Africa is becoming an important policy agenda due to an increase in the frequency and magnitude of shocks and stresses resulting from significant changes not only in biophysical factors but also in socio-economic factors (Nkonya et al. 2018). Climate change and variability have added a significant layer in as far as these stresses and shocks are concerned, posing a major challenge for African agriculture. This chapter brings together aspects of livelihood resilience (LR) and climate risk management (CRM) towards analysing agricultural systems' sensitivities in the semi-arid mid-Zambezi

A. Nyamwanza (✉)
Economic Performance and Development Unit, Human Sciences Research Council, Cape Town, South Africa
e-mail: anyamwanza@hsrc.ac.za

Y. Bamutaze et al. (eds.), *Agriculture and Ecosystem Resilience in Sub Saharan Africa*, Climate Change Management,
https://doi.org/10.1007/978-3-030-12974-3_21

valley area of northern Zimbabwe and the capacities of these agricultural systems to respond to climatic change and variability. The two LR and CRM constructs are briefly discussed in the following two subsections.

1.1 Livelihood Resilience

Livelihood resilience can be defined as a process linking the capacities of households and communities to anticipate, respond to, recover and learn from changes, disturbances and uncertainties (Marschke and Berkes 2006; Nyamwanza 2012). This includes reinstating, renewing and reinvigorating earnings and livelihood patterns disturbed and/or compromised by changes and challenges in the social and/or physical environment. It involves the ability to reconfigure livelihood systems without significant declines in critical functions, particularly those related to primary productivity, sustainability of the natural resource base, social relations and well-being in the face of stresses and shocks.

The main pillars of the construct revolve around three aspects (cf. Folke et al. 2003; Berkes 2007). The first aspect is the capacity of livelihood systems to live with change and uncertainty, which involves the ability to undertake livelihood diversification, learn from crisis and build rapid feedback mechanisms for responding to various livelihood constraints. The second aspect of the livelihood resilience construct is the ability of livelihood systems to nurture processes of social learning and adapting. Social learning is mainly associated with Bandura's (1986) social cognitive theory, whose core aspects are observation learning, imitation and modelling. In the context of this discussion, the social learning aspect involves people's willingness and ability to tap into social memory, rectify mistakes from past experiences and enhance useful livelihood strategies created during previous periods of crises. The third aspect of the livelihood resilience construct refers to the ability of livelihood systems to self-organize, particularly with respect to creating and strengthening collaborative institutional structures in as far as livelihood activities in an area are concerned. Livelihood resilience, in this chapter, therefore links with ideas and practices that promote and conserve options and opportunities for adaptation, novelty, innovation and renewal with respect to monitoring, interpreting, anticipating and responding to dynamic changes and challenges.

1.2 Climate Risk Management

Climate risk management is an approach to climate-sensitive decision-making to deal with climate variability and change (Hellmuth et al. 2007). The approach focuses on coordinated responses for addressing climate risks centred on a committed engagement with farmers, agricultural support services, relevant institutions and enabling policy towards resilient livelihoods (Selvaraju 2012). In the agricultural sector, it involves such initiatives as changing crop and/or livestock enterprises, using more or less of certain inputs, implementing new resource management practices, diversifying farming systems and/or even diversifying into non-farm activities (Mulwa et al. 2017). It is also widely recognized that climate risk management, particularly for agricultural systems, should revolve around the availing and use of effective climate information services that are sufficient to support agricultural and general livelihood decision-making, promote adaptive management and improve preparedness to disaster risks (Hellmuth et al. 2007; Selvaraju 2012). Broadly put therefore, climate risk management in the context of livelihood resilience and agriculture refers to different aspects of the risk management process, which include risk reduction (involving planning and preparation), risk assessments for informed decision-making and risk sharing, pooling and transfer in the context of adaptation (UNFCCC 2011).

This chapter is structured as follows: following this introduction section is the methodology section, which outlines the research techniques used and methods of data analysis. This is followed by Sect. 3 which presents results on the agricultural system, climate risks and risk responses in the case study area. Section 4 discusses the results centering mainly on challenges to and opportunities for enhancing climate risk management in the area towards a resilient agricultural and livelihood system. The chapter wraps up with a conclusion in Sect. 5.

2 Methodology

This discussion is a culmination of a 6-year (2010–2016) doctoral and postdoctoral longitudinal primary fieldwork research focused on livelihood resilience, climate change adaptation and farmer climate information needs in northern Zimbabwe. A case study of the mid-Zambezi valley area, which forms part of the Zimbabwean lowveld situated north of the Zambezi Escarpment bordered by Mozambique to the north and east and Zambia to the north-west, was used. The area is listed in the country's agroecological zones IV and V. Agroecological zones (AEZs) are geographical areas which exhibit similar climatic conditions that determine their ability to support rainfed agriculture (HarvestChoice 2010). Zimbabwe is divided into five AEZs, with AEZs IV and V, respectively, categorized as characterized by dry and very dry rainfall conditions. The mid-Zambezi valley area is characterized by temperatures of up to 40 degrees Celsius in summer and low, increasingly irregular and unpredictable rainfalls averaging 450–650 mm annually (Mupangwa et al. 2006).

The study employed qualitative methods, with analysis of relevant published and unpublished secondary documents on livelihoods, agriculture and climate variability and change in the area; key informant interviews; life histories; and focus group discussions being used in data collection. Purposive sampling of individuals who had continuously stayed in the area at least since 1990 was used to select respondents for life histories, as the study sought to cover livelihoods and agricultural dynamics as well as climatic challenges experienced in the area at least since 1990. Key informant interviews were conducted with selected officials working in various relevant subnational (state and non-state) entities involved in livelihood and agricultural activities in the area, whilst focus group discussions were conducted with selected adult male and female community members in the area. All data was analysed using thematic analysis.

3 Results

3.1 The Agriculture System in the Mid-Zambezi Valley

The main agricultural activities in the mid-Zambezi valley area revolve around crop and livestock production. These are also the major sources of livelihoods in the area. There are two crop production systems in the area, which are upland crop production and riverbank crop production.

Upland Crop Production

Upland fields consist mainly of shallow sandy clay soils and are used for dryland crop production involving mainly cotton, sorghum and groundnut farming though a variety of other crops particularly cowpea, sunflower and millet are also grown there. Upland crop activity begins with the clearing of fields and preparations for planting in the months of September and/or October just before and continuing after the falling of the first rains. It becomes intensive from November to March with the planting and weeding periods and the protecting of crops from wild animals, birds and other pests, before the commencing of harvesting from end of March/beginning of April onwards. With the exception of cotton, which is exclusively for selling on the cotton market, respondents during fieldwork noted that most upland crops are for subsistence purposes, i.e. family consumption, barter trade for accessing other foodstuffs and essential services such as health, transport and education for children.

Riverbank Crop Production

Riverbanks form the most arable lands in the area, and the major crop grown there is maize. Green vegetables, tomatoes and onions are also cultivated in these fields, albeit in small quantities. The mid-Zambezi valley area has a long history of riverbank cultivation going back to the fifth-century farming communities. This is therefore an old practice carried over and evolving through generations. Riverbank fields are characterized by rich alluvial soils which store residual moisture from the rainy season into the dry season; hence villagers are able to conduct farming there the whole year round – planting crops around October/November and harvesting March/April and then planting again around May and harvesting September/October. The latter crop depends on the amount of moisture held in the riverbank soils after the rainy season. Riverbank fields are inherited and passed on from generation to generation, and they are treated very much as private properties. As noted by the interviewed community members during fieldwork, kinship and lineage are very important factors in the ownership of these fields.

Livestock Production

Livestock raised in the area include cattle, goats, sheep, pigs and poultry, with cattle, goats and poultry forming the larger numbers of animals. Livestock in the area is important as a source of draught power, meat, milk, manure and is essential in other important social processes (e.g. paying bride price, settling serious societal conflicts and payment of fines in traditional courts) (Nyamwanza 2012). As in many other rural communities in Africa, cattle are the most highly ranked form of livestock; and the more cattle a household owns, the wealthier and/or financially stable they are (perceived to be). The number of cattle one has also determines the area that can be planted in upland fields and how fast this can be done, which is important given the fact that timely planting is a major factor in crop success in these marginal areas (Nyamudeza 1999). From observations made during the fieldwork period, cattle are also a local moneymaking business (e.g. ploughing other villagers' fields for cash and transporting other people's commodities to the market). It also helps in spreading social networks since owning more cattle in the area generally means more social recognition and being 'culturally anchored'.

3.2 Climatic Sensitivities

The main climatic sensitivities in the mid-Zambezi valley are linked to inadequate rainfall and increasing drought cycles. Rains and rainfall patterns are central to the agricultural system in the area. Upland crop production is exclusively dependent on the rains. Good rains ensure the availability of water in major rivers for continued moisture around the riverbanks important for easier all-year-round maize (and other crop) farming in the riverbank fields. Livestock water points and grazing lands in the villages also thrive on rainwater. Being a semi-arid zone, the area is characterized by low and erratic rainfall. Inter-seasonal dry spells are also common. In the listing of livelihood constraints hierarchy in focus group discussions, low rainfalls and increased drought cycles came out as the top livelihood challenges. Respondents noted that rains are generally unreliable every year, and that rainfall patterns are increasingly becoming erratic in recent decades. A normal rainy season in the area begins late November and ends in March. It was also noted that there is a lull in rainfalls in the month of January almost every year, which leads to crop stress at the critical tasselling stage. A season may therefore receive normal rainfall amounts; however, these in-season lull periods coming at a time when crops are at a critical stage of growth have many a time led to reduced quantity and quality of crop yields.

Droughts are broadly defined here as the deficiency of precipitation, or water deficit over an extended period of time, resulting in serious adverse effects on water supplies and agricultural production. Respondents in fieldwork interactions stated that droughts occur in the area every 3–5 years. This trend in Zimbabwe's drylands is also noted by scholars such as Illiffe (1990) and Bird and Shepherd (2003). This often results in total crop failure (especially upland crops), the drying up of rivers and boreholes and livestock deaths.

3.3 Climate Risk Management and Responses

The management of and responses to climate risks vis-à-vis the agricultural system in the mid-Zambezi valley area have revolved around a number of strategies initiated at the household and subnational levels. Strategies at the household level have included diversification, temporary migration and increased use of and reliance on indigenous knowledge. From the subnational level, climate risk management initiatives to assist local communities have mainly revolved around the provision of inputs for cotton and sorgum (which are the major crops in the area), and food aid.

Household-Level Responses to Climate Risks

Diversification

Diversification reflects the livelihood resilience aspect of the ability to live with change and uncertainty. In livelihoods, it refers to the process by which households construct a diverse portfolio of activities and social support capabilities in their struggle against various stresses and shocks as well as to improve their standards of living (Ellis 1998). Respondents during fieldwork interactions noted that diversification in as far as the agricultural system in the area is concerned has mainly involved crop diversification i.e. mixing a variety of crops in general, and diversifying more into a specific drought-tolerant major crop, sorghum, over the years. As noted earlier, crops grown in upland fields have included cotton, sorghum, groundnuts, cowpea, sunflower and millet, whilst maize and vegetables are the dominant crops in riverbank fields. An examination of general trends in upland crop production over the years showed a shift towards diversification into the drought-tolerant sorghum, particularly from maize. This has assisted in managing the progressively drought conditions obtaining in the area.

Temporary Migration

Two forms of temporary migration associated with households in the area are (a) temporary outmigration to commercial farms and/or urban areas for employment and to supplement household income and food supplies and (b) temporary outmigration beyond the country's borders mainly to nearby Zambia and

Mozambique, again for employment purposes and in search of food supplies, particularly during times of severe droughts. These forms of migration exhibited local residents' endeavours in nurturing social learning processes (i.e. especially with respect to the second type of temporary migration), as well as to live with change and uncertainty.

Discussions with respondents revealed that the first type of migration was popular throughout the 1990s, before various macroeconomic and political challenges set in. These challenges include the effects of Economic Structural Adjustment Programmes (ESAPs) in the 1990s, which closed off employment opportunities in the commercial farming areas and led to the virtual collapse of activity in surrounding commercial farms in the 2000s. The type of migration involving commercial farm working had been very important in the area, and most households had had at least one household member who migrated to commercial farms at one point or the other in the 1990s. Respondents mentioned that whole families, in years characterized by severe drought, particularly during the 1992 drought, would leave to temporarily settle with other household members in commercial farms and urban areas. This type of migration, particularly to the farms, was therefore among the most valued livelihood diversification activities, and it appeared to be an essential part of living with change and uncertainty in the pre-2000 period. After 2000, the government-led Fast Track Land Reform Programme dissolved most commercial farms, and this represented the closure of one of the most important household adaptation portfolios.

Migration to Zambia and Mozambique was popular during periods of food insecurity. Respondents however noted that this type of migration has always been part of the mid-Zambezi valley communities' interaction with neighbouring communities in these countries during difficult periods. For example, during the Zimbabwe's war of liberation in the 1970s, local community members would temporarily seek refuge in these two countries as the war intensified; then during the days of Mozambique's civil war throughout the 1980s, residents of nearby communities in that country would cross over into Zimbabwe as hostilities intensified. This type of migration may therefore be said to be a long-created strategy during periods of previous crises and therefore a product of local people's nurturing of social learning processes, modified according to the type and intensity of crisis being faced.

Increased Reliance on Indigenous Knowledge

Indigenous knowledge is knowledge of a community accumulated over generations of living in a particular environment, and it covers all forms of knowledge – technologies, know-how, skills, practices and beliefs that enable people to achieve stable livelihoods (Mwaura 2008). It is part of a community's institutional memory; and in the mid-Zambezi valley, it involves social, intellectual, technological and ecological knowledge. It includes dealing with and/or predicting rainfall and droughts, crop production and responding to challenges of food shortage, particularly in drought years and during the November to February 'food-gap' period for most households. The increased reliance on indigenous knowledge showed households' and the community's ability to live with change and uncertainty, as well as to nurture social learning processes within livelihood systems in the area. This is because besides being based on community members tapping into social memory, it also comprised of specific strategies (mentioned below) meant to deal with particular livelihood challenges.

Among the foremost forms of indigenous knowledge gaining prominence in the area vis-à-vis climate risk management within the local agricultural system is the use of indigenous seasonal early warning systems. This relates to the prediction of seasonal rainfall and temperature trends, which then assists communities in activating appropriate responses to inadequate rainfall and drought periods. The system involves social learning processes of recalling signs of previous years as told by others (normally elders and/or autochthonous residents) or as experienced by oneself, to predict the future. Respondents mentioned three main ways through which these indigenous early warning systems are manifest. These include:

(a) Atmospheric indicators – Especially the appearance of a particular seasonal star, called '*nyamatsatse*' in Shona vernacular, around the beginning of November each year which signifies good rains for that season; and each time it has not appeared, it becomes a signal that the season will be characterized by erratic rains.
(b) Meteorological indicators – Particularly wind direction and temperature trends. Respondents noted that fast-blowing wind from the east at the beginning of the summer season (i.e. early September) represents heavy and good rains, whilst that from the west signifies imminent dry conditions and a bad farming season. In the same vein, extremely high temperatures beginning mid-August right into the middle of summer (i.e. October/November) without any intermittent showers in-between were said to generally point to a dry and bad season.
(c) Insect indicators – Especially wild insects and the blooming cycle of particular wild trees. The abundance of butterflies (*Danaus plexippus*) at the beginning of and during the farming season has always locally signified good and continuing rains in the season. At the same time, the failure of particular local wild trees (e.g. the *muhurongwa* and the *musiga* trees) to bloom, signals drought and a bad season according to respondents.

These indigenous indicators assist households to plan, for example, on when to plant and which crops to concentrate on (and which ones to leave out during a particular season) and in making labour-allocation decisions in the home especially with respect to the uptake of off-farm activities. These indicators have therefore been integral in climate risk management and for living with change and uncertainty. Apart from indigenous seasonal early warning systems, some indigenous knowledge-based crop farming activities have also been employed as useful strategies in responding particularly to the challenge of low rainfalls in the area. These mainly came up in two areas, namely, (a) the riverbank farming practice and (b) intercropping.

The riverbank farming practice is essentially undertaken to maximize maize and vegetable production in the rich alluvial riverbank soils long identified as the most fertile lands in this drought-prone area. Whilst this is a long-established strategy of people in the area to live with drought and low rainfall challenges, it also exhibits the nurturing of social learning processes as the practice has been passed on for generations. Households in the area also practise intercropping in both riverbank and upland fields. They fill remaining spaces within cotton, sorghum and maize fields with indigenous drought-resistant crops such as the pumpkin (*manhanga*) – which not only provides a special kind of nutritious vegetable (*muboora*) before and after it fully produces the actual pumpkin but also a vegetation canopy within these fields thereby protecting the soil as well as controlling weed growth. Other indigenous crops filling the remaining spaces between major crops in the fields and revered for their nutritious value in the area include wild okra (*derere renyenje*), spider flower (*nyevhe*), black nightshade (*musungusungu*) and pigweed (*mowa*).

Lastly, indigenous knowledge in wild vegetation and plants has also enabled villagers to pick fruits, berries, vegetables, roots and tubers from the wild (i.e. surrounding forests) as substitutes and supplements for cereals and other food crops during the food-gap in the year and during drought periods over the years. Examples of these wild fruits, tubers and plants include *mawuyu*, *maroro*, *nhengeni*, *tsvanzva*, *mhande*, *katunguru*, *manyanya* and *pfumvudza*.

Subnational Level Initiatives in Assisting Local Climate Risk Responses

The subnational level comprises institutional entities originating from the national level and supporting local livelihoods in various roles. In the mid-Zambezi valley area, the main active entities vis-à-vis climate risk management and the agricultural system include cotton companies involved primarily in the availing of cotton inputs to local farmers, as well as non-governmental organizations (NGOs) involved in the availing of food aid to the most vulnerable in the community.

Cotton Companies

Cotton companies have emerged as critical in the mid-Zambezi valley agricultural system given their role in the availing of cotton inputs to enable the profitable farming of this drought-tolerant cash crop in the area. Cotton companies enter into contracts with individual farmers at the beginning of the farming season whereby the companies provide seed, chemicals and fertilizers on credit during the farming season, as well as technical advice to farmers since all companies have their own extension officers who are supposed to offer farming advice to their contracted farmers. On their part, farmers are supposed to sell their harvest to the particular company that they would have entered into contract with. Cotton companies have therefore supported cotton farming in the area in a huge way and enabled poorest subsistence farmers to participate in cash crop farming in this drought-prone area.

NGOs

Whilst many NGOs have operated in the mid-Zambezi valley area over the years, two are specifically singled out because of their continued, uninterrupted and consistent operations in the area since the 1990s. The first one is Christian Care, an ecumenical international NGO that has been involved in delivering food aid to the area. During major droughts and every year during the 'food-gap' November to March period, Christian Care parcels out food handouts to local residents, carefully targeting the poorest, the needy and those who are considered to be the most vulnerable in the community, such as the elderly, the disabled, orphans and widows. The foods distributed include cooking oil, maize meal, rice, salt, beans and soap. This aid has been very important for communities vis-à-vis climate risk management and livelihood resilience in this drought-prone area.

A second prominent NGO operating in the area is the Lower Guruve Development Association (LGDA). This is an NGO that is involved in such initiatives as the establishment of such community projects as nutrition gardens. LGDA has worked closely with Christian Care to mobilize food for local communities. A notable programme by LGDA was the introduction of the now popular drought-tolerant *macia* sorghum variety in 2002, which gradually changed the crop production system in upland fields, with sorghum becoming the second major crop after cotton in the uplands and displacing maize in that regard. According to residents, the *macia* crop has extensively helped in managing food insecurity as the crop can withstand conditions of drought and seasonal dry spells prevalent in the area. The two NGOs have therefore been at the centre of livelihood resilience and climate risk management in the mid-Zambezi valley area.

4 Discussion

From the foregoing discussions and from fieldwork experiences in the case study area, there are a number of challenges to and opportunities for enhanced climate risk management towards building a more resilient agricultural and livelihood system in the area. This section discusses these challenges and opportunities thereby positioning research results presented in Sect. 3 within a broader conversation of how to adequately deal with barriers to and stimulate the enablers of livelihood resilience and climate risk management within rural agricultural and livelihood systems in Africa. Among the main challenges are lack of proper government institutional support to farmers, lack of appropriate climate services in the area and lack of an effective national climate risk management policy framework to capacitate and support agricultural systems in such marginal areas as the mid-Zambezi valley. The main opportunities include the prominent role of indigenous knowledge in the area, which provides some foundation for the kind of climate risk management information needed, as well as the presence of collaborative arrangements among some key institutions in the area vis-à-vis livelihood resilience and farmer agricultural support. These factors are discussed in detail in the following subsections.

4.1 Challenges

Lack of Proper Government Institutional Support

There has not been proper government institutional support to farmers for climate risk management and livelihood resilience in the mid-Zambezi valley particularly in recent years. Agrawal (2010) mentions four ways through which institutions (in this case government institutions) might be helpful in climate risk management in rural communities. These include the provision of credible and reliable weather and climate information, financial and material support, technological interventions that help in increasing productivity and leadership efforts that promote collective action for risk management. Whilst there is a local government meteorological office in the area, it has largely been dysfunctional in recent years, and has not been helpful in the production of useful weather and climate information. The local government agricultural extension (Agritex) office, which is supposed to spearhead the provision of advice to local farmers on proper agricultural and related activities, is understaffed and poorly resourced, and the few officers that are there are not able to cover all communities in the area. A local Agritex officer who has worked in the area since the 1980s pointed out that they are supposed to be extensively mobile in visiting farmers around the area yet they have not been able to do so. Since the turn of the 2000s, the provision of motorcycles and fuel from central government, which had greatly enabled them to go round and effectively cover their catchment area, had been stopped in the late 1990s. This was also the same story with the local Veterinary Services Department.

Between 1982 and 1999, the government through the local rural district council also used to roll out drought relief programmes in the area to mitigate against the effects of droughts and food insecurity given the local climatic conditions. These programmes had however long stopped by the time of first fieldwork contact in the area in 2010 due to macroeconomic challenges. As discussions in the previous section have shown, much of meaningful institutional activity in as far as climate risk management and livelihood resilience in the mid-Zambezi valley is concerned has been spearheaded by private and non-governmental organizations.

Lack of Proper Climate Information Services

In managing climate risk, information, information services and their transmission to target audience is vital (Egeru 2016). Climate information services refer to the provision of this (climate) information in such a way as to assist decision-making. Climate information services must be based on scientifically credible information and expertise, have appropriate engagement from users and providers, have effective access mechanisms and meet user needs (Graham et al. 2015). In the mid-Zambezi valley, all this is lacking. Scientific weather and climate information is generally not trusted, and farmers mostly rely on indigenous knowledge for making their agricultural and other decisions. There is also no engagement between farmers, climate information producers and intermediaries on the type of information needed by farmers in the area as well as the communication channels preferred. This has therefore resulted in weak climate information services in the area, despite the critical role of climate information services in providing answers to a range of questions related to climate risk management, livelihood resilience and adaptation in marginal rural semi-arid areas.

Lack of an Effective National Climate Risk Management Policy Framework

Zimbabwe is yet to come up with a clear integrated climate risk management policy framework, yet this is of huge importance in managing risk and building resilience for agricultural systems in such rural areas as the mid-Zambezi valley. A climate risk management policy framework which seeks to understand past successful efforts to reduce climate risk through the integration of social, economic, scientific and technological research and action should aid and result in effective climate risk management. This will help in the reframing of issues from policy initiatives that are reactive vis-à-vis responding to climate risk in such drought-prone areas as the mid-Zambezi valley to ones that allow proactive planning.

4.2 Opportunities

Prominent Role of Indigenous Knowledge

As shown in Sect. 3, indigenous knowledge systems occupy a prominent role with respect to climate risk management and livelihood resilience in the mid-Zambezi valley. They are used in early warning and enhancing food security and crop production. Indigenous knowledge provides opportunities for understanding historical and current risk management processes, facilitating community-based learning, co-production of knowledge and increasing the usability of climate science information – all of which are critical aspects of effective climate risk management. The integration of indigenous and scientific knowledge is also essential in enhancing climate risk management, and with such an extensive base of indigenous knowledge as existing in the mid-Zambezi valley, the integration process may become easier.

Presence of Collaborative Institutional Arrangements

Collaboration of institutions in climate risk management is important in harmonizing institutional interventions, effectively deploying resources where they are most needed, preventing overlaps and sharing knowledge and ideas. In the mid-Zambezi valley, there is already some collaboration of particularly non-governmental organizations specifically in food aid distribution. Government and cotton company agricultural extension officers have also, in some instances, collaborated in delivering agricultural advisories to farmers. Whilst current collaborative institutional arrangements in the area are clearly not extensive, they provide a foundation for broader institutional engagement and collaboration around climate risk management and livelihood resilience.

5 Conclusion

Climate risk management is integral in building the resilience of agricultural and livelihood systems in such marginal semi-arid areas as the mid-Zambezi valley. From discussions in this chapter, it is apparent that whilst farmers in such rural communities as the mid-Zambezi valley area can develop a suite of strategies in managing and responding to climate risks, there is a need for more integrated and proactive approaches for effective climate risk management in rural African agricultural systems. Such integrated and proactive approaches should be anchored on clear support of farmers by both government institutions and non-governmental organizations, collaboration of these institutions and the availing of effective climate services.

References

Agrawal A (2010) Local institutions and adaptation to climate change. In: Mearns R, Norton N (eds) Social dimensions of climate change: equity and vulnerability in a warming world. New Frontiers of social policy. The World Bank, Washington D.C

Bandura A (1986) Social foundations of thought and action: a social cognitive theory. Prentice-Hall, Englewood Cliffs

Berkes F (2007) Understanding uncertainty and reducing vulnerability: lessons from resilience thinking. Nat Hazards 41:283–295

Bird K, Shepherd A (2003) Chronic poverty in Semi-arid Zimbabwe, CPRC working paper no. 18

Egeru A (2016) Climate risk management information, sources and responses in a pastoral region in East Africa. Climate Risk Manag 11:1–14. https://doi.org/10.1016/j.crm.2015.12.001

Ellis F (1998) Household strategies and rural livelihood diversification. J Dev Stud 35(1):1–38

Folke C, Colding J, Berkes F (2003) Synthesis: building resilience and adaptive capacity in socio-ecological systems. In: Berkes F, Colding J, Folke C (eds) Navigating social-ecological systems: building resilience for complexity and change. Cambridge University Press, Cambridge

Graham R, Visman E, Wade S, Amato R, Bain C, Janes T, Leathes B, Lumbroso D, Cornforth R, Boyd E, Parker D (2015) Scoping, options analysis and design of a 'Climate Information and Services Programme' for Africa (CIASA): literature review, UK Met Office

HarvestChoice (2010) Agro-ecological zones of sub-Saharan Africa. International Food Policy Research Institute, Washington, DC., and University of Minnesota, St. Paul, MN. Available online at http://harvestchoice.org/node/8853

Hellmuth ME., Moorhead A, Thomson MC and Williams J. (Eds) 2007. Climate risk management in Africa: learning from practice. International research Institute for Climate and Society (IRI), Columbia University, New York, USA

Illiffe J (1990) Famine in Zimbabwe. Mambo Press, Gweru

Marschke MJ, Berkes F (2006) Exploring strategies that build livelihood resilience: a case from Cambodia. Ecol Soc 11(1). Available online at: http://www.ecologyandsociety.org/vol11/iss1/art42/

Mulwa C, Marenya P, Rahut DB, Kassie M (2017) Response to climate risks among smallholder farmers in Malawi: a multivariate probit assessment of the role of information, household demographics, and farm characteristics. Climate Risk Manag. Available online at. https://doi.org/10.1016/j.crm.2017.01.002

Mupangwa JF, Nyabadza T, Mberengwa I, Chaipa I (2006) Problem animal control strategies in human-elephant conflict areas of Dande Communal Lands, Zimbabwe: a study in sustainability. J Sustain Dev Africa 8(1):53–69

Mwaura P (ed) (2008). Indigenous knowledge in disaster management in Africa, UNDP

Nkonya E, Koo J, Kato E, Johnson T (2018) Climate risk management through sustainable land and water management in Sub-Saharan Africa. In: Lipper L, McCarthy N, Zilberman D, Asfaw S, Branca G (eds) Climate smart agriculture. Natural Resource Management and Policy, Springer, Cham

Nyamudeza P (1999) Agronomic practices for the low rainfall natural regions of Zimbabwe. In: Manzungu E (ed) Water for agriculture in Zimbabwe: policy and management. University of Zimbabwe, Harare

Nyamwanza A (2012) Resilience and livelihoods inquiry in dynamic vulnerability contexts: insights from northern Zimbabwe. PhD Thesis, The University of Manchester, Manchester, United Kingdom

Selvaraju R (2012) Climate risk assessment and management in agriculture, climate, energy and tenure division. FAO, Rome

UNFCCC (2011) Report on the workshop to identify challenges and gaps in the implementation of risk management approaches to the adverse effects of climate change. Subsidiary body for implementation. Thirty-fifth session. Durban, 28 November–3 December 2011

A Synthesis of Determinants of Urban Resilience in Sub-Saharan Africa

Oriangi George

Abstract Urban resilience to natural and human-induced shocks and stresses has become an important issue in the contemporary world. Several reviews exist on urban resilience, but limited attempts have been made to critically review studies that shed light on determinants of urban resilience in sub-Saharan Africa (SSA). This study synthesizes both peer-reviewed and grey literature on the determinants of urban resilience to natural and human-induced shocks and stresses in SSA. A considerable number of studies that shed some light on the determinants of urban resilience in SSA have been conducted since the year 2000, but limited attempts have been made to synthesize and integrate them into the pool of knowledge. In this study, the preferred reporting items for systematic reviews and meta-analysis (PRISMA) protocol was followed. Findings indicate that urban resilience in SSA is understood, firstly, as a social and organizational construct and, secondly, as a social, organizational, and ecological construct. The most reported determinants of resilience in cities of SSA are access to basic services, social networks, employment, ownership of productive assets, involvement in none-agricultural activities, building flood retention facilities, and environmental preservation. In conclusion, necessity exists to conduct more studies in secondary cities in SSA while considering the social, institutional, economic, and ecological aspects of resilience so as to understand the multidimensional, location-specific dynamics of the determinants of resilience, given the growing role that secondary cities will play in the strong urban growth trajectories projected over the next decades.

Keywords Urban resilience · Determinants · Sub-Saharan Africa · Shocks · Stresses

---

O. George (✉)
Department of Geography, Geo-Informatics and Climatic Sciences, Makerere University, Kampala, Uganda

Department of Geography, Gulu University, Gulu, Uganda

Department of Physical Geography and Ecosystems Science, Lund University, Lund, Sweden

Y. Bamutaze et al. (eds.), *Agriculture and Ecosystem Resilience in Sub Saharan Africa*, Climate Change Management,
https://doi.org/10.1007/978-3-030-12974-3_22

1 Introduction

As natural and human-induced shocks and stresses continue to pose increasing risks to the urban areas of sub-Saharan Africa (SSA) (Boyd and Juhola 2014a, b; NCEA 2015), it becomes important to establish the factors that contribute to urban resilience, to enable urban areas prepare, recover, and adapt (Bozza et al. 2015). The increasing risks from natural and human-induced shocks and stresses on urban areas of SSA are often attributed to inadequate levels of preparedness, limited ability to accommodate and recover, and low adaptation capabilities by urban communities and governments (Pharoah and Ross 2016). This has been exacerbated by under-resourced socio-economic infrastructure, poor service delivery, and high levels of poverty (Pharoah and Ross 2016; Thompson et al. 2010). These often make urban areas in SSA not only hotspots to risks posed by natural and human-induced shocks and stresses but also potential sources of options to build resilience (Greenwalt and Raasakka 2018; Pharoah and Ross 2016; Solecki 2015). Urban areas are hotspots because they have large concentrations of population and socio-economic infrastructure and they consequently continuously degrade the surrounding ecosystems (Bozza et al. 2015; Dobson et al. 2015; Dobson 2017; Solecki et al. 2015; Yuen and Kumssa 2011). However, urban areas are also potential sources of options to build resilience because they are engines of social and economic development (Bottazzi et al. 2018; Greenwalt and Raasakka 2018; Saghir and Santoro 2018). Hence, policymakers and practitioners in the urban areas of SSA need to identify the factors that have been found as important determinants of urban resilience and learn lessons in order to enhance the resilience of urban areas. To date, paucity of studies that bring together several findings on what determines urban resilience in SSA exists and yet over the last few years, there has been a rapidly expanding body of knowledge from studies that give some light on what determines urban resilience to natural and human-induced shocks and stresses among the scientific community (Chun et al. 2017; Dhar and Khirfan 2017; Dobson 2017; Lee and Kim 2017; Suárez et al. 2016; Tawodzera 2012; Tippens 2016). This has been mainly a response to the global calls by the United Nations International Strategy for Disaster Reduction (UNISDR 2015), the New Urban Agenda (2016), and the IPCC Fifth Assessment Report (IPCC 2015), which emphasize the need for more research undertaking on urban resilience to multiple shocks and stresses.

This review investigates the determinants of urban resilience in SSA by synthesizing studies that have been undertaken on the subject. Studies that review the subject of urban resilience exist (e.g. Ayyoob 2016; Elena et al. 2017; Juan-García et al. 2017; Meerow et al. 2016). However, paucity exists in reviewing studies that shed light on the determinants of urban resilience in the context of SSA. Reviews that have been undertaken on the subject of urban resilience can be summarized as follows: Meerow et al. (2016) synthesized 25 definitions of urban resilience across studies for the period 1973–2013 and came up with a new definition of urban resilience. Ilmola (2016), on the other hand, reviewed five resilience frameworks developed for different purposes, categorizing them into survey based, secondary data based, and both survey and secondary data based. Furthermore, Juan-García et al. (2017) reviewed urban resilience frameworks specific to waste water and concluded that there existed inadequacy of key elements and no agreement on the conceptualization of the concept. Morestill, Elena et al. (2017) focused on urban energy resilience studies and concluded that there is need for worthwhile energy plans that can achieve climate change resilience. Additionally, Tyler and Moench (2012) focused on the concept and theories of resilience to come up with how the concept can further be developed and operationalized. To date, one of the most detailed reviews is by Ayyoob (2016), who examined 36 community resilience assessment tools to validate their contentions and possibilities in community resilience.

Analysis of the above reviews indicates that none of them focuses on reviewing the determinants of urban resilience, particularly in the context of sub-Saharan Africa. Thus, reviewing studies with the purpose of assessing what determines urban resilience to natural and human-induced shocks and stresses in SSA needs consideration. This is because reports indicate that over the last four decades, natural and human induced shocks and stresses are increasingly affecting urban areas of SSA (Nsubuga et al. 2014; Waithaka et al. 2013). Thus a need to integrate findings on determinants of urban resilience to create a body of knowledge. This review contributes to urban resilience research by integrating a pool of knowledge that can be useful for urban planners, academia, and civil society organizations interested in enhancing urban resilience. Urban resilience is also still a recent and a growing field (Ayyoob 2016), and thus this synthesis can aid further understanding of what constitutes urban resilience in the context of SSA and thus facilitating future studies in this field. This review is divided into five sections, and the following sections present the background of the resilient concept, methods, results and discussions, and conclusions.

2 Resilience Background

The word resilience was used in physics as early as the 1960s (Peng et al. 2017), where it meant the "reaction of a substance to its external forces". The concept was then operationalized in ecology where Holling (1973:14) in his seminal work "resilience and stability of ecological systems" defined resilience as "the measure of the persistence of systems and of their ability to absorb change and disturbance and still maintain the same relationships between populations or state variables". In this respect, an ecosystem will flip into another equilibrium position while still maintaining its functioning. Thus resilience is viewed as a dynamic process. Later, Holling (1996) in his work "engineering resilience versus ecological resilience" had a different interpretation of the concept, where he defined resilience as the capacity of material objects and physical structures to withstand disturbances without undergoing structural change. In this conceptualization, resilience was taken to mean maintenance of pre-disaster status and hence maintenance of equilibrium state (Ronak and Nosal 2016). The concept was later adopted in socioecological sciences, psychology, anthropology, organizational management, economics, and adaptation to climate change (Batica and Gourbesville 2016; Béné et al. 2017; De Boer 2016; Ronak and Nosal 2016). In these fields, the general meaning of resilience has been the capacity of a system or community that is exposed to hazards to resist, absorb, accommodate, and recover from the effects of shocks and stresses in a timely and efficient way, including the preservation and restoration of its essential basic structures and functions (UNISDR 2015; Elena et al. 2017). The application of the concept of resilience in several disciplines ushers contentions in its framing and conceptualization. Nevertheless, consensus exists that communities must build their resilience in order to prepare for future uncertainties, recover, and adapt when faced with shocks and stresses (Boyd and Juhola 2014a, b).

2.1 Urban Resilience

This paper focuses on urban resilience which has been defined by the United Nations University and Centre for Policy Research (UNU-CPR) (2016 p. 6) as "the ability to activate protective qualities and processes at individual, community, institutional, and systems level to engage with hazards and stress and cooperate with each other in order to maintain or recover functionality and prosper while adapting to a new equilibrium and minimizing the accumulation of pre-existing or additional risks and vulnerabilities". The definition upholds that resilience involves the acquisition of a new equilibrium, arising from complex interaction between individuals, institutions, and government. Resilience is viewed more so as a social construct. Furthermore, Rockefellers Foundation/Arup (2016) defines urban resilience as the ability of an urban area to continue functioning such that its inhabitants, more so the poor and vulnerable, can be able to survive and thrive in the midst of shocks and stresses. In their view, resilience is a dynamic process and a function of health and well-being, economy and society, infrastructure and ecosystem, and leadership and strategy. They further emphasize that particular qualities should be observed while measuring resilience, i.e. the quality of being inclusive, reflective, integrated, resourceful, robust, redundant, and flexible. More still, Abdrabo and Hassaan (2015 p. 556) define urban resilience as "the ability of an urban system in all its dynamism to support, in the face of hazards or pressure, the provision and accessibility to services and functions essential for the wellbeing of all residents, especially those lacking means to buffer stress". This definition views resilience as a dynamic process encapsulating socio-economic, physical, environmental, institutional, emergency preparedness, and climate change hazards. In addition, Batica and Gourbesville (2016 p. 812) whose study was focused on flood resilience defines urban resilience as "… acceptable level of flooding that an urban system can tolerate e.g. a system is able to function during and after flooding". This conceptualization looks at the capacity to accept and resist, recover, and learn so as to improve, thus moving to another equilibrium position. Resilience is viewed as an interaction between social, economic, physical, natural, and institutional factors.

3 Methods

This review was conducted following the preferred reporting items for systematic reviews and meta-analysis (PRISMA) protocol (Shamseer et al. 2015). While using the PRISMA protocol, the search strategy, inclusion and exclusion criteria used, the screening and eligibility criteria employed to select relevant articles for this synthesis, the design, and the analysis method are shown in the next subsections of this section. Studies were synthesized to assess how urban resilience is being understood in SSA and what determines urban resilience in the context of SSA, at the same time exploring the scope of the studies, the shocks and stresses assessed, the type of data used, and the type of analysis used by the studies that shed light on the determinants of urban resilience.

3.1 Search Strategy (Table 1)

Table 1 The search terms, databases searched, and articles generated in the search

Search terms	Databases searched	Articles generated, included, and excluded in the review
Urban resilience*SSA	Web of Science Core Collection	95 articles were generated, 3 were included, and 92 were excluded
	Personal knowledge and Google Alerts	Five articles were generated and included in the review
Urban or city resilience determinants*SSA	Google Scholar	17,200 articles were generated, 3 were included, and 17,197 were excluded
	ScienceDirect	614 articles were generated, and 3 studies were included, while 611 were excluded
	Personal knowledge and Google Alerts	Seven articles were generated and included

3.2 Inclusion and Exclusion Criteria (Table 2)

Table 2 Inclusion and exclusion criteria of the different categories of studies searched

Category of literature	Inclusion	Exclusion and remarks
Types of studies	Empirical studies Conceptual Theoretical	Was not an empirical study on urban resilience in SSA Did not focus on the concept of urban resilience in the context of SSA Did not focus on the theory of urban resilience in SSA
Focus	Understanding urban resilience in SSA Determinants of urban resilience in SSA	Did not define urban resilience and if the study was not conducted in SSA Did not focus on what contributes to urban resilience in SSA
Publication type	Journal articles, books, book chapters, reports, thesis, working papers, conference proceedings, and grey literature	Editorials, newspaper articles, encyclopedia articles, magazine articles, statutes, hearings, cases, patents, and television broadcasts
Language	English	Literature in other languages
Time period (publication year)	2000–2018	Papers published earlier were excluded

3.3 Screening and Eligibility Criteria (Fig. 1)

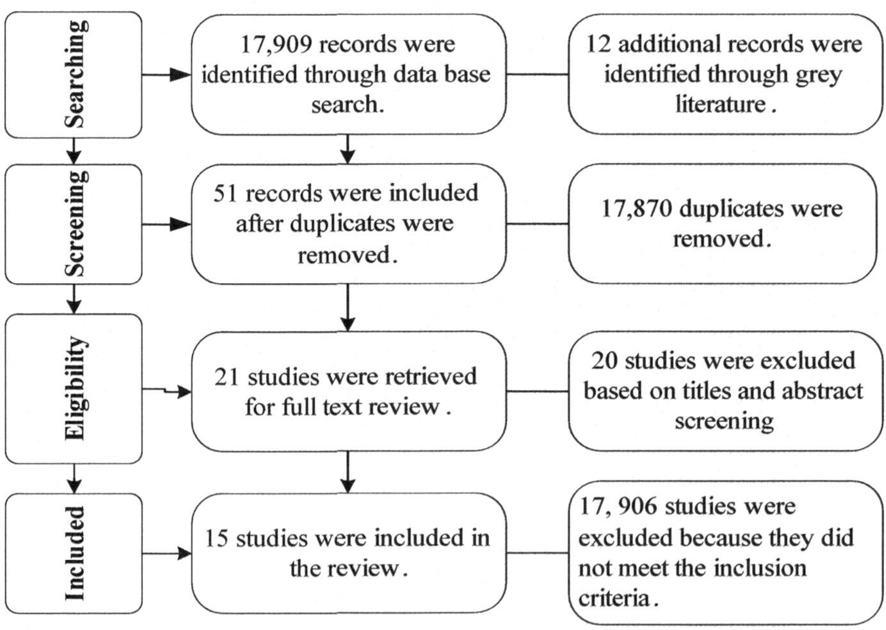

Fig. 1 Eligibility and screening criteria for the studies searched on how urban resilience is understood and the determinants of urban resilience in SSA

3.4 *Design and Analysis*

The study considered articles that applied qualitative, quantitative, and mixed methods. The reviewed articles were analysed using summative content analysis (Potter and Levine-Donnerstein 1999). Content analysis is a method which is embedded in systematic narrative synthesis where data is analysed narratively (Snilstveit et al. 2012). The analysis involved developing categories priori, coding using both open and vivo codes (Corley and Gioia 2004), and creating frequency counts to identify key findings. The analysis follows purely an inductive approach where prior knowledge does not influence the process of coding and generation of categories (Braun and Clarke 2006). This method of analysis has been popularized by many scholars to be used in literature review studies e.g., by Dixon-Woods et al. (2005), Popay et al. (2006), and Snilstveit et al. (2012). Furthermore, this method of analysis was selected based on its suitability for synthesizing a mix of qualitative and quantitative studies (Snilstveit et al. 2012). The questions that addressed how urban resilience is being understood and the determinants of urban resilience while exploring the geographical scope of the reviewed studies, the shocks and stresses assessed by studies, and the methodology used were systematically applied to each article. The results of the study were presented in tables, texts, and figures to summarize the findings of the reviewed articles.

4 Results and Discussion

4.1 *Urban Resilience in Sub-Saharan Africa*

Understanding resilience to shocks and stresses is now an important issue in the global policy agenda (New Urban Agenda 2016; UNISDR 2015). Urban areas in SSA play a significant role in driving economic growth and poverty reduction, 472 million people live in urban areas of SSA, 50 percent of the gross domestic product (GDP) in SSA is generated in urban areas, and most of the non-agricultural jobs are located in these urban areas (Saghir and Santoro 2018). However, urban areas in SSA are being threatened by several shocks and stresses due to the limited infrastructure, high poverty levels, and low service provision (Pharoah and Ross 2016). These can significantly reverse the development trajectory of countries in SSA. Therefore, understanding what constitutes urban resilience in SSA could guide policymakers and practitioners on where to put more resources so as to enhance urban resilience. In addition, it can guide researchers in the formulation of relevant research questions. This section is divided into two parts, i.e. studies specific to a country and those that cover more than one country in SSA.

Country-Specific Studies

FAO-UNICEF-WFP (2014) conducted a study in Dolow, Somalia, a region that has suffered from several calamities ranging from wars, prolonged droughts, and dysfunctional government. The purpose of their study was to evaluate the performance of help given by international agencies in building household resilience and the important building blocks of household resilience in both the urban and rural context. The study employed the Resilience Index Measurement and Analysis (RIMA) framework while defining resilience as "the ability to withstand, accommodate, and transform in the face of shocks and stresses". In this essence, resilience is understood as a dynamic processes built over time. They explain that resilience of a household depends on five factors, i.e. access to basic social services, productive assets, adaptive capacity, social safety networks, and sensitivity. The authors further explain that more access to basic social services such as schools, health, water, electricity, markets, sanitation, etc. by the household would contribute to more resilience. More productive assets owned by a household such as land, livestock, business, and other durables would enable a household to respond, recover, and adapt easily when faced with shocks and stresses, thus becoming more resilient. Furthermore, adaptive capacity constituting things like diversified income-generating activities can increase the capacity of a household to adapt to new situations and develop new livelihoods when faced with shocks and stresses, hence contributing to more resilient households. Additionally, households that derive their income from shock-affected activities are highly sensitive, thus less resilient.

Another study by D'Errico and Di Giuseppe (2018) was conducted in Northern, Eastern, and Central Uganda, focusing on what drives household resilience to food security from a low-to a high-profile resilience in both urban and rural context. The authors define household resilience as "the capacity that ensures that stressors and shocks do not have long lasting adverse development consequences". The study understands resilience as dynamic and time-dependent concept. The authors further explain that resilience is a concept which is multidimensional and context specific. Thus the drivers of resilience may vary from one region to another. The multidimensional components of resilience that the authors considered in their study were drawn from the RIMA framework (FAO 2012), i.e. income and food access, access to basic needs, assets, social safety nets, and adaptive capacity similar to the aforementioned study by FAO-UNICEF-WFP (2014). The study understands that the availability and access of the variables in these four dimensions by a household can boost the ability of a household to move from a low to a high level of resilience.

Furthermore, Dobson et al. (2015) studied the local and participatory approaches to resilience building in the informal settlements in Uganda's urban areas. They postulate that building resilience in the urban areas where the majority of the population lives in slums will depend on government support and on the active participation of the urban poor and their partnership with their local leaders. According to Dobson et al., urban resilience can best be understood as being a function of the role of resilient agents, which should include the local participation of urban residents from low-income groups living in informal settlements, alongside urban systems (infrastructure), and institutions (Tyler and Moench 2012). In essence, resilience is viewed mostly from societal and organizational point of view.

More still, Kotzee and Reyers (2016) conducted a study in Eden District in South Africa and piloted a socio-ecological index to measure flood resilience. In their study, resilience can best be understood by assessing the interaction between the social, organizational, and ecological systems. The authors indicate that the ability of a socio-ecological system to adapt and be able to take advantage of the shocks and stresses is based on the characteristics of vulnerability (the degree of harm owing to exposure and sensitivity to a specific hazard and absence of the capacity to adapt) and adaptive capacity (the ability of actors within the system to influence resilience via collective action and learning) (Walker-Springett et al. 2017). In this view, therefore, resilience is understood from the social, organizational, and ecological point of view.

Additionally, Mugume et al. (2015) conducted a study to evaluate the performance of Kampala's drainage systems when subjected to a wide range of structural failure scenarios, emanating from random cumulative link failure. The study adopts the engineering view of understanding resilience (Holling 1996) as "system's ability to maintain its basic structure and patterns of behaviour...". In this essence, resilience is meant to ensure continuity and efficiency of system's function during and after failure (Juan-García et al. 2017; Mugume et al. 2015). Resilience, from this perspective, is the maintenance of the status quo.

Finally, Woolf et al. (2016) developed and piloted a framework to assess resilience-related projects in Kibera in Nairobi, Kenya. Here urban resilience of slum dwellers is understood as an interaction between the asset base, external resources, capacities (resourcefulness, adaptive and flexible, and learn), and qualities (robust, well located, diverse, redundant, and equitable).

Studies Covering More Than One Country in SSA

Dobson (2017) conducted a study on community-driven pathways for urban resilience building in SSA. He employs the City Resilience Index (CRI) to frame his study while emphasizing the importance of organized communities and community-based slum upgrading in building urban resilience. The author defines resilience as a technical and social construct which can be explained through infrastructure design, household capacities, civic organizations, government, and business. He further explains that much as the physical, financial, and social assets accumulated over time provide the foundational ability for disaster response, the social capital is what always provides the flexibility and support to build back better after shocks. In this essence, resilience is understood as a dynamic process build over time by organized communities and slum upgrading. Thus, viewing resilience as being more of a social and organizational construct. The author argues that in order to understand the resilience of urban areas in SSA, the following three dimensions of resilience drawn from the City Resilience Index (CRI) are important, i.e. health and well-being, economy and society, and leadership and society.

Furthermore, Pharoah and Ross (2016) conducted a study in three cities of SSA, i.e. Dakar in Senegal, Banjul in the Gambia, and Harare in Zimbabwe. They focused on identifying common urban hazards to understand the underlying vulnerabilities that drive risks at national, metropolitan, and community level and to identify processes that can build resilience of the people living in urban areas. The authors defined resilience as "the ability to recover from disasters and adapt to changing conditions. Building resilience reduces vulnerabilities and involves helping people to mitigate risks before, during and after disasters". Similarly, this definition looks at resilience as a social construct, dynamic in nature, and built over time.

Urban resilience studies in SSA understand resilience severally, i.e. (a) those that understand resilience as a social construct and emphasizing that resilience of urban communities is a result of social and institutional factors. However, Kotzee and Reyers (2016) indicate that this understanding of resilience has a potential limitation of ignoring the biophysical component, which has been shown to play an important role in determining resilience (Joerin et al. 2014; Dhar and Khirfan 2017; Kamh et al. 2016; Hung et al. 2016; Abdrabo and Hassaan 2015). (b) Those that understand resilience as a socio-ecological construct (Kotzee and Reyers 2016; D'Errico and Di Giuseppe 2018) and (c) those that understand resilience from the engineering perspective as maintenance of pre-disaster status, i.e. Mugume et al. (2015) and Mugume and Butler (2017). In addition, most studies take on the current popular understanding of resilience as a dynamic process while a few from the engineering perspective understand resilience as more of maintenance of the status quo (Mugume et al. 2015; Mugume and Butler 2017).

4.2 Determinants of Urban Resilience in Sub-Saharan Africa

More than half of the 7.7 billion world's population now lives in urban areas, making the globe experience the largest urban growth in its history (Saghir and Santoro 2018; Pharoah and Ross 2016). It is projected that, by 2020, the number of people that live in urban areas will increase to approximately 5 billion with a greater proportion of the increase experienced in Africa and Asia (Pharoah and Ross 2016). This will particularly affect the resilience of the poor population; and thus there is a need to understand the factors that determine urban resilient to natural and human-induced shocks. This can guide the selection of the most effective combination of short and long - term strategies that can be beneficial in delivering urban areas from being vulnerable (FAO-UNICEF-WPF 2014).

A synthesis of studies on urban resilience to natural and human-induced shocks and stresses in SSA (Table 3) reveals the following determinants of urban resilience (Fig. 2): access to basic services represents 25%, social networks (25%), employment (5.1%), ownership of productive assets (5.1%), none involvement in agricultural activities (5.1%), building flood retention facilities in urban areas (5.1%), knowledge about disasters (5.1%), environmental preservation (5.1%), preparedness (2.5%), per capita expenditure (2.5%), government commitment to

Table 3 Studies that give light on determinants of urban resilience in SSA

Author	Spatial scope	Shocks/stresses	Data	Methods	Key determinants of resilience
D'Errico and Di Giuseppe (2018)	Northern, Eastern, and Central Uganda	Food security	Objective data	Multinomial logit regression	Self-employment, Non-involvement in agricultural activities Low dependency ratio Household type Education Per capita expenditure
Dobson (2017)	More than two region in SSA	Natural and human induced	Objective	Document analysis	Community-driven slum upgrading Partnership with organized communities
Dobson et al. (2015)	Uganda	Climate change induced	Subjective	–	Savings Manufacturing environmentally friendly building and cooking materials Urban profiling Improving access to clean water and drainage systems Partnership with organized communities
Tutu (2012)	Accra, Ghana	Natural and human induced	Subjective	Analysis of variance, chi-square, and ordinal regression	Employment Social capital Medication Number of stresses experienced
Tippens (2016)	Nairobi, Kenya	Psychological stress	Subjective	Thematic content analysis	Establishing borrowing networks Trust in religious communities and faith in God
Tawodzera (2012)	Zimbabwe	Food security	Subjective	Thematic content analysis	Multiple income-generating activities Strong social links with rural households International remittances
Pharoah and Ross (2016)	Senegal, Gambia, and Zimbabwe	Natural and human induced	Objective and subjective	Descriptive statistics and content analysis	Increasing commitments by governments Petty and cross-border trade Networks at different levels Mobile phones

(continued)

Table 3 (continued)

Author	Spatial scope	Shocks/stresses	Data	Methods	Key determinants of resilience
FAO-UNICEF-WPF (2014)	Dolow, Somalia	Natural and human induced	Subjective	Factor analysis	Productive assets Access to basic needs Adaptive capacity
Kotzee and Reyers (2016)	Eden District, South Africa	Floods	Objective	Principle component analysis and geographical information techniques (GIS)	Social factor (housing, health, education, and networks)
Mugume et al. (2015)	Kampala, Uganda	Flash floods	Objective	Modelling	Building more flood storage tanks Temporary storage areas or increasing spare capacity Building more centralized detention ponds upstream
Mugume and Butler (2017)	Kampala, Uganda	Flash floods	Objective	Modelling	Improving urban drainage systems to accommodate floods during heavy rainfall events occurring at short durations
Lwasa et al. (2011)	East and West Africa	Food security and climate change	Documentary review	Content analysis	Supportive infrastructure Reforms on institutional architecture and policy International support
Jones and Samman (2016)	Tanzania	Floods	Subjective	Ordinal logistic regression and descriptive statistics	Previous knowledge about floods

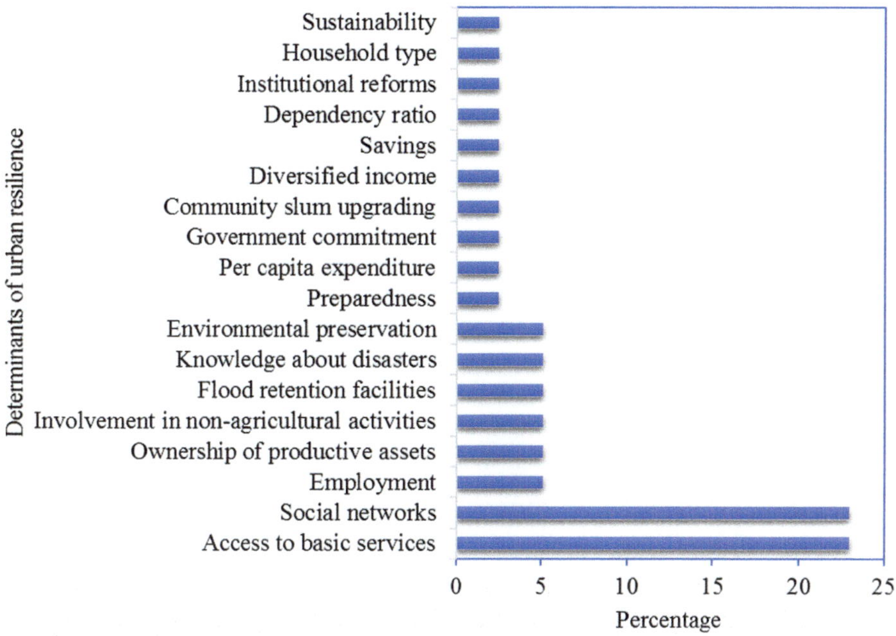

Fig. 2 Determinants of urban resilience to natural and human-induced shocks and stresses in SSA

enhance urban resilience (2.5%), community slum upgrading (2.5%), diversified income-generating activities (2.5%), savings (2.5%), dependency ratio (2.5%), institutional reforms (2.5%), household type (2.5%), and sustainability (2.5%). Access to basic services, social networks, employment, ownership of productive assets, none involvement in agricultural activities, building flood retention facilities in urban areas, knowledge about disasters, and environmental preservation are the most reported determinants of urban resilience in SSA. The importance of these most reported factors in determining urban resilience can further be explained in the next paragraphs.

Access to Basic Services

Access to robust, flexible, and inclusive basic services such as water, electricity, health, education, housing, sanitation, and transportation can facilitate the well-functioning of an urban area and can enhance the ability of urban residents to respond, absorb, and recover easily when faced with natural and human-induced shocks and stresses (The Rockefeller Foundation/Arup 2014; Dobson et al. 2015; FAO-UNICEF-WFP 2014; Kotzee and Reyers 2016). For example, Tutu (2012) reported that access to medication significantly contributed to the resilience capacity of young migrants living in the slum of Fadama in Accra in Ghana.

FAO-UNICEF-WFP (2014) found out that access to basic services (i.e. water, electricity, toilets, waste disposal, education, means of transport, and markets) significantly contributed to resilience more than other dimensions of resilience in Dolow in Somalia. Similarly, Kotzee and Reyers (2016) reported that social factors such as access to housing, education, and health significantly contributed to community resilience in Eden District in South Africa.

Social Networks

The importance of social networks such as networks with relatives, friends, non-governmental organizations, charity organizations, etc. has severally been reported as important determinants of resilience (Albrecht 2018; Aldrich and Meyer 2015; Sadri et al. 2017). A strong network of urban households with the rural community was shown to have a significant contribution towards the resilience of urban households to food in security resilience in Harare in Zimbabwe (Tawodzera 2012). One of the factors that will drive urban resilience in SSA Africa is networks with organized communities such as Slum Dwellers International (SDI), Rockefeller Foundation International, Cities Alliance, and United Cities and Local Governments of Africa (UCLG-A) (Dobson 2017). SDI, for example, which is the largest federation of slum dwellers in the world, and Slum Dwellers Federation (SDF) have improved over 13,000 houses in South Africa; they have also given household access to solar energy in the East, West, and Southern parts of Africa. In Malawi, the SDF has provided access to over 800 Ecosan (dry composting) toilets, serving over 14,400 people (Dobson 2017). The same author reports that SDI are also operating in Kenya, Namibia, and Malawi, and they provide similar services to enhance household resilience. Tippens (2016) reported that, in Kenya, the resilience of urban refugees has been enhanced through establishing borrowing networks with religious communities. Therefore, the importance of social networks in enhancing the resilience of urban population in SSA can never be underestimated.

Employment

Employment has been proved to be able to enhance the resilience of urban people in SSA especially employment in non-agricultural activities (D'Errico and Di Giuseppe 2018; Tutu 2012). Tutu (2012) found that employment significantly improved the resilience of youth in Accra in Ghana. Similarly, D'Errico and Di Giuseppe (2018) found that employment in non-agricultural activities significantly contributed to household resilience to food security in Northern, Eastern, and Central Uganda. The importance of employment is because it provides people with a source of income which is important in providing access to basic needs of life, thus enhancing the people's capacity to thrive during shocks and stresses.

Productive Assets

Productive assets such as land, livestock, and durables are important determinants of resilience because they are key elements of livelihood that facilitate production of consumable and tradeable commodities (FAO-UNICEF-WFP 2014; The Rockefeller Foundation/Arup 2014). Households who had more productive assets were more resilient than those with limited productive assets in Dolow in Somalia (FAO-UNICEF-WFP 2014).

Building Flood Retention Facilities

Because of the highly humanized nature of the urban environment with many surfaces compacted, much of the run-off during heavy precipitation events rushes to low-lying areas, presenting a challenge of flash floods and floods to those communities. Mugume et al. (2015) report that such challenges can be reduced by establishing flood retention facilities such as dams upstream.

Findings on shocks and stresses assessed by studies reveal that 38.5% of studies were on determinants of urban resilience to climate change-induced shocks and stresses, 38.5% were on resilience to both natural and human-induced shocks and stresses, 15.4% were on resilience to food insecurity, and 7.7% were on resilience to psychological stress (Fig. 3). The large number of studies on urban resilience to climate-related shocks and stresses (i.e. heavy precipitation events, flash floods, floods, and prolonged droughts) confirms reports that climate change is one of the key challenges that cities in the world are contending with (Bozza et al. 2015; Forino et al. 2016; Joerin et al. 2014; Kamh et al. 2016). And in an attempt to search for better options to enable urban areas anticipate and prepare, absorb and recover, and adapt when faced with shocks and stresses, scholars are becoming more and more interested in the subject of urban resilience. Furthermore, many studies also assessed determinants of urban resilience to both natural and human-induced

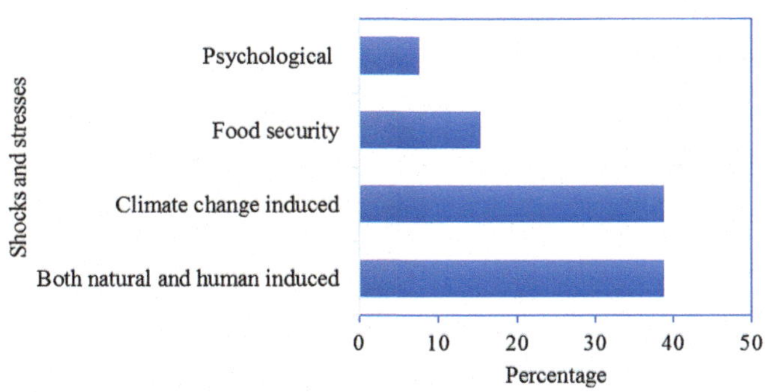

Fig. 3 Shocks and stresses focused by studies that shed light on the determinants of urban resilience in SSA

shocks and stresses. Kotzee and Reyers (2016) state that in order to account for the varied ecosystem characteristics and processes related to a particular shock or stress, resilience studies which are specific to a particular shock or stress would enable the selection of relevant indicators and parameters; at the same time, they would correctly capture the role, location, and condition of ecosystem services.

Results on the spatial scope of studies that shed light on the determinants of urban resilience to natural and human-induced shocks and stresses (Fig. 4) reveal that 53.3% of the studies were conducted in the cities of East African region (i.e. Nairobi in Kenya, Kampala in Uganda, Dar es Salaam in Tanzania, and Dolow in Somalia), 20% in South African cities (i.e. Eden in South Africa and Harare in Zimbabwe), 20% in West African cities (i.e. Accra in Ghana, Dakar in Senegal, and Banjul in The Gambia), and 6.7% were conducted in SSA in general. This finding shows that much of the focus was in the capital cities with less attention to secondary urban area.

Results on the type of data used by studies that shed light on determinants of urban resilience in SSA (Fig. 5) show that 46% of the studies used subjective data, 39% used objective data, while only 15% used both subjective and objective data.

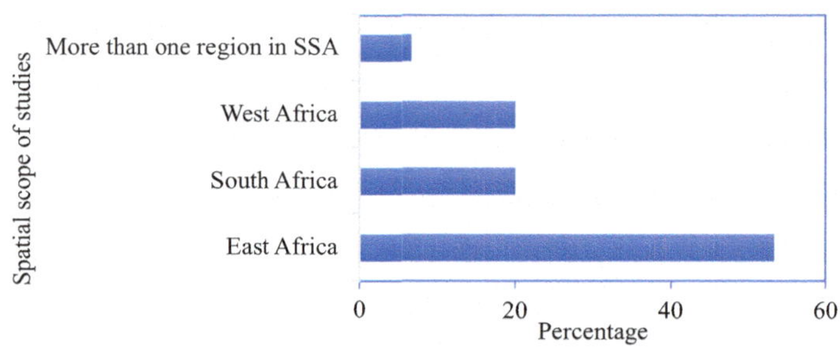

Fig. 4 The spatial scope of studies that shed light on the determinants of urban resilience to natural and human-induced shocks and stresses in SSA

Fig. 5 Type of data used by studies that shed light on urban resilience to natural and human-induced shocks and stresses in SSA

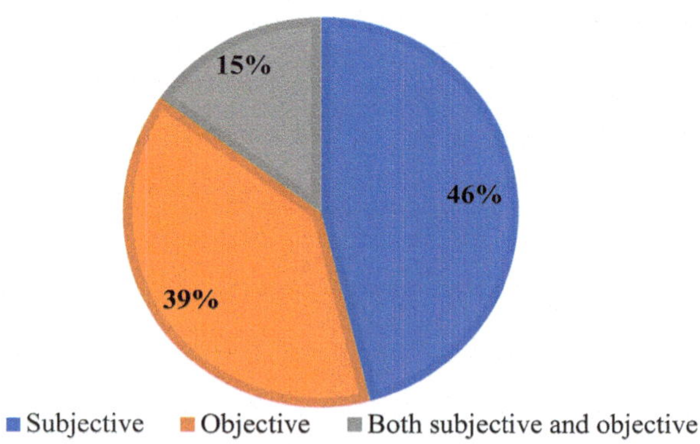

Subjective data is data obtained from conducting surveys, while objective data is secondary data that requires robust socio-economic and biophysical data sets. Using either objective or subjective data has both strengths and weaknesses. For example, subjective methods can help a researcher to understand the intangible livelihood assets such as social networks, sense of place, beliefs, and cultural identity which are difficult to measure using objective means (Jones and Tanner 2015; Wong-Parodi et al. 2015). Subjective methods also help in identifying improvements in resilience over time by using consistent assessors, measuring the marginal effect of a resilience enhancement strategy, and promoting the use of context specific knowledge (Choptiany et al. 2017; Rader et al. 2013). However, subjective methods may be influenced by inter-assessor bias, hence limiting generalization to different locations or between different organizations. On the other hand, objective data is drawn from robust socio-economic and biophysical data sets, making resilience assessment faster, does not need much resources, reduces inter-assessor bias, and offers a high level of generalization. However, objective data sets are often lacking in most developing countries, making subjective methods suitable in such localities. Because of the weaknesses embedded in both objective and subjective methods, scholars now emphasize the use of both objective and subjective methods in a single study to alleviate the weaknesses of using a single method (Rader et al. 2013; Jones and Tanner 2015; The Rockefeller Foundation/Arup 2014).

Results on methods of data analysis used by studies that shed light on determinants of urban resilience in SSA (Fig. 6) indicate that 31% of the studies used thematic content analysis, 25% used regression analysis, 12.5% used factor analysis, 12.5% used descriptive statistics, and 12.5% used modelling techniques, while 6.5% used geographical information system techniques. Most studies therefore used regression analysis and content analysis. The prevalent use of regression analysis could be because of its strength to tell significant predictors of resilience as an outcome variable and thus able to show which parameters are important in predicting urban resilience to natural and human shocks and stresses.

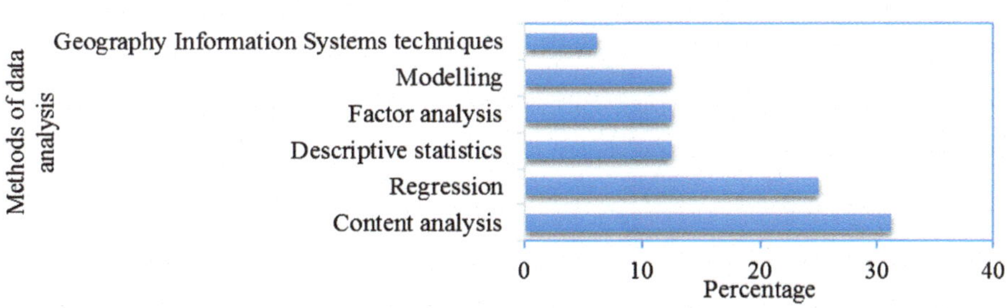

Fig. 6 Methods of data analysis used by studies that shed light on urban resilience to natural and human-induced shocks and stresses in SSA

5 Conclusion

Studies that shed light on the determinants of urban resilience to natural and human-induced shocks and stresses in cities of SSA can be categorized into two classes: (i) those that understand resilience as a function of social and organizational systems and (ii) those that view resilience as a function of the social, organizational, and ecological systems.

Findings suggest that several factors determine the resilience of urban areas to natural and human-induced shocks and stresses in SSA. The most important determinants include access to basic services, social networks, employment, ownership of productive assets, non-involvement in agricultural activities, building flood retention facilities in urban areas, knowledge about disasters, and environmental preservation. The majority of the studies pay limited attention to resilience to specific shocks or stresses, and yet this would enable accurate selection of indicators and parameters that are relevant to a particular shock or stress. Most of the studies were carried out in big cities of East Africa, South Africa, and West Africa with limited attention to secondary cities. Most studies either used objective data or subjective data and those that use both are limited. Regression analysis and thematic content analysis were the most used analysis methods employed by studies with limited use of GIS and modelling techniques.

Based on the above conclusions, this review suggests that more studies on the determinants of urban resilience need to consider a multi-dimensional approach to resilience, covering: the social, economic, organizational, and ecological dimensions of resilience. There is also a need to focus more on assessing resilience to specific shocks and stresses, more so on secondary urban areas of SSA while employing both subjective and objective methods in any single study since these help to complement each other.

Acknowledgements This book chapter represents the views of the named author only. The guide, motivation, and supervision from Assoc. Prof. Yazidhi Bamutaze, Prof. Petter Pilesjö, Assoc. Prof. Jonas Ardö, and Prof. Giuliano Di Baldassarre are acknowledged. The author also acknowledges the generous funding provided by Makerere-SIDA bilateral program phase IV (2015–2020) under the Building Resilient Ecosystems and Livelihoods to Climate Change and Disasters Risks (BREAD) project, grant number: 331 in the Department of Geography, Geo-Informatics and Climatic Sciences at Makerere University in Uganda in collaboration with Lund, and Uppsala Universities in Sweden. The project has provided funding for my university fees, travels, fieldwork activities and the living allowance for my PhD studies.

References

Abdrabo MA, Hassaan MA (2015) An integrated framework for urban resilience to climate change - case study: sea level rise impacts on the Nile Delta coastal urban areas. Urban Climate 14(2015):554–565. Available at: https://doi.org/10.1016/j.uclim.2015.09.005

Albrecht F (2018) Natural hazard events and social capital: the social impact of natural disasters. Disasters 42(2):336–360

Aldrich PD, Meyer M (2015) Social capital and community resilience. Am Behav Sci 59(2):254–269

Ayyoob S (2016) A critical review of selected tools for assessing community resilience. Ecol Indic 69(2016):629–647. Available at: https://doi.org/10.1016/j.ecolind.2016.05.023

Batica J, Gourbesville P (2016) Resilience in flood risk management - a new communication tool. Procedia Engineering 154:811–817. Available at: https://doi.org/10.1016/j.proeng.2016.07.411

Béné C et al (2017) Resilience as a policy narrative: potentials and limits in the context of urban planning. Climate Dev 5529(January):1–18

Bottazzi P et al (2018) Measuring subjective flood resilience in suburban Dakar : a before – after evaluation of the "live with water" project. Sustainability 10(2018):12–35

Boyd E, Juhola S (2014a) Adaptive climate change governance for urban resilience. Urban Stud J Limited 2014:1–31. Available at: usj.sagepub.com

Boyd E, Juhola S (2014b) Adaptive climate change governance for urban resilience. Urban Stud 52(7):1234–1264. Available at: http://usj.sagepub.com/cgi/doi/10.1177/0042098014527483

Bozza A, Asprone D, Manfredi G (2015) Developing an integrated framework to quantify resilience of urban systems against disasters. Nat Hazards 78(3):1729–1748

Braun V, Clarke V (2006) Using thematic analysis in psychology. Qual Res Psychol 3(2):77–101

Choptiany JMH et al (2017) SHARP: integrating a traditional survey with participatory self-evaluation and learning for climate change resilience assessment. Climate Dev 9(6):505–517

Chun H, Chi S, Hwang BG (2017) A spatial disaster assessment model of social resilience based on geographically weighted regression. Sustainability 9(12):2222–2238. Available at: http://www.mdpi.com/2071-1050/9/12/2222

Corley KJ, Gioia DA (2004) Identify ambiguity and change in the wake of corperate spin-off. Adm Sci Q 49(2004):173–208

De Boer J (2016) Conceptualizing city fragility and resilience. Available at: https://i.unu.edu/media/cpr.unu.edu/attachment/2227/WP05.02_Conceptualizing_City_Fragility_and_Resilience.pdf

D'Errico M, Di Giuseppe S (2018) Resilience mobility in Uganda: a dynamic analysis. World Dev 104:78–96. Available at: http://linkinghub.elsevier.com/retrieve/pii/S0305750X17303820

Dhar TK, Khirfan L (2017) A multi-scale and multi-dimensional framework for enhancing the resilience of urban form to climate change. Urban Climate 19:72–91. Available at:. https://doi.org/10.1016/j.uclim.2016.12.004

Dixon-Woods M et al (2005) Synthesysing qualitative and quantitative evidence: a review of possible methods. J Health Serv Res Policy 10(1):45–53. https://doi.org/10.1258/1355819052801804e

Dobson S (2017) Community-driven pathways for implementation of global urban resilience goals in Africa. Int J Disaster Risk Reduct 26(September):78–84. Available at: https://doi.org/10.1016/j.ijdrr.2017.09.028

Dobson S, Nyamweru H, Dodman D (2015) Local and participatory approaches to building resilience in informal settlements in Uganda. Environ Urban 27(2):605–620. Available at: http://journals.sagepub.com/doi/10.1177/0956247815598520

Elena C, Mariella DF, Fabio F (2017) Energy resilience of historical urban districts: a state of art review towards a new approach. Energy Procedia. Elsevier B.V.:426–434. Available at:. https://doi.org/10.1016/j.egypro.2017.03.204

FAO (2012) Resilience index: measurement and analysis model, Fiat

FAO-UNICEF-WFP (2014) Household resilience in Dolow, Somalia

Forino G, MacKee J, von Meding J (2016) A proposed assessment index for climate change-related risk for cultural heritage protection in Newcastle (Australia). Int J Disaster Risk Reduct 19(2016):235–248. Available at:. https://doi.org/10.1016/j.ijdrr.2016.09.003

Greenwalt J, Raasakka N (2018) Building urban resilience to address urbanization and climate change. In: Raasakka N, Alverson K, Greenwalt J (eds) . Elsevier Inc.. Available at: https://www.sciencedirect.com/science/article/pii/B9780128118917000128

Habitat III (2016) New Urban Agenda. In: Habitat III Conference. Quito, p 24. Available at: http://www.eukn.eu/news/detail/agreed-final-draft-of-the-new-urban-agenda-is-now-available/

Holling CS (1973) Resilience and stability of ecological systems. Annu Rev Ecol Syst 4(1973):1–23

Holling CS (1996) Engineering resilience versus ecological resilience. Eng Ecol Constraints 31(1996):31–44

Hung HC et al (2016) Building resilience: mainstreaming community participation into integrated assessment of resilience to climatic hazards in metropolitan land use management. Land Use Policy 50:48–58

Ilmola L (2016) Approaches to measurement of urban resilience. Springer, Switzerland, pp 207–237

IPCC (2015) Climate change 2014 synthesis report, Switzerland. Available at: http://www.ipcc.ch

Joerin J et al (2014) The adoption of a climate disaster resilience index in Chennai, India. Disasters 38(3):540–561

Jones L, Samman E (2016) Measuring subjective household resilience. Overseas Development Institute (ODI), London, UK

Jones L, Tanner T (2015) Measuring "subjective resilience" using people' s perceptions to quantify household resilience, (July)

Juan-García P et al (2017) Resilience theory incorporated into urban wastewater systems management. State of the art. Water Res 115(2017):149–161. Available at:. https://doi.org/10.1016/j.watres.2017.02.047

Kamh YZ, Khalifa MA, El-Bahrawy AN (2016) Comparative study of community resilience in mega coastal cities threatened by sea level rise: the case of Alexandria and Jakarta. Procedia Soc Behav Sci 216(October 2015):503–517. http://linkinghub.elsevier.com/retrieve/pii/S187704281506187X

Kotzee I, Reyers B (2016) Piloting a social-ecological index for measuring flood resilience: a composite index approach. Ecol Indic 60:45–53

Lee EH, Kim JH (2017) Development of resilience index based on flooding damage in urban areas. Water (Switzerland) 9(6):1–15

Lwasa S et al (2011) The role of urban and peri-urban agriculture in enhancing food security and climate change resilience in East and West African cities. Available at: https://www.google.com/#q=The+Role+of+Urban+and+Peri-Urban+Agriculture+in+Enhancing+Food+Security+and+Climate+Change+Resilience+in+East+and+West+African+Cities+Abstract

Meerow S, Newell JP, Stults M (2016) Defining urban resilience: a review. Landsc Urban Plan 147(2016):38–49. Available at:. https://doi.org/10.1016/j.landurbplan.2015.11.011

Mugume SN, Butler D (2017) Evaluation of functional resilience in urban drainage and flood management systems using a global analysis approach. Urban Water J 14(7):727–736. Available at:. https://doi.org/10.1080/1573062X.2016.1253754

Mugume SN et al (2015) A global analysis approach for investigating structural resilience in urban drainage systems. Water Res 81:15–26. Available at:. https://doi.org/10.1016/j.watres.2015.05.030

NCEA (2015) Climate change profile Uganda. Retrieved from climatehelpdesk@minbuza.nl

Nsubuga FWN et al (2014) The nature of rainfall in the main drainage sub-basins of Uganda. Hydrol Sci J 59(2):278–299

Peng C et al (2017) A review of the theory and practice of regional resilience. Sustain Cities Soc 29:86–96. Available at:. https://doi.org/10.1016/j.scs.2016.12.003

Pharoah R, Ross S (ed) (2016) Strengthening climate resilience in African cities, Stellenbosch. Available at: http://cdkn.org/wp-content/uploads/2014/05/CDKN_ACC_WP_final_web-res.pdf

Popay J et al (2006) Guidance on the conduct of narrative synthesis in systematic reviews. In: Katrina R, Duffy S (eds). https://doi.org/10.13140/2.1.1018.4643

Potter WJ, Levine-Donnerstein D (1999) Rethinking validity and reliability in content analysis. J Appl Commun Res 27(3):258–284. https://doi.org/10.1080/00909889909365539

Rader et al (2013) Measuring the resilience of Cities: the role of big data. In: J.A. Rader and S. Cole (eds) Proceedings of the conference measuring the resilience of cities: the role of big data. Royal united services institute for defence and security studies, Whitehall, London, UK, pp 1–143, 25 Oct 2013. Available at: www.stfc.ac.uk, www.rusi.org

Ronak P, Nosal L (2016) Defining the resilient city: United Nations University Centre for policy research working paper 6. United Nations University
Sadri AM et al (2017) The role of social capital, personal networks, and emergency responders in post-disaster recovery and resilience: a study of rural communities in Indiana. Nat Hazards 90(3):1–30
Saghir J, Santoro J (2018) Urbanization in sub-Saharan Africa: meeting challenges by bridging stakeholders. Available at: https://en.oxforddictionaries.com/definition/sharing_economy
Shamseer L et al (2015) Preferred reporting items for systematic review and meta-analysis protocols (prisma-p) 2015: elaboration and explanation. BMJ (Online) 349(January):1–25. Available at: https://doi.org/10.1136/bmj.g7647
Snilstveit B, Sandy O, Martina V (2012) Narrative approaches to systematic review and synthesis of evidence for international development policy and practice. J Dev Effect 4(3):409–429
Solecki W (2015) ScienceDirect Hurricane Sandy in New York, extreme climate events and the urbanization of climate change : perspectives in the context of sub-Saharan African cities. Curr Opin Environ Sustain 13(2015):88–94. Available at: https://doi.org/10.1016/j.cosust.2015.02.007
Solecki W et al (2015) A conceptual framework for an urban areas typology to integrate climate change mitigation and adaptation. Elsevier. 14:116–137. https://doi.org/10.1016/j.uclim.2015.07.001
Suárez M et al (2016) Towards an urban resilience index: a case study in 50 Spanish cities. Sustainability 8(8):774. Available at: http://www.mdpi.com/2071-1050/8/8/774
Tawodzera G (2012) Urban household survival and resilience to food insecurity in crisis conditions: the case of Epworth in Harare, Zimbabwe. J Hunger Environ Nutr 7(2012):293–320. http://www.tandfonline.com/loi/when20%0AUrban
The Rockefeller Foundation/Arup (2014) City resilience index. Research report volume 3 urban measurement report, 3(May). Available at: http://publications.arup.com/Publications/C/City_Resilience_Framework.aspx
The Rockefeller Foundation/Arup (2016) Measuring city resilience, New York
Thompson HE, Berrang-Ford L, Ford DJ (2010) Climate change and food security in sub-Saharan Africa: a systematic literature review. Sustainability 2(8):2719–2733
Tippens JA (2016) Urban Congolese refugees in Kenya : the contingencies of coping and resilience in a context marked by structural vulnerability Urban Congolese Refugees in Kenya : The Marked by Structural Vulnerability
Tutu RA (2012) Self-rated resilience among young migrants in old Fadama, Accra, Ghana. GeoJournal 78(4):709–725
Tyler S, Moench M (2012) A framework for urban climate resilience. Climate Dev 4(4):311–326
UNISDR (2015) Making development sustainable: the future of disaster risk management. Global assessment report on disaster risk reduction, Tokyo, Japan, New York, U.S.A.
UNU-CPR (United Nations University Centre for Policy Research) (2016) Defining the resilient city. Stanford University, Stanford, USA and Washington D.C., USA
Waithaka M et al (2013) East African agriculture and climate change: a comprehensive analysis. Internal Food Policy Research Institute (IFRI)
Walker-Springett K, Butler C, Adger WN (2017) Wellbeing in the aftermath of floods. Health and Place 43(March 2016):66–74. Available at: https://doi.org/10.1016/j.healthplace.2016.11.005
Wong-Parodi G, Fischhoff B, Strauss B (2015) Resilience vs. adaptation: framing and action. Climate Risk Manag 10(2015):1–7. Available at: https://doi.org/10.1016/j.crm.2015.07.002
Woolf S et al (2016) Towards measurable resilience: a novel framework tool for the assessment of resilience levels in slums. Int J Disaster Risk Reduc 19:280–302. Available at: https://doi.org/10.1016/j.ijdrr.2016.08.003
Yuen B, Kumssa A (2011) In: Yuen B, Kumssa A (eds) Climate change and sustainable development in Africa and Asia First. Springer, Nairobi

Determinants of Strategies that Enhance Farmers' Resilience to Rainfall Variability in Mt. Elgon Region, Eastern Uganda

Narisi Mubangizi, Florence Birungi Kyazze, and Paul Isolo Mukwaya

Abstract Smallholder farmers whose livelihoods depend on rain-fed agricultural production need to be resilient to sustain their wellbeing in a context of highly variable rainfall patterns. This study set out to establish the strategies that build household resilience to rainfall variability and the conditions that influence the use of such strategies. The study employed a descriptive cross-sectional design. It had 12 focus group discussions with 123 participants, semi-structured interviews with 255 farmers, and in-depth interviews with 5 purposively selected households. Results indicate that only 30% of the smallholder farmer households were resilient to the effects of abnormal rainfall seasons. Five strategies including use of physical and financial reserves, agronomic practices, alteration of livelihood asset portfolio, diversification into off-farm ventures, and uptake of soil fertility and water conservation were associated with households that were resilient to effects of abnormal rainfall seasons. The use of these strategies was enhanced by livestock ownership, group membership, access to land, amount of regular financial savings, and use of climatic forecast information. Overall, these findings suggest that building smallholder farmers' resilience to rainfall variability requires a multipronged approach to improve farmers' agricultural production as well as expanding their asset base and off-farm livelihood options.

Keywords Smallholder farmers · Rainfall variability · Household wellbeing · Resilience building strategies

N. Mubangizi (✉) · F. B. Kyazze
College of Agriculture and Environmental Sciences, Makerere University, Kampala, Uganda
e-mail: mubnarisi@caes.mak.ac.ug

P. I. Mukwaya
Department of Geography, Geo-informatics and Climatic Sciences, College of Agricultural and Environmental Sciences, Makerere University, Kampala, Uganda

Y. Bamutaze et al. (eds.), *Agriculture and Ecosystem Resilience in Sub Saharan Africa*, Climate Change Management,
https://doi.org/10.1007/978-3-030-12974-3_23

1 Introduction

The majority of countries in Africa have land that is over 1500 m above sea level and therefore classified as mountainous (United Nations Environment Programme [UNEP] 2014). It is estimated that about 20% of Africa's surface area is mountainous (Alweny et al. 2014). East Africa is one of the most mountainous areas in Africa and is home to a number of mountains including Kilimanjaro, Mount Kenya, Rwenzori and Mount Elgon (UNEP 2014). The mountain ecosystems in East Africa just like elsewhere in the world are among the most fragile but significant ecosystems providing vital services and products to human beings (Mugagga et al. 2012; Jiang et al. 2014). These mountainous areas favor production of a range of agricultural commodities because of their fertile volcanic soils, cooler temperatures, and more stable rainfall patterns (Food and Agriculture Organization [FAO] 2013, 2014). As a result, rain-fed agriculture is the main source of livelihood for the inhabitants of the mountainous areas of East Africa.

However, due to favorable agricultural conditions, mountainous areas in East Africa are very densely populated. The Mt. Elgon which is shared between Eastern Uganda and Western Kenya is one of the most densely populated rural areas in East Africa. The population density in Mt. Elgon area is estimated at 900 persons per square kilometer (UNEP 2014). The high population density has led to shrinkage of farm sizes to as low as 0.5 acres in some areas (Kaggwa et al. 2010) and unsustainable land use practices such as deforestation (Jiang et al. 2014). The unsustainable land use practices have in turn led to severe land degradation processes including disastrous landslides and high rates of soil erosion (Jiang et al. 2014; Bamutaze 2015). As a direct consequence of high rates of soil erosion, the Mt. Elgon area experiences negative soil nutrient balances and declining agricultural yields (Bamutaze 2015).

In addition to the land degradation challenges, mountainous areas in East Africa are also among the areas most affected by climate change and climate variability. Mountainous areas are reported to be warming at a faster rate than the low-lying areas, and the warming trend is projected to continue within the foreseeable future. For instance, the Mbale subregion (Bududa, Manafwa, and Mbale districts) within Mt. Elgon region experienced an increase of between 0.4 and 1.2 °C in monthly temperatures during the 2001–2011 period when compared to the 1961–1990 (Bomuhangi et al. 2016). With regard to precipitation, the Mbale subregion has experienced a shift from April to May as the wettest month implying delayed onset of the rainfall season (Mbogga 2012). The increases in temperature and changes in the rainfall patterns have had adverse impacts on the agricultural production in the area. Already, substantial proportions of the population in the Mt. Elgon region have experienced food shortages due to climate-related challenges. For instance, the Uganda National Agricultural Census of 2008/2009 reported that about 89% and 80% of the agricultural households from Manafwa and Bududa districts in Mt. Elgon region, respectively, experienced food shortage during the month of April 2009 due to rainfall variability (Uganda Bureau of Statistics [UBOS] and Ministry of Agriculture, Animal Industry and Fisheries [MAAIF] 2010).

The adverse impacts of climatic changes in the Mt. Elgon region are likely to continue in the foreseeable future, given that climatic projections based on two emission scenarios (A1b and A2) from at least five general circulation models indicate an increase in temperature and more rainfall in the 2010–2039 period (Mbogga 2012; Ministry of Water and Environment [MWE] 2013; Bomuhangi et al. 2016). This implies that farmers in Mt. Elgon region whose livelihoods depend on rain-fed agricultural production need to have capacity to sustain their wellbeing in a context of highly unreliable climatic conditions. This capacity of farmers to maintain and/or bounce back to their normal wellbeing, when faced with change including rainfall variability, is what is referred to as resilience.

The concept of resilience was first applied in the engineering field by Tredgold in 1818 to explain why certain types of timber were able to accommodate sudden and severe loads without breaking (McAslan 2010). The resilience concept then spread to other subject areas including ecology where it was introduced by Holling in 1973 (McAslan 2010; Barrett and Constas 2014). Holling labeled the new concept as "ecological resilience" and used it as a measure of the ecosystem's ability to absorb changes and still continue to exist and function normally (Allinovi et al. 2010; McAslan 2010; Barrett and Constas 2014). Holling's use of the resilience concept in ecology provided the basis for using it in other fields including disaster management and dealing with impacts of climate change and climate variability (Barret and Constas 2014).

In this chapter, the term resilience is operationally defined as the ability of a household to maintain and/or bounce back to its normal wellbeing situation after being affected by rainfall variability. Currently, there is limited knowledge about smallholder farmers' resilience to rainfall variability because most studies have focused on their vulnerability to rainfall variability and its effects (Rwenzori Think Tank 2011; Mbogga 2012; Mubiru et al. 2012; Caffrey et al. 2013). Due to their emphasis on the vulnerability perspective, past studies have generally inventoried strategies used by smallholder farmers to cope with and adapt to changes in rainfall patterns without delving into the contribution of such strategies to resilience. This chapter therefore focuses on establishing the strategies associated with smallholder farmer households that are resilient to rainfall variability and the conditions that influence the use of such strategies. This chapter specifically (i) categorizes farmer households in terms of their resilience to the effects of rainfall variability, (ii) identifies resilience building coping and adaptation strategies employed by households in Mt. Elgon region, and (iii) presents the factors that influence the likelihood of use of resilience building strategies.

2 Materials and Methods

2.1 Description of Study Area

The study was conducted in Bududa and Manafwa districts located in Mt. Elgon region in Eastern Uganda. Bududa district lies at the foot of the southwestern slopes of Mt. Elgon in Eastern Uganda. It is geographically bound by latitudes 2° 49'

North and 2° 55' North, longitude 34° 15' East and 34° 34' East (National Environment Management Authority [NEMA] 2010). The district has a total of 211,683 people of whom 204,953 people, representing about 97%, are in the rural areas (UBOS 2014). Manafwa district is located between latitude 01° 01' North and longitude 34° 21' East. The district has a total population of 352,864 people of whom 330,103 people, representing about 94%, are rural based (UBOS 2014). The majority of the people in the two districts belong to the Gishu tribe whose local language is Lumasaba.

Both Bududa and Manafwa districts are described as medium to high altitude and are characterized by steep terrain. They both have high population densities of over 590 persons per km^2 (Mbogga 2012). The high population density results into people cultivating on very steep slopes and other marginal lands. The mean minimum and maximum temperatures are 15 °C and 23 °C, respectively, while annual rainfall is just above 1500 mm (Mugagga et al. 2012). The rainfall in the area follows a bimodal pattern, with the first and main season running from March to May and the second one running from September to November. However, the rainfall patterns in the area have reportedly become variable in terms of the timing of onset and cessation of rainfall seasons, distribution, and intensity of rainfall (Mbogga 2012; Kansiime et al. 2013). Specifically, the timing of the seasons has increasingly become unpredictable with the onset of the first season tending to shift from March to April, while the September to November season has increasingly tended to receive more rain concentrated over a short spell (Mbogga 2012).

The main source of livelihood is subsistence agricultural production of crops and livestock. The agricultural production is rain-fed since the area receives over 1500 mm of rainfall per year over two seasons which is adequate for the production of a variety of crops and rearing of animals. Bananas, Arabica coffee, and Irish potatoes are the major crops at higher elevations, while maize, millet, cassava, beans, sweet potatoes, and vegetables are dominant at lower elevations (UBOS & MAAIF 2010; Mbogga 2012).

2.2 Data Sources and Processing

Research Design

This study employed a descriptive cross-sectional design to gather data about farmers' wellbeing before, during, and after the perceived abnormal rainfall seasons and the strategies used to cope with and/or adapt to the effects of rainfall variability. The study consisted of three logically linked phases. The first phase was qualitative employing focus group discussions (FGDs) with selected representatives of smallholder farmer households. The FGDs provided insights about farmers' characterization of resilient and non-resilient households with respect to the effects of abnormal rainfall seasons. The information generated from the FGDs was used to fine-tune the data collection tool for the second phase and to identify households to participate in the third phase of the study. The second phase of the study was a household survey to assess household resilience to the effects of rainfall variability and identify strategies used to build such resilience. The third phase involved in-depth interviews with smallholder farmer households that were identified during FGDs in phase 1 as being fully resilient. The in-depth interviews focused on obtaining deeper insights about how and why smallholder farmer households differ in terms of their resilience to effects of rainfall variability.

Sampling

Bududa and Manafwa districts were purposively selected as cases from Mt. Elgon region because they are the most affected by rainfall variability in the region and Uganda at large. For instance, as a result of prolonged heavy rains interacting with the steep slopes, soil type, and human activities, Bududa district has had five major landslides between 1933 and 2010. These landslides have destroyed property and killed over 500 people (NEMA 2010; Staudt et al. 2014). Furthermore, both Bududa and Manafwa districts frequently experience household food shortages due to rainfall variability (UBOS and MAAIF 2010).

Within each of the two districts, three sub-counties that were considered to be the most affected by rainfall variability were purposively selected, namely, Bushika, Bukalasi, and Bumasheti from Bududa district and Khabutoola, Nalondo, and Bugobero from Manafwa district (Fig. 1). The selection of sub-counties was done with guidance from the respective District Production and Marketing Officers (DPMOs). In each sub-county, three villages were also purposively selected with guidance from the respective sub-county agricultural officers. Preference was given to villages that were perceived to have recently been affected by changes in rainfall patterns and their associated shocks.

A total of 123 farmers (62 males and 61 females) were purposively selected to participate in the FGDs. These were purposively selected based on sex, age, and implementation of strategies to mitigate, cope with, and adapt to the effects of rainfall variability shocks experienced in the study area. In terms of sex, efforts were made to select an equal number of male and female farmers so as obtain perspectives from both sexes about the occurrence and effects of rainfall-related shocks and how farmers are resilient to them. With regard to age, preference was given to farmers aged 40–60 years because these were considered to have more knowledge and experience with the rainfall variability shocks experienced in the study area. Additionally, preference was given to those farmers considered to be exemplary with regard to implementation of strategies to mitigate, cope with, and adapt to rainfall variability shocks experienced in their respective areas. The 123 farmers that met the above three criteria were identified with guidance from the sub-county agricultural officers and respective village local council chairpersons. On average, seven people were selected from each village to participate in the FGDs.

Fig. 1 Location of study sub-counties in Bududa and Manafwa districts

For households to be involved in the survey, the sampling frame comprised of all households in all the 18 selected villages in the 6 sub-counties. Lists of all households in the villages were generated by the respective village chairpersons and three other members of the local council 1 committee. The names of household heads from all the villages were compiled to form a sampling frame of 2124 households. Basing on this sampling frame, the sample size was computed using the formula by Krejcie and Morgan (1970). The calculated sample size was 236. However, to cater for the non-responses, the estimated sample size was increased by 10%, resulting into a sample of 255. The number of households drawn from each village was then obtained through proportionate stratified random sampling. For the third phase of the study (in-depth interviews), five households that were earmarked by FGD participants as being the most resilient to effects of rainfall variability were purposively selected.

Data and Methods

The study employed focus group discussions (FGDs), in-depth interviews, and face-to-face semi-structured interviews (SSIs) as the main methods for collecting qualitative and quantitative data, respectively. A total of 12 FGDs were conducted as the first phase of data collection. The FGDs were facilitated using a checklist developed with input from two scholars from Makerere University and reviewed by peers for content validity. The FGDs focused on gathering in-depth data from farmers on effects of rainfall variability, strategies used to cope with and adapt to the effects of rainfall variability, and the local indicators of resilient and non-resilient households. The data from the FGDs provided the indicators for household wellbeing that were included in the semi-structured interview tool.

The second phase of data collection involved conducting semi-structured interviews with 255 household heads or their spouses. The interviews focused on collecting household data on socioeconomic, demographic characteristics and household wellbeing. In operationalizing household wellbeing, this study drew insights from Narayan et al. (2000). Narayan et al. (2000) highlight five dimensions of wellbeing, namely, material wellbeing, bodily wellbeing, social wellbeing, security and freedom of choice, and action in all aspects of life. Among these five dimensions of wellbeing highlighted by Narayan et al. (2000), this study focused on material, bodily, and social wellbeing because they can be directly influenced by the outcomes of households' livelihood activities. Thus these three wellbeing dimensions were expected to be influenced by shocks such as rainfall variability which affect the household livelihood activities and their outcomes.

In order to capture the household status with regard to the three dimensions of wellbeing, the semi-structured interviews collected data on the following aspects in the households:

- Food consumption focusing on frequency of consumption of; adequate quantity of food, preferred foods, and having three or more meals per day. These specific indicators for food consumption were developed based on insights from the Household Food Insecurity Access Scale (HFIAS) developed by Coates et al. (2007).
- Access to nonfood basic items (salt, soap, and Vaseline).
- Children's school attendance.
- Access to healthcare for sick household members.
- Socialization and leisure. This was assessed in terms of attendance of meetings and activities of groups that the household was involved in and attendance of social gatherings like prayers, ceremonies, and going out for leisure.

Data on each of the above items was collected over three periods: before the most recent perceived abnormal rainfall season, 1–2 months after the most recent perceived abnormal rainfall season, and at the time of study (July 2016). Each household identified the most recent season that they considered to have been abnormal, and this was used as the reference point for the subsequent parts of the interview. To minimize challenges related to recall, respondents were restricted to focus on seasons in 3 years before the time of the study in July 2016 (2013, 2014, and 2015). The period for assessing the effects of abnormal rainfall season on household wellbeing was set at 1–2 months after its end not during the season because the food and income from such a season are consumed immediately after its end. For each of the three periods, data on all the indicators were collected in terms of frequency within a week and later computed to per month. For example, if a respondent reported that his/her household consumed an adequate quantity of food twice a week, this would be computed as 2/7*30, and this would be equivalent to 9 days in a month.

The SSI questionnaire was developed with input from two scholars based at Makerere University for clarity of the questions and content validity. The questionnaire was also pretested on 16 farmers from Ikaali parish in Bukhofu sub-county (not part of the study sub-counties) in Manafwa district to avoid spillover of information and contamination of the sample. Prior to data collection, research assistants were trained and involved in the pretesting to acquaint them with the tool and to ensure quality of the data collected. The third and last phase of data collection involved conducting in-depth interviews with five farmers who were purposively selected during the FGDs to generate deeper understanding on their resilience to the effects of rainfall variability and the associated shocks.

Data Analysis

Content analysis was used to analyze the qualitative data that was generated through FGDs and in-depth interviews. Survey data were analyzed using Statistical Package for Social Sciences (SPSS) version 21 to generate frequencies and percentages of farmers' sociodemographics. Means were also obtained for continuous attributes of households and wellbeing indicators. After this initial analysis, the households were then characterized in terms of resilience to the effects of abnormal rainfall seasons and strategies used to cope with and/adapt to the effects of abnormal rainfall seasons.

(i) *Categorization of Households in Terms of Resilience to Effects of Rainfall Variability*

Given that resilience is a latent variable which cannot be directly observed, principal component analysis (PCA) was used to identify the observed wellbeing indicators that could be aggregated into a proxy measure for resilience. To ascertain whether the sample met the minimum requirements for PCA, the Kaiser-Meyer-Olkin (KMO) measure of sampling adequacy test was performed. All principal components (factors) with eigenvalues of more than one were considered; and in accordance with Keil et al. (2008), Allinovi et al. (2010), and Banda et al. (2016), the first factor which extracted the highest variance was used as the proxy measure for household resilience to the effects of abnormal rainfall seasons. The specific wellbeing indicators that loaded highest onto this first factor were then used to categorize the households into three categories.

Resilience includes both the ability to maintain and/or "bounce" back to the normal wellbeing situation after experiencing shocks. Based on this understanding, resilient households were broken into two categories, namely, the fully resilient and the partly resilient. The fully resilient households were those that were able to maintain their normal wellbeing on more than half of the indicators that loaded highest on the proxy for resilience (factor 1). Partly resilient households were those that had been able to recover to the normal wellbeing situation on more than half of indicators that loaded highest onto factor 1. The households were categorized as follows:

Suppose W_a is the household's perceived normal wellbeing, W_b is the wellbeing 1–2 months after abnormal rainfall season, and W_c is the wellbeing at the time of the study (July 2016). Then for a household to be in a given category below, the following expression must be true on more than half of the wellbeing indicators that loaded highest on factor 1:

$$\text{Fully resilient households}: W_a = W_b = W_c \tag{1}$$

$$\text{Partly resilient households}: W_b < W_a = W_c \tag{2}$$

$$\text{Non resilient households}: W_b \langle W_a \rangle W_c \tag{3}$$

(ii) *Identification of Resilience Building Strategies and Factors that Influence Their Use at Household Level*

The three categories of households were then used as the basis for identifying the short-term coping and long-term adaptation strategies that build resilience. Principal component analysis (PCA) was used to objectively aggregate the specific short-term coping strategies and long-term adaptation strategies into broad categories. A chi-square test was then used to identify which broad categories of both short-term coping and long-term adaptation strategies had a statistically significant association with households that were fully resilient and partly resilient to the effects of abnormal rainfall seasons.

(iii) *Factors Influencing Smallholder Farmer Households' Use of Resilience Building Strategies*

Binary logistic regression model was used to identify the factors that influenced the likelihood of the household to use each of the resilience building strategies (i.e., those coping and adaptation strategies associated with households that were fully or partly resilient to effects of rainfall variability). The model was chosen so as to analyze factors associated with use of each broad category of resilience building strategies. Such information about the factors influencing use of specific strategies is essential for promoting specific strategies, either singly or as part of an intervention package.

The binary logistic regression model had each of the identified resilience building strategies as the dependent variable and the different sociodemographic, economic, and institutional factors as predictors.

Suppose Y is the household's use of a coping/adaptation strategy associated with fully and partly resilient households and X are the sociodemographic, economic, and institutional factors presumed to predict the likelihood of using the strategy. Then, the probability of Y can be predicted from the range of predictor variables through Eq. 4 given by Field (2009):

$$P(Y) = \frac{1}{1 + e^{-(\beta_{0+}\beta_1 X_1 + \beta_2 X_2 + \beta_3 X_3 + \dots\dots\dots \beta_n X_n + \mu)}}. \quad (4)$$

Where:

P(Y) is the probability of a farmer using the coping/adaptation strategy associated with fully and partly resilient households.

X_1 to X_n are independent variables presumed to predict the probability of a farmer using the coping/adaptation strategy associated with fully and partly resilient households.

β_1 to β_n are the coefficients (weights) attached to each of the predictors x_1 to x_n.

μ is the error term.

The above coefficients are only used to show the direction of the relationship between the dependent variable and the specific predictor variable. In order to indicate the magnitude of the influence that a specific predictor variable has on the dependent variable, odds ratios are used. The odds ratio corresponding to a given predictor variable represent the change in odds of using a given resilience building strategy that occurs as a result of a unit change in the predictor variable divided by the original odds. Thus an odds ratio of more than one indicates that as the predictor variable increases, the odds of using the coping/adaptation strategy associated with fully and partly resilient households increase, while a value of less than one shows that as the predictor variable increases, the odds of the outcome occurring decrease (Field 2009).

Given that the probability of using a given resilience building strategy is *P(Y)*, the probability of not using such a strategy = 1- *P(Y)*. Then the odds of using the resilience building strategy under consideration can be computed through equation 5 below:

$$\frac{P(Y)}{1 - P(Y)} \quad (5)$$

Combining Eqs. (4) and (5), the odds of using the resilience building strategy under consideration can be given by:

$$\frac{\frac{1}{1+e^{-z}}}{1-\frac{1}{1+e^{-z}}} \tag{6}$$

where $z = \beta_{0+}\beta_1 X_1 + \beta_2 X_2 + \beta_3 X_3 + \ldots \ldots \ldots \beta_n X_n + \mu$.

The above equation can be simplified by finding the natural logarithm

$$\text{Ln}\left(\frac{\frac{1}{1+e^{-z}}}{1-\frac{1}{1+e^{-z}}}\right) = Z = \beta_{0+}\beta_1 X_1 + \beta_2 X_2 + \beta_3 X_3 + \ldots\ldots\ldots \beta_n X_n + \mu \tag{7}$$

The odds ratio corresponding to a given predictor variable (X_1) was computed as the odds of using resilience building strategy under consideration after a unit change in X_1 divided by the original odds. Table 1 shows how the predictors included in the model were operationalized. The specific variables (Table 1) included within each category of factors depended on theory, data availability, and bivariate correlation results.

Table 1 Description of the predictor variables entered into the empirical model

Predictor variables	Definition (how the predictor was defined in this chapter)
Sociodemographic and biophysical factors	
Household type	1 for male headed and 0 for female
Household size	Total number of members of the household
Household head's education	Number of years of schooling
Age of the household head	Age in years
Location of the household	Altitude (meters above sea level)
Economic factors	
Farm size	Total land (acres) accessed by household
Livestock owned	Tropical livestock units (TLU)[a]
Engagement in production of perennial crops	1 if a household farmer has one or more perennial crops, 0 for otherwise
Financial savings	Monthly savings (UGX per month)
Institutional factors	
Group membership	Number of groups in which household is represented
Access to extension services	Number of development organizations to which household is affiliated
Rental land tenure system	1 if a farmer rented any land, 0 for otherwise
Use of scientific climate forecast (SCF) information	1 if a farmer used any SCF to make production decisions and 0 for otherwise
Use of indigenous indicators for rainfall forecasting	1 if a farmer used any indigenous indicators to make production decisions and 0 for otherwise

[a]Irrespective of weight and age, cattle, 0.70 TLU; goat and sheep, 0.10TLU; pig, 0.20TLU; chicken, 0.01TLU (Otte and Chilonda 2002)

3 Results and Discussion

3.1 Household Sociodemographic Characteristics

The household is the primary unit in which decisions about production, consumption choices, and resource allocation are made (Niehof 2011). The household therefore provides the basic context for understanding the effects of rainfall variability and what actions are taken to cope with and/or adapt to them. Table 2 shows the socioeconomic and demographic characteristics of the households.

Results in Table 2 show that majority of the households (83%) were male headed. Over 75% of the households involved in the study had one or more members belonging to a community-based group. Most of the community groups had a savings and credit component, and this partly explains why about 70% of respondents reported to be savings in groups. The groups are avenues for pooling and accumulating financial and social capitals which are essential for resilience building.

The mean age and years of schooling of the respondents were 43 and 7, respectively. The average years of schooling indicate that most of the respondents had completed primary school education and could therefore read and write. This implies that the respondents had the potential to comprehend written information about changes in rainfall patterns and strategies to cope with and/or adapt to them. The average household size was seven people, and this is noticeably higher than the national average of 4.7 (UBOS 2014). Table 2 further shows that out of the seven people per household, only two members per household provided farm labor. This

Table 2 Sociodemographic and economic characteristics of households

Socioeconomic characteristics	**Percentage (n = 255)**	
Proportion of male-headed households	83.3	
Proportion of female-headed households	16.7	
Proportion of households with members in farmers' groups	78.0	
Proportion of respondents saving in groups	70.2	
Demographic characteristics	**Mean (n = 255)**	
Age of respondent (years)	42.73	
Education level of respondent (years in school)	7.08	
Total household size (number of household members)	7.30	
Number of household members who contribute farming labor	1.92	
Economic characteristics	**Percentage (n = 255)**	**Mean (n = 255)**
Total farm size (acres)	100	2.82
Land owned (acres)	100	2.37
Land rented (in) at a fee (acres)	41.2	0.42
Land borrowed (in) for free (acres)	3.1	0.03

translates into a high dependency ratio, besides having adverse implications on the uptake and use of labor-intensive agricultural productivity-enhancing practices.

Land is a core natural resource for agricultural production; and therefore, its size and security of tenure influence the kind of production activities that farmers engage in. Table 2 shows that the average farm size was approximately 3 acres. This average total farm size of approximately 3 is less than the 5 acres upper limit used by FAO to classify a farmer as smallholder (Lowder et al. 2014). This implies that the farmers involved in the study were typical smallholders. Out of the total farm size of about 3 acres, at least 2 acres were owned by the household, while the rest of the land was either rented or borrowed. Slightly over 40% of the respondents rented land at a fee. Discussions with farmers during FGDs revealed that on average, the rental fee ranged from US$ 20 to 50 per year, depending on the plot size and location. The high proportion of farmers (more than 40%) renting land at fee indicates scarcity of arable land for most farmer households in the study area.

3.2 *Rainfall Variability in Mt. Elgon Region*

Farmers noted that their respective areas had been experiencing rainfall variability since the early 1980s. Rainfall variability in this case was about any deviations from the normal rainfall patterns in terms of onset and cessation of seasons, amount and intensity of rain, temporal distribution of rains during the season, and frequency of extremes such as hailstorms. The farmers involved in the study revealed that incidences of seasons that deviated from the normal ones had been increasing ever since the early 1980s. Table 3 shows the most recent abnormal rainfall seasons from the farmers' perspective.

Results in Table 3 above show that for all the 3 years, higher proportions of farmers considered the first season (March to May) to have been abnormal. In particular, the first season for 2014 and the one for 2015 were the most frequently mentioned as having been abnormal over the last 3 years. These two seasons were then characterized by farmers in terms of how they deviated from the normal seasons regarding the timing of onset and cessation, distribution and amount of rains, and occurrence of extremes. The results about how the first season of 2014 and 2015 deviated from the normal are presented in Fig. 2.

Figure 2 shows that over 60% of the farmers reported that the first season of 2014 and 2015 started later and ended earlier than expected. The late onset and early

Table 3 Abnormal rainfall seasons over the last 3 years (2013–2015)

	Percentage of farmers perceiving season as abnormal	
Year	First season (March–May)	Second season (September–November)
2013	14.1	2.7
2014	31.8	4.7
2015	37.3	9.4

cessation of rains translates into shorter seasons with adverse effects on agricultural production and general wellbeing of farmers. Farmers involved in the study reported that in addition to the delay of onset of rain, it rained nonstop throughout the season before it ceased earlier than expected. The farmers interviewed further noted that even the amount of rainfall was more than normal, and it was associated with heavy winds and hailstones (Fig. 2).

3.3 Trends in Household Wellbeing in Relation to Abnormal Rainfall Seasons

In this study, the quantity and quality of food, nonfood basic goods (salt, soap, Vaseline), and services (health, education, and leisure) that a household uses were used as an appropriate proxy for the household wellbeing. Given that households in the study area relied on rain-fed agriculture for their food and income, changes in rainfall seasons were expected to affect their wellbeing. Figure 3 presents a comparison of the households' welling after a normal rainfall season (i.e., one with good rains), immediately after an abnormal rainfall season, and at the time of this study (July 2016). The wellbeing situation after a normal rainfall season was considered to be the normal wellbeing situation for each household and was used as a reference point for assessing household resilience to effects of abnormal rainfall seasons.

Results in Fig. 3 show that even during the normal situation, households were on average unable to achieve their perceived normal wellbeing situation for all the 30 days in a month on any of the nine indicators. The results further show that the abnormal rainfall season had resulted into declines on all the nine wellbeing indicators. Food consumption-related household wellbeing indicators exhibited the most drastic declines compared to nonfood basic household items (salt, soap, Vaseline) and access to social services (school attendance and healthcare). Farmers

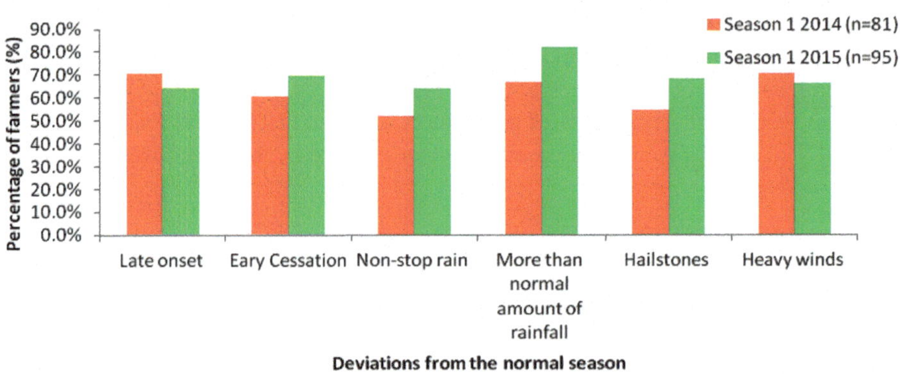

Fig. 2 Characteristics of the most recent abnormal rainfall seasons

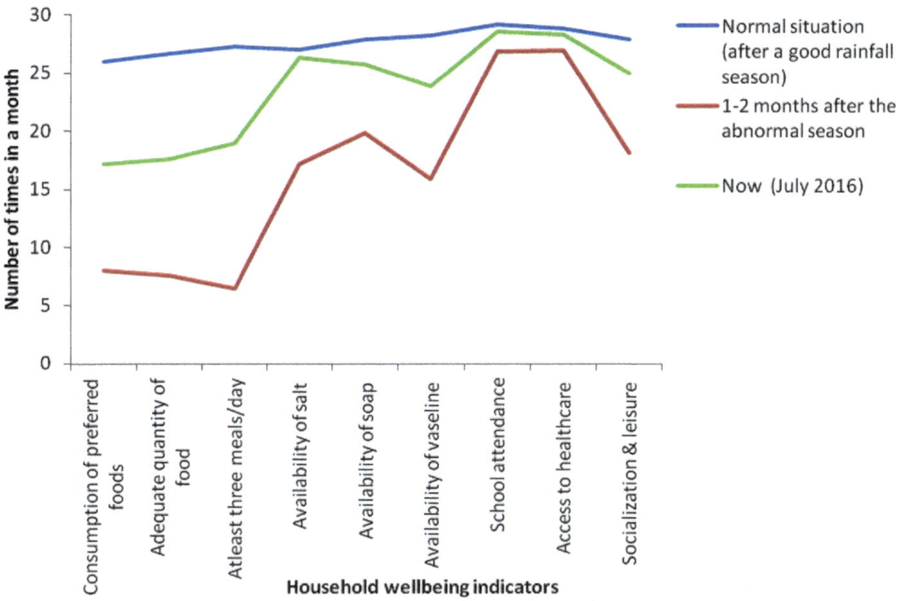

Fig. 3 Trends in household wellbeing in relation to abnormal rainfall season

largely attributed the drastic decline in food consumption-related indicators after an abnormal rainfall season to food price hikes that normally happen when their harvests have been poor and most of the food being sold in their areas has been brought in from other areas. On the contrary, they noted that the consumption of nonfood basic household items was not drastically affected by abnormal rainfall seasons because the prices of these items are generally stable irrespective of the season. The marginal change in the frequency of access to healthcare and school attendance across the three periods was attributed to the existence of free social services (universal primary and secondary education and free primary healthcare) from the government facilities.

The wellbeing situation at the time of this study (July 2016) showed improvements on all the nine indicators. However, households had generally not yet fully recovered to their normal situation on any of the indicators. The difference between the normal and the situation at the time of study was more apparent for the indicators related to food consumption mainly because these indicators had suffered the largest decline as a result of abnormal rainfall season.

3.4 Determining Proxy Measure for Household Resilience to Effects of Abnormal Rainfall Seasons

In this study, household wellbeing was used as a proxy measure for resilience. Principal component analysis was used to aggregate the nine different indicators of wellbeing into two principal components with eigenvalues of more than one as indicated in Table 4.

According to Keil et al. (2008), Allinovi et al. (2010), and Banda et al. (2016), the first principal component which extracts the highest variance is used as a proxy measure for resilience. Therefore in this study, the first factor which extracted the highest variance and was assigned the label "consumption of food and nonfood basic items" was used as a proxy for household resilience to the effects of abnormal rainfall seasons.

3.5 Categorization of Households in Terms of Resilience to Abnormal Rainfall Seasons

The six indicators that loaded highest on the first factor (consumption of food and nonfood basic household items) were used to categorize households in terms of their resilience to the effects of abnormal rainfall seasons. Specifically, these indicators were consumption of preferred types of food, consumption of adequate quantity of food, having at least three meals per day, and availability of salt, soap, and Vaseline. The households were categorized into three categories, namely, the fully resilient, partly resilient, and non-resilient, based on Eqs. (1), (2), and (3) seen earlier in this chapter. Thus, fully resilient and partly resilient households were

Table 4 Aggregation of wellbeing indicators into resilience proxy measure

Broad wellbeing indicators and their constituent-specific indicators	Component loadings
Consumption of food and nonfood basic items	
Availability of salt	0.886
Consumption of adequate quantity of food per meal	0.845
Consumption of preferred foods	0.818
Availability of soap	0.813
Availability of Vaseline	0.812
Having at least three meals per day	0.760
Access to social services and socialization	
School attendance	−0.781
Going for socialization	0.686
Access to healthcare	0.594

those for which equation (1) and equation (2) respectively held true. The non-resilient households were those for which equation (3) held true.

Results in Table 5 show that only about 8% of the households could be categorized as fully resilient in that they had been able to maintain their normal wellbeing on four or more of the six wellbeing indicators when they were faced by the most recent abnormal rainfall season. The table further shows that an additional 22% of the households could be categorized as partly resilient to the effects of abnormal rainfall season because they had been able to restore their perceived normal wellbeing situation on four or more of the six indicators at the time of this study in July 2016. The rest of the households (about 70%) belonged to the non-resilient category, because they were neither able to maintain nor recover to their normal wellbeing situation on four of the six indicators after being affected by abnormal rainfall seasons.

Table 5 Categorization of households in terms of resilience to effects of abnormal rainfall seasons

Number of indicators	Specific indicator combinations	Percentages of households ($n = 255$)		
		Fully resilient ($w_a = w_b = w_c$)	Partly resilient ($w_a < w_b = w_c$)	Non-resilient ($w_b < w_a > w_c$)
4	Consumption of at least three meals per day and availability of all three nonfood basic items (salt, soap, and Vaseline)	3.9	6.5	10.2
	Consumption of adequate quantity of food and availability of all the three nonfood basic items	2.8	5.5	9
	Consumption of preferred foods, availability of all three nonfood basic items	0.0	0.8	11
5	Consumption of at least three meals per day, consumption of adequate quantity of food, and availability of all the three nonfood basic items	0.8	1.6	10.2
	Consumption of preferred types of foods, consumption of adequate quantity of foods, and availability of all the three nonfood basic items	0.4	2.4	11
6	All three food consumption-related indicators and availability of all three nonfood basic items	0	5.1	18.8
Total		**7.9**	**21.9**	**70.2**

3.6 *Household Strategies for Resilience to Abnormal Rainfall Seasons*

Households employed several short-term coping strategies in attempt to maintain their wellbeing when faced with the effects of the most recent abnormal rainfall season. The specific strategies were aggregated into five broader categories using principal component analysis. The five broader categories accounted for about 54% of the variance in the specific strategies from which they were generated. Table 6 below indicates the variance that was extracted by each of the factors (broader categories of short-term coping strategies).

Kaiser-Meyer-Olkin (KMO) measure of sampling adequacy of 0.621 was obtained for the short-term coping strategies, and this satisfies the threshold of 0.50 for suitability of the data for dimension reduction. Table 6 shows that the first and second component extracted 18% and 13% of the variance, respectively. Components 3, 4, and 5 extracted 9%, 8%, and 6% of the variance, respectively. The five components were assigned labels based on the specific short-term strategies that loaded highest onto them as indicated in Table 7.

Results in Table 7 show that reliance on social capital and depletion of assets needed to sustain future consumption were predominant among both the resilient and non-resilient households. However, there were variations between the resilient and non-resilient households in terms of the specific strategies related to reliance on social capital and depletion of assets that were used. For instance, in terms of reliance on social capital, fully resilient and partly resilient households relied on remittances, while the non-resilient ones relied on sale of labor for cash and for food. Regarding depletion of assets needed for future consumption, while over half of the households in category of the fully resilient and partly resilient households relied on income from sale of livestock and their products, over 70% of the non-resilient households relied on harvesting and consuming premature foods such as bananas, cassava, and beans.

Coping through consumption reduction (i.e., by reduction of number of meals per day and resorting to less preferred foods) was a more predominant short-term strategy in partly resilient and non-resilient households than in the fully resilient households. In particular, approximately 60% of the partly resilient households and more than 80% of non-resilient households employed coping strategies related to

Table 6 Aggregation of the short-term coping strategies

Component	Eigenvalues	Percentage of variance extracted	Cumulative percentage of variance extracted
1	3.63	18.26	18.26
2	2.42	12.64	30.90
3	1.88	9.17	40.07
4	1.78	7.72	47.79
5	1.54	6.48	54.27

Table 7 Categories of short-term strategies to cope with effects of abnormal rainfall seasons

	Household resilience category			
Categories of short-term coping strategies and their constituent-specific strategies	Fully resilient (n = 19)	Partly resilient (n = 55)	Non-resilient (n = 181)	Component loading
1. *Desperate measures to obtain food and nonfood basic items*				
Consumption of stored seed	21.4	26.0	54.8	0.908
Gathering food from the wild	14.3	18.0	30.9	0.934
Begging for food and nonfood items	0	1.8	23.2	0.439
Illegally harvesting forest products for sale and food	0	8.0	16.0	0.737
2. *Reliance on social capital*				
Selling labor for cash	5.3	36.4	72.9	0.701
Selling labor in exchange for food	0	10.9	33.2	−0.469
Receiving remittances from h/h and non-h/h members	52.6	50.9	32.7	0.635
Receipt of relief from agencies and individuals	21.4	18.0	13.8	−0.400
3. *Depletion of assets needed to sustain future consumption*				
Harvesting and consumption of premature food	36.8	45.5	73.1	0.459
Selling of livestock and their products	57.9	54.5	47.0	0.792
Acquiring of food and nonfood basic items on credit	14.3	42.0	29.8	0.746
Acquiring of a financial loan	15.8	27.3	19.9	0.549
4. *Use of physical and financial reserves*				
Consuming stored food	84.2	78.2	48.6	0.720
Using income from off-farm business	52.1	35.5	14.4	0.362
Withdrawing own financial savings	42.1	23.6	11.6	−0.528
Renting out land	31.6	20.0	5.5	−0.554
5. *Reduction of consumption*				
Reducing number of meals	5.3	58.2	85.1	0.775
Eating less preferred foods	31.6	69.1	80.1	0.834
Sending away some household members	0	1.8	6.1	0.425

reduction of food consumption. On the contrary, only about one third of fully resilient households employed coping strategies related to reducing food consumption. Results further indicate that use of physical and financial reserves (mainly consumption of stored food and use of income/operating capital of off-farm business) was more predominant among the fully and partly resilient households. On the contrary, use of desperate measures to obtain food, especially eating grain stored as seed for the next season, was common among households that were not resilient to the effects of the most recent abnormal rainfall season.

Long-Term Strategies Used by Smallholder Farmer Households to Build Resilience to Effects of Abnormal Rainfall Seasons

The long-term strategies to adapt to abnormal rainfall seasons are aimed at reducing the negative impacts onto the household while also harnessing any opportunities presented by the effects of the abnormal rainfall seasons. Households were employing several long-term adaptation strategies, and these strategies were aggregated using principal component analysis into five broader categories, which extracted a total of 51% of the variance. Table 8 indicates the variance that was extracted by each of the five broader categories of long-term adaptation strategies.

KMO measure of sampling adequacy of 0.593 was obtained, and this satisfies the threshold of 0.50 for suitability of the data for dimension reduction. The first and second components, respectively, accounted for 15% and 13% of the variance in all the specific long-term adaptation strategies from which they were obtained. Components 3, 4, and 5 extracted 10%, 8%, and 6% of the variance, respectively. The labels assigned to each of the components and proportions of households in each of the resilience categories that were using the specific strategies are presented in Table 9.

Results in Table 9 show that altering livelihood asset portfolio and diversifying income sources and use of soil fertility and water conservation measures were more predominantly used by fully and partly resilient households than the non-resilient ones. The use of agronomic practices and diversification of agro-enterprises were the predominant adaptation strategies among both the resilient and non-resilient households. Intercropping was the most predominantly used agronomic practice by all the three categories of households, while acquisition of livestock and expansion of land under perennial crops were the most predominant specific strategies related to agro-enterprise diversification among all the three categories of households. Intercropping, acquisition of livestock, and engagement in perennial crops production were common strategies among all categories of households because farmers considered them to be less prone to climatic risks.

Table 9 further shows that changing planting dates, tree planting, application of manure, and making soil and water conservation structures were more predominant among fully and partly resilient households. Such strategies like renting out land, buying more land, and tree planting were exclusively predominant among the fully resilient households. These results generally show that fully resilient households

Table 8 Aggregation of long-term strategies to adapt to abnormal rainfall seasons

Component	Eigenvalues	Percentage of variance extracted	Cumulative percentage of variance extracted
1	3.34	15.12	15.12
2	2.25	12.64	27.76
3	1.98	9.52	37.29
4	1.76	7.86	45.15
5	1.62	5.81	50.96

Table 9 Long-term strategies to adapt to effects of abnormal rainfall seasons

	Household resilience category			
Categories of long-term adaptation strategies and their constituent-specific strategies	Fully resilient (*n* = 19)	Partly resilient (*n* = 55)	Non-resilient (*n* = 181)	Component loading
Use of agronomic practices				
Practicing intercropping	68.4	69.1	56.4	0.858
Changing planting dates	57.9	56.4	14.4	0.737
Widening crop spacing	31.6	34.3	9.4	0.468
Practicing crop rotation	15.8	30.9	22.1	0.639
Use of mulching	34.2	13.6	0.0	0.405
Staggered planting	0.0	15.2	13.3	0.401
Narrowing crop spacing	15.8	21.2	33.3	0.421
Altering the household livelihood asset portfolio				
Tree planting	73.7	69.1	35.9	0.450
Starting to save/increasing on savings	56.8	34.5	28.7	0.502
Renting out land	51.6	12.7	5.0	0.750
Buying land	56.8	33.7	4.5	0.418
Selling land	26.3	10.9	9.4	0.755
Agro-enterprise diversification				
Acquisition of livestock	52.6	65.5	53.0	0.676
Taking on/expanding on perennial crops	57.9	58.2	53.0	0.450
Taking up fast maturing crops	31.6	50.9	56.4	0.759
Renting in more land	0.0	12.7	21.0	−0.335
Growing food security crops	5.3	7.7	22.0	0.419
Diversification of income sources				
Initiation of off-farm business activities	52.6	41.8	29.8	0.676
Brick making	15.8	16.4	9.9	0.654
Sand mining/stone quarrying	5.3	13.0	14.4	0.647
Soil fertility and water conservation				
Making soil and water conservation structures	57.9	56.4	36.5	0.796
Use of manure/fertilizers	52.6	52.7	17.7	0.661

had a wider range of adaptation strategies, and this could be attributed to their larger asset base. On the contrary, non-resilient households had fewer adaptation options; and their predominant strategies – specifically intercropping, acquisition of livestock, more involvement in production of perennial crops, and growing fast maturing crops –were all still sensitive to rainfall variability and other climatic shocks. This implies that the wellbeing of the non-resilient households is exposed and sensitive to climatic shocks now and within the foreseeable future. This finding concurs with Antwi-Agyei et al. (2013) who noted that poor households were less likely to be resilient to climate variability because they lack the capital assets to adapt.

Our findings, however, indicate that the resilience to future shocks for all categories of households might be endangered given the minimal attention to the improvement and protection of soil by the non-resilient households, who are about 70% of all the farming households in the area. This is attributed to the fact that planning for and implementation of soil fertility and water conservation measures was done at individual household rather than landscape level. Planning and implementing such measures like soil and water conservation at individual household level results in haphazard adoption of such measures, and this compromises their effectiveness (Kato et al. 2011; FAO 2014). There is need for soil fertility and water conservation to be planned and implemented at a level higher than the farm; this is particularly critical in the Mt. Elgon area given the steep terrain, severe land fragmentation phenomenon, and episodes of heavy and prolonged rains in the area.

3.7 Identification of Strategies that Build Household Resilience to the Effects of Abnormal Rainfall Seasons

Resilience building strategies were considered as those that were predominantly used by the fully and partly resilient households. Therefore, this study used a chi-square test to identify the broad categories of strategies that had a statistically significant association with fully resilient and partly resilient households. Table 10 shows the percentages of households using the different broad categories of strategies and their statistical association with the resilient household categories.

Based on this study's criterion of identifying resilience building strategies as those only predominantly used by fully resilient and partly resilient households, results in Table 10 show that reliance on physical and financial reserves was the only resilience building short-term coping strategy. Among the long-term adaptation strategies, use of agronomic practices, alteration of household livelihood asset portfolio, diversification into off-farm income options, and use of soil fertility and water conservation measures were identified as the resilience building strategies. These findings about the strategies associated with household resilience concur with Harris and Orr (2014), who identified agronomic practices, diversification into off-farm income options, soil improvement, and protection as having positive influence on household food and income security. Furthermore, most of the long-term adaptation strategies identified by this study as being associated with resilient households are part of the practices within climate smart agriculture which is considered key for building resilience to current and future climatic shocks (Alliance for a Green Revolution in Africa [AGRA] 2014).

Table 10 Strategies used by different categories of households

Strategies		Percentages by household resilience category			Chi-test statistics (df = 2)
		Category 1 (*n* = 19)	Category 2 (*n* = 55)	Category 3 (*n* = 181)	X^2
Short-term coping	Desperate measures to obtain food	15.8	25.5	54.2	10.600***
	Reliance on social capital	78.9	80.0	74.6	0.772
	Depletion of assets needed for future consumption	89.5	89.1	81.8	2.145
	Use of physical and financial reserves	89.5	89.1	62.9	11.003***
	Reduction in consumption	45.2	64.8	100.0	17.768***
Long-term adaptation	Use of agronomic practices	94.7	92.7	68.5	9.587***
	Altering household livelihood asset portfolio	100	92.7	70.2	9.745***
	Agro-enterprise diversification	68.4	87.3	80.1	0.181
	Income diversification	57.9	56.4	34.8	10.462***
	Soil fertility and water conservation	57.9	61.8	37.0	12.161***

***Significant at $P < 0.05$, ***Significant at $P < 0.01$*

3.8 Factors that Influence the Use of Strategies Associated with Household Resilience to Effects of Abnormal Rainfall Seasons

The type of strategies used by a household to cope with and adapt to climatic shocks is dependent on several factors within and beyond the household. This study employed binary logistic regression to identify the different sociodemographic, economic, and institutional factors that influence the likelihood of a household to use a specific strategy associated with households that were resilient to the effects of rainfall variability (see Table 11).

Sociodemographic and Biophysical Factors

Results in Table 11 indicate that among the sociodemographic factors, only household type and household size had a statistically significant influence on the likelihood of a household to take up any of the resilience building strategies.

Table 11 Factors influencing use of short-term coping and long-term adaptation strategies associated with resilience to effects of abnormal rainfall seasons

	Coping and adaptation strategies and their corresponding inferential statistics									
	Use of physical and financial reserves		Use of agronomic practices and food storage		Alteration of livelihood asset portfolio		Diversification of income sources		Soil fertility and water conservation	
Predictors	B	Exp (B)	B	Exp (B)	B	Exp (B)	B	Exp (B)	B	Exp (B)
Sociodemographic and biophysical factors										
Household type	0.231	1.260	−0.293	0.746	−0.111	0.895	0.223	1.250**	−0.244	0.894
Years of schooling of household head	0.041	1.042	−0.038	0.963	0.037	1.037	0.064	1.067	0.078	1.094
Household size	−0.453	0.567**	0.049	1.050	−0.062	0.976	−0.329	0.759**	−0.032	0.0968
Age of the household head	0.162	1.031	0.153	1.023	0.111	1.035	0.123	1.061	−0.134	0.693
Location of the household (altitude)	0.273	1.084	0.229	1.028	0.290	1.092	0.128	1.00	0.398	1.232**
Economic factors										
Tropical livestock units	0.512	1.641**	−0.319	0.727**	0.896	2.426***	0.492	1.596**	0.134	1.158**
Total land accessed by household	0.424	1.528**	0.215	1.124**	0.513	1.662**	0.028	1.013	0.244	1.383**
Savings (UGX per month)	0.806	2.968***	0.318	1.012	−0.001	0.999	0.493	1.523**	0.212	1.101
Institutional factors										
Number of groups in which household has members	0.506	1.659**	0.318	1.214**	−0.001	0.999	0.493	1.098	0.537	1.712***
Rental land tenure system	0.136	1.0095	0.416	1.344**	−0.215	0.806	0.038	1.039	−0.467	0.766**
Use of scientific climate forecasts	0.256	1.292	0.328	1.219**	0.490	1.101	0.328	1.048	0.473	1.184
Use of indigenous rainfall forecast indicators	−0.606	0.545	0.412	1.304**	0.200	1.222	0.066	1.006	0.560	1.015
Constant	0.631	5.032	−0.580	0.560	0.948	2.581	0.453	1.573	−1.891	0.151

**Significant at $P < 0.05$, ***Significant at $P < 0.01$

Specifically, household type positively influenced the likelihood of a household to diversify into off-farm income options. The odds ratio corresponding to household type indicates that male households were about 1.25 times more likely to diversify into off-farm income options compared to female-headed households. This result could be attributed to the social cultural constraints where female-headed households carry a huge burden of productive, reproductive, and community service roles that reduce their mobility and time needed to start and operate an off-farm income venture (Blackden and Wodon 2006). Female household heads may also have difficulties in obtaining capital to initiate and/or sustain off-farm ventures because of their limited ownership and control over resources that can be sold or used as collateral security to obtain business capital (Blackden and Wodon 2006; Nabikolo et al. 2012).

Household size had a negative influence on the likelihood of a household to rely on physical and financial reserves and diversify into off-farm income options. Results show that when household size increased by one member, the odds to rely on reserves (physical and financial) and diversify into off-farm income options reduced by about 43% and 24%, respectively. This result concurs with Obayelu et al. (2014) who also found that household size had a negative influence on households' use of adaptation strategies. This finding however contradicts with Bryan et al. (2009) and Apata (2011) who noted that household size positively influences adaptation because it reflects the availability of family labor. The argument about household size enhancing adaptation is valid in contexts where key production resources, especially land, are not a constraint. On the contrary, households in Mt. Elgon region face an acute shortage of farming land. Therefore in large households, the acute land shortage increases the dependency ratio since only few people within the household will be needed to work on the land and the rest will largely be underutilized. The high dependency ratio in large households implies that such households have minimal surpluses to accumulate reserves or initiate off-farm income ventures. This argument is especially validated by Kibet et al. (2009) and Kiran and Dhawan (2015), who found that large households had low savings to build any substantial reserves.

Location of the household in terms of altitude had a positive influence on the likelihood of a household to use soil fertility and water conservation measures. Specifically, the results indicate that an increase of 1 meter in the altitude at which the household is located translates into a 23% increase in the likelihood of the household using any soil fertility and water conservation measures. This finding is consistent with those of Kabede (2014) whose study in Ethiopia found out that altitude and slope had a positive influence on adoption of soil and water conservation measures. This finding further implies that interventions employing a watershed logic which requires adoption of soil fertility and water conservation commencing from uphill progressing downslope should be possible in the study area.

Economic Factors

Tropical livestock units as a measure of number and value of livestock owned by the household had a positive influence on the likelihood of a household to engage in all the strategies except use of agronomic practices for which the influence was negative. The odds ratios show that when the tropical livestock units owned by the household increase by one, the odds to rely on reserves, alter livelihood asset portfolio, diversify income sources, and use soil fertility and water conservation measures increase by about 1.4, 2.4, 1.6, and 1.2 times, respectively. The positive influence of tropical livestock units on the likelihood of households to use most of the resilience building strategies is attributed to the ability of livestock to easily be sold to raise cash that can be used to facilitate the initiation and operation of the different strategies. The positive influence of tropical livestock units on household's likelihood to use soil fertility and water conservation measures can be attributed to the increased supply of animal manure resulting from more livestock. Additionally, increase in tropical livestock units also encourages farmers to establish grass bunds and fodder trees to provide feeds for the livestock that are dominantly stall fed. On the contrary, a unit increase in the tropical livestock units owned by the household results into a reduction of about 30% in the odds of the household to use agronomic practices. This could be attributed to the fact that increase in livestock units results into additional labor requirements to mobilize livestock feeds and water, thereby translating into labor constraints for implementing agronomic practices. The positive influence of livestock ownership on uptake of adaptation measures has also been reported by Nhemachena and Hassan (2007) who found that farmers who had mixed crop-livestock enterprises were more likely to change their planting dates, venture into nonfarm activities, and adopt varieties of crops.

Land ownership as an economic factor had a positive influence on the likelihood of the household to cope through use of physical and financial reserves and adapt through use of agronomic practices and alter asset portfolio and soil fertility and water conservation. The odds ratios show that when land owned by a household increases by 1 acre, its odds to rely on reserves (physical and financial), use agronomic practices, alter livelihood assets, and use soil fertility and water conservation measures increase by 1.5, 1.1, 1.7, and 1.4 times, respectively. Increase in land owned enables households to rely on reserves because land could be rented out at fee. Interactions with farmers during FGDs and in-depth interviews revealed that farmers with more land could adopt agronomic practices, soil fertility, and water conservation because they had more space on which to experiment such measures. These findings concur with those of Knowler and Bradshaw (2007) and Bryan et al. (2009) who reported that farm size positively influenced adoption of soil and water conservation measures.

The amount of financial savings that a household made per month had a positive and statistically significant influence on the likelihood of the household to rely on reserves and diversify into other income sources. Specifically, when the monthly savings of the household increase by one shilling, the likelihood of the household to rely on reserves and diversify into other income sources increases by about three and two times, respectively. This implies that more regular savings enable the household to build reserves on which it can draw to maintain and/or recover to the normal wellbeing situation or even invest in off farm livelihood options. This therefore suggests the need to encourage and support households to save as much as possible during good seasons as way of enhancing their resilience to effects of rainfall variability.

Institutional Factors

The number of groups in which a household had at least one member positively influenced its likelihood to rely on reserves, use agronomic practices and soil fertility and water conservation measures. When the number of groups in which household is represented increases by one, its odds to use agronomic practices and rely on reserves and soil fertility and water conservation measures increase by 1.2, 1.7, and 1.7 times, respectively. The positive influence of number of groups is attributed to the fact that groups provide avenues for households to accumulate their reserves through pooling financial savings while also sharing information and learning about farming practices. Groups also enhance household's access to material and technical support from NGOs and government programs, which often target people organized in groups. The positive influence of membership in groups on use of adaptation strategies has also been reported by Shongwe et al. (2014).

Renting land by a household influenced the use of agronomic practices and soil fertility and water management measures as strategies to adapt to rainfall variability. On the one hand, renting in land positively influenced household use of agronomic practices, while on the other hand, it had negative influence on use of soil fertility and water conservation measures. The odds of a household to use agronomic practices are about 1.3 times higher for a household which rents in land compared to one that does not. This implies that farmers who use rented land were motivated to use agronomic practices to optimize production. This concurs with Idoma and Ismail (2014) whose study in Benue State of Nigeria reported that farmers who rented land had higher yields than those who owned land under freehold and communal land tenure systems.

However, households that rented land had lower likelihood of using soil fertility and water conservation measures. Specifically, a household that rented land was about 20% less likely to take up soil fertility and water conservation than one which did work on its own land. This is attributed to the fact that short-term rental tenure (usually 1 year) discourages farmers from investing in long-term improvements like soil fertility improvements and soil erosion control structures. During FGDs in both districts, farmers revealed that there were instances where tenants in the land-renting arrangements destroyed existing soil and water conservation structures such as trenches and terraces, in order to increase on the area available for crop production so as to recoup the rental investment. The importance of secure land tenure as a driver for investment in long-term adaptation measures has been documented by Nhemachena and Hassan (2007) and Bryan et al. (2009).

Results further show that utilization of scientific climatic forecast information and indigenous indicators for forecasting rainfall onset cessation to make production decisions increases the likelihood of a household adapting to rainfall variability through use of agronomic practices. When a household utilizes scientific climate information and indigenous indicators for forecasting rains, its odds to apply agronomic practices are about 1.2 and 1.3 times higher than those of a household that does not utilize any of the two knowledge bases, respectively. These results on the positive influence of scientific climate information corroborate with Nhemachena and Hassan (2007) and Deressa et al. (2009), and they justify the need for investments in climatic information generation and dissemination to farmers to enhance their resilience.

4 Conclusion and Recommendations

The empirical results suggest that rainfall variability has adverse impacts on smallholder farmer households' wellbeing in Mt. Elgon region. Food consumption-related wellbeing indicators are the most sensitive to rainfall variability, largely because smallholder farmer households derive the livelihoods from rain-fed agriculture. Similar findings about severe negative impacts of rainfall variability on smallholder farmers' food consumption have also been documented by UBOS and MAAIF (2010) and AGRA (2014).

More worrying, however, is the finding that about 70% of the farmers do not recover from the decline in their wellbeing after being affected by rainfall variability. This points to the likelihood of rainfall variability pushing most of the households into a vicious wellbeing trap, from which they may never disentangle themselves without external support. Given that rainfall variability is projected to continue in the foreseeable future, it is critical to implement interventions that prevent the households from descending into the wellbeing trap when affected by rainfall variability. The empirical results from this study demonstrate that interventions to prevent households from sliding into the wellbeing trap should be multipronged focusing on enhancing agricultural productivity, protecting and improving the available land as well as widening the asset base for smallholder farmer households, and expanding the nonfarm sector to relieve the pressure on land. The need to expand smallholder farmers' asset base is especially critical given that the likelihood of taking up most of the resilience building strategies was enhanced by the amounts of such assets like livestock, land, financial savings, and group membership that the households had access to.

Much as the empirical results from this study have demonstrated that soil fertility and water conservation measures were associated with resilience to effects of rainfall variability, the fact that they are planned at individual household level has adversely affected their widespread use. Only about one third of the non-resilient households who form the majority in the study area were using any soil fertility and water conservation measures largely because of lack of capacity. The limited and haphazard use of such soil fertility and water conservation measures within mountainous landscapes compromises their effectiveness (Kato et al. 2011; FAO 2014). Therefore, the planning and implementation of such measures should be done at landscape level targeting watersheds. This may call for establishment of mechanisms such as bylaws, sanctions, and incentives to ensure that as many people as possible within a watershed take up the appropriate soil fertility and water conservation measures.

Acknowledgments This study was funded by the Norwegian Agency for Development Cooperation (NORAD) through the Regional Capacity Building for Sustainable Natural Resource Management and Agricultural Productivity under Climate Change (CAPSNAC) Project Grant No: UGA 13/0020.

References

AGRA (2014) Africa agriculture status report 2014: climate change and smallholder agriculture in Sub-Saharan Africa. Nairobi

Allinovi L, D'Errico M, Mane E, Romano D (2010) Livelihoods strategies and household resilience to food insecurity: an empirical analysis to Kenya. Paper prepared for the Conference on Promoting Resilience through Social Protection in Sub-Saharan Africa, organised by the European Report of Development in Dakar, Senegal, 28–30 June, 2010

Alweny S, Nsengiyumva P, Gatarabirwa W (2014) African Mountains status report technical report no. 1. Albertine Rift Conservation Society (ARCOS). Available: www.fao.org/mountain-partnership/publications/publication-detail/en/c/278290/. Accessed 18 Apr 2018

Antwi-Agyei P, Dougill AJ, Stringer LC (2013) Barriers to climate change adaptation in sub-Saharan Africa: evidence from Northeast Ghana & systematic literature review. Centre for Climate Change Economics and Policy Working Paper No. 154

Apata TG (2011) Factors influencing the perception and choice of adaptation measures to climate change among farmers in Nigeria. Evidence from farm households in Southwest Nigeria. Environ Econ 2(4):74–83

Bamutaze Y (2015) Revisiting socio-ecological resilience and sustainability in the coupled mountain landscapes in Eastern Africa. Curr Opin Environ Sustain 14:257. https://doi.org/10.1016/j.cosust.2015.06.010

Banda TF, Phiri MAR, Mapemba LD, Maonga BB (2016) Household resilience to drought: The case of Salima district in Malawi. International Food Policy Research Institute [IFPRI] Working Paper 14, May 2016

Barrett CB, Constas MA (2014) Toward a theory of resilience for international development applications. PNAS 11(40):14625–14630. https://doi.org/10.1073/pnas.1320880111

Blackden CM, Wodon Q (2006) Gender, time use and poverty in sub-Saharan Africa. World Bank Working Papers 73. World Bank, Washington DC

Bomuhangi A, Nabanoga G, Namaalwa JJ, Jacobson MM, Abwoli B (2016) Local communities' perceptions of climate variability in the Mt. Elgon region, eastern Uganda. Cogent Environ Sci 2:1168276. https://doi.org/10.1080/23311843.2016.1168276

Bryan E, Deressa TT, Gbetibouo GA, Ringler C (2009) Adaptation to climate change in Ethiopia and South Africa: options and constraints. Environ Sci Pol 12(4):413–426

Caffrey P, Finan T, Trzaska S, Miller D, Laker-Ojok R, Houston S (2013) Uganda climate change vulnerability assessment report. United States Agency for International Development [USAID] African and Latin American Resilience to Climate Change [ARCC] August, 2013

Coates J, Swindale A, Bilinsky P (2007) Household food insecurity access scale (HFIAS) for measurement of household food access: Indicator guide (version 3). Food and Nutrition Technical Assistance (FANTA). United of America, Washington, D.C.

Deressa TT, Hassan RM, Ringler C, Alemu T, Yusuf M (2009) Determinants of farmers' choice of adaptation methods to climate change in the Nile Basin of Ethiopia. Glob Environ Chang 19(2):248–255

FAO (2013) Mountain farming is family farming. Food and Agriculture Organization of the United Nations, Rome

FAO (2014) Adapting to climate change through land and water management. Results of pilot projects in Kenya, Ethiopia and Tanzania. Food and Agriculture Organization, Rome

Field A (2009) Discovering statistics using SPSS. Third Edition. SAGE publications, London

Harris D, Orr A (2014) Is rainfed agriculture really a pathway from poverty? Agric Syst 123:84–96

Idoma K, Ismai'l M (2014) The effects of land tenure practices on agricultural output in Agatu local government area of Benue State, Nigeria. J Dev Agric Econ 6(5):212–219

Jiang B, Bamutaze Y, Pilesjö P (2014) Climate change and land degradation in Africa: a case study in the Mount Elgon region, Uganda. Geo-spatial Inf Sci 17(1):39–53

Kabede W (2014) Effect of soil and water conservation measures and challenges for its adoption: Ethiopia in focus. J Environ Sci Technol 7(4):185–199

Kaggwa R, Hogan R Gowa H (2010) Climate change, the environment and human welfare: lessons learned from the Lake Kyoga catchment area. Policy brief on Demographic change and its impacts on ecosystems and human welfare – UNDP-UNEP Poverty-Environment Initiative. Nairobi

Kansiime MK, Wambugu SK, Shisanya CA (2013) Perceived and actual rainfall trends and variability in Eastern Uganda: implications for community preparedness and response. J Nat Sci Res 3(8):179–195

Kato E, Ringler C, Yesuf M, Bryan E (2011) Soil and water conservation technologies: a buffer against production risk in the face of climate change? Insights from the Nile basin in Ethiopia. Agric Econ 42(2011):593–604

Keil A, Zeller M, Wida A, Sanim B, Birner R (2008) What determines farmers' resilience towards ENSO related drought? An empirical assessment in Central Sulawesi, Indonesia. Climate Change 86:291–307. https://doi.org/10.1007/s10584-007-9326-4

Kibet LK, Mutai BK, Ouma DE, Ouma SA, Owuor G (2009) Determinants of household saving: case study of smallholder farmers, entrepreneurs and teachers in rural areas of Kenya. J Dev Agric Econ 1(7):137–143

Kiran T, Dhawan S (2015) The impact of family size on savings and consumption expenditure of industrial workers: a cross-sectional study. Am J Econ Bus Adm 7(4):177–184

Knowler D, Bradshaw B (2007) Farmers' adoption of conservation agriculture: a review and synthesis of recent research. Food Policy 32(1):25–48

Krejcie RV, Morgan DW (1970) Determining sample size for research activities. Educ Psychol Meas 30:607–610

Lowder SK, Skoet J, Singh S (2014) What do we really know about the number and distribution of farms and family farms worldwide? Background paper for The State of Food and Agriculture 2014. ESA Working Paper No. 14-02. Rome, FAO

Mbogga MS (2012) Climate profiles and climate change vulnerability assessment for the Mbale region of Uganda. UNDP Consultancy report. Kampala

McAslan A (2010) The concept of resilience. Understanding its origins, meaning and utility. A Strawman paper. Torrens Resilience Institute, Adelaide, 14 March 2010

MWE (2013) Integrated territorial climate plan 2014–2029 for Mbale region Uganda (Bududa, Manafwa and Mbale districts). Kampala

Mubiru DN, Komutunga E, Agona A, Apok A, Ngara T (2012) Characterizing agrometeorological climate risks and uncertainties: crop production in Uganda. S Afr J Sci. https://doi.org/10.4102/sajs.v108i3/4.470

Mugagga F, Kakembo V, Buyinza M (2012) Land use changes on the slopes of Mount Elgon and the implications for the occurrence of landslides. Catena 90(2012):39–46

Nabikolo D, Bashasha B, Mangheni MN, Majaliwa JGM (2012) Determinants of climate change adaptation among male and female headed farm households in eastern Uganda. Afr Crop Sci J 20(S2):203–212

Narayan D, Chambers R, Shah MK, Petersch P (2000) Voices of the poor: crying out for change. Oxford University Printing Press, Oxford

NEMA (2010) Landslides in Bududa district, their causes and consequences. March 2010. Available on http://www.nemaug.org/reports/Current_reports/Bududa_report.pdf. Accessed on 10 Oct 2017

Nhemachena C, Hassan R (2007) Micro-level analysis of farmers' adaptation to climate change in Southern Africa. International Food Policy Research Institute. [IFPRI] Discussion Paper 00714. August 2007

Niehof A (2011) Conceptualizing the household as an object of study. Int J Consum Stud 35(2011):488–497

Obayelu OA, Adepoju AO, Idowu T (2014) Factors influencing farmers' choices of adaptation to climate change in Ekiti State, Nigeria. J Agric Environ for Int Dev 108(1):3–16

Otte MJ, Chilonda P (2002) Cattle and small ruminant production systems in sub-Saharan Africa - a systematic review. Food and Agriculture Organization, Rome

Rwenzori Think Tank (2011) Small holder farmers' knowledge and adaptation to climate change in the Rwenzori region. Research report No. 001, March 2011

Shongwe P, Masuku MB, Manyatsi AM (2014) Factors influencing the choice of climate change adaptation strategies by households: a case of Mpolonjeni Area Development Programme (ADP) in Swaziland. J Agric Stud 2(1):86–98

Staudt M, Kuosmanen E, Babirye P, Lugaizi I (2014) The Bududa landslide of 1 March 2010. Geol Surv Finland Spec Pap 56:373–384

UBOS & MAAIF (2010) Uganda census of agriculture 2008/2009. Volume III: Agricultural household and holding characteristics report, December 2010, Government of Uganda, Kampala

UBOS (2014) National Population and Housing Census 2014 Provisional Results November 2014: Revised Edition. Kampala

UNEP (2014) Africa Mountains Atlas. United Nations Environment Programme (UNEP), Nairobi

Adoption of Climate Smart Agricultural Technologies and Practices in Drylands in Uganda: Evidence from a Microlevel Study in Nakasongola District

Rhoda Nakabugo, Isolo Paul Mukwaya, and Sabiiti Geoffrey

Abstract Agriculture is the most important sector in Uganda's economy, but it continues to experience challenges of erratic rainfall variability and environmental degradation. This paper is situated within post-structuralist geographical thought to (i) classify CSA practices and (ii) empirically quantify the relative importance of household socio-economic factors that structure the adoption of CSA practices in drylands in Uganda. The study was carried out in Nakasongola District in Central Uganda, and 143 geo-referenced questionnaires were used to collect relevant farming household and CSA data. Results indicated that timely planting, crop rotation, intercropping, and proper spacing were the most prevalent CSA practices, while rotational grazing, mulching, fertilizer use, and use of pesticides and herbicides were the least prevalent practices. Principal component analysis (PCA) generated a factor solution, and the components were clustered into three CSA practices: crop management, conservation agriculture, and land management practices. There are important differences in the propensity of households living in village settings to adapt. Parameter estimates indicated that the size of household, household income diversity index, access to pesticide uses, fertilizers, extension services, domestic water sources, improved seeds, credit, main decision-maker in the household, and education levels of the head of the household significantly influence ($p < 0.05$) the adoption of CSA practices.

Keywords Climate-smart agriculture · Adoption · Farming households · Nakasongola · Uganda

R. Nakabugo (✉) · I. P. Mukwaya
Department of Geography, Geo-informatics and Climatic Sciences, School of Forestry, Environment and Geographical Sciences, Makerere University, Kampala, Uganda

S. Geoffrey
Department of Geography, Geo-informatics and Climatic Sciences, School of Forestry, Environment and Geographical Sciences, Makerere University, Kampala, Uganda

IGAD Climate Prediction and Applications Centre, Nairobi, Kenya

Y. Bamutaze et al. (eds.), *Agriculture and Ecosystem Resilience in Sub Saharan Africa*, Climate Change Management,
https://doi.org/10.1007/978-3-030-12974-3_24

1 Introduction

Agriculture constitutes the backbone of the economies of most developing countries, including Uganda (Bernard and Dulle 2014). Most of the agriculture is rain-fed and thus susceptible to the impacts of climate variability and change. For most African countries, agriculture remains the main source of livelihoods and therefore the basis for subsistence and the largest employer of any sector in the economy with about 80% of the employment (MAAIF 2011; Majaliwa et al. 2015). Agriculture employs about 65% of the total work labor force in sub-Saharan Africa (SSA). Though there has been a decline in the relative number of agriculture workers, it still accounts for the majority of the working population in the region (AGRA 2015). A 1% increase in agricultural per capita gross domestic product (GDP) reduced the poverty gap five times more than a 1% increase in GDP per capita in other sectors, mainly among the poorest people because it is a pro-poor, income-generating, and employment-creating sector for most SSA economies (AGRA 2015). Recent evidence suggests that global climate change is likely to increase the incidence of natural hazards, as well as the variability of rainfall, temperature, and other climatic parameters (IPCC 2012), and yet studies by Cline (2008) showed that most countries in sub-Saharan Africa are at risk because of the scarce resources for adaptation. The effects of future climate change will manifest themselves in more expensive foods, a shortage of water, and more extreme weather. Those who will suffer (and already are suffering) as a result are ordinary working- and middle-class people and peasant farmers (Burrows 2012). Given that agricultural production remains the main source of income for most rural communities, the increased risk of crop failure, associated with increased frequency of extreme climate events, poses a major threat to food security and poverty reduction (FAO 2015). The threat of climate change in agriculture has been acknowledged by the Comprehensive Africa Agriculture Development Programme (CAADP), the overarching framework and Africa's policy framework for agricultural transformation, wealth creation, food security and nutrition, economic growth, and prosperity for all. [Are all these bodies/programs like the first? Clarify!] The African Union's New Partnership for Africa's Development (AU-NEPAD) has set a goal of supporting at least 25 million farm families to adopt and practice climate-smart agriculture (CSA) by the year 2025. As a result, there is an urgent need for responses that contribute to building resilience and developing adaptation (Aldunce, Handmer, Beilin & Howden 2016). The adoption of climate-smart agricultural (CSA) practices by farmers in the developing world has been viewed as one of the approaches to achieve food security, reduce greenhouse gas emissions, and adapt to climatic risks.

CSA is not a single specific agricultural practice that can be applied universally (Carr 2016), but rather, it is a package that involves changing the way in which farmers manage their agricultural resources, and it recognizes the urgent need to reduce both greenhouse gas emissions and the vulnerability of agriculture and thus of food supplies to the impacts of climate change. It therefore requires location-specific assessments to identify how given practices endure at particular sites. CSA generally aims to achieve three main objectives: sustainably increasing agricultural productivity and incomes, adapting and building resilience to climate change, and reducing and/or removing greenhouse gas emissions relative to conventional practices (El-Fattal 2012; FAO 2013). A detailed description of CSA is given by the FAO (2013), but the main argument is that agricultural practices that constitute a CSA approach in most cases are not new and largely coincide with conservation agriculture, sustainable agriculture, and sustainable intensification (Kaczan et al. 2013). In practice, it means having access to agricultural technologies such as crop varieties and livestock breeds that are more adapted to a changing climate, improved soil and water management techniques to use water more efficiently (Blanco and Lal 2008; FAO 2013; Giller, Witter, Corbeels, & Tittonell 2009; McCarthy et al. 2011; Sturm et al. 2009; Teshome, Rolker, & de Graaff 2013), agroforestry (Ajayi et al. 2006; FAO 2013; Franzel et al. 2004), crop rotation and intercropping (Adiku et al. 2009; Blanco and Lal 2008; Chiputwa et al. 2011; Chomba 2004; Stringer et al. 2009), mulching (Acharya et al. 1998; Erenstein 2002), and integrated crop-livestock management (James et al. 2012), among others. In many cases, CSA involves investment and long-term commitment to improved land management (El-Fattal 2012).

A number of CSA technologies and practices have been promoted and used by different stakeholders across Uganda in order to support climate-smart agriculture, but studies on their exact nature and factors that influence their adoption are patchy and far from clear especially for dryland areas. Despite clear benefits, adoption of CSA by smallholder farmers has been slow, piecemeal, and largely unsustained (Carr 2016). Reference can be made to other studies in the sub-Saharan Africa (SSA) region that have underscored the importance of socio-economic factors in influencing the adoption of CSA practices and technologies. These household socio-economic characteristics include (Tizale 2007; Birungi 2007), state of education of household members (Uaiene 2009; Uematsu and Mishra 2010), and gender of household heads (Bayard et al. 2007; Bekele and Drake 2003; Dolisca et al. 2006 and Marenya and Barrett 2007). Other factors that have received tremendous attention across sub-Saharan Africa include presence or absence of a critical resource base for such land (Uaiene et al. 2009), credit (Doss 2006; Mohamed and Temu 2008; and Teferi 2013), human resource base and management intensity associated with CSA (Doss 2006; Mohamed and Temu 2008; Senyolo et al. 2017; and Teferi 2013), level of community cohesion and kinds of social support available among community members (Anderson 1980; Gandure, Walker and Botha 2013), availability of agricultural inputs and initial CSA investment costs (Bekele and Drake 2003; Chomba 2004; Grainger-Jones and Rydén 2011; Kurukulasuriya and Mendelsohn 2006; Long, Blok and Coninx 2016; Mignouna et al. 2011; Pender and Gebremedhin 2007; Rainforest 2016; Senyolo et al. 2017; Tadesse 2001; Uaiene et al. 2009; Zegeye et al. 2001), and access to water sources (Kassa et al. 2014; Teklewold et al. 2016.

Other scholars such as Biazin et al. (2012) underscore the importance of geographical and location-specific attributes of place. Biazin et al. (2012) argue that micro-catchment and *in situ* rainwater harvesting techniques are more common than rainwater irrigation techniques from macro-catchment systems and these are dependent on rainfall patterns and local soil characteristics. The result of these micro-catchment techniques is that they could improve the soil water content of the rooting zone by up to 30% and up to sixfold crop yields would be realized through combinations of rainwater harvesting and fertilizer use, as compared to traditional practices (Biazin et al. 2012).

Nakasongola District, a dryland zone in Uganda, is one of the target districts for the climate-smart agriculture (CSA) and National Adaptation Programmes of Action (NAPA) intervention projects in Uganda focused on reducing food insecurity through drought management (Nyasimi, et al. 2016). The district also hosts the "Enabling Environment for Sustainable Land Management (SLM) to overcome Land Degradation in the Uganda Cattle Corridor" project to provide land users and managers with the enabling policy, institutional, and capacity environment for effective adoption of SLM within the complexity of the cattle corridor production system (FAO 2016). The district was selected to (1) classify the existing CSA practices existing in dryland areas and (2) empirically quantify the relative importance of household socio-economic factors in influencing the adoption of CSA practices. It was hypothesized that (1) the CSA practices across farming households in the district were not heterogeneous and (2) the variance in the adoption of CSA practices across the district was not influenced by household socio-economic factors.

The rest of the paper is structured as follows: Section 2 presents the theoretical framework against which the whole paper leans; Section 3 describes the materials and methods with emphasis on the location and socio-economic characteristics of Nakasongola District, research design, empirical model parameters and related literature, and data analytical methods; Section 4 presents the detailed description of CSA adopters and results of the empirical model; and the paper concludes with emerging issues and key recommendations in Sect. 5.

2 Post-structuralist Geographies and Adoption of CSA Practices

The paper is situated broadly within post-structuralist epistemology whose roots in critical geography can be traced to the 1960s. Created by *Jacques Derrida (REF)*, post-structuralist epistemology was a movement within the French philosophy. It describes how the ontological and epistemological characterization of something as a "problem" establishes and justifies proper objects and subjects for its "solution," thereby constituting an understanding of what practices and rationales are considered as legitimate in order to resolve a problem (Foucault 2003).

The variable adoption of CSA practices speaks to post-structuralist worldviews and causal beliefs (Roberts and Parks 2007). Whatever the particular course of action, ideas about how the world works "put blinders on people" and "[reduce] the number of conceivable alternatives" that they choose from (Goldstein and Keohane 1993). Goldstein and Keohane (1993) define worldviews as ideas that "define the universe of possibilities for action." For example, culture, religion, rationality, emotion, ethnicity, race, class, gender, and identity all shape the way that humans (including policy-makers) perceive the opportunities and challenges facing them. In a similar way, farming households have to contend with climatic risks and change ways of living or adopt behavioral and structural measures to withstand the climatic challenges. As such, having a worldview implies "[limited] choice because it logically excludes other interpretations of reality, or at least suggests that such interpretations are not worthy of sustained exploration" (Goldstein and Keohane 1993). By limiting one's menu of available options, worldviews and causal beliefs have an instrumental impact on how adoption decisions are implemented. They also influence the very way in which farming households identify and select CSA practices. Therefore, understanding the environment in which households operate and the motivation of household members to adopt CSA practices informs the type of agricultural decisions at microlevel and facilitates planning for the adoption of associated practices across the country.

Post-structuralists use hierarchies, arguing that, for example, households can move up and down the CSA ladder, given the inherently unequal household socio-economic environment (REF). The volatility and variability of climatic patterns may according to post-structuralism explain why many households across many farming communities are still currently locked into precarious, unpredictable, and unsustainable agricultural practices (REF). As we argue in greater detail in the next sections, the environment in which households operate to make their agricultural decisions shapes their worldviews and beliefs, which in turn significantly influence and shape the decisions to adopt CSA practices. We argue that understanding household socio-economic environment is a useful starting point because it directs our attention to the root causes of variance in adoption of CSA technologies and practices in rural areas and understanding pathways to increased adoption of climate-smart agriculture in Uganda.

3 Materials and Methods

3.1 *Location and Socio-economic Characteristics of Nakasongola District*

Nakasongola District is located in Central Uganda, 114 km north of Kampala City, the administrative capital of Uganda, and covers an area of 3424 square kilometers (Fig. 1). The district is located on latitudes 05° 5′N and 1° 40′N and longitudes 31° 55′E and 32° 50′E (Nakasongola District Local Government 2009). Administratively Nakasongola District comprises of 1 county and 11 sub-counties, namely, Lwampanga, Wabinyonyi, Nakasongola Town Council, Kakooge Town Council, Migeera Town Council, Nakitoma, Nabiswera, Kakooge, Kalongo, Kalungi, and Lwabyata (Fig. 2). According to the Uganda 2014 National Population and Housing Census, the population of Nakasongola District was estimated at 181,863 persons (UBOS 2014).

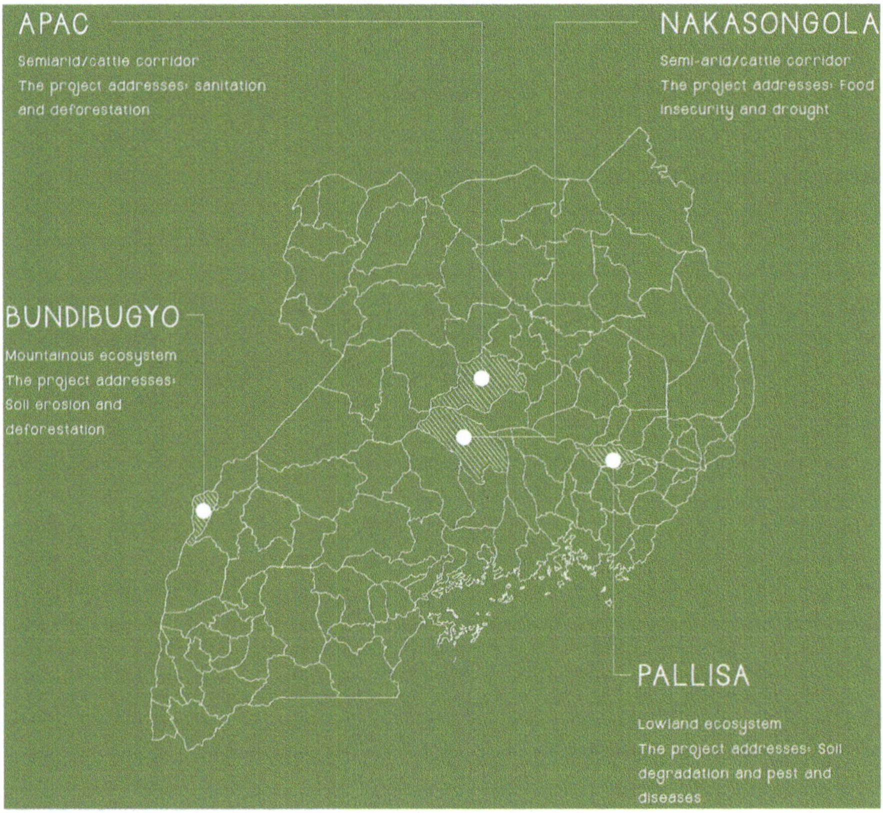

Fig. 1 NAPA implementation areas in Uganda. (Source: CCU 2014)

The district lies within the pastoral rangelands agroecological zone (AEZ), one of the AEZs in the cattle corridor of Uganda (Mubiru et al. 2017). It experiences variable climatic conditions such as unpredictable rainfall onsets and withdrawals, floods, as well as prolonged droughts among other climate variability-related phenomena (Kisamba-Mugerwa 2001). The district also experiences a bimodal rainfall pattern like most parts across Uganda, and annual rainfall totals received range between 500 and 1000 mm. The maximum temperature ranges between 25 °C and 35 °C, while the minimum diurnal range is between 18 °C and 25 °C (Nakasongola District Local Government 2009). The prolonged droughts, sometimes stretching more than a year, have intensified in the district over the last 10 years (Ninsiima 2016); and this has put people's livelihoods increasingly at risk, especially due to the inadequate local capacities to adapt.

Nakasongola is one of the districts in Uganda to implement NAPA activities and the main NAPA interventions that have been introduced in the district include excavation of valley tanks for crop irrigation, livestock watering, and domestic use and water harvesting including construction of 11,110,000 liters household water tanks

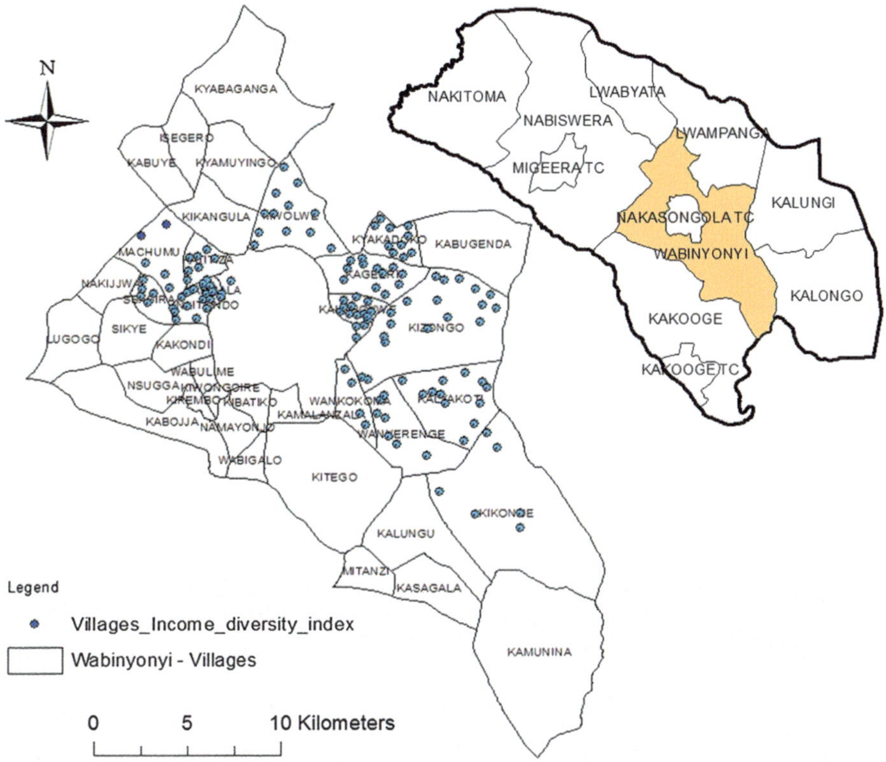

Fig. 2 Location of the study parishes and households in Wabinyonyi sub-county, Nakasongola District, Uganda

and installation of community drip irrigation schemes. Other interventions include provision of assorted seeds, agro-inputs, and fruit tree seedlings to farmers, provision of a pair of oxen and ox plows, provision of pigs and goats, and skills and knowledge development for community through trainings in records management, group dynamics, procurement, agronomy, enterprise selection, agriculture, livestock feeding, and management (CCU 2014).

3.2 Research Design and Sampling Procedure

Semi-structured questionnaires were used to collect relevant cross-sectional data from a randomly selected set of farming households in Nakasongola District. We selected representative parishes from Wabinyonyi sub-county; and then from each selected parish, farming households were targeted. Wabinyonyi sub-county is one of those sub-counties across the district that have been the target for the introduction and promotion of CSA practices under the central government program on agricultural improvement. The outcomes of the agricultural improvement program are unclear today due in part to inadequate information on adoption rates of CSA practices.

By 2014, Wabinyonyi sub-county had 3370 households, and these households are found in 8 parishes of Wabigalo, Sikye, Ssasira, Kiwongoire, Kyamuyingo, Kageeri, Wampiti, and Kamunina. The details of the sampling procedure are given in Table 1.

Out of the 3370 farming households in Wabinyonyi sub-county, we employed Equation 1 given by Krejcie and Morgan (1970), to arrive at a sample size of 311. Three parishes, Ssasira, Kageeri, and Wampiti, were purposively selected in the sub-county given that they were the main targets for the CSA interventions. A proportional sampling procedure was then used to select 120, 107, and 85 farming households from the 3 parishes, respectively. The location information for all households that responded were mapped, and these are displayed in Fig. 2. One hundred forty-three farming households were able to respond to the survey tool. The reasons for non-response ranged from outright refusal to participate in the survey to inability to spend the required time on the questionnaire. It should be emphasized that the study was conducted at a time when the government of Uganda was rolling out the "Operation Wealth Creation" program, whose major aim was supplying agricultural equipment, seedlings, and equipment to rural farming households. Therefore, reports of household socio-economic conditions and available assets at household level may have influenced the responses that were received.

$$S = \frac{X^2 NP(1-P)}{d^2(N-1) + X^2 P(1-P)} \tag{1}$$

where:

S = required sample size

X^2 = table value of chi-square for 1 degree of freedom at the desired confidence level

N = population size

p = the population proportion (assumed to be 0.5)

q = 1-p

d = the degree of accuracy expressed as a proportion (0.05)

Table 1 Summary of sampling procedures

Target geographical scale		Sampling procedure
Uganda	Nakasongola	Purposive sampling
Nakasongola	Wabinyonyi	Purposive sampling
Wabinyonyi	Ssasira, Kageeri, and Wampiti	Purposive sampling
Ssasira, Kageeri, and Wampiti	Households	Proportional sampling
Households	Individual farming households	Random sampling

Data Analysis

We used Rogers (1995) innovation adoption curve to describe the categories of adopters in the sub-county. The innovation adoption curve is a model that distinguishes between different types of adopters of innovations, based on the idea that certain individuals are more open to adoption than others. Past research has generally shown that the adoption of an innovation follows a normal, bell-shaped curve when plotted over time on a frequency basis. If the cumulative number of adopters is plotted, the result is an s-shaped curve. Figure 3 shows that the same adoption data can be represented by either a bell-shaped (frequency) or an s-shaped (cumulative) curve to give rise to five categories of adopters: innovators (2.5%), early adopters (13.5%), early majority (34%), late majority (34%), and laggards (16%) (Rogers 1995). These categories, based on standard deviations from the mean of the normal curve, provide a common language for innovation researchers (Rogers 1995).

Factor analysis was further used to classify the climate-smart agricultural (CSA) practices in the selected sub-county. According to Yong and Pearce (2013), factor analysis is a method of data reduction that helps to summarize data so that relationships and patterns can be easily interpreted and understood. Principal component analysis was used to generate clusters of CSA practices into homogeneous sets that contained most of the information in the CSA data. We employed factor analysis to condense and cluster the 12 CSA practices into new factors/clusters of CSAs; and it was these new factors for which the logistic regression modeling was employed. Prior to the application of factor analysis, two checks of internal reliability and validity were undertaken. The application of Cronbach's alpha test to the 12 CSA items yielded a score of three components whose reliability according to George and Mallery (2003) was good. The Kaiser (1974) criterion and Bartlett's test of sphericity were employed to inspect the acceptability and reliability of the identity matrix, and it was deemed appropriate to pursue factor analysis. The factor-loadings criterion level of 0.35 (Stevens, in Field, 2000) was used to identify the structure of relationships among the variables. This technique allowed the factors and their relative explanatory power to be identified (Cattell, in Field, 2000).

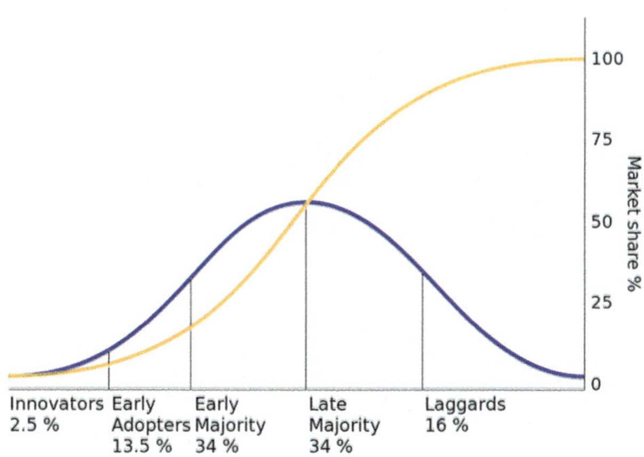

Fig. 3 The innovation adoption curve (the bell-shaped frequency curve and the s-shaped cumulative curve for an adopter distribution). (Source: Rogers 1995)

Emphasis was put on socio-economic factors that influence farming household's decision to adopt CSA practices. As indicated in Table 2, most of the empirical studies have focused on socio-economic variables as having an influence on the decision of farming households to adopt CSA practices. Before the logistic regression analysis was conducted, there was a need to test for multicollinearity. Two approaches were used to test this, including the variance inflation factor (VIF) and tolerance.

Tolerance is given as 1-R^2 (coefficient of determination) (Field 2009). Menard (2002) and other scholars suggest that a tolerance value less than 0.1 should be investigated further, as it indicates a serious collinearity problem. If a low tolerance value is accompanied by large standard errors and non-significance, multicollinearity may be an issue. The variance inflation factor (VIF) is 1/tolerance, and it is always greater than or equal to 1. Like tolerance, there is no formal cutoff value to use with VIF value lies between 1 and 10, then there is no multicollinearity. Values of VIF that are less than 1 or exceed 10 are often regarded as indicating multicollinearity, but in weaker models, values above 2.5 may be a cause for concern (Allison 2001). Equation 2 given by Simon (2003) was used to empirically quantify the relative importance of household socio-economic factors in influencing the adoption of CSA practices in the sub-county.

$$Y = \beta_0 + \beta_1 X_1 + \beta_2 X_2 + \beta_3 X_3 + \beta_4 X_4 + \beta_5 X_5 + \beta_6 X_6 + \beta_7 X_7 \ldots\ldots + \beta_{19} X_{19} + \mu \ldots\ldots\ldots\ldots \quad (2)$$

where:

Y = climate-smart agricultural (CSA) practices
X_1 = gender of the head of household
X_2 = age of the household head
X_3 = education level of the household head
X_4 = household size
X_5 = availability of labor force
X_6 = size of the land (acres)
X_7 = type of ownership of land
X_8 = main decision-maker in the household
X_9 = access to credit
X_{10} = access to extension services
X_{11} = access to agricultural tools
X_{12} = access to fertilizer use
X_{13} = access to pesticides and herbicides
X_{14} = membership to farmers group
X_{15} = access to improved seeds
X_{16} = marital status of the head of household
X_{17} = access to domestic water sources
X_{18} = access to irrigation water and other facilities

Table 2 Model predictors and summary of related literature

Variables	Variable type	Definition and relationship		Author(s)
1. Gender of household head	Binary	1 = male 2 = female	±	Marenya and Barrett (2007), Bayard et al. (2007), Bekele and Drake (2003), and Dolisca et al. (2006)
2. Age of household head	Discrete	Years	±	Mignouna et al. (2011), Marenya and Barrett (2007), and Mauceri et al. (2005)
3. Educational level	Categorical	1 = none, 2 = secondary, 3 = certificate, 4 = primary, 5 = degree	±	Uaiene (2009) and Uematsu and Mishra (2010)
4. Marital status	Binary	1 = married, 0 = unmarried	±	Ani et al. (2004) and Nnadi and Akwiwu (2008)
5. Household size	Discrete	Number of household members	±	Tizale (2007) and Birungi (2007)
6. Incomes sources (income diversity index)	Categorical	1 = livestock, 2 = trading, 3 = fishing, 4 = boda-boda, 5 = crop farming, 6 = forestry, 7 = civil servants, 8 = unemployed, 9 = others	±	Zhao and Barry (2013), Mehta (2009), Mathenge et al. (2015), Pender and Gebremedhin (2007), Davis, Winters, Reardon, and Stamoulis (2009), Maguza-tembo and Mangisoni (2016)
7. Size of land	Discrete	Total farm size (acres)	+	Kassa et al. (2014), Parvan (2011), and Uaiene et al. (2009)
8. Type of land tenure system – system of landholding	Categorical	1 = traditional/customary, 2 = freehold, 3 = mailo, 4 = leasehold	±	Doss (2006), Doss and Morris (2001), and Tizale (2007)
9. Main decision-maker	Categorical	1 = household head, 2 = spouse, 3 = children, 4 = relatives and others	+	Bekele and Drake (2003)
10. Household labor force	Binary	1 = yes 2 = no	+	Bonabana-Wabbi (2002) and Nanyeenya et al. (1997)
11. Household access to credit	Binary	1 = yes 2 = no	±	Mohamed and Temu (2008), Doss (2006), and Teferi (2013)
12. Access to extension services	Binary	1 = yes 2 = no	+	Mignouna et al. (2011) and Uaiene et al. (2009)
13. Agricultural tools (assets)	Binary	1 = yes 2 = no	+	Bekele and Drake (2003) and Kurukulasuriya and Mendelsohn (2006)
14. Access to fertilizer	Binary	1 = yes 2 = no	±	Tadesse (2001), Zegeye et al. (2001), Bekele and Drake (2003), and Chomba (2004)
15. Membership in farmers groups	Binary	1 = yes 2 = no	+	Anderson (1980)
16. Access to pesticides and herbicides	Binary	1 = yes 2 = no	±	Grainger-Jones and Rydén (2011), Rainforest (2016), WorldBank et al. (2014), and Grainger-Jones and Rydén (2011)

(continued)

Table 2 (continued)

Variables	Variable type	Definition and relationship		Author(s)
17. Access to improved seeds	Binary	1 = yes 2 = no	±	Pender and Gebremedhin (2007)
18. Access to domestic water sources	Binary	1 = yes 2 = no	+	Teklewold et al. (2016)
19. Access to irrigation water and other facilities	Binary	1 = yes 2 = no	±	Kassa et al. (2014)

X_{19} = household income diversity index[1]
β_0 up to β_{19} = the coefficients of the above variables
μ = the error term

[1] It has been realized for some time that rural people no longer remain confined to crop production, fishing, forest management, or livestock rearing but combine a range of occupations (Khatun and Roy 2012). This kind of practice, sometimes referred to as livelihood diversification, is one of the most remarkable characteristics of rural livelihoods (Gautam and Andersen 2016). Given as a process by which rural families construct a diverse portfolio of activities and social support capabilities in their struggle for survival and improvement in their standards of living (Ellis 1998), it was important to measure the level of diversification of farming households or the overall degree of income diversity/diversification using a single index.

There are several income diversification and household labor allocation indices, and some of these are given by Gebreyesus (2016), McNamara and Weiss (2001), and Zhao and Barry (2013). They include the Herfindahl index, ogive index, inverse Simpson index, Shannon index (H), Shannon equitability index, entropy index, modified entropy index, composite entropy index (Gebreyesus 2016), USA Today Diversity Index (Meyer and McIntosh 1992), Berger-Parker index, Shannon-Weaver index, and Margalef index (FAO 2015), among others. These indices or approaches can be divided into two groups. One group contains one-dimensional indices, which include indicators that count the number of activities or evaluate changes in the volumes of different divisions. The other group measures diversification based on two dimensions considering both the number of areas of activities and their relative volumes of turnover (Zhao and Barry 2013). Most of these indices require data on the proportion of income from one activity relative to all other activities. Given that, during the data collection exercise, we found difficulties in estimating incomes from individual activities that the farming households engaged in, except for reports of the number and types of activities, we constructed the income diversity index for this study using the Margalef index (MI) given as:

$$\mathrm{D}i = \frac{[S_i - 1]}{In[N_i]} \quad (3)$$

where:
N_i = total count of income sources that support household livelihoods
S_i = number of income sources for each individual household i (namely, a count)
The index has a lower limit of zero if only one unit of diversity is observed.

4 Results and Discussion

4.1 Identification and Classification of Climate-Smart Agricultural Practices

This section presents the adoption profile of farming households in the study parishes. Table 3 shows the mean age of the head of household, size of household, household income diversity index, and size of land were 42 years, 9 persons, 0.4, and 15 acres, respectively. The highest education for over one-half of the heads of household was primary education. Indeed, in Fig. 4, over 77% of the household head has education below lower secondary school. About 50% of the household head were in the cohabiting marriage category; this is an informal marriage category, but it dominates the area of the study. Figure 4 indicates that farming households in the study area engage in a diverse portfolio of activities including activities in the farm sector such as crop growing and livestock rearing and activities in the nonfarm sector such as seeking wage employment, self-employment, boda-boda riding, fishing, and small-scale trading, among others.

It is clearly shown that Timely planting was the most prevalent CSA practice followed by the rest of the practices as indicated in Table 4. This is because most farmers apply these practices basing on the traditional knowledge and experience from their grandparents as well as basing on the size of land they have. Intercropping, crop rotation, and proper spacing have been shown as the prevalent practices with their respective percentage cases of 41, 54, and 55. These practices help in diversifying crop production and reduce soil erosion and risks of total crop failure which is similar to studies done by Adiku et al. (2009), Laube et al. (2012), and Stringer et al. (2009).

Table 3 Descriptive statistics of the model variables

	N	Min	Max	Mean	Std. deviation
Age of the head of household (years)	143	19	83	41.50	14.417
Size of the household	143	1	6	6.804	1.245
Household income diversity index	143	0.00	1.82	0.387	0.387
Size of land (acres)	143	1	200	15.26	28.317
Access to household labor	143	1	2	1.03	.184
Access to credit	143	1	2	1.21	.409
Access to extension services	143	1	2	1.22	.418
Access to domestic water sources	143	1	2	1.03	.184
Access to irrigation water and other facilities	143	1	2	1.98	.144
Access to agricultural tools	143	1	2	1.01	.084
Access to fertilizers	143	1	2	1.74	.439
Access to pesticides and herbicides	143	1	2	1.71	.454
Membership in farmers group	143	1	2	1.71	.457
Access to improved seeds	143	1	2	1.36	.483

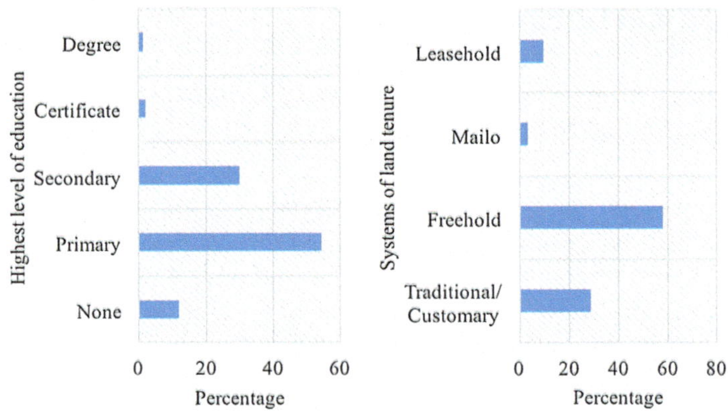

Fig. 4 Socio-economic and demographic characteristics of selected households

Table 4 The prevalence of the CSA practices in Wabinyonyi sub-county

	Parish			Total/percent of cases
	Wampiti	Kageeri	Ssasira	
1. Timely planting	42 (34.4)	31 (25.4)	29 (23.8)	102 (83.6)
2. Proper spacing	27 (22.1)	13 (10.7)	27 (22.1)	67 (54.9)
3. Crop rotation	29 (23.8)	19 (15.6)	18 (14.8)	66 (54.1)
4. Intercropping	20 (16.4)	14 (11.5)	16 (13.1)	50 (41.0)
5. Agroforestry	8 (6.6)	7 (5.7)	11 (9.0)	26 (21.3)
6. Improved crop and animal varieties	11 (9.0)	6 (4.9)	6 (4.9)	23 (18.9)
7. Fertilizer use	7 (5.7)	4 (3.3)	4 (3.3)	15 (12.3)
8. Use of pesticides and herbicides	7 (5.7)	3 (2.5)	4 (3.3)	14 (11.5)
9. Contour planting	6 (4.9)	5 (4.1)	2 (1.6)	13 (10.7)
10. Mulching	4 (3.3)	5 (4.1)	2 (1.6)	11 (9.0)
11. Rotational grazing	5 (4.1)	0 (0.0)	0 (0.0)	5 (4.1)
12. Water harvesting	0 (0.0)	1 (0.8)	1 (0.8)	2 (1.6)
Total	47 (38.5)	36 (29.5)	39 (32.0)	122 (100)

Finally, the least prevalent practices were water harvesting, rotational grazing, mulching, contour planting, use of fertilizer, access to pesticides and herbicides, improved crop and animal varieties and then agroforestry. This is because water harvesting reduce crop and livestock loss from unreliable rainfall as reported by Below et al. (2010) and Sturm et al. (2009). Agroforestry prevents soil and water erosion as well as improves water management which is in line with Ajayi et al. (2006) and FAO (2013). During the study, household farmers mentioned that they are advised to plant more trees but they do not have access to tree seedlings; and the few who have received them plant them in seasons without enough rain, and they end up drying.

When we classified CSA adopters and fitted the household adoption of CSA technologies and practices on the innovation adoption curve, our results indicated that they fall in similar categories of innovators, early adopters, early majority, late majority, and laggards, but the percentage values associated with each varied (Fig. 6). Although the actual figures and innovation adoption curve departed from what Rogers (1995) observed, the shape and direction of the curve were similar. Figure 5 shows a cumulative percentage of adopters over time – slow at the start, more rapid as adoption increases, and then leveling off until only a small percentage of laggards have not adopted. A further exploration of the variability in types of adopters and the CSA diversity index in Fig. 7 shows that not all villages are diversified in a similar pattern. There is spatial heterogeneity in the level of diversification in CSA practices with Kizongo village as the most diversified village, while Kikonge and Wakonkoma villages are the least diversified. The performance of Kizongo village, according to MAAIF (2015) and UNDP (2014), is attributed to being one of those villages where the national CSA program was started in the district, with the hope that associated activities would spread to other villages in the district such as Kikonge and Wakonkoma accordingly.

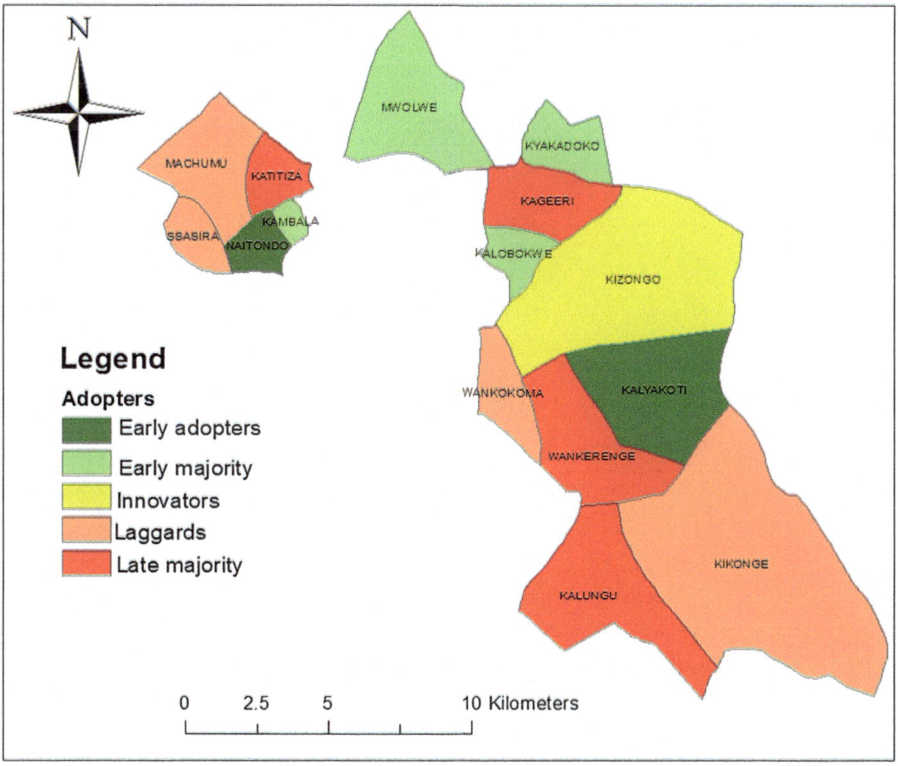

Fig. 5 Distribution of CSA adopters by parish

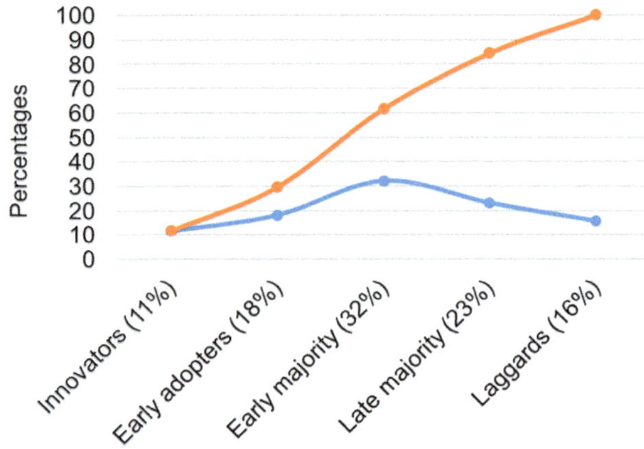

Fig. 6 The types of CSA adopters in Wabinyonyi sub-county along the innovation adoption curve

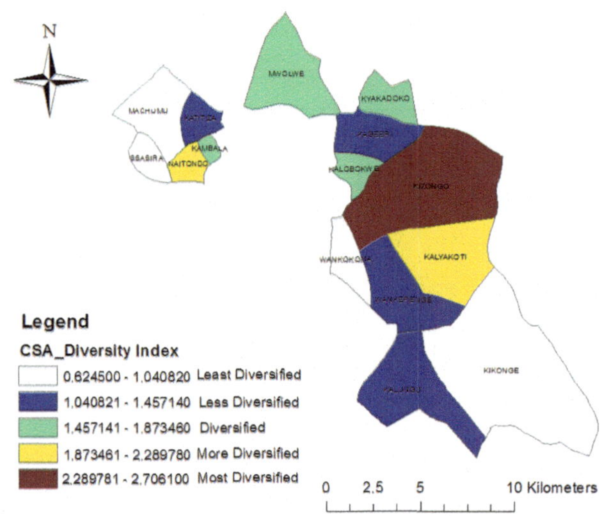

Village	CSA Diversity Index
Kizongo	2.7061
Kalyakoti	2.0816
Naitondo	2.0816
Kalobokwe	1.8734
Molwe	1.8734
Kambala	1.8734
Kyakadoko	1.6653
Kalungu	1.2490
Kageeri	1.2490
Wankerenge	1.2490
Katitiza	1.2490
Ssasira	1.0408
Macumu	0.8326
Kikonge	0.6245
Wankokoma	0.6245

Fig. 7 The level of diversity of CSA technologies and practices in Wabinyonyi sub-county

The application of Cronbach's alpha test to the 12 CSA practices items yielded a score of 3 components. Bartlett's test of sphericity was also statistically significant ($p < 0.05$). In Table 5, the three components extracted explained 51% of the variance in adoption of CSA practices with a sampling adequacy of 0.667 ($p < 0.001$).

A further inspection of the factor loadings was done using the rotated component matrix. For a substantive interpretation of the factors, only significant loadings were considered. As far as factor one was concerned, a close inspection of the loadings

Table 5 Total variance explained

	Initial eigenvalues			Extraction sums of squared loadings			Rotation sums of squared loadings		
Component	Total	% of Variance	Cumulative %	Total	% of variance	Cumulative %	Total	% of variance	Cumulative %
1	2.65	22.093	22.093	2.65	22.093	22.093	2.46	20.487	20.487
2	2.10	17.497	39.590	2.10	17.497	39.590	1.92	16.030	36.517
3	1.39	11.603	**51.193**	1.39	11.603	51.193	1.76	14.676	51.193
4	.979	8.156	59.349						
5	.890	7.416	66.766						
6	.816	6.801	73.567						
7	.723	6.027	79.595						
8	.711	5.927	85.521						
9	.575	4.795	90.316						
10	.501	4.174	94.490						
11	.374	3.119	97.609						
12	.287	2.391	100.000						

Extraction method: principal component analysis

and a number of items indicates that items were highly correlated with the use of pesticides and herbicides, fertilizer use, water harvesting, and rotational grazing. We considered these as land management practices. This component is primarily a measure of pesticides, herbicides, and fertilizers. However, the inclusion of rotational grazing within this component will add a bit of redundancy within our results. Four elements including agroforestry, mulching, contour planting, and crop and animal varieties, loaded onto component 2, which we referred to as conservation agriculture. It should be noted that crop and animal varieties enter with a negative sign to indicate that mulching, agroforestry, and contour plowing increase with decreasing crop and animal varieties. The elements loading onto component 3 included crop rotation, intercropping, proper spacing, and timely planting, and we labeled them broadly as crop management practices. If any of these variables goes up, so do the remaining ones. They are all positively related as they all have positive signs (Table 6).

4.2 *Model Estimation Results*

Before the logistic regression model was conducted, we tested for multicollinearity. There was no collinearity among all the independent factors since their VIFs and tolerance values were all above 1 and 0.1, respectively. The values for testing multicollinearity are given in Table 7. Three models were estimated and the estimation results are given in Table 8.

The most significant factor in the model across the three categories of climate-smart agriculture was the age of the head of the household. It was expected that as age increases, household probability of adopting climate-smart agricultural prac-

Table 6 Rotated component matrix of CSA practices

	Component			
	1	2	3	Construct
Crop rotation			.638	Crop management practices (model 1)
Intercropping			.671	
Proper spacing			.549	
Timely planting			.672	
Crop and animal varieties		–.560		Conservation agriculture (model 2)
Mulching		.671		
Agroforestry		.668		
Contour planting		.708		
Use of pesticides and herbicides	.842			Land management practices (model 3)
Fertilizer use	.836			
Water harvesting	.621			
Rotational grazing	.377			

Principal component analysis. Rotation method: varimax with Kaiser normalization

Table 7 Collinearity diagnostics

	Eigenvalue	Condition index	Tolerance	VIF
Gender of household head	17.089	1.000	.770	1.299
Age of household head	.825	4.550	.680	1.470
Marital status of the head of the household	.521	5.727	.714	1.400
Education level of the head of the household	.368	6.813	.886	1.128
Size of the household	.223	8.749	.751	1.331
Household income diversity index	.171	10.009	.776	1.288
Size of land	.152	10.590	.781	1.280
Main type of landownership	.129	11.498	.816	1.226
Main decision-maker in the household	.101	13.023	.704	1.420
Access to household labor	.077	14.928	.868	1.152
Access to credit	.074	15.224	.777	1.287
Access to extension services	.068	15.853	.583	1.716
Access to domestic water sources	.049	18.757	.871	1.148
Access to irrigation water and other facilities	.044	19.659	.760	1.317
Access to agricultural tools	.040	20.555	.863	1.158
Access to fertilizers	.030	23.730	.514	1.944
Access to pesticides and herbicides	.020	29.170	.544	1.838
Membership of farmers groups	.012	37.312	.785	1.274
Access to improved seeds	.005	59.329	.599	1.669

tices would decrease, where younger farmers were expected to adopt unlike elder farmers. Our estimation results indicate that an increase in the age of the head of household by 1 year on average increased the probability of adoption of crop management, land management, and conservation agricultural practices by about 1% in all the three categories. This is consistent with studies conducted by Mignouna et al. (2011), Marenya and Barrett (2007), and Mauceri et al. (2005) who reported that increases in the age of household heads increased the likelihood for adoption of climate-smart agricultural practices.

The size of land from our estimation results indicates that it influences the three broad categories of climate-smart agriculture, i.e., crop management, land management, and conservation agricultural practices. An increase of a household's land area by 1 acre, on average, increased the probability of adoption of crop management, land management, and conservation agricultural practices by 0.4%, 1%, and 0.4%, respectively. The size of land can be considered as a proxy for many other wealth-related household variables, and all climate-smart agricultural practices need a considerable amount of land for them to be successful. Those households with large land sizes could engage in proper spacing, planting of trees/woodlots, and rotational grazing. Farming households with big pieces of land would essentially be in position to afford experimenting across a number of climate-smart agricultural practices, and as several scholars have indicated, this makes them likely to engage in and adopt a variety of practices, because the extra land means they can also stand short-term and long-term shocks while having enough buffer, in comparison to small landowners.

Table 8 Marginal effects

	Model 1 Crop management practices				Model 2 Conservation agriculture				Model 3 Land management practices			
	Coef.	Std. Err.	*z*	95% C.I.	Coef.	Std. Err.	*z*	95% C.I.	Coef.	Std. Err.	*z*	95% C.I.
Gender of household head	–.0802	.0999	–0.80	–.27596	–.1928	.1612	–1.20	–.5087	.1797	.181	0.99	–.1756
Age of household head	.00035	.0036	0.10	**.0068****	.0018	.0059	0.30	**.0098****	–.0004	.007	–0.06	**.0134****
Marital status of the head of the household	.02306	.1085	0.21	–.1896	.1143	.1751	0.65	–.2289	.0647	.197	0.33	–.3214
Size of the household	–.0747	.0428	–1.75	–.1585	.0518	.0690	0.75	**.0835***	.096	.078	1.24	**.0562***
Household income diversity index	–.3535	.1346	–2.63	–.6173	.9803	.2174	4.51	.554	–.119	.244	–0.49	–.5978
Size of land (acres)	–.0004	.0018	–0.25	**.0039****	–.0039	.0029	–1.37	**.0095****	.0019	.0032	0.59	**.0044****
Access to household labor	–.1609	.2609	–0.62	–.6723	.4624	.4211	1.10	–.363	–.212	.475	–0.45	–1.141
Access to credit	.1379	.1244	1.11	–.1060	.2057	.2008	1.02	–.1879	–.3427	.226	–1.52	–.785
Access to extension services	–.0224	.1413	–0.16	–.2992	–.4538	.228	–1.99	–.900	.2870	.256	1.12	–.215
Access to domestic water sources	.3648	.26	1.40	–.1448	1.060	.4196	2.53	.238	.0577	.472	0.12	–.867
Access to irrigation water and other facilities	–.174	.3363	–0.52	–.8335	–.5199	.5427	–0.96	–1.58	.0716	.610	0.12	–1.125
Access to agricultural tools	.140	.5749	0.24	–.9867	–.3459	.9278	–0.37	–2.16	–.9374	1.04	–0.90	–2.983
Access to fertilizers	–.9847	.140	–7.02	–1.259	.16591	.2262	0.73	–.277	.3104	.255	1.22	–.1884
Access to pesticides and herbicides	–.978	.133	–7.34	–1.24	–.3634	.2152	–1.69	–.785	–.4819	.242	–1.99	–.956
Membership of farmers group	–.080	.114	–0.70	–.3042	–.1128	.1842	–0.61	–.48	–.0035	.207	–0.02	–.4096
Access to improved seeds	–.2057	.1201	–1.71	–.4412	.4701	.1939	2.42	**.090***	–.2486	.218	–1.14	–.676
Education level of the head of the household	–.237	.1471	–1.61	–.5252	.07325	.2374	0.31	–.392	.523	.267	1.96	**.0002****
Land tenure system – system of landholding	.1526	.1251	1.22	**.093***	.08697	.2019	0.43	–.309	–.0042	.227	–0.02	–.4493
Main decision-maker in the household	.0188	.1518	0.12	–.2787	–.2375	.2449	–0.97	–.718	–.6112	.276	–2.22	–1.151
Cons.	.595	.769	0.77	–.9276	–1.430	1.241	–1.15	–3.89	1.508	1.396	1.08	–1.255

Significant ***$p < 0.001$, **$p < 0.05$, *$p < 0.1$

The study results are in tandem with what Kassa et al. (2014) and Uaiene et al. (2009) found, to the effect that land size has a positive effect on adoption, because farmers with more cultivatable land are more likely to set aside a piece of land for climate-smart agricultural practices without impacting heavily on other agricultural practices. This is, however, in contrast with two studies done in Kenya. Gelder and Kerkholf (1984) observed that despite the pressure of land, trees were grown in 5–10% of the agricultural land, while Collier (1989) showed that 80% of the rural households had planted trees on 25% of their farms despite the small household land sizes in the district. This is because there was high awareness on tree planting in the district and that tree growing awareness through extension services was related positively to tree growing in the fields (Githiomi et al. 2012).

The size of household was significant for the adoption of conservation agriculture and land management practices. An increase in the size of the household by one person, on average, increased the probability of adoption of conservation agriculture and land management practices by 8% and 6%, respectively. The size of the household can also be considered as a proxy for many other labor- or manpower-related variables; and because climate-smart agricultural practices are labor-intensive, they require a considerable amount of labor resources for them to be successful. Tizale (2007) and Birungi (2007) conducted similar studies on the influence of household size on agricultural technologies adoption, and our results are in tandem with theirs.

The system of landownership was found to be statistically significant in determining the adoption of crop management practices. An increase in the security of tenure increased the probability of adoption of crop management practices by 9%. Households with secure land tenure, keeping other variables constant, had 9% higher probability of adopting crop rotation, intercropping, proper spacing, and timely planting than their counterparts with insecure land tenure regime. If farming households do have landownership rights, they would tend to engage more in crop management practices; on the other hand, if they do not have landownership rights, they become reluctant to adopt and incur a cost for crop management practices. It was a common practice in Nakasongola District that farming households were more interested in employing climate-smart agricultural practices on lands that they had full control over than those lands that they shared or communally owned. These results are also similar to what Doss (2006), Doss and Morris (2001), Kassa et al. (2014), and Tizale (2007) observed in their studies across several parts of the world. While it does not come out clearly in our research, it is possible that when households have secure systems of landownership, they can also lease out part of their landholdings for monetary returns to purchase and engage in some climate-smart agricultural practices.

Access to improved seeds was also found to be statistically significant in determining the adoption of conservation agriculture. An increase in the access to improved seeds increased the probability of adoption of improved animal varieties, mulching, agroforestry, and contour planting by 9%. This result is also consistent in similar semiarid areas of Tigray, Ethiopia, where Pender and Gebremedhin (2007) observed that farmers who had access to better and large chunks of land may be able to purchase improved seeds using the available credit which enables them to adopt CSA practices. The education level of the head of household was significant in influencing the adoption of land management practices. An increase in the education level of the head of household increased the probability of adoption of pesticides and herbicides, fertilizer use, water harvesting, and rotational grazing marginally by 0.02%.

5 Conclusions and Policy Implications

Our results indicated that timely planting, crop rotation, intercropping, and proper spacing were the most prevalent CSA practices. On the one hand, rotational grazing, mulching, fertilizer use, and use of pesticides and herbicides were the least prevalent practices. The PCA generated a three-factor solution and the components which clustered the CSA practices into three categories: crop management, conservation agriculture, and land management practices. Post-structuralist geographical thought, the context in which the paper is situated, is relevant for understanding adoption studies. A farming household is never just a farming household, and a farmer is never a farmer, according to post-structuralist geographers. Their actions are interwoven in a mesh of intersecting variables, and it is these factors which, according to post-structuralist geographical thought, provide both limitations and opportunities to adopt CSA practices across the selected households. Parameter estimates further indicated that households are not motivated by the same factors; but the most significant socio-economic factors influencing the adoption of CSA practices were the size of household, access to fertilizers, pesticides/herbicides, credit, seed, water sources and extension services, income diversity index, and the main household decision-maker. All the significant independent variables are given in Fig. 8.

Fig. 8 CSA practices and their determinants summarized

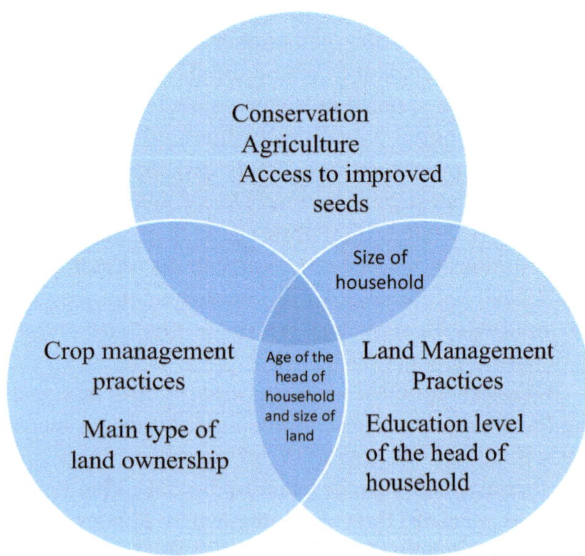

This study recommends that to realize the benefits of CSA practices across the study villages, there is a need to pay greater attention to innovators, early adopters, and early majority of adopters by:

1. Increasing the size of landholdings, together with securing the systems of land-ownership, would enhance further adoption of CSA practices. There is a need for the central government working with District Land Boards and Area Land Committees spread across Nakasongola District to take urgent and appropriate steps to increase households' legal claims and security over their land. Although larger land sizes may not be possible in the future with a growing population, increasing access to land through land rental market to enable land-constrained smallholders to acquire additional farmland would be appropriate.
2. Strengthening the existing Operation Wealth Creation (OWC), a national program and an offshoot of a conventional agricultural advisory service. This should include interventions that:

   (a) Build an elaborate agricultural education and training program for farmers to make better and informed agricultural decisions in the face of climatic risks and uncertainties.
   (b) Increase the supply of improved seeds that withstand the changing climatic conditions and improve crop security and agricultural productivity. The improved crop seeds should be made available to the poor rural farmers through the programs implemented in the district.

Acknowledgments This research forms part of the Climate and Water Resources Centre, Department of Geography, Geo-informatics and Climatic Sciences (http://www.mak.ac.ug/geography), research agenda. We would like to thank the Regional Capacity Building for Sustainable Natural Resource Management and Agricultural Productivity under Climate Change (CAPSNAC) for funding this study. Any errors herewith are the responsibility only of the authors, and this paper reflects the opinions of the authors and not the institutions which they represent or with which they are affiliated.

References

Acharya CL, Kapur OC, Dixit SP (1998) Moisture conservation for rainfed wheat production with alternative mulches and conservation tillage in the hills of north-west India. Soil Tillage Res 46(3–4):153–163. https://doi.org/10.1016/S0167-1987(98)00030-0

Adiku SGK, Jones JW, Kumaga FK, Tonyigah A (2009) Effects of crop rotation and fallow residue management on maize growth, yield and soil carbon in a savannah-forest transition zone of Ghana. J Agric Sci 147(3):313–322. https://doi.org/http://dx.doi.org.libproxy.mit.edu/10.1017/S002185960900851X

AGRA (2015) Youth in agriculture in Sub-Saharan Africa. Climate change and smallholder agriculture in Sub-Saharan Africa. Nairobi, Kenya. https://doi.org/http://hdl.handle.net/10568/42343

Ajayi OC, Masi C, Katanga R, Kabwe G (2006) Typology and characteristics of farmers testing agroforestry based soil fertility management in Eastern Zambia. Zambian J Agric Sci 8:1–5

Aldunce P, Beilin R, Handmer J, Howden M (2016) Stakeholder participation in building resilience to disasters in a changing climate. Environ Hazard 15(1):58–73. https://doi.org/10.1080/17477891.2015.1134427

Allison PD (2001) Missing data. In: Santoyo D, Journey K (eds) Sage University papers series on quantitative applications in the social sciences. Sage Publications, Thousand Oaks
Anderson P (1980) Conservation and agriculture. Biol Conserv 18. https://doi.org/10.1016/0006-3207(80)90070-1
Ani OA, Ogunnika O, Ifah SS (2004) Taking farm decisions and socio-economic characteristics of rural women farmers in Southern Ebonyi State, Nigeria. Int J Agric Biol 6(13):645–649
Augustina DS, Pârvan L, Popescu I (2011) Research on forestry recultivation of sterile dumps within the Jil ţ basin. Bull UASVM Agric 68(1):89–94. Retrieved from https://scholar.google.com/scholar?hl=en&as_sdt=0%2C5&q=parvan+2011+and+size+of+land&btnG=
Bayard B, Jolly CM, Shannon DA (2007) The economics of adoption and management of alley cropping in Haiti. J Environ Manag 84(1):62–70. https://doi.org/10.1016/j.jenvman.2006.05.001
Bekele W, Drake L (2003) Soil and water conservation decision behavior of subsistence farmers in the Eastern Highlands of Ethiopia: a case study of the Hunde-Lafto area. Ecol Econ 46(3):437–451. https://doi.org/10.1016/S0921-8009(03)00166-6
Below T, Artner A, Siebert R, Seiber S (2010) Micro-level practices to adapt to climate change for African small-scale farmers: a review of selected literature. IFPRI Discus Pap 0953(February):28
Bernard R, Dulle FW (2014) Access and use of mass media by small-scale farmers in accessing agricultural information for poverty alleviation in Tanzania: a case study of Kilombero district. Proceedings of the 1st COTUL annual conference November (pp 55–77). Sokoine University of Agriculture, Tanzania
Biazin B, Sterk G, Temesgen M, Abdulkedir A, Stroosnijder L (2012) Rainwater harvesting and management in rainfed agricultural systems in sub-Saharan Africa - a review. Phys Chem Earth 47–48:139–151. https://doi.org/10.1016/j.pce.2011.08.015
Birungi PB (2007) The linkages between land degradation, poverty and social capital in Uganda. University of Pretoria, South Africa
Blanco H, Lal R (2008) Principles of soil conservation and management. Wind erosion. Retrieved from http://books.google.com/books?id=Wj3690PbDY0C&pg=PA57
Bonabana-Wabbi J (2002) Assessing factors affecting adoption of agricultural technologies: the case of integrated pest management (IPM) in Kumi district, Eastern Uganda. Thesis. Virginia Polytechnic Institute and State University
Burrows C (2012) Global-warming-marxist-perspective.. Retrieved February 27, 2019, from https://www.marxist.com/global-warming-marxist-perspective.htm
Carr E (2016) Pathways to adoption of climate-smart agriculture. Retrieved from https://www.wilsoncenter.org/sites/default/files/csa_carr_presentation_-_wwc_-_final_0.pdf
CCU (2014) In: Nyasimi M, Radeny M, Mungani C, Kamini C (eds) Uganda' s national adaptation programme of action: implementation, challenges and emerging lessons, CGIAR research program on Climate Change, Agriculture and Food Security (CCAFS), Copenhagen, Denmark. Retrieved from www.ccafs.cgiar.org
Chiputwa B, Langyintuo AS, Wall P (2011) Adoption of conservation agriculture technologies by smallholder farmers in the Shamva District of Zimbabwe: a Tobit application. Proceedings of the 2011 meeting of the Southern Agricultural Economics Association
Chomba GN (2004) Factors affecting smallholder farmers' adoption of soil and water conservation practices in Zambia. Chemistry & …. Michigan State University. Retrieved from http://onlinelibrary.wiley.com/doi/10.1002/cbdv.200490137/abstract
Cline, W. R. (2008). Global warming and agriculture. Finance Dev 45. Southestern Zimbabwe. https://doi.org/10.1002/9780470752630.ch18
Collier P (1989) Contractual constraints on labour exchange in rural Kenya. Int Labour Rev 128(6):745–768
Davis B, Winters P, Reardon T, Stamoulis K (2009) Rural nonfarm employment and farming: household-level linkages. Agric Econ 40(2):119–123. https://doi.org/10.1111/j.1574-0862.2009.00374
Dolisca F, Carter DR, McDaniel JM, Shannon DA, Jolly CM (2006) Factors influencing farmers' participation in forestry management programs: a case study from Haiti. Forest Ecol Manag 236(2–3):324–331. https://doi.org/10.1016/j.foreco.2006.09.017

Doss (2006) Analysing technology adoption: challenges and limitations of micro-studies. Agric Econ 34:207–219
Doss CR, Morris ML (2001) How does gender affect the adoption of agricultural innovations ? the case of improved maize technology in Ghana. Agric Econ 25:27–39
El-Fattal L (2012) Climate-smart agriculture is "Smarter" when informed by a gender perspective, WOCAN policy brief. Women organising for change in agriculture and natural resource management. Bangkok, Thailand
Ellis F (1998) Household strategies and rural livelihood diversification. J Dev Stud 35(1):1–38. https://doi.org/10.1080/00220389808422553
Erenstein O (2002) Crop residue mulching in tropical and semi-tropical countries: an evaluation of residue availability and other technological implications. Soil Tillage Res 67(2):115–133. https://doi.org/10.1016/S0167-1987(02)00062-4
FAO (2013) Climate-smart agriculture sourcebook. Sourcebook on climate-smart agriculture, forestry and fisheries. Retrieved from http://www.fao.org/docrep/018/i3325e/i3325e00.htm
FAO (2015) Food security impact of agricultural technology adoption under climate change: micro - evidence from Niger climate variability and adoption of farming practices in Niger, Rome
FAO (2016) Eastern Africa climate-smart agriculture scoping study: Ethiopia, Kenya and Uganda. By Njeru E, Grey S, Kilawe E Addis Ababa, Ethiopia
Field DA (2000) Qualitative measures for initial meshes. Int J Numer Methods Eng 47(4):887–906. https://doi.org/10.1002/(SICI)1097-0207(20000210)47:4<887::AID-NME804>3.0.CO;2-H
Field A (2009) Discovering statistics using spss. SAGE Publications, New Delhi, p 110. https://doi.org/10.14359/51686441
Franzel S, Denning GL, Lillesø JPB, Mercado AR (2004) Scaling up the impact of agroforestry: lessons from three sites in Africa and Asia. Agrofor Sys 61–62(1–3):329–344. https://doi.org/10.1023/B:AGFO.0000029008.71743.2d
Foucault M (2003) Society must be defended: Lectures at the Collège de France. Picador, New York, pp 1975–1976
Gandure S, Walker S, Botha JJ (2013) Farmers' perceptions of adaptation to climate change and water stress in a South African rural community. Environ Dev 5(1):39–53. https://doi.org/10.1016/j.envdev.2012.11.004
Gautam Y, Andersen P (2016) Rural livelihood diversification and household well-being: insights from Humla, Nepal. J Rural Stud 44:239–249. https://doi.org/10.1016/j.jrurstud.2016.02.001
Gebreyesus B (2016) Determinants of livelihood diversification: the case of Kembata Tambaro zone, Southern Ethiopia. J Poverty Invest Dev 23(0):1–10. Retrieved from https://www.iiste.org/Journals/index.php/JPID/article/view/30208
George D, Mallery P (2003) SPSS for windows step by step. A simple guide and reference fourth edition. Allyn and Bacon, Boston. https://doi.org/9780335262588
Githiomi JK, Mugendi D, Kung'u J (2012) Household tree planting and its related constraints in meeting woodfuel production in Kiambu, Thika and Maragwa Districts of Central Kenya. J Hortic For 4(7):120–125. https://doi.org/10.5897/JHF12.006
Giller KE, Witter E, Corbeels M, Tittonell P (2009) Conservation agriculture and smallholder farming in Africa: the heretics' view. Field Crop Res 114(1):23–34. https://doi.org/10.1016/j.fcr.2009.06.017
Grainger-Jones E, Rydén P (2011) Climate-smart smallholder agriculture : what' s different ? IFAD OCCASIONAL PAPER, 1–23
Goldstein, Keohane (1993) Ideas and foreign policy beliefs, institutions, and political change. Cornell University Press, Ithaca
IPCC (2012) In: Field CB, Barros V, Stocker TF, Qin D, Dokken DJ, Ebi KL, Mastrandrea MD, Mach KJ, Plattner G-K, Allen SK, Tignor M (eds) Managing the risks of extreme events and disasters to advance climate change adaptation: a special report of working groups I and II of the intergovernmental panel on climate change. Cambridge University Press, Cambridge, UK/New York, NY, 582 pp

James B, Manyire H, Tambi E, Bangali S (2012) Barriers to scaling up/out climate smart agriculture and strategies to enhance adoption in Africa

Kaczan D, Arslan A, Lipper L (2013) Climate smart agriculture? a review of current practice of agroforestry and conservation agriculture in Malawi and Zambia

Kaiser HF (1974) An index of factorial simplicity. Psychometrika 39:31–36

Kassa BH, Abrha K, Aregawi KW (2014) Adoption and impact of agricultural technologies on farm income: evidence from Southern Tigray, Northern Ethiopia. I J Food Agric Econ ISSN 2(4):2147–8988

Khantun D, Roy BC (2012) Rural livelihood diversification in West Bengal: determinants. Agric Econ Res Rev 25(1):115–124

Kisamba-Mugerwa W (2001) Rangelands management policy in Uganda (Vol. 2001). Kampala. Retrieved from https://www.rmportal.net/framelib/rangeland-uganda.pdf

Krejcie RV, Morgan DW (1970) Determining sample size for research activities Robert. Educ Psychol Meas 38(1):607–610. https://doi.org/10.1177/001316447003000308

Kurukulasuriya P, Mendelsohn R (2006) A Ricardian analysis of the impact of climate change on African, 2(8): 1–62. Retrieved from http://www.researchgate.net/profile/Pradeep_Kurukulasuriya/publication/23550305_A_ricardian_analysis_of_the_impact_of_climate_change_on_African_cropland/links/0912f5084b3098b1c3000000.pdf

Laube W, Schraven B, Awo M (2012) Smallholder adaptation to climate change: dynamics and limits in Northern Ghana. Clim Change 111(3):753–774. https://doi.org/10.1007/s10584-011-0199-1

Long TB, Blok V, Coninx I (2016) Barriers to the adoption and diffusion of technological innovations for climate-smart agriculture in Europe: evidence from the Netherlands, France, Switzerland and Italy. J Clean Prod 112:9–21. https://doi.org/10.1016/j.jclepro.2015.06.044

MAAIF (2011) Ministry of agriculture, animal industry and fisheries sector strategic plan for statistics

MAAIF (2015) Uganda climate smart-agriculture country program 2015–2025 jointly implemented by Ministry of agriculture animal industry and fisheries, Ministry of water and environment. http://canafrica.com/publication/uganda-climate-smart-agriculture-country-program-2015-2025

Maguza-tembo F, Mangisoni J (2016) Impact of soil and water conservation improvement on the welfare of smallholder farmers in Southern Malawi. J Econ Sustain Dev 7(22):118–125

Majaliwa JG, Tenywa MM, Bamanya D, Majugu W, Isabirye P, Nandozi C, Nampijja J, Musinguzi P, Nimusiima A, Luswata KC, Rao KP (2015) Characterization of historical seasonal and annual rainfall and temperature trends in selected climatological homogenous rainfall zones of Uganda. Global J Sci Front Res 15(4):21. Retrieved from https://globaljournals.org/GJSFR_Volume15/3-Characterization-of-Historical-Seasonal.pdf

Marenya PP, Barrett CB (2007) Household-level determinants of adoption of improved natural resources management practices among smallholder farmers in western Kenya. Food Policy 32(4):515–536. https://doi.org/10.1016/j.foodpol.2006.10.002

Mathenge MK, Smale M, Tschirley D (2015) Off-farm employment and input intensification among smallholder maize farmers in Kenya. J Agric Econ 66(2):519–536. https://doi.org/10.1111/1477-9552.12093

Mauceri M, Alwang J, Norton G, Barrera V (2005) Adoption of integrated pest management technologies: a case study of potato farmers in Carchi, Ecuador

Mcnamara K, Weiss C (2001) On- and Off-farm diversification. AAEA-CAES meeting in Chicago copyright. Retrieved from http://ageconsearch.umn.edu/record/20520/files/sp01mc01.pdf

Mehta R (2009) Rural livelihood diversification and its measurement issues. Focus India, Rome

Menard S (2002) An introduction to logistic regression diagnostics In: Applied logistic regression analysis, 2nd edn. SAGE Publications, Thousand Oaks. https://doi.org/10.4135/9781412983433.n4

Meyer P, McIntosh S (1992) The USA today index of ethnic diversity. Int J Public Opin Res, Oxford Acad 4(1):51–58. https://doi.org/10.1093/ijpor/4.1.51

Mignouna DB, Manyong VM, Rusike J (2011) Determinants of adopting imazapyr-resistant maize technologies and its impact on household income in Western Kenya. AgBioForum 14(3):158–163

Mohamed KS, Temu AE (2008) Access to credit and its effect on the adoption of agricultural technologies: the case of Zanzibar. Bergamo, Italy. Retrieved from http://www.jstor.org/stable/41410533

Nakasongola District Local Government (2009) Higher local government statistical abstract. Retrieved from http://www.ubos.org/onlinefiles/uploads/ubos/2009_HLG

Nanyeenya WN, Mutetikka MM, Mwangi W, Verkuij H (1997) An assessment of factors affecting adaption of Maize production technologies in Iganga district, Uganda. Addis Ababa, Ethiopia. Retrieved from cimmyt.org/xmlui/bitstream/handle/10883/956/65998.pdf

Ninsiima R (2016) FAO pushes for climate change technology in Nakasongola. Retrieved April 8, 2016, from http://observer.ug/business/38-business/43492-fao-pushes-for-climate-change-technology-in-nakasongola

Nnadi FN, Akwiwu CD (2008) Determinants of youths' participation in rural agriculture in Imo State, Nigeria. J Appl Sci 8(2):328–333

Pender J, Gebremedhin B (2007) Determinants of agricultural and land management practices and impacts on crop production and household income in the highlands of Tigray, Ethiopia. J Afr Econ 17(3):395–450. https://doi.org/10.1093/jae/ejm028

Rainforest A (2016) Summary of climate-smart agriculture in the 2017 SAN Sustainable Agriculture Standard

Roberts JT, Parks BC (2007) A climate of injustice: global inequality, north-south politics, and climate policy. In: Global environmental accord (vol 1). The MIT Press, London, UK/ Cambridge, MA. https://doi.org/10.1017/CBO9781107415324.004

Rogers EM (1995) Elements of diffusion - what is diffusion. Diffusion of innovations, New York. https://doi.org/citeulike-article-id:126680

Senyolo MP, Long TB, Blok V, Omta O (2017) How the characteristics of innovations impact their adoption: an exploration of climate-smart agricultural innovations in South Africa. J Clean Prod 172:3825–3840. https://doi.org/10.1016/j.jclepro.2017.06.019

Simon G (2003) Multiple regression basics. Science (New York, N.Y.). Newyork. Retrieved from http://people.stern.nyu.edu/wgreene/Statistics/MultipleRegressionBasicsCollection.pdf

Stringer LC, Dyer JC, Reed MS, Dougill AJ, Twyman C, Mkwambisi D (2009) Adaptations to climate change, drought and desertification: local insights to enhance policy in southern Africa. Environ Sci Policy 12(7):748–765. https://doi.org/10.1016/j.envsci.2009.04.002

Sturm M, Zimmermann M, Schütz K, Urban W, Hartung H (2009) Rainwater harvesting as an alternative water resource in rural sites in central northern Namibia. Phys Chem Earth 34(13–16):776–785. https://doi.org/10.1016/j.pce.2009.07.004

Tadesse B (2001) Differential adoption of technologies and its implications for policy choice between equity and growth

Teferi ET (2013) Adoption of improved sorghum varieties and farmers' varietal trait preference in Kobo District, North Wolo Zone, Ethiopia. Haramaya University

Teklewold H, Mekonnen A, Di Falco S (2016) Impact of multiple climate smart practices in the climate resilient green economy: empirical evidence from the Nile Basin of Ethiopia. Jeju International Convention Center, Republic of Korea

Teshome A, Rolker D, de Graaff J (2013) Financial viability of soil and water conservation technologies in northwestern Ethiopian highlands. Appl Geogr 37(1):139–149. https://doi.org/10.1016/j.apgeog.2012.11.007

Tizale CY (2007) The dynamics of soil degradation and incentives for optimal management in the Central Highlands of Ethiopia. Retrieved from http://repository.up.ac.za:8080/handle/2263/25333

Uaiene RN (2009) Determinants of agricultural technology adoption in Mozambique (Vol. 10)

Uaiene RN, Arndt C, Masters WA (2009) Determinants of agricultural technology adoption in Mozambique. National Directorate of Studies and Policy Analysis. Retrieved from http://196.46.4.208/gest/documents/67E_AgTechAdoptionMoz.pdf

UBOS (2014) National population and housing census. Uganda Bureau of Statistics. Kampala, Uganda. Retrieved from http://www.ubos.org/onlinefiles/uploads/ubos/NPHC/NPHC2014PROVISIONAL RESULTSREPORT.pdf

Uematsu H, Mishra AK (2010) Can education Be a barrier to technology adoption ? Agricultural & Applied Economics Association 2010 AAEA. Retrieved from http://ageconsearch.umn.edu/bitstream/61630/2/11158AAEAEducationPaper.pdf

UNDP (2014) Uganda strategic investment framework for sustainable land management 2010–2020, (March), 104. Retrieved from www.agriculture.go.ug

Van Gelder B, Kerkholf P (1984) The agroforestry survey in Kakamega District. Kenya woodfuel development programme, working paper No. 3, Nairobi, Kenya. Workshop held in Arusha, Tanzania. The African Academy of Sciences (AAS), pp 156–170

WorldBank, CIAT, CATIE. Climate-Smart Agriculture in Sinaloa, M. C. C. P. for L. A. S. W. D. C. T. W. B. G (2014) Climate-smart agriculture in Sinaloa, Mexico

Yong AG, Pearce S (2013) A beginner's guide to factor analysis: focusing on exploratory factor analysis. tutorials in quantitative methods for psychology (Vol. 9). Ottawa, Ontario, Canada

Zegeye T, Taye G, Tanner D, Verkuijl H, Agidie A, Mwangi W (2001) Adoption of improved bread wheat varieties and inorganic fertilizer by small-scale farmers in Yelmana Densa and Farta Districts of Northwestern Ethiopia. Ethiopian Agricultural Research Organization (EARO) and International Maize and Wheat Improvement Center (CIMMYT), Mexico, DF. Independent Science and Partnership Council. Ethiopia

Zhao J, Barry P (2013) Implications of different income diversification indexes: the case of rural China. Econ Bus Lett 2(1):13. https://doi.org/10.17811/ebl.2.1.2013.13-20

Part IV
ICT for Ecosystems and Human Resilience

The Efficacy of ICT in Weather Forecast Information Dissemination: Evidence from Farming Communities in Mbale and Rakai Districts, Uganda

Doreen Tuheirwe-Mukasa, Magne Haveraaen, Julianne Sansa-Otim, Benjamin Kanagwa, and Godfrey Rwamahe Mujuni

Abstract Information and communication technology (ICT) has pervaded all spheres of life from the upscale rich to the rural poor in developing countries. ICT is increasingly being harnessed universally to better the quality of life of communities. This chapter highlights the need to utilize ICT tools to improve livelihoods of farmers in Uganda, in the face of climate change and variations in seasonal weather. We investigated the use of ICT tools and services in enhancing farmers' access to weather forecast information to improve agricultural productivity in Uganda. The ICT tools in question included mobile phones and computers/laptops, while the services included the use of emails, websites, and social network sites. We used focus group sessions with farmers in Mbale and Rakai districts to (1) capture their perception of the use of ICT tools and (2) establish the mode of ICT-supported dissemination that would be most effective and efficient for relevant weather forecast information dissemination. Extra information was sourced from key informant interviews with agricultural extension workers and personnel from Uganda National Meteorological Authority. We transcribed the information gathered into descriptive narratives, used thematic analysis and coding with spreadsheets for analysis. We found the mobile phone to be the ICT tool that most farmers have access to, and we found them open to solutions designed around the mobile phone. We establish and recommend using ICT tools to complement existing and conventional weather information dissemination strategies such as mass media. ICT tools allow for customized information to be sent to farmers in text, graphic, audio, or visual formats.

D. Tuheirwe-Mukasa (✉) · J. Sansa-Otim · B. Kanagwa
College of Computing and Information Sciences, Makerere University, Kampala, Uganda
e-mail: sansa@cit.ac.ug

M. Haveraaen
Department of Informatics, University of Bergen, Bergen, Norway
e-mail: Magne.Haveraaen@ii.uib.no

G. R. Mujuni
Uganda National Meteorological Authority, Kampala, Uganda

Y. Bamutaze et al. (eds.), *Agriculture and Ecosystem Resilience in Sub Saharan Africa*, Climate Change Management,
https://doi.org/10.1007/978-3-030-12974-3_25

We also found that ICT tools enhance user control and interaction, allow farmers to query and receive specific responses of weather forecast information they are interested in, provide a mechanism to collect crowdsourced feedback that can be used to improve the weather information services and products provided by UNMA, and promote timely and easy access to weather information. We noted that if fully exploited, ICT tools have the potential to contribute to effective dissemination of tailored weather information, which would help to improve the livelihoods of the farmers in Uganda.

Keywords ICT efficacy · ICT tools · Farmer perception · Weather information dissemination · Mobile phone

1 Introduction

The use of information and communication technology (ICT) is on the increase the world over. There are several global problems that can be alleviated and supported by the application of digital technologies (Heeks 2008). ICT provides great potential for accelerating human progress (National Research Council and others 2007). According to Agenda 2030, one of the greatest challenges facing humanity today is climate change. The Sustainable Development Goals (SDGs), particularly goal 13, are geared toward taking action on climate change. Handling climate change will foster the achievement of some of the other SDGs such as ending poverty and hunger and making human settlements safe, resilient, and sustainable. Furthermore, the United Nations General Assembly 2015 envisions a world where there is development and application of technology that are climate sensitive and resilient (United Nations General Assembly and others 2015).

Uganda is among the countries adversely affected by climate change (Barihaihi and Mwanzia 2017; Barros et al. 2014). This is because most of its economic sectors are weather-climate dependent. Four out of five people in Uganda depend on agriculture for income and food security, and any threat to agricultural production degrades Uganda's socioeconomic status and puts 80% of the population at risk of poverty and hunger. Erratic rainfall seasons and the increased frequency of droughts and floods have led to failed harvests, food insecurity, insufficient intake of food for children and adults, and low household incomes, among others (Bamanya 2014; Barihaihi and Mwanzia 2017). These situations can be further aggravated by a lack of weather information. While considerable progress has been made in the collection, archiving, and analysis of weather and climate data, their transformation into information that can be readily used has lagged behind, especially in developing countries where such information needs are the greatest (Sivakumar 2006). The farmers, particularly, are interested in seasonal weather information, especially the onset and cessation of the rain and dry periods. Seasonal rainfall forecasts are particularly suited for rain-fed farming systems (Klopper et al. 2006). Farmers use this information to decide on when to plant their crops, weed, and harvest. They can also decide what type of crops to plant and what soil conservation practices to put in place, in case of flooding (Kanagwa et al. 2015). Barihaihi and Mwanzia report that "empirical studies among African farmers have shown that usable climate forecasts can help them reduce their vulnerability to drought and climate extremes, while also allowing them to maximize opportunities when favorable rainfall conditions are predicted" (Barihaihi and Mwanzia 2017). Reliable weather information is therefore vital for informed decision-making (Sivakumar 2006; Weiss et al. 2000; Rijks and Baradas 2000).

ICT tools offer solutions for easing communication of information. For this study, we define ICT tools broadly as encompassing computing, communications, and digital technologies (Heeks 2008). These range from computer-based devices such as computers/laptops and mobile telephones to software systems including Internet services such as email, websites, and social network platforms, among others (Weiss et al. 2000; Ministry of Agriculture, Animal Industry and Fisheries n.d.). Globally, the most common means of dissemination are the mass media, mainly radio, television, and newspapers. Telephone and pagers are used to target specific users. The Internet is also increasingly gaining importance in dissemination (World Meteorological Organization n.d.; Uganda Communications Commission 2015). Specifically, these tools can be harnessed for the dissemination of weather information to farmers to improve agricultural productivity. Dissemination allows for clear communication of the results of climate science efforts (Barihaihi and Mwanzia 2017). However, despite the advances made in the field of ICT that have made the task of data manipulation, analysis, interpretation, and preparation of information easier, the challenge of communicating the right kind of information to meet the user needs remains even today (Sivakumar 2006; Weiss et al. 2000). There are challenges in getting weather forecast information to farmers (Barihaihi and Mwanzia 2017). User awareness of what information is available, where it can be found, and how it can be used in effective decision-making needs to be enhanced (Sivakumar 2006; Weiss et al. 2000).

This chapter presents results from a qualitative survey carried out to establish the level of ICT usage in accessing weather information by farmers in Uganda and to establish what modes of ICT-supported dissemination would be most effective and efficient to convey relevant weather information to them. We sought insights on the behavior of the farmers, their perceptions, and thoughts to using ICT tools and whether these tools would improve their livelihoods. The study is part of a larger project geared to addressing the problem of access to reliable weather information that meets the varying needs of stakeholders (Kanagwa et al. 2015). The project aims at improving the farmers' experience in accessing timely and personalized weather information to enhance their decision-making through the use of suitable ICT tools (Reuder and Sansa-Otim 2013).

This chapter is structured in seven sections. Section 2 presents the conventional methods of conveying weather information from the national meteorological authority in Uganda, the weather services provider, to rural farmers. We highlight some challenges to this flow of information. Section 3 presents related work that describes ICT tools, their advantages, and examples of projects where ICT has been used in Uganda. We then present our focus group and survey approach, tailored to the design science methodology, in Sect. 4; this is followed by Sect. 5 in which we present the results from the survey, structured according to themes, as well as the limitations. We discuss the results from the survey in Sect. 6, presenting possible usage scenarios arising from the use of ICT tools; and we conclude in Sect. 7.

2 Challenges of Weather Information Dissemination in Uganda

Uganda is divided into 121 districts (Uganda National Bureau of Statistics 2009), with 13 climatological zones (Basalirwa 1995). The languages of the people in Uganda overlap across districts, but a total of 52 local dialects are spoken, with English as the official language. Swahili is being promoted as a national language in the spirit of regional integration within the East African Community (Uganda National Bureau of Statistics 2009; Barihaihi and Mwanzia 2017). Uganda has nine known farming systems (Mwebaze 1999). These farming systems, like the climatological zones, do overlap. The Uganda National Meteorological Authority (UNMA) currently disseminates weather information to different stakeholders in Uganda through electronic and print media, including news bulletins on radio and television (TV), email, newspapers, bulletins, magazines, websites, as well as tailored mobile weather alerts in the form of short message service (SMS). Information is also relayed through stakeholder workshops, the National Media Centre (NMC), non-governmental organizations (NGOs) charged with facilitating the access of weather information by relevant stakeholder groups, the Ministry of Agriculture early warning unit, and more recently through the Internet, specifically using social media (Facebook, Twitter, YouTube) (UNMA n.d.; Bamanya 2014; Barihaihi and Mwanzia 2017). The Ministry of Agriculture uses the extension service to disseminate agricultural and weather information to farmers. In this dissemination avenue, farmers acquire knowledge of new developments, technologies, and practices to improve their current practices through extension workers based at district local governments (Barungi et al. 2016; Gakuru et al. 2009). The weather information in Uganda ranges from weather alerts/warnings, daily weather updates, dekadals, and monthly weather updates to seasonal weather forecasts (UNMA n.d.). The weather information disseminated is further simplified by translating it into some of the local dialects to make it easier to understand by the relevant stakeholders (Bamanya 2014).

Most farming communities in Uganda are interested in seasonal weather information that indicates the duration of the rain and dry periods and the amount and distribution of rainfall to be expected. Given that most of the farming carried out in Uganda is seasonal, UNMA's main product for weather dissemination is the seasonal weather forecast, which typically contains a forecast for 3 months periods with accompanying advisories for different sectors. It shows the rainfall conditions for the four major regions of the country. The seasonal forecast is prepared for periods March–April–May (MAM), June–July–August (JJA), and September–October–November–December (SOND) (UNMA n.d.). The farmers can then use this information to guide their farming calendar and related agricultural decisions.

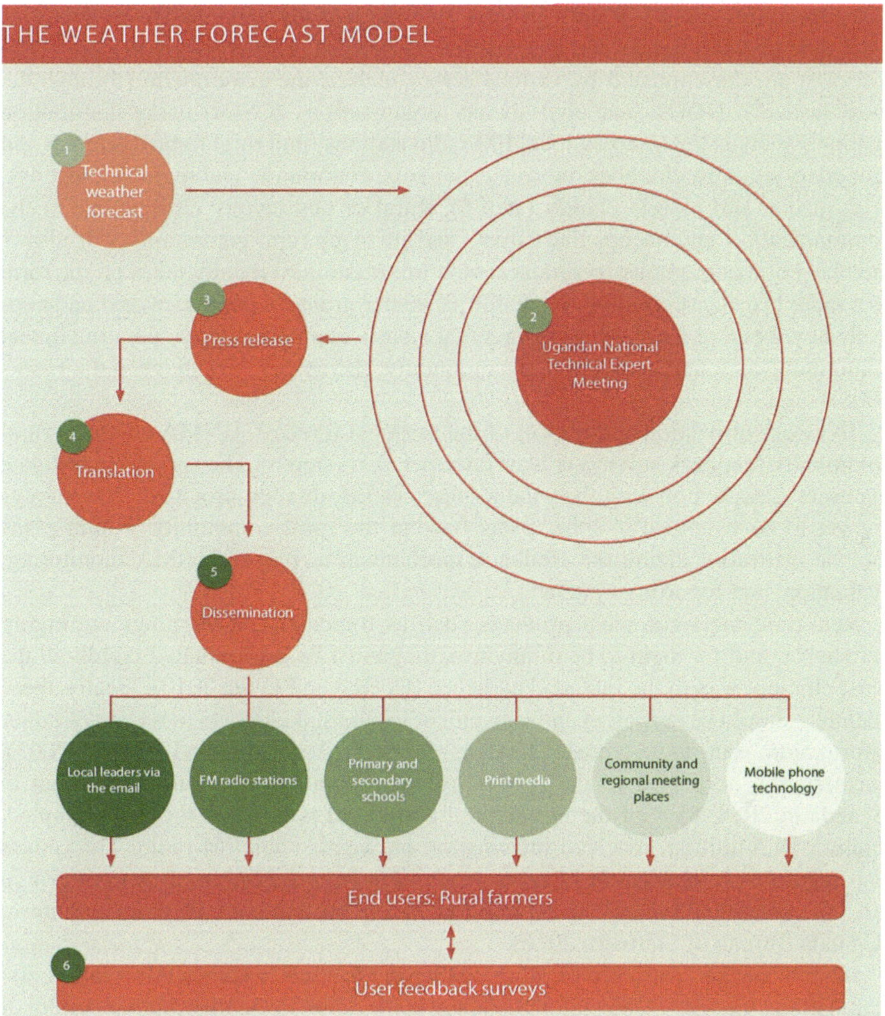

Fig. 1 Weather forecast model

Figure 1 shows the information flow from UNMA to farmers in the weather forecast model (Bamanya 2014).

In step 1, UNMA produces the seasonal weather forecast in relation to the regional consensus with other countries in the Greater Horn of Africa. In step 2, a meeting is held, where weather stakeholders representing various economic sectors discuss, simplify, and work out advisories for the seasonal forecast. At step 3, a press release is issued, and the forecast with advisories is also published in the national newspaper, the New Vision, and other newspapers. The forecast with advisories is then translated into local dialects by specialized translators trained in meteorology terminology. The translators work with UNMA to clarify any uncertain issues.

The translation takes 3–4 days in total, and the various versions are produced in electronic text soft copies and prerecorded audio CDs. This is shown in step 4. The forecasts are disseminated by various actors such as the government (through the local leaders), NGOs, and civil society organizations (CSOs) using distribution channels such as the Internet, local FM radio stations (and rural radio), primary and secondary schools, churches, mosques, markets, community and regional meetings, print media, and mobile phones (step 5). Rural or community radio is one of the communication approaches that targets and involves rural communities. It allows members of a community to gain access to information. It mainly takes on the form of a radio broadcast specially made for particular group of people, owned and controlled by them, or a radio station targeting a particular community operating in that community and managed by that community (Biraahwa Nakabugu n.d.; Heeks 2008).

To assess utilization and impact of the seasonal forecast, UNMA conducts rural community feedback surveys at least once per year (step 6). The surveys investigate the users' satisfaction with the weather forecast information provided. The surveys are yet to be extended to other users beyond the rural community. Future plans include institutionalizing the feedback mechanism as part of UNMA monitoring, evaluation, and reporting system.

Real-time meteorological products such as forecasts and warnings are highly perishable, and for them to be of any use, they must be disseminated rapidly in the most efficient way to the intended audience (Coelho and Costa 2010). Ideally, these channels should be sufficient and convenient for the stakeholders to receive weather information, but their uptake remains limited (Ziervogel and Calder 2003). Furthermore, according to the results from the population census carried out in Uganda in 2014, 85% of the agricultural household population that was sampled, totaling 19.3 million, received information on weather via FM radio. The census also reported an increase in Internet penetration from 29.5% in 2014 to 39.5% in 2015. The national literacy levels stand at 72.2% for adults of 18 years and above (Uganda Bureau of Statistics 2016).

In spite of this, there are existing challenges to these weather information dissemination strategies. One major challenge is the delay in the distribution system of the weather information. This challenge arises at points 4 and 5 of Fig. 1. In the dissemination approach, local leaders receive the weather information and use agricultural extension workers to disseminate the information to farmers. According to the census report of 2014, 19% of the agricultural households were visited by an extension worker, and 63.3% of them were at a distance of 5 km to receive extension services (Uganda Bureau of Statistics 2016). Considering statistics from a district such as Rakai district, there are 22 sub-counties, with a total of 752 villages. In 2009, there were 25 registered agricultural extension workers for Rakai (Uganda National Bureau of Statistics 2009). That would roughly approximate to one extension worker for each sub-county, expected to cover an estimate of 35 villages. This makes it practically impossible to effectively disseminate relevant weather information to the farmers on time (Barungi et al. 2016; Gakuru et al. 2009). Some of the weather information is therefore received when there is no value for it. For instance, seasonal weather information is first received in English by meteorological experts, translated into local dialects, and disseminated to the user communities (Bamanya 2014). Considering the time taken from the translation to final dissemination and the challenges of dissemination logistics, sometimes the end user receives it when the season it pertains to is either half gone or has ended, and sometimes he/she may not receive the information at all. This makes this information irrelevant to the users and not useful for decision-making (Barihaihi and Mwanzia 2017).

Another challenge is that the user has no control over the formats or of information they access or receive. For example, when weather forecast information is provided at the end of a news bulletin on radio or TV or presented in print media, it is clear that the user has not requested for the information. They cannot change the format in which they receive it. For weather information that is disseminated using print media, some of the consumers of this information may read it but not understand it and fail to apply it.

Owing to the diversity of the people in Uganda, in terms of dialects spoken, and the variations in farming systems, UNMA faces challenges in providing specific weather information that meets individual stakeholder needs due to a lack of computing resources.

A survey carried out to establish the status of the weather information dissemination system in Uganda found that many stakeholders do not receive timely and relevant weather information, with some stakeholders claiming that the weather information they receive is still complex in wording, making it difficult to understand. The nature of the forecasts provided, in terms of being probabilistic and using meteorological terms, means that they are not appropriate for all users (Ziervogel and Calder 2003; Barihaihi and Mwanzia 2017). The survey concluded that the current modes of communication used are not effective, and stakeholders have limited access to weather information (Kanagwa et al. 2015). This impedes the stakeholders' ability for making decisions that would lead to improved productivity. Other studies have identified similar challenges (Coelho and Costa 2010). To support the agricultural activities of farmers, the weather information needs to be timely, relevant, and packaged appropriately so as to lead to increased productivity (Rijks and Baradas 2000; Kanagwa et al. 2015).

3 ICT and Dissemination

Efforts are being increasingly made to provide disadvantaged people in rural areas with access to digital content and services using ICT tools in many developing countries (Pitula and Radhakrishnan 2011). These ICT tools collect, process, store, retrieve, and disseminate data and information (Weiss et al. 2000). According to a survey carried out to establish the access and usage of communication services across Uganda, ICT tools included radio, TV, cassette/DVD/CD, mobile phones, laptops, and desktop computers. Radio was the most widely owned form of communication device for both households and individuals, followed by mobile phones for individuals, while desktop computers were the least owned by both (Uganda Communications Commission 2015). For our study, we investigated the use of mobile phones, laptops, and computers specifically.

Several pilot projects have been attempted over the past decades to integrate ICTs into the dissemination of information to stakeholders in developing countries, but few have managed to bring long-term sustained benefits to the people that they target. This is mainly because the projects tend to focus on technical success and not on the specific end users' needs. Most initiatives around rural ICT in agriculture are donor (or at least externally) driven. The majority of such projects are implemented by, or in partnership with, international organizations with local or national nongovernmental organizations (NGOs), private companies, or government institutions that play a large part in several projects. They remain at pilot or "proof-of-concept" phase which limits their impact, as the projects cannot be financially sustained beyond the funding of the donors (Barihaihi and Mwanzia 2017; Gakuru et al. 2009). Another challenge these initiatives meet is that they are largely institutional-based and product- and platform-specific (Gakuru et al. 2009). This tends to have the effect of several isolated and replicated "small" pilot projects that cannot be fully sustained as the resources to do so are scattered over a number of them.

One such project targeted the fishing community in Kalangala, an island of Lake Victoria in the south of Uganda. In May 2011, the Uganda National Meteorological Authority (UNMA) partnered with Mobile Telecommunications Network (MTN), World Meteorological Organization (WMO), Ericsson, and the National Lake Rescue Institute (NLRI) to provide tailored SMS information to fisherfolk about the weather conditions over Lake Victoria, so as to improve their safety. The information provided included daily weather forecasts and real-time warnings in the Luganda language. The fisherfolk received a color-coded message, using the traffic light color scheme, that described the daily weather condition over the lake, and provided advisories on how to proceed. The green color code indicated to them that the lake was safe, orange communicated that they should proceed to carry out fishing activity on the lake with caution, and red communicated dangerous weather conditions over the lake. The red code warned the fisherfolk not to venture out onto the lake for fishing activities. This information would also be displayed on the MTN website (UNMA n.d.; World Meteorological Organization 2012; Ericsson Corporate Public, and Media Relations, World Meteorological Organization, MTN Group, National Lake Rescue Institute, Grameen Foundation 2012; MetOffice 2012). The project demonstrated that mobile telephone solutions were suitable for disseminating weather and climate information directly to end users in Uganda (World Meteorological Organization 2012).

Another pilot project targeted farmers in Kasese district by providing weather information through SMS. The project provided the dekadal, monthly, and seasonal forecast with related agricultural advisories to farmers, to answer their questions on when the rains would come. The project covered a total of 22 sub-counties in the district and covered two crop growing seasons (World Meteorological Organization 2012; Byarugaba 2014). Over a single-year period, from November 2011 to December 2012, the project showed that providing timely and accurate weather information to farmers could mitigate the increased risks of weather and climate variability that they faced and by boosting agricultural production, food security, and poverty reduction would be enhanced (Byarugaba 2014).

The effective dissemination of weather data and forecasts also demands close coordination between the meteorological and user communities – cooperation that notably has often been lacking (Mass 2006). The Fourth Assessment Report of the Intergovernmental Panel on Climate Change (IPCC) noted that one of the determinants of adaptation efforts to cope with the impacts of climate change is the need to increase access to resources (including information and technology) (Yohe et al. 2006). Furthermore, the challenges of climate change in Africa can be addressed by more effective engagement with climate variability through better integrating climate science and user communities (including in early-warning and disaster management systems), improving the dissemination and communication of information, and developing seasonal and intraseasonal information (Washington et al. 2006).

Other studies have investigated dissemination of information using ICT, especially over the World Wide Web and the Internet, and shown the impact of economic development as a result (Sivertsen 2000; Aker and Mbiti 2010). These have shown that ICT has the ability to support graphics, animations, and simulations (demonstrated live and/or in three-dimensional aspects), which is not possible with presentations on print media. Computer-based ICT tools are able to filter relevant information for a user and deliver it in a timely and appropriate manner (Sivakumar 2006). This would contribute to efficient dissemination of relevant information to stakeholders.

4 Data and Methods

To optimize the efficiency of the weather forecast information, the presentation techniques of the information including content, format, language, and style must fit the chosen means of dissemination as well as the target audience (World Meteorological Organization n.d.). There is need to take advantage of the respective features of the various dissemination channels so as to present a product in a way that will be appreciated by users and encourage positive response. Specifically, aspects of weather forecast information that appeal to stakeholders' interests need to be noted, so that the relevant content is prepared. The presentation of the weather information should attempt to arrest and retain the users' attention. It should be concise and easy to understand and interpret; but when necessary, emphasis should be placed on the most important meteorological phenomena and their impacts (Sivle et al. 2014).

One approach to identifying appropriate content for dissemination is to elicit forecast information that agricultural decision-makers desire, often done in the context of exploratory surveys designed to characterize perceptions and perspectives on a range of related issues. For this reason, we chose a qualitative strategy (Miles et al. 2013) which is person-centered to understand the process involved in weather information dissemination. We set out to understand the weather information services which agricultural stakeholders in Uganda are interested in, so as to allow us to generate a list of requirements for an effective and efficient weather information dissemination system.

First, we carried out a stakeholder analysis to identify the key stakeholders. We identified these as the farmers, agricultural extension workers who aid in weather information dissemination to farmers, and the weather service providers (UNMA). For the information gathering, we chose to interview the farmers first, as the immediate end users, then agricultural extension workers, and finally UNMA personnel. Farmers provide a representative sample of stakeholders that consume weather information. The targeted farmers were involved in either or both of crop production and animal rearing for subsistence and commercial purposes. UNMA is the only institution mandated to provide weather forecasts and advisories in Uganda.

We used focus group discussions (FGDs) with the farmers. We chose focus groups because of the resources available to us, and they are productive. To improve the responses, the FGDs were conducted in the local dialects of the farmers to ensure that they understood and answered the questions appropriately. We followed up with interviews using guided questionnaires with the agricultural extension workers and UNMA personnel, to address arising follow-up questions.

The information gathering techniques were chosen in line with the design science methodology. The main goal of design science is to create artefacts from an understanding of an environment (Von Alan et al. 2004). Tremblay, Hevner, and Berndt propose focus groups as an effective technique to improve artefact design and to provide evidence of its ability to solve a real problem (Tremblay et al. 2010). Focus groups provide a cost-effective and fast way to obtain experiences (Kontio et al. 2004) and obtain specific type of information from a clearly identified set of individuals (Stewart and Shamdasani 2014).

4.1 Setup of the Focus Group Sessions

We conducted one pretest focus group discussion to practice the discussion and evaluate if the questions were clear (Stewart and Shamdasani 2014). The feedback from the pretest session was also used to improve the clarity of some questions (Kontio et al. 2004). The study focused on two districts in Uganda: Mbale and Rakai. Mbale and Rakai are two of the districts in Uganda that have been affected by the weather changes. Choosing Mbale and Rakai provides us with varied user input because the districts are in different climatological zones (Basalirwa 1995) and the farmers in the two districts grow similar crops. This provides a basis for comparison, for instance, in terms of the weather requirements for the crops and arising decisions that can be taken.

To focus our research, we considered farmers dealing with two categories of crops. The National Development Plan (NDP) II 2015/2016 to 2019/2020 identifies these as subsistence and cash crops (Ministry of Agriculture, Animal Industry and

Fisheries n.d.). In Rakai, we conducted four focus group sessions and, in Mbale, six focus group sessions. Mbale is divided into three major regions the upstream, midstream, and downstream areas. We conducted two focus group sessions in each of these areas so as to enrich our findings. As we progressed with the focus group discussions, we realized we were getting the same responses. This confirmed that the ten focus groups were enough. This has also been supported by literature, as the point of saturation (Tremblay et al. 2010).

The extension workers and UNMA personnel completed structured questionnaires with the researcher(s) present or had an oral interview. The questionnaires were used as tools to guide the interviews. The extension workers were found at district level and represented different sub-counties in the two districts.

Selection of Participants

The district leaders involved in agriculture supported us in identifying the specific farmers. These officials are local experts and can identify farmers with suitable profiles (Meyer and Booker 2001). In Mbale, these were the Chief Agricultural Officer (CAO), Community Development Officer (CDO), and the District Environment Officer/Natural Resources Officer and extension officers. In Rakai, the District Production Officer (DPO), the District Natural Resources Officer, the Sub-county Chief, the Local Council Chairman, and the District Police Commander were contacted. We selected farmers based on their experience and insight into weather information for farming activities (Kontio et al. 2004). We considered both farmers who have experienced the effects of fluctuating weather conditions and those who were knowledgeable of the changes in the weather conditions and had taken measures to adjust accordingly. The farmers selected were part of farmer groups and were familiar with the farming domain. We sent out invitations to 12 farmers per focus group (Tremblay et al. 2010; Stewart and Shamdasani 2014), targeting the preferred 6–8 (Kontio et al. 2004). For nine of the ten focus groups, there were at least ten participants. One focus group however registered 15 participants because some of the invited farmers invited other colleagues. The farmers chosen were of similar socioeconomic standing, as similarities in abilities, knowledge, and intelligence facilitate communication (Stewart and Shamdasani 2014). After the sessions, we discovered that some (at most three in one of the groups in Rakai) of the farmers belonged to the same farming group and were familiar to each other, but most were unfamiliar. This, in our view, did not affect the group cohesiveness.

We observed during the pretest that there were cultural hindrances attached to mixing the farmers by gender (if a husband and his wife were in the same session, the interviewers had to coerce the wife to respond), so we disaggregated the sessions by gender, to allow for free interactions. Some studies have shown that men and women behave differently in group settings (Stewart and Shamdasani 2014). Of the four sessions in Rakai, two were composed of female-only farmers and the other two of male-only farmers. The six in Mbale were equally divided into three male and three female groups (one group of male-only and another of female-only farmers from each of the upstream, midstream, and downstream areas).

We also selected farmers by age with ranges of 15–25, 26–35, 36–45, and above 46. The age ranges begin at 15–25 because most of the people in Uganda are the youth, with 21.13% within this age group (Uganda Bureau of Statistics 2016). This was done to include farmers of all age groups and determine if ICT tools are more proficient for a particular age group. The National Analytical Report shows that generally ICT usage increases as an individual's age advances (Uganda Bureau of Statistics 2016). We considered exposure (none, some, high) to ICT tools and mechanisms of information dissemination ranging from radios, televisions, computers/laptops, and mobile telephones; and the applications that run on these, including Internet services such as email, and social media platforms (WhatsApp and Facebook); and the use of SMS.

The district leaders also identified the extension workers who filled out the questionnaires and participated in interviews. These were randomly selected from the sub-counties they oversee in working with the farmers. The UNMA personnel were also randomly selected from the directorates in UNMA. A total of 17 questionnaires were given to the extension workers from both Mbale and Rakai districts and ten questionnaires to UNMA personnel. The structured questionnaires and key informant interviews for the extension workers and UNMA personnel were all administered in English.

FGD Session Management

A moderator, an assistant, and an observer (part of the authors) were in charge of each session. We recorded each session using an audio recorder, and the participants were notified about the recording before the sessions. The average session time was 1.5 hours, with the longest taking 2 hours. The moderator provided ground rules including assuring the participants of privacy and confidentiality, respect for differing viewpoints of participants, and handling of mobile telephone interruptions. Pictures were also taken at each of the sessions, as a way of data capture. The pictures show participants raising their hands to answer some of the questions, for instance, to answer "how many people have a smart mobile phone?" and some showing the mobile phone they have. We obtained consent from the participants for both the audio recording of the sessions and taking of pictures.

Meeting Venues

The district leaders identified meeting venues easily accessible to the farmers. These included homes of village Local Council Chairpersons, churches, and schools. These sites turned out to be excellent as they were comfortable for both the farmers and researchers and allowed proper recording of sessions. Participants from hard-to-reach areas of the upstream region of Mbale traveled to the midstream areas for their sessions. The researchers also traveled by car to one such meeting.

Question Guides and Themes

We categorized the questions in four themes: agricultural activities of the farmers; agro-meteorological information pertaining to the farmers; ICT aspects and dissemination channels; and indigenous knowledge forecasts. Each of the themes was designed to capture a different aspect of weather information dissemination. Each theme consisted of three open-ended questions structured from general to specific (Stewart and Shamdasani 2014). We translated the questions from English into the farmers' local dialects using translators. We conducted the Rakai sessions in Luganda and Mbale sessions in Lugisu, the local dialects for the districts.

4.2 Analysis

We conducted the focus group sessions with a cross section of farmers chosen from the sub-counties in the districts of Rakai and Mbale. Focus groups are not self-contained by themselves, so we used a limited content analysis (Morgan 1996). We targeted thematic analysis since the interviews were structured into themes. Each focus group session and interviews was transcribed in form of a simple descriptive narrative (Stewart and Shamdasani 2014). The transcribing was done verbatim, and no corrections were made to the participants' grammar to preserve the responses. The results from the pretest session formed part of the analysis. Questionnaires from the agricultural extension workers and UNMA personnel were coded using spreadsheets.

4.3 Limitations

Some of the participants were not as active as others, particularly those younger in age, and the moderator directed individual questions to them to capture their responses (Stewart and Shamdasani 2014). The sessions were conducted in the participants' local dialects, and the participants tackled complex questions, rephrasing them for counterparts, with the guidance of the moderator. However, we cannot rule out that there may have been some information lost in translation. One limitation of focus group research is generalization (Tremblay et al. 2010) arising from the way participants are recruited and the number of groups sufficient for a correct sample. Basing on the fact that after the ten sessions, we reached a point of saturation and received the same responses, we suggest that our findings are useful for further research.

5 Results

This study sought to establish the level of ICT usage in accessing weather forecast information by farmers in Uganda and establish what modes of ICT-supported dissemination would be most effective and efficient to convey relevant weather information to them. We sought insights on the behavior of the farmers in light of climate change and existing mechanisms of information dissemination, their perceptions, and thoughts to using ICT tools and whether these tools would improve their livelihoods. We present the results from the focus group sessions according to the themes in the survey.

5.1 Agricultural Activities

We established that the farmers engage in either or both of subsistence or commercial farming. Most farmers practice crop production, with livestock rearing done for purposes of supplementing household incomes and food supply. The farmers in Rakai and Mbale noted that the pricing of agricultural produce is not consistent. This leads to unanticipated losses from their sales.

The crops grown can be categorized as either short-maturing crops such as market fruits and vegetables (onions, carrots, tomatoes) or long-maturing crops such as coffee, bananas (matooke), maize, and cassava. Animals reared included cows, pigs, poultry, and goats. Mbale farmers grow more market fruits and vegetables than Rakai farmers. Farmers in both districts reported that they rarely have food left over from one planting season for storage and use in the next season, as any surplus is sold for upkeep. One of the participants from a male group in Mbale proposed that each household stores about 50 kg of food with government, which could be later redistributed in hunger time.

The farming activities are carried out on average on an acre of land. In Mbale, the land is fragmented according to the size of one's family, and this determines the kind of farming activity undertaken. The more children in a family, the more land is fragmented. For this reason, farmers in Rakai rear more animals compared to the farmers in Mbale, as the land is less fragmented. The farmers in Mbale rear a maximum of five cows, most of them zero-grazed, while the cows in Rakai are kept free-range. Some farmers in Mbale pay for grazing their animals in areas with pasture.

The farmers in both districts make comparable decisions including when to prepare the land, plant, apply fertilizers (mostly from the animal waste of cows, goats, and chicken) irrigate, harvest rainwater, and apply pesticides to both crops and animals (cows). Notably, only farmers who grow market vegetables invest effort in irrigating their crops. Water for irrigation comes from streams, swamps, rivers, harvested rainwater, boreholes, and piped water from the National Water and Sewerage Corporation (NWSC) and Lake Wagagai, a crater lake in Mbale. Residents who have NWSC water installed sell a 20 L jerrycan on average at 0.2USD (600UGX).

To be able to pay for the water bills, these residents sell water to their neighbors who come from different places to fetch the water. Irrigation is generally not practiced because the distances the farmers have to travel for water are prohibitive. The decision to use fertilizers is also dependent on the type of soil; participants in Mbale claimed that fertilizer is not helpful on sandy soil. This information is useful for understanding the factors that farmers consider before planting, during planting, harvesting, and post-harvest stages. In order to disseminate weather forecast information that is relevant to the farmers, their farming decisions must be considered.

5.2 Agro-meteorological Information

Most of the farmers receive weather information, especially via the radio (handset or mobile phone radio) from farmer program broadcast at designated times. Radio has been found to be the most widely owned form of communication device (Uganda Communications Commission 2015). Some farmers pointed out that they are not available at broadcast times and miss out on the disseminated information. Some areas of Mbale have poor radio signal strength, and participants here do not have access to radio. Other areas do not have access to electricity, so the participants do not have televisions. For the areas where electricity is available, televisions are prohibitive because of costs attached to buying set-top boxes after digitization of broadcasting.

The farmers acknowledged the change in seasons and unreliability of weather information received, because it is general information, noting that it would rain in one village, for instance, and not in the neighboring village. Most farmers claimed they would plant their seed after experiencing one shower of rain in a season when the rains were expected and then wait expectantly for the crop to grow. In this particular season (January), the rains had not come as expected resulting in a severe scarcity of water to the extent that many animals died in Rakai. Some farmers noted that when the rains are heavy (above normal), there are more pests to deal with.

The government provides the farmers with agricultural inputs such as seeds and fertilizers, but most farmers complained that these arrived late in season to be useful. Some NGOs have given the farmers nonlocal heifers, and the farmers found that they have no pastures to feed the animals. The animals are exotic and require more specialized care than local breeds, as they are not adapted to the local conditions, which makes them more prone to diseases.

The extension workers chose information on crop development, plant disease reports, and food security reports as parameters relevant to aid farmers in making decisions. Crop development included information on planting seasons, rain periods, prolonged droughts, and harvesting aspects, while food security encompasses food usage, storage, and hunger forecasts. They cited that some weather element parameters are unpopular, including soil temperature and moisture, relative humidity, and air pressure, and yet these have been cited to be some of the main variables of interest for an effective growth season in farming (Coelho and Costa 2010).

The prepared weather forecast does not aid the farmers in understanding this particular information, and therefore they do not have any motivation toward understanding and using it. There is a need therefore to bridge this information for the farmers.

The extension workers also cited challenges in accessing and distributing agro-meteorological information including unavailability of the information, low literacy levels among the farmers, and high poverty levels (farmers cannot afford modern equipment such as radios, mobile phones, computers). Some of the farmers are in hard-to-reach areas and therefore inaccessible. Many of the farmers do not believe in the accuracy of the seasonal weather forecast information and rely more on their own experiences. Agro-meteorological information here provides insights into what channels of dissemination to use, appropriate times for availing weather information, providing location-specific information, and structuring meaningful advisories for the expected weather conditions.

5.3 ICT and Dissemination Channels

We found that over 95% of the farmers own a mobile telephone, those that did not have one were of school-going age and therefore unable to afford one. These, however, expressed the desire to own a mobile phone. Despite the high poverty levels, the farmers prioritize acquiring a mobile phone, also as Heeks points out (Heeks 2008). The 2017 National Analytical Report reports that 52.3% of the population owns a mobile phone, and this is projected to increase as the population gets more exposure to ICT tools (Uganda Bureau of Statistics 2016). The majority of the farmers had the simple feature phone (the "button" type), while a few of the farmers had smartphones (commonly referred to as "touch"). Many of the farmers argued that the smartphones were not practical considering the nature of their work. "You cannot be in the garden digging and use your dirty hands to slide on the phone!" This can be considered a needs assessment for improving the quality of smartphones, for instance, by designing more resilience in waterproofing phones.

One male focus group in Mbale explained that there was an ongoing project that was teaching the farmers how to use computers. Most of the farmers, however, did not have exposure to a laptop or computer. These have been shown to be the least owned ICT devices (Uganda Communications Commission 2015; Ministry of Agriculture, Animal Industry and Fisheries n.d.). Those who could access the Internet used their mobile phones to do so. These were the minority. The more youthful participants had knowledge and access to social media (particularly WhatsApp). 77.2% of the population is signed up for social media (Uganda Communications Commission 2015).

The farmers had good command of how to operate their mobile phones and could use short codes, e.g., for loading airtime, checking account balance, accessing mobile money, and calling customer care lines, among others. This shows that the farmers would know how to utilize various applications on a smartphone. If, for example, farmers ask for SMS information with parameters of language, district, and farming practices, they could get back a text message, audio recording, or a toll-free line to call for related weather information.

Overwhelming majority were willing to pay up to 0.14USD (420UGX) for weather information services, as long as they initiated the request and received pertinent information. This is due to the volume of unsolicited for messages telecommunication companies normally send, particularly for advertisement. Some of the farmers felt that they should meet part of the costs for receiving weather information as individuals, and other companies such as telecommunication companies top up as part of their corporate social responsibility. The majority voted to receive weather information using their mobile phones, via SMS, because they have access to their mobile phones all of the time. This is positive since the farmers realize that the service has to be funded somehow.

One focus group of male farmers in Mbale suggested the introduction of a weather call center, sponsored by the government, where they can call in for the weather information at no cost. Another group proposed that they receive audio phone messages directly from UNMA, just as they received audio political campaign messages.

All of the women groups vouched to receive weather information in communal settings such as places of worship, at weddings, or funerals, organized demonstration sessions among others. They reported that radios, televisions, and mobile phones are controlled by the men, and the women have to listen or watch programs prioritized by the men, such as football. A greater percentage of men own mobile phones and radios, listened to radio, watched TV, and used a computer and Internet than women. Men have been shown to have more ownership and control of ICT tools than women (Uganda Communications Commission 2015; Ministry of Agriculture, Animal Industry and Fisheries n.d.; Barihaihi and Mwanzia 2017).

The farmers in Mbale vouched for receiving weather forecast information in Lugisu, and those in Rakai in Luganda, or be given the option to choose a preferred language, with a bias for audio messages than any other format. They reasoned that the government had carried out mass registration of citizens, so one's local language could be determined through their identity. However, English and Swahili should also be maintained as general national languages.

The extension workers advocated for the use of mobile phone capabilities, especially SMS (as most of the farmers own or at least have access to a mobile phone) and Internet resources, particularly social media be used for dissemination. However, radio and the services of extension workers must be maintained, to cater for those who cannot read but can listen. Aspects of the Internet such as email and websites are currently not being used for dissemination to the farmers. This is in tandem with the results from the Uganda population census report that showed that the Internet stands at 7.3% among the sources of information for households in Uganda (Ministry of Agriculture, Animal Industry and Fisheries n.d.). Acceptable formats of information chosen for dissemination of weather forecast information were audio, text, and pictorial/graphical. The extension workers also advocated for trainings for themselves, to be able to provide better services to the farmers, and sensitization workshops for the farmers for increased uptake of meteorological products.

We established that UNMA mainly provides seasonal weather forecasts, accompanied with advisories, to the public. These take the form of PDF (portable document format) reports mainly. They also provide alerts in form of warnings and updates which may be daily, dekadal, and monthly. All of the services they provide are public (common good) services. For the services provided to the farmers, there is a demand for location-specific advisories for specific crops, which they are unable to cater for at the moment. UNMA is currently unable to provide regular and timely updates on seasonal climate outlook (due to late collection of the data to generate the product), and forecast of dry spells, due to a low computing capacity. Information that is not freely available to the public is weather and climate information tailored to particular transactions, historical data, and raw climate data.

UNMA personnel also agree that there is a need to understand the weather information needs of the farmers, so as to create tailored products and services. They concur that to complement existing dissemination efforts, SMS should be fully adopted and promoted, and a strategy for a crowdsourced feedback mechanism established. One of their recommendations to improve the weather information dissemination strategy is to mainstream forecast dissemination in government programs, so that effective policies can be structured, and to promote integrating traditional weather forecasts in order for the stakeholders to appreciate scientific forecasts.

From this information, we can deduce the proficiency of the farmers in the use of ICT tools such as the mobile phone, computers, and Internet services. We also noted the availability and access to these tools; the need for a demand-driven service where the farmers initiate the request for weather information; and willingness to pay for the services provided.

There was also a clear need for language-specific weather information services, packaged in different formats for different people.

6 Discussion

This study allowed us to interact closely with farmers and enabled us to understand their weather information needs, as well as their impression in using ICT tools for dissemination of this information. The farmers expressed interest in receiving weather forecast information because it affects the productivity of their small-scale agricultural efforts. More targeted weather information will lead to improved productivity. They unanimously agreed that the mobile phone be harnessed in the dissemination strategy.

We believe that an ICT-based dissemination system will provide the best solution to complement the existing weather information dissemination strategies, as reaching a majority of farmers requires special targeted communication efforts. ICT can provide user-friendly access and cost-effective solutions. ICT tools can effectively manage information delays, foster interactivity, and provide services palatable for all.

ICT however cannot be a stand-alone solution in the effort to ease dissemination of weather forecast information but can be effectively used to boost existing solutions. Gakuru et al. Argue that since users cannot be forced to accept a product, they are to be enticed and convinced by the benefits of the product. If they are satisfied with the product, they will give it better publicity than anyone else (Gakuru et al. 2009). If the farmers fully appreciate the benefits of using ICT tools, they will embrace their usage and will convince their counterparts to do the same. Training sessions can be held for all the stakeholders to ease the uptake of enhanced dissemination.

We found that the mobile phone is the most prevalent ICT device the farmers have access to, and there is some access to computers and laptops. Mobile penetration rates continue to increase, as the costs of handsets decrease (approaching 10USD per unit). The mobile subscribers also continue to increase as shown by the increase in number of active subscriber identification module (SIM) ownership. Many mobile subscribers own one or multiple SIM cards without actually owning a phone set (Ndiwalana and Tusubira 2012). Mobile phones require basic literacy, making them accessible to a large portion of the population. This means that more people have access to mobile phones, including women, and the number will continue to increase. Availability of Internet-enabled mobile handsets (smartphones) is also enabling access to mobile Internet (Ndiwalana and Tusubira 2012). The current state shows a progression from a period when there was a lack of access to mobile phones to date when mobile phone possession is widespread. We anticipate that in the near future, smartphones will be more common among the farmers. We assume that progressively, more farmers will own or have access to a mobile device, and this will reduce the gender gap over ownership of the mobile phone. This should also open up dissemination of information in a nondiscriminatory manner. There is ready access to ICT tools, and the weather forecast information to be disseminated is also available. What remains is to find seamless ways to use these existing ICT tools to disseminate the information.

Internet coverage is also on the increase, with access and speeds projected to keep increasing as the infrastructure to support is harnessed (Uganda Communications Commission 2015). This implies that ICT usage and use of social media will increase. Mobile phones can therefore be harnessed to disseminate weather information and reduce on the proverbial delays by eliminating any third-party actors. The information can be provided in the audio and SMS formats.

The farmers are somewhat familiar with mobile phone user interface aspects such as the use of short codes, and this familiarity can be harnessed to enhance usability. The recent introduction of tax on over the top (OTT) services in Uganda (MTN Uganda 2018), that include the popular social media platforms Facebook and WhatsApp, may humper the uptake of weather forecast information. However, availing a nonsocial network weather information application for download would promote dissemination. UNMA can take the lead in the development of such applications. ICT tools support such a venture for farmers with smartphones.

Using the mobile/smartphone, the farmers can subscribe to receive SMS weather forecast information at a small fee. They can also make a call to listen to an audio recording, in a language of their choice, or download an application that provides the forecast information. In responding to a query made by a farmer, the information to be sent back can be filtered appropriately to fit the request. Using these channels allow tailored and relevant information to be sent to the farmers. Because mobile phones offer mobility, weather forecast information can be accessed anytime, anywhere.

The requests made by farmers for weather forecast information can be studied, and corresponding profiles can be derived. For instance, if several farmers make requests for information on soil temperature and moisture in their locality, this can provide an indication to UNMA of what other services and/or products to provide to the public, aside from the seasonal forecast. UNMA can also receive feedback on the services they offer directly from the farmers, for example, on simplifying the meteorological terminology used. This will foster understanding and application of the weather forecast information.

ICT tools can offer UNMA the needed computing resources to be able to offer and organize tailored and relevant information needed by the farmers to meet their individual needs. ICT tools can also offer relief to the overstretched and under-resourced extension service. If every extension worker is equipped with a smartphone, they can receive the weather forecast information directly. They can then either pass on the information by SMS to farmers or organize communal gatherings to share the information. We have also learned that attention should be paid both to the technical solution and to the workflow as information trickles down to the farmer. Some of the processes involved are not computerizable. They need to be organized more efficiently for a smooth flow of information from one human being to another. For instance, the need to provide an extension service that can positively contribute to the weather information dissemination. The nontechnical solution requires a change in attitude of the farmers, policy changes at the level of weather information service provision, for instance, in instituting a designated team of stakeholders that participate in weather information break down into meaningful advisories. Also, the incorporation of the farmers' feedback, and the further refinement of the disseminated information, would create more ownership of the resultant system.

7 Conclusion

This study presented the need to utilize ICT tools to improve livelihoods of farmers in Uganda, in the face of climate change, and the variations in seasonal weather. We listened to farmers voice their concerns and give ideas and suggestions for weather information dissemination using ICT tools, particularly the mobile phone.

We found that mobile phones are generally available to the farmers, and the farmers are open to the use of them for accessing weather forecast information. We posit that solutions built around the mobile phone will receive more ownership and have greater impact, as long as they are developed in close interaction with the farmers.

We also presented the several capabilities of ICT to enhance weather forecast information dissemination, particularly in targeting a larger population than the conventional dissemination channels. ICT tools allow for tailoring of the weather forecast information so that it is relevant for the farmers, including parameters such as the language of interest, the locality of the farmer, and any other related constraints, such as specific advisories. The farmers can receive the information in text form on SMS or in audio form from a recording.

However, ICT solutions cannot be stand-alone; linking them to other dissemination channels, especially those which are more accessible to farmers, will create a multiplier effect. A multichannel weather dissemination system which must be accessible, anytime, anywhere, with requirements structured from the interactions with the farmers needs to be developed. The envisaged system is an integrated system online (web) and mobile (USSD, SMS-based) platform that provides voice, textual, graphical, and analytical weather information. The information provided by the system can also be used to create advisories for the farmers, and be language-specific, making it simple to read and interpret.

To provide an efficient weather forecast information dissemination system, issues of technology, location, policy, pricing of services, politics, and training/sensitization, among others, cannot be ignored. We anticipate that the dissemination system to be developed should be of low cost, as a number of existing dissemination channels already are in place, and can lead to successful implementation.

A deeper investigation should explore how to foster the use of services such as email and websites in weather forecast information dissemination to farmers. An open challenge to address is assessing the impact on agricultural productivity after disseminated weather forecast information has been used by the farmers to take agricultural-related decisions. We also propose structuring a business model to ensure sustainability of the weather information dissemination system, so that its impact among stakeholders can be fully realized.

Acknowledgments This work was funded by the WIMEA-ICT Project (http://wimea.mak.ac.ug/), and it is part of the PhD research work of the corresponding author. The authors thank UNMA for their collaboration on project activities. We thank all the respondents (farmers) who were willing to be interviewed. We thank the WIMEA-ICT project research team, particularly Andrew Kagwa, Sophie Nabukenya, and Gift Mwesigwa, who engaged in conducting the focus group sessions, interviews, and transcribing the gathered information. We thank Maximus Byamukama for making possible our presentation at the International Conference on Ecosystem Resilience and Agricultural Productivity 2017.

References

Aker JC, Mbiti IM (2010) Mobile phones and economic development in Africa. J Econ Perspect 24(3):207–232

Bamanya D (2014, October) The climate forecast model (Tech. Rep.). Uganda National Meteorological Authority

Barihaihi M, Mwanzia J (2017) Enhancing farmers' resilience and adaptive capacity through access to usable weather information: a case study from Uganda. In: Climate change adaptation in Africa. Springer, Switzerland, pp 617–627

Barros V, Field C, Dokke D, Mastrandrea M, Mach K, Bilir TE et al (2014) Climate change 2014: impacts, adaptation, and vulnerability. Part B: regional aspects-contribution of working group II to the fifth assessment report of the intergovernmental panel on climate change. Cambridge University Press, New York

Barungi M, Guloba M, Adong A (2016, March) Uganda's agricultural extension systems: how appropriate is the single spine structure? Economic Policy Research Centre (EPRC). (Research report no. 16). https://ageconsearch.umn.edu/record/253558/files/16%20Ugandas%20Agricultural%20Extension%20Systems.pdf

Basalirwa C (1995) Delineation of Uganda into climatological rainfall zones using the method of principal component analysis. Int J Climatol 15(10):1161–1177

Biraahwa Nakabugu S (n.d.) The role of rural radio in agricultural and rural development translating agricultural research information into messages for farm audiences. http://www.fao.org/docrep/003/x6721ex6721e31.htm. Accessed 30 Jan 2018

Byarugaba GA (2014) Mobile weather alert and advisories, building resilience through innovation. In: Proceedings of CRS 2014 ICT4D Conference, March 25–28, 2014, Nairobi, Kenya

Coelho CA, Costa SM (2010) Challenges for integrating seasonal climate forecasts in user applications. Curr Opin Environ Sustain 2(5):317–325

Ericsson Corporate Public & Media Relations, World Meteorological Organization, MTN Group, National Lake Rescue Institute, Grameen Foundation (2012) Mobile weather service improves safety of fishermen in Uganda. https://public.wmo.int/en/media/news/mobile-weather-service-improves-safety-of-fishermen-uganda. Accessed 21 Feb 2019

Gakuru M, Winters K, Stepman F (2009) Innovative farmer advisory services using ICT. Documento presentado en el taller de W3C Africa perspective on the role of mobile technologies in fostering social development, Maputo, 1

Heeks R (2008) ICT4D 2.0: the next phase of applying ICT for international development. Computer 41(6):26–33

Kanagwa B, Tuheirwe-Mukasa D, Muwembe K (2015) The need for an integrated effective weather dissemination system for Uganda. In: Proceedings of the 2015 international conference on frontiers in education: computer science and computer engineering. CSREA Press, Athens, pp 330–336

Klopper E, Vogel CH, Landman WA (2006) Seasonal climate forecasts – potential agricultural-risk management tools? Clim Chang 76(1–2):73–90

Kontio J, Lehtola L, Bragge J (2004) Using the focus group method in software engineering: obtaining practitioner and user experiences. In: Empirical software engineering, 2004. ISESE'04. Proceedings: 2004 international symposium on, pp 271–280

Mass C (2006) The uncoordinated giant: why US weather research and prediction are not achieving their potential. Bull Am Meteorol Soc 87(5):573–584

MetOffice (2012) Saving lives on Lake Victoria. https://www.metoffice.gov.uk/binaries/content/assets/mohippo/pdf/m/r/14_0500_saving_lives_on_lake_victoria.pdf. Accessed 20 Feb 2019

Meyer MA, Booker JM (2001) Eliciting and analyzing expert judgment: a practical guide. SIAM, Philadelphia

Miles MB, Huberman AM, Saldana J (2013) Qualitative data analysis. Sage, Los Angeles

Ministry of Agriculture, Animal Industry and Fisheries (n.d.) Local government budget framework paper FY 2016/17 preparation workshops

Morgan DL (1996) Focus groups as qualitative research, vol 16. Sage, Thousand Oaks

MTN Uganda (2018, July) OTT services. https://www.mtn.co.ug/en/products/internet/Pages/OTT-Services-.aspx. Accessed 21 Aug 2018

Mwebaze SMN (1999) Country pasture/forage resource profiles. http://www.fao.org/ag/agp/agpc/doc/counprof/uganda/uganda.htm. Accessed 30 Nov 2017

National Research Council and others (2007) Improving disaster management: the role of IT in mitigation, preparedness, response, and recovery. National Academies Press, Washington, DC
Ndiwalana A, Tusubira F (2012) Understanding what is happening in ICT in Uganda: a supply and demand-side analysis of the ICT sector, Research ICT Africa. Available from https://researchictafrica.net/publications/Evidence_for_ICT_Policy_Action/Policy_Paper_8_-_Understanding_what_is_happening_in_ICT_in_Uganda.pdf
Pitula K, Radhakrishnan T (2011) On eliciting requirements from end-users in the ICT4D domain. Requir Eng 16(4):323
Reuder J, Sansa-Otim J (2013, November) WIMEA-ICT: Improving Weather Information Management in East Africa for effective service provision through the application of suitable ICTs. Geophysical Institute, University of Bergen, Bergen
Rijks D, Baradas M (2000) The clients for agrometeorological information. Agric For Meteorol 103(1):27–42
Sivakumar M (2006) Dissemination and communication of agrometeorological information – global perspectives. Meteorol Appl 13(S1):21–30
Sivertsen T (2000) Weather information, site information and a system for dissemination of information on the Worldwide Web from a network of 52 automatic agrometeorological stations. EPPO Bull 30(1):77–81
Sivle AD, Kolstø SD, Kirkeby Hansen PJ, Kristiansen J (2014) How do laypeople evaluate the degree of certainty in a weather report? A case study of the use of the web service yr.no. Weather Clim Soc 6(3):399–412
Stewart DW, Shamdasani PN (2014) Focus groups: theory and practice, vol 20. Sage, Los Angeles
Tremblay MC, Hevner AR, Berndt DJ (2010) The use of focus groups in design science research. In: Design research in information systems. Springer, Boston, pp 121–143
Uganda Bureau of Statistics (2016) The national population and housing census 2014 – Main report, Kampala, Uganda. Available from https://www.ubos.org/onlinefiles/uploads/ubos/NPHC/NPHC%202014%20FINAL%20RESULTS%20REPORT.pdf
Uganda Communications Commission (2015, January) Access and usage of communication services across Uganda. http://www.ucc.co.ug/files/downloads/20150130-UCC-access-usage-dissemination.pdf. Accessed 30 Jan 2018
Uganda National Bureau of Statistics (2009, June) Higher local government statistical abstract Rakai district. http://www.ubos.org/onlinefiles/uploads/ubos/2009_HLG_%20Abstract_printed/Rakai%20district%20statistical%20abstract.pdf. Accessed 10 Nov 2018
United Nations General Assembly and others (2015) Resolution adopted by the general assembly on 25 September 2015. United Nations, Washington, DC. http://www.ipu.org/splz-e/unga16/2030-e.pdf.
UNMA (n.d..) https://www.unma.go.ug/. Accessed 16 Sept 2016
Von Alan RH, March ST, Park J, Ram S (2004) Design science in information systems research. MIS Q 28(1):75–105
Washington R, Harrison M, Conway D, Black E et al (2006) African climate change. Bull Am Meteorol Soc 87(10):1355
Weiss A, Van Crowder L, Bernardi M (2000) Communicating agrometeorological information to farming communities. Agric For Meteorol 103(1):185–196
World Meteorological Organization (2012) Reaching the last mile with mobile weather alert: the user-interface platform in action. Bulletin 61(1). https://library.wmo.int/pmb_ged/bulletin_61-1_en.pdf
World Meteorological Organization (n.d.) Dissemination and presentation of weather information. https://www.wmo.int/pages/prog/amp/pwsp/communicationdissemination\en.htm
Yohe G, Lasco R, Ahmad QK, Cohen S, Janetos T, Perez R et al (2006) Perspectives on climate change and sustainability 3. Change 25(48):49
Ziervogel G, Calder R (2003) Climate variability and rural livelihoods: assessing the impact of seasonal climate forecasts in Lesotho. Area 35(4):403–417

Empowering Rural Farmers to Improve Livelihoods Through Environmental Risk Communication: A Case Study of Uganda

Goretti L. Nassanga

Abstract The chapter examines how effective environmental risk communication can empower rural farmers in Africa to improve their livelihoods, using Uganda as a case study. Specifically looking at climate change that is considered a global risk, the chapter explores how media are domesticating the climate issues to meet the grassroots' right to relevant information, thereby assessing how media are carrying out their Responsibility to Protect (R2P) the vulnerable rural farmers. The analysis is based on primary findings from a survey of rural farmers that assessed their access to environment and climate change information, the extent they perceived these as "risks" at the personal, community and national levels and the relevancy of the information received. Findings showed that although farmers perceived climate change as a big risk, they did not go beyond the risk identification and risk assessment stages to the last stage of taking action for risk reduction/protection as the information received from the media was not solution-oriented. Rather than engage in "risk communication" or "risk dialogue" that is a participatory process, media engaged more in "risk information giving". Gaps to effective risk communication are highlighted, and strategies presented to address them to ensure media carry out their R2P to the rural farmers. These findings provide a useful guide in effective environmental risk communication to rural farmers that should result into better adaptation to CC and improved production, thus contributing to a reduction in poverty levels and overall development in Africa.

Keywords Climate change communication · Responsibility to Protect · Risk communication · Risk information giving · Solution-oriented communication

G. L. Nassanga (✉)
Makerere University, Kampala, Uganda
e-mail: nassanga@chuss.mak.ac.ug

Y. Bamutaze et al. (eds.), *Agriculture and Ecosystem Resilience in Sub Saharan Africa*, Climate Change Management,
https://doi.org/10.1007/978-3-030-12974-3_26

1 Introduction

This chapter is premised within the precincts of the "Responsibility to Protect" (R2P) principle that recognizes our shared environment. The R2P citizens is a UN institutional framework that was adopted by heads of state and government at the 2005 UN World Summit. It is an outgrowth of the "right of humanitarian intervention", and it was originally meant to be protection against genocide, war crimes, ethnic cleansing and crimes against humanity, taken as a responsibility of the state, with the international community intervening if the state failed (Bellamy 2010; Global R2P 2015; The Responsibility to Protect 2001; UN General Assembly Security Council 2013). However, this has expanded to the R2P "people suffering from avoidable catastrophe … deliberate starvation and exposure to disease" (The Responsibility to Protect 2001).

The R2P has become a prominent feature in international debates about preventing genocide, mass atrocities and about protecting potential victims (Bellamy 2010). The International Crisis Group[1] points out that although states have the primary R2P and, where necessary, the international community coming in, civil society groups and individuals also have a responsibility, which is to force the attention of policymakers on what needs to be done, by whom and when. It is within this framework that media is obliged to carry out the R2P, given their key role as the major source of information to citizens (Banda 2013; BBC World Service Trust 2010; Bodt 2007; Boykoff and Roberts 2007; Carvalho 2007, 2008; Engesser 2017; Hellmuth et al. 2007; Klinsky 2007; Nassanga 2010; Olausson 2011; Okello 2011; Shanahan 2009; Zamith et al. 2012). Although globalization processes have expanded media's scope to target audiences beyond the local/national level, media still have a responsibility to meet the local information needs.

The analysis in the chapter specifically examines environmental risk communication of climate change that is considered a "disaster risk" by the IPCC, who defines it as constituting severe alterations in the normal functioning of a community or a society due to hazardous physical events interacting with vulnerable social conditions, leading to widespread adverse human, material, economic or environmental effects that require immediate emergency response to satisfy critical human needs and that may require external support for recovery (IPCC 2012, p. 3).

The chapter explores how media are domesticating the climate issue to meet the grassroots' right to relevant climate information. By implication, the study will assess how media are carrying out their R2P of the vulnerable group of the rural farmers, basing on the perceptions of the farmers on various aspects of environmental risk communication.

This chapter makes a case for specifically targeting rural farmers in environmental risk communication. As Hellmuth et al. (2007) observe, community-level stakeholders, especially farmers, are the most important group to engage in the drive to

[1] International Crisis Group. The Responsibility to Protect. http://www.crisisgroup.org/en/publication-type/key-issues/thematic/responsibility-to-protect.aspx.

incorporate climate information into decision-making because apart from being the most climate-vulnerable group, they also comprise the largest group of decision-makers within the hugely important agriculture sector. They further observe that climate presents a risk not only to livelihoods and subsistence of the farmers at the individual level but to the economy and infrastructure at the national and regional levels as well.

It is noted that although in parts of Africa, Asia and Latin America, the quality and quantity of climate change articles are growing fast in the newspapers read by urban elite, there is little research available on how much information is reaching the poorer communities (Shanahan 2007). Further, Olausson (2011) notes that studies that empirically explore media's role in shaping citizens' understanding of climate issues are scarce, compared to those that analyse media content alone. It is further noted that most of these studies focus on the United States, the United Kingdom and Sweden, with few studies from Africa (Engesser 2017). The lack of media research on climate change in Africa is confirmed by the findings of a BBC World Service Trust (2010) study that explored public understanding of climate change in Africa. It was observed that the fundamental shortage of relevant information for African audiences hampered African citizens' response to climate change.

Applied to Uganda, while several studies have been done on climate change (ACCRA 2010; Busingye 2010; Hepworth and Goulden 2008; Kasimbazi 2013; McGrath 2008; Mwiturubani 2010; Namanya 2009; Tadesse 2010; Zizinga et al. 2017), these have not focused on media's role. The area of how climate change issues are communicated, particularly through the media, has not yet become a priority for research in Uganda as evidenced by just a handful of studies that have been conducted in the recent years (BBC World Service Trust 2010; Berglez and Nassanga 2015; Corner 2011; Nassanga 2013; Nassanga et al. 2016; Okello 2011; Rowling 2008; Shanahan 2009; Semujju 2013, 2015).

Hellmuth et al. (2007) submit that if climate information and services are to support rural farmers, their involvement is crucial; he explains that much of the climate discourse so far has been at the national and international levels. They caution that this mismatch needs to be addressed if successful practices and policies are to be developed that make the best possible use of climate information. This information gap is echoed by Anderson (2017) in his study on media coverage of climate change, where he observes that not all sources have equal opportunity of getting heard in the media as news routines tend to systematically favour the voices of elites and exclude those who lack social dominance. This situation of information dualism very much applies to Uganda's situation, where the voices of rural farmers do not feature as much as those of the urban elite in climate change news (Nassanga 2013; Nassanga et al. 2016; Semujju 2013). The information gap is also pronounced at the gender level as Semujju (2015) finds in his study, where he explored the positioning of farmers as sources on climate change news. His analysis showed that the gender gap in Uganda was highly pronounced, with women as sources ranking third in importance after "men" and "anonymous sources".

While analysing the reporting of risk and uncertainty by the media, Painter (2013) posits that scientists and politicians are increasingly turning from the language of uncertainty to the language of risk in their descriptions of likely future climate change impacts. He views this as a positive development, explaining that uncertainty is a difficult concept to communicate and is prone to being misunderstood or misinterpreted. It is thus imperative to carry out studies like this one that investigate how the climate change risk is being communicated in specific contexts or interpretive communities.

The findings as analysed in this chapter are a valuable resource in contributing to filling the above identified gaps, and they enhance better understanding of the dynamics in environment and climate risk communication targeting rural farmers from an African perspective.

The question of interest in this analysis is: How effective is the environment and climate change risk communication targeted at the rural farmers in Uganda? Basing on the different perspectives of the rural farmers, it was possible to analyse how effective the environment and climate change risk communication has been, thus indirectly assessing media performance of their R2P the rural farmers.

2 Context for the R2P Against Climate Change Impacts in Uganda

Climate change may be taken to be outside the "relatively narrow interpretation of 'crimes against humanity' that excludes crimes not associated with the deliberate killing and displacement of civilians" (Bellamy 2010, p. 151). However, due to the devastating impacts which climate change has inflicted on the livelihoods of people in African that are dependent on rain-fed agriculture (ACCRA 2010; Busingye 2010; Hepworth and Goulden 2008; Kasimbazi 2013; McGrath 2008; Mwiturubani 2010; Namanya 2009; Saleh 2012; Tadesse 2010; Zizinga et al. 2017), it is now viewed as a crisis that warrants intervention based on the R2P. This means going beyond the state responsibility to extend to other actors, one of these is the media, given its function as the major source of environment and climate change information for the public and policymakers (BBC World Service Trust 2010; Bodt 2007); Boykoff and Roberts 2007; Carvalho 2007, 2008; Hellmuth et al. 2007; Klinsky 2007; Nassanga 2010; Okello 2011; Olausson 2011; Shanahan 2009; Zamith et al. 2012). Depending on the way media frame issues, media transmit information that shape how people understand climate change as well as the actions they are ultimately willing to support to address the problem (Bolsen and Shapiro 2017; Saleh 2012).

At the global level, the UN Sustainable Development Goals (SDG) that replaced the Millennium Development Goals, 13 requires countries to "take urgent action to combat climate change and its impacts". In the Uganda National Development Plan 2 (2015–2020), under the Section on "International, Regional and National Development Obligations", this has been domesticated with a commitment to "Mainstream Climate Change Adaptation and Mitigation" (Uganda Government 2015, p. 91). Media is amongst the non-state actors tasked with the responsibility of monitoring and reporting both achievements and setbacks countries make on these global agendas (UNEP 2017).

This chapter assesses how media have mainstreamed climate information and communicated it to the rural farmers to enhance adaptation and mitigation to climate change impacts. In other words, how have media carried out this obligation or the R2P the farmers?

The concerns over climate change impacts are very pertinent to Uganda's situation. The National Development Plan 2 acknowledges that climate change impacts pose considerable environmental and development challenges for the country (Uganda Government 2015). Given that over 80% of Ugandans are engaged in agriculture as the primary source of livelihood, a shock to agricultural production not only affects the basic livelihoods of majority of Ugandans but also undermines economic performance (Uganda Debt Network 2017). Climate change has proved to be a major threat to the country's development through impacts it has had including increased food insecurity, shifts in the spread of diseases like malaria, soil erosion and land degradation, flood damage to infrastructure and settlements and shifts in the productivity of agricultural and natural resources, with the rural poor feeling these impacts most (Banda 2013; Hepworth and Goulden 2008; Kasimbazi 2013; McGrath 2008; Tadesse 2010; Uganda Debt Network 2017; UNEP 2008; Zizinga et al. 2017).

Amongst the factors identified that constrain the goal of poverty reduction, climate change is predicted to have substantive impacts on Africa's environment and the livelihoods of people as it will affect food production, health and security (UNEP 2008). For Uganda, a report by the DFID Uganda scoping mission (Hepworth and Goulden 2008) warns that climate change will not only exacerbate poverty and trigger migration, it will also heighten competition over strategic water resources, which could lead to regional insecurity. The country has already witnessed many cases of conflicts over land, including the "balalo" (cattle keepers) from the Western part of Uganda who had migrated with their cattle to the Northern part, being expelled from the area by the local leaders in these districts.

Uganda is reported to have one of the fastest-changing climates in the world, and the rapidly changing climate is projected to have an impact on the rural population who are dependent on rain-fed agriculture. Climate-related crop failure and income loss are likely to affect food security and trap millions of children in poverty (Uganda Government and UNICEF 2017).

While the effects of climate change will affect people differently, it is the rural poor farmers who are most hit as they are dependent on subsistence farming that is subject to the weather hazards (ACCRA 2010; Banda 2013; Cox 2010; Hepworth and Goulden 2008; Kasimbazi 2013; UNEP 2008; Zizinga et al. 2017). Oftentimes, farmers lack the requisite information and other resources to be able to mitigate the negative effects or adapt to these climate changes (Zizinga et al. 2017). This often results in poor crop yields and food insecurity, which threaten their survival, given the low incomes and the hand-to-mouth existence that does not allow them to make savings. As Escobar (1995) contends, massive poverty in the modern sense appeared when the spread of the market economy broke down community ties and deprived many people from access to land, water and other resources, which along with the consolidation of capitalism brought systemic pauperization. The IPCC (2007) report projected that by 2020, yields from rain-fed agriculture could be reduced by up to 50%, which would further adversely affect food security and exacerbate malnutrition.

The concern over the impacts of climate change is an issue that has become very prominent in the sustainable environment management debates. In a global survey by the Pew Research Centre (2017), it was found that most publics believe global climate change to be the second-highest threat by 61%, closely following behind ISIS that was considered as the top most threat by 62%. In Africa, the climate change problem was rated slightly lower at 58%.

In the Fifth Assessment Report (AR5) by the UN Expert body of the Intergovernment Panel on Climate Change (IPCC), it is affirmed that climate change is a threat to sustainable development, and it is predicted that "climate change will amplify existing risks and create new risks for natural and human systems", explaining that risks are unevenly distributed and are generally greater for disadvantaged people and communities (IPCC 2014, p. 13). For Africa, as Tadesse (2010) observes, the productivity and sustainability of her environment are heavily dependent on how climate change is managed, and if sustainable development is to be realized, one of the imperatives should be accessibility to climate change information in Africa.

Although the climate change crisis has no borders (Eide and Kunelius 2010), it is noted that the adverse impacts of climate change present a formidable threat that already limits the economic progress and the fight against poverty in many African nations (Banda 2013; Tagbo 2010; Uganda Debt Network 2017; Uganda Government 2015; UNECA 2011; UNEP 2008). As a means to improve climate change adaptation in Uganda, Kasimbazi (2013) calls for the strengthening of research that will enhance climate monitoring and effective response, in order to provide regular climate change information that supports sustainable socio-economic development. It is thus very valuable to conduct studies like this one that focus on how the environment and climate risk can be communicated effectively, thus serving as a mechanism of monitoring how media are carrying out their R2P the rural farmers.

3 Methodological Approach

This chapter specifically analyses environmental risk communication targeting rural farmers because this is the most climate-vulnerable group, as they are dependent on rain-fed agriculture, with few resources to mitigate or adapt to climate change impacts.

3.1 Data Collection Methods

The respondents ($n = 451$) were drawn from four representative districts (Nakaseke, Luweero, Mayuge and Iganga), around the Lake Victoria basin, where farming is a predominant activity. The respondents were selected with the assistance of the respective District Agricultural officers, using cluster sampling of four villages in each of the four districts. Efforts were made to have a gender representative sample; women were slightly more ($n = 235$ or 52%) than men ($n = 216$ or 48%). This gender representation is similar to the structure of Uganda's population, where women are slightly higher than men, and it can also be partly attributed to the traditional gender roles, where subsistence farming is largely an occupation for women.

The primary data was collected using a questionnaire, after which the data was analysed quantitatively and qualitatively. Due to the low literacy levels amongst the rural farmers, the questionnaire was administered to the farmers by the author/principal researcher, with the assistance of two research assistants. This approach helped to ensure a high response rate.

3.2 *Framework for Effective Risk Communication*

In order to have successful risk communication, where people take protective action, three stages in the risk communication process have been identified. (i) The first stage is risk identification that involves identifying or becoming aware of the risk. (ii) The second stage involves risk assessment, where the person assesses the probability and the likely severity of the consequences that the risk could cause. (iii) The third stage is risk reduction or protection that requires identifying and choosing an action(s) that can eliminate or reduce the risk (Lindell and Perry 2004).

Guided by this framework, this analysis is based on findings from a survey conducted to assess various aspects relating to environment and climate change risk communication to rural farmers. When assessing how a risk is gauged, several factors that influence the perception of the risk involved have to be taken into consideration. Covello and Sandman (2001) opine that risks from activities viewed by people to place them (or their families) personally and directly at risk are judged to be greater than risks from activities that appear to pose no direct or personal threat. Further, for people to make the change from unsustainable to sustainable behaviour, it is important for them to have sufficient information regarding the issue and also to have the capacity to evaluate this information within their personal context, in order to make the right decisions (Bodt 2007).

In this study, farmers were asked about their understanding of climate change; perceptions on how climate change had affected them at the individual, community and national levels as well as the degree to which they perceived environment and climate change as "risks" to them at individual, community and national level. The yardstick for measuring the "risk" was the degree of intensity as reflected on a five-point Likert scale ranging from 0% to 100% (very low = 0–24%; low = 25–49%; average = 50%; high = 51–75%; very high = 76–100%).

Another important consideration in climate communication for adaptation in Africa is the accessibility to weather forecast information. It is noted that despite being projected to face the greatest impact of climate change, vulnerable people in Africa have inadequate access to the support services and the information they need to build their adaptive capacity to an effective level. Real-time meteorological products such as forecasts and warnings that are highly perishable are not disseminated efficiently to the intended audience and are therefore hardly of any use (Shaka 2013).

While media have a responsibility to avail climate information, the way the message is framed is also equally important if it is to achieve the desired response (Bolsen and Shapiro 2017; Eide and Kunelius 2012). Particular attention thus needs to be given to the type of frames used and the perceived risk. Lindell and Perry (2004) contend that risk communication involves an examination of the perception of personal risk and protective action, so the perceived relevancy of the message is a key variable influencing the response in risk communication. They explain that if the individual feels the message is relevant to him/her, this will motivate the person to continue to think about a message, to seek additional information and to follow through with taking the desired action. So, the way media frame climate change will affect how audiences respond (Bolsen and Shapiro 2017; Eide and Kunelius 2012; Shanahan 2007). The messages or information should provide access to a full range of interpretive frameworks that would facilitate the individual to convert the raw information into explanations on how particular decisions would impact his/her life (Murdock 2005).

This means that while the information given by the media is not the only key factor, the content or the way the message is packaged is an important consideration. The information or the stories should provide solutions or should be action-oriented, thereby fulfilling media's R2P. Media's role in climate change communication should not only be geared towards raising awareness of climate issues or alerting people when specific threats arise, media should also convey the risks and the recommended responses (Eide and Kunelius 2012; Tadesse 2010). This analysis sought to find out perceptions on the type of information media gives and whether some direction on the solutions or necessary action is provided.

The study also sought to find out the farmers' access to environment and climate change information and how this was utilized. Of particular interest was how they benefitted from weather forecast information. Views were also sought to establish the level of satisfaction with the status of the environment in Uganda and the views of the farmers on possible solutions.

4 Results and Discussion

4.1 Grassroots Perception of Environment and Climate Change as Risks

In order to gauge the significance of farming in the communities, respondents were asked to indicate whether they engaged in farming and to indicate their major source of income. Farming was found to be the main livelihood activity for 442

(98%) of respondents and the major source of income for most of them 356 (80%). With almost all the respondents dependent on farming and considering the negative impacts of climate change, media has the R2P this vulnerable group by availing them with relevant environment and climate change information to enhance their adaptation. As Eide and Kunelius (2012) posit, ideal journalism practice should not only compel journalists to strike a balance and give fair coverage to all, but it also calls for defending the vulnerable, in this case, the rural subsistence farmers.

Almost all respondents 408 (95%) were aware and had heard of CC. Moving beyond the point of awareness, when analysing the concept of "risk", it is necessary to distinguish between different perspectives as there is often a gap between "real" risk as formulated by scientists and the "perceived" risk as interpreted by lay people (Bussotti 2014). As reflected in a study that explored perceptions of communities at fish-landing sites in Uganda, the interpretation of "good environment management" practices around Lake Victoria by these communities was different from that as defined by government and other actors (Nassanga 2010). This was largely attributed to the differences in the coding and encoding in the communication process as well as issues of differences in ideological horizons.

A related challenge here is how media interpret the complex climate science (Anderson 2017; Corner 2011; Fahn 2009; Painter 2013; Tagbo 2010; Shanahan 2009, 2011; UNEP 2017; Wihbey and Ward 2016) or translate the information for the public (Saleh 2012) to make it comprehensible. In the study, farmers were asked to indicate what they understood by climate change. They described it differently, including changes in weather pattern, limited rainfall, changing planting seasons, excessive heat with low crop yields, prolonged drought, heavy rain followed by floods and new crop and animal pests, amongst others. The farmers were asked to rate how they had been affected by climate change at the various levels. It was indicated that at the personal level, 394 (93%); 402 (96%) at community level and 347 (88%) at national level had been affected.

Asked to rank the degree to which climate change was considered a risk at the various levels, at the personal level, 285 (67%) considered climate change to be either of average risk, high risk or very high risk; at community level, 284 (68%) considered climate change to be either of average risk, high risk or very high risk; and at national level, 256 (62%) considered climate change to be either of average risk, high risk or very high risk. From the gender perspective, there wasn't much significant difference in the rating of climate change as a personal risk between male 139 (33%) and female 146 (35%) respondents.

It is important to note the concept of proximity as it relates to the perceived risk of climate change. The farmers reported almost similar degrees of having been affected at personal and community level, while they ranked the effect at national level (which is more distant) to be lower. Similarly, they ranked the risk at personal and community level at almost the same intensity, while the perceived risk at national level was lower.

4.2 *Information Sources, Nature and Utility*

Knowledge and access to information are essential for effective environmental management, and these have significant impacts on the economy and the livelihood choices people make (Tadesse 2010). The survey sought to explore the farmers' access to environment and climate change information, their information sources, the nature and utility of the information they received.

Radio was found to be the major source 274 (68%), with males having more access 153 (40%) than female farmers 121 (27%). TV and newspapers were not considered important sources, with only a few having access to these. This is contrary to what Painter (2016) finds in his analysis of the IPCC TV coverage of the Working Group Reports in 2013–2014. He found that for most countries in Europe, television was the most used and trusted source of information for news, including news about science.

Interpersonal channels were also found to be important sources of environment and climate information. Half of the respondents – 165 (50%) – belonged to a farmers group, where they reported getting information during their meetings. The majority of these were females 102 (31%), with fewer males 65 (20%) belonging to the farmers groups. A similar pattern was reflected for those who got information through LC1 meetings 262 (66%), where more female farmers attended such meetings 158 (40%) than male farmers 104 (26%). District officials, NGOs and seminars were also cited as important interpersonal sources.

On the use of mobile phones, 310 (71%) had access, most of them being male farmers 180 (41%) and female farmers were 131 (30%). This difference in access reflects a gender digital gap that confirms the pattern where women tend to have less access to digital sources of information than men. Only 124 (33%) of the respondents had ever received environment and climate change information on them. However, most of the respondents – 317 (82%) – thought this was a good channel that could be used to receive information quickly, which shows that mobile phones potential as a source of environment and climate change has not been fully exploited.

Analysing the content of the environment and climate change information received from radio, most of the respondents reported that programs do not give solutions or what action to take. Specifically focusing on weather information, 346 (85%) said they get weather information, but most of them indicated that it was not very useful often as it was found to be largely inaccurate, particularly as pertains to rainfall predictions that usually don't tally with the actual patterns forecast. The farmers complained that the weather forecasts were very general and were not correct most times, so people tended to disregard forecast information completely. The distrust is confirmed in another study by Kasimbazi (2013), where he noted that most farmers in Uganda were planting at the onset of rains and they were largely guided by traditional season calendars rather than forecast information. He calls for the strengthening of the meteorological station to facilitate the generation and dissemination of relevant information required in the design of adaptive response mechanisms.

Only about a quarter 108 (26%) of the respondents were either satisfied or very satisfied with the information received. Further, although most respondents 296 (66%) found information helpful, one-third 155 (34%) did not find the information helpful. The implication of these figures is that a big percentage of the respondents are not satisfied with the information received and a relatively big number did not find the information helpful with females being 87 (19%) and males 71 (16%). This means that the communication of environment and climate change information is not very effective and there is room for improvement on how this is communicated so that the farmers can utilize and benefit from such information. As Shanahan (2011) contends, effective communication is essential for the success of both climate mitigation and adaptation, explaining that it is like the lubricant that helps the whole engine of climate response to turn and keep running. The little climate change information received by farmers is collaborated by similar findings from a study by Zizinga et al. (2017) who investigated farmers' choices on climate change adaptation practices in Uganda. They identify lack of adequate information on adaptation methods as one of the main barriers to farmers' adaptation to climate change.

4.3 Nationalization and Individualization in Coverage of Environment Management and Climate Action

Various stakeholders have different expected roles to play for proper environment and climate action, starting from the individual, extending to the community, nation and global level. For media, they have a responsibility to highlight the different roles and draw attention to government, individuals and the respective stakeholders.[2] Previously, risk communication was largely based on the traditional model, which was influenced by "technical" meanings of risk and involved dissemination of one-way technical information from experts to lay people. This has now opened up to have "risk dialogue", which involves exchange of information amongst various concerned stakeholders, including the affected communities. Thus new risk communication has stressed public participation or what is termed as the cultural model of risk communication (Cox 2010; Covello and Sandman 2001; Herber 2004). Applying this observation to climate change, Anderson (2017) cautions that the climate problem will not be solved without mass participation by the general public in countries around the globe, which makes the concept of participation key to successful environment and climate change risk communication.

To assess perceptions on who is responsible for environment and climate change action, respondents were asked about possible solutions to deal with environment and climate change. Most attributed the responsibility for this to the government. This could partly be attributed to the way media package the messages, coming out as largely having diagnostic frames, rather being solution-oriented. In other words,

[2] International Crisis Group. The Responsibility to Protect. http://www.crisisgroup.org/en/publication-type/key-issues/thematic/responsibility-to-protect.aspx.

the action that the individual or the audience is supposed to take is not highlighted. This tallies with what Shanahan (2007) observed that amongst the most common frames in climate communication is one that inspires inaction or what he terms the "catastrophe" frame, elsewhere described as a disaster frame (Painter 2016) or time bomb (Semujju 2013). Similarly, other researchers (Covello and Sandman 2001; Cox 2010; Eide and Kunelius 2012; Nassanga 2013; Nassanga et al. 2016; UNEP 2017) found that climate coverage tends to be events-driven and focussing on drama and victims rather than focussing on the processes, while overlooking the need to highlight the desired action or behavioural change required.

Applying these observations to Uganda's media, there were similar findings from this analysis where the farmers revealed that the media programs most times did not provide them with the required action in their stories. Additionally, the majority of respondents were not satisfied with the level of climate information they get from the media. This can be explained partly by the tendency to have more of "risk information giving" from the media, rather than "risk communication" or risk dialogue that should offer opportunity for public participation bringing technology and scientific expertise together with the knowledge held by other stakeholders including indigenous knowledge (Herber 2004).

Contrary to the Ugandan coverage where responsibility for climate change action is largely attributed to the government, the Swedish mainstream news media tends to "individualize" climate change by emphasizing the responsibility of individual Swedish citizens as potential solvers of the global warming problem. In addition, the "nationalization" of climate change approach is also reflected, whereby Sweden is depicted as a "green" role model for other nations, as well as an individualization discourse that highlights the duty of the individual to go green, thus promoting "climate smart" ideas and communication (Berglez and Lidskog 2017; Berglez et al. 2009; Berglez and Nassanga 2015; Olausson 2011).

Amongst the solutions respondents gave were for government to supply them with seedlings for tree planting and fruit trees, provide irrigation equipment and water provision for cattle and other animals, arrange sensitization workshops and provide fertilizers, pesticides, valley dams and disease-resistant crops (cassava, bananas, maize, etc).

5 Conclusions

Risk communication is effective when the planned goal or objective is achieved, which could be behavioural change or taking specific action to mitigate or remove the risk (Herber 2004). The analysis in this study showed that although farmers perceived climate change as a big risk, they did not go beyond the risk identification and risk assessment stages to the last stage of taking action for risk reduction/protection as the information received from the media was largely diagnostic and not solution-oriented, so the public did not know the expected action to take.

This type of communication is what Lundgren and McMakin (2004) describe as "technical communication" rather than "risk communication". They clarify that the latter involves motivating the audience to action, but the "technical communication" has a lot of jargon that lay people find incomprehensible, which leads to perceived apathy towards environmental risks and the desired change will not be achieved.

Another observation is that the farmers viewed environment and climate change as higher risks at the personal and community level, but did not consider the risk as high at the national level, implying a degree of detachment from what goes on at the national level. The coverage also reflected the nationalization approach mostly leading to the farmers perceiving the responsibility for environment and climate action being for government, given that the responsibilities for the individuals were not highlighted. So the individualization approach was missing from the coverage.

The packaging of the risk messages was thus inadequate as a good risk message should not only incorporate the type of hazard but should also have the expected time of the hazard and the recommended protective actions to prevent or mitigate the impacts (Abunyewah et al. 2016). The implication is that in environment and climate change risk communication in Uganda, media are not paying due attention to the third stage of giving information to motivate the public to take protective action. One can term this as an information gap, which reflects an oversight in the performance of their R2P.

This chapter has highlighted several aspects relating to the effectiveness of environment and climate change risk communication in Uganda, specifically targeting the rural farmers. Gaps in the performance of media's R2P were identified, and solutions suggested to address these. It is hoped that this article has contributed to better understanding of the dynamics involved in risk communication in Africa, using Uganda as a case study. The study has provided useful guides to formulating strategies for effective environment and climate change risk communication, which should enhance better access to environment and climate change information targeting rural farmers. The increased access to relevant information should translate into better agricultural production and ultimately contribute to increased income and better standards of living for the rural farmers.

6 Recommendations

A key factor that has been emphasized as vital for successful risk communication is the concept of risk dialogue and public participation (Anderson 2017; Covello and Sandman 2001; Cox 2010; Herber 2004; Lundgren and McMakin 2004). Media in Uganda should engage in "risk communication" or dialogue, rather than "risk information giving" that flows one way and does not allow for public participation.

This is in line with one of the recommendations at the Climate and Development conference (UNECA 2011), where media were called upon to engage in coordinated and sustainable awareness campaigns on the causes, impacts and course of action needed to prepare African populations for the devastating effects of climate change. This implies engaging in consistent coverage of environment and climate change issues instead of being event-driven. In addition, there should be better audience segmentation and targeting different audience groups so that the rural farmers get to benefit from the available information on how to mitigate and adapt to environment and climate change risks.

Journalists, editors and media managers should put environment and climate change information into the context of sustainable development for the country and globally. Equally, environment and climate change issues need to be given due priority on media's agenda. Given the media's capacity to influence public perceptions, this should increase public awareness on different environment and climate change issues and motivate moving to the third stage of taking risk reduction or protection measures.

The three stages for effective risk communication each require seeking relevant information to facilitate getting to the last stage of taking protective action. Farmers should be encouraged to use existing interpersonal communication channels like village LC1, farmers groups, women and youth groups to facilitate better information seeking and information sharing, habits amongst the communities. Such interpersonal and group communication channels are very conducive for behavioural change and go to supplement information from the mass media.

When considering effective environment and climate change risk communication, there is an assumption that journalists are well prepared to engage in this. However, several studies have shown that journalists in the developing world have not been equipped to deal with the extent of the climate change challenge and to make sense of the complex climate science (Anderson 2017; Corner 2011; Engesser 2017; Fahn 2009; Painter 2013, 2016; Shanahan 2009, 2011; Tagbo 2010; UNEP 2017; Wihbey and Ward 2016). More concerted efforts should be put in training journalists in environment and climate change issues so they can be empowered to communicate these effectively to their audiences. The training would enhance the performance of their R2P vulnerable groups like rural farmers against climate change impacts.

In the packaging of environment and climate change messages, there is a need to ensure that both the individualization and nationalization approaches are applied so as to encourage more audience participation in taking environment and climate change action.

Acknowledgement The research for this chapter was funded by the collaborative Research Project "Strengthening Media in Post-Conflict Societies through Education and Research - Bridging Gaps, Building Futures in Uganda, South Sudan, Nepal and Norway", under the support to Makerere University Kampala by the Norwegian Agency for Development Cooperation (NORAD).

References

Abunyewah M, Gajendran T, Maund K (2016) Influence of risk communication on intention to prepare for flood hazards in informal settlements. In: 6th international conference on building resilience. Paper no 49, Meassy University, New Zealand

ACCRA (2010) Preparing for the future in Uganda: understanding the influence of development interventions on adaptive capacity at the local level. Uganda synthesis report, Kampala

Anderson A (2017) Source influence on journalistic decisions and news coverage of climate change. Oxford Research Encyclopedia of Climate Science [Online]. Available: http://climatescience.oxfordre.com/view/10.1093/acrefore/9780190228620.001.0001/acrefore-9780190228620-e-356. Accessed 7 Feb 2018

Banda F (2013) Climate change in Africa: a guide book for journalists. UNESCO, Paris

BBC World Service Trust (2010) Africa talks climate. Research report Uganda. British Council, London

Bellamy AJ (2010) The responsibility to protect – five years on [Online]. Available: http://responsibilitytoprotect.org/Bellamy.pdf. Accessed 7 Feb 2018

Berglez P, Lidskog R (2017) Foreign, domestic, and cultural factors in climate change reporting: Swedish media's coverage of wildfires in three continents. Environ Commun [Online]. Available: https://doi.org/10.1080/17524032.2017.1397040. Accessed 7 Feb 2018

Berglez P, Nassanga GL (2015) What is the difference between cross-national comparisons and semi-comparative work? Example of Swedish-Ugandan climate change communication research. J Dev Commun Stud 4(1):33–48

Berglez P, Hoijer B, Olausson U (2009) Individualization and nationalization of the climate issue. In: Boyce T, Lewis T (eds) Climate change and the media. Peter Lang, New York, pp 211–223

Bodt T (2007) Role of the media in achieving a sustainable society [Online]. Available: http://www.bhutanstudies.org.bt/pubFiles/M-21.pdf. Accessed 24 Nov 2017

Bolsen T, Shapiro MA (2017) The US news media, polarization on climate change, and pathways to effective communication. Environ Commun [Online] Available: https://research.iit.edu/publications/2017/us-news-media-polarization-climate-change-and-pathways-effective-communication. Accessed 7 Feb 2018

Boykoff MT, Roberts JT (2007) Media coverage of climate change: current trends, strengths, weaknesses. Human development report 2007 background paper. United Nations Development Program, New York

Busingye G (2010) The link between climate change, law, sustainable development and livelihoods in Uganda. In: Mwiturubani AD, Wyk VJ (eds) Climate change and natural resources conflicts in Africa. Institute for Security Studies, Pretoria. [Online]. Available: http://www.iss.co.za/uploads/Mono170.pdf. Accessed 7 Feb 2018

Bussotti L (2014) Environmental risk management and communication in an African context: the case of the Mozal bypass in Mozambique. In: Multinational enterprises in Africa: corporate governance, social responsibility and risk management, pp 93–118 {Online]. Available: https://journals.openedition.org/cea/1692?lang=en. Accessed 19 February 2019

Carvalho A (2007) Ideological cultures and media discourses on scientific knowledge: re-reading the news on climate change. Public Underst Sci 16(2):223–243

Carvalho A (2008) Communicating climate change: discourses, mediations and perceptions. Centro de Estudos de Comunicação e Sociedade, Universidade do Minho, Braga

Corner A (2011) Hidden heat communicating climate change in Uganda: challenges and opportunities. Panos Eastern Africa, Kampala

Covello V, Sandman PM (2001) Risk communication: evolution and revolution. In: Wolbarst A (ed) Solutions to an environment in peril. John Hopkins University Press, Baltimore, pp 164–178

Cox R (2010) Environmental communication and the public sphere. Sage, Thousand Oaks

Eide E, Kunelius R (2010) Domesticating global moments – a transnational study on the coverage of the Bali and Copenhagen climate summits. In: Eide E, Kunelius R, Kumpu V (eds) Global climate – local journalisms: a transnational study of how media make sense of climate summits. Pojektverlag, Bochum/Freiburg, pp 11–51

Eide E, Kunelius R (2012) Epilogue. Challenges for future journalism. In: Eide E, Kunelius R (eds) Media meets climate: the global challenge for journalism. NORDICOM, Gothenburg, pp 331–339

Engesser S (2017) Impact of journalistic background, professional norms, and culture on climate change coverage. Clim Sci [Online]. Available: http://climatescience.oxfordre.com/view/10.1093/acrefore/9780190228620.001.0001/acrefore-9780190228620-e-353. Accessed 7 Feb 2018

Escobar A (1995) Encountering development, the making and unmaking of the third world. Princeton Academic Press, Princeton

Fahn J (2009) Rescuing climate reporting in the south. Media Dev (2):21–24

Global R2P (2015) The responsibility to protect: a background briefing. Global Centre for the Responsibility to Protect [Online]. Available: http://www.globalr2p.org/media/files/r2p-backgrounder.pdf. Accessed 7 Feb 2018

Hellmuth ME, Moorhead A, Thomson MC, Williams J (eds) (2007) Climate risk management in Africa: learning from practice. International Research Institute for Climate and Society (IRI). Columbia University, New York

Hepworth N, Goulden M (2008) Climate change in Uganda: understanding the implications and appraising the response. LTS International, Edinburgh

Herber MW (2004) Talking to me? Risk communication to a diverse public. Orebro University, Orebro

IPCC (2007) Climate Change 2007: Synthesis Report. IPCC, Geneva, Switzerland

IPCC (2012) Summary for policymakers. In: Managing the risks of extreme events and disasters to advance climate change adaptation. Cambridge University Press, Cambridge, pp 1–19

IPCC (2014) Climate change 2014 synthesis report summary for policymakers [Online]. Available: https://www.ipcc.ch/pdf/assessment-report/ar5/syr/AR5_SYR_FINAL_SPM.pdf. Accessed 24 Nov 2017

Kasimbazi E (2013) Uganda's economic development: the challenges and opportunities of climate change. Konrad-Adenauer-Stiftung, Kampala

Klinsky S (2007) Mapping emergence: network analysis of climate change media coverage. The integrated assessment. Bridg Sci Policy 7(1):1–24

Lindell MK, Perry WR (2004) Communicating environmental risks in multiethnic communities. Sage, Thousand Oaks

Lundgren RE, McMakin A (2004) Risk communication. Battelle Press, Columbus

McGrath J (2008) Turning up the heat: climate change and poverty in Uganda. Oxfam GB, London

Murdock G (2005) Building the digital commons. Public service broadcasting in the age of the internet. In: Lowe GF, Jauert P (eds) Cultural dilemmas in public service broadcasting. Gothenberg, NORDICOM

Mwiturubani AD (2010) Climate change and access to water resources in the Lake Victoria basin. In: Mwiturubani AD, Wyk VJ (eds) Climate change and natural resources conflicts in Africa. Institute for Security Studies, Pretoria. [Online]. Available: http://www.iss.co.za/uploads/Mono170.pdf. Accessed 7 Feb 2018

Namanya DBM (2009) An assessment of the impact of climate change on the health sector in Uganda: a case of malaria and cholera epidemics and how to improve planning for effective preparedness and response. Ministry of Health, Kampala

Nassanga GL (2010) Button-less on the information superhighway: issues of ideological horizons in environmental communication amongst communities at fish landing sites along Lake Victoria in Uganda. Communicatio 36(3):327–342

Nassanga GL (2013) African media and the global climate change discourse: implications for sustainable development. In: Open Science Repository Communication and Journalism, Online (open-access), e23050451. https://doi.org/10.7392/openaccess.23050451

Nassanga G, Eide E, Hahn O, Rhaman M, Sarwono B (2016) Climate change and development journalism in the Global South. In: Risto K, Eide E, Tegelberg M, Dmitry Y (eds) Media and global climate knowledge: journalism and the IPCC. Palgrave Macmillan US, New York, pp 213–234

Okello OF (2011) Media coverage of climate change in East Africa: importance, challenges, opportunities and needs [Online]. Available: ej.msu.edu/.../Okello_Media_Coverage.ppt. Accessed 24 Nov 2017

Olausson U (2011) We're the ones to blame: citizens representations of climate change and the role of the media. Environ Commun 5(3):281–299

Painter J (2013) Climate change in the media: reporting risk and uncertainty. I.B. Tauris & Institute for the Study of Journalism, Oxford

Painter J (2016) Disaster, uncertainty, opportunity or risk? Key messages from the television coverage of the IPCC's 2013/2014 reports. Metode Sci Stud J (6):81–87. University of Valencia. https://doi.org/10.7203/metode.85.4179

Pew Research Centre (2017) Globally, people point to ISIS and climate change as leading security threats [Online]. Available: http://www.pewglobal.org/2017/08/01/globally-people-point-to-isis-and-climate-change-as-leading-security-threats/. Accessed 7 Feb 2018

Rowling M (2008) Who's helping media in developing countries tackle climate change? Reuters AlertNet [Online]. Available: http://www.alertnet.org/db/blogs/20316/2008/07/15-173925-1.htm. Accessed 7 Feb 2018

Saleh I (2012) Ups and Downs From Cape to Cairo: The journalistic practice of climate change in Africa. In: Eide E, Kunelius R (eds) Media meets climate the global challenge for journalism. NORDICOM, Gothenburg, pp 49–67

Semujju B (2013) Climate change in Ugandan media: a "global warming" of journalism ethics. J Afr Media Stud 5(3):337–352

Semujju B (2015) Frontline farmers, backline sources: women as a tertiary voice in climate change coverage. Fem Media Stud 15(4):658–674

Shaka A (2013) Dissemination of climate information using radio in Kenya. Jotoafrika: adapting to climate change in Africa. Issue 12, Nairobi, June 2013

Shanahan M (2007) Talking about a revolution: Climate change and the media. An IIED Briefing. [Online]. Available: http://dlc.dlib.indiana.edu/dlc/bitstream/handle/10535/6263/Talking%20about%20a%20revolution.pdf?sequence=1&is%20allowed=y. Accessed 19 February 2019

Shanahan M (2009) Time to adapt? Media coverage of climate change in non-industrialised countries. Chapter 12 [Online]. Available: http://pubs.iied.org/pdfs/G02512.pdf. Accessed 24 Nov 2017

Shanahan M (2011) Why the media matters in a warming world: a guide for policy makers in the global South. Climate change media partnership Policy Brief. IIED, Internews & Panos

Tadesse D (2010) The impact of climate change in Africa. ISS paper 220, Institute for Security Studies, November [Online]. Available: https://www.africaportal.org/documents/2753/Paper220.pdf. Accessed 24 Nov 2017

Tagbo E (2010) Media coverage of climate change in Africa: a case study of Nigeria and South Africa. Reuters Institute for the Study of Journalism & University of Oxford, London

The Responsibility to Protect (2001) Report of the International Commission on Intervention and State Sovereignty, co-chaired by Gareth Evans and Mohamed Sahnoun [Online]. Available: https://www.idrc.ca/en/book/responsibility-protect-report-international-commission-intervention-and-state-sovereignty. Accessed 24 Nov 2017

Uganda Debt Network (2017) Climate change could undermine Uganda's economic performance [Online]. Available: https://www.udn.or.ug/udn-media/news/112-climate-change-could-undermine-uganda-s-economic-performance.html. Accessed 7 Feb 2018

Uganda Government (2015) National development plan 2 (2015–2020). National Planning Authority, Kampala

Uganda Government and UNICEF (2017) Emerging global challenges: climate related hazards and urbanization – protecting Uganda's children [Online]. Available: https://reliefweb.int/

report/uganda/emerging-global-challenges-climate-related-hazards-and-urbanization-protecting-uganda. Accessed 7 Feb 2018

UN General Assembly Security Council (2013) Veto code of conduct [Online]. Available: https://www.una.org.uk/sites/default/files/Briefing%20%20Veto%20code%20of%20conduct.pdf. Accessed 7 Feb 2018

UNECA (2011) First conference on climate change and development in Africa (CCDAI), Addis Ababa, Ethiopia [Online]. Available: http://www1.uneca.org/ccda1/home_ccda1.aspx. Accessed 7 Feb 2018

UNEP (2008) Mainstreaming environment and sustainability in African universities partnership 2004–2008 report: supporting universities to respond to environment, sustainable development and climate change challenges. UNEP, Nairobi

UNEP (2017) UN environment supports capacity building environmental journalists in Africa [Online]. Available: http://www.unep.org/africa/news/un-environment-supports-capacity-building-environmental-journalists-africa. Accessed 7 Feb 2018

Wihbey J, Ward B (2016) Communicating about climate change with journalists and media producers. Oxford Res Encycl Clim Sci. [Online]. Available: http://climatescience.oxfordre.com/view/10.1093/acrefore/9780190228620.001.0001/acrefore-9780190228620-e-407. Accessed 7 Feb 2018

Zamith R, Pinto J, Villar ME (2012) Constructing climate change in the Americas: an analysis of news coverage in U.S. and South American newspapers. Sci Commun 35(3):334–357

Zizinga A, Kangalawe RYM, Ainslie A, Tenywa MM, Majaliwa J, Saronga NJ, Amoako EE (2017) Analysis of farmer's choices for climate change adaptation practices in South-Western Uganda, 1980–2009. Climate 5(4):89. [Online]. Available: http://www.mdpi.com/2225-1154/5/4/89. Accessed 7 Feb 2018

Effectiveness of Communication Channels on Level of Awareness and Determinants of Adoption of Improved Common Bean Technologies Among Smallholder Farmers in Tanzania

Charles B. Lugamara, Justin K. Urassa, Paul M. Dontsop Nguezet, and Cargele Masso

Abstract Increased legume productivity contributes to nutritional security as they are a source of cheap proteins. However, there is limited access to information on improved legume technologies among smallholder farmers in resource poor countries such as Tanzania. This chapter is aimed at assessing the effectiveness of communication channels (i.e. demonstration plots, farmer field days, technological briefs) on level of awareness and the determinants of adoption of improved common bean technologies among smallholder farmers in Tanzania. The study on which the chapter is based used a cross-sectional design on 400 households in Gairo and Mvomero districts, Tanzania. Results show that more than a half of the farmers were aware of all the seven improved legume technologies assessed. However, the level of awareness on all the technologies differed across the treatments, with a high level of the awareness recorded in areas with interventions. Among others, intervention included sharing information with farmers on land preparation, legume variety selection, use of quality seed, fertiliser application at planting, planting and spacing, weeding, control of insect and storage pests and diseases, harvesting and storage

C. B. Lugamara (✉)
Department of Development Studies, Sokoine University of Agriculture (SUA), Morogoro, Tanzania

J. K. Urassa
Department of Policy Planning and Management, College of Social Sciences and Humanities, Sokoine University of Agriculture, Morogoro, Tanzania
e-mail: urassa@suanet.ac.tz

P. M. Dontsop Nguezet
International Institute of Tropical Agriculture (IITA), Bukavu, Democratic Republic of Congo
e-mail: P.Dontsop@cgiar.org

C. Masso
International Institute of Tropical Agriculture (IITA), Nairobi, Kenya
e-mail: C.Masso@cgiar.org

Y. Bamutaze et al. (eds.), *Agriculture and Ecosystem Resilience in Sub Saharan Africa*, Climate Change Management,
https://doi.org/10.1007/978-3-030-12974-3_27

and safe use of chemicals. The awareness was low in areas without intervention. Nonetheless, there was a low level of adoption of the improved legume technologies. This could be due to the fact that the intervention was at its initial stage of implementation; but it was expected to increase with time as knowledge diffuses to the communities. In addition, as pointed out in the focus group discussions, low adoption could be because of difficulties in accessing improved bean technologies (high costs associated), unavailability of improved seeds and absence of seed dealers nearby villages. The factors significantly ($p \leq 0.05$) associated with smallholder farmers' adoption of improved legume technologies were visits by extension officers, age of household head, being member of a farmers' association, revenue from other income-generating activities and household size. Therefore, it can be concluded that a combination of demonstration plots, farmer field days and technological briefs (leaflets and brochures) accounted for the effective communication and awareness creation. Thus, it is recommended that the government and non-governmental organisations should invest more in awareness creation approaches in order to make sure that all smallholder farmers are sensitised on the improved legume technologies. In addition, the government and non-governmental organisations should insist more on visits by extension officers, formation of/joining farmers association and participating in other income-generating activities to enhance adoption of improved legume technologies.

Keywords Effectiveness · Communication channels · Smallholder farmers' adoption · Bean technologies · Tanzania

1 Introduction

Food legumes play important and diverse roles in the farming systems and in the diets of poor people around the world such as reducing poverty, improving human health and nutrition and enhancing ecosystem resilience (Katungi et al. 2010). In sub-Saharan Africa (SSA), legume crops play an important role economically, socially and environmentally by providing jobs, providing the cheap protein consumed mostly at the household level, improving health and nutrition and improving soil fertility through ground cover, weed suppression and nitrogen fixation (Akibode 2011; Sanginga and Bergvinson 2015).

In Tanzania, most small-scale farmers, especially women, participate in legume production. Generally, legumes act as a good and inexpensive protein source compared to meat and fish (Malema 2006; ProFound and Mugenyi 2012). Other significant roles of legumes in Tanzania include their early maturity compared to other staple food crops, being a quick source of income at every stage of their growth such as green leaves, fresh pods and dry grains (Birachi 2012). Generally, legumes excel in human and livestock nutrition, soil fertility improvement and foreign currency earning through export.

The importance of legumes to communities has led to a need for development and dissemination of various improved technologies. As a result of the above, extension agents have used a variety of ways to reach farmers. Generally, a number of delivery approaches and communication channels exist in legume producing areas. These include, but are not limited to, conventional approaches (agricultural extension officers visiting farmers), multimedia approaches (radio, television, mobile phones, newspapers, leaflets, brochures, etc.) and other extension methods such as demonstration plots, farmer field days, etc. However, the conventional agricultural extension services commonly used in Tanzania Morogoro included visits by extension officers to farmers in order to disseminate agricultural technologies. This method is important since it helps extension workers to provide technical assistance directly to farmers. However, the method faces a number of challenges, which include insufficient numbers of extension officers and inadequate resources (finances and transport) (Sanga et al. 2013).

In Tanzania and beyond, multimedia methods such as radio, television, mobile phones, newspapers, leaflets and brochures are also used in disseminating agricultural technologies. A prominent example is the Farmer Voice Radio project which was launched in 2009 and implemented in some of the districts in Tanzania. The project linked extension officers and farmers with a radio-based system. Generally, the multimedia method helps to reach many farmers within a short time in disseminating agricultural technologies (Sanga et al. 2013). Other research conducted in different parts of Africa found that multimedia methods are effective in awareness creation to smallholder farmers pertaining agricultural technologies because less time and costs are incurred while covering large areas (Ango et al. 2013; Ariyo et al. 2013; Kakade 2013; Chapota et al. 2014; Sam and Dzandu 2012). Despite their importance, multimedia methods have limitations. For example, duration of the programme tends to be short for farmers to capture all necessary information. Another limitation is that of language barrier, most facilitators are not fluent in local languages, and there may be lack of communication skills to communicate with the audience (Sam and Dzandu 2012).

Apart from the above extension methods, demonstration plots have been another avenue through which agricultural technologies are disseminated. Generally, demonstration plots help farmers to learn more by seeing and doing/practising. Also they are among the best methods to improve yield and help extension workers to effect desirable changes to smallholder farmers. Demonstration plots are arranged at the best learning locations (rural setting); and they provide opportunities through which useful communication and interaction can take place between extension workers and smallholder farmers (Khan et al. 2009). Nonetheless, the method also has limitations, as only a few farmers can be made to learn at a time.

Tanzania has for a long time been making efforts to scale up crop productivity (legumes included) under the Agricultural Sector Development Strategy 2001. These efforts have included financing agriculture and promoting research activities, improving extension services provided to smallholder farmers, training for updating skills and knowledge of farmers, improving agricultural mechanisation and improving agricultural information systems (URT 2001). Despite the efforts made to increase food productivity, legume yields are low (below a ton per hectare) (Malema 2006; URT 2012). In addition, the cost of obtaining such crops for food is high

(ProFound and Mugenyi 2012). Moreover, poor productivity may either be a result of the ineffective awareness creation approaches used or farmers' unwillingness to adopt improved legume technologies. Generally, the low productivity could be a consequence of farmers' low access to the legume technologies due to the shortfalls of awareness creation approaches.

In the 2015/2016 cropping season, a project on Scaling up Improved Legume Technologies (SILT) through Sustainable use of Input Supply and Information Systems was implemented in Morogoro Region with the support of the International Development Research Centre (IDRC) and the Canadian International Food Security Research Fund. Through the project there was sharing of knowledge with smallholder farming families using multimedia approaches such as technological briefs (leaflets and brochures) and other extension approaches, especially demonstration plots and farmer field days (MLE 2016). The knowledge disseminated included positive contribution of legumes to human and livestock nutrition, livelihoods, soil fertility and the environment, land preparation, legume variety selection, use of quality seed, fertiliser application at planting, planting, spacing and weeding. Others were control of insects and storage pests and diseases, harvesting and storage, and safe use of chemicals.

According to the Productivity Commission (2013), effectiveness is the extent to which stated objectives are met. Indicators of the effectiveness of programmes generally focus on measuring the changes in outcomes that reflect the objectives of the programme. According to SCRGSP (2006) cited in Productivity Commission (2013), the performance of any programme can be measured at two levels: cost-effectiveness performance indicators can be used to estimate unit cost of producing certain output, and programme effectiveness performance indicators can be used to look at agreed measures of access, appropriateness and quality. Therefore, the chapter adopted the definition of Production Commission 2013 by measuring effectiveness of communication channels in terms of awareness raised and determinants of adoption of improved legume technologies as the aim of the study.

2 Methodology

The study was conducted in Gairo and Mvomero districts in Morogoro Region, Tanzania, from February to March 2017. Morogoro Region lies between latitudes 5° 58″ and 10° 0″ South of the Equator and longitudes 35° 25″ and 30° 30″ to the East. Its climate is highly influenced by the Indian Ocean. The Nguru, Uluguru and Udzungwa Mountains as well as the Mahenge Hills form part of the Eastern Arc Mountains (URT 2016a, b).

Gairo District constitutes different agroecological zones with different climatic conditions. Generally, rainfall varies between 600 mm and 1200 mm and between altitudes of 1100 to 2200 metres above sea level (m.a.s.l). Land is characterised by moderately fertile well-drained soils, comprising sandy/clay loam soils. Agriculture is the mainstay of the district residents employing 90% of the households. The main subsistence crops cultivated include maize and beans (URT 2016a).

Mvomero District's climate varies from semi and warm tropical to cool high altitude. The district is characterised by high rainfall between March and May and from October to December. Annual rainfall is between 600 mm and 2000 mm and highest between the altitudes 400 and 2000 m.a.s.l. (URT 2016b). The land is very fertile, and about 90.1% of the district's total population is engaged in agriculture and agricultural-related occupations for their livelihood (URT 2016b). The above districts were purposively selected due to the fact that multimedia approaches (i.e. technological briefs such as leaflets and brochures) and other extension approaches (i.e. demonstration plots and field days) were used to scale up improved legume technologies in the last (i.e. 2015/2016) cropping season in these particular areas.

3 Research Design

The study used a cross-sectional research design, and two wards (Kinda Ward in Mvomero District and Rubeho Ward in Gairo District) that received improved common bean technologies (from November 2015 to April 2016) were purposively selected for data collection. The sampling units were households within the villages (i.e. Ndole and Makate in Kinda Ward as well as Ikenge and Rubeho in Rubeho Ward) with and without intervention. A structured questionnaire was administered to 400 respondents, about two thirds (66.25%) were from the area of intervention (Ndole and Ikenge villages) and the rest were from areas with no intervention (Makate and Rubeho villages). Based on the proportion of the number of respondents in the particular intervention, the research needed to capture at least minimum representation from each group, i.e. out of 66.25% respondents interviewed in the area of intervention, 63% received two interventions (demonstration and farmer field days), while 37% received three interventions (demonstration, farmer field days and technological briefs). Qualitative data were collected using focus group discussions (FGDs) and key informant interviews, which were conducted at the ward level (Table 1).

4 Sample Size

In calculating the sample size, it was assumed that 50% of smallholder farmers in both control and intervention areas are willing to adopt legume technology. This is because, from the reviewed literature, no reference was obtained showing percentage of smallholder farmers who were willing to adopt improved legume technologies; thus in calculating the sample size, the generic proportion, i.e. 50%, was used.

Table 1 Number of respondents selected in Mvomero and Gairo districts

District	Ward	Village	Intervention	People received intervention	Sample
Mvomero	Kinda	1 = Ndole	Farmer field days	82	47
			Farmer field days +Technological briefs	44	30
		2 = Makate	No intervention (control village)	00	69
Gairo	Rubeho	1 = Ikenge	Farmer field days	215	120
			Farmer field days +Technological briefs	120	68
		2 = Rubeho	No intervention (control village)	00	66
Total				**461**	**400**

NB: The non-response rate was 0.7% (3 respondents) which means, out of 403 targeted respondents, authors managed to interview 400 respondents. In addition, the types of data collected from these respondents included socio-demographic characteristics, types of communication channels used to create awareness on legume technologies in the area of study and types of legume technologies

Therefore, using Eq. 1 given by Cochran (1977), the sample size of 403 smallholder farmers was arrived at:

$$n = Z2 \propto /2P\left(1 - P / e2\right) \tag{1}$$

where

n = Sample size
$Z2\alpha/2$ = The probability distribution with level of significance $\alpha = 5\%$
"P" = Proportion of smallholder farmers adopted legume technologies
$(1 - P)$ = Proportion of smallholder farmers not adopted legume technologies
"e" = The level of marginal error

4.1 Data Analysis

The primary data collected through the questionnaire was coded and entered into the SPSS software (version 20) and checked for accuracy, and the anomalies found were corrected. The data was then analysed, computing descriptive statistics including frequencies, percentages, mean and standard deviations (SD). In addition, the binary logistic regression model was used to determine the factors predicting adoption of improved legume technologies. As given by Agresti (2002), it is specified as:

$$\text{Logit}(\text{Pi}) = \log(\text{Pi}/1-\text{Pi}) = b_0 + b_1x_1 + b_2x_2 + b_1x_1 + \cdots b_kx_k \ldots\ldots \quad (2)$$

Logit (Pi) = In odds (event), that is natural log of the odds of an event (adoption of technologies) occurring
Pi = Prob (event), that is the probability that the event will occur
1−Pi = Prob (no-event), that is the probability that the event will not occur
b_0 = Constant of the equation
b_1–b_k = Coefficient of the independent (predicator, response) variables
k = Number of independent variable
x_1 to x_k = Independent variables entered in the model
x_1 = Household size (total number of people in a household)
x_2 = Sex of household head (male 1, 0 female)
x_3 = Age of household head measured in years
x_4 = Marital status of household head (married 1, 0 otherwise)
x_5 = Education level of household head (primary and above 1, 0 otherwise)
x_6 = Type of intervention (with intervention 1, 0 otherwise)
x_7 = Total income from other income-generating activities (IGA)
x_8 = actual land in hectares cultivated
x_9 = Belonging to farmers association (1 Yes, 0 No)
x_{10} = Access to extension service (1 Yes, 0 No)
x_{11} = Access to credit (ever received credit 1, 0 otherwise)

5 Results and Discussion

5.1 Respondents Socio-demographic Characteristics

The results in Table 2 show that more than three quarters of the households were headed by males. The household head ages ranged from 18 to 79 years. Nevertheless, the majority of household heads were in the age range of 36–60 years (middle-aged household heads) and 18–35 years (youthful heads), which means the majority of household heads were in the economic active group (URT 2015). Study results further show that more than three quarters of the household heads had attained primary school education. This means the level of literacy in the study areas was high to the extent that programmes intended to create awareness in the particular area can be easily delivered and understood by the smallholder farmers through use of different communication channels/methods. Study results also show that almost all of household heads depend on agricultural production as their main occupation. The above is supported by Gairo and Mvomero districts socio-economic profiles which show that agriculture employs over 90% of the district's residents (URT 2016a, b).

Table 2 Demographic and socio-economic characteristics of the respondents ($n = 400$)

Variable	Category	Overall	Mvomero	Gairo
Household size	1–6	343(85.8)	125(85.6)	218(85.6)
	>6	57(14.2)	21(14.4)	36(85.8)
	Mean = 4.75, SD = 1.74			
Sex	Male	348(87)	123(84.2)	225(88.6)
	Female	52(13)	23(15.8)	29(11.4)
Age	18–35	171(42.8)	49(33.60	122(48)
	36–60	209(52.2)	87(59.6)	122(48)
	>60	20(5)	10(6.8)	10(3.9)
	Mean = 39.79, SD = 12.17			
Education level	None	68(17)	20(13.7)	48(18.9)
	Primary education	326(81.5)	120(82.2)	206(81.1)
	Above primary education	6(1.5)	6(4.1)	0(0)
Marital status	Single	11(2.8)	6(4.1)	5(2)
	Married	334(83.5)	114(78.1)	220(86.6)
	Divorced	26(6.5)	13(8.9)	13(5.1)
	Separated	10(2.5)	6(4.1)	4(1.6)
	Widow/er	19(4.8)	7(4.8)	12(4.7)
Main occupation	Crop production	394(98.5)	143(97.9)	251(98.8)
	Others	6(1.5)	3(2.1)	3(1.2)

NB: Number in brackets indicates percentage

Above primary includes secondary education; tertiary (certificate and diploma)

Others refers to livestock production; salaried employment (government); and casual labour (off-farm activities)

5.2 Communication Channels Used to Create Awareness

A number of communication channels are used in dissemination of agriculture technology in Tanzania. In the last production season (2015/2016), the African Fertilizer Agribusiness Partnership (AFAP) established demonstration plots in Gairo and Mvomero districts on common bean production covering one ward and one village in each district (AFAP 2016). In the same cropping season, 299 farmers attended farmer field days on common bean husbandry of which 217 farmers were from Ikenge village Gairo District and 82 farmers were from Ndole village Mvomero District (AFAP 2016). In addition, in the last production season (2015/16), the Centre for Agricultural Biosciences International (CABI) distributed leaflets to 205 farmers which contain all agronomic practices concerning common bean production. One hundred twenty of farmers who received the leaflets were from Ikenge village in Gairo District, and 85 farmers were from Ndole village in Mvomero District (CABI 2016).

5.3 *Levels of Awareness of Smallholder Farmers on Improved Legume Technologies*

Study results in Table 3 show that more than three quarters of the respondents were aware of improved common bean technologies. Results also show that more than two thirds of the respondents were aware of new planting methods (time of planting and proper spacing). Table 3 further shows that more than a half of the respondents were aware of the type, rate and time of using fertilisers (basal and boosting fertilisers). In addition, the results show that under two thirds of the respondents were aware of the weeding methods (stage, when and number of times to weed).

The results in Table 3 also show that more than a half of the respondents were aware of harvesting methods (stage of maturity and proper time to harvest). In addition to the above, results in Table 3 show that more than a half of the respondents were aware of the type, rate, time and safe use of chemicals. Lastly, the results show that more than a half of the respondents were aware of postharvest and storage management technologies. Generally, the results seem to suggest that there has been an impact of the interventions availed through the SILT project which is dealing with improved legume technologies. The above explanation is mainly based on the observation that the levels of respondents' awareness were high, suggesting communication channels used were effective. Moreover, awareness was very high in areas with demonstration plots + farmer field days and demonstration plots + farmer field days + technological briefs (leaflets and brochures), while for the areas with no intervention, there was low awareness. Further, the study results suggest that most

Table 3 Awareness of improved legume technologies among smallholder farmers in the study area ($n = 313$)

Technology	Aware of technology	No intervention	Demo/ FFD	Demo/ FFD/tech. briefs	Chi-squire	*P*-value
1. Improved common bean varieties	313 (78)	57 (43)	158 (94)	98 (100)	152.29	$p < 0.001$
2. New planting methods	271 (68)	26 (19)	150 (89)	95 (97)	217.23	$p < 0.001$
3. Type, rate and time of use of fertilisers	231 (58)	11(8)	132 (79)	88 (90)	205.89	$p < 0.001$
4. Weeding method (when and times)	241 (60)	13(10)	141 (84)	87 (89)	215.59	$p < 0.001$
5. Harvesting method (stage and when)	219 (55)	11(8)	124 (74)	84 (86)	179.72	$p < 0.001$
6. Type, rate, time and safe use of chemicals	224 (56)	11(8)	126 (75)	87 (89)	191.55	$p < 0.001$
7. Postharvest and storage management	217 (54)	10(8)	125 (74)	82 (84)	179.87	$p < 0.001$

NB: Number in bracket indicates percentages

of the smallholder farmers are in good position to raise their legume production and productivity based on the fact that they have a high level of awareness particularly on improved legume technologies. However, productivity can only be raised through adoption and proper application of the same.

It was also pointed out in the FGDs that farmer field days and demonstration plots are the two most important communication channels for raising awareness on improved legume technologies. One of the participants in the FGDs said:

> Despite the bad weather (drought) which occurred in the last cropping season (2015/2016), common beans planted in the demonstration plot continued to be good, the seeds were of high quality and we saw the required spacing practically and the yields were high. Generally, it was encouraging.

The above views are supported by the feedback from the project implementers: African Fertilizer Agribusiness Partnership (AFAP) who said that farmer field days take less time to deliver information and demonstration plots lessons are easily understood; the other implementers were the Centre for Agricultural Biosciences International (CABI), who said that technological briefs are less expensive and take less time to prepare. Similarly, the results from key informants (District Council Extension Officers) who said farmer field days and demonstration plots take less time to deliver information, and lessons are easily understood. In addition, many people are taught at a time through technological briefs.

In addition, the results in Table 3 conform to those reported by Ariyo et al. (2013), which hold that 90% of smallholder farmers confirmed multimedia methods to be effective in creating awareness of improved agricultural technologies. Moreover, Khan and Akram's (2012) found that farm/home visits, farmer field days and demonstration plots are the most effective communication channels in disseminating agricultural technologies. Generally, results of the study imply that farmer field days + demonstration plots + technological briefs if combined could be effective in disseminating improved legume technologies.

5.4 Farmers Adoption of Improved Legume Technologies

The study results (Table 4) show that only a few farmers adopted improved legume technologies. Generally, it was revealed that the improved legume technologies mostly adopted were weeding methods (proper time and number of weedings) (7.5%) and new planting methods (proper spacing and timely planting) (6.2%). The above figures seem to be low in relation to the respondents using the technologies; this could be due to the fact that the intervention was in its initial stage of implementation. Therefore, the figure may increase with time as knowledge diffuses to the communities in the study area.

During the FGDs participants pointed out that access to improved technologies in particular seeds was difficult due to the associated high costs. It was also pointed out in other FGDs that improved seeds were not available, and there were no seed dealers nearby their villages, hence poor adoption or their dependence on local seeds. The quote below emphasises the above:

Table 4 Smallholder farmers' adoption of improved legume technologies (n = 18)

Technology	Total adopted	Without intervention	With demo/ FFD	With demo/FFD/ Tech. briefs
Improved common bean varieties	11(3.5)	2(3.5)	5(3.2)	4(4.1)
New planting methods	17(6.2)	2(7.7)	2(1.3)	13(13.7)
Type, rate and time of use of fertilisers	7(3)	1(8.3)	4(3)	2(2.3)
Weeding method (when and times)	18(7.5)	2(15.4)	9(6.4)	7(8)
Harvesting method (stage and when)	9(4.1)	1(8.3)	3(2.4)	5(6)
Type, rate, time and safe use of chemicals	7(3.1)	0(0)	4(3.2)	3(3.4)
Postharvest and storage management	8(3.7)	0(0)	2(1.6)	6(7.3)

NB: Number in bracket indicates percentages

> Local seeds are very cheap hence most farmers rely on these. Moreover, nowadays farming is like gambling you may incur huge costs and end up harvesting nothing, like what happened to most of us in the last cropping season (2015/2016), because of the unreliable rains. (FGD participant, Ndole village, Mvomero, 20 March 2017)

The above is supported by Ngwira et al. (2012) who hold that adoption of the best legume technologies requires well-established innovation platforms with multiple stakeholder involvements, sufficient supply of high-quality legume seeds together with farmer training or access to extension services; otherwise adoption or actual use of the technologies is likely to remain low.

5.5 *Factors Associated with Smallholder Farmers' Adoption of Improved Legume Technologies by Type of Intervention*

Study results (Table 5 and Appendices 1, 2, 3, 4, 5, 6, and 7) show that there was a significant (p = 0.032) association between visits by extension officers and smallholder farmers' adoption of improved common bean seeds. Similarly, the results show there was a significant association between visit by extension officers (p = 0.001), household head's age (p = 0.021) and smallholder farmers' proper use of planting method (timely planting and proper spacing). Study results further show existence of a significant association between visits by extension officers (p = 0.011) and farmers' proper use of the type, rate and time to use of fertiliser. The results above conform to those of FAO (2015) and Pan et al. (2015) that access or visits by extension services influences the use of improved crop technologies or modern inputs. The results in Table 5 further show that there was a significant (p = 0.033) association between being a member of a farmer's association, revenue from IGA (p = 0.034) and smallholder farmers' use of proper harvesting methods (stage of harvesting and proper time

Table 5 Factors associated with smallholder farmers' adoption of improved legume technologies by type of intervention

Factor/determinants	Improved common bean varieties	New planting methods	Type, rate and time of use of fertilisers	Weeding method (when and times)	Harvesting method (stage and when)	Type, rate, time and safe use of chemicals	Postharvest and storage management
Household size	0.085 (0.205)	−0.356* (0.208)	−0.052 (0.029)	−0.203 (0.182)	−0.487* (2.850)	−0.522 (2.348)	−0.756** (4.660)
Sex of household head	0.957 (1.389)	−1.868 (2.033)	–	0.797 (1.538)	−1.845 (0.578)	−1.22 (0.313)	−2.542 (1.057)
Age of household head	−0.027 (0.035)	0.060** (0.026)	0.027 (0.457)	0.012 (0.025)	0.033 (0.995)	−0.004* (0.01)	0.110*** (7.314)
Marital status	−0.925 (1.094)	2.110 (2.056)	–	0.074 (1.236)	1.232 (0.247)	1.933 (0.739)	1.095 (0.193)
Education	0.542 (1.093)	1.690 (1.171)	–	1.122 (1.088)	–	–	1.992 (1.882)
Availability of technology intervention (Yes)	−0.093 (0.816)	0.451 (0.937)	0.280 (0.026)	−0.528 (0.886)	−0.07 (0.003)	–	–
Total income from IGA	−0.000 (0.000)	0.000 (0.000)	0.000* (3.227)	0.000 (0.000)	0** (4.477)	0.000** (4.940)	0.000 (1.848)
Total area cultivated	0.199 (0.190)	0.091 (0.205)	0.248 (0.392)	−0.052 (0.219)	0.141 (0.244)	−0.335 (0.478)	−0.765 (0.715)
Being member farmers association	−0.307 (1.102)	0.804 (0.822)	0.784 (0.418)	0.434 (0.724)	1.926** (4.528)	–	–
Visit by extension officer (Yes)	1.680** (0.784)	1.955*** (0.609)	2.413** (6.525)	1.241* (0.646)	0.299 (0.086)	–	–
Borrowing money for farming (Yes)	0.330 (0.730)	0.005 (0.657)	−0.923 (0.543)	0.383 (0.585)	−0.045 (0.002)	–	−0.104* (0.011)

Constant	−3.366 (2.660)	−6.663*** (9.778)	−6.397** (6.131)	−3.733* (3.703)	−3.097 (2.104)	−2.094 (1.549)	−5.304** (4.288)
−2 Log likelihood	88.753	102.903	46.948	117.009	61.310	54.234	50.250
Number of observations	313	272	232	241	220	225	218
Cox & Snell R^2	0.021	0.085	0.066	0.045	0.061	0.035	0.081
Nagelkerke R^2	0.079	0.229	0.278	0.108	0.211	0.147	0.299
Chi-square	6.969	4.011	5.698	8.562	25.804	10.405	8.133
P-value	0.54	0.856	0.681	0.381	0.001	0.238	0.421

The specific logistic regression model results are presented as Appendices 1, 2, 3, 4, 5, 6, and 7
NB: Number outside the bracket refers to B values, while number in bracket indicates Wald statistics
***, ** and * are significance levels at 1%, 5% and 10%, respectively
The '–' are omitted because of collinearity

of harvesting). In addition to the above, the study results (Table 5) show that there was a significant association between a household head's age ($p = 0.007$), household size ($p = 0.031$) and proper use of postharvest and storage management. The results in Table 5 conform to those reported by Ainembabazi et al. (2017), to the effect that smallholder farmers' membership to associations and extension services provided significantly influence smallholder farmers' use of improved legume technologies. The study's observation conforms to Uaiene et al. (2009), Abate et al. (2011) and Katengeza et al. (2015) who have reported that farmers' membership to associations has an impact on their use of improved technologies. The results in Table 5 further conform to those reported by Kasirye (2013) that low education and land holding or small area cultivated does not influence smallholder farmers' adoption of improved agricultural technologies (especially improved seeds and fertilisers).

6 Conclusions and Recommendations

The chapter has assessed the effectiveness of communication channels on level of awareness and adoption of improved common bean technologies among smallholder farmers in Gairo and Mvomero districts, Morogoro Region. Based on the findings, it can be concluded that people in the study area are generally aware of all the improved legume technologies assessed. Nevertheless, the level of awareness was high in areas with the intervention and low in the area with no intervention. It is also concluded that a combination of demonstration plots + farmer field days + technological briefs (leaflets and brochures) was the most effective communication channels in creating awareness, followed by a combination of demonstration plots + farmer field days. It is further concluded that smallholder farmers' adoption of improved legume technologies is influenced by visits by extension officers, age of household head, being member of farmers association, revenue from other income-generating activities and household size.

Based on the study findings and conclusions, the following are recommended:

(i) The government and non-governmental organisations should invest more in awareness creation approaches in order to make sure that all smallholder farmers are sensitised on the improved legume technologies.
(ii) The government and non-governmental organisations should insist more on visits by extension officers, farmers association and participating in other income-generating activities so as to enhance adoption of improved legume technologies. Doing the above will highly impact on the farmer's productivity, food security, incomes and general well-being.

Acknowledgements We are grateful to the International Development Research Centre (IDRC) and the Canadian International Food Security Research Fund (CIFSRF) for supporting this study and the Sokoine University of Agriculture (SUA) and the International Institute of Tropical Agriculture (IITA) for the field and technical assistance. We also acknowledge the smallholder common bean farmers in Gairo and Mvomero districts, Tanzania who wholeheartedly provided important information for the study.

Appendices

Appendix 1 Factors associated with adoption of improved common bean seeds

Factor/determinants	B	Std. Err.	Wald	df	Sig.	Exp(B)	95% CI Lower	95% CI Upper
Household size	0.085	0.205	0.171	1	0.679	1.089	0.728	1.627
Sex	0.957	1.389	0.475	1	0.491	2.604	0.171	39.648
Age	−0.027	0.035	0.599	1	0.439	0.974	0.910	1.042
Marital status	−0.925	1.094	0.715	1	0.398	0.396	0.046	3.385
Education	0.542	1.093	0.246	1	0.620	1.720	0.202	14.659
Availability of technology intervention (yes)	−0.093	0.816	0.013	1	0.910	0.911	0.184	4.516
Total income from IGA	0.000	0.000	0.485	1	0.486	1.000	1.000	1.000
Total area cultivated	0.199	0.190	1.092	1	0.296	1.220	0.840	1.771
Being member of farmers association (yes)	−0.307	1.102	0.078	1	0.780	0.736	0.085	6.376
Visit by extension officer (yes)	1.680	0.784	4.598	1	0.032	5.367	1.155	24.930
Borrowing money for farming (yes)	0.330	0.730	0.205	1	0.651	1.392	0.333	5.822
Constant	−3.366	2.064	2.660	1	0.103	0.035		

Variable(s) entered on step 1: V1_QNB1, V2_QNB2IBb, V3_QNB2IC, V4_QNB2IEb, V5_QNB2IFb, V6_Intervtn_status, V7_TOTALINCOME, V8_TOTAREACULT, V9_FAMERASSOC, V10_EXTOFVISIT, V11_L01

Appendix 2 Factors associated with planting method (timely planting and proper spacing)

Factor/determinants	B	Std. Err.	Wald	df	Sig.	Exp(B)	95% CI Lower	95% CI Upper
Household size	−0.356	0.208	2.926	1	0.087	0.701	0.466	1.053
Sex	−1.868	2.033	0.844	1	0.358	0.154	0.003	8.300
Age	0.060	0.026	5.294	1	0.021	1.062	1.009	1.118
Marital status	2.110	2.056	1.053	1	0.305	8.244	0.147	463.767
Education	1.690	1.171	2.084	1	0.149	5.419	0.546	53.749
Availability of technology intervention (yes)	0.451	0.937	0.231	1	0.631	1.569	0.250	9.850
Total income from IGA	0.000	0.000	1.015	1	0.314	1.000	1.000	1.000
Total area cultivated	0.091	0.205	0.196	1	0.658	1.095	0.732	1.638
Being member of farmers association (yes)	0.804	0.822	0.958	1	0.328	2.235	0.447	11.191
Visit by extension officer (yes)	1.955	0.609	10.303	1	0.001	7.064	2.141	23.306
Borrowing money for farming (yes)	0.005	0.657	0.000	1	0.993	1.005	0.278	3.641
Constant	−6.663	2.131	9.778	1	0.002	0.001		

Variable(s) entered on step 1: V1_QNB1, V2_QNB2IBb, V3_QNB2IC, V4_QNB2IEb, V5_QNB2IFb, V6_Intervtn_status, V7_TOTALINCOME, V8_TOTAREACULT, V9_FAMERASSOC, V10_EXTOFVISIT, V11_L01

Appendix 3 Factors associated with adoption of proper type, rate and time to apply fertiliser

Factor/determinants	B	Std. Err.	Wald	df	Sig.	Exp(B)	95% CI Lower	Upper
Household size	−0.052	0.302	0.029	1	0.865	0.95	0.525	1.718
Age	0.027	0.04	0.457	1	0.499	1.028	0.95	1.112
Marital status	0.28	1.74	0.026	1	0.872	1.323	0.044	40.073
Total income from IGA	0	0	3.227	1	0.072	1	1	1
Total area cultivated	0.248	0.257	0.932	1	0.334	1.282	0.774	2.123
Being member of farmers association (yes)	0.784	1.213	0.418	1	0.518	2.191	0.203	23.633
Visit by extension officer (yes)	2.413	0.945	6.525	1	0.011	11.168	1.754	71.131
Borrowing money for farming (yes)	−0.923	1.252	0.543	1	0.461	0.397	0.034	4.626
Constant	−6.397	2.584	6.131	1	0.013	0.002		

Variable(s) entered on step 1: V1_QNB1, V3_QNB2IC, V6_Intervtn_status, V7_TOTALINCOME, V8_TOTAREACULT, V9_FAMERASSOC, V10_EXTOFVISIT, V11_L01

Appendix 4 Factors associated with weeding methods (when to weed and number of times to weed)

Factor/determinants	B	Std. Err.	Wald	df	Sig.	Exp(B)	95% CI Lower	Upper
Household size	−0.203	0.182	1.248	1	0.264	0.816	0.571	1.166
Sex	0.797	1.538	0.269	1	0.604	2.219	0.109	45.201
Age	0.012	0.025	0.218	1	0.640	1.012	0.963	1.063
Marital status	0.074	1.236	0.004	1	0.952	1.077	0.096	12.139
Education	1.122	1.088	1.063	1	0.303	3.070	0.364	25.904
Availability of technology intervention (yes)	−0.528	0.886	0.355	1	0.551	0.590	0.104	3.347
Total income from IGA	0.000	0.000	0.104	1	0.747	1.000	1.000	1.000
Total area cultivated	−0.052	0.219	0.057	1	0.812	0.949	0.618	1.457
Being member of farmers association (yes)	0.434	0.724	0.359	1	0.549	1.543	0.373	6.372
Visit by extension officer (yes)	1.241	0.644	3.715	1	0.054	3.458	0.979	12.210
Borrowing money for farming (yes)	0.383	0.585	0.429	1	0.513	1.467	0.466	4.621
Constant	−3.733	1.940	3.703	1	0.054	0.024		

Variable(s) entered on step 1: V1_QNB1, V2_QNB2IBb, V3_QNB2IC, V4_QNB2IEb, V5_QNB2IFb, V6_Intervtn_status, V7_TOTALINCOME, V8_TOTAREACULT, V9_FAMERASSOC, V10_EXTOFVISIT, V11_L01

Appendix 5 Factors associated with harvesting methods (stage of harvesting and proper time of harvesting)

Factor/determinant	B	Std. Err.	Wald	df	Sig.	Exp(B)	95% CI Lower	Upper
Household size	−0.487	0.288	2.850	1	0.091	0.615	0.349	1.081
Sex	−1.845	2.428	0.578	1	0.447	0.158	0.001	18.407
Age	0.033	0.033	0.995	1	0.319	1.033	0.969	1.102
Marital status	1.232	2.481	0.247	1	0.620	3.428	0.026	443.431
Availability of technology intervention (yes)	−0.070	1.398	0.003	1	0.960	0.932	0.060	14.440
Total income from IGA	0.000	0.000	4.477	1	0.034	1.000	1.000	1.000
Total area cultivated	0.141	0.285	0.244	1	0.622	1.151	0.658	2.012
Being member of farmers association (yes)	1.926	0.905	4.528	1	0.033	6.865	1.164	40.478
Visit by extension officer (yes)	0.299	1.020	0.086	1	0.769	1.349	0.183	9.958
Borrowing money for farming (yes)	−0.045	0.928	0.002	1	0.962	0.956	0.155	5.893
Constant	−3.097	2.135	2.104	1	0.147	0.045		

Variable(s) entered on step 1: V1_QNB1, V2_QNB2IBb, V3_QNB2IC, V4_QNB2IEb, V6_Intervtn_status, V7_TOTALINCOME, V8_TOTAREACULT, V9_FAMERASSOC, V10_EXTOFVISIT, V11_L01

Appendix 6 Factors associated with adoption of the type, rate, time and safe use of chemicals

Factor/determinants	B	Std. Err.	Wald	df	Sig.	Exp(B)	95% CI Lower	Upper
Household size	−0.522	0.340	2.348	1	0.125	0.594	0.305	1.157
Sex	−1.220	2.180	0.313	1	0.576	0.295	0.004	21.190
Age	−0.004	0.039	0.010	1	0.922	0.996	0.923	1.076
Marital status	1.933	2.248	0.739	1	0.390	6.907	0.084	565.961
Total income from IGA	0.000	0.000	4.940	1	0.026	1.000	1.000	1.000
Total area cultivated	−0.335	0.484	0.478	1	0.489	0.716	0.277	1.848
Constant	−2.094	1.683	1.549	1	0.213	0.123		

Variable(s) entered on step 1: V1_QNB1, V2_QNB2IBb, V3_QNB2IC, V4_QNB2IEb, V7_TOTALINCOME, V8_TOTAREACULT

Appendix 7 Factors associated with postharvest and storage management

Factor/determinants	B	Std. Err.	Wald	df	Sig.	Exp(B)	95% CI Lower	95% CI Upper
Household size	−0.756	0.350	4.660	1	0.031	0.470	0.237	0.933
Sex	−2.542	2.472	1.057	1	0.304	0.079	0.001	10.003
Age	0.110	0.041	7.314	1	0.007	1.117	1.031	1.210
Marital status	1.095	2.494	0.193	1	0.661	2.989	0.023	396.496
Education	1.992	1.452	1.882	1	0.170	7.327	0.426	126.057
Total income from IGA	0.000	0.000	1.848	1	0.174	1.000	1.000	1.000
Total area cultivated	−0.765	0.905	0.715	1	0.398	0.465	0.079	2.741
Borrowing money for farming (yes)	−0.104	0.975	0.011	1	0.915	0.902	0.133	6.095
Constant	−5.304	2.562	4.288	1	0.038	0.005		

Variable(s) entered on step 1: V1_QNB1, V2_QNB2IBb, V3_QNB2IC, V4_QNB2IEb, V5_QNB2IFb, V7_TOTALINCOME, V8_TOTAREACULT, V11_L01

References

Abate T, Shiferaw B, Gebeyehu S, Amsalu B, Negash K, Assefa K, Eshete M, Aliye S, Hagmann J (2011) A systems and partnership approach to agricultural research for development lessons from Ethiopia. Outlook Agric 40(3):213–220

AFAP (African Fertilizer Agrobusiness Partnership) (2016) SILT project progress report 2016

Agresti A (2002) Categorical data analysis, 2nd edn. Wiley, Hoboken, 710 pp

Ainembabazi JH, Asten PV, Vanlauwe B, Ouma E, Blomme G, Birachi EA, Nguezet PMD, Mignouna DB, Manyongi VM (2017) Improving the speed of adoption of agricultural technologies and farm performance through farmer groups: evidence from the Great Lakes region of Africa. J Int Assoc Agric Econ 48(2):241–259

Akibode CS (2011) Trends in the production, trade, and consumption of food-legume crops in Sub-Saharan Africa. Dissertation for Award of MSc Degree at Michigan State University, 76 pp

Ango AK, Illo AI, Yakubu AA, Yelwa FJ, Aliyu A (2013) Radio agricultural programmes: a means of bridging research findings – rural farmers gap. a case of Zaria Metropolitan area, Kaduna State, North Western, Nigeria. Int J Sci Nat 4(3):538–545

Ariyo OC, Ariyo MO, Okelola OE, Aasa OS, Awotide OG, Aaron AJ, Oni OB (2013) Assessment of the role of mass media in the dissemination of agricultural technologies among farmers in Kaduna North Local Government Area of Kaduna State, Nigeria. J Biol Agric Healthc 3(6):19–28

Birachi EA (2012) Value chain analysis of beans in eastern and southern Africa: building partnerships for impact through research on sustainable intensification of farming systems. International Institute of Tropical Agriculture, 21 pp

CABI (Centre for Agricultural Biosciences International) (2016) SILT project progress report 2016

Chapota R, Fatch P, Mthinda C (2014) The role of radio in agricultural extension and advisory services – experiences and lessons from farm radio programming in Malawi – modernising extension and advisory services (MEAS). http://www.farmradiomw.org/publications/MEAS-CS-Malawi-Farm-Radio-Chapota-Fatch-Mthinda-Feb-2014.pdf. Site visited on 26/9/2016

Cochran WG (1977) Sampling Techniques, 3rd edn. Wiley, New York. http://hbanaszak.mjr.uw.edu.pl/StatRozw/Books/Cochran_1977_Sampling%20Techniques.pdf. Site visited on 30/9/2016

FAO (Food and Agriculture Organization of the United Nations) (2015) Food security impact of agricultural technology adoption under climate change: micro-evidence from Niger. FAO, Rome, 139 pp
Kakade O (2013) Credibility of radio programmes in the dissemination of agricultural information: a case study of air Dharwad, Karnataka. IOSR J Humanit Soc Sci 12(3):18–22
Kasirye I (2013) Constraints to agricultural technology adoption in Uganda: evidence from the 2005/06–2009/10 Uganda National Panel Survey. Kampala, Uganda, pp 90–107
Katengeza S, Kankwamba H, Mangisoni JH (2015) Maize and legume technology adoption in Malawi: gender, social networks and SIMLESA effects. Adoption Pathways project discussion paper 7. Malawi, 15 pp
Katungi E, Farrow A,Mutuoki T, Gebeyehu S, Karanja D, Alamayehu F, Sperling L, Beebe S, Rubyogo JC, Buruchara R (2010) Improving common bean productivity: an analysis of socioeconomic factors in Ethiopia and Eastern Kenya. Baseline Report Tropical Legumes II. Centro Internationale Agricultura Tropical-CIAT. Cali, Colombia, 126 pp
Khan A, Akram M (2012) Farmers' perception of extension methods used by extension personnel for dissemination of new agricultural technologies in Khyber Pakhtunkhwa, Pakistan. Sarhad J Agric 28(3):511–520
Khan A, Pervaiz U, Khan NM, Ahmad S, Nigar S (2009) Effectiveness of demonstration plots as extension method adopted by AKRSP for agricultural technology dissemination in District Chitral. Sarhad J Agric 25(2):313–319
Malema BA (2006) An overview of soyabean promotion in Tanzania. In: Malema BA, Laswai H, Myaka FA (eds) Production and sustainable use of soyabean in Tanzania. Soya bean production and utilization in Tanzania. Proceedings of the second national soyabean stakeholders workshop, Morogoro, Tanzania, 21–22 December, 2006, 57 pp
MLE (Monitoring Learning and Evaluation Strategy) (2016) Scale up of improved legume technologies through sustainable input supply and information systems in Tanzania (SILT). MLE working Group, 19 pp
Ngwira AR, Kabambe VH, Kambauwa G, Mhango WG, Mwale CD, Chimphero L, Chimbizi A, Mapfumo P (2012) Scaling out best fit legume technologies for soil fertility enhancement among smallholder farmers in Malawi. Afr J Agric Res 7(6):918–928
Pan Y, Smith SC, Sulaiman M (2015) Agricultural extension and technology adoption for food security: evidence from Uganda. IZA Discussion Paper No. 9206. Bonn, Germany, 45 pp
Productivity Commission (2013) On efficiency and effectiveness: some definitions. Staff research note: Canberra. http://www.pc.gov.au/research/supporting/efficiency-effectiveness/efficiency-effectiveness.pdf. Site visited on 18/8/2016
ProFound and Mugenyi F (2012) Organic kidney beans: potential for certified producers in Tanzania. Trade for Development Centre – BTC (Belgian Development Agency), Brussels, 29 pp
Sam J, Dzandu L (2012) The use of radio to disseminate agricultural information to farmers: the Ghana Agricultural Information Network System (GAINS) experience. In: Proceeding of third IAALD Africa chapter conference, e-agriculture for improved livelihoods and food security in Africa, Johannesburg, South Africa, 21–23 May, 2012, 23 pp
Sanga C, Kalungwizi VJ, Msuya CP (2013) Building an agricultural extension services system supported by ICTs in Tanzania: Progress made, challenges remain. Int J Educ Dev Inf Commun Technol 9(1):80–99
Sanginga N, Bergvinson D (2015) Oilseeds and cowpeas. An action plan for African agricultural transformation. Background paper. http://www.afdb.org/fileadmin/uploads/afdb/Documents/Events/DakAgri2015/Oilseeds_and_Cowpeas.pdf. Site visited on 28/04/2016
Uaiene RN, Arndt C, Masters WA (2009) Determinants of agricultural technology adoption in Mozambique. Discussion papers No. 67E. Mozambique, 29 pp
URT (United Republic of Tanzania) (2001) Agricultural sector development strategy. United Republic of Tanzania, Dar es Salaam, 73 pp

URT (United Republic of Tanzania) (2012) National sample census of agriculture 2007/2008. Volume Ve: regional report. Morogoro, 317 pp

URT (United Republic of Tanzania) (2015) Economic activity monograph. National Bureau of Statistics, Ministry of Finance Dar es Salaam and Office of Chief Government Statistician Ministry of State, President's Office, State House and Good Governance, Dar es Salaam, 117 pp

URT (United Republic of Tanzania) (2016a) Regional administration and local government. Gairo District Council: The Council Profile, Morogoro, 4 pp

URT (United Republic of Tanzania) (2016b) Regional administration and local government. Mvomero District Council: The Council Profile, Morogoro, 5 pp

Implications of Media-Scientists' Relationship on Crop Biotechnology Debate in Uganda

Ivan Nathanael Lukanda

Abstract Individuals often turn to the media for information about science and to track developments in their chosen fields of science, including medicine, climate change, biotechnology, and ecosystems resilience. Thus, media are key gateways to belief and doubt in knowledge, including science. Indeed, media houses have noted this trend and indexed it by establishing pages in their print versions, airtime on their electronic platforms, publishing science articles online, and sharing relevant information on social media. In order to achieve the desired visibility, science institutions have reciprocated by adopting and adapting training for scientists in public relations and providing guidelines for their researchers interested in going public or facing the media. Yet, findings from interviews with four scientists and ten science journalists show that there is still friction between journalists and scientists in what should have been a mutual relationship in sharing information about ecosystems risk science, especially in the crop biotechnology debate, with the non-expert society in Uganda. The implication of media playing the dual roles of being watchdogs and supporting scientific developments is that the awareness created through the various platforms greases and sustains the debate on issues of GMOs, in a way similar to the debate on climate change and ecosystems resilience.

Keywords Media for information about science · Ecosystems resilience · Science journalists · Scientific developments · Scientists in public relations

1 Introduction

Science is a vital component of society. The use of some scientific means to improve food quality and quantity, however, is controversial as reflected by the media. The controversy extends to the "revolutionary" science of biotechnology (Chen et al. 2016:1). Studies show that mass media are the dominant sources of scientific

I. N. Lukanda (✉)
Department of Journalism, Stellenbosch University, Stellenbosch, South Africa

Y. Bamutaze et al. (eds.), *Agriculture and Ecosystem Resilience in Sub Saharan Africa*, Climate Change Management,
https://doi.org/10.1007/978-3-030-12974-3_28

information for the general public, including discoveries and debates on science (Bucchi 2016; Claassen 2011). Thus, media are key gateways to belief and doubt in knowledge, including science. Indeed, media houses have noted this trend and indexed it by establishing pages in their print versions, airtime on their electronic platforms, publishing science articles online, and sharing relevant information on social media. As such, news media cannot be ignored as carriers of scientific information and debates on the adoption of the contested science of modern biotechnology to commercialise GMOs in Uganda (Ashwell 2014). Biotechnology is the use of living materials or their products to generate or modify other products and to enhance the quality of plants, animals, and other organisms for specific purposes (Okafor and Okafor 2017). The science can also be referred to as the "genetic manipulation", genetic modification (GM), or genetic engineering (GE) of living organisms to produce useful products for people (Rodriguez and Lee 2016:102). Biotechnology has been exploited commercially, at times contentiously, in the medical and pharmaceutical fields to make products such as insulin and antibiotics; in industry to make bread, wine, beer, yogurt, and juice and for the extraction of cobalt; in forensics to identify DNA at crime scenes and determining parentage; and in agriculture to breed crops and animals (Giorno and Drioli 2000). The contention, however, is most pronounced in agriculture, where it can be used to produce GMOs, often simply referred to as GM food or GM crops (Hicks 2017; Ventura et al. 2017). For the purposes of this study, the terms biotechnology and GM[Os] will be used interchangeably.

Considering the centrality of food and science in human lives, therefore, media (re)presentation of scientific information is of keen interest to policymakers, scientists, and the public in modern democracies, where accountability is an integral part (Hicks 2017). The connection between science and the media is best understood by analysing the relationship between scientists and science journalists who pave the way for entry of scientific information into society on issues such as medicine, climate change, ecosystems resilience, and biotechnology.

This chapter is based on the premise that the link between science and society is not obvious and must be triggered, "nurtured, and sustained" through science journalism and other communication methods (Suldovsky 2016:417). Underlying such attempts is the thinking that sharing scientific knowledge in socially stimulating and pleasurable settings will make science more acceptable, especially if it is controversial. Such endeavours are aimed at waving science into the fabric of citizens as a form of social responsibility and accountability to the taxpayers who directly or indirectly fund research (Lukanda 2018). By journalists linking science with society, it appears the formerly considered boundaries between the cultures of science institutions, and the broader society of non-scientists becomes porous. Such porosity should be considered in agricultural biotechnology, which results into the making and commercialisation of GMOs. This chapter assesses whether the relationship between the two groups could have a significant impact on how the Ugandan society understands the controversial science of biotechnology and GMOs.

1.1 Method

This chapter draws on interviews with four scientists and ten journalists covering biotechnology issues to assess whether the relationship between the two groups could have an impact on how the Ugandan society understands the science of GMOs. Of the four scientists interviewed for this study, three were biotechnologists and had appeared in the media several times supporting the adoption of GMOs. The other was a soil scientist and had appeared in the media opposing the adoption of GMOs. All the ten journalists interviewed had had by-lines either in the mainstream newspaper hardcopy or its online version. The results were analysed using Atlas.ti, and they depict a complex but promising relationship between the scientists and the journalists.

1.2 Results

The relationship between scientists and journalists can be described in two parts – before and after familiarisation. Journalists described biotechnologists as an arrogant, mean, and secretive group of experts, who think journalists are ignorant and inaccurate and misreport everything they tell them. Thus, the relationship was labelled as that of "traditional enemies" (Journalist G), "hide and seek, and blame game" (Journalist B), "confrontational" (Journalist I), "mistrust" (Journalist H), "murky, tricky" (Journalist G), and "hate and love" (Journalist C). Scientists B and C described their relationship with journalists as a "complexity". Scientist D described the association as an "opportunistic interaction", a "fight" and a "tough task", owing to the journalists' need to sell to the public information produced through rigorous scientific processes. The need to sell often leads to sensationalism and consequent distortion of laboratory findings as a result of recontextualisation. The respondents' description of the friction in the relationship is consistent with a study that argues that the initial interface between scientists and journalists is tense, strange, and stained by a "fault line" between the laboratory and the newsroom (Franklin 2010:145). Such a clouded relationship is fragile and makes reporting difficult, for fear of annoying the few available sources. A delicate relationship of this kind could explain why activists are infuriated by the rapport between biotechnologists and some science journalists who "love" science (Lukanda 2018).

Yet, after familiarisation, on one hand, science journalists describe the relationship as "cordial" (Journalist H), "collaborative" (Journalist I), "partners" (Journalist B), and "good" (Journalist J). On the other hand, the biotechnologists described the journalists as their "allies" (Scientist A), "friends" (Scientist B), and "cooperative" (Scientist C) in the dissemination of scientific knowledge. Nonetheless, Scientist D asserted that the science journalists are "unquestioning" and are bound to regurgitate the views from biotechnologists in Uganda. The subsequent analysis sections indicate that the multifaced relationship between scientists and journalists is historical, partly influenced by orientation, and the practitioners face challenges in establishing a mutual relationship. Besides, the relationship between biotechnologists and journalists has implications for public understanding science in Uganda.

2 Previous Studies on the Relationship Between Scientists and Journalists

The relationship between scientists who produce knowledge and journalist who share it with the public is critical to the public understanding of science in Uganda and other parts of the world. Journalists mirror the scientific knowledge to the general public. This relationship, however, has been described as complex, owing to the cooperation and misunderstandings between scientists and the journalists on how the evidence is presented to the public. From the findings in the previous sections, it appears that the results from laboratories are biased, hyped, or sensationalised, and out of context for Uganda, as journalists try to make the stories sellable. Uganda's case can only be understood by referring to the history of science communication in other countries.

Although studies about the development of science communication are rare, Bucchi (2016), Dunwoody (2008b), and Govoni (2010) observe that there were traces of science communication in the nineteenth century through magazines and newspapers. The content of the publications commonly composed of lecture notes and opinions on natural occurrences such as earthquakes and meteorites. The scholars explain that scientists, then, supposed that they had a moral duty to inform and educate the public using the communication platforms of the time about their discoveries as part of garnering societal support and approval. The Royal Society of England was central to improving the relations between experts and non-scientists through engagement especially on issues of medicine, evolution, and exploration, among others, by rewarding excellence in science. Thus, popularisation reached a peak during the Victorian age, a period marked by industrial revolution, colonial expansion, and creation of modern states, as scientists intensified their "dialogue" in different languages with politicians, industrialists, the middle class, workers, artisans, women, and the youth as their audience (Govoni 2010:24). It was through such engagements that "Newtonian physics" and "chemistry" became popular in England, much the same way "Pasteurisation" became endeared in France (Govoni 2010:23).

Dunwoody (2008a) notes that scientists' public participation started declining in the United States in the first decades of the twentieth century when they started specialising. At the same time, scientists started professionalising, developing, coining exclusive terms, and preferring to share information with fellow scientists through journals and conferences rather than with the general public. The systems of professional development included stringent rules that barred scientists from "popularising" their knowledge and denied them affiliation if they defied the establishment (Peters 2013). By the 1980s, scientists afraid of ruining their careers abandoned public engagements, and some journalists have since filled the gap. The practice, however, has been riddled with a serious shortage of specialist reporters, often requiring editors to interchange the "generalists" not only to ensure that the beat is covered but also to "prevent the pitfalls of reporter/source intimacy" (Dunwoody 2008a:16), which sometimes limits investigative reporting (Einsiedel and Thorne 2008). Although there are currently many Ugandan science journalists belonging to various national and international associations, observers admit that science remains under-reported because many media houses lack specialists (Unesco 2011). Indeed, even in the digital age, science communication has largely remained a "goodwill exercise" (Bucchi 2016:905). It is a voluntary activity for researchers because institutions seem to regard it as secondary to research and teaching.

In Italy, science was used as a political tool to demonstrate support for or against Catholicism through "indoctrination [researcher's emphasis]" of the public (Govoni 2010:25). This approach allowed Catholic fundamentalists to attack scientists who toed with the Darwinian theory of evolution. A notable victim of Italian scientific propaganda was astronomer Galileo Galilei, who maintained that the earth rotated around the sun contrary to Catholic teachings that the earth was at the centre of the solar system. In 1616, Galileo was consequently banned from popularising his thinking as it contradicted the Church's interpretation of the Bible. "So the sun stood still, and the moon stopped...." (The Bible, Joshua 10:31). Galileo was later tried and sentenced to home imprisonment for 8 years in 1633 for heresy. It was as late as 1992 that Pope John Paul II apologised for the church's mistake.

Bucchi and Trench (2014) chronicle the story of science communication from 1800 to 2014 under six research themes: 1800–1900s (deficit model paid attention to increasing public knowledge on ready-made science); 1930s–1940s (selling science as redemption of societal problems); 1960s (science journalism began to question science); 1980s (public relations in science are born); late 1900s (dialogue gave rise to the science-in-society model pushing the House of Lords in United Kingdom to document the need for sharing information with the citizens at all levels in 2000); and 2006 to date (crisis of traditional mediators [media] of science communication as social media entrenched itself). Science communication has been globalised raising the issue of democratisation of expertise. The question of integrating science with democracy seems to be at the core of GMO debate in Uganda.

Based on the above history, the relationship between scientists and journalists is of fundamental interest to this chapter for three major reasons – (1) attracting and justifying funding; (2) legitimising one's work before fellow scientists and the public; and (3) building partnerships between scientists and commercial enterprises, beyond advertising (Bauer and Gaskell 2002; Claassen 2011; Malyska et al. 2016). The scientists have the information, but they need journalists to share this information in a language understandable by the ordinary people for visibility of their work and brand. The public's understanding of science, more so, social debates on subjects such as biotechnology, is usually through the lenses of the media, which transmit and popularise the knowledge (Peters 2013; Maille et al. 2010; Bauer and

Gaskell 2002). Once information sinks into people's minds, it becomes difficult to change the public's perception especially if such perception has been built based on information got through the media (Malyska et al. 2016). The relationship is based on the premise that scientific findings can only be beneficial to society if they are accessible to the public. Widest dissemination is best achieved through the media (Bucchi 2016; Claassen 2011; Miller 2010; Tran 2013). However, this symbiosis has sometimes been antagonised by the suspicion that has plagued science communication for decades and recently in the case of Uganda and biotechnology.

Peters (2013:14102) uses six different metaphors – "distance, gap, barriers, fence, oil and water, and creative tension" – to describe the divergence between internal scientific communication and public communication. The use of different metaphors is an indicator that scientists and journalists probably have different agendas. Peters (2013:14103) opines that scientists and journalists are like "strangers, not able to understand each other's language" and how the professions vary in principle and practice. The antagonism seems to be rooted in the struggle between the scientists and journalists for power to communicate to the public. On the one hand, the scientists generate the information, but on the other hand, journalists have more contact with the public in filling the "structural hole" (Maille et al. 2010:76). In Uganda, journalists use newspapers, radio, television, blogs, and other social media platforms.

Maille et al. (2010:71) explain that journalists accuse scientists of confining themselves to the "ivory towers" and lacking the expertise to explain the results of their routine work to ordinary people. Journalists criticise scientists for wanting to check the stories where they have appeared as sources for purposes of checking accuracy before they are published, a point the journalists object to as an infringement on editorial independence (Claassen 2011; Maille et al. 2010). Stocking (2008) lists seven concerns of the scientists: (1) journalists often omit caveats on scientific claims; (2) journalists often use a single source for their stories, thereby locking out other scientists from interpreting the results; (3) journalists lack understanding of context (many stories do not connect to the past, ongoing, and the future) to related studies; (4) journalists focus on product triumph over process (stories often omit the method used in the study) thereby "solidify[ing], mystify[ing]" scientific claims or "obscuring scientific uncertainties"; (5) journalists sometimes make science appear uncertain yet knowledge will have settled on some issues such as the causes of HIV/AIDS and global warming; (6) journalists engage in unexpected and unexplained contradictions in linking new stories to the initial story with similar results; and (7) journalists give equal weight to minority and fringe scientists. Scientists complain that sometimes journalists allocate the same space and airtime to scientists who have published widely and those working for think tanks, and this creates doubt in mainstream science or gives equal weight to victims who are non-scientists and the scientists who cannot find a link between the effect and the available scientific evidence (cf. Davies 2009; Oreskes and Conway 2010; Pigliucci 2010).These concerns apply to Uganda, where the relationship is still budding.

Thus, many scientists have a generally negative perception about the way media cover science, especially controversial subjects such as climate change and biotechnology in ecosystems resilience. Scientists have argued that they resent the media because journalists misrepresent science, leading to misinformation among the public. Scientists feel that in many cases, scientific data are deliberately ignored when reporting science (Bhatta and Misra 2016; Caple and Bednarek 2013; Carver 2014; Malyska et al. 2016; O'Brien and Pizmony-levy 2015). The Ugandan scientists interviewed for this project criticised journalists for seeking views from anti-GMO activists, weighing scientific work with non-laboratory research to create a false balance.

On the other hand, there is also a lack of understanding on the part of the two actors of what makes news, resulting from differences in training (Basu and Leeuwis 2012; Berglez 2011; Brants and Van Praag 2017; Rodriguez and Lee 2016). To the scientists, their findings are news enough, but to the journalists the results become news after adding the social, political, economic, and environmental context and recontextualisation. For the scientists, this leads to sensation and hence distortion. The distortion sometimes emanates from the difference in training, which breeds terminologies hard to understand by journalists. Maille et al. (2010) found that journalists think that scientists are not mindful of the time element in responding to their questions. Often journalists want scientists to give them timely comments for the next by-line, yet the scientists want to first read, understand, and contextualise before commenting. In an attempt to beat deadlines and save space, journalists omit the methods used in conducting the research, to the chagrin of scientists, who think methodology is important in contextualising the results. To the scientists, the method is as good as the results. Maille and coresearchers further argue that science communication is complicated by the fact that some scientists think that because of the complex methods they use, the results they generate are secondary to public interests. The Ugandan journalists interviewed for this project objected to this, arguing that science must be understandable to the laypeople.

Some scientists suggest that the general public should be left out of the science production and validation process altogether, but the scientists can share their already published research in the public arena according to the "Ingelfinger rule" (Toy et al. 2002:195). The rule is a general principle developed by Franz Ingelfinger of *The New England Journal of Medicine* in the 1970s. The principle is used by editors of scientific journals to reject any article whose results have already appeared in the journalistic media, for purposes of exclusivity. In Peters (2013) study, scientists revealed that it is realistic to talk to journalists about results which have already been approved by their peers. According to Peters (2013:10402–9), 60% of the scientists believed that science should be treated as "specialist" knowledge and the general public should be treated as an external "audience". In this case the public is not the primary target of the research, but the findings can be used to educate them about issues such as biotechnology as the deficit model or scientific literacy model suggests (Lewenstein 2003; Trench 2008; Secko et al. 2013). The literacy model approach is a significant departure from the humanities and social sciences, where research results tend to focus on the ordinary people as an "active" primary audience (Fairclough 2008:812).

Therefore, social scientists are inclined to have more interaction with the media than researchers from natural sciences, law, archaeology, and philosophy. For the scientists, it becomes hard to communicate results to the unintended group, unlike for the social scientists who produce knowledge with the vision that the general public have to validate the results. On this issue, Peters (2013:14103) concludes that "the gap may be a steep canyon in the sciences but a smooth valley in the humanities and social sciences". Indeed, studies involving Ugandan politicians tend to attract more media attention than those involving scientists only.

So, as democracy deepens, public participation in science should also increase. The right to know should extend to what scientists do, since, as already discussed, most of the scientific work is maintained by public funds to find solutions to problems the citizenry may be faced with. Countries therefore need to craft ways of adopting "democracy for science" as a knowledge area, which has been dictated "by individual curiosity" for millennia to bring it to conformity with democracy, which survives principally on consent of the governed (Priest 2008:98). The working of democracy is certainly contrary to science, largely driven by systematic measurement, peer review, caution, reason, evaluation, and replication (Pigliucci 2010; Davies 2009; Priest 2008). But public accountability is necessary if leaders are to answer a vital question: how can the world feed the current population and ensure a sustainable biodiversity for the future generations?

In terms of frequency of contacts between scientists and journalists, Peters (2013) explains that sometimes the difference in coverage is a result of individual interests of journalists and their audiences for some topics, but not others. Hence, the interests of researchers in some fields and the contacts such scientists build with journalists play a critical role in interesting journalists and the public to follow their fields of research. Public information activities such as annual events, exhibitions and fairs draw scientists and journalists close (Broom and Dozier 1986; Carver 2014; Gunter et al. 1999; Lewenstein 2003). The other reason is the scientific journal in which the research appears. Publishing in journals such as *Nature*, *Science* or *The Lancet* is likely to draw media attention to scientists' research (Bhatta and Misra 2016; Ji-kun and Bo-wen 2015; Vilella-Vila and Costa-Font 2008). The last reason is the willingness of the researcher to talk to media will either draw journalists close or keep them away. This point also explains why there is "heavy reliance" on some individuals or organisations as sources of quotes on similar stories (Reul et al. 2018:6), providing a possibility for "neglect of sociological, cultural, ethical, historical and educational contexts" bolstering science, specifically subjects such as biotechnology (Petersen 2001:1258). Journalists complained that only a few scientists avail themselves to talk about biotechnology in Uganda as most scientists are still "media averse".

The popularity of the organisation and rate at which they publish are normally associated with frequency of media coverage. The relationship becomes even clearer if such organisations have strong public relations personnel who take the opportunity to publish as "advertising" or have "principal investigators" associated with them (Peters 2013:14105). Such organisations tend to drive coverage (Dunwoody 2008a, b). However, journalists have to be cautious in such engagements and when using press releases to write stories, because some organisations package pseudo-science and want to use the media as conduits to the public (Pigliucci 2010; Davies 2009). The over-reliance on the information provided by the science institutions could be the reason Scientist D described Uganda's journalists as "unquestioning" when covering the science of GMOs.

Also, critical in understanding the frequency of coverage is the ratio of journalists to scientists; the number of local publications vis-à-vis international publications; and routines of journalists (news releases, scientific publications, or interviews with scientists). Generally, there are few science journalists compared to the number of science streams to be covered, hence the under-reporting of science. Journalists tend to report streams of science such as medicine, which are popular to ordinary people than astronomy, which the audience hardly encounters in everyday life (Dunwoody 2008a). However, the interests of the public vary from society to society. Island countries like Japan may be interested in understanding biotechnology and marine science, but desert countries like Sudan would be better off concentrating on understanding biotechnology and desert life. In the same way, tropical countries like Uganda should prioritise understanding biotechnology in the context of tropical agriculture. Reporters who depend on news releases are more likely to cover science only when the research organisation is ready to release the information rather than enterprising journalists who take initiatives to investigate and report using angles the organisations may not have anticipated. Noteworthy is that disciplines like medicine and the environment, where contact with scientists is a routine activity, tend to get more coverage than others. Sometimes, journalists use public relations events to give their decaying information a "timely edge" (Dunwoody 2008b:66). Relatedly, scientists occupying high offices stand higher chances of interacting with the media than others, because their prominence drives journalists to seek their opinions (Peters 2013:14102), especially after an event or when there are social issue media want to report on. In Uganda, it is common for directors of agricultural departments and ministers of agriculture to appear in the media commenting about biotechnology.

Therefore, in the scientist-journalist relationship, scientists are able to drive the agenda because they choose what to tell the journalists (Dunwoody 2008b). In this, they are able to directly and indirectly dictate the story angles through news releases, press conferences, video releases, flyers, posters, and choosing the journalists to give the information. They also choose or recommend the journals to which media can have access. By Altheide and Snow's (1979) media logic and Entman's (1993) principle of framing, journalists can only be creative when using accessible material. Considering that most journalistic stories are reactive, scientists are more powerful in the relationship. Journalists frame the stories following the media logic, but scientists are asked to interpret the stories often creating uncertainty in the process for the audience. The ensuing debate culminates into a controversy, as it has happened in the case of Uganda and GMOs (Lukanda 2018).

The uncertainties involved in science could be another possible reason for avoidance by journalists. Moreover, training journalists and doing investigative stories on science are very expensive. This scenario can be attributed to astronomical costs involved in investigative reporting on the part of the media organisation and the journalist. For that reason, investigative stories are "scarce indeed" (Dunwoody 2008b:65), but it is worse in science reporting and possibly worst in biotechnology. Yet, when journalists investigate science, "they don't check the past papers of the scientists for mistakes, check raw data, or ask for source of funding" (Dunwoody 2008b:65). Often journalists select stories "which are quick to cover and safe to publish" (Davies 2009:114). Therefore, it is important to emphasise that what the general public reads as news is largely at the convenience of the journalists and their respective organisations.

In closing this section, it is imperative to assert that the "diffusion of scientific culture" requires the public to use their "knowledge and skills" to face their lived challenges, beyond knowledge acquired through the school curriculum to achieve an attitude of "participation and scepticism". Such a combination can be achieved through collaboration among science communicators, journalists, scientists, and politicians through the education system or orientation outside the academic curricula.

3 Differences in Orientation Between Scientists and Journalists

Despite all the antagonism, recent studies have suggested that there are factors that influence scientists to engage the journalists and the public, including subjects such as GMOs, whose adoption is debatable in the context of Uganda. Peters (2013:14102–9) found out that scientists' interaction with the media is based more on orientation – benefits, costs, and moral obligation. The interaction is usually linked to the leadership role of contemporary scientists rather than psychological factors such as scientists' intrinsic motivations, perception of moral duty, extroversion, presumed media influence, stereotypes of journalists, and motivations based on the perception of costs and benefits. When scientists take up leadership positions in some organisations, it is expected that they will face the media to explain key issues, regardless of their character as individuals. In other words, psychological factors are secondary to the orientation of a scientist. Indeed, this is nowadays encouraged by many scientific journals, which include the contacts of the principal investigators for follow-up stories. Such stories, especially in the media, increase the visibility of the scientist, the organisation they work for, and the journal where the findings were published. Many scientists have taken this as a moral duty. Another study found that race and gender play a role in one's ability to nurture a relationship with journalists. Joubert (2017) finds male and white scientists working in university settings more willing to engage the media than female and black scientists working outside university settings. Such engagement usually includes addressing policymakers, defending science, and promoting values of science in everyday life. No local studies have been done to ascertain the relevance of the above findings to Uganda.

Rödder (2012) argues that scientists who choose to interact with the media must be competent on the subject matter; the content of their statements must be substantive; they must be confident in their presentation; they must ensure that scientific communication is a priority; and they must further ensure that they speak to only reputable news organisations. To achieve this, it is important that scientists are trained in packaging their findings to accommodate the interests of the media, since journalists' understanding of science is crucial to the social contexts in which science applies and is debated. In the case of biotechnology, it is important that scientists know the "bias", if any, of respective journalists and the media houses they work for before addressing them. Such knowledge, on the part of the scientists, is important for purposes of anticipating the likelihood of being misquoted and emphasising certain issues while "facing" the media. In the case of Uganda, this could involve translating the message from English to local language to allow journalists capture the right information about GMOs for their audiences in "vernacular".

Of recent, public relations officers of organisations are facilitating the interaction between scientists and journalists for purposes of increasing visibility, securing public support and legitimacy, attracting attention of sponsors, and increasing competitiveness for contracts (Peters 2013:14102–9). Hence public relations tools such as press releases, blogs, newsletters, and wikis have been adopted in the management of organisations doing scientific research for enjoyment, visibility, attracting opinion, and understanding of science (Jarreau 2016; Carver 2014). Such publicity is linked to the growing commercialisation of science as the field expands to compete, maintain, and expand opportunities for government and international funding (Jasinsk 2010; Mackenzie et al. 2003; Townson et al. 2016). For this reason, in many cases, publishing of scientific research requires clearance from the organisation or by senior colleagues where the individual disseminating is not the principal investigator. This is true in the case of Uganda.

Further, Peters (2013:14105–6) notes that many scientists have come to embrace publishing their research in the media, though significant percentages have remained "ambivalent" about popularising science. He argues that institutions are using subtle means to encourage their scientists to face the media by offering them rewards, crafting press releases for researchers, and training them in dealing with the media in the long run. Often, the role of clearance is to ensure that the interaction with the media is anchored within the organisation's strategic plan, since adoption of GMOs is on a case-by-case basis. Indeed, Uganda is testing only those crops such as maize, bananas, cassava, and rice, which are popular in the country.

On the part of journalists, the way they cover science is sometimes skewed towards edging their "competitors" (Stocking 2008:67). Stories have to sell, hence the spicing with political and socio-economic angles for occasional sensationalism, even when the scientific facts may not have been interesting (Berglez and Nassanga 2015; Maeseele and Schuurman 2008; Takens et al. 2013). In fact, editors always argue that "what customers want drives the editorial judgement" (Davies 2009:135).

This strategy is aimed at increasing revenue for the media houses. In contrast, the focus on revenue makes media outlets ignore stories that would require them to invest much. Instead, they opt for sports and entertainment where the content is almost readily available and has a captive audience. This is far preferable to biotechnology which requires training journalists in science communication and being patient with scientists to compose their responses. In short, profits are central to the political economy of media houses (Susen 2011; Unesco 2011; Rodriguez and Lee 2016), and they take precedence over covering biotechnology and ecosystems resiliency in Uganda.

It should also be stressed that for journalists, it is very important to balance the story by including many voices. This is because sometimes scientific claims turn out to be faulty and have to be retracted (Dunwoody 2008b:66), as demonstrated in the case of Mbeki's basis for rejecting ARVs (antiretrovirals) for South African HIV patients (Duesberg and Rasnick 1998). Journalists seek protection in the principle of balance, especially where they are not sure of the facts. Covering a story from different angles portrays neutrality and allows media houses to avoid apologising to aggrieved parties. This very principle, however, according to Davies (2009:131), sometimes leads journalists "to abandon their primary purpose – truth telling". In the 1980s and 1990s, for example, when scientists warned that tobacco is linked to lung cancer, journalists kept seeking voices from tobacco companies, which vigorously responded with counter claims which raised doubts about a claim that had scientific data backing. Journalists have also given audience to oil companies to punch holes in scientific evidence on global warming. In other cases, journalists go about their work by selecting "safe facts". This way, they steer clear of individuals with power to influence the news or cause their sacking from the job. By so doing, they only give the public what it believes in and succeed in playing with the moral panic of society (Davies 2009:109–153). Such journalism practices, however, complicate the public's ability to understand biotechnology in Uganda and other parts of the world.

Stocking (2008:23–24) divides Davies' production rules into four parts – (1) individual factors; (2) media routines; (3) organisational demands; and (4) ownership patterns. Individual factors such as education and experience are important in covering science with accuracy and completeness in reporting science; experience in number of years has been proven to improve the ability to report science. At the same time, individual journalists' consideration for scientists' values and journalistic ethics tend to put the burden of responsibility on the journalists to report well. In addition, journalistic desire for significance entices them to ignore the caveats put by the scientists. These factors combine with organisational demands, especially audience interests, pressure from advertisers, professional norms, level of competition, the ownership patterns, and editorial policy, to determine what is published, which sources, and how they should be quoted (Caple and Bednarek 2013; Reul et al. 2018); Sarrimo 2016:1–16). In their article, *The case of the media by the media, The New York Times* (2016: n.p.), they concluded that social media have exposed traditional media as "partisan, mendacious, lazy, sloppy, and shrill", different from the "angels" the media purport to imitate. However, Bucchi describes this attitude as "simplistic and idealized" (2004:108–109):

> On the one hand, it legitimates the social and professional role of the 'mediators' – popularizers, and scientific journalists in particular – who undoubtedly comprise the most visible and the most closely studied component of the mediation. On the other hand, it authorizes scientists to proclaim themselves extraneous to the process of public communication so that they may be free to criticize errors and excesses – especially in terms of distortion and sensationalism. There has thus arisen a view of the media as a 'dirty mirror' held up to science, an opaque lens unable adequately to reflect and filter scientific facts.

It is apt to conclude that scientists and journalists see reality using different lenses, hence aggravating the controversies in science, more so on already controversial topics such as biotechnology and climate change. Having noted the above views, it is important to explain the challenges in the scientist-journalist's relations in light of the biotechnology and GMO debate in Uganda.

4 Challenges in Establishing a Mutual Relationship

Some scientists consider their knowledge special and not for everyone. They consider engagements with the media as arenas that lead to oversimplification of their findings. They also think that any strategy to reach the ordinary people through the media "erodes" the facts (Dunwoody 2008b:60), especially if it involves "fostering excitement" (Besley et al. 2016:370). In the case of Uganda, this excitement could involve incorporation of music, dance, and comedy into messages relating to biotechnology and science generally. The challenge of entwining entertainment and science or biotechnology for that matter is outside the scope of this chapter and could be the subject of another study.

Indeed, the need for increased public engagement between scientists and journalists faces the challenge of the push by some scholars who demand that scientists should also face the public directly, in addition to the occasional media appearance (Howard 2012; Miller and Fahy 2010; O'Brien and Pizmony-Levy 2015; Von Roten 2011). Certainly, this will most likely be time-consuming since scientists would have to meet small groups of people, but it may be more efficient. For it to be efficient, it will require biotechnologist and other scientists to be re-oriented from their structural exclusivity during training to embracing approaches that consider disseminating information to the ordinary person from project design to implementation. Importantly, this may have the effect of reducing the time available to stay in laboratories but could provide the opportunity to get more ideas from many people. Ideas can also be received through the Internet-powered platforms, which are becoming popular with scientists in Uganda and globally.

However, the reduced cost of uploading and maintaining a website have meant that many science organisations have put a lot of their content online. Cheap online technology enables institutions to open websites, run blogs, wikis, Twitter handles, and Facebook accounts. Peters (2013) regards this as direct competition with the media, in an age where sales of print media are dwindling. The use of social media to publish scientific information directly to the public seems to be slowly but steadily limiting face-to-face contact between journalists and scientists; on the other hand, it is also accelerating the flow of information (Bell 2016). Moreover, *The New York Times* (2016: n.p.) online affirms that social media are "bringing new voices to the fore and connecting established ones to new audiences". The connections are becoming easier with the development of non-commercial investigative technology such as The Intercept and ProPublica that select content depending on the readers' historical interest in related information and post them. Indeed, many science organisations in Uganda have websites, though the platforms are rarely updated.

The emergence of social media as a phenomenon in science communication may be a threat to processes of authenticating facts, but it should also be seen as an opportunity for journalists to get information readily from the websites rather than looking for scientists to interview. It also provides an opportunity for sustained discussions on topics from which ordinary people have been traditionally excluded. The live(ly) exchange and debate on several topics on Facebook, Twitter, and WhatsApp are evidences of this. In other words, the new media platforms are balancing the flow of information to the citizens, thereby reducing the journalists' privilege to exclusive information born out of their contacts with principal investigators. Scientists can now discuss new innovations, implications for funding, and legal, moral, and ethical implications directly with the public. But there are scholars who think that the right to information encompasses "non-journalistic" media and continue to agitate for bidirectional or multidimensional approaches beyond the media (Dudo et al. 2011:756) and should include intrapersonal, interpersonal, and public discussions for a healthy debate (Besley et al. 2016). The sustainability of this approach to science communication is subject to further research.

As it appears, science communicators now want to minimise mediated communication with the public. Lessening mediation should allow scientists to address issues of "misinformation" (Besley et al. 2016:370), which they have complained about for many years. Such interactivity will democratise science and may increase public "support" for specific streams of science, and attract potential funders (Ceccoli and Hixon 2012:301; Geary et al. 2016:740), who may strengthen the biotechnology movement. Some scientists have already chosen platforms (e.g. radio, television, newspapers, community outreach, science fairs, and exhibitions) where they are "engaging" the ordinary people (Besley et al. 2016; Von Roten 2011). However, it is not clear how the other scientists, who belong to the old school that argues that science is beyond the understanding of the ordinary people, will adapt to this wave. These scientists continue to hold paternalistic views about ordinary people's competence to understand scientific findings and to use the results to engage in meaningful discussions about scientific topics even through social media (Besley et al. 2016; Dunwoody 2008a). In the case of Uganda, it would be prudent for science institutions to support their researchers to increase their online presence as a way of facilitating knowledge flow in society. Certainly, the incorporation of the component of online presence is subject to availability of funds and the interests of scientists as indicted in their project proposals.

It should be noted that the technological developments are not a panacea to the structural challenges of the media. Journalists seem to corroborate this structural problem when they allude to:

> … the pressure for speed and easy hits that squeezes the nuance out of the complicated stories, editors who knowingly simplified stories past the point of accuracy, and publishers who spent resources on subjects they believed were trivial rather than those they felt were important. (*The New York Times* 2016: n.p.)

While it is hard to predict the future of science journalism and communication, it is clear that social media present new platforms for sharing scientific findings (Bucher and Helmond 2017; Bucchi 2016; Smailhodzic et al. 2016). However, the issue of authenticity of what is published on social media is likely to keep the audience with the mainstream media as a more trusted source than the latter. Indeed, it is a worthwhile consideration that the new media such as blogs and social networks open up new opportunities for science communication, which Uganda can benefit from immensely. If the expectations of social media defenders were to come true, the "gap" between internal scientific communication and public science communication would be narrowed. Journalist Goujard (2016:n.p.) proposes five ways reporters can engage scientists and their audiences in meaningful storytelling: (1) engage the scientist and audience before publishing by opening up the newsroom to allow them to contribute ideas for the newsrooms to follow up; (2) create a direct relationship with the public by contacting them directly on phone or email to verify the content they share with the journalists; (3) reward the scientists and the members of the general public who contribute by quoting them in the stories, without demanding too much from them since they are not professional journalists; (4) newsrooms should adapt to the scientific community and the audience by analysing whether their contributors want to remain anonymous or to feature; and (5) show the methods used to developing the story to enable them appreciate the cost of news production and the price the audience pay for the news. Further research in this area will be crucial.

5 Implications for Public Understanding of Ecosystem Resilience

This relationship and the resulting coverage suggest that the media platforms are playing both the "watchdog role" of allowing critical issues to trickle to the public and the "development role" in supporting scientists to enhance their knowledge (De Beer et al. 2016). While this dual role is healthy in politics, it has the effect of permeating controversy in science by slanting the debate through cherry-picking the issues to focus on. Nevertheless, the awareness created through the press greases and sustains the debate on issues of GMOs, in a way similar to the debate on climate change.

It is worth noting that there is a reciprocal relationship between scientists and science journalists. Science journalists need the news that results from the trips and conferences, on the one hand. On the other hand, scientists need coverage. As Eveland and Cooper (2013:14089) report, only "0.5% of science journal articles get media coverage", and most of the coverage is commanded by health and medicine. Chances are other stakeholders, especially anti-GMO activists and policymakers, involved in the biotechnology debate will strengthen their network of journalists to cover their events. The implication is that all stakeholders will benefit, except the consumers or the general public who will be manipulated by the various stakeholders through the press.

Pairing journalists with scientists will allow mutual learning. The pairing will help the scientists and journalists to understand one another's work and strengthen their relationship. This may work to improve science reporting, but may not take away the controversies that are heightened when a biotechnology story is moved from a science section to the front page or national or international news sections where stories draw attention to themselves.

Although a symbiotic relationship between scientists and journalists is necessary to soften the perception of the science behind biotechnology, such moves have been opposed by GMO sceptics (including some scientists). True, the closeness of scientists to journalists is likely to lessen the antagonism between the two groups, but there is fear that science organisations may use the unquestioning media to portray biotechnology as a risk-free science in combating climate change and enhancing ecosystems resilience. Journalists may also be forced to continue publishing what scientists want for fear of losing valuable sources of information and to maintain the close relationship. The implication is that public mistrust of the media as avenues of receiving information may increase. The demand from the public to engage individual biotechnologists in face-to-face dialogues away from the media may increase in Uganda, if trust in the media reduces.

There is also the fear that the dangers associated with GMOs can be extended to journalists who report about genetic engineering of food in Uganda. The stigmatisation of the science and the researchers involved may force more journalists to abandon the beat altogether for fear of being accused of being Monsanto agents.

The future relationship of scientists and journalists remains in balance as reporting on biotechnology is not limited to science journalism since the subject tends to arise in political arenas where non-specialised journalists generally apply the principle of balance. Balancing allows them to seek opinions of activists, politicians and other non-experts, whose views are erroneously weighed on the same scale as those of the scientists to create a false balance. Basing on such "balance as bias" (Boykoff and Boykoff 2004:125), rather that basing on "weight of evidence" (Dunwoody and Kohl 2017:1–20), escalates the controversy around GMOs, and widens the divide between scientists and journalists in Uganda. Hence, while a close relationship between scientists and science journalists can improve accuracy and smoothen the relationship, it cannot completely take away the controversy and global stigma of the risk associated with genetic engineering.

6 Conclusion

The relationship between scientists and journalists remains a point of contention in the public understanding of biotechnology and ecosystems resilience in Uganda. It is important to note that in countries like Uganda where media houses are small and can hardly afford to facilitate their journalists, partnerships between them and science institutions remain the most feasible way of enabling reporters get scientific information to share with the public. While reconciliation between the two groups seems to be the most feasible option, such moves have been opposed by anti-GMO activists who stigmatise the allied journalists as "renegades" who want to harm the country's citizens and its environment. This accusation from the activists and the counter accusation of activists being "ignorant" about GMOs make the debate on biotechnology and ecosystems resilience foggy. Yet, attempts by media houses to "balance" the views of the stakeholders only breed a false balance between the pro- and the anti-GMO activists. While the new media platforms, such as WhatsApp, Facebook, and Twitter, where scientists can post relevant information as part of both internal and public communication, can improve engagements between science institutions and the public, access is still limited. The implication is that discussing GMOs will be limited to the elite who can afford the gadgets and data bundles. Yet, it is general knowledge that in sub-Saharan Africa, the majority of people involved in agriculture are the rural poor, who can hardly afford Internet or access electricity/power to charge them. Moreover, the multiple platforms are likely to stir the debate further and confuse the audience. When the audience gets confused by the multiple sources of information, they resort to their partisan political, cultural, or religious identity to sieve information, with the likely consequence of reinforcing their exiting perceptions.

References

Altheide DL, Snow RP (1979) Media logic. Sage, California

Ashwell DJ (2014) The challenges of science journalism: the perspectives of scientists, science communication advisors and journalists from New Zealand. Public Underst Sci 25(3):379–393

Basu S, Leeuwis C (2012) Understanding the rapid spread of System of Rice Intensification (SRI) in Andhra Pradesh: exploring the building of support networks and media representation. Agric Syst 111:34–44

Bauer MW, Gaskell G (2002) In: Bauer MW, Gaskell G (eds) Biotechnology: the making of a global controversy. Cambridge University Press, Cambridge, UK

Bell E (2016) Facebook is eating the world. Columbia Journal Rev:1–7

Berglez P (2011) Inside, outside, and beyond media logic: journalistic creativity in climate reporting. Media Cult Soc 33(3):449–465

Berglez P, Nassanga GL (2015) What is the difference between cross-national comparisons and semicomparative work? Example of Swedish-Ugandan climate change communication Research. J Dev Commun Stud 4(1):33–48

Besley JC, Dudo AD, Yuan S, AbiGhannam N (2016) Qualitative interviews with science communication trainers about communication objectives and goals. Sci Commun 38(3):356–381

Bhatta A, Misra KD (2016) Biotechnology communication needs a rethink. Curr Sci 110(4):573–578
Boykoff MT, Boykoff JM (2004) Balance as bias: global warming and the US Prestige Press. Glob Environ Chang 14(2):125–136
Brants K, Van Praag P (2017) Beyond media logic. In: Journalism studies. Routledge, pp 1–14. https://www.tandfonline.com/doi/pdf/10.1080/1461670X.2015.1065200?needAccess=true
Broom GM, Dozier DM (1986) Advancement for public relations role models. Public Relat Rev 12(1):37–56
Bucchi M (2004) Science in society: an introduction to social studies of science, 5th edn. Routledge, London
Bucchi M (2016) Science communication and science in society: a conceptual review in ten keywords. Ital J Sci Technol Stud 7(2):151–168
Bucchi M, Trench B (eds) (2014) Routledge handbook of public communication of science and technology, 2nd edn. Routledge, London
Bucher T, Helmond A (2017) The affordances of social media platforms. In: Burgess J, Poell T, Marwick A (eds) The Sage handbook of social media. Sage, London
Caple H, Bednarek M (2013) Delving into the discourse: approaches to news values in journalism studies and beyond. The Reuters Institute for the Study of Journalism, Oxford
Carver RB (2014) Public communication from research institutes: is it science communication or public relations? J Sci Commun 13(3):1–4
Ceccoli S, Hixon W (2012) Explaining attitudes toward genetically modified foods in the European Union. Int Polit Sci Rev 33(3):301–319
Chen SY, Chu YR, Lin CY, Chiang TY (2016) Students' knowledge of, and attitudes towards biotechnology revisited, 1995–2014: changes in agriculture biotechnology but not in medical biotechnology. Biochem Mol Biol Educ 44(5):475–491
Claassen G (2011) Science and the media in South Africa: reflecting a "dirty mirror". Communicatio 37(3):351–366
Davies N (2009) Flat earth news. Vintage, London
De Beer AS, Malila V, Beckett S, Wasserman H (2016) Binary opposites – can South African journalists be both watchdogs and developmental journalists? J Afr Media Stud 8(1):35–53
Dudo A, Brossard D, Shanahan J, Scheufele DA, Morgan M, Signorielli N (2011) Science on television in the 21st century. Commun Res 38(6):754–777
Duesberg P, Rasnick D (1998) The AIDS dilemma: drug diseases blamed on a passenger virus. Genetica 104(2):85–132
Dunwoody S (2008a) Science journalism. In: Bucchi M, Trench B (eds) Handbook of public communication of science and technology. Routledge, London, pp 15–26
Dunwoody S (2008b) Scientists, journalists and the meaning of uncertainty. In: Friedman S, Dunwoody S, Rogers C (eds) Communicating uncertainty: media coverage of new and controversial science. Lawrence Erlbaum Associates, Publishers, London, pp 59–80
Dunwoody S, Kohl PA (2017) Using weight-of-experts messaging to communicate accurately about contested science. Sci Commun 39(3):338–357
Einsiedel E, Thorne B (2008) Public responses to uncertainty. In: Friedman S, Dunwoody S, Rogers C (eds) Communicating uncertainty: media coverage of new and controversial science. Lawrence Erlbaum Associates, Publishers, London, pp 43–58
Entman R (1993) Framing: toward clarification of a fractured paradigm. J Commun 43(4):51–58
Eveland WP, Cooper KE (2013) An integrated model of communication influence on beliefs. Proc Natl Acad Sci 110(Suppl. 3):14088–14095
Fairclough N (2008) The language of critical discourse analysis: reply to Micheal Billig. Discourse Soc 6(2):185–206
Franklin J (2010) The end of science journalism. In: Bauer MW, Bucchi M (eds) Journalism, science and society. Routledge, New York, pp 143–156
Geary J, Camicioli E, Bubela T (2016) DNA barcoding in the media: does coverage of cool science reflect its social context? Genome 59(9):738–750

Giorno L, Drioli E (2000) Biocatalytic membrane reactors: applications and perspectives. Trends Biotechnol 18(8):339–349
Goujard C (2016) Five ways journalists can engage their audiences in storytelling: International Journalis' Network. [Online]. Available: https://ijnet.org/en/blog/five-ways-journalists-can-engage-their-audiences-storytelling. Accessed 1 July 2016
Govoni P (2010) The rise and fall of science communication in late nineteenth century Italy. In: Bauer MW, Bucchi M (eds) Journalism, science and society. Routledge, New York, pp 21–29
Gunter B, Kinderlerer J, Beyleveld D (1999) The media and public understanding of biotechnology: a survey of scientists and journalists. Sci Commun 20(4):373–394
Hicks DJ (2017) Scientific controversies as proxy politics. Issues Sci Technol 33(2):67–73
Howard A (2012) Connecting with communities: how local government is using social media to engage with citizens. ANZSOG Institute for Governance at the University of Canberra and Australian Centre of Excellence for Local Government, Canberra. [Online]. Available: https://opus.lib.uts.edu.au/bitstream/10453/42107/3/Connecting-With-Communities-Social-Media.pdf. Accessed 4 June 2016
Jarreau PB (2016) Using Twitter to interact, but science communication to preach. SciLogs [Online]. Available: http://www.scilogs.com/from_the_lab_bench/using-twitter-to-interact-but-science-communication-to-preach/ [2017, 27 Feb]
Jasinsk AH (2010) Public relations as a tool of science communication with society. Revista CENIC. Cienc Biol 41:1–10
Ji-kun H, Bo-wen P (2015) Consumers' perceptions on GM food safety in urban China. J Integr Agric 14(11):2391–2400
Joubert M (2017) Who and where are the visible scientists in South Africa? Available: https://www.sun.ac.za/english/Lists/news/DispForm.aspx?ID=5307
Lewenstein BV (2003) Models of public communication of science and technology. Public Underst Sci 96(3):288–293
Lukanda IN (2018) From lab to fork? Press coverage and public (mis)perception of crop biotechnology in Uganda. Unpublished PhD thesis, Stellenbosch University, Stellenbosch
Mackenzie R, Burhenne-Guilmin F, La Viña AGM, Werksman JD, Kinderlerer J, Kummer K, Tapper R (2003) An explanatory guide to the cartagena protocol on biosafety. Environ Bus 46:1–13
Maeseele PA, Schuurman D (2008) Biotechnology and the popular press in northern Belgium: a case study of hegemonic media discourses and the interpretive struggle. Sci Commun 29(4):435–471
Maille ME, Saint-Charles J, Lucotte M (2010) The gap between scientists and journalists: the case of mercury science in Quebec's press. Public Underst Sci 19(1):70–79
Malyska A, Bolla R, Twardowski T (2016) The role of public opinion in shaping trajectories of agricultural biotechnology. Trends Biotechnol 34(7):530–534
Miller JD (2010) Civic scientific literacy: the role of the media in the electronic era. In: Kennedy D, Overholser G (eds) Science and the media. American Academy of Arts and Sciences [Online], Cambridge, MA. Available: https://www.amacad.org/content/publications/pubContent.aspx?d=1093 [2016, 29 Apr]
Miller S, Fahy D (2010) Can science communication workshops train scientists for reflexive public engagement? Sci Commun 31(1):116–126
O'Brien TLO, Pizmony-levy O (2015) Going public, gaining credibility: student perceptions of publicly engaged scholars. Sociol Perspect:1–24
Okafor OE, Okafor PI (2017) Membership of cooperative society and adoption of agricultural technology in Awka North LGA of Anambra State, Nigeria. J Agric Biol Res 6:1–11
Oreskes N, Conway EM (2010) Merchants of doubt: how a handful of scientists obscured the truth on issues from tobacco smoke to global warming. Bloomsbury, London
Peters HP (2013) Gap between science and media revisited: Scientists as public communicators. Proc Natl Acad Sci U S A 110(Suppl):14102–14109

Petersen A (2001) Biofantasies: genetics and medicine in the print news media. Social Science and Medicine 52(8):1255–1268
Pigliucci M (2010) Nonsense on stilts: how to tell science from bunk. University of Chicago Press, Chicago
Priest SH (2008) Popular beliefs, media, and biotechnology. In: Friedman S, Dunwoody S, Rogers C (eds) Communicating uncertainty: media coverage of new and controversial science. Lawrence Erlbaum Associates, London, pp 95–112
Reul R, Paulussen S, Van der Steen L, Maeseele P (2018) Professional journalistic routines and the protest paradigm: the Big Potato Swap in traditional and alterative media. Journalism 19(7):899–916
Rödder S (2012) The ambivalence of visible scientists. In: Rödder S, Franzen M, Weingart P (eds) The sciences' media connection – public communication and its repercussions. Springer, Dordrecht, The Netherlands, pp 155–177
Rodriguez L, Lee S (2016) What can be gleaned from news coverage to improve science reporting and enhance public literacy about agricultural biotechnology in Ghana? J Agric Food Inf 17(2–3):91–109
Sarrimo C (2016) The press crisis and its impact on Swedish arts journalism: autonomy loss, a shifting paradigm and a 'journalistification' of the profession. Journalism 18(6):1–16
Secko DM, Amend E, Friday T (2013) Four models of science journalism. Journal Pract 7(1):62–80
Smailhodzic E, Boonstra A, Langley D (2016) Towards new social media logic in healthcare and its interplay with clinical logic. In: Twenty-fourth European conference on information systems (ECIS), pp 1–11
Stocking SH (2008) How journalists deal with scientific uncertainty. In: Friedman S, Dunwoody S, Rogers C (eds) Communicating uncertainty: media coverage of new and controversial science. Lawrence Erlbaum Associates, Publishers, London, pp 23–42
Suldovsky B (2016) In science communication, why does the idea of the public deficit always return? Exploring key influences. Public Underst Sci 25(4):415–426
Susen BS (2011) Critical notes on Habermas's theory of the public sphere. Sociol Anal 5(1):37–62
Takens J, Van Atteveldt W, Van Hoof A, Kleinnijenhuis J (2013) Media logic in election campaign coverage. Eur J Commun 28:277–293
The New York Times (2016) The case of the media against the media. By the media. The New York Times (New York). 25 July. [Online]. Available: http://nymag.com/daily/intelligencer/2016/07/case-against-media.html. Accessed 28 July 2016
Townson C, Brewer PR, Ley BL (2016) Public responses to forensic DNA testing backlogs: media use and understandings of science. Bull Sci Technol Soc 35(5–6):1–8
Toy J, Vandenbroucke JP, Journal E, Holden E, Franz J (2002) The Ingelfinger rule: Franz Ingelfinger at The New England Journal of Medicine 1967–77. Science 25(6):195–198
Tran H (2013) Does exposure to online media matter? The knowledge gap and the mediating role of news use. Int J Commun 7:831–852
Trench B (2008) Towards an analytical framework of science communication models. In: Cheng D, Claessens M, Gascoigne NRJ, Metcalfe J, Schiele B, Shi S (eds) Communicating science in social contexts: new models, new practices. Springer Netherlands, pp 119–135
Unesco (2011) Media coverage of science and technology (October), pp 1–40
Ventura V, Frisio DG, Ferrazzi G, Siletti E (2017) How scary! An analysis of visual communication concerning genetically modified organisms in Italy. Public Underst Sci 26(5):547–563
Vilella-Vila M, Costa-Font J (2008) Press media reporting effects on risk perceptions and attitudes towards genetically modified (GM) food. J Socioecono 37(5):2095–2106
Von Roten FC (2011) Gender differences in scientists' public outreach and engagement activities. Sci Commun 33(1):52–75

Pathways for Addressing Gender-Based Constraints for Effective Participation in Profitable Crop Value Chains in Tanzania

Tatu Said Mnimbo, Joyce Lyimo-Macha, Justin Kalisti Urassa, and Khalmadin D. Mutabazi

Abstract Gender-based constraints in crop value chains are important considerations for equitable and sustainable participation of men, women and the youth. Women and youth make vital contributions to the agricultural sector despite the many gender-based constraints (GBCs) they face in accessing resources. The study on which the chapter is based aimed at analysing constraints that men, women and youth face, which hinder their participation in the profitable crop value chains (CVCs). Specifically, it analysed intra-household decision-making, assets associated with gender-based constraints and socio-economic factors influencing participation in profitable CVC and determined the pathways for addressing GBC. The study adopted a cross-sectional design whereby data was collected from 594, i.e. 295 and 299, from Chamwino and Kilosa Districts, respectively. Study results show that women use more time in performing agricultural activities such as planting, harvesting and post-harvesting activities, except for post-harvesting in Chamwino. Results further show that lack of wage labour, gender norms and household responsibilities negatively and significantly ($p \leq 0.05$) influence one's participation in the

T. S. Mnimbo (✉)
Department of Development Studies, College of Social Sciences and Humanities, Sokoine University of Agriculture, Morogoro, Tanzania

J. Lyimo-Macha
Institute of Continuing Education, Sokoine University of Agriculture, Morogoro, Tanzania
e-mail: joyce@sua.ac.tz

J. K. Urassa
Department of Policy, Planning and Management, Sokoine University of Agriculture, Morogoro, Tanzania
e-mail: urassa@sua.ac.tz

K. D. Mutabazi
Department of Food and Resource Economics, School of Agricultural Economics and Business Studies, Sokoine University of Agriculture, Morogoro, Tanzania
e-mail: kmutabazi@sua.ac.tz

Y. Bamutaze et al. (eds.), *Agriculture and Ecosystem Resilience in Sub Saharan Africa*, Climate Change Management,
https://doi.org/10.1007/978-3-030-12974-3_29

CVC. Generally, an increase in income was associated with participation in the profitable nodes of the value chain. The chapter concludes that the current gender inequalities and stereo types perpetuate an ‘exploitative status quo’ which is depriving women and youth of opportunities to properly engage in the more profitable nodes of the CVC. To address the above, the study recommends the adoption of gender-transformative strategies.

Keywords Gender · Gender-based constraints · Crop value chain · Participation · Pathways

1 Introduction

Value chains have become a key concept in international discussions on development, especially when one considers the effects of globalization on employment and poverty reduction (Carayannis et al. 2017). In the context of gendered economies, women and men participate at multiple levels in food crop value chains, often in different tasks, and with different opportunities for upgrading (Barrientos et al. 2003). In Africa, the participation of both women and men in agriculture is critical to production and growth; however, there is a limited understanding of the gender dynamics related to crop value chains (Shackleton et al. 2011). While men and women may face similar constraints to upgrading in crop value chains, their capability and incentives to overcome them often differ (Barrientos et al. 2010). Therefore, understanding these gender dynamics can help to get the right incentives to the right actors to promote better positions of men, women and the youth in crop value chains (Njuki et al. 2011). In Tanzania, women in rural areas have one thing in common across regions; they have less access than men to productive resources and opportunities in agriculture (Mnimbo et al. 2017). These gender gaps and constraints are found in different dimensions: assets, inputs, education, extension and financial services. Furthermore, they impose costs on the agricultural sector, the broader economy and society, as well as on the women themselves (Maertens and Swinnen (2012); Mnimbo et al. 2017).

Over the past few years, the question of how to promote increased gender-equitable agricultural development has emerged as an explicit component of value chain development efforts (Rubin et al. 2009; Chan et al. 2010; Bullock et al. 2017). Socio-economic researchers are increasingly analysing interventions to be able to achieve the dual objectives of economic efficiency and increasing gender equality which is goal number five of the Sustainable Development Goals (SDG).

Ensuring gender issues are considered in value chain-related interventions is vital for facilitating the development of inclusive value chains that equally benefit both women and men (Norell et al. 2016). According to CARE (2015), the above is quite essential for effective participation in crop value chains; thus, the paper sees effective participation in crop value chains as having more actors further along the value chains (CARE 2015). According to CARE effective value chains are good for reliability and sustainability of crop value chains.

The value chain, as a concept, describes the full range of activities that firms, farms and workers do to bring a product from its conception to its end use and beyond (Kaplinsky and Morris 2001). This includes activities such as production, processing, marketing, distribution and support to the final consumer. Value chain activities can produce goods or services and can be contained within a single geographical location or spread over wider areas (Coles and Mitchell 2011).

Gender is conceptualized as the socially constructed roles associated with being male or female (Morgan et al. 2017). Thus, understanding men's and women's position in a value chain, how changes in a value chain might affect gender inequality and the main constraints in terms of gaining from value chain participation requires one to place gender in the context of intra-household bargaining power (Dolan 2001). Generally, women and men enter value chains for commercial purposes (Chagomoka et al. 2014). However, their opportunities in the value chains are shaped by their physical, financial and human assets of which access to land and other productive assets (e.g. land, credit, extension, inputs) are key enabling factors (Mnimbo et al. 2017). Moreover, social assets and norms can also expand or limit the character and extent of men's and women's involvement (David 2015). Men and women stand to benefit in a number of ways from their participation in value chains through employment, wages or other income and empowerment, all of which can accrue to an individual or a household (Quisumbing et al. 2014). Generally, accessing value chain benefits is determined by the type of participation (e.g. as a wage worker or unpaid family worker) and the gender dynamics and power relations at multiple levels of the value chain. The above determines who gains and how these benefits are accessed and distributed (Meaton et al. 2015). According to Coles and Mitchell (2011), gendered patterns of benefit distribution are such that participation in the value chain does not always translate into gains. For example, in Kenya, despite women providing 72 percent of the labour, they only got 38 percent of the food crop value chain total income (Dolan 2001). At the same time, non-participation does not equate to a lack of benefit (Norell et al. 2016). What matters is not simply the level of income derived from value chain activities but a combination of factors related to the perception of ownership or management of a particular commodity, the scheduling of payment and the point of entry into the chain (Maertens and Swinnen 2012).

Agricultural value chains are equally important to men and women as a source of employment; however, gender inequalities run through agricultural systems; hence, action is required at all levels from the household and community up to the national level (MAFAP 2013). In this chapter the pathways to remove gender-based constraints (GBCs) take into account the daunting constraints that prevent women and men from engaging more equitably with the agricultural systems. The main objective of the chapter is to identify the key gender-based constraints that stand in the way of achieving an equitable and profitable participation in the value chains. Specifically, the study looked at sources of gender-based constraints and socio-economic constraints for participation in crop value chains. The chapter recommends pathways for reducing gender-based constraints in the crop value chains in the study villages. While both men and women can be disadvantaged or excluded from value chains (Bullock et al. 2017), the chapter focuses more on women's general participation in crop value chains, since they are more disadvantaged than men in the context of value chain operations (Quisumbing et al. 2014).

2 Conceptual Framework

The chapter adopts the Community Capitals Framework (CCF) to provide a holistic perspective of the interaction between the capitals required by men and women farmers for effective engagement in agriculture (Gutierrez-Montes et al. 2009). In this chapter, the gendered constraints and opportunities are grouped into seven capitals of the Community Capitals Framework (CCF), i.e. natural, human, financial, cultural, social, physical and political capitals as reported in literature (Bhandari 2013). These assets, especially the natural, human, social, physical and financial, play a major role in determining the extent to which smallholder farmers may take or could be willing to take the risks associated with participation markets (Sheck et al. 2013). According to Flora and Flora (2013), community capitals are assets or resources that can be utilized to produce additional resources. A holistic perspective helps to determine the actions to improve women's access to and control over resources to effectively engage in agriculture (Bullock et al. 2017). Human capital describes the skills, knowledge and abilities of people which they can deploy to mobilize other resources. It also enables individuals to strengthen their understanding, identify promising technologies and practices and obtain information to enhance their mobilization of other resources. At the household level, human capital includes the amount and quality of labour available and the ability to command this labour (Flora and Flora 2013). Financial capital refers to the monetary resources available for investment. For monetary resources to become capital, they must be invested to create new resource.

According to Voora and Venema (2008), natural capital is the spectrum of physical assets within the natural environment that deliver economic value through ecosystem services. Physical capital refers to the basic infrastructure and production inputs needed to produce wealth for purposes of supporting livelihoods (UNDP 2017). Social capital includes social networks and norms such as trust, which have an impact on the potential for coordination and cooperation for mutual benefit (LI et al. 2015). And according to Bourdieu (1986), cultural capital exists in three forms: in the embodied state, i.e. in the form long-lasting dispositions of the mind and body; in the objectified state (objects that are available in physical form like drums, tombs, etc. but which can also be harnessed into a production asset); and in the form of cultural goods (pictures, books, dictionaries, instruments, machines, etc.). Political capital according to Casey (2005) *is generally ill-defined. However, using* Pierre Bourdieu's interconvertibility theory, Casey defines political capital as 'an amalgamation of capital types combined in various ways for specific political markets'.

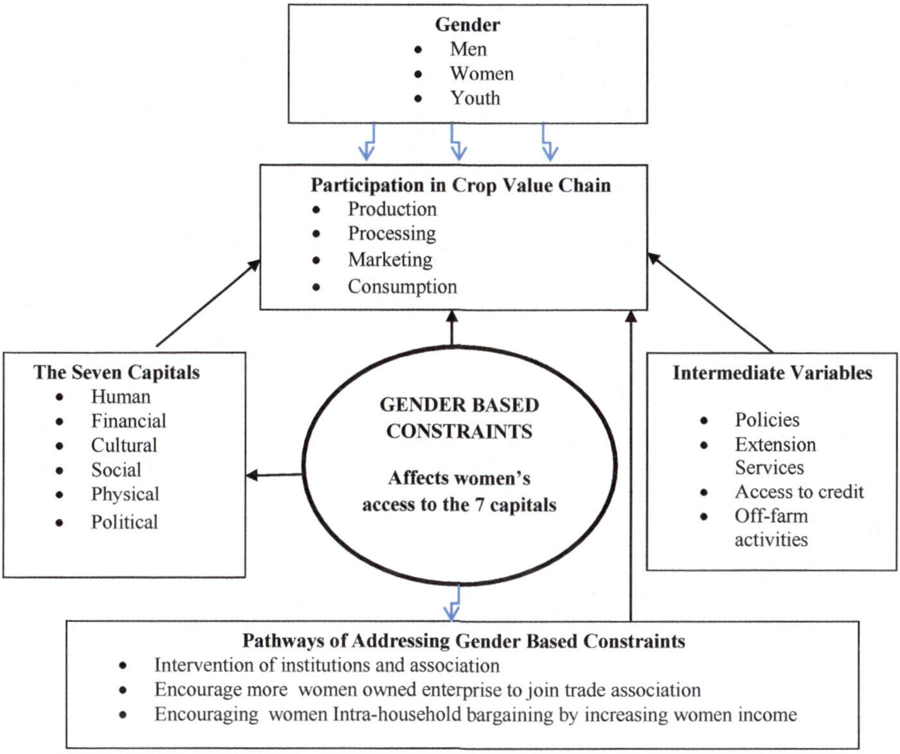

Fig. 1 Conceptual framework for pathways to address gender-based constraints in crop value chains

In this study, value chain is defined based on the definition by Kaplinsky et al. (2002), which is the full range of activities which are required to bring a product or service from conception through the different phases of production, transformation and delivery to final consumers and eventual disposal after use. In the Kaplinsky et al. (2002) approach, a value chain should have economic viability and sustainability as one of its core values and should aim at enhancing win-win outcomes for all participants (Coles and Mitchell 2011) (Fig. 1).

3 Theoretical Framework

The study on which this chapter is based adopted the social cognitive theory (SCT) of gender development. The theory looks at people and the conceptions they hold of themselves and others and the socio-structural opportunities and constraints they encounter (Bussey and Bandura 1999). The theory focuses on the social life and occupational paths people pursue and how they are prescribed by their society

(Giddens 1984). The social cognitive theory explains that people are producers as well as products of social systems. Furthermore, social structures are created by human activity (Giddens 1984; Bandura et al. 1996, 1999). The structural practices, in turn, impose constraints and provide resources and opportunity structures for personal development and functioning (Bandura et al. 1999). Viewed from this sociological perspective, the pattern of opportunity structures and formal and informal constraints shape gendered styles of behaviour and channel men and women into different life paths (Epstein 1997). People cannot derive similar benefits from life, even when they are of the same socio-economic status, live under the same opportunity structures, are subject to the same social and familial controls, are exposed to the same educational opportunities and draw from the same community resources and face the same normative climate. A host of personal and others come in to determine differences in the way some people benefit more from the same circumstances than others. Major among these factors is gender.

According to the SCT, gender constraints and the difference between men and women take on added importance because many of the attributes and roles selectively promoted in males and females tend to be valued differentially, with those ascribed to males generally being regarded as more desirable, effectual and of higher status (Berscheid 1993). Although some gender differences are biologically founded, most of the stereotypic attributes and roles linked to gender arise more from cultural design than from biological endowment (Bandura and Jeffery 1973; Beall and Sternberg 1993; Epstein 1997).

As regards gender development and functioning, the SCT integrates psychological and socio-structural determinants within a unified conceptual framework (Bandura and Jeffery 1973). In this perspective, gender conceptions and role behaviour are the products of a broad network of social influences operating both at the family level and in the many societal systems encountered in everyday life (Bandura et al. 1996). Many gender differences in social behaviour are viewed as products of division of labour between the sexes that get replicated through socio-structural practices governed by gender status and power (Eagly 1987; Geis 1993).

4 Research Methodology

4.1 Description of Chamwino and Kilosa Districts

The current study was conducted in two districts, Chamwino (Dodoma Region) and Kilosa (Morogoro Region). The former is semiarid with flat plains, and the food system is primarily based on sorghum, millet and maize. In addition, there is a deep attachment to livestock keeping (Mnenwa and Maliti 2010). Chamwino District, just like other districts in Dodoma Region, receives an average rainfall of about 500 mm annually, with 85% of the rain falling between December and March. The District experiences unreliable rainfall, leading to poor harvests of the traditional crops, particularly maize and millet, leading to recurrent shortages of food and culminating into extreme poverty among communities (PMO-RALG/TOA 2012).

Kilosa District, on the other hand, is predominantly semi-humid with flat plains, highlands and dry alluvial valleys, hence more diverse. Kilosa District experiences an average of 8 months of rainfall (October to May), the highest levels being between February and March. Generally, the rainfall distribution is bimodal in good years, with the short rains falling between October and January and the long rains falling between mid-February and May. The district's mean annual rainfall ranges between 1000 and 1400 mm in the southern flood plain, while further north towards Gairo District, it ranges from 800 to 1100 mm (Kajembe et al. 2013). The crops cultivated in the district include maize, sorghum, legumes, paddy and horticulture. In addition, some households keep livestock, but to a lesser extent when compared to Chamwino District.

The study areas were selected based on their diverse food value chains. Semiarid Chamwino is particularly sensitive to food insecurity as pointed above. On the other hand, Kilosa has both food-insecure and food-secure areas. The study villages in Kilosa are closer to the market, compared to those in Chamwino; they thus have differing commercialization opportunities and challenges (Mnenwa and Maliti 2010). Chamwino District is generally more patriarchal in comparison to Kilosa, which is mostly populated by the *Luguru* who are matrilineal (Beidelma 2017), even though it has several other ethnic groups. Because of the patriarchal and matrilineal systems practised in the above districts, one would expect access to resources and participation in important decisions to be different between gender groups. According to Paul and Meena (2016), access and control over productive resources is critical when ones considers agricultural productivity. In matrilineal communities, women tend to have better chances to asset ownership (Beildelman 1967); on the other hand in patriarchal communities such as Chamwino, which is dominated by the Gogo ethnic group, men dominate decision-making at the household level. Women seldom participate in decision-making, especially in matters related to income, asset control or duties that men are also do. The study on which the chapter is based sets out to assess whether there would be differences in the participation in the different food value chain nodes based on whether one came from a matrilineal or patriarchal community (Fig. 2).

4.2 Research Design

The study adopted a cross-sectional research design whereby data were collected at a single point and time. This design was adopted because information collected can be used to show association between variables and the dependent variable (Harvey et al. 2016). For example, the way the respondents' socio-economic characteristics influence their participation in the food value chain.

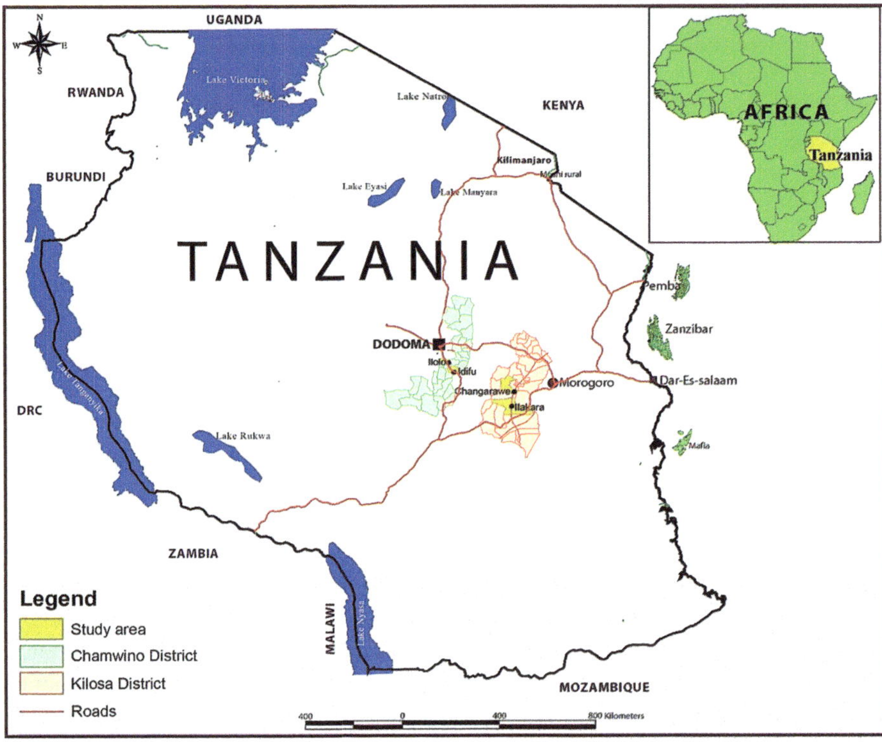

Fig. 2 Map of Tanzania showing the study villages (i.e. Ilolo and Idifu-Chamwino District and Ilakala and Changarawe-Kilosa District)

Sampling Techniques and Sample Size

The study areas were purposively selected, while the households were randomly sampled from village registers. The village registers contained information of the names of household heads in their villages. In each village, 150 households were selected from each of the 4 villages to make a total of 600 households from the two study areas. However, data was only collected from 594 households as 6 of the household heads were not available during actual data collection. In this study the youth are defined based on the Tanzanian youth policy (2007), as male and female individuals aged between 15 and 35 years (Table 1).

4.3 Data Collection

Primary data from 594 randomly selected households was collected using a pre-structured questionnaire with open- and close-ended questions.

To allow for triangulation, a mixed methods approach was used in data collection whereby both quantitative and qualitative data were collected. Qualitative data on

Table 1 Sampling per district and village

District	Villages	Sampled households
Chamwino	Ilolo	144
	Idifu	150
Kilosa	Changarawe	150
	Ilakala	150
Total		594

gender-based constraints (GBCs) at the different nodes of the food crop value chain, consequences of the GBCs for value chain participation and benefits derived from the food value chain (FVC)[1, 2] participation were collected through focus group discussions (FGDs).

A total of 12 FGDs were conducted, 6 in each of the 2 study districts: the FGD's involved separate groups of men, women and the youth (males and females). Each group comprised of 12 participants. Data on gender-based constraints (GBCs) at different nodes of the food crop value chain, consequences of the GBCs for value chain participation and benefits derived from value chain participation were collected using the Gender Dimensions Framework (GDF) (Rubin et al. 2009). The GDF has four major elements or dimensions that are critical in the intersecting dimensions of social life: (i) observed practices and patterns of participation, (ii) existing patterns of access to productive assets, (iii) social beliefs and perceptions and (iv) laws, policies and institutions. Rubin et al. argue that although overlapping in real life, the above dimensions are conceptually distinct.

What data did the authors capture on participation? What is a profitable crop value chain and how different is it from an unprofitable value chain? What does participation in a crop value chain mean? How was participation captured? What variables measure participation? Is participation uniform across crop value chains? What data did the authors capture on crop value chains?

[1] The agriculture and food value chain is complex spanning from (i) input companies, (ii) farmers/producers, (iii) traders, (iv) food companies and (v) retailers, all of whom must ultimately satisfy the varying demands of the consumer (KPMG International 2013). However, Cucagnaa ME and Goldsmith PD (In press) do present a shorter version with only four main stages, i.e. Stage 1, inputs; Stage 2, production; Stage 3, processing; and Stage 4, retail/selling to the consumers.

[2] Food crops considered by the study include maize (*Zea mays*), bulrush millet (*Pennisetum glaucum*), sesame (*Sesamum indicum*) and sunflower (*Heliunthus* spp.).

4.4 *Data Analysis*

Intra-household Decision-Making

A 1*2 cross-tabulation was done using Statistical Package for Social Sciences (SPSS) to determine intra-household decision-making on what to produce, sell and spend on. Weights were assigned as '1' meaning joint decision, '2' meaning women made the decision and '3' meaning men made the decision in the household. The household decisions were analysed based on the three aspects (decision on what to produce, where to sell and how much to spend).

Determination of Gender-Based Constraints (GBCs) in the Crop Value Chain

The GBCs were modelled using the multinomial logistic regression shown below:

$$\log\left[p/(1-p)\right] = \beta_0 + \beta_1 X_1 + \beta_2 X_2 + \beta_3 X_3 + \beta_4 X_4 + \beta_5 X_5 + \beta_6 X_6 + \beta_7 X_7 + \beta_8 X_8 + \beta_9 X_9 + \varepsilon$$

Where:

Log $[p/(1-p)]$ = Natural logarithm of the odds of probability of participation in the more profitable nodes of the crop value chain
β_1, β_2, β_3, β_4, β_5, β_6, β_7, β_8, and β_9 are parameters to be estimated and
ε is the error term.
The explanatory variables X_1, X_2, X_3, X_4, X_5, X_6, X_7, X_8, and X_9 were as follows:

X_1 = access to education
X_2 = access to farmland
X_3 = access to productive assets
X_4 = access to fertilizers
X_5 = availability of wage labour
X_6 = access to market information
X_7 = women's household responsibilities
X_8 = presence of segregative societal gender norms
X_9 = availability of water sources
ε = an error term

These explanatory variables are specified in Table 2 with their expected sign of influence.

Table 2 Definition of hypothesized effects of explanatory on gender-based constraints in the participation in crop value chain (CVC)

Variable name	Variable definition	Variable type	Hypothesized effect on participation in the CVC
Access to education	Ability to read and write yes = 1, no = 0	Dummy	+
Ownership of farmland	Ownership of agricultural land yes = 1, no = 0	Dummy	+
Access to productive assets	Yes = 1, no = 0	Dummy	+
Access to fertilizer	Yes = 1, no = 0	Dummy	+
Access to wage labour	Respondent being hired as a wage labour yes = 1, no = 0	Dummy	+
Access to market information	Availability of market info yes = 1, no = 0	Dummy	+
Women's household responsibilities	Number of hours used to work in the HH	Continuous	–
Presence of discriminatory societal gender norms in relation to household decision-making	If society rules and regulation hinders participation in decision-making yes = 1, no = 0	Dummy	–
Availability of water sources	Yes = 1, no = 0	Dummy	+

4.5 Results and Discussion

Intra-household Decision-Making in the Crop Value Chains

Generally, as presented earlier in Subsection 2.1, Chamwino and Kilosa Districts have different crop production potentials and market access. In addition, crop production and food preferences do differ between the districts too. For example, according to Mnimbo et al. (2017), maize (*Zea mays*) and sesame (*Sesamum indicum*) are widely grown in Kilosa, while bulrush millet (*Pennisetum glaucum*) and groundnut (*Arachis hypogaea*) are grown in Chamwino. Many of the food crops are consumed locally, with the exception of bulrush millet, groundnut and sunflower (*Helianthus* spp.), whereby about 50% of what is produced is sold. Further to the above, gender differences do exist with regard to crop production preferences between various gender groups within the above-mentioned districts (Mnimbo et al. 2017). For example in Chamwino District, the youth and women preferred maize to sorghum as a second food crop when compared to elderly men. On the other hand, in Kilosa District, the priority crops were maize for food and sesame for cash; however, when it came to the second most preferred food and cash crops in Kilosa District, there was a mismatch between men and women. The factors responsible for the mismatch included differences in biophysical and agroclimatic conditions and economic and labour needs (ibid).

Table 3 Decision-making on household crop production ($n = 594$)

How are decisions made		Chamwino (%)		Kilosa (%)	
		Male	Female	Male	Female
What to produce	Joint	31.1	26.4	38.7	31.0.3
	Women	2.7	40.2	1.9	35.8
	Men	66.2	33.3	59.4	32.8
What to sell	Joint	28.1	25.3	39.0	29.9
	Women	5.4	41.4	0.9	37.3
	Men	66.5	33.3	60.1	32.8
What to spend on	Joint	30.8	27.6	42.7	32.8
	Women	2.7	40.2	0.9	38.8
	Men	66.5	32.2	56.3	28.4

Generally, findings from the study (Table 3) show that men are the major decision-makers when it comes to what crops to be produced, what and how much to sell and how much to spend on the production process (Table 3). In the FGDs, women in Kilosa argued that women's decision-making power increases when they earn more than the men or just as much as men do. Literature (Bullock et al. 2017) concurs with the study's finding that decision-making within households has to do with bargaining, and this bargaining depends on the endowments of the parties and that a woman's ability to bargain in the household is usually augmented by the increase in her income, which leads to greater equity in the dispensation of household resources. Furthermore, a study by Ngome (2003) found that lack of income affects men's decision-making power, in that, if the man cannot afford to cater for family needs and the woman takes charge of that responsibility, then the man tends to involve the woman more in the decisions which he could have made alone. Cultural and religious beliefs and practices do influence participation in the value chain. This was evident in the Kenyan French bean value chain, where women were escorted to the market place in order for the husbands to verify the prices paid (Dolan 2001).

Time Use in the Crop Value Chain

Female and male farmers' time use in performing agriculture activities in the study area is presented in Table 4. In general, results show that in Chamwino District, female farmers use more time (51%) (more than 6 hours) in planting and spend 52% of the time in harvesting and post-harvesting activities compared to males. On the other hand in Kilosa, male famers use more hours (51%) in planting and in harvesting than females (50%). This implies that, since post-harvesting activities include crop product processing and marketing and since Chamwino is mostly patriarchal, then men could be more attracted to the activity as it involves getting cash. In Kilosa, more females (50%) spent more time in post-harvesting activities than males. However, during the FGDs, male farmers in Kilosa reported that female farmers are

Table 4 Distribution of respondents by hours spent in the crop value chain ($n = 594$)

		Chamwino ($n = 295$)		Kilosa ($n = 299$)	
Activity	Working hours	Male (%)	Female (%)	Male (%)	Female (%)
Planting	1–5	44.4	55.6	46.4	53.6
	More than 6 hours	49.1	50.9	51.0	49.0
Harvesting	1–5	45.0	55.0	39.5	60.5
	More than 6	47.9	52.1	50.4	49.6
Post-harvesting	1–5	39.7	60.3	46.8	53.2
	More than 6	50.2	49.8	49.7	50.3

involved in many household chores, which require them to leave the farm earlier than male farmers; hence, males stay longer in the farms. Nonetheless, during FGDs in Kilosa, female famers reported that time use in the farm between male and female varies by crop, activity and age. It was reported that for cash crops, males tend to use more time compared to females. Further, male farmers during FGDs in Kilosa pointed out that activities such as weeding or planting are to be females' work. This was attributed to the fact that weeding involves a lot of bending which men desist from and therefore leave women to perform this activity.

The above results are similar to what has been reported by Seymour et al. (2016) indicating that female farmers use more time in undertaking agricultural activities compared to male farmers. On the contrary, FAO (2011) suggests that women's contribution in agriculture is slightly less than half because of their involvement in household chores. In addition, Doss (2014) and Palacios-Lopez et al. (2017) have reported that females' share of agriculture labour as opposed to time use is 40% on average in Tanzania and Uganda, respectively, which is substantially less than the 60–80% cited by most literature on gender (e.g. Dolan (2001) and Barrientos et al. (2003)). Thus, this study shows that females use more time than males in undertaking the mentioned agricultural activities, because females participate in performing activities in all the important nodes in the value chain. It was evident that women participate in the production node where they plant, prepare the seeds and weed and in the processing node where they process the harvested crop by winnowing, drying and storing.

Gender-Based Constraint Participation in the Crop Value Chain

A multinomial logistic regression model was used to estimate the selected constraints related to participation in value chain. The overall model in determining the GBC was significant at ($p = 0.05$), indicating that the independent variables had satisfactory explanatory power in determining men and women's participation in the crop/food value chains. Table 5 shows gender-based constraints faced by male and female respondents in the study area for participation in the crop value chains.

Generally, the study results show that education is negatively associated with the probability of participation in the food value chain. This implies that persons with

Table 5 Gender-based constraints and their influence on participation in the crop value chain

Gender-based constraint	B	S.E	Wald	df	Sig	Exp (B)
1. Access to education	−4.899	5.031	0.948	1	0.330	0.007
2. Ownership of farmland	−1.485	0.177	70.824	1	0.000***	0.006
3. Access to productive Assets	4.643	5.029	0.852	1	0.356	103.878
4. Access to fertilizer	1.649	0.357	21.359	1	0.000***	5.200
5. Wage labour.	−1.972	0.260	57.530	1	0.000***	0.139
6. Lack of Market information	−0.279	0.417	0.447	1	0.504	0.757
7. Household responsibilities	−1.072	0.316	11.535	1	0.001***	0.342
8. Gender norms	−1.869	0.421	19.739	1	0.000***	0.154
9. Water scarcity	0.066	0.052	1.622	1	0.203	1.068

NB: *** significant at the $p = 0.001$ level

lower levels of education are less likely to participate in the profitable nodes of the food value chain. Further, wage labour was found to be negatively associated with one's participation in the food value chain, and this association was very significant ($p = 0.05$). Generally, wage labour involves doing work for pay which increases income. According to Palacios-Lopez et al. (2017), there is an association of increased income and participation in the profitable node of agricultural value chains (AVC), which implies that if work is paid for, an individual is able to participate in the profitable nodes of the value chains.

Gender norms negatively and significantly influence participation in AVC. This may be attributed to women's limited mobility (e.g. moving outside the village to commercialize crops). Therefore, this sometimes excludes them from participating in the marketing node which is considered one of the profitable parts of AVC. In line with these results, observations from the FGDs showed that in Chamwino, female participants explained that it was normal in their societies for males to be in charge and in control of assets. On the contrary, the society sees them as 'out of the ordinary' as shown in the quote below:

> When a household is comprised of a wife and a husband, and then somehow the wife controls the important assets such as land, money and makes all the important decisions, then that husband (man) is considered to be weak and to be under a spell (i.e. under control). (Female FGD participant, Chamwino District, March, 2014)

It is believed that the more assets people have, the less vulnerable they are (De Mel et al. 2009). Palacios-Lopez et al. (2017) also argue that women and the youth are more vulnerable because tradition gives them less control over assets than men, while at the same time, their opportunities to engage in remunerative activities to acquire their own assets are limited. Results in Table 5 show that land ownership is negatively and significantly ($p = 0.05$) associated with participation in the food value chain. These results are in line with views of the female FGD participants that being deprived of land ownership makes it difficult when making important decisions such as which crops to grow or the possibility of using the land to obtain credit as ownership of land is synonymous to managing resources. According to Grabe (2010) land remains an important livelihood resource in many societies and is likely to increase in times of economic crisis; it is emblematic of social belonging and is a highly gendered phenomenon. Grabe (2010) further reports that, women are often excluded or marginalized from access to land relative to men in similar social positioning. In addition, reports show that women face disadvantages compared to men in accessing the basic assets and resources needed to participate fully in realizing their growth potential and that gender-based differences affect supply response, resource allocation within households and labour productivity (Palacios-Lopez et al. 2017). Thus, this study suggests that women need to have access and control over land in order to participate in the important and profitable nodes of the AVC/CVC/FVC. This will enable them to choose the types and quantities of crops to produce based on the available land.

Pathways in Addressing Gender-Based Constraints (GBCs)

The current section tries to recommend or propose some pathways to address the GBC to crop value chains observed in the study areas. Given the diversity of communities, there is no single way of mitigating or removing GBC. It is therefore useful to adopt a continuum of different strategies. In this study 'continuum' is defined based on the gender manual of 2009 (FAO 2016), which uses gender elements in a continuous sequence that are adjacent but not clearly different from each other, with the extremes being quite distinct on the intended issue to be resolved. Thus, this study uses the continuum of three categories for identifying pathways to removing gender-based constraints and ensuring an equitable participation in value chain, namely, gender-exploitative, gender-accommodating and gender-transformative pathways. The study opted to use the gender continuum of strategies after observing the GBC and how they influence participation in the food value chain (Table 5).

Based on Table 5, which also reports on the gender norms which negatively affect participation, there is a likelihood of the communities and households being faced by the *exploitative status quo* whereby there are deliberate existing gender inequalities and stereotypes that stand in the women's way when it comes to pursuit of economic activities/income generation. For example, in the area of decision-making, women are left to use most of their time on tedious and time-consuming less profitable agricultural-related activities compared to men who are mostly found in the profitable nodes of the crop value chain (Mnimbo et al. 2017). In addition, women are involved more in household labour as 'free labour'. Barrientos et al. (2010) and Milberg and Winkler (2011) showed that *exploitative status quo* exists in value chains, and women are the ones being exploited relative to men.

The gender-accommodating strategy is the pathway which focuses on addressing specific gender inequalities such as decision-making (Table 5) by addressing the isolated issues that may create more dynamic change in a broad range of activities (Gereffi 2014). Pretty and Bharucha (2014) show the importance of looking at policy as a driver of change. For example, the 1990 community development gender policy and the children and the women and gender development policy of 2000 were observed to have many challenges. In many cases, implementation proved difficult when it came, to the patriarchal systems, customs and tradition, which still discriminate against women (Bullock et al. 2017). Hence, there is a need for policies that will ensure the GBC observed in the study area and in Tanzania generally. Generally, equitable and sustainable participation of men and women in the value chains can significantly contribute to the achievement of the SDGs such as reduction of poverty, food security (zero hunger) and gender equality (Allen and Sachs 2012; Maestre et al. 2017; Oduol et al. 2017).

Lastly, the gender-transformative strategy focuses on achieving mutually supportive goals in achieving the 'win-win' situation by finding synergies between gender relations and equitable participation in crop value chains (Mnimbo et al. 2017). In addition, due to the results on time use in agriculture activities (Table 5), there is a need to introduce labour-saving technologies that reduce women's load while at the same time increasing men's involvement (Johnson et al. 2016). For example, the introduction of winnowing and maize processing technology has changed the roles from an activity performed by women to one performed by men (change of gender role) (Mnimbo et al. 2017).

5 Conclusions and Recommendations

The chapter aimed at analysing constraints that men, women and the youth face which hinder their participation in the profitable nodes of crop/food value chains. Specifically, it aimed at analysing intra-household decision-making, community capital associated with gender-based constraints and household decision-making influencing participation in the above-mentioned value chains and pathways for addressing gender-based constraints. Based on the study, the following conclusion are made: (i) GBCs are negatively associated with women's participation in food crop value chains; (ii) ownership and access to assets such as land are vital for participation in the food value chains; (iii) of the two gender groups, women are more constrained when it comes to participation in the food value chains; (iv) household responsibilities, gender norms and not getting employed as a wage labourer have a strong negative association with participation in food crop value chains; and (v) men make the important decisions in the household and that women are only involved in decisions already made.

Based on the study's findings and conclusions, the study recommends introduction of technologies that will, where possible, reduce household duties for women and involve men more in the household activities. Due to the societal gender norms, men can only get involved in certain domestic activities if new innovations/technologies simplify and de-stigmatize the work and attract them to do it. In the policy arena, it is important to understand that without changing the 'business as usual' mindset of considering women as the 'weaker farmer' in all activities surrounding agricultural production, sustainable participation will continue to fail. Furthermore, there is a need to be careful in the adoption of a package of sustainable agricultural practices, to avoid the possibility of increasing women's workload by increasing time spent, which may affect their decisions to adopt and take part in the food crop value chains.

References

Allen P, Sachs C (2012) Women and food chains: the gendered politics of food. Taking food public: redefining food ways in a changing world. Int J Sociol 5:23–40

Bandura A, Jeffery RW (1973) Role of symbolic coding and rehearsal processes in observational learning. J Pers Soc Psychol 26:122–130

Bandura A, Barbaranelli C, Caprara CV, Pastorelli C (1999) Efficacy beliefs as shapers of aspirations and occupational trajectories. J Pers Soc Psychol 76:285–298

Bandura A, Barbaranelli C, Caprara GV, Pastorelli C (1996) Mechanisms of moral disengagement in the exercise of moral agency. J Pers Soc Psychol 71:364–374

Barrientos S, Dolan C, Tallontire A (2003) A gendered value chain approach to codes of conduct in African horticulture. World Dev 31(9):1511–1526

Barrientos S, Gereffi G, Rossi A (2010) Economic and social upgrading in global production networks: developing a framework for analysis. Int Labor Rev 150(4):319–340

Beall AE, Sternberg RJ (eds) (1993) The psychology of gender. Guilford Press, New York. 147pp

Beidelma TO (2017) The matrilineal peoples of Eastern Tanzania (Zaramo, Luguru, Kaguru, Part 16. Routledge, 10 Feb 2017 – Social Science – 89 pages

Beildelman TO (1967). The matrilineal peoples of eastern Tanzania. Ethnographic survey of Africa, East Central Africa, Part XIV. International African Institute, London

Berscheid E (1993) Forward. In: Beall AE, Sternberg RJ (eds) The psychology of gender. Guilford Press, New York, pp 7–17

Bhandari PB (2013) Rural livelihood change? Household capital, community resources and livelihood transition. J Rural Stud 32:126–136

Bourdieu P (1986) The Forms of capital. In: Richardson J (ed) Handbook of theory and research for the sociology of education. Greenwood, New York

Bullock R, Gyau A, Mithoefer D, Swisher M (2017) Contracting and gender equity in Tanzania: using a value chain approach to understand the role of gender in organic spice certification. Renewable agriculture and food systems. Cambridge University Press, London. 13pp

Bussey K, Bandura A (1999) Social cognitive theory of gender development and differentiation. Psychol Rev 106:676–713

Carayannis EG, Meissner D, Edelkina A (2017) Targeted innovation policy and practice intelligence: concepts and implications for theory, policy and practice. J Technol Transf 42(3):460–484

Casey KL (2005) Defining political capital: A reconsideration of Bourdieu's interconvertibility theory [WWW document]. http://lilt.ilstu.edu/

CARE (2015) Adding value to value chains how to unlock the poverty-fighting potential of value chains. https://www.care.org/sites/default/files/documents/Adding-value-to-value-chains.pdf. 25 Step 2018

Chagomoka T, Afari-Sefa V, Pitoro R (2014) Value chain analysis of traditional vegetables from Malawi and Mozambique. Int Food Agribu Man Rev 17(4):59–86

Chan KW, Yim CK, Lam SS (2010) Is customer participation in value creation a double-edged sword? Evidence from professional financial services across cultures. J Mark 74(3):48–64

Coles C, Mitchell J (2011) Gender and agricultural value chains: a review of current knowledge and practice and their policy implications. Working Paper No. 5. Food and Agriculture Organization of the United Nations, London. 29pp

Cucagnaa ME, Goldsmith PD (In press) Value adding in the agri-food value chain. International Food and Agribusiness Management Review. file:///C:/Users/Public/Documents/Downloads/IFAMR 2017.0051valueaddingproof%20(1).pdf. 27 Sept 2018

David S (2015) Getting a piece of the pie: an analysis of factors influencing women's production of sweet potato in Northern Nigeria. J Gend Agric Food Sec 1(1):1–19

De Mel S, McKenzie D, Woodruff C (2009) Are women more credit constrained? Experimental evidence on gender and microenterprise returns. Am Econ J Appl Econ 1, 1(3):–32

Doss C (2014) If women hold up half the sky, how much of the world's food do they produce? In: Quisumbing A, Meinzen-Dick R, Raney T, Croppenstedt A, Behrman J, Peterman A (eds)

Gender in agriculture and food security: closing the knowledge gap, Springer and FAO. Chapter 4. FAO, Rome Italy

Dolan D (2001) The 'Good Wife': struggles over resources in the Kenyan horticultural sector. J Dev Stud 37(3):39–70

Eagly AH (1987) Reporting sex differences. Am Psychol 42:755–756

Epstein CF (1997) The multiple realities of sameness and difference: ideology and practice. J Soc Issues 53:259–278

FAO (Food and Agriculture Organization) (2011) The state of food and agriculture 2010–2011. Bull Econ Stat 65(3):283–328

FAO (2016) The state of food insecurity in the world 2015. Meeting the 2015 international hunger targets: taking stock of uneven progress. Food and Agriculture Organization Publications, Rome, pp 1–35

Flora CB, Flora JL (2013) Rural communities: legacy and change. Soc Nat Resour 26(1):5–20

Geis FL (1993) Self-fulfilling prophecies: a social psychological view of gender. In: Beall AE, Sternberg RJ (eds) The psychology of gender. Guilford Press, New York, pp 9–54

Gereffi G (2014) Global value chains in a post-Washington Consensus world. Rev Int Polit Econ 21(1):9–37

Giddens A (1984) The constitution of society: outline of the theory of structuration. Polity Press, Berkley, University of California Press, Cambridge. 402pp

Grabe S (2010) Promoting gender equality: the role of ideology, power, and control in the link between land ownership and violence in Nicaragua. Anal Soc Issues Public Policy 10(1):146–170

Gutierrez-Montes I, Emery M, Fernandez-Baca E (2009) The sustainable livelihoods approach and the community capitals framework: the importance of system-level approaches to community change efforts. Community Dev 40(2):106–113

Harvey CR, Liu Y, Zhu H (2016) The cross-section of expected returns. Rev Financ Stud 29(1):5–68

Johnson NL, Kovarik C, Meinzen-Dick R, Njuki J, Quisumbing A (2016) Gender, assets, and agricultural development: lessons from eight projects. World Dev 83:295–311

Kajembe GC, Silayo DSA, Mwakalobo ABS, Mutabazi K (2013) The Kilosa District REDD+ pilot project, Tanzania. A socio-economic baseline study. IIED, London. http://pubs.iied.org/pdfs/G03624.pdf. 25 Sept 2018

Kaplinsky R, Morris M (2001) A handbook for value chain research. Report Prepared for International Development Research Center. Institute of Development Studies, Brighton. 6pp

Kaplinsky R, Morris M, Readman J (2002) The globalization of product markets and immiserizing growth: lessons from the South African furniture industry. World Dev 30(7):1159–1177

KPMG International (2013) The agricultural and food value chain: Entering a new era of cooperation. https://assets.kpmg.com/content/dam/kpmg/pdf/2013/06/agricultural-and-food-value-chain-v2.pdf. 25 Sept 2018

LI Y, Wang X, Westlund H, Liu Y (2015) Physical capital, human capital, and social capital: the changing roles in China's economic growth. Growth and Change 46(1):133–149. file:///C:/Users/Public/Documents/Downloads/grow12084publication.pdf. 27 Sept 2018

Maertens M, Swinnen JFM (2012) Gender and modern supply chains in developing countries. J Dev Stud 48(10):1412–1430

Maestre M, Poole N, Henson S (2017) Assessing food value chain pathways, linkages and impacts for better nutrition of vulnerable groups. Food Policy 68:31–39

MAFAP (2013) Review of food and agricultural policies in the United Republic of Tanzania. Monitoring African Food and Agricultural Policies Country Report Series. Food and Agriculture Organization of the United Nations, Rome, Italy. 42pp

Meaton J, Abebe B, Wood AP (2015) Forest spice development: the use of value chain analysis to identify opportunities for the sustainable development of Ethiopian Cardamom (Korerima). Sustain Dev 23(1):1–15

Milberg W, Winkler D (2011) Economic and social upgrading in global production networks: problems of theory and measurement. Int Labour Rev 150(4):341–365

Mnenwa R, Maliti E (2010) A comparative analysis of poverty incidence in farming systems of Tanzania. Special Paper No.10/4REPOA, Dar-es-Salaam, Tanzania

Mnimbo TS, Lyimo-Macha J, Urassa JK, Mahoo HF, Tumbo SD, Graef F (2017) Influence of gender on roles, choices of crop types and value chain upgrading strategies in semi-arid and sub-humid Tanzania. Food Sec 2017:1–15

Morgan T, Williams LA, Gott M (2017) A feminist quality appraisal tool: exposing gender bias and gender inequities in health research. Crit Public Health 27(2):263–274

Ngome AN (2003) Gender division of labour and women's decision- making power in rural households: the Case of Mbalangi, Ediki and Mabonji Villages of Meme Division. Dissertation for Award of M.Sc. at University of Buea, 71pp

Njuki J, Kaaria S, Chamunorwa A, Chiuri W (2011) Linking smallholder farmers to markets, gender and intra-household dynamics: does the choice of commodity matter? Eur J Dev Res 23(3):426–443

Norell D, Janoch E, Kaganzi E, Tolat M, Lynn ML, Riley EC (2016) Value chain development with the extremely poor: evidence and lessons from CARE, Save the Children, and World Vision. Enterp Dev Microfinanc 28(2):44–62

Oduol JBA, Mithöfer D, Place F, Nang'ole E, Olwande J, Kirimi L, Mathenge M (2017) Women's participation in high value agricultural commodity chains in Kenya: strategies for closing the gender gap. J Rural Stud 50:228–239

Palacios-Lopez A, Christiaensen L, Kilic T (2017) How much of the labor in African agriculture is provided by women? Food Policy 67:52–63

Paul P, Meena BS (2016) A study on access to and control over resources; gender perspective. Int J Sci Envir Technol 5(5):2982–2988

PMO-RALG/TOA (2012). Best practices hand book. https://www.jica.go.jp/tanzania/english/activities/c8h0vm00008doll0-att/handbook_en.pdf. 25 Sept 2018

Pretty J, Bharucha ZP (2014) Sustainable intensification in agricultural systems. Ann Bot 114(8):1571–1596

Quisumbing AR, Rubin D, Manfre C, Waithanji E, van den Bold M, Olney D, Meinzen-Dick R (2014) Closing the gender asset gap – Learning from value chain development in Africa and Asia. Discussion Paper No. 01321. International Food Policy Research Institute, Washington, DC. 32pp

Rubin D, Manfre C, Barrett K (2009) Promoting gender equitable opportunities in agricultural value chains handbook. United States Agency for International Development, Washington, DC. 31pp

Seymour CW, Liu VX, Iwashyna TJ, Brunkhorst FM, Rea TD, Scherag A, Rubenfeld G, Kahn JM, Shankar-Hari M, Singer M, Deutschman CS (2016) Assessment of clinical criteria for sepsis: for the Third International Consensus Definitions for Sepsis and Septic Shock (Sepsis-3). JAMA 315(8):762–774

Shackleton S, Paumgarten F, Kassa H, Husselman M, Zida M (2011) Opportunities for enhancing poor women's socioeconomic empowerment in the value chains of three African non-timber forest products. Int For Rev 13(2):136–151

Sheck R, Donovan J, Stoian D (Eds.) (2013) Assessing the impacts of value chain development on poverty: a case study companion to the 5Capitals tool. Technical Report No. 396, Rural Enterprise Development Collection No. 8. CATIE: ICRAF: Bioversity International, pp. 63

UNDP (United Nations Development Programme) (2017) Guidance note; application of the sustainable livelihoods framework in development projects. http://www.latinamerica.undp.org/content/dam/rblac/docs/Research%20and%20Publications/Poverty%20Reduction/UNDP_RBLAC_Livelihoods%20Guidance%20Note_EN-210July2017.pdf. 27 Sept 2018

Voora and Venema (2008) The natural capital approach a concept paper. International Institute for Sustainable Development (IISD). https://www.iisd.org/pdf/2008/natural_capital_approach.pdf. 27 Sept 2018

Impacts of Climate Change on Small Holder Households in Mt. Elgon Region of Uganda: Does Gender Matter?

Nabanoga N. Gorettie, Namaalwa J. Justine, and Bomuhangi Allan

Abstract Agriculture, the main livelihood activity for several communities in Uganda, is threatened by short- and long-term changes in temperatures and precipitation. The increasing involvement of women in agriculture has attracted a myriad of gender-climate studies. However, much of the focus has concentrated on the usual gender dichotomy that assumes homogeneity within gender identities. This study is based on the premise that while an evaluation of male- and female-headed households is important, it only forms an initial stride in understanding climate change impacts' and adaptation analyses. Using focus group discussions and household surveys, this study unveils the climate change adaptation dynamics created by the different positions that men and women hold across defined household typologies in the Mt. Elgon region. The results indicate that male divorced/separated/widowed households are more impacted by crop failure than female divorced/separated/widowed households. Across the households, adult male and female vulnerability was attributed mainly to a limited asset ownership portfolio. Due to ownership of more diverse assets, coupled households had more flexibility to engage in a number of adaptive/coping strategies compared to any other household type. Women in coupled households were also more likely to adapt to crop failure compared to women in other household types, given that they have some access to and use rights of their spouse's assets. Given these dynamics, it is concluded that issues of gender and climate change are multifaceted and that meaningful design and implementation of adaptation strategies should not view "male," "female," and "household" as homogeneous categories but rather recognize their variation in adaptation process.

Keywords Climate change · Gender · Household type · Adaptation · Crop failure · Mt. Elgon · Uganda

---

N. N. Gorettie (✉) · N. J. Justine · B. Allan
School of Forestry, Environmental and Geographical Sciences, Makerere University, Kampala, Uganda
e-mail: nabanoga@caes.mak.ac.ug; namaalwa@caes.mak.ac.ug

Y. Bamutaze et al. (eds.), *Agriculture and Ecosystem Resilience in Sub Saharan Africa*, Climate Change Management,
https://doi.org/10.1007/978-3-030-12974-3_30

1 Introduction

Agriculture continues to be the major contributor to Uganda's economy in terms of the country's gross domestic product (GDP) (UBOS 2017). It provides about 23.7% of the total GDP, employs about 72% of the labor force, and accounts for 43.2% of the country's total export (UBOS 2017). There are over 5.2 million agricultural households in the country, over 97% of whom are engaged in smallholder subsistence farming of mixed crop and livestock production (Okonya et al. 2013; NEPAD 2015; UBOS 2017).

Uganda has, however, over that last decades experienced climate-induced shocks such as droughts, floods, and mudslides in several regions of the country (Markandya et al. 2015). The impacts of climate change and variability are barriers to development, with immense affects to agricultural production, food security, and livelihoods in sub-Saharan Africa (e.g., Deressa et al. 2011; Nhemachena 2009). The Mt. Elgon region has experienced extreme climate change, with the length of the rainfall seasons varying, with some seasons becoming shorter and heavier, while others become drier and recording warmer temperature with more hot days (Mbogga 2012; MWE 2013; Bomuhangi et al. 2016). The impacts of climate change have created several challenges and imposed severe losses and hardships on the poorest communities whose livelihoods are largely dependent on rain-fed agricultural practices, thus leaving most of the households in precarious situations (e.g., Okonya et al. 2013).

Scholarly investigations with general focus on climate change effects, as well as gender interests, have increased, with the latter prompted by the observations of women's involvement in agricultural production. However, several studies on gender in the realm of climate change have provided sex disaggregated information with regard to vulnerability (e.g., Kisauzi et al. 2012; Jost et al. 2016; Denton 2002; Kakota et al. 2011) and adaptation (e.g., Perez et al. 2015; Denton 2002). These studies assume that the binary categories male and female are homogeneous within a community, while some suggest that with climate-induced shocks, women are more likely to be disproportionately affected (e.g., Twyman et al. 2014; Jost et al. 2015). Other studies have considered the characteristics of the household heads that influence the choice of adaptation strategies (e.g., Bomuhangi et al. 2016; Masuku and Manyatsi 2014). There is a paucity of literature on the nature of the household with regard to the lead decision-maker in adapting to climate change. Some of the available literature has investigated male- versus female-headed households in climate variability adaption (Nabikolo et al. 2012; Ngigi et al. 2017; Van Aelst and Holvoet 2015).

We postulate that, while an evaluation of male-headed and female-headed households is important, it is a partial contribution to the discussion of climate change impacts and adaptation strategies. We find it informative to interrogate the dynamics created by the different positions men and women hold in the different types of households. Thus, this study demonstrates that it is important to consider the heterogeneity that exists among "women," "men," or "households headed by different types of individuals," given that these constitute individuals with differentiated levels of access to climate change adaptation strategies.

This study aimed at contributing to the gender-climate information by understanding the climate change adaptation dynamics created by the different positions men and women hold across defined household typologies of small scale farmers in the Mt. Elgon region.

1.1 Past Studies on Farmer's Perceptions on Climate Change in the Mt. Elgon Region

Globally, it has been argued by various scholars that climate change or risk, knowledge, and experience are important factors at the individual and household level in determining whether and how climate change adaptation takes place (Ferguson and Bargh 2004; Adger et al. 2009). O'Brien et al. (2007) reported that in situations where no changes or risk is perceived, there is little or no adaptation action undertaken Ferguson and Bargh (2004), and Adger et al. (2009) articulate that choices are shaped by whether local impacts are known or anticipated and by the cognitive–behavioral gap that exists in individuals between knowledge of impacts, values, beliefs, norms, and action.

Small holder farmers in the Mt. Elgon region, eastern Uganda, have over time noticed changes in their climate. According Mubangizi et al. (2018), farmers reported that the agricultural seasons were reportedly becoming shorter due to late onset and early cessation of rains. They further observe that the amount of rain was also considered more than normal and was concentrated in a short period at the beginning or end of the agricultural seasons. Similar results were reported by Mugagga (2017) who noted that farmers generally perceived rainfall and temperature to have moderately increased, while winds and droughts were noted to have significantly increased.

Despite the general acknowledgment that climate in the Mt. Elgon region was variable, small holder farmers had diverging opinions regarding the trends of climate change in their community. A study by Bomuhangi et al. (2016) revealed that despite both the male and female farmers having identical perceptions of changes in climate indicated by the increase in temperature and rainfall, there were variations on the exact nature and magnitude of change across gender. While the male and female farmers aged between 15 and 29 years were more confident in reporting observed changes in climate, their counterparts aged 30 years and above expressed skepticism to the concept of climate change given their experience in practicing farming amidst rainfall variability over many years. Women farmers in the region more often reported variability in climate, as opposed to their male counterparts, and this was mainly attributed to their daily engagement in farming activities (Bomuhangi et al. 2016).

1.2 Women and Men's Roles and Responsibilities in Mt. Elgon, Uganda

Globally, addressing the threat of climate change has emerged as a key priority, and in this context debates regarding identification of gender perspectives in climate change adaptation have arisen (Carvajal et al. 2008). However, insufficient attention has been paid to understanding the contexts, roles, and responsibilities of men and women that shape climate change adaptation. This failure not only generates concerns in terms of respect for gender equity; it also leads to shortcomings in the efficiency and efficacy of climate-related adaptation. To address this, there is therefore a need to understand the rules that determine gender identity.

In the most of the communities of the Mt. Elgon region, cultural rules determine the gender identity, expression, and roles of its members (Otiso 2006). The behavioral patterns and activities carried out by women and men, young and old, and rich and poor in this region therefore differ. This is due to the social characteristics of the society as shaped by the gender relations in the region. The Bagisu and Sabiny, who form the largest ethnicities in the region, are mostly patrilineal; and women relocate to their husband's family dwelling at marriage. When marriage is constituted, the man will often be the breadwinner/household head and often takes decisions on behalf of the household. In the event of loss, separation or divorce of any of the spouses, the household would then be headed by the surviving member. Women are only heads of the household when the man is away for a long period of the year or when widowed (Balikoowa et al. 2018). There are also households headed by individual males or females that have never been married (Otiso 2006).

In the Gisu and Sabiny culture, women are often responsible for sustaining the household's welfare by ensuring that the household is food secure (Otiso 2006). Because of this distinctive role, women with the help of their children are always engaged in reproductive chores, while men take on more productive roles.

2 Study Area and Methodology

The study was carried out in Kapchorwa and Manafwa districts in the Mt. Elgon region of eastern Uganda. These districts are characterized by mountainous terrain which influences its climate (NEMA 2008). The region has a bimodal rainfall pattern (Mbogga 2012) with subsistence agriculture and livestock farming as the major occupations (MFEP 2014). The districts are also characterized by poor infrastructure and basic services delivery, as well as natural resource degradation (Mbogga 2012; MWE 2013). The districts of Kapchorwa and Manafwa and the focus parishes (Kapnarwaba, Chemangang, Maalo, and Bunasambi) were selected purposively based on their acute vulnerability to climate change (MWE 2013).

Focus group discussions (FGD) and a household survey were used to elicit information on the impact of climate change shocks on small holder farmers in different types of households and how they have adjusted their farming practices to cope with the changes in climate and whether the adopted strategies are gendered. A total of eight focus group discussions, two in each parish, disaggregated by male and females were conducted in the parishes. The FGD sessions had an average of 12 persons each, representing the youth, elderly, single, divorced/separated, or widows. A checklist was used to generate discussions and solicit information from the group participants.

Four hundred and twenty (420) households were randomly selected from a list of households in the study parishes for interviews. Within each household, data was collected from any adult (above 18 years) available at the time of the visit.

3 Data Analysis

The data was disaggregated by household type, based on the marital status of the household head at the time of the investigation. The reported effects of climate change were ranked and the most prominent considered for subsequent analyses across the household types. This study considered the following household categories: (i) the single male-headed (with individual never married), (ii) male-headed separated or widower, (iii) male-headed couple/married or cohabiting, (iv) single female-headed (with individual never married), and (v) female-headed divorced or widowed. Cross-tabulations were used to generate Pearson chi-square ($\chi 2$) values to test the association between the household types and the severity of climate change impacts, reasons for household vulnerabilities, household coping strategies, and type of assets/ownership of asset used to cope/adapt to crop failure.

4 Results and Discussion

4.1 Climate Change Impacts on Different Household Types

Crop failure, increased price of food products, decreased animal production, and decreased family health were the most commonly experienced impacts of climate change by both men and women. It is however observed that while crop failure and decreased animal production are primary level impacts, increased price of food products and decreased family health are secondary level impacts. The fact that increased price of food products was reported with primary level effects could explain the magnitude of the effect on the communities in the study area. While the incidence of decreased family health was reported, it was more commonly reported by the female respondents (Table 1). The least reported impacts by both men and women farmers included loss of wage employment and increased wildlife predation. There was no statistical variation between the impacts of climate change

Table 1 Climate change impacts on smallholder farmers

Climate impacts	Male (%)	Female (%)	$\chi 2$ value	*P*-value
Primary level impacts				
Crop failure	100	100		
Loss of farming land	10	8	0.581	0.276
Decreased animal production	50	49	0.039	0.460
Increased wildlife predation	3	4	0.285	0.385
Secondary level impacts				
Decreased family health	55	60	2.232	0.327
Loss of wage employment	7	4	1.076	0.299
Increase price of food products	100	99	1.049	0.305
Poor access to markets	27	26	0.146	0.701

reported by male and female respondents. The findings provide an alternative and perhaps a less frequently documented opinion that the impacts of climate change may be similar for men in more homogeneous environments.

To further understand, if the impacts of climate change were gendered, data was disaggregated by the types of household. The results reveal that across all the household types, crop failure and increased price of food products are the predominantly experienced impacts (Table 2). The results also indicate that female and male divorced/separated or widowed households are more vulnerable to crop failure impacts in comparison to the single and coupled households. When female and male divorced/separated or widowed household are compared, results indicate that male divorced/separated or widowed households are more impacted by changes in climate than female divorced/separated or widowed households. The results also reveal that across all the household categories, the single male and female households were least impacted. This could be attributed to the fact that single men and women command several resources on their own, compared to those that are either married or widowed individuals.

For further analysis, crop failure as the most experienced primary level effect was considered. Crop failure was defined as a situation whereby farmers experienced greatly diminished crop yield relative to their expectation (Challinor et al. 2014). The participants in the FGDs reported that crop failure was mainly caused by adverse weather conditions which in this case were inadequate and poorly distributed rainfall as well as high temperatures. The result of crop failure affects farmers' income and also decreases the amount of food available for consumption. At the secondary level, crop failure negatively affect household economy(ies) in the selected parishes. This seems to confirm what other scholars (e.g., Mubangizi et al. 2018) reported for the Mt. Elgon region, where farmers perceived that the shocks associated with changes in rainfall patterns resulted into reduced crop and livestock production as well as depleting the livelihood assets on which they depended.

While all households reported crop failure as a climatic impact, households registered differences in their perception of the severity of impact and in the category of individuals within the households that were affected most by the impact. With the

Table 2 Climate change impacts by household type

Impact	Type of household				
	Male single ($n = 52$)	Female single ($n = 39$)	Coupled (married/cohabiting) ($n = 255$)	Female divorced/separated/widowed ($n = 34$)	Male divorced/widowed/separated ($n = 40$)
Primary level impacts					
Crop failure	100	100	100	100	100
Loss of farming land	8	5	10	3	15
Decreased animal production	42	23	50	65	68
Increased wildlife predation	2	3	4	3	0
Secondary level impacts					
Decreased family health	56	51	56	65	74
Loss of wage employment	8	6	4	16	25
Increase price of food products	98	97	100	79	100
Poor access to markets	21	21	24	38	47

Table 3 Perceived vulnerability of community members to crop failure

Household type	Level of effect	Most affected (%)				$\chi 2$ value	*P*-value
		Adult male	Adult female	Youth	Elderly		
Single male ($\underline{n}$ = 32	Somewhat severe	25	9	3		3.793	0.705
	Severe	25	13	6	3		
	Very severe	6	6		3		
Single female (*n* = 25)	Somewhat severe	44	4			11.355	0.023***
	Severe	20	8	4			
	Very severe	4	16				
Coupled(married/ cohabiting) (*n* = 169)	Somewhat severe	30	13	1		15.967	0.014***
	Severe	23	8	4	1		
	Very severe	12	4	2	2		
Divorced female (*n* = 25)	Somewhat severe	20	28	4		6.617	0.358
	Severe	16	8	4			
	Very severe	12	4		4		
Divorced male (*n* = 27)	Somewhat severe	30	7		4	10.861	0.093**
	Severe	19	15	4	7		
	Very severe		15				

***Significant at $p < 0.05$; ** Significant at $p < 0.1$

exception of single female-headed household category, male adults were reported to be impacted the most (Table 3), which could largely be associated with the patriarchal nature of the society in which most of men bear the burden of bread weaning. Within the category of single-headed households, the respondents perceived the adult female to be more vulnerable, while male singles perceived the male adults to be the most vulnerable.

Considering all household types, males (61%) were more affected by crop failure, while the elderly (4%) were the least affected. Among the single male households, both males and females (each 6%) were severely affected. Also, among the single female and divorced male households (16% and 15%, respectively), females were reported to be the most severely affected. However, among the divorced households, men were reported to be the ones very severely affected. The results suggest that vulnerability to climate change may not be straightforward and that the line between who is bound to be more vulnerable and therefore impacted most is not clear. However, across all the household types, the elderly persons were considered the least vulnerable, given the fact that they are often dependent on the male and female heads of households and also receive support from their children and relatives outside the household. They thus experience secondary effects. This conforms to what other scholars (e.g., Mutongi 1999) reported for Kenya, where elderly parents, widows, and widowers usually perceived it as their right to receive support from their children, both financially and in kind. It can therefore be assumed that this entitlement for support from other relatives, specifically children, can partially cover the would-be vulnerability gap. This coping/adaptive strategy is unique to the elderly in the African sociocultural context. These findings suggest that when understanding climate change impacts, studies should not start with preconceived connections between identities (men and women), household type, and vulnerability as generalizing identities may hinder the ability to address the climate-related needs of the most vulnerable in each peculiar context.

To understand the causes of gendered vulnerability to climate impacts, the reasons for perceived vulnerability across the different household types were analyzed. The study found that reasons for the perceived vulnerability for male, female adults, and the elderly within the different household types differ (Table 4).

Across all household types, male and female adults' vulnerability was attributed mainly to limited assets (34% & 16%) and more male responsibilities (18% & 10%), while the vulnerability of the elderly was mainly attributed to their health status (4%). The results also revealed a significant difference between household types and the reasons sighted for their vulnerabilities ($\chi 2$ = 137.535, df = 9, P = 0.000). The single males had very limited livelihood options, and this could be attributed to the fact that farming is the major livelihood activity in the area, with limited alternative options. This finding further seems to link perceived vulnerability to climate impacts with the different gender roles.

4.2 Gendered Adaptation Strategies to Crop Failure

Given the perceived impacts and vulnerabilities, the coping and adaptive strategies to crop failure were investigated. The findings revealed that men and women farmers demonstrated different preferences for adaptation to crop failure. Overall, the adaptive/coping strategies identified included horizontal diversification of crops and livestock (43%), spending cash savings on food items (17%), selling off assets (17%), obtaining support from friends/relatives or loans from financial institutions (10%), seeking alternative employment such as providing wage labor or trading (10%), and reducing the number of meals consumed per day (1%). Further, the findings revealed that men adapted by diversifying crops and livestock farming practices (21%) and disposing of assets (10%), while women mainly adapted by diversifying crops and livestock farming practices (22%) and spending cash savings on food items. However, the chi-square test did not reveal any significant differences between the coping strategies of both men and women ($\chi 2$ = 7.400, df = 6, P = 0.285). This is contrary to what other studies found out. For example, Ngigi et al. (2017) and Kristjanson et al. (2015) reported significant differences in the adaptation strategies between men and women, which was due not only to differences in their perception of change in climate but also differences in access to and control over resources and decision-making authority.

Table 4 Reasons for perceived vulnerability to crop failure due to climate change

Type of household	Most affected	Reason for perceived vulnerability				χ2 value	*P-value*
		Limited assets	Men have more responsibilities	Limited livelihood options	Poor health		
Male single	Adult male	38	16	3		37.926	0.000***
	Adult female	19	6	3			
	Youth		6	3			
	Elderly				6		
Female single	Adult male	36	24	8		4.643	0.326
	Adult female	24	4				
	Youth		4				
Coupled (married/cohabiting)	Adult male	36	20	7	2	62.265	0.000***
	Adult female	13	8	2	2		
	Youth	2	1	2	1		
	Elderly			1	3		
Female divorced	Adult male	28	8	12		34.812	0.000***
	Adult female	12	28				
	Youth	4		4			
	Elderly				4		
Male divorced	Adult male	19	11	19		24.673	0.003***
	Adult female	22	15				
	Youth	4					
	Elderly			4	7		

***Significant at $p < 0.05$

The study further revealed that adaptation strategies may differ depending on the household type (Table 5). Despite these differences, diversification of crop and livestock is the most predominant strategy taken on by all, regardless of the household type.

Coupled (married/cohabiting) households were observed to have more flexibility to engage in a number of adaptive/coping strategies compared to any other household type. Divorced/separated female households were found to have the least coping strategies. The single women, like the divorced/widows, also depended mostly on diversifying crops and livestock farming practices and to a lesser extent received support from their relatives as well as financial institutions, which helped their

Table 5 Adaptation to crop failure by type of HH

Crop failure coping strategy	Type of household (%) Male single ($n = 48$)	Female single ($n = 38$)	Coupled (married/cohabiting) ($n = 236$)	Female divorced ($n = 34$)	Male divorced ($n = 38$)
Diversified crops and livestock farming practices	6	4	25	4	5
Spend cash savings on food items	2	2	10	1	1
Sell assets	3	1	10	1	1
Obtain support from friends or relatives or loan from financial institution	1	2	5	1	2
Seek alternative employment	1	2	6	1	1
Reduced number of meals per day			1		
Do nothing			2		

capacity to cope with crop failure. This could be explained by the fact the patriarchal nature of this society limits women's access to resources, given that women majorly access resources based on their relationship with their male counterparts. This means that women will most likely lose their access to resources in the event of divorce or separation from husbands. This finding is in line with the study in Tanzania by Van Aelst and Holvoet (2016) who reported that female divorcees and widows were most likely to face challenges in accessing valley land to cope with climate change because they could not depend on a husband to secure their land rights. Although women are the ones who mostly implement farm work in any household, the divorcees and widowed females lack independent access to or control over the household resources. The male mostly remained the legal owners of land and other household production resources.

4.3 *Assets Used for Adaptation to Crop Failure*

The study also explored the assets used to cope with crop failure resulting from climate change in the different households. The findings revealed that both male and female respondents used mostly livestock (large and small) as well as trees/woodlots to cope with crop failure. The large livestock included cattle and donkeys, small livestock included goats, sheep, and poultry, while communication equipment included mobile telephones and radios. Men were more likely to dispose of large livestock and trees, while women were more likely to dispose of plots of land,

Table 6 Assets used in coping with crop failure

Sex of respondent	Asset used for crop failure grouped	Owner of the asset(s) used to cope with crop failure				Chi-square test	
		Myself	My spouse	Jointly with spouse	Jointly with other household members	$\chi2$ value	*P*-value
Male	Large livestock ($n = 45$)	62	4	31	2		
	Small livestock ($n = 47$)	51	30	19			
	Trees/woodlots ($n = 21$)	43	10	43	5	22.983	0.084
	House furniture ($n = 9$)	44	33	22			
	Communication equipment ($n = 4$)	25		75			
	Plot of land ($n = 3$)	67		33			
Female	Large livestock ($n = 35$)	77		23			
	Small livestock ($n = 54$)	54	19	28		24.651	0.006
	Trees/woodlots ($n = 19$)	58	11	32			
	House furniture ($n = 12$)	50	8	42			
	Communication equipment ($n = 6$)		33	67			
	Plot of land ($n = 8$)	100					

communication equipment, and house furniture. Further, the owner of the assets used to cope with crop failure was also considered (Table 6).

Generally, there was a significant difference ($\chi2 = 38.966$, $df = 15$, $P = 0.001$) in the ownership of assets used by men and women in coping with crop failure. Although use of assets jointly owned with other household members was rarely reported, it was interesting to note that none of the female respondents reported using assets jointly owned with other household members to cope with crop failure. This was most likely due to the sociocultural norm in the study area, which limits women's joint ownership of assets to mainly males and usually their husbands. The relationship between asset ownership and type of assets used for crop failure was very significant in the females category ($\chi2 = 24.651$, $df = 15$, $P = 0.0024.651$) compared to that of the males ($\chi2 = 22.983$, $df = 15$, $P = 0.084$). The study also revealed that women did not report on using large livestock solely owned by males, while males reported using large livestock that were owned by their spouses. This could be explained by the fact that most of the women do not own land and therefore rear their livestock on male owned land. This, thus, could entitle males to claim "ownership" of the livestock by virtue of owning the land on which women's animals graze. This finding reinforces the findings of Bomuhangi et al. (2011) and Doss et al. (2012) who found that land in Uganda is owned by men and that women gain use rights by virtue of their relationship with the men. They further found that both men and women report a relatively high degree of joint ownership of land, even though women's names are rarely on the documents, and women may lose rights to land if their marriage dissolved. They also reported that women had fewer recognized decision-making rights than men, especially for alienation of (to sell, bequeath, or rent) land.

The results also suggest that more men, as compared to their women counterparts, reported more frequently using their spouses' asset to cope with crop failure. Further, none of the female respondents reported sole ownership of communication assets although a good proportion reported their use as a coping strategy. This could be attributed to the fact that most of the radios and telephone sets in the household are owned by the males and rarely do women own such assets. This finding reinforces the findings of Doss et al. (2012), who revealed the asset ownership portfolio of males and females in households in Uganda. This finding suggests that within coupled households, women are more likely to adapt to climate change impacts compared to women in other household types given that they have some access and use rights of their spouses assets.

The assets used by each types of household to cope with crop failure were also explored and are presented in Table 7. Coupled households, where you find both husband and spouse, compare to all other household types, were found to have more diverse assets in coping with crop failure.

Single male and divorced female households did not report having used land to cope with crop failure. This could be attributed to that fact that single/unmarried males are less likely to own land, while divorced females are more likely to lose land when the relationship/marriage with their husbands is terminated or dissolved, as land is more often owned by the men in this society. A relatively large proportion of single female households (27%) reported using land to cope with crop failure. This could be attributed to the fact that single women, though few, can independently

Table 7 Assets used for adaptation/cope with crop failure by household type

Assets used to cope with crop failure	Type of household (%)				
	Male single ($n = 31$)	Female single ($n = 25$)	Married/ cohabiting ($n = 163$)	Female divorced ($n = 24$)	Male divorced ($n = 26$)
Large livestock	12		72	7	8
Small livestock	13	10	57	11	10
Trees/woodlots	18	15	45	13	10
Household furniture	9	14	64	5	9
Communication equipment	10	10	60	10	10
Plot of land		27	55		18

own land especially through individual purchase (Doss et al. 2012). It could also be further explained by the fact that since single women have limited inheritance land rights from their parents in this society, able single females acquire land mainly through purchase for their independent use as means to guard against any vulnerability. These findings conform to those of Van Aelst and Holvoet (2016) who found that while women in female-headed households often have to endure the problem of a limited asset base, women in male-headed ones have less access to control over the assets in the household. They further note that female-headed households are occasionally able to reap the benefits of their greater independence.

There were no reports of use of large livestock in single female household types in coping with crop failure. This could be mainly attributed to the fact that most of the cattle are owned by men and women in this society, while the youthful male and female/not yet married commonly own the small livestock which include goats, sheep, and poultry. Respondents (18%) in divorced male-headed households reported using plots of land as the main asset to cope with crop failure, while 18% of single male-headed households mainly used trees/woodlots. It was also noted that female divorced households reported using trees/woodlots (13%) as well as small livestock (10%) to cope with crop failure. Cross-tabulations were used to generate Pearson chi-square ($\chi 2$) values which were used to test any association between the household types and assets used to cope/adapt to crop failure. The analysis revealed no significant association ($\chi 2 = 20.334$; $df = 20$; $P = 0.437$).

4.4 Challenges of Adaptation to Crop Failure

The challenges reported by respondents included inadequate financial resources to purchase inputs and labor (50%), population increase (22%), shortage of household labor (18%), poor quality seeds (5%), and limited awareness on copping strategies (4%). Considering the household types, inadequate financial resources and population increase were the most frequently reported challenges by single male, single female, and divorced female households (Table 8). Coupled households and

Table 8 Challenges in coping/adapting to crop failure

Challenge in coping/ adapting to crop failure	Type of household (%)					Total
	Male single	Female single	Married/ cohabiting	Female divorced	Male divorced	
Inadequate financial resources to purchase inputs and labor	38	49	55	47	43	50
Population increase	44	23	18	24	20	22
Shortage of household labor	13	18	19	18	25	18
Poor quality seeds	2	10	4	6	10	5
Limited awareness on coping strategies	2		4	6	3	4

divorced male households reported inadequate financial resources and shortage of household labor as the major challenge. The analysis also revealed a relatively significant relationship ($\chi2 = 25.657$; $df = 16$; $P = 0.059$) at 10% significance level, for challenges encountered by the different household types.

5 Conclusions

This study reveals the interaction between the decision-maker in a given household type and reports impacts and adaptation strategies to the impacts of climate change. Crop failure is used to demonstrate the adaptation strategies and assets used in the coping strategies. The household types depicted varying coping strategies, vulnerability levels of different individuals within the community and assets used to cope to crop failure.

Crop failure and decreased animal production were the primary level impacts, while increased price of food products and decreased family health are secondary level impacts most reported. The study has revealed that although both female and male divorced/separated or widowed households were more vulnerable to crop failure impacts, male divorced/separated or widowed households were found to be more impacted than female divorced/separated or widowed households. Across all the household types, adult males were more affected by crop failure, while the elderly were the least affected. Across all of household types, male and female adults' vulnerability was attributed to mainly limited asset and more male responsibilities, while the elderly attributed their vulnerability to poor health. Diversification of crop and livestock was the most predominant strategy taken on by all regardless of the household type. Households mostly used livestock (large and small) trees/woodlots plots of land to cope with crop failure. Coupled households, compare to all other household types, were found to have more diverse assets in coping with crop failure. Single women (whether married or divorce/widows) depend mostly on diversifying crops and livestock farming practices and to a lesser extent received support from their relatives as well as financial institutions which helped their capacity to cope with crop failure. The assets used were either individually owned, jointly owned with spouse, or very few jointly owned with other household members. None of the female respondents reported using assets jointly owned with other household members to cope with crop failure. Coupled households and divorced male households had inadequate financial resources and shortage of household labor as most frequently reported challenges.

In this article, it is highlighted that the household type in which farmers belong had great influence on how they are impacted by effects of climate variability as well as their coping/adaptive strategies. For instance, the single male, never married; single male, either divorced, separated, or widowed; single female, never married; single female, either divorced, separated, or widowed; and married or cohabiting husband and wife living together are each impacted differently and each encounter different challenges and opportunities as they attempt to cope/adapt to climate variability. It is therefore important to note that, male divorced households were more vulnerable to crop failure impacts than female divorced households; across all the household types, adult males were more affected by crop failure, while the elderly were the least affected; male and female adults' vulnerability was attributed to mainly limited asset and more male responsibilities, while the elderly attributed their vulnerability to poor health; coupled households have more diverse assets in coping with crop failure; women never used assets jointly owned with other members of the household. It is therefore concluded that the issues of gender and climate change are multifaceted; and therefore, we should not view "male," "female," and "household" as homogeneous categories but rather recognize their variation in adaptive capacity.

These findings are relevant for informing the growing need for gender mainstreaming in national climate change policies and implementation plans. It therefore guards against the traditional limited view of gender mainstreaming that only deals with sex disaggregation, which may lead to the development of ineffective policies that will further marginalize of some categories of women and men.

References

Adger WN, Dessai S, Goulden M, Hulme M, Lorenzoni I, Nelson DR, Naess LO, Wolf J, Wreford A (2009) Are there social limits to adaptation to climate change? Clim Change 93:335–354

Balikoowa K, Nabanoga G, Tumusiime DM (2018) Gender stereotyping: Evidence from gender differentiated household vulnerability to climate change in Eastern Uganda. Cogent Environ Sci 4(1):1512838

Bomuhangi A, Nabanoga G, Namaalwa JJ, Jacobson MG, Abwoli B (2016) Local communities' perceptions of climate variability in the Mt. Elgon region, eastern Uganda. Cogent Environ Sci 2(1):1168276

Bomuhangi A, Doss C, Meinzen-Dick R (2011) Who owns the land?: perspectives from rural Ugandans and implications for land acquisitions (No. 1136). International Food Policy Research Institute (IFPRI)

Carvajal-Escobar Y, Quintero-Angel M, García-Vargas M (2008) Women's role in adapting to climate change and variability. Adv Geosci 14:277–280

Challinor AJ, Watson J, Lobell DB, Howden SM, Smith DR, Chhetri N (2014) A meta-analysis of crop yield under climate change and adaptation. Nat Clim Change 4(4):287

Deressa TT, Hassan RM, Ringler C (2011) Perception of and adaptation to climate change by farmers in the Nile basin of Ethiopia. J Agric Sci 149(1):23–31

Doss C, Truong M, Nabanoga G, Namaalwa J (2012) Women, marriage and asset inheritance in Uganda. Dev Policy Rev 30(5):597–616

Denton F (2002) Climate change vulnerability, impacts, and adaptation: Why does gender matter? Gend Dev 10(2):10–20

Ferguson MJ, Bargh JA (2004) How social perception can automatically influence behavior. Trends Cogn Sci 8:33–39

Jost C, Kyazze F, Naab J, Neelormi S, Kinyangi J, Zougmore R, Aggarwal P, Bhatta G, Chaudhury M, Tapio-Bistrom ML, Nelson S, Kristjanson P (2015) Understanding gender dimensions of agriculture and climate change in smallholder farming communities. Clim Dev 8(2):133–144. https://doi.org/10.1080/17565529.2015.1050978

Jost C, Kyazze F, Naab J, Neelormi S, Kinyangi J, Zougmore R, Nelson S (2016) Understanding gender dimensions of agriculture and climate change in smallholder farming communities. Clim Dev 8(2):133–144

Kakota T, Nyariki D, Mkwambisi D, Kogi-Makau W (2011) Gender vulnerability to climate variability and household food insecurity. Clim Dev 3(4):298–309

Kisauzi T, Mangheni MN, Sseguya H, Bashaasha B (2012) Gender dimensions of farmers' perceptions and knowledge on climate change in Teso sub-region, eastern Uganda. Afr Crop Sci J 20(2):275–286

Kristjanson P, Bernier Q, Bryan E, Ringler C, Meinzen-Dick RS, Ampaire E, Rufino M (2015) Gender and climate change adaptation in Uganda: insights from Rakai. International Food Policy Research Institute (IFPRI).–Project note, (03)

Ngigi MW, Mueller U, Birner R (2017) Gender differences in climate change adaptation strategies and participation in group-based approaches: An intra-household analysis from rural Kenya. Ecol Econ 138:99–108

Markandya A, Cabot-Venton C, Beucher O (2015) Economic assessment of the impacts of climate change in Uganda. Climate and Development Knowledge Network (CDKN)

Masuku M, Manyatsi A (2014) Factors influencing the choice of climate change adaptation strategies by households: a case of Mpolonjeni Area development Programme (ADP) in Swaziland. J Agric Stud 2(1):86

Mbogga MS (2012) Climate profiles and climate change vulnerability assessment for the Mbale region of Uganda. UNDP Consultancy report. Kampala, Uganda

MFEP (2014) Housing and population census. Ministry of Finance and Economic Planning, Kampala, Uganda

Mubangizi N, Kyazze FB, Mukwaya PI (2018) Smallholder farmers' perception and adaptation to rainfall variability in Mt. Elgon region, eastern Uganda. Int J Agric Ext 5(3):103–117

Mugagga F (2017) Perceptions and response actions of smallholder coffee farmers to climate variability in montane ecosystems. Environ Ecol Res 5(5):357–366

Mutongi K (1999) Worries of the heart: widowed mothers, daughters and masculinities in Maragoli, western Kenya, 1940–60. J Afr Hist 40(1):67–86

MWE (2013) Integrated territorial climate plan 2014–2029 for Mbale region Uganda (Bududa, Manafwa and Mbale district), Ministry of Water and Environment, Kampala, Uganda

Nabikolo D, Bashaasha B, Mangheni M, Majaliwa JGM (2012) Determinants of climate change adaptation among male and female headed farm households in eastern Uganda. Afr Crop Sci J 20(2):203–212

NEMA (2008) State of environment report for Uganda. National Environment Management Authority. Kampala Uganda

NEPAD (2015) New partnership for Africa's development annual report 2015

Nhemachena C (2009) Agriculture and future climate dynamics in Africa: Impacts and adaptation options. Doctoral dissertation, University of Pretoria

O'Brien K, Eriksen S, Nygaard LP, Schjolden A (2007) Why different interpretations of vulnerability matter in climate change discourses. Clim Policy 7(1):73–88

Okonya JS, Syndikus K, Kroschel J (2013) Farmers' perception of and coping strategies to climate change: Evidence from six agro-ecological zones of Uganda. J Agric Sci 5(8):252

Otiso KM (2006) Culture and customs of Uganda. Greenwood Publishing Group

Perez C, Jones EM, Kristjanson P, Cramer L, Thornton PK, Förch W, Barahona C (2015) How resilient are farming households and communities to a changing climate in Africa? A gender-based perspective. Glob Environ Chang 34:95–107. https://doi.org/10.1016/j.gloenvcha.2015.06.003

Twyman J, Green M, Bernier Q, Kristjanson P, Russo S, Tall A, Ampaire E, Nyasimi M, Mango J, McKune S, Mwongera C, Ndourba Y (2014) Gender and climate change perceptions, adaptation strategies, and information needs preliminary results from four sites in Africa. CCAFS working paper no. 83. CGIAR research program on climate change, agriculture and food security (CCAFS). Copenhagen, Denmark. Available online at: www.ccafs.cgiar.org

UBOS (2017) Uganda Bureau of Statistics. Statistical abstract of 2017
Van Aelst K, Holvoet N (2015) Intersections of gender and Marital Status in Accessing Climate Change Adaptation: Evidence from Rural Tanzania. World Dev 79:40–50, 2016 0305-750X/2015 Elsevier. https://doi.org/10.1016/j.worlddev.2015.11.003
Van Aelst K, Holvoet N (2016) Intersections of gender and marital status in accessing climate change adaptation: evidence from rural Tanzania. World Dev 79:40–50

Using Indigenous Knowledge to Enhance Rainfall Forecasts Among Smallholder Farmers in Mt. Elgon Region, Eastern Uganda

Florence Birungi Kyazze, Narisi Mubangizi, Paul Isolo Mukwaya, and Samuel Kyamanywa

Abstract The frequency and severity of uncertain rainfall and climate extremes are projected to increase across many parts of the world. Access to rainfall forecasting information becomes an essential and critical resource that smallholder farmers should use to take advantage of good rains and avoid its adverse effects. In many smallholder farming communities, the reliability and accuracy of the scientific information is questionable and therefore not adequately used to make informed farming decisions. Amidst this dilemma, smallholder farmers rely heavily on indigenous knowledge to comprehend rainfall patterns in their day-to-day and seasonal farming calendar. A study carried out among smallholder farmers in the Mt. Elgon region indicated that a large proportion of farmers used a wide range of indigenous indicators to predict rainfall patterns. The indicators used by farmers were largely celestial objects and/or animal/plant behaviour to forecast onset and cessation of rains. While this is true, the type of indicators used to forecast the rainfall patterns were site specific, made prediction over a short temporal scale (days to a few weeks) and did not provide adequate information on rainfall amount, intensity and distribution which are key parameters for making evidence-based farming decisions.

Keywords Indigenous knowledge · Rainfall forecasts · Farming decisions · Mt. Elgon · Uganda

1 Introduction

Access to rainfall forecasting information is considered essential in enabling farmers to prepare to take advantage of good rains and avoid its adverse effects (Kaggwa et al. 2009; Hansen et al. 2011; Mwatu et al. 2016; Okonya and Kroschel 2013, AGRA 2014). Rainfall forecasting information is available from

---

F. B. Kyazze (✉) · N. Mubangizi · P. I. Mukwaya · S. Kyamanywa
College of Agriculture and Environmental Sciences, Makerere University, Kampala, Uganda
e-mail: fbirungikyazze@caes.mak.ac.ug

Y. Bamutaze et al. (eds.), *Agriculture and Ecosystem Resilience in Sub Saharan Africa*, Climate Change Management,
https://doi.org/10.1007/978-3-030-12974-3_31

both science-based meteorological forecasts and indigenous knowledge systems (Ziervogel and Opere 2010) that are widespread across many parts of the world. Smallholder farmers in sub-Saharan Africa (SSA) and elsewhere in the world have, because of their livelihoods closely associated with the natural environment (United Nations Environment Programme (UNEP) 2007), traditionally relied on indigenous knowledge to comprehend weather and climate patterns that influence their farming decisions (Ziervogel and Opere 2010; Nganzi et al. 2015). This kind of indigenous knowledge is obtained through observing the surrounding natural environment and is often based on cumulative experience handed down from one generation to the next (Pareek and Trivedi 2011).

While it is critical that local, national and international agencies involved in generating and disseminating seasonal forecasts recognize that farmers have a rich indigenous knowledge base for weather forecasting (Nganzi et al. 2015), their forecasts are solely based on scientific knowledge. It must, however, be stated that most of these agencies are largely national in character (Orlove et al. 2010), and therefore most of the weather forecasting information doesn't reach farmers easily. It is further reported that exclusion of indigenous knowledge in the existing weather forecasting systems and structures compromises their acceptability and subsequent utilization by farmers in taking agricultural-related decisions and actions (Masinde and Bagula 2011). Even in circumstances where such scientific forecast information is available, few smallholder farmers tend to integrate and use it in their agricultural decision-making processes (Cooper et al. 2008; AGRA 2014). This study set out to explore the underlying reasons that explain the inadequate integration of indigenous knowledge practices and use of scientific information in making agricultural-related decisions by farmers. In most cases, the scientific information is often reported in complicated formats that cannot be comprehended by smallholder farmers (Cooper et al. 2008; Hansen et al. 2011; AGRA 2014).

2 Methodology

2.1 Description of Study Area

The study was conducted in two districts, Bududa and Manafwa districts, which are located in Mt. Elgon region in Eastern Uganda. Bududa district lies on south-western slopes of Mount Elgon in Eastern Uganda. It is geographically bounded by latitude 2° 49′ North and 2° 55′ North and longitude 34° 15′ East and 34° 34′ East (Knapen et al. 2006). Bududa district has 1 town council (Bududa) and 15 sub-counties, namely, Bududa, Bubiita, Bukibokolo, Bukigai, Bulucheke, Bumayoka, Bukalasi, Bushika, Bumasheti, Bushiribo, Bushiyi, Buwali, Nabweya, Nakatsi and Nalwanza (UBOS 2014). By 2014, the district had a total of 211,683 people, and 97% of them lived in rural areas (UBOS 2014). On the other hand, Manafwa district is located between latitude 01° 01' North and longitude 34° 21′ East, and it has 2 town councils (Manafwa and Lwakhakha) and 28 sub-counties, namely, Bubutu, Bugobero, Bumbo, Bumwoni, Bupoto, Butiru, Buwabwala, Buwagogo, Bukhabusi, Bukhaweka, Bukhofu, Bukiabi, Bukhokho, Bukusu, Bukusuya, Bunabwana, Butta, Kaato, Khabutoola, Magale, Mukoto, Nalondo, Namabya, Namboko, Sibanga, Sisuni, Tsekululu and Weswa (UBOS 2014). The district had a total population of 352,864 people, and about 94% were rural based (UBOS 2014). The majority of the people in the two districts belong to the Gishu tribe whose local language is Lumasaba.

The Gishu people traditionally depend on rain-fed agricultural for their livelihoods. As a result of their dependency on rain-fed agricultural production for their livelihoods, farmers in two districts have accumulated a wealth of indigenous knowledge about their local climatic conditions. Besides the indigenous knowledge that the people possess on their local climatic conditions, Uganda National Meteorological Authority (UNMA) disseminates seasonal climate forecasts, monthly, decadal climate outlooks, daily weather forecasts and warnings about likelihood of climatic disasters through electronic and print media.

2.2 Research Design

The study was conducted from May to August 2016 to assess farmers' awareness and use of indigenous knowledge on forecasting rainfall. The study employed a descriptive cross-sectional design to gather data about farmers' awareness and use of indigenous knowledge for forecasting onset and cessation of rains. The main data collection method was a survey that was complemented by Focus Group Discussions (FGDs) to obtain in-depth understanding about farmers' perceived reliability and use of indigenous knowledge on forecasting onset and cessation of rainfall.

2.3 Sampling and Subject Selection

Bududa and Manafwa districts were selected as case study districts in Mt. Elgon region because the two districts are reported as some of the climate risk hotspots in Uganda. Both districts frequently experience landslides and household food shortages due to rainfall variability. For instance, the Uganda National Agricultural Census of 2008/2009 reported that over 89% and about 80% of the agricultural households from Manafwa and Bududa districts, respectively, experienced food shortage during the month of April 2009 due to rainfall variability (UBOS, 2010).

Within each of the two districts, three sub-counties that were considered by the district local governments to be the most affected by rainfall variability were purposively selected, namely, Bushika, Bukalasi and Bumasheti from Bududa district and Khabutoola, Nalondo and Bugobero from Manafwa district (Fig. 1). The selection

Fig. 1 Location of study sub-counties in Mt. Elgon area, Eastern Uganda

of sub-counties was done with guidance from the respective district local government production coordinators. In each sub-county, three villages were also purposively selected with guidance from the respective local council 1 chairpersons and sub-county agricultural officers. Preference was given to villages that were perceived to have recently been affected by changes in rainfall patterns and their associated shocks.

A total of 12 FGDs (2 in each of the 6 study sub-counties) were conducted. The FGDs involved 123 participants (62 males and 61 females) who were purposively selected based on age, gender and possession of knowledge on indigenous indicators of forecasting rains. In terms of age, preference was given to both male and female farmers aged 40–60 years because these were considered to be more knowledgeable about indigenous indicators of forecasting rainfall patterns. The FGD participants were also identified with guidance from the sub-county agricultural officers and respective village local council chairpersons. On average, seven people were selected from each village to participate in the FGDs that were held at the respective sub-county headquarters.

For households to be involved in the study, the sampling frame comprised of all households in all the 18 selected villages in the 6 sub-counties. Lists of all households in the villages were generated by the respective village chairpersons and three other members of the local council 1 committee. The names of household heads from all the villages were compiled to form a sampling frame of 2124 households. Basing on this sampling frame, the sample size was computed using the formula by Krejcie and Morgan (1970). The calculated sample size was 236. However, to cater for the non-response, the estimated sample size was increased by 10% resulting into a sample size of 255. The number of households drawn from each village was then obtained through proportionate stratified random sampling.

2.4 Instrumentation and Data Collection

Data collection was conducted in two phases. The first phase involved 12 FGDs which were facilitated using a checklist that had been developed with input from two scholars from Makerere University and reviewed by peers for content validity. The FGDs focused on gathering in-depth data on awareness and use of indigenous climate forecast information. The data from the FGDs helped to focus the study, develop the survey tool and provide detailed explanations for the study results.

The second phase involved a survey where semi-structured interviews were administered to 255 farmers. The semi-structured interviews (SSIs) focused on collecting quantitative data on household socio-economic and demographic characteristics and awareness about indigenous indicators for onset and cessation of rains. The interviews also collected data on farmers' perceived reliability of the indigenous indicators for onset and cessation of rains, use of such indicators in making agricultural decisions and challenges faced in the process. In assessing the reliability of the indigenous indicators for forecasting onset and cessation of rains, the study drew insights from the concept of hit rate normally used to assess quality of forecasts over a long time (Sultan et al. 2010). The hit rate refers to the proportion of forecasts that agree with what eventually happens in reality (Sultan et al. 2010). In line with this concept, reliability indigenous indicators were assessed by capturing farmers' perceptions about the extent to which they correctly projected the actual weather conditions they forecasted. The reliability was measured using a rating scale with the following scores: 1 = very low; 2 = low; 3 = high; and 4 = very high.

The survey questionnaire used in the SSIs was developed with input from two scholars from Makerere University for clarity of the questions and content validity. The questionnaire was also pretested on 16 farmers in Ikaali parish, Bukhofu sub-county, in Manafwa district to enhance its suitability. Prior to data collection, research assistants were trained and involved in the pretesting to acquaint them with the tool and ensure quality of the data collected.

Data from the FGDs was analyzed through content analysis. Data from the SSIs was analyzed using Statistical Package for Social Sciences (SPSS) version 21 to produce frequencies and means to describe the characteristics of the study respondents. Frequencies and percentages were also used to describe farmers' awareness and perception about reliability of indigenous indicators for onset and cessation of rains as well the types of agricultural decisions made based on the different indigenous indicators known.

A binary logistic regression model was used to establish the factors that influence the likelihood of farmers to use indigenous indicators on onset and cessation of rains in making agricultural decisions. The dependent variable in the binary logistic regression model was whether a farmer used any of the indigenous indicators for onset and cessation of rains to make agricultural decisions. The decision to or not to use the indigenous indicators was presumed to be dependent on several socio-economic factors including farmer characteristics and perceived reliability of the indigenous indicators of onset and cessation of rains known. The binary choice was dummied as 1 if a farmer used any of the indigenous indicators for onset or cessation or rains to make any agricultural decision and 0 for otherwise. The selection of predictor variables included in the model was based on literature, theory and data availability.

Suppose Y is the decision to use any indigenous indicator for onset and cessation of rain to make an agricultural decision which is a random variable and X are the socio-economic factors, farm characteristics and perceptions about the reliability of the indigenous indicators presumed to predict the decision to use or not to use the indicator.

Then, the probability of Y can be predicted from the range of predictor variables through Eq. 1 given by Field (2009):

$$P(Y) = \frac{1}{1 + e^{-(\beta_0 + \beta_1 X_1 + \beta_2 X_2 + \beta_3 X_3 + \ldots\ldots\ldots \beta_n X_n + \mu)}} \tag{1}$$

where:

$P(Y)$ is the probability of a farmer using the indigenous indicators for onset and cessation of rains

X_1 to X_n are independent variables presumed to predict the probability of a farmer using the indigenous indicators for onset and cessation of rains

β_1 to β_n are the coefficients (weights) attached to each of the predictors X_1 to X_n

μ is the error term

The above coefficients are only used to show the direction of the relationship between the dependent variable and the specific predictor variable (Deressa et al. 2009). In order to indicate the magnitude of the influence that a specific predictor variable has on the dependent variable, odds ratios are used. The odds ratio corresponding to a given predictor variable represents the change in odds of using information that occurs as a result of a unit change in the predictor variable divided by the original odds. Thus an odds ratio of more than one indicates that as the predictor variable increases, the odds of using the indigenous indicators for onset and cessation of rains to make agricultural decisions increase, while a value of less than one shows that as the predictor variable increases, the odds of the outcome occurring decrease (Field 2009).

Given that the probability of using any indigenous indicator for onset or cessation of rains is $P(Y)$, the probability of not using the indigenous indicator = $1 - P(Y)$. Then the odds of using the indigenous indicator are given by.

$$\frac{P(Y)}{1-P(Y)} \tag{2}$$

Combining Eqs. (1) and (2), the odds of using the indigenous indicator can be given by.

$$\frac{\frac{1}{1+e^{-Z}}}{1-\frac{1}{1+e^{-Z}}} \tag{3}$$

where $z = \beta_0 + \beta_1 X_1 + \beta_2 X_2 + \beta_3 X_3 + \ldots\ldots\ldots \beta_n X_n + \mu$

The above equation can be simplified by finding the natural logarithm:

$$\text{Ln}\left(\frac{\frac{1}{1+e^{-Z}}}{1-\frac{1}{1+e^{-Z}}}\right) = Z = \beta_0 + \beta_1 X_1 + \beta_2 X_2 + \beta_3 X_3 + \ldots\ldots\ldots \beta_n X_n + \mu \tag{4}$$

The odds ratio corresponding to a given predictor variable (X_1) was computed as the odds of using indigenous indicator for onset and cessation of rains after a unit change in X_1 divided by the original odds. Table 1 shows how the predictors included in the model were operationalized and their hypothesized influence on farmers' use of climatic information to make production decisions.

Table 1 Description of the predictor variables entered into the empirical model

Predictor variables	Definition (how the predictor was defined in this chapter)	Expected sign / relationship
X_1 = sex of respondent	1 for male and 0 for female	+
X_2 = group membership	1 for yes and 0 for no	+
X_3 = age	Years	+
X_4 = highest education level in the household	Years of schooling	+
X_5 = farm size accessed	Acres	+
X_6 = off-farm income sources	Count of off -farm income sources	–
X_7 = access to scientific weather info	1 for yes and 0 for no	+
X_8 = perceived reliability of indigenous indicator	Reliability score	+

Table 2 Farmers' socio-economic and demographic characteristics

Socio-economic characteristics	Categories	Percentage (n = 255)	Mean
Sex of the respondent	Male	64.3	
	Female	35.7	
Household type	Male-headed	83.1	
	Female-headed	16.9	
Highest level of education in the household	No formal education	3.5	
	At least primary	48.8	
	At least secondary	38.3	
	Post-secondary	9.3	
Age of the respondent (years)			42.7
Household size (number of members)			7.3
Farm size (acres)			2.8

3 Results

3.1 Farmers' Socio-Economic and Demographic Characteristics

The respondents to this study were drawn from Manafwa and Bududa in the Mt. Elgon area where people's livelihood is derived largely from agricultural production. In these districts, approximately 80% of households which participated in the study were headed by men and only about 17% by women (see Table 2). Also note from Table 2 the large proportions of male respondents (64.3% as compared to 35.7% of the females). By and large Uganda is a patriarchal society, and therefore it is very common for households to have a man as the head as well as the "rightful" respondent to issues that concern livelihood and agricultural production. While this may be a gender issue, it could be put to good use as male-headed households are more likely to be aware of technologies and more likely to take risks than female-headed households (Asfaw and Admassie 2004).

The predominantly male-headed households are coupled with a diverse of education status ranging from no formal education to post-secondary school education. The study respondents were on average 43 years. Family sizes in the study sites were relative large and averaged to at least seven persons per household. Farmers in the respective districts grew a variety of crops and also reared some animals. The most common animals reared included cattle, pigs, goats, sheep and chicken. Though this was the case, the farming communities heavily relied on crop production to earn a living.

3.2 Description of the Farming System

Farmers involved in the study were typically smallholder farmers engaged in both crops and livestock farming. Table 3 shows the major crops grown and livestock reared by the farmers involved in the study.

Table 3 Description of the farming system in the study area

Farming system		Response (%)
Type of crops	Beans	93.6
	Maize	92.4
	Bananas	72.8
	Coffee	64.8
	Cassava	35.7
Type of livestock	Chicken	75.7
	Cattle	67.1
	Goats	54.3
	Pigs	18.6
	Sheep	2.9
Off-farm income sources	Sale of labour	35.7
	Brick laying	21.4
	General merchandize shops	16.4
	Remittances	15.7
	Sand mining	14.3
	Agricultural produce trade	12.1
	Livestock trading	6.4
	Formal employment	3.6

Results in Table 3 show that most of the households grew and kept a diversity of crops and livestock, respectively. Approximately 70% of the farmers kept cattle and chicken as the major livestock though other livestock including goats (54%), sheep (3%) and pigs (19%) were also reared on the farm. Respondents reported that they largely kept the animals in small numbers to supplement household nutrition and income. Other than crop and livestock production, some of the farmers participated in other off-farm income such as sale of labour, brick making, small general merchandize shops, remittances, sand mining and trading in agricultural produce. The small proportions of farmers involved in off-farm income generating activities however indicate that the livelihoods of the households in the area largely depended on agriculture.

4 Trends of Rainfall Structure and Seasonality on Mt. Elgon: Farmers perspectives

Farmer experience in the Mt. Elgon zone indicates that two rainfall seasons exist with peaks in April and October. The rainfall seasons are also preceded by two dry seasons that run between December to February and June to August. For these smallholder farmers, any deviation from the anticipated rainfall peaks and dry season, respectively, is referred to as an abnormal rainfall season. The trends in Fig. 2 show mixed perceptions of farmers towards (i) timing of onset and cessation of rainfall seasons and (ii) the amount and distribution of rainfall. From Fig. 2 more

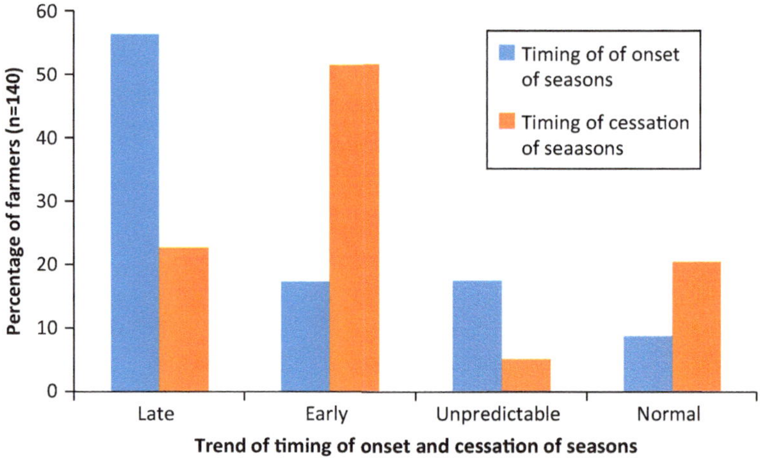

Fig 2 Farmers' perceived changes in the timing of the rainfall seasons

than half of the respondents recognize that both the onset and cessation of rains have changed from the perceived normal season. The general consensus for these farmers is that the onset of the rains is late and the cessation is early. While this is true, approximately 20% of the respondents acknowledge that the onset and cessation of the rains are both early. Other farmers on the other hand noted that the rainfall is unpredictable and could be received early, as expected or late. The overall impression of the results implies a tendency towards shorter growing seasons than normal (Mbogga 2012; Mubiru et al. 2012; Kansiime et al. 2013; Bomuhangi et al. 2016), and this has adversely altered the farmers' cropping calendars. This, coupled with the increasing mid-season droughts, makes smallholder farming a risky business. The risk averseness of the farming industry is further aggravated by the lack of proper prediction on the amount and distribution of rainfall for an entire season that farmers cannot foresee. Within the context of rainfall variability, farmers are prone to a number of climatic risks.

4.1 Rainfall Variability-Related Risks and Related Impacts on Crop and Livestock Production

In the past 5–30 years, farmers have recognized major climate-related risk that has and will continue to threat the agricultural industry. Table 4 shows the rainfall variability risks in the Mt. Elgon zone. As mentioned earlier, rainfall variability-related risks can have devastating impacts on crop and livestock production. Table 5 shows the farmers' perceived impacts of these risks.

The impacts on crop farming ranged from crop failure, crop damage and pest infestation to disease prevalence with crop failure being the heaviest impact on crop

Table 4 Rainfall variability-related risks in the Mt. Elgon zone

Rain variability-related risk	% of farmer perceiving the risk as a risk ($n = 255$)
Drought (food insecurity)	49. 0
Floods and flooding	39.6
Hail storms	36.0
Pests and disease prevalence	29.7

Table 5 Major impacts of climate-related risk on crops and livestock

Type of impact		Percentage of respondents ($n = 255$)
Major impact on crops	Crop failure (droughts)	93.3
	Crop damage (flooding/destructive storms)	45.0
	Pest infestation	48.3
	Disease prevalence	46.7
Major impacts on livestock	Lack of feeds (in dry season)	85.2
	Water shortage	66.7
	Low milk production	24.1
	Small grazing areas	35.2
	Disease prevalence	48.1
	Vector prevalence	9.3

production among farmers in the Mt. Elgon zone. Other impacts such as crop damage, pest infestation and disease prevalence were reported by over 40% of the respondents. Similarly, impacts in the livestock sector ranged from lack of feeds, water shortage, reduced milk yields to increased vector prevalence.

Table 6 shows that there is a diversity of technologies that farmers are using either on their own or through advice from other sources including community farming groups, local government extension staff and staff from development agencies to cope with the changing rainfall patterns in the region. While some of the technologies are popular among the farming communities, others are only practised a few of the farmers. It should be noted that the popular technologies included mulching (89%), intercropping (85.2%) and planting of food security crops (78.8%). Notably, these three practices have been in existence for a long time and still feature as important practices. Water harvesting for domestic consumption (89.6%) is among the fast-rising local innovation in the area. This is not to underestimate the importance of other technologies such as the use of soil bunds, use of plant and animal manure and use of storage facilities. Using the aforementioned adaptation strategies effectively requires that farmers receive adequate weather and/or advisory climatic information in order to make informed production decisions. Farmers therefore need access to climatic information in order to maintain and/or improve

Table 6 Rain variability adaptation technologies

Rainfall variability adaptation technologies		% of farmers using the technology ($n = 255$)
Soil moisture conservation	Mulching	89.1
	Use of cover crops	1.8
	Agroforestry	40.0
	Use of grass strips	7.3
	Strip cropping	1.8
	Soil bunds	45.5
	Irrigation	10.9
	None	0.0
Soil fertility enhancement	Use of backyard manure	55.4
	Use of compost manure	51.8
	Use of inorganic fertilizer	39.3
	Use of green manure	10.7
	No technology	2.1
Diversification technology	Intercropping	85.2
	Integration of livestock and crops	22.2
	Use of new crop varieties	38.9
	Planting of fruit tree	29.6
	No technology	1.9
Use of improved technology	Improved seed	48.1
	Proper spacing	63.0
	Timely planting	66.7
	Timely weeding	61.1
	Postharvest handling	40.7
Food storage	Planting of food security crops	78.8
	Use of storage facilities	56.7
Local innovations	Kitchen gardens	25.0
	Rainwater harvesting for domestic use	89.6
	Rainwater harvesting for production	2.1
	Organic pesticides	6.2
	Micro-irrigation	8.3
	Nonconventional organic fertilizers, e.g. urine	12.5

their agricultural production amidst increasing climatic uncertainties. Survey results indicate that majority (84%) of the farmers in the study area received scientific rainfall forecast information. Despite the fact that the majority of farmers received scientific information, over 60% of the respondents criticized it for its unreliability and untimeliness. Due to the above challenges, farmers have over the years learned to predict climate patterns especially for onset and cessation of the rain seasons using local indicators.

5 Use of Indigenous Indicators for Onset and Cessation of Rainfall

Heavy reliance on rain-fed agriculture has increased the vulnerability of farmers to rainfall variability, and this has increased the need for precise weather information using indigenous or local indicators. Table 7 shows the different indicators that farmers used to predict the onset of the rains. Farmers used various meteorological, animal and plant indicators as good forecasters for the onset of rains.

Certain meteorological conditions were commonly used by old folks to predict the onset of rains; among these were appearance of dark clouds and the direction of the wind from west to the east. Farmers further noted that whenever the wind started blowing from the west to the east (i.e. up the Mt. Elgon), rain came within a week. Half of the respondents (50%) also reported that night sweating among humans indicated the likelihood of onset of rain.

Other farmers observed the behaviour of selected animals and plants to obtain clues about the onset of rains. About 75% of the farmers mentioned sounds made by specific local bird species (*Makumeti*, *Namutwitwi*, *Lisoso* and *Tsikhu*), while about 50% mentioned arrival of migratory birds from east to west as indicators for onset of rains. The main plant behaviour-related indicator for onset of rains was the sprouting of Mvule (*Milicia excelsa*) tree which was mentioned by 49% of the farmers. Likewise, farmers relied on meteorological, plant and animal indicators to predict the cessation of rains as indicated in Table 8. Table 8 shows the different indicators that farmers used to predict the onset of the rains. Farmers used various meteorological, animal and plant indicators as good forecasters for the onset of rains.

Table 7 Indigenous indicators for forecasting onset of rains

Specific indicators	% of farmers who perceive this as an indicator ($n = 255$)
Meteorological indicators	
Appearance of dark clouds	90.6
Wind blowing from west to east	67.7
High temperatures during the night	49.6
Presence of lightning and thunder at night without rain	20.8
Position of the moon	
Plant and animal indicators	
Sounds made by some specific birds	74.9
Movement of migratory birds from east to west	49.4
Sprouting of young shoots of Mvule (*Milicia excelsa)* tree	48.6
Appearance of many Safari ants	18.8
Croaking of frogs during day time	5.9

Table 8 Indigenous indicators for forecasting cessation of rains

Specific indicators	% of farmers who perceive this as an indicator ($n = 255$)
Meteorological indicators	
Presence of very clear sky	70.3
Presence of white clouds	61.8
Wind blowing from the east to west	54.5
Prevalence of frequent drizzles	48.6
Incidences of cold mornings and evenings	34.7
Appearance of a lot of mist in the morning and at night	32.9
Appearance of many stars in the sky during the night	28.9
Presence of a deep red colour during the setting of the sun	23.9
Appearance of a bright/clear moon	21.8
High frequencies of lightening when it rains	15.0
Appearance of the rainbow	11.3
Plant and animal indicators	
Migratory birds moving from west to east	42.1
Noise made by birds	28.9
Increased incidences of butterflies	11.8

Meteorological indicators such as appearance of a clear sky (70%), prevalence of white clouds (62%) and direction of wind moving from east to west (55%) were highly associated with cessation of rain. Animal behaviour related to cessation of rains was migratory birds moving from the west to the eastern direction, and this was mentioned by 42% of the farmers that were interviewed. Farmers reported that appearance of a clear sky and white clouds was an indicator of cessation of rains over a shorter time (1–2 days), while departure of migratory birds from the communities indicated cessation of rains over the next 1–2 weeks.

6 Reliability of Indigenous Indicators for Forecasting Onset and Cessation of Rains

Farmers involved in the study were asked to rate the different indicators they used in terms of reliability (i.e. the extent to which they correctly predict the climatic conditions they purported). Results in Table 9 show that farmers considered meteorological indicators to be more reliable than plant and animal behaviour-related indicators of onset of rains.

Table 9 Perceived reliability of indigenous indicators for forecasting rainfall

Indicator category	Specific indicator	Number of farmers who know this as an indicator	Perceived reliability of the indicator (%) Very low	Low	High	Very high
Onset	*Meteorological indicators*					
	Presence of dark clouds	232	1.3	8.2	40.3	50.2
	Direction of wind – West to east	173	0.0	1.7	38.3	60.0
	High temperature	126	0.0	10.4	25.0	64.6
	Plant and animal behaviours					
	Sounds made by specific birds	191	0.0	6.3	31.3	62.4
	Movement of migratory birds from east to west	125	22.8	39.0	12.7	25.5
	Sprouting of Mvule trees	124	26.0	39.2	15.4	19.5
Cessation	*Meteorological indicators*					
	Presence of very clear sky	179	5.0	18.1	44.6	32.2
	Presence of white clouds	158	1.4	20.0	38.6	40.0
	Direction of wind – East to west	139	12.7	39.6	23.3	24.5
	More frequent drizzles	124	4.6	62.2	26.5	6.7
	Plant and animal indicators					
	Departure of migratory birds	107	0.0	33.3	29.2	37.5

For instance, over 80% of the farmers perceived the appearance of dark clouds, wind blowing from west to east and high temperatures especially at night to have high to very high reliability. On the other hand, over 60% of the farmers perceived animal behaviour (movement of migratory birds from east to west) and plant behaviour (sprouting of trees) to have low to very low reliability. Farmers' perceptions regarding the reliability of indigenous indicators for cessation of rains showed mixed feeling. For instance, on one hand, over 75% of the farmers perceived the presence of clear sky and white clouds as having high to very high reliability. On the other hand, about 67% and 52% of the farmers reported that wind blowing from east to west and more frequent drizzles, respectively, had low to very low reliability. Farmers reported that appearance of a clear sky, white clouds and departure of migratory birds was a reliable indicator of cessation of rains within 1 week and 1–2 weeks, respectively. These findings about the perceived high reliability of meteorological indicators agree with those of Gyampoh et al. (2011), Nganzi et al. (2015) and Soropa et al. (2015) in Ghana, Uganda, and Zimbabwe, respectively. Furthermore, the unreliability of the behaviour of trees such as *Milicia excelsa* and *Erythrina abyssinica* as indicators of rainfall forecasting has also been documented in Teso region of Uganda by Egeru (2012).

Table 10 Decisions made based on indigenous indicators of onset and cessation of rains

Type of indicator	Type of production decision	Percentage of farmers using the indicator to make a production decisions		
		Dark clouds ($n = 231$)	Sounds made by birds ($n = 191$)	Wind blowing west to east ($n = 173$)
Onset of rains	Prepare land for planting	78.0	73. 6	73.4
	Plant appropriate crops immediately	62.4	68.0	63.3
	Purchase of farm inputs, e.g. seed	39.9	45.1	18.1
	Select types of crops to grow	2.6	19.4	11.0
	Make no decision	9.6	9.2	9.2
		Clear sky ($n = 179$)	White clouds ($n = 158$)	Wind blowing east to west ($n = 139$)
Cessation of rains	Delay planting	81.0	74.1	73.4
	Planning to harvest	42.4	49.5	45.8
	Types of crops to be grown and how to manage them	11.7	18.6	13.5
	Decide on whether and when to store food	10.6	5.3	13.5
	Make no decision	3.8	5.9	23.3

7 Use of Indigenous Indicators for Onset and Cessation of Rains in Agricultural Decision Making

Farmers used different meteorological and animal behaviour indicators that forecasted onset and cessation of rains to make several production decisions. Results in Table 10 show the commonly used indicators by farmers to make informed decisions on production choices. Results in Table 10 indicate that farmers relied heavily on indigenous indicators for onset and cessation of rains to take agricultural production decisions and actions. Farmers mainly used indigenous indicators for onset of rains to make decisions on land preparation and planting crops. At least 73% and 62% of the farmers decided to prepare land and plant crops, respectively, based on appearance of dark clouds, direction of wind from west to east as well as sounds made by birds. While this was true, at least 40% of the farmers also used the appearance of dark clouds and the sound made by birds to purchase farm inputs to use during the cropping season.

On the contrary, the main agricultural production decisions that farmers made based on indigenous indicators of cessation of rains were delaying to plant and

Table 11 Incentives for use of indigenous indicators to forecast onset and cessation of rainfall

Predictor variables	B	S.E.	Wald	Sig.	Exp (B)
Sex of respondent	1.983	1.105	3.217	0.073	7.261
Age of respondent	−0.003	0.024	0.018	0.894	0.997
Education level of respondent	0.088	0.106	0.692	0.405	1.092
Diversity of off-farm income	−1.031	0.314	10.753	0.001	0.357***
Membership to a farming group	−1.361	0.789	2.971	0.085	0.256
Total land available for farming	−0.232	0.141	2.693	0.101	0.793
Access to scientific information	0.375	0.137	7.536	0.006	1.456***
Reliability score of ITK	1.637	0.843	3.771	0.05	5.138**
Constant	3.178	1.783	3.179	0.075	24.004

** p<0.05
***p<0.01

planning to harvest. At least 73% and 42% of the farmers had, respectively, decided to delay to plant and to harvest their crops immediately based on one or more of the commonly known indigenous indicators of cessation of rains. Furthermore, results show that less than 20% based on the indigenous indicators for onset and cessation of rains to decide on which crops to grow and how to manage them. Farmers revealed that they rarely used indigenous indicators of forecasting rains to decide on types of crops to grow because such indicators were unable to clearly forecast the amount and distribution of rain over the entire season. The model in Table 11 predicts what would incentivize a farmer to use indigenous knowledge for forecast rainfall variability and/or make production decision based on the indicator in equation.

Results in Table 11 indicate that farmers' likelihood to use indigenous indicators to forecast the onset and cessation of rains was dependent on the diversity of off-farm income, the access to scientific weather information and the reliability score for the particular indigenous indicator. Surprisingly, farmer characteristics including age, sex and education level within the household did not influence the likelihood of using an indigenous indicator to predict rainfall patterns. Nonetheless, the diversity in off-farm income influenced the likelihood to use or not to use indigenous information to predict rainfall patterns at $p < 0.01$. The odd ratio (0.357) implies that the farmers investing more in off-farm income are less likely to use ITK to predict rainfall patterns. This is because the larger proportion of off-farm income does not rely on rainfall patterns for their implementation. On the other hand, farmers who accessed scientific information were one and half times more likely to use ITK to predict the onset and cessation of rain than those that did not access scientific information. Results further indicate that the perceived reliability of the indigenous indicator for rainfall positively influenced farmers' likelihood to use it to predict rainfall onset and cessation. Farmers who believed that an indicator was reliable were more likely to use it to predict onset and/or cessation of rainfall than those that did not.

8 Discussion

The farmers in the Mt. Elgon zones depended on rain-fed agriculture for both income and food security. Farmers owned very small land holdings, and on these holding grew a wide range of crops and kept livestock as livelihood options. The major crops included bananas and beans, while the livestock included cattle, chicken and goats. Within this farming system, rainfall variability was the largest challenges that these smallholder farmers faced and was characterized by various risks and shocks. The most prevalent climatic shock was drought, and this had adverse impacts on both crop and livestock production. Devastating impacts including severe crop and livestock losses were observed, and these culminated into reduction in other physical and production resources that ultimately reduced the production potential of the farming system. Heavy rains were also a predominant rain variability shocks, and this led to secondary effects including increased prevalence of pests and diseases, destruction of road networks and other social infrastructures. Within these shocks however, farmers have over time built their resilience potential by using several adaptation strategies to cope with the rain variability.

Farmers used a wide ranges of strategies to curb the impacts of rainfall variability. Among these were (i) the use of recommended agronomical practices, (ii) soil enhancement technologies, (iii) diversified livelihoods and (iv) use of storage infrastructure. Though this is true, these strategies can only be effective if an adequate weather and rainfall information system is made available to farmers at local level.

Farmers' reliance on indigenous indicators of rainfall onset and cessation to make agricultural production decisions is consistent with Gyampoh et al. (2011), Egeru (2012), Nganzi et al. (2015) and Soropa et al. (2015). Gyampoh et al. (2011) and Nganzi et al. (2015) noted that farmers in Ghana and Uganda, respectively, relied on indigenous knowledge to make agricultural decisions even when they had access to scientific seasonal climatic forecasts. Farmers had a wide range of indigenous indicators that were used to forecast both the onset and cessation of rainfall. Among the commonest indicators for onset and/or cessation of rain was the appearance of dark clouds. Since time immemorial, clouds have been associated with rain. Two categories of clouds, namely, are nimbostratus, the dark, rain-carrying cloud of bad weather, and cumulonimbus, the cloud that produces showers and thunderstorms. The rain comes and goes with these clouds. The appearance of these two types of clouds makes a high probability of rainfall. While this is true, farmers also associated white clouds and clear skies as indicators for cessation of rain. Frequently farmers also used the positioning of the moon as an indicator of rainfall onset. This fits well with an old proverb that goes, "a ring around the sun or moon brings rain upon you soon". Change in animal behaviour was also indicative of whether the rains would come or stop. When sensing an upcoming rain, animal may behave in a certain way. Their behaviour is believed to be triggered by odours/aromas that become intense than usual when rain clouds develop as air density drops (Galacgac and Balicasan 2009). Other biological factors that might trigger behavioural change in animals due to upcoming rainfall may include increasing humidity and air pressure. Humidity and pressure are also important parameters that science uses to give precise weather forecasts. Many climate/weather watchers have recognized that some plants and trees are more sensitive to changes in atmospheric conditions than others. Some bloom or bear fruit during rainy seasons while others during the dry season. Some well-known plant indicators in Uganda include the sprouting of young shoots of the *Mvule* tree and blooming in coffee. The blossoming of the *Mvule* often predicts the onset of the rain season, while the shedding of the old leaves signals the beginning of the dry season. According to scientific literature, rising/falling humidity associated with the beginning of the rainy/dry season triggers these phonological behaviours in plants.

Study results on indigenous indicators for forecasting onset and cessation of rainfall are consistent with other studies by Orlove et al. (2010), Egeru (2012), Risiro et al. (2012), Okonya and Kroschel (2013), and Nganzi et al. 2015). Orlove et al. (2010), for instance, found out that increased night-time temperatures, shifts in the direction of prevailing winds (from West to East) and plant and animal behaviour predicted the onset of rains. Other scholars including Gyampoh et al. (2011), Rusinga et al. (2014) and Soropa et al. (2015) found similar results regarding farmers' use of plant and animal behaviour to forecast rains in other areas in SSA. However, the specific plants and animals mentioned in each of the case studies differed largely because of the difference in agroecological conditions. The question to date is how reliable the indigenous indicators are in predicting or forecasting rainfall patterns across the region. The findings indicate that farmers perceive a number of indigenous indicators to be reliable for forecasting onset and cessation of rains but over a very short time (up to 2 weeks). When these signs appear, some of the farmers make important production decisions that enable them to cope in the future. Some of the farmers (i) start field preparations, (ii) carry out dry planting, (iii) purchase seed for planting, (iv) decide on the type of crops to grow, (v) plan for type of agronomic practices to use and (vi) plan for labour redistribution and allocation. Though farmers can make some production decisions, they considered the short time over which the indigenous indicators forecasted to be insufficient for them to adequately prepare for the seasons. Results further show that there were more indicators of onset of rains that farmers considered to be reliable compared to those for cessation. This is largely because farmers closely monitor indicators for onset of rains because onset shapes most of the agricultural production decisions and actions which once taken are irreversible. In essence, indicators for cessation of rains only enable farmers not to be misled by intermittent rains to make irrational production decisions.

Farmers will, however, use the indigenous indicator if they believe that they precise forecasters. These findings corroborate with those of Stone and Meinke (2006), Gyampoh et al. (2011), Lemos et al. (2012) and Mpandeli and Maponya (2013) who observed that farmers were likely to use information if they considered it to be credible, reliable and timely. Given that farmers who accessed scientific weather information were likely to use indigenous knowledge as well, points to a new direction of integrating both scientific and indigenous knowledge to forecast rainfall patterns. Various scholars including Orlove et al. (2010) report that integration of scientific and indigenous knowledge related to climate forecasting can enhance the success of efforts aimed at supporting farmers, thus building resilience to climate extremes. Orlove et al. (2010) and Ziervogel and Opere (2010), for example, argued that indigenous knowledge can enhance the customization and communication of scientific climate predictions to farmers. Indigenous knowledge is highly localized, and as such farmers can easily relate to the spatial scale at which it is collected. However, efforts to pave way for the integration of indigenous knowledge and scientific on forecasting rainfall patterns have been curtailed by limited documentation of the indigenous knowledge (Orlove et al. 2010).

9 Conclusion and Recommendations

The Mt. Elgon zone experiences a bimodal climatic pattern. The communities in the Mt. Elgon zone were predominantly agriculturalists and rely on rain-fed agriculture for both food and income security. A wide variety of crops and livestock are grown and kept, respectively, on approximately 3 acres of land. The households were headed predominantly by men who also controlled most of the production resources in the home. Within these largely subsistence farming system, rain variability was one of the biggest challenges that farmers faced. Rainfall variability was characterized with several shocks including droughts and heavy rains, and these adversely affected their livelihoods through reducing agricultural production and depleting different types of livelihood assets. Because agriculture is a livelihood option, farmer had to plan and implement different strategies to cope with and adapt to the effects of the changes in rainfall patterns.

A wide range of adaptation strategies were used though these were on individual farm and therefore resulted in less impact at a landscape level. This partly explains the farmers' perceptions about the continued declines in agricultural production and depletion of various livelihood capitals resulting from the shocks triggered by changes in rainfall patterns. Weather and climate information therefore becomes a critical resource for the farmers who need to make informed production decision on a daily basis. While this is true, there is limited appreciation of scientific climatic forecasts as they are assumed unreliable and inaccurate. In this dilemma, farmers relied heavily on indigenous knowledge to forecast the onset and cessation of rain. Several astronomical and biological indicators were used by farmers to predict both onset and cessation of rain. Even within the same locality, farmer described different indicators as predictors of the same scenario. Ultimately, no single indicator was adequate to predict a weather scenario, so farmers had to use several indicators to precisely predict the rainfall pattern. An indicator would only be useful if a farmer can make a meaningful production decision after its appearance.

In this case, farmer noted several indicators that they considered precise in predicting either onset or cessation of rain and used them to make production decisions. However, farmers could only make short-term decisions ranging from 2 to 14 days.

In addition, the indicators could not make any prediction on the amount of rain or the distribution of the rain during the entire cropping season. This meant that indigenous indicators though important were not sufficient to predict rainfall patterns over a long period of time. This therefore calls for an integrated approach where scientific and indigenous knowledge are used in combination to predict weather patterns. Such a tendency for farmers to integrate scientific climate forecast information and indigenous indicators is driven by their recognition that both knowledge bases have limitations and complementary strengths. Scientific climate forecast information had longer lead time and wider scope in terms of rainfall characteristics forecasted despite being less reliable and inappropriately timed. On the other hand, they considered most indigenous indicators to be reliable, but they had short lead times, and they could only forecast the timing of onset and cessation of rains. To this effect, development practitioners and various government entities can facilitate the integration process by documenting information on indigenous climatic forecast indicators in a form in which it can be preserved, shared and updated over generations.

Acknowledgements The writing of this paper arises out studies funded by the Regional Capacity Building for Natural Resources and Agricultural Productivity in view of Climate Change (CAPSNAC) project, to which the authors are greatly indebted.

References

Asfaw A, Admassie A (2004) The role of education on the adoption of chemical fertiliser under different socioeconomic environments in Ethiopia. Agricultural Economics 30(3):215–228

African Green Revolution Alliance [AGRA] (2014) Africa agriculture status report 2014: climate change and smallholder agriculture in Sub-Saharan Africa. Nairobi, Kenya

Bomuhangi A, Nabanoga G, Namaalwa J, Jacobson M, Abwoli B (2016) Local communities' perceptions of climate variability in the Mt. Elgon region, eastern Uganda Cogent Environmental Science, 2

Cooper PJM, Dimes J, Rao KPC, Shapiro B, Shiferaw B, Twomlow S (2008) Coping better with current climatic variability in the rain-fed farming systems of sub-Saharan Africa: an essential first step in adapting to future climate change? Agric Ecosyst Environ 126(2008):24–35

Deressa TT, Hassan RM, Ringler C, Alemu T, Yusuf M (2009) Determinants of farmers' choice of adaptation methods to climate change in the Nile Basin of Ethiopia. Glob Environ Chang 19(2):248–255

Egeru A (2012) Role of indigenous knowledge in climate change adaptation: a case study of Teso sub-region, eastern Uganda. Indian J Tradit Knowl 11(2):217–224

Field A (2009) Discovering statistics using SPSS, 3rd edn. SAGE publications, London

Galacgac ES, Balisacan CM (2009) Traditional weather forecasting for sustainable agroforestry practices in Ilocos Norte Province, Philippines. Forest Ecology and Management 257:2044–2053

Gyampoh BA, Asante WA, La Rose DJ, Adu-Acheampong G, Assimeng T, Opoku AG (2011) Mapping and documenting indigenous knowledge in climate change adaptation in Ghana. Recovery Program Support to Northern Ghana-UNDP/NADMO and African Adaptation Program – UNDP/Environmental Protection Agency, https://www.undp-aap.org/sites/undp-aap.org/files/Ghana_Mapping%20and%20Documenting%20Indegenous%20Knowledge%20in%20CCA%20in%20Ghana_2011.pdf.

Hansen JW, Mason SJ, Sun L, Tall A (2011) Review of seasonal climate forecasting for agriculture in sub-Saharan Africa. Exp Agric 47(2):205–240. https://doi.org/10.1017/S0014479710000876
Kaggwa R, Hogan R, Hall B (2009) Enhancing the contribution of weather, climate and climate change to growth, employment and prosperity. UNDP/NEMA/UNEP Poverty Environment Initiative, Kampala, Uganda
Kansiime MK, Wambugu SK, Shisanya CA (2013) Perceived and actual rainfall trends and variability in eastern Uganda: implications for community preparedness and response. J Nat Sci Res 3(8):179–195
Knapen A, Kitutu MG, Poesen J, Breugelmans W, Deckers J, Muwanga A (2006) Landslides in a densely populated county at the foot slopes of Mount Elgon (Uganda): characteristics and causal factors. Geomorphology 73(2006):149–165. https://doi.org/10.1016/j.geomorph.2005.07.004
Krejcie RV, Morgan DW (1970) Determining sample size for research activities. Educ Psychol Meas 30:607–610
Lemos MC, Kirchhoff CJ, Ramprasad V (2012) Narrowing the climate information usability gap. Nat Clim Chang 2(2012):789–794. https://doi.org/10.1038/NCLIMATE1614
Masinde M, Bagula A (2011) ITIKI: bridge between African indigenous knowledge and modern science of drought prediction. Knowl Manag Dev J 7(3):274–290.. Available at. https://doi.org/10.1080/19474199.2012.683444
Mbogga MS (2012) Climate profiles and climate change vulnerability assessment for the Mbale region of Uganda. UNDP Consultancy report. Kampala, Uganda
Mpandeli S, Maponya P (2013) The use of climate forecasts information by farmers in Limpopo Province, South Africa. J Agric Sci 5(2):47–55. https://doi.org/10.5539/jas.v5n2p47
Mubiru DN, Komutunga E, Agona A, Apok A, Ngara T (2012) Characterizing agrometeorological climate risks and uncertainties: crop production in Uganda. S Afr J Sci 108. https://doi.org/10.4102/sajs.v108i3/4.470
Mwatu MM, Recha CW, Kumba JK (2016) Perception of Seasonal Climate Forecast Information among Smallholder Farmers in Semi-Arid Southeastern Kenya: A Case of Voi Sub-County. Journal of Environment and Earth Science 6(4)
Nganzi P, Kajumba TC, Barihaihi M, Bataze J, Mujuni G (2015) Use of indigenous knowledge in weather forecasting in Uganda National survey. Africa Climate Change Resilience Alliance (ACCRA), Kampala, Uganda
Okonya JS, Kroschel J (2013) Indigenous knowledge of seasonal weather forecasting: a case study in six regions of Uganda. Agric Sci 4(12):641–648
Orlove B, Roncoli C, Kabugo M, Majugu A (2010) Indigenous climate knowledge in southern Uganda: the multiple components of a dynamic regional system. Clim Chang 2010(100):243–265. https://doi.org/10.1007/s10584-009-9586-2
Pareek A, Trivedi PC (2011) Cultural values and indigenous knowledge of climate change and disaster prediction in Rajasthan, India. Indian J Tradit Knowl 10(1):183–189
Risiro JD, Mashoko D, Tshuma T, Rurinda E (2012) Weather forecasting and indigenous knowledge systems in Chimanimani district of Manicaland, Zimbabwe. J Emer Trends Educ Res Policy Stud 3(4):561–566
Rusinga O, Chapungu L, Moyo P, Stigter K (2014) Perceptions of climate change and adaptation to microclimate change and variability among smallholder farmers in Mhakwe communal area, Manicaland province, Zimbabwe. Ethiop J Environ Stud Manage 7(3):310–318. https://doi.org/10.4314/ejesm.v7i3.11
Soropa G, Gwatibaya S, Musiyiwa K, Rusere F, Mazima GA, Kasasa P (2015) Indigenous knowledge systems weather forecasts as a climate change adaptation strategy in smallholder farming systems of Zimbabwe: case study of Murehwa, Tsholotsho and Chiredzi districts. Afr J Agric Res 10(10):1067–1075. https://doi.org/10.5897/AJAR2013.7205.
Stone RC, Meinke H (2006) Weather, climate, and farmers: an overview. Meteorol Appl (Supplement) 13:7–20. https://doi.org/10.1017/S1350482706002519

Sultan B, Barbier B, Fortilus J, Mbaye SM, Leclerc G (2010) Estimating the potential economic value of seasonal forecasts in West Africa: a long-term ex-ante assessment in Senegal. Weather Clim Soc 2010(2):69–86. https://doi.org/10.1175/2009WCAS1022.1
Uganda Bureau of Statistics [UBOS] (2010) Uganda Census of Agriculture 2008/2009. Crop area and production report. Volume IV. Available at https://www.ubos.org/wp-content/uploads/publications/03_2018UCACrop.pdf
Uganda Bureau of Statistics [UBOS] (2014) National population and housing census 2014 provisional results. November 2014: Revised Edition. Kampala, Uganda
United Nations Environment Programme [UNEP] (2007) Nature conservation and natural disaster management: the role of indigenous knowledge in Kenya. UNEP, Nairobi, Kenya
Ziervogel G, Opere A (2010) Integrating meteorological and indigenous knowledge-based seasonal climate forecasts in the agricultural sector. International Development Research Centre, Ottawa, Canada. Climate Change Adaptation in Africa learning paper series

Gender Norms, Technology Access, and Women Farmers' Vulnerability to Climate Change in Sub-Saharan Africa

M. N. Mangheni, H. A. Tufan, L. Nkengla, B. O. Aman, and B. Boonabaana

Abstract The challenges of climate change are defined by biophysical unpredictability and the sociocultural context in which communities adjust to these challenges. Gender norms to which women and men generally conform influence women smallholder farmers' vulnerability to climate change. Understanding the social context within which an intervention is introduced can therefore greatly influence its transformative capacity. This review addresses the evidence on the influence of gender norms on climate-smart agricultural systems in sub-Saharan Africa through the dual lenses of equitable system productivity and women's empowerment. It makes a case for inclusive strategies to enhance equitable access to improved seed and other technologies as an adaptation option. We conclude that challenging gender norms around seed systems and extension services in SSA will increase our chance of success in mitigating climate disasters.

Keywords Climate-smart agriculture · Sub-Saharan Africa · Gender norms · Seed systems · Extension services

M. N. Mangheni (✉) · B. Boonabaana
Makerere University, Kampala, Uganda

H. A. Tufan
Cornell University, Ithaca, NY, USA

L. Nkengla
ICRISAT, Bamako, Mali

B. O. Aman
Bioversity International, Cotonou, Benin

Y. Bamutaze et al. (eds.), *Agriculture and Ecosystem Resilience in Sub Saharan Africa*, Climate Change Management,
https://doi.org/10.1007/978-3-030-12974-3_32

1 Introduction

Climate-smart agriculture which emphasizes enhanced agricultural productivity and climate change adaptation and mitigation employs a wide array of approaches, ranging from natural resource management, soil conservation practices to more technology-focused interventions such as crop and livestock diversification (Lipper et al. 2014). Changing crop varieties is a top adaptation strategy employed by both men and women smallholder farmers to mitigate climate change (Twyman et al. 2014). Many climate-smart agriculture practices therefore require access and utilization of quality improved seeds and other agricultural inputs and extension services. The improved seeds and inputs enhance productivity, while agricultural extension services support farmers to access information, link with other actors and services, and form viable institutions through which they can access improved seeds and complementary technologies and services. We argue that in all these interventions, equal attention should be devoted to gender because besides the biophysical factors, social dynamics further contribute to system vulnerability and resilience. Better understanding of gender is particularly vital to cope with the prevailing complex and highly challenging environment characterized by population explosion, fragile natural resources, unfavorable social systems, and gender norms.

Smallholder farming in sub-Saharan Africa (SSA) is intimately intertwined with the social fabric of society. Gender defines how men and women relate within the household, the roles they perform, how they interact with the factors of production, the enterprises they manage, and benefits they enjoy from these investments. The sociocultural construction of men and women, boys and girls manifesting across key institutions at household, local, national, and international level, has usually subordinated women and girls (Kabeer 1999). This has resulted in gender-based constraints and privileges in most cases to the disadvantage of women and girls. For example, studies have revealed disparities in women and men smallholder farmers' access to adoption and benefit from new knowledge and technologies (Doss and Morris 2001; Quisumbing and Pandolfelli 2010, Mudege et al. 2017). Resource-poor farmers, the majority of whom are rural women, often have very limited access to land and improved technologies such as inputs, credit, and extension needed for improving livelihoods (FAO 2011). Resource allocation may also reflect competition, investment trade-off, and gender-linked orientation within the household. Consequently, a holistic understanding of the vulnerability of smallholder agriculture in SSA that embodies the social physical and economic dimensions should underpin climate response interventions. Similarly, there is need to tap into the potential and capacity of women as key players to strengthen resilience such as drawing on their local knowledge, creativity, and existing networks. There is evidence that where women were aware of climate adaptation options, they were likely to adopt, suggesting that targeting women with climate and agricultural information is effective in promoting uptake of new agricultural practices for adaptation (Twyman et al. 2014).

This review aims at highlighting the growing body of evidence regarding the influence of gender norms on sustainable agricultural systems in sub-Saharan Africa through the dual lenses of equitable system productivity and women's empowerment. It assesses how gender norms contribute to the vulnerability of women smallholder farmers in SSA by limiting their access and utilization of improved seeds, complementary technologies, and extension services. While this topic has been widely examined and acknowledged, it has not been satisfactorily integrated into thinking about and planning for climate change. We start by defining gender norms discussing how they affect agricultural technology utilization. In order to attain optimum returns from adoption of improved seeds, farmers often have to access complementary technologies, inputs, and extension services. The chapter therefore expounds how gender norms influence seed and agro-input systems and extension services. This review reveals potential biases engraved in gender norms and important entry points for a more gender inclusive application of improved seeds and other agricultural technologies toward a resilient system in the face of climate change.

2 Gender Norms Defined

Gender norms have been variously defined. The United Nations Statistics Division (2015) defined gender norms as the standards and expectations to which women and men generally conform, within a range that defines a particular society or group, culture, and community at a particular point in time. Kabeer (1994) defined gender norms as rules embedded in the communities, activities or behaviors, resources available, and decision-making which influence the roles and activities of men and women differently. Norms are context specific and time bound as they may change from place to place and over time. Members of a group or community tend to adhere to the norms or rules of that particular group or society. Gender norms are therefore a type of sociocultural regulation, a social control mechanism, and structural elements providing a sense of direction to men and women (Spencer et al. 2015).

Gender norms influence the roles played by men and women in production activities. In some communities, certain agriculture-related roles are assigned to men and women differently based on the type of crop (Doss 2002). Consequently, men and women are engaged in different activities in the production value chains, and this impacts on their productivity and climate response. Ajambo et al. (2018) found that in nine banana bunchy top disease pilot sites in SSA, gender division of labor in banana production differed across sites. For instance, in Nigeria, men were involved in all banana production activities, while women and children were mostly involved in processing and marketing. By contrast, in Burundi and Gabon, men and women jointly did most banana production activities with the children mainly assisting in activities such as weeding and transporting of planting materials. The different roles performed by men and women were linked to community gendered perceptions. In some cultural contexts, men are perceived as physically stronger than women and are required to take up strenuous and labor-intensive tasks such as field preparation, while women are expected to engage in "soft" tasks that require keenness such as seed sourcing and processing. In Nigeria, for instance, in Idologun community, men who allowed their wives to engage in labor-intensive banana production activities were regarded as lazy and incapable of providing for their families and were not respected in the community (Nkengla et al. 2018 in press).

3 Gender and Agricultural Technology: A Case of Improved Seeds

Gender norms explain and help to understand the root causes of smallholder women farmers' vulnerability, a useful starting point to developing inclusive strategies for enhanced access to improved seeds and other technologies for climate change response. Besides roles, the norms underpin gender-differentiated needs, priorities, and access to productive resources for men and women (Kristjanson et al. 2014; Meinzen-Dick et al. 2012; Quisumbing and Pandolfelli 2010). Prevailing social and gender norms also shape the way men and women behave, interact, and react toward agricultural technologies including agricultural inputs and extension services as well as benefit sharing from participation and adoption of agricultural technologies (Ajani 2008; Mudege et al. 2015;Twyman et al. 2015; Padmanabhan 2002; Jeckoniah et al. 2013). Several studies in Uganda on climate change showed that gender and class can shape and influence the processing of climate-related information (Roncoli 2006 cited in World Bank Group, FAO and IFAD 2015). Societal norms and beliefs related to resource access and control in developing countries have led to differential vulnerabilities and impacts of climate change for men and women, thus affecting their willingness and capacity to adopt climate-related technologies (Huyer et al. 2015; Kristjanson et al. 2014; Nyasimi et al. 2017). For instance, women's role as collectors of firewood influenced higher adoption of agroforestry systems such as improved tree fallows in Zambia and Uganda (Phiri et al. 2004). Other studies in Nigeria (Sanginga et al. 2007) and Malawi (Gilbert et al. 2002) found higher adoption rate of seeds and fertilizers, respectively, among men. Hence, ignoring gender norms in the dissemination of agricultural technology can impact adoption as it is dependent on access to and decision-making rights over resources such as land and labor which in sub-Saharan Africa is unequal, mostly favoring men (Doss and Morris 2001).

A mainstay in strategies to improve agricultural productivity in SSA has long been the development and delivery of improved seeds, which has had limited success mainly due to the complexity of SSA seed systems. Informal seed systems have undergone restructuring over the past decades, moving from formalization under government control in the 1970s to privatization starting in the 1980s, finally with the 1990s characterized by NGO and relief organizations becoming involved (Rubyogo et al. 2007). Current trends in SSA seed system development include a mixture of formalization of seed systems through links with the private sector, subregional organizations, and NGOs while also strengthening existing networks and farmer seed multipliers (McGuire and Sperling 2016). Recent investments also focus on integration of formal and informal seed systems, which remain mostly ad hoc and localized (Sperling et al. 2014). Formal seed systems in SSA largely remain undeveloped, with most farmers obtaining seed from informal channels: farm-saved seeds, exchanges, or gifts characterizing the bulk (90%) of seed movement (Maredia et al. 1999; McGuire and Sperling 2016). Adoption rates of improved seeds differ by crop, country, context, but most interestingly also by the sex of the household head (Smale et al. 1991). Technology adoption decisions are informed by access to resources, and with a demonstrated large gender gap in assets, it means improved varieties are often not "gender neutral" but depend on the context within which the technology is released (Doss and Morris 2001). Besides gender, wealth has been found to be another major structuring factor in local seed circulation in Cameroon, with wealthy households having access to wider diversity of seed sources (Wencélius et al. 2016). Smale et al. (2018) also found intersectional variables such as marital status, education, and age, in combination with gender, strongly explained if improved seed was grown in households examined. Though women may have less access to formal seed systems, this is not insurmountable. Women's access to improved seed and input in variety development decision-making can be strengthened through participatory plant breeding approaches, for example (Galiè 2013).

The social context within which an intervention is introduced to communities under threat of climate change can greatly influence adaptive capacity, especially for disadvantaged groups (Kristjanson et al. 2017). In Ethiopia, households most vulnerable to extreme weather events (drought) chose landraces over improved sorghum varieties to mitigate risk, whereas households with only moderate risk chose to adopt improved varieties (Cavatassi et al. 2011), putting into question the use of improved varieties as a means to mitigate climate stress. The same study questioned the suitability of different varieties to handle different kinds of climate stresses (catastrophic vs chronic), which should be considered in developing varieties. On the other hand, new seeds may in fact benefit marginalized groups and their ability to weather climate shocks. Randomized control trials (RCTs) with flood-tolerant rice varieties in India showed that the highest gains in productivity were in regions where flooding was most severe, where lower caste social groups tended to cultivate land due to historical land allocation practices (Dar et al. 2013).

Barriers to women's access to farm inputs including improved seeds exacerbate the gender gap in agriculture (World Bank 2008) and, without amelioration, will only widen the gender gap in climate change resilience (Nyasimi and Huyer 2017). Men are more likely than women to adopt climate-stress-tolerant seeds (Kristjanson et al. 2017), in line with the observation that adoption rates of agricultural technologies remain low among women in SSA, likely due to gender norms (Peterman et al. 2014; see section on how gender norms influence extension services). In Ghana, for example, men mentioned adoption of improved varieties as a climate change mitigation strategy, while women did not (Naab and Koranteng 2012). Similarly, in Tanzania 95% of farmers used improved crop varieties as climate-smart agriculture (CSA) practices, yet when the data was disaggregated by sex, improved varieties were mentioned three times more by men than by women, indicating key differences in CSA practices of choice among men and women (Nyasimi et al. 2017). Decisions around crop diversification (including use of new varieties) in response to climate stress in maize production systems in Benin varied significantly by gender (Yegbemey et al. 2013). This may be explained by the higher land ownership enjoyed by men in the region, which was found to be highly correlated to any climate change adaptation strategies in the same study. Lastly in Uganda, studies by Fisher and Carr (2015) revealed that women farmers have much lower adoption rates of drought-resistant varieties of maize than men. The cumulative results of these studies indicate that despite the availability of climate-smart seeds and best intentions to distribute them, gender-based barriers may constrain women's access to these as a mitigation strategy.

From a holistic standpoint, agricultural technology can affect farmers positively or negatively. Some innovations that aim at achieving increased productivity may worsen gender (and other social) inequity. For instance, introduction of agricultural technologies may affect the well-being of women farmers resulting in labor drudgery, increased time for farm activities or household chores, and lower access to agricultural inputs and technology (Doss 2001; Ragasa 2012). To ensure that vulnerable groups such as women have equal access to proposed technologies, it is important to identify and consider the impediments embedded in gender norms in the society. It is now apparent that agricultural innovation that does not consider the socio-technical aspects often different for men and women (Jacobson 2011) is likely not to acquire the desired results.

There are no silver bullets in combating climate change, including improved seed. What is clear though is that any proposed intervention and practice should include women as part of the solution through researchers and communities working to overcome gender-based constraints to women's adoption of climate resilience practices (Kristjanson et al. 2017). This is part of the "transformative change" vision put forth by CCAFS to overcome gender norms and empower women to become agents of change in the face of climate change (Jost et al. 2016). Gender-responsive CSA practices and technologies provide an opportunity to close the gender gap as well as bring women into the forefront in the fight against climate change (Nyasimi and Huyer 2017). There is great potential in empowering women to lead the drive against climate change in SSA.

4 Gender and Agricultural Extension Services

Agricultural extension services play a key role in farmers' uptake of improved technologies. The traditional role of such services has been to provide information on reliable sources of quality inputs (including seeds) to farmers coupled with unbiased training and advice on proper utilization and agronomic practices. More recently, the role of extension is evolving to embrace emerging needs to organize farmers into institutions that promote collective action, empowerment, voice, and efficient access to productivity enhancing technologies and services. This is in response to the increased recognition of the complexity involved as practitioners of commercializing SSA agriculture have began to adopt a value chain focus. The prevailing setup necessitates linking farmers with a range of actors in the innovation system through multi-stakeholder innovation platforms and other institutions where extension's capacity building, brokering, and convening roles are central.

Smallholder women farmers in SSA are not able to fully benefit from extension services that support climate response despite their crucial role illustrated in the above sections. Many obstacles to gender equitable extension services are rooted in the local culture of various SSA countries which shapes women's roles and position in society. Gender norms which manifest in extension organizations, farmer organizations, and households largely explain this status quo. They shape the extension staffing, methods used, extension packages, or messages promoted all of which disadvantage women. There are bottlenecks at the level of access whereby women have been found to have less contact with services compared to men; and participation in farmers' institutions is often used as a vehicle for service delivery.

Women often have to travel to access information and training and purchase seeds and other technologies, yet they are often more constrained than men in their movements. Gender norms that vary across societies determine where, when, for how long, and for what reasons women should leave their homes. These restrictions are more prevalent in rural areas (Mandel 2004; Porter 2011). Constraints related to mobility influence women's access to information, training, and adoption of agricultural technologies (Bergman Lodin et al. 2018 in press). However, in some regions, unmarried women are often able to move more freely to secure a living since they do not need to seek approval from husbands or other family members. Several case studies (Uganda, Kenya, Malawi) from the GENNOVATE project indicate that the inability of women to move freely limits their learning and exposure to agricultural related information. Reasons postulated for this restriction include withholding of permission by families and communities, household drudgery, and jealousy by husband for fears of promiscuity (GENNOVATE 2017). Restrictions on interaction with men extension staff further limit opportunities to participation in extension activities. Other cultural prohibitions include women not being allowed to leave home alone, use public transport, or ride a motorbike all of which effectively prevent women from attending trainings in neighboring villages or work as extensionists (GIZ 2013).

Women need resources to access improved seed, complementary technologies, and supporting extension services to cope with climate change. Lack of decision-making powers on certain agricultural enterprises and lack of access to and control over land and other factors of production demotivate women from investing time and other resources needed to participate in extension activities related to such enterprises. Gender norms around roles and responsibilities in productive, reproductive, and community spheres place a heavy burden on women. Women's disproportionately heavier workload also rooted in the norm that women are expected to take care of domestic work, childcare, elder care, and care of the sick among other domestic tasks, leaves them with limited time to earn an income and seek services available in the public sphere which would promote technology utilization.

Besides social norms in the wider community, organizational gender-based barriers within the extension service organizations and organizations offering complementary services (e.g., microfinance and marketing) further limit capacity to effectively serve women farmers. Within extension organizations, women extension staff face challenges attributed to traditional, male-dominated organizational dynamics and cultural barriers to women's education in science and agriculture, professional performance, and career advancement (Mangheni et al. 2010; FARA and AFAAS 2015). These challenges have partly led to women's minority status in agricultural professions in general and in many extension organizations in particular, especially in leadership positions (Mangheni et al. 2010). This in turn contributes to gender-biased decision-making and priority setting within extension organizations (Manyire and Apekey 2010). This is particularly important in cultures that restrict interactions between men extension providers and women farmers. Other gender issues within extension organizations include extension messages that are not responsive to strategic agricultural activities, interests, and responsibilities of women small-scale farmers' choice of advisory methods and approaches that exclude women, for instance, those requiring high literacy levels and commitment of much time (which women don't have) also exclude women. In Malawi, Mudege et al. (2015) reported that gender and cultural norms influence access to training and agricultural information. It was revealed that with men mostly identified as household heads, extension officers tend to be biased in selecting more men than women when running their training programs. Inadequate capacity within agricultural extension service organizations creates a key barrier to gender-responsive services. There is inadequate gender awareness and capacity in most organizations (Chipeta 2013), particularly the public sector which has the mandate for policy guidance and quality assurance within the pluralistic extension systems consisting of multiple providers in the public and private sector.

Addressing the above issues calls for gender-responsive extension and advisory services designed and implemented in a manner that effectively addresses needs (practical and strategic), interests, and issues affecting men and women beneficiaries, with guiding principles of promoting gender equity and women's agency (Chipeta 2013). Such services would by necessity take into consideration the complex sociocultural aspects of the target communities and other relevant institutions including the implementing extension organizations in order to deliver gender-equitable agricultural extension that empowers women (GIZ 2013). The packages disseminated to farmers and the delivery approaches and methods used to reach them should incorporate gender-specific targeting of women by providing them with their preferred types of seed varieties, technologies, information, and knowledge, in a form they understand and can use, within an organizational and institutional context guided by the principles of gender equality and women's empowerment. For example, in instances where women farm smaller plots and have low purchasing power, extension can promote marketing of seeds in smaller packs suited to the women's needs. The organizational and institutional context is key because institutions (rules, attitudes, routines) and policies form the enabling environment that largely determines practices (Rasheed and Davis 2012). Transformation at this level is a key driver of gender equity and women's empowerment.

A synthesis of findings from seven country scoping studies (Benin, Ethiopia, Ghana, Malawi, Nigeria, Sudan, and Uganda) conducted by the Forum for Agricultural Research in Africa (FARA) and the African Forum for Agricultural Advisory Services (AFAAS) via desk reviews and key informant interviews documented case studies and identified best practices for effectively targeting women and youth (FARA and AFAAS 2015). The study recommended that gender-responsive extension approaches should address the formal and informal exclusion and/or unfavorable inclusion of women in the development process. In addition, best practices for gender-responsive rural advisory services should embody characteristics along the continuum of targeting deliberate participation, inclusion, and empowerment of women and, ultimately, transformation of the gender status quo. Tackling the root causes of women's subordination calls for questioning the status quo which is often sensitive, with a risk of eliciting resistance from the community. This requires long-term interventions, a lot of funds, and expertise which many extension service providers do not ordinarily possess. Such interventions also fall outside the conventional mandate of most extension organizations particularly those in government ministries of agriculture. Effective interventions would therefore call for establishment of partnerships with other organizations and/or review of mandates, structures of extension organizations, as well as training curricula for extension service providers. Since the concerned organizations fall under different ministries/organizational settings, coordinated harmonized action is quite challenging. This may explain why most innovations in gender-responsive extension often make no serious attempts to address cultural root causes of women's marginalization. In order to achieve women's empowerment goals, advisory services will have to include practical services that address issues related to women's rights and reduce obstacles to women's participation in extension services, for example, services that enhance women's voice in households and society and secure property rights. However, for greater impact, these need to be combined with or linked to other types of advocacy efforts at higher strategic and policy levels.

5 Conclusion

Climate change is a complex challenge threatening both current and future agricultural productivity in SSA. It is not only the biophysical unpredictability that defines the complexity of climate change and its potential impact on smallholders in SSA but also the sociocultural context in which communities face these challenges. Gender norms have enormous implications on the "who's" and "how's" that could underpin climate mitigation strategies. CSA practices offer a diverse basket of options to communities impacted by climate change, shaping how they may adapt to changing patterns of rainfall and temperature. This chapter considered one of these CSA options: improved seeds. We argue that effective and efficient interventions in seed systems that foster resilience to climate stress should be gender responsive. This applies from the varietal design stage, where design principles for considering gender throughout the crop breeding cycle should be applied to the dissemination stage through seed and extension systems. There is a need to understand the gender norms governing the target communities to identify the constraints and opportunities for targeted seed interventions. Known gender-based constraints including access to labor, land, inputs, and training may all impact the adoption and utilization of CSA seeds and must therefore be carefully mapped out and identified through gender analysis and foresight studies. This diagnosis should anticipate potential negative impacts on women such as increase in workload and reduced access to seed arising from privatization of seed systems and commercialization.

Extension services that support CSA, seed, and technology dissemination programs should be designed to positively impact women. However, overcoming gender bias requires an understanding of the root causes of inequitable extension service provision, rather than simplistic strategies that address symptoms by targeting women. This calls for rigorous studies to map out and understand negative gender norms affecting extension which are often grounded in the wider societal and policy environment and rural development norms. Such studies can also uncover positive gender norms, on which one can build more effective CSA.

The vision for transformative change to empower women to become more effective agents of change in the face of climate change is a powerful narrative that aims to empower women by challenging norms and practices that perpetuate their disadvantaged status. CSA practices, including seeds and systems to disseminate these, should be coupled with these transformative approaches to improve on the norms that hinder or limit the participation of women. Moreover, understanding gender norms surrounding agricultural practices and agricultural technology uptake is a vital step toward appropriate and gender-responsive strategies for technology adoption. Gender equity is a key determinant in success of CSA practices, and expansion to scale is only possible if gender norms are challenged in partnership with communities to ensure that women have the same opportunities to combat climate change. We conclude that challenging gender norms around seed systems and extension systems in SSA will increase our chance of success in mitigating climate disasters. Women are affected by climate change as much as if not more than men, so they must be given a place at the table and a voice to help shape the solutions that will have such a large impact on their future as farmers in SSA. It is clear that considering gender in CSA technology design and dissemination is not only smart but necessary if success of a technology is to be determined by the number of lives it improves on adoption.

Acknowledgment The authors acknowledge the contribution of Dr. Carol Colfer (Cornell University) and Dr. Peace Musiimenta (Makerere University) for their input into the chapter as internal reviewers.

References

Ajambo S, Rietveld A, Nkengla L, Omondi A, Niyongere C, Dhed'a D, Olaosebikan DO, Nitunga E, Toengaho J, Lava Kumar P, Hanna R, Sufo Kankeu R (2018) Recovering banana production in banana bunchy top affected areas in Sub-Saharan Africa: developing gender-responsive approaches. ActaHortic. 1196. ISHS 2018. https://doi.org/10.17660/ActaHortic.2018.1196.27

Ajani OI (2008) Gender dimensions in agriculture, poverty, nutrition and food security in Nigeria. Nigeria Strategy Support Program (NSSP). IFPRI. pp. 5–7

Bergman Lodin J, Tegbaru A, Degrande A, Nkengla-Asi L, Gaya HI, Bullock R (2018) Gendered mobilities and immobilities and women's and men's capacities for agricultural innovation. Insights from Kenya and Nigeria. Gend Place Cul, in press

Cavatassi R, Lipper L, Narloch U (2011) Modern variety adoption and risk management in drought prone areas: insights from the sorghum farmers of eastern Ethiopia. Agric Econ 42:279–292. https://doi.org/10.1111/j.1574-0862.2010.00514.x

Chipeta S (2013) Gender Equality in Rural Advisory Services. Towards a common understanding: A working document. GFRAS Working Group on Gender Equality in Rural Advisory Services. Global Forum for Rural Advisory Services (GFRAS), Lindau, Switzerland. http://www.g-fras.org/en/activities/gender-equality-in-ras-3.html. Accessed 11 May 2018

Dar MH, de Janvry A, Emerick K, Raitzer D, Sadoulet E (2013) Flood-tolerant rice reduces yield variability and raises expected yield, differentially benefitting socially disadvantaged groups. Sci Rep 3. https://doi.org/10.1038/srep03315

Doss C (2001) How does gender affect the adoption of agricultural innovations? The case of improved maize technology in Ghana. Agric Econ 25:27–39. https://doi.org/10.1016/S0169-5150(00)00096-7

Doss C (2002) Men's crops? Women's crops? The gender patterns of cropping in Ghana. World Dev 30(11):1987–2000

Doss CR, Morris ML (2001) How does gender affect the adoption of agricultural innovations: the case of improved maize technology in Ghana. Agric Econ 25(1):27–39

Fisher M, Carr ER (2015) The influence of gendered roles and responsibilities on the adoption of technologies that mitigate drought risk: the case of drought-tolerant maize seed in eastern Uganda. Glob Environ Chang 35:82–92

FARA and AFAAS (2015) Scoping study of gender-responsive agricultural services for rural poverty reduction in Africa. Accra, Ghana, FARA

FAO (2011) The state of food and agriculture 2010–2011: women in agriculture: closing the gender gap for development. Food and Agriculture Organization of the United Nations, Rome, Italy

Galiè A (2013) Governance of seed and food security through participatory plant breeding: empirical evidence and gender analysis from Syria. Nat Res Forum 37:31–42. https://doi.org/10.1111/1477-8947.12008

Gilbert RA, Sakala WD, Benson TD (2002) Gender analysis of a nationwide cropping system trial survey in Malawi. Available at http://agris.fao.org/agrissearch/search.do?recordID=GB2013201770. Accessed 11th May 2018

GENNOVATE RTB-HT team (2017) Gender in agricultural change: towards more inclusive innovation in farming communities. GENNOVATE Report to the CGIARResearch Programs on Roots, Tubers and Bananas and Humidtropics. GENNOVATE Research Paper, RTB, Lima, Peru

GIZ (2013) Gender and agricultural extension. A report produced for the Federal Ministry for Economic Cooperation and Development (BMZ)

Huyer S, Twyman J, Koningstein M, Ashby J, Vermeulen SJ (2015) Supporting women farmers in a changing climate: five policy lessons. CCAFS Policy Brief 10. CGIAR Research Programme on Climate Change, Agriculture and Food Security (CCAFS), Copenhagen, Denmark

Jacobson JP (2011) The role of technological change in increasing gender equity with a focus on information and communications technology. Wesleyan university, Middletown

Jeckoniah J, Mdoe N, Nombo C (2013) Mapping of gender roles and relations along onion value chain in Northern Tanzania. Int J Asian Soc Sci 3(2):523–541

Jost C, Kyazze F, Naab J, Neelormi S, Kinyangi J, Zougmore R, Aggarwal P, Bhatta G, Chaudhury M, Tapio-Bistrom ML, Nelson S (2016) Understanding gender dimensions of agriculture and climate change in smallholder farming communities. Clim Dev 8(2):133–144

Kabeer N (1994) Reversed realities. Gender hierarchies in development thought. London, Verso

Kabeer N (1999) Resources, agency, achievements: reflections on the measurement of women's empowerment. Dev Chang 30:435–464

Kristjanson P, Waters-Bayer A, Johnson N, Tipilda A, Njuki J, Baltenweck GD, MacMillan D (2014) Livestock and women's livelihoods: a review of the recent evidence. In: Quisumbing A, Meinzen-Dick R, Raney T, Croppenstedt A, Behrman JA, Peterman A (eds) Gender in agriculture closing the knowledge gap. Springer with FAO, New York, NY, pp 209–233

Kristjanson P, Bryan E, Bernier Q, Twyman J, Meinzen-Dick R, Kieran C, Ringler C, Jost C, Doss C (2017) Addressing gender in agricultural research for development in the face of a changing climate: where are we and where should we be going? Int J Agric Sustain 15:482–500. https://doi.org/10.1080/14735903.2017.1336411

Lipper L, Thornton P, Campbell BM et al (2014) Climate-smart agriculture for food security. Nat Clim Chang 4:1068–1072. www.nature.com/natureclimatechange

Mandel J (2004) Mobility matters: women's livelihood strategies in Porto Novo, Benin. Gend Place Cult 11(2):257–287

Mangheni MN, Ekirikubinza TL, Forsythe L (2010) Gender issues in agricultural education within African Universities. Gender background paper. Ministerial conference on higher education in agriculture in Africa. Accessed 10/8/2018 from https://www.nri.org/images/documents/development-programmes/gender_soc_dif/publications/Gender_Issues_AgrHrhEd.pdf

Manyire H, Apekey AD (2010) Mainstreaming gender equality in African agricultural research and development: a study of constraints and opportunities. Forum for Agricultural Research in Africa (FARA), Accra, Ghana

Maredia MK, Howard JA, Boughton D, Naseem A, Wanzala MN, and Kajisa K (1999) Increasing Seed System Efficiency in Africa: Concepts, Strategies and Issues, Food Security International Development Working Papers 54578, Michigan State University, Department of Agricultural, Food, and Resource Economics.

McGuire S, Sperling L (2016) Seed systems smallholder farmers use. Food Sec 8:179–195. https://doi.org/10.1007/s12571-015-0528-8

Meinzen-Dick R, Quisumbing A, Behrman J, Biermayr-Jenzano P, Wilde V, Noordeloos M, Beintema N (2012) Engendering agricultural research, development and extension. International Food Policy Research Institute (IFPRI), Washington, DC. Retrieved from http://www.ifpri.org/sites/default/files/publications/rr176.pdf. Accessed, 10th May 2018

Mudege NN, Mdege N, Abidin PE, Bhatasara S (2017) The role of gender norms in access to agricultural training in Chikwawa and Phalombe, Malawi. Gend Place Cul 24(12):1689–1710. https://doi.org/10.1080/0966369X.2017.1383363

Mudege NN, Chevo T, Nyekanyeka T, Kapalasa E, Demo P (2015) Gender norms and access to extension services and training among potato farmers in Dedza and Ntcheu in Malawi. J Agric Educ Ext 22(3):291–305. https://doi.org/10.1080/1389224X.2015.1038282

Naab JB, Koranteng H (2012) Using a gender lens to explore farmers' adaptation options in the face of climate change: results of a pilot study in Ghana. Working Paper No. 17, CGIAR Research Programme on Climate Change, Agriculture and Food Security. Nairobi, Kenya

Nkengla L, Olaosebikan DO, CheVicent S, Rachid H, Lava Kumar P (2018) Gender norms and implications for Banana recovery and Banana Bunchy Top Disease containment in West Africa. J Gend Agric Food Sec, in press

Nyasimi M, Huyer S (2017) Closing the gender gap in agriculture under climate change. Agric Dev 30:37–40

Nyasimi M, Kimeli P, Sayula G, Radeny M, Kinyangi J, Mungai C (2017) Adoption and dissemination pathways for climate-smart agriculture technologies and practices for climate-resilient livelihoods in Lushoto, Northeast Tanzania. Climate 5:63. https://doi.org/10.3390/cli5030063

Padmanabhan MA (2002) Trying to grow: gender relations and agricultural innovations in Northern Ghana. Rurale Geschlechterforschung 3:255. Munster: LIT Verlag

Peterman A, Behrman JA, Quisumbing AR (2014) A review of empirical evidence on gender differences in non-land agricultural inputs, technology, and services in developing countries. In: Gender in agriculture. Springer, Dordrecht, pp 145–186

Phiri D, Franzel S, Mafongoya P, Jere I, Katanga R, Phiri S (2004) Who is using the new technology? The association of wealth status and gender with the planting of improved tree fallows in Eastern Province, Zambia. Agr Syst 79:131–144

Porter G (2011) "I think a woman who travels a lot is befriending other men and that's why she travels": mobility constraints and their implications for rural women and girls in sub-Saharan Africa. Gend Place Cul 18(1):65–81

Quisumbing AR, Pandolfelli L (2010) Promising approaches to address the need of poor female farmers: resources, constraints, and interventions. World Dev 38(4):581–592

Rasheed SV, Davis K (2012) The "new extensionist": Roles, strategies and capacities to strengthen extension and advisory services. Global Forum for Rural Advisory Services (GFRAS), Lindau, Switzerland. http://www.g-fras.org/en/activities/the-new-extensionist.html. Accessed 11 May 2018

Ragasa C (2012) Gender and institutional dimensions of agricultural technology adoption: a review of literature and synthesis of 35 case studies. http://ageconsearch.umn.edu/bitstream/126747/2/IAAE.2012.gender.pdf. Accessed 08, 2018

Rubyogo JC, Sperling L, Nasirumbi L, Kasambala SP (2007) Developing seed systems with and for the marginalized : Case of common bean (Phaseolus vulgaris L.) in East, Central and Southern Africa. In: Farmer First Revisited Conference (2007, Sussex, England). Papers presented. Future Agricultures Consortium; Institute of Development Studies (IDS), Sussex, Great Britain. https://hdl.handle.net/10568/56162

Sanginga PC, Adesina AA, Manyong VM, Otite O, Dashiell KE (2007) Social impact of soybean in Nigeria's southern Guinea savanna Ibadan, Nigeria: International Institute of Tropical Agriculture, Ibadan, Nigeria

Smale M, Assima A, Kergna A, Thériault V, Weltzien E (2018) Farm family effects of adopting improved and hybrid sorghum seed in the Sudan savanna of West Africa. Food Policy 74:162–171. https://doi.org/10.1016/j.foodpol.2018.01.001

Smale M, Kaunda ZHW, Makina HL, Mkandawire MMMK, Msowoya MNS, Mwale DJEK, Heisey PW (1991) Chimanga cha makolo, hybrids, and composites: an analysis of farmers' adoption of maize technology in Malawi, 1989–91. Economics Program Working Paper 91–04. CIMMYT, Mexico, DF

Spencer RA, Rehman L, Kirk SF (2015) Understanding gender norms, nutrition, and physical activity in adolescent girls: a scoping review. Int J Behav Nutr Phys Act 12(1):–6

Sperling L, Boettiger S, Barker I (2014) Integrating seed systems, Planning for Scale Brief # 3: AgPartnerXChange. https://seedsystem.org/wp-content/uploads/2014/03/Integrating-Seed-Systems-.pdf

Twyman J, Green M, Bernier Q, Kristjanson P, Russo S, Tall A, Ndourba Y (2014) Adaptation actions in Africa: evidence that gender matters (CCAFS Working Paper no. 83). Copenhagen: CGIAR Research Program on Climate Change, Agriculture and Food Security (CCAFS). Retrieved from https://cgspace.cgiar.org/handle/10568/51391

Twyman J, Muriel J, García M (2015) Identifying women farmers: informal gender norms as institutional barriers to recognizing women's contributions to agriculture. J Gend Agric Food Sec 1(2):1–17

Wencélius J, Thomas M, Barbillon P, Garine E (2016) Interhousehold variability and its effects on seed circulation networks: a case study from northern Cameroon. Ecology and Society 21. https://doi.org/10.5751/ES-08208-210144

United Nations Statistics Division (UNSD) (2015) Global gender statistics programme. Available at: http://unstats.un.org/unsd/genderstatmanual/Glossary.ashx. Accessed on, 09th May 2018
World Bank, Food and Agriculture Organization, International Fund for Agricultural Development (2008) Gender in agriculture sourcebook. The World Bank. https://doi.org/10.1596/978-0-8213-7587-7
World Bank Group, FAO and IFAD (2015) Gender in climate-smart agriculture: module 18 for the gender in agriculture sourcebook. http://www.fao.org/3/a-i5546e.pdf. Accessed 27 May 2018
Yegbemey RN, Yabi JA, Tovignan SD, Gantoli G, HarollKokoye SE (2013) Farmers' decisions to adapt to climate change under various property rights: a case study of maize farming in northern Benin (West Africa). Land Use Policy 34:168–175. https://doi.org/10.1016/j.landusepol.2013.03.001

Identification of Optimal Agricultural Development Strategies in the West African Sahel Mékrou Transboundary River Basin

Marco Pastori, Céline Dondeynaz, Bernard Minoungou, Angel Udias, Iban Ameztoy, Mohamed Hamatan, Luigi Cattaneo, Abdou Ali, César Carmona Moreno, and Paolo Ronco

Abstract Mékrou is a transboundary river basin across Bénin, Burkina Faso, and Niger. In this area, agricultural production plays a major strategic role in guaranteeing household food security for the sustainable development of the basin. Increasing agricultural productivity is crucial for meeting the growing population demands. Climate change (CC) is already affecting agricultural production due to increased frequency and intensity of extreme weather events. In this paper, a systematic approach to identify the main factors that influence observed yield gaps and to evaluate the impacts of agricultural intensification on the environment and support local development plans is presented. By using a newly developed decision support system tool (E-Water) linking a geographic information system with the biophysical model EPIC, we demonstrate that the agricultural sector has a growing potential by adopting more integrated and efficient agricultural production systems. The analysis of the crop yield gap shows that the main limiting factor for crop production is low soil fertility, while water is limiting during dry years. We predict that the adoption of more intensive farming techniques and sustainable irrigation practices can increase the capacity of agriculture to efficiently mitigate the impacts of climate change and other external environmental stress factors. We confirm that irrigation

M. Pastori · P. Ronco (✉)
GFT Italia, Milano, Italy
e-mail: Paolo.RONCO@ec.europa.eu

C. Dondeynaz · A. Udias · C. C. Moreno
European Commission, DG Joint Research Centre, Ispra, Italy

B. Minoungou · M. Hamatan · A. Ali
Centre Régional AGRHYMET, Niamey, Niger

I. Ameztoy · L. Cattaneo
Piksel Italia, Milano, Italy

Y. Bamutaze et al. (eds.), *Agriculture and Ecosystem Resilience in Sub Saharan Africa*, Climate Change Management,
https://doi.org/10.1007/978-3-030-12974-3_33

can substantially increase yield and that it will be required to adapt to expected climate and increase of rainfall variability.

Keywords Sustainable agriculture · Yield gap · Food security · Water management · Climate variability

1 Introduction

FAO reports that after declining for over a decade, global hunger appears to be on the rise, especially in rural Sub-Saharan Africa where agriculture is the main source of livelihood (FAO et al. 2017). In this region, agriculture is the predominant economic activity; it contributes to income generation and employment, and it is critical for the identification of optimal development strategies. In sub-Saharan Africa, agriculture is characterized by low productivity with a low level of intensification, and its development is considered vital for enhancing growth and for poverty alleviation (Conceição et al. 2016; Dzanku et al. 2015). Poverty reduction (and country growth) and agricultural productivity are strongly correlated (Thirtle and Piesse 2007). Unfortunately, climate change (CC) is already hampering agricultural growth due to increased frequency and intensity of extreme events such as droughts and flooding (Cohn et al. 2017; FAO 2016; Lesk et al. 2016; Lipper et al. 2014). It has been observed that climate change and variability have negatively impacted Western African Sahel rain-fed agriculture (Zougmoré et al. 2016), although agronomic studies suggest that low soil fertility management is the major limiting factor to agricultural production (Van Keulen and Breman 1990). High rainfall variability and climate change can severely impact the management of water resources, particularly in water-stressed transboundary river basins.

A 3-year cooperation project (funded by the European Commission DG DEVCO and starting in 2014) implemented by the Joint Research Centre of the European Commission and Global Water Partnership in West Africa addressed the issue of agricultural development in the Mékrou transboundary river basin (shared among Benin, Burkina Faso, and Niger). Food insecurity in this basin is perceived as a major issue, largely explained by very low agricultural productivity (Markantonis et al. 2017). There is a need to enhance resilience of local agricultural systems to cope with increasing food demands, climate risks, and competition for land. Agriculture is affected by a set of biophysical stressors, including climate change, which need to be spatially identified and quantified in order to identify effective measures and strategies that enhance agricultural resilience. Indeed, agricultural productivity is not only affected by climate change but also by other important factors including soil degradation, soil fertility management, soil conservation practices for water and crop management, etc.

Consequently, there is a need to identify the key factors that explain observed yield gaps (differences between current and maximum obtainable crop yields) to evaluate the impact of potential intensification on the environment and to support the development of sustainable local management plans. Biophysical models and decision support systems (DSSs) are computer software packages that allow the development of site-specific recommendations for different purposes such as agricultural pesticide management (Beck et al. 1989; Strassemeyer et al. 2017), management of livestock enterprises (Stuth and Lyons 1993), and crop and land management systems (Basso et al. 2013; Rossi et al. 2014; Xiong et al. 2010). DSSs have mainly been used by farm advisors and other specialists working with farmers and policymakers (Fraisse et al. 2015; Nelson et al. 2002), although some may be used directly by farmers. In addition to this type of farm-level decision-making support, agricultural system models are increasingly being used for various types of local, regional, national, and global modeling and analysis to provide support for policy design and implementation and investment decision (Jones et al. 2017).

In the context of the Mékrou transboundary river basin project, the E-Water DSS tool (Udias et al. 2018) was developed and applied in order to assess the WEFE (Water Energy Food Ecosystem) Nexus, providing optimal management solutions at river basin level for local managers. The local stakeholders were intensely involved in the development and the application of the DSS and in the formulation and testing of alternative catchment management plans. The Nexus approach allows the simultaneous consideration of several factors that all together contribute to the development of integrated agricultural systems strategies. Such integration limits reliance on external input and indirectly increases agricultural system resilience (Gil et al. 2017; Mapfumo et al. 2010; Salton et al. 2014).

2 The DSS System: E-Water

E-Water is a DSS conceived to allow users to manage and analyze data and scenarios at river basin and/or regional level. Given the large amount and distinct nature of datasets required in the context of the Nexus approach, the tool gathers all of them into a common dataset in order to provide a direct overview of suitable inputs for further operations. The software framework enables decision-makers to identify policy-relevant solutions, quantify trade-offs between alternative strategies in different scenarios, and identify key system factors to be monitored as triggers for future actions or additional planning (Fig. 1). In addition it allows decision-makers to easily share and visualize their datasets. The framework is flexible, and it can be extended with customized scripts in the R programming language. In more details, E-Water integrates several tools (SWAT and EPIC models, Refran-CV, L-Moments, Optimization routines, etc.) in order to cope with heterogeneity of data, analysis, and issues required by the Nexus approach such as (i) the management of pressures on environmental resources in the context of climate change and variability, (ii) the preservation of ecosystems and associated services and benefits, and (iii) the management and mitigation of floods and droughts.

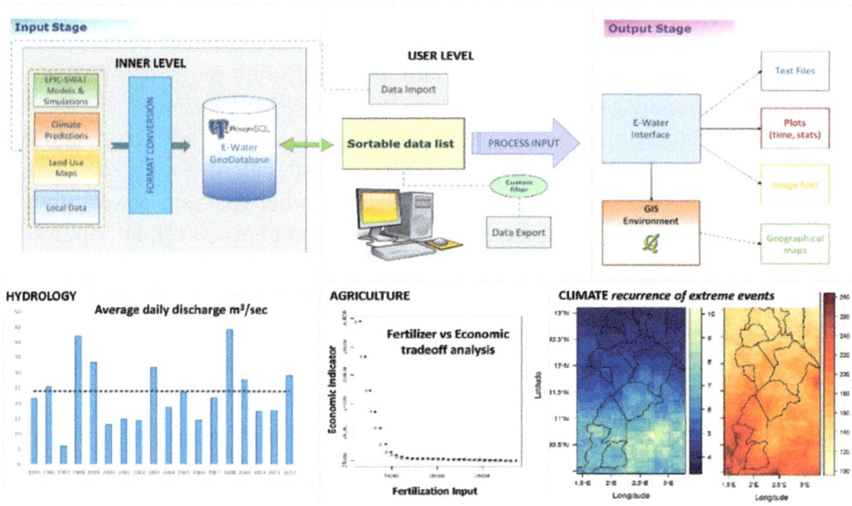

Fig. 1 The E-Water open-source framework

2.1 The Thematic Components

Several components have been developed within the DSS system. Each of these components can be used to assess a specific issue. All module capabilities (hydrological, multi-objective optimization for food security, climate variability), functionalities, and methodology can be found online in the AquaKnow platform (JRC 2018). This paper will focus more on the agricultural components of the tool.

The Agricultural Modeling Component

The biophysical EPIC model was integrated in the DSS in order to assess crop productivity and variability and nutrient/water requirements across an area of interest and develop alternative management strategies for nutrient and irrigation water.

EPIC is a biophysical, time-continuous field-scale agriculture management model (Williams 1995). It simulates crop growth and production, crop water requirements, and the fate of nutrients and pesticides as affected by farm management activities such as the timing of agrochemical application, tillage, crop rotation, irrigation strategies, etc. The main components can be divided into the following items: hydrology, weather, erosion, nutrients, and plant growth. The model estimates crop growth development on a daily time step based on light interception and conversion of CO_2 into biomass. The phenological development is based on daily heat units accumulation (PHU 2007). Water and nutrient stresses are based on the departure from the optimal supply on each day. Crop yield is calculated via harvest index as a fraction of aboveground biomass. The hydrological model is based on the water balance in the soil profile where the processes simulated include surface runoff and infiltration, evapotranspiration, lateral subsurface flow, and percolation. EPIC takes into account nitrogen and phosphorus cycles (mineralization, denitrification, volatilization, and fixation processes). For a detailed and complete description of the model and the simulated processes, see Williams (1995) and Wang et al. (2012). A specific geodatabase was developed to support the analysis and application of EPIC for the whole area of study. It includes all data required by EPIC (such as meteorological daily data, soil data, land use data with crop distribution, and agriculture management information) to simulate different agronomic management strategies and all data required to define the optimization analysis (such as population density, food calorie requirements per habitant, diet habits for each region and total food demands, future population growth trends, and agricultural product selling prices).

3 Description of Study Area

The Mékrou river basin is a sub-basin of the Niger River Basin that covers an area of 10,635 km^2 (about 3% of the total Niger River Basin area) (Fig. 2). The transboundary Mékrou river basin is distributed across Benin (80%), Burkina Faso (10%), and Niger (10%). It is characterized by underdeveloped infrastructure and weak socioeconomic conditions. In the context of this project, we defined a wider area of interest, named Mékrou Area of Influence (Fig. 2), corresponding to the 12 administrative communes involved in the management of water and natural resources belonging to the river basin. The arable land accounts for about 22% of the total area and is mainly used for food crop production and cattle breeding. Agriculture is the key sector of the economy in all the three riparian countries, and it is critical for poverty alleviation and food security. The current extension of the cultivated area of the whole region for the current analysis is about 730,000 ha.

The water resources of the Mékrou river basin are shared for different usages including domestic consumption, crop irrigation, animal production, fishing and fish farming, recreation, and religious practices. The need for clean drinking water in 2014 is estimated at 3.1×10^6 m^3 (2.0×10^6 m^3 for urban areas and 1.1×10^6 m^3 for rural areas), serving a total population of 280,000 inhabitants. The main pressures on water resources come from production activities linked to agriculture, livestock, fisheries, forests, hunting, mining, industry, and energy.

3.1 Development of Simulation Scenarios for Analysis of the Impact of Alternative Strategies on the Agricultural Sector

In the Mékrou river basin, the agricultural sector continues to face several constraints that hinder its development: cropping practices that do not maintain/protect soil fertility, drought periods during rain-fed cropping system seasons, difficulty in

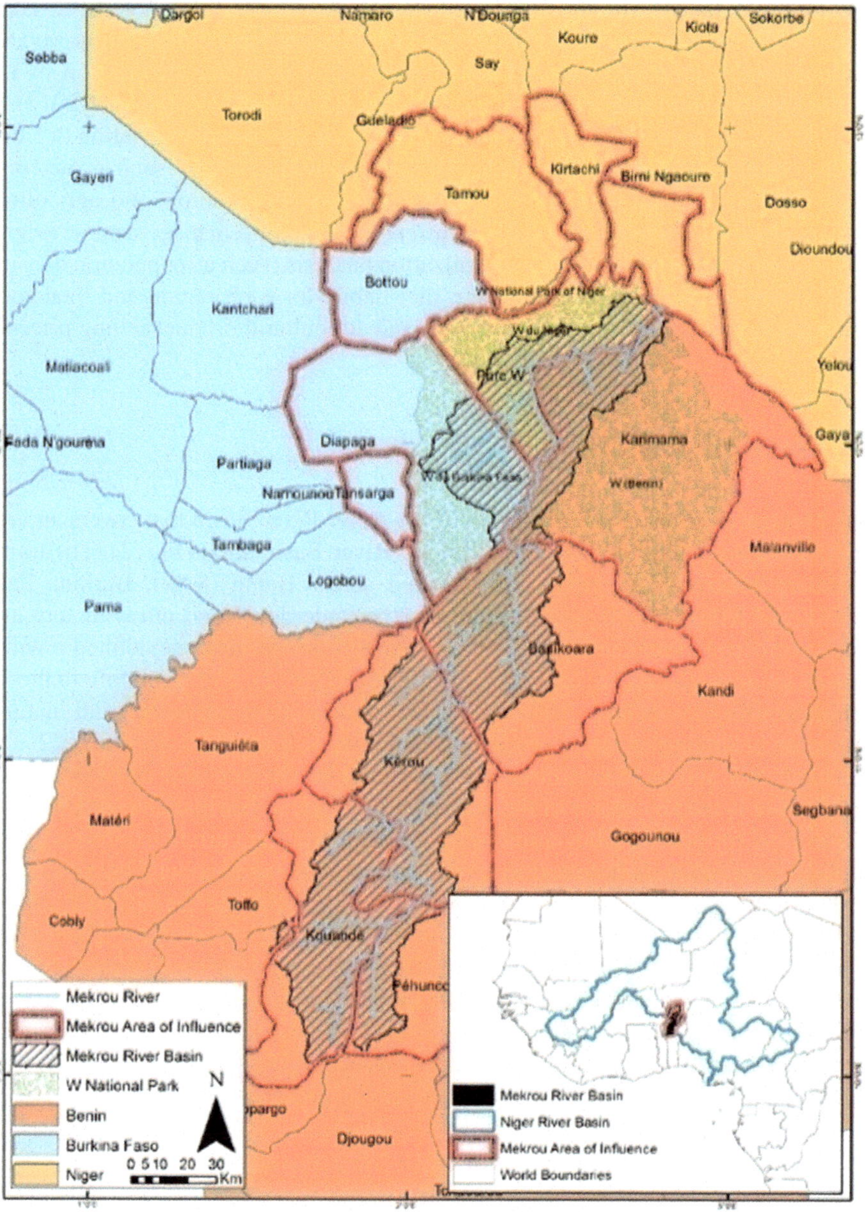

Fig. 2 Location of the Mékrou river basin and area of influence

accessing synthetic fertilizers, management of livestock manure, low mechanization, low seed quality, poor crop storing systems, competition with livestock and pastoral systems, abandonment of low fertility soil increasing erosion risks, etc. (IPCC 2007; Pastori and Bouraoui 2017). In addition low yields and high dependence on cropland expansion are viewed as a major threat to sustainability of African food and nutrition security and ecosystems. Actual fertilizer applications in the Mékrou river basin are generally low (3–12 kg·ha^{-1} average for all crops). This low input practice is reflected in crop productivity, which is the lowest in the region. Insufficient nutrient fertilizer inputs have been identified as a major cause for the extremely low crop production in Africa (Brams 1971; Folberth et al. 2012; Pastori et al. 2017b; Payne 2010). Agroecosystems with long-term low fertilization inputs can lead to serious soil nutrient depletion (Weight and Kelly 1998) and to loss of soil organic matter and consequently to loss of soil functionalities, thereby reducing resilience of agricultural systems.

In this paper, the EPIC model, specifically set up with local data by integrating the information discussed and provided during validation workshops (models and data validation with local experts and stakeholders within the context of the Mékrou Project, https://aquaknow.jrc.ec.europa.eu/mekrou-water4growth), was firstly used to simulate the impact of increasing fertilizer and water irrigation inputs so as to identify possible strategies to increase agricultural productivity and mitigating negative impacts of climate change. Other important analyses include the quantification of nutrient demands and water requirements for specific crops (rice, vegetables, and cereals). All these analyses were carried out in producing supporting information for land and water planning (and specifically in supporting local strategic planning such as "*CaSSE*" and "*SDAGE*" plans) in the river basin.

A Yield Gap Analysis for Identification of Factors Limiting Crop Productivity

A yield gap analysis was performed for each region to identify the major factors limiting crop growth and calculating nutrient and water requirements to bring crop production to the optimum level. Yield gap is defined in our study as the difference between potential crop yield under no water and nutrient constraints and the actual yield. In order to analyze the potential variability of crop production under different management strategies, EPIC is set up to run under alternative scenarios. The baseline (baseline management scenario BSL) was derived from current management practices. The scenarios are mainly based on two controlling factors, i.e., fertilization and irrigation. The first scenario (N1) is conservative with respect to water use and is characterized by a maximum fertilization strategy (very low nutrient crop stress allowed). In this scenario, the E-Water modeling system was set to run EPIC in a free auto-fertilization mode, while irrigation was identical to the baseline. With this configuration, the model applies fertilizer in order to maximize the yield according to crop nitrogen requirements. This scenario is focused on nitrogen fertilization, which has been identified as the main limitation in restricting yield production, usually more than water availability (Van Keulen and Breman 1990). This results in a water-limited potential yield that is a good benchmark for rain-fed agriculture (van Ittersum et al. 2013). The second scenario (W1) is conservative with respect to nutrient use and is characterized by a maximum irrigation strategy (very low crop water stress allowed). In this scenario the fertilization is set to that of the baseline, while the model runs in an auto-fertilization mode. The third scenario (O1) is the "high production potential" with no limitation of fertilizers and irrigation. This scenario aims at simulating the higher range of potential production of agriculture in the area under actual land use. Under these conditions, crops can always obtain sufficient water and nitrogen when stress occurs.

The EPIC model is applied for the most dominant crops of each grid cell (3 km resolution) for the time period extending from 1990 to 2016. The crops used for this analysis are maize, sorghum, millet, and rice for cereals; cassava, yam, and potatoes for tubers; a vegetable crop, cowpea, and soybeans for leguminous crops; and peanuts, oil crops, fruit crops (banana), and cotton for cash crops.

As already pointed out, under the current management, one of the most dominant factors reducing productivity is low levels of nutrient availability. This is reflected by the simulated total number of stress days due to lack of nitrogen. Nitrogen limitation is more marked in cereal crops such as maize, sorghum, millet, and rice, as well as in cotton, sugarcane, and vegetables (Table 1). In the case of maize and cotton, for example, about 45–55 days of stress is estimated (with baseline management plans), and this corresponds to a very long period of stress that would significantly affect the harvest. Grain legumes such as peanuts, cowpeas, and soybeans are good nitrogen fixers and can normally fix all of their nitrogen needs other than that absorbed from the soil. Indeed, these legumes are able to fix up to 280 kg of nitrogen per ha, and for this reason, they are not usually fertilized.

Table 1 Average stress day by crop type for nutrient and water availability under different scenarios

	Baseline	N1		W1		O1		
	Total number of stress days for nutrient and water availability							
Crop type	Nut.	Wat.	Nut.	Wat.	Nut.	Wat.	Nut.	Wat.
Cassava	10	9	8	9	11	5	9	5
Maize	45	17	14	26	59	4	23	11
Cotton	57	38	19	52	86	8	33	21
Cowpea	0	7	0	7	0	4	0	4
Peanuts	0	31	0	31	0	17	0	17
Millet	26	2	17	2	26	1	17	2
Potato	18	20	12	21	22	8	16	9
Rice	28	20	15	22	39	6	22	8
Sorghum	28	23	15	26	39	6	22	9
Sugar crops	31	21	19	22	46	6	29	10
Vegetables	30	7	0	0	34	2	1	7
Yam	23	16	7	36	29	5	11	16

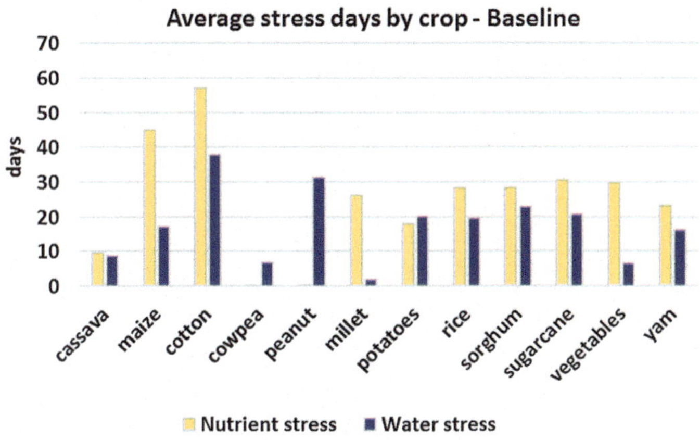

Fig. 3 Average number of simulated days of crop stress according to nutrient and water requirements for dominant crops in the Mékrou river basin under current agricultural management strategy

In the following section, the analysis of yield gap is used to identify the major factors limiting growth for each crop and to calculate nutrient and water requirements to bring production to near optimal levels. For instance, in the case of maize, all the areas in the Mékrou river basin produce yield significantly lower than optimal. More specifically, in the Banikoara community, the current yield gap is about 75%, while adopting a scenario with higher fertilization, the yield gap can be potentially reduced to 20% and only to 72% when increasing irrigation (Fig. 3).

It is clear that the most limiting factor influencing maize production is nutrient availability. Table 2 reports the ratio among percentage differences in Baseline vs Scenario Fertilization F1, and the percentage difference in Baseline vs Scenario Irrigation W1 (where values over 1 mean the increase related to added water is more important than the increase for added fertilizer) is close to 0.

Therefore, in general, most of the crops and for all the river basin areas, the main limiting factors are nutrient fertilization and low soil fertility. For potatoes, cassava, and rice, there is a more complex situation; and in this case, the limiting factors can be nitrogen and/or water, depending on the region; for example, results in Kérou commune (Benin) show that potato production is mostly controlled by irrigation. Crop yields can significantly increase with an increase in nitrogen and water applications. A more detailed analysis can be performed to assess if the estimated additional water requirements are available at river basin level (hydrological module and multi-objective optimization (MOO)) by considering the requirements of competing sectors as well, including the environment, through the additional potential losses of nutrients to water bodies.

This first assessment highlights the need to increase soil fertility management across the river basin through appropriate nutrient and crop management. To this end, the potential of a better integration of agricultural and livestock systems needs to be analyzed by including manure as a source of fertilizer and a mean to increase soil fertility.

Table 2 Impact on yield productivity for different scenarios and some example crops typically produced in the Mékrou river basin: yield gap is calculated versus optimal productivity scenario O1

	Yield tons/ha				Ratio	Yield gap (vs optimal O1)		
Commune	Bas	N1	W1	O1	[(N1 – Bas)/N1]/ [(W1 – Bas)/W1]	Bas.	N1	W1
Maize								
Banikoara	1.3	3.9	1.4	4.9	0.04	−74	−20	−72
Karimama	2.3	3	2.5	4.3	0.33	−48	−32	−43
Kerou	2.2	6.2	2.2	6.7	0.01	−68	−8	−67
Kouande	2.6	6.1	2.6	6	0.01	−57	2	−56
Pehunco	3.3	6.4	3.3	6.3	0.01	−47	2	−47
Bottou	1.4	3.8	1.5	4.3	0.03	−67	−11	−65
Diapaga	1.6	4.1	1.7	4.5	0.03	−65	−9	−63
Tansarga	1.8	5.9	1.9	6.1	0.01	−71	−4	−70
Kirtachi	1.2	4.2	1.3	6.5	0.03	−82	−35	−80
Tamou	1.1	4.7	1.2	6	0.02	−82	−20	−80
Falmey	1.5	4.7	1.6	6.2	0.03	−76	−25	−74
Potatotes								
Banikoara	13.4	14.1	14	15.3	0.8	−13	−8	−9
Karimama	10.5	15.9	11.4	17.1	0.17	−39	−7	−33
Kerou	16.1	16.3	16.4	17.1	1.53	−5	−5	−4
Kouande	12.6	24.3	12.6	23.7	0	−47	2	−47
Pehunco	14.7	22.9	14.7	21.8	0	−33	5	−32
Bottou	13.4	15.2	13.9	15.5	0.26	−13	−2	−10
Diapaga	13.4	16.3	13.9	17.3	0.18	−22	−6	−19
Tansarga	16.3	17.5	16.9	18.2	0.5	−10	−4	−7
Kirtachi	8.8	20.5	9.3	20.6	0.04	−57	0	−55
Tamou	10.5	18.7	11.1	20.4	0.07	−49	−8	−46
Falmey	8.5	18.4	8.9	18.5	0.04	−54	0	−52
Vegetables								
Banikoara	7.7	17.9	8.2	23.3	0.05	−67	−23	−65
Karimama	6.1	13.8	6.8	19.5	0.08	−68	−29	−65
Kerou	7.9	22.6	9.6	23.6	0.11	−66	−4	−59
Kouande	9.9	28.5	9.9	29.1	0	−66	−2	−66
Pehunco	10.1	28.4	10.1	28.9	0	−65	−1	−65
Bottou	7.8	20.1	7.8	24.6	0	−68	−18	−68
Diapaga	7.9	20.4	8	24.8	0	−68	−18	−68
Tansarga	9.6	24.8	9.6	27	0	−65	−8	−65
Kirtachi	8.4	18.6	8.5	24.9	0.01	−66	−25	−66
Tamou	8.3	20.2	8.4	24.5	0.01	−66	−18	−66
Falmey	8.1	18.9	8.1	24.4	0	−67	−23	−67

(continued)

Table 2 (continued)

	Yield tons/ha				Ratio	Yield gap (vs optimal O1)		
Commune	Bas	N1	W1	O1	[(N1 – Bas)/N1]/ [(W1 – Bas)/W1]	Bas.	N1	W1
Rice								
Banikoara	3.3	3.3	3.5	3.5	221.27	−5	−5	0
Karimama	4	4	4.6	4.7	924.63	−14	−14	−1
Kerou	3.7	3.7	3.7	3.8	3.54	−3	−3	−2
Kouande	2.3	5.1	2.3	5	0	−54	0	−54
Pehunco	2.7	5.6	2.7	5.6	0	−52	0	−52
Bottou	1.8	2.6	1.9	3.8	0.14	−52	−31	−49
Diapaga	2.1	3	2.2	3.4	0.12	−38	−11	−35
Tansarga	2.1	3.1	2.2	3.5	0.08	−40	−12	−38
Kirtachi	1.3	3	1.5	4.1	0.08	−67	−27	−64
Tamou	1.3	3.2	1.4	4	0.05	−67	−21	−65
Falmey	1.4	3.1	1.5	4	0.07	−65	−22	−62

Irrigation Strategies to Mitigate Climate Variability Impact on Crop Productivity

The spatial and temporal variability of crop production depends on several factors. Climatic factors are more important in the case of rain-fed agriculture, where crop growth is affected by temperature and rainfall. In rain-fed agriculture, the most important climatic variable affecting crop growth is the rainfall distribution during the year. In order to analyze the impact of climate change and variability on crop production, regional climatic scenarios (provided by the Agrhymet Regional Centre, Niamey) were identified, analyzed and bias-corrected for integration and use within the E-Water agricultural module (EPIC modeling).

The climatic scenarios used are based on climate models AFR-44 CORDEX (Giorgi et al. 2009) and two different scenarios of climate change (RCP4.5 and RCP8.5). The CORDEX Africa domain ("*AFR-44*") has a resolution of 0.44°. The precipitation trend is expected to remain stable on average for both scenarios on the Mékrou (Fig. 4). While annual precipitation quantities remain stable or slightly more abundant at horizon 2100, the interannual variability is expected to increase from 2030: this aspect is particularly pronounced for RCP8.5 scenario.

The EPIC model specifically allows assessment of the annual variability of rain-fed cropping systems as affected by dry years. For the Mékrou river basin, dry years mean cropping seasons characterized by late start or anticipated end of the rainy season. The analysis focused on maize as the dominant crop for food security.

The impact of climate risk in controlling crop productivity is shown by variations of the simulated annual crop yield. Crop productivity can be very different from year to year (Fig. 5). Under the baseline scenario, maize productivity fluctuates between −15% and +15%. Precipitation ranges between 768 and 1612 mm, while

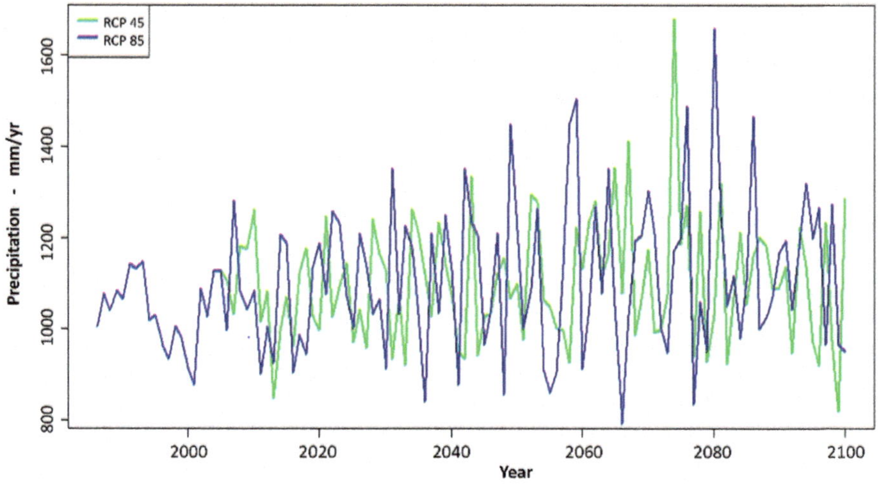

Fig. 4 Rainfall variability in the Mékrou river basin under climate scenarios RCP45 and RCP85 (Agrhymet)

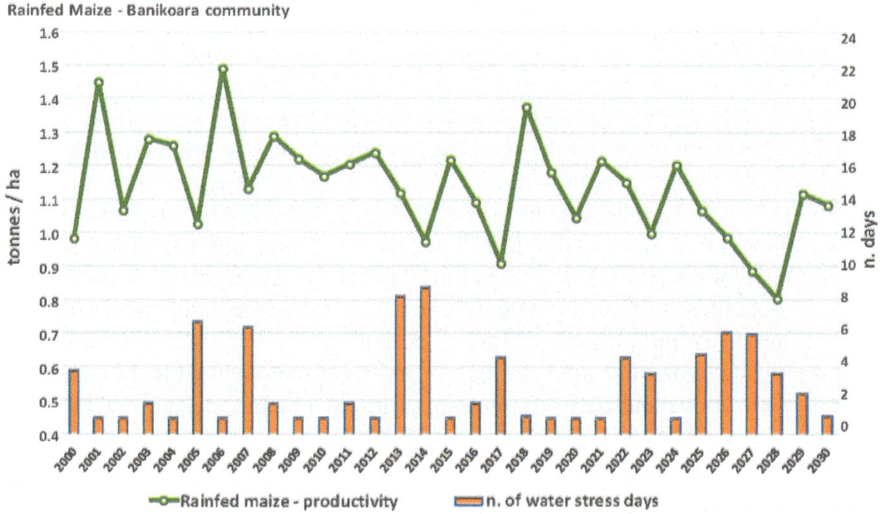

Fig. 5 Annual maize productivity in an agricultural area of the Mékrou river basin (Banikoara) estimated by assuming a rain-fed agriculture (baseline) for the period 2000–2030 (The green line is the productivity of maize; the blue line is the number of days with water stress during the growing season)

the annual average is about 1060 mm for the simulated period 2000–2030. The fertilization rate applied is very low (3 kg/ha), and therefore, the dominant limiting factor is again nutrient availability. Nevertheless, an important effect of water availability (rainfall) on crop productivity can be clearly observed, at least for drier years (Fig. 5, Table 3). Years with higher water stress (characterized by the highest num-

Table 3 Mean dry yields, max, min, and coefficient of variation (CV) simulated in a region within the Mékrou river basin under a management scenario with rain-fed and medium fertilized agriculture

	Maize	Cotton	Rice	Yam	Vegetables	Sorghum
Average yield (tons/ha)	3.6	0.8	4.4	2.8	1.5	2
Max (tons/ha)	5.1	0.7	5.6	4	1.6	2.7
Min (tons/ha)	2	1	2.1	1.6	0.6	1.2
CV (%)	17.7	6.6	16.4	21.1	28.7	17.8

bers of total water stress days as calculated by EPIC model) correspond with lower crop yields. This is the case for 2005, 2007, 2013, 2014, 2017, and 2025–2028.

Similar analyses have been performed for other crops, and more details are given by Pastori et al. (2017a). In the case of cotton, annual variability can be observed in all the communes, even though it is more pronounced in Bottou and Banikoara regions (−16%, +13%), while it is less important in Kerou and Pehunco. Bottou is the commune facing the highest water stress. Vegetable annual variability is very high in the regions of Karimama, Bottou, and Kirtachi (it can reach −60% of the yield reduction). Vegetable crop can benefit from additional "irrigation." Consistent annual crop yield variability is estimated for the remaining dominant crops in the region (specifically sorghum, yam, and rice; see Table 3).

3.2 *Irrigation*

Irrigation is expected to increase in the whole region to mitigate the impact of the climate change. In this region rice is the highest water-demanding crop. According to the model simulation, an increase in irrigated surface area dedicated to rice would result in an increase of rice production by 15%. In addition, the adoption of an integrated crop management combining more irrigation and increased fertilization input (e.g., by livestock manure management) would increase the productivity of rice to 7 tons/ha/year (corresponding to an increase of about +110% versus current levels). Similar results are found for vegetables, characterized by a potential increase in productivity of about +9%, with irrigation extension and about +220%, with an optimal combination of irrigation and fertility management strategies. It is also interesting to note that crops usually grown under rain-fed conditions, such as sorghum and millet, can significantly benefit from irrigation (+10%), a solution particularly of interest when dealing with climate change and necessity to increase food production self-sufficiency and security (Saxena et al. 2018).

As previously mentioned, the nutrient depletion is a major problem in this area. Within EPIC, this constraint was evaluated by allowing the model to apply all the nutrients required for optimal crop growth. In this case, a potential productivity scenario can be estimated under both rain-fed and irrigated conditions. With this hypothesis, it is possible to isolate the contribution of irrigation to limit annual variability of maize productivity (Fig. 6).

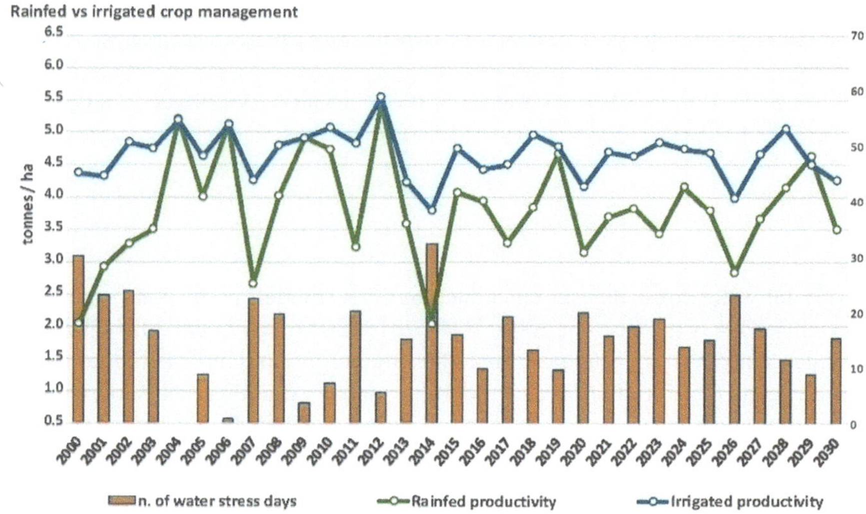

Fig. 6 Annual crop maize productivity in an agricultural region in the Mékrou region best estimated by assuming a rain-fed agriculture (with no stress for nutrient – green line). Blue line represents crop yield harvest (tons/ha) under optimal irrigation conditions. Bars (secondary axis) refer to the total days of crop water stress calculated by the model during the growing season (for rain-fed scenario)

The simulated rain-fed annual crop variability (green line) is much stronger if compared with the irrigated one (blue line).With irrigation (blue line), the yield loss in dry years is very much reduced. The importance of irrigation is shown in the box plots of annual variability in case of rain-fed and irrigated agriculture (Fig. 7): the variability of irrigated scenario (in blue) is reduced as illustrated by the inter-quantile (average productivity changes from 3.8 to 4.7 ton/ha and CV from 23% to 8%).

Irrigation timing is also an important factor to be taken into account for future development and planning. Climate change and variability can potentially introduce important changes in precipitation patterns and distribution, thereby causing delays in the onset of the rainy season. In order to effectively cope with these changes and mitigate their impacts on crop productivity, it is important to ensure water availability when needed. Crop water requirements are more important during dry periods when surface water resources are scarce. In the Mékrou river basin, daily crop water needs were estimated by using the E-Water tool and compared with water flow in the river (Fig. 8), pointing out how irrigation needs are much important when there is no water or when there is very limited water available in the river. Satisfying such demands would require other water sources (groundwater, small dams, lakes, etc.).

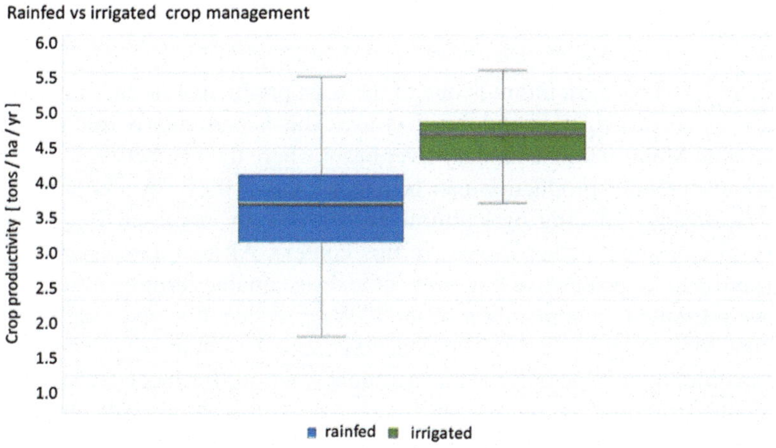

Fig. 7 Annual maize productivity variability in an agricultural region in the Mékrou river basin, estimated by assuming a rain-fed versus an irrigated agriculture (with no nutrient stresses)

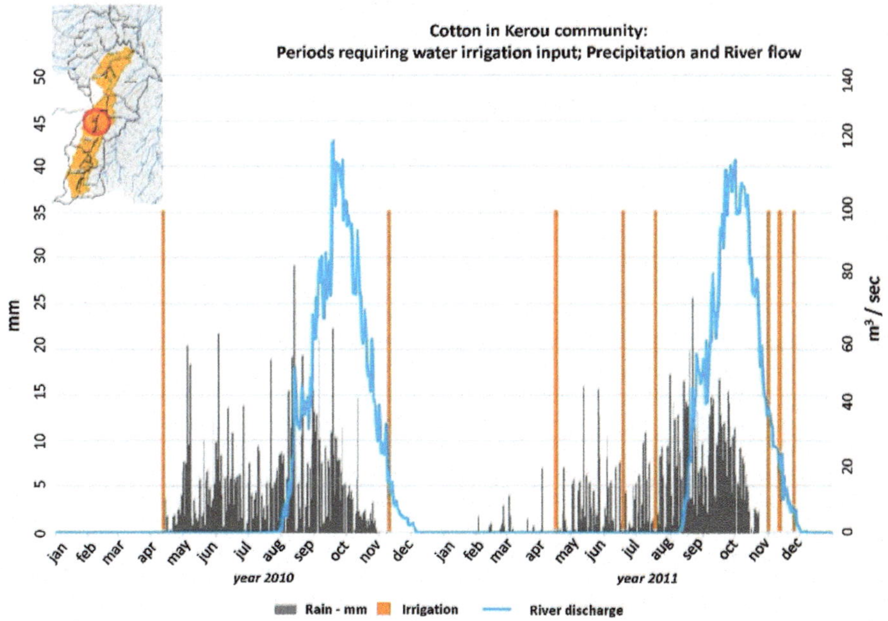

Fig. 8 Simulation of daily irrigation crop water requirements for cotton in the Mékrou river basin: irrigation water amounts are expressed in mm (first axis) as orange bars, daily rain is in dark gray (first axis), and river water flow is in blue (second axis)

4 Conclusion

Sub-Saharan African agriculture is one of the least productive in the world because of the low levels of irrigation and fertilizer use. This is particularly true in the West African Sahel Mékrou transboundary river basin where the gap between actual production and potential production can be greater than 100%. In addition, climate change is already reducing agricultural growth due to increased frequency and intensity of extreme weather events. In this context, African agriculture needs to shift toward a more productive but efficient and sustainable farming system. In this paper, we presented an application of the E-Water decision support system (DSS) tool in the Mékrou river basin. The decision support system was developed and applied in order to provide optimal management solutions for local decision-makers in the context of food insecurity and increasing competition with other sectors struggling for the same resources (primarily land and water).

The study showed that agriculture has the potential to mitigate the impact of climate change by adopting more integrated and efficient practices. The main limiting factor for crop production is low soil fertility, while water limitation is restricted to specific dry years. The adoption of more intensive farming and introduction of irrigation can increase the capacity of agriculture to efficiently respond and adapt to climate change and other external stress factors. Infrastructure for water harvesting and storage, together with best agricultural practices for increasing fertility, is the way forward in improving productivity and meeting food security goals.

References

Basso B, Cammarano D, Fiorentino C, Ritchie JT (2013) Wheat yield response to spatially variable nitrogen fertilizer in Mediterranean environment. Eur J Agron 51:65–70

Beck HW, Jones P, Jones JW (1989) SOYBUG: an expert system for soybean insect pest management. Agric Syst 30:269–286

Brams EA (1971) Continuous cultivation of West African soils: organic matter diminution and effects of applied lime and phosphorus. Plant Soil 35:401–414. https://doi.org/10.1007/BF01372671

Cohn AS, Newton P, Gil JDB, Kuhl L, Samberg L, Ricciardi V, Manly JR, Northrop S (2017) Smallholder agriculture and climate change. Annu Rev Environ Resour 42:347–375. https://doi.org/10.1146/annurev-environ-102016-060946

Conceição P, Levine S, Lipton M, Warren-Rodríguez A (2016) Toward a food secure future: ensuring food security for sustainable human development in sub-Saharan Africa. Food Policy 60:1–9. https://doi.org/10.1016/j.foodpol.2016.02.003

Dzanku FM, Jirström M, Marstorp H (2015) Yield gap-based poverty gaps in rural sub-Saharan Africa. World Dev 67:336–362. https://doi.org/10.1016/j.worlddev.2014.10.030

FAO (2016) The state of food and agriculture: climate change, agriculture and food security. ISBN: 978-92-5-107671-2

FAO, IFAD, UNICEF, WFP, WHO (2017) The state of food security and nutrition in the world 2017. Building resilience for peace and food security. FAO, Rome

Folberth C, Gaiser T, Abbaspour KC, Schulin R, Yang H (2012) Regionalization of a large-scale crop growth model for sub-Saharan Africa: model setup, evaluation, and estimation of maize yields. Agric Ecosyst Environ 151:21–33. https://doi.org/10.1016/j.agee.2012.01.026

Fraisse CW, Perez N, Andreis JH (2015) Smart strawberry advisory system for mobile devices. U.S. Department of Agriculture, UF/IFAS Extension Service, University of Florida, Gainesville, USA.

Gil JDB, Cohn AS, Duncan J, Newton P, Vermeulen S (2017) The resilience of integrated agricultural systems to climate change. Wiley Interdiscip Rev Clim Chang 8:e461. https://doi.org/10.1002/wcc.461

Giorgi F, Jones C, Asrar GR (2009) Addressing climate information needs at the regional level: the CORDEX framework. World Meteorol Organ Bull 58(3): 175–83

IPCC (2007) Climate Change 2007: Impacts, Adaptation and Vulnerability. Contribution of working group II to the fourth assessment report of the intergovernmental panel on climate change, Edited by M.L. Parry, O.F. Canziani, J.P. Palutikof, P.J. van der Linden, and C.E. Hanson. New York, USA: Cambridge University Press, Cambridge, UK

Jones JW, Antle JM, Basso B, Boote KJ, Conant RT, Foster I, Godfray HCJ, Herrero M, Howitt RE, Janssen S et al (2017) Toward a new generation of agricultural system data, models, and knowledge products: state of agricultural systems science. Agric Syst 155:269–288

JRC (2018) E-WATER: technical documentation of E-Water software | Aquaknow (WWW document). https://aquaknow.jrc.ec.europa.eu/6583/documents/e-water-technical-documentation-e-water-software. Accessed 5 Apr 2018

Lesk C, Rowhani P, Ramankutty N (2016) Influence of extreme weather disasters on global crop production. Nature 529:84–87. https://doi.org/10.1038/nature16467

Lipper L, Thornton P, Campbell BM, Baedeker T, Braimoh A, Bwalya M, Caron P, Cattaneo A, Garrity D, Henry K, Hottle R, Jackson L, Jarvis A, Kossam F, Mann W, McCarthy N, Meybeck A, Neufeldt H, Remington T, Sen PT, Sessa R, Shula R, Tibu A, Torquebiau EF (2014) Climate-smart agriculture for food security. Nat Clim Chang 4:1068–1072. https://doi.org/10.1038/nclimate2437

Mapfumo P, Chikowo R, Mtambanengwe F (2010) Lack of resilience in African smallholder farming: exploring measures to enhance the adaptive capacity of local communities to pressure climate change: final technical report – Zimbabwe (October 2010). Department of Soil Science and Agricultural Engineering, University of Zimbabwe, Harare

Markantonis V, Farinosi F, Dondeynaz C, Ameztoy I, Pastori M, Marletta L, Ali A, Carmona Moreno C (2017) Assessing floods and droughts in the Mékrou River basin (West Africa): a combined household survey and climatic trends analysis approach. Nat Hazards Earth Syst Sci 18:1279–1296. https://doi.org/10.5194/nhess-18-1279-2018

Nelson RA, Holzworth DP, Hammer GL, Hayman PT (2002) Infusing the use of seasonal climate forecasting into crop management practice in North East Australia using discussion support software. Agric Syst 74:393–414

Pastori M, Bouraoui F (2017) Modeling the impact of water and nutrient management in African agriculture. Afr J Water Conserv Sustain 5:206–220

Pastori M, Dondeynaz C, Carmona Moreno C (2017a) Simulation et analyse scientifique-technique des scenarii de developpement CaSSE sur le bassin versant de la Mekrou. JRC Report N 106603

Pastori M, Udías A, Bouraoui F, Bidoglio G (2017b) A multi-objective approach to evaluate the economic and environmental impacts of alternative water and nutrient management strategies in Africa. J Environ Inform 29:16–28. https://doi.org/10.3808/jei.201500313

Payne WA (2010) Farming systems and food security in sub-Saharan Africa. In: Lal R, Stewart BA (eds) Food security and soil quality. CRC Press, Boca Raton, p 416

PHU (2007) Potential Heat Unit (PHU) program. Download at http://swat.tamu.edu/software/potential-heat-unit-program/

Rossi V, Salinari F, Poni S, Caffi T, Bettati T (2014) Addressing the implementation problem in agricultural decision support systems: the example of vite.net®. Comput Electron Agric 100:88–99. https://doi.org/10.1016/j.compag.2013.10.011

Salton JC, Mercante FM, Tomazi M, Zanatta JA, Concenço G, Silva WM, Retore M (2014) Integrated crop-livestock system in tropical Brazil: toward a sustainable production system. Agric Ecosyst Environ 190:70–79. https://doi.org/10.1016/J.AGEE.2013.09.023

Saxena R, Vanga SK, Wang J, Orsat V, Raghavan V (2018) Millets for food security in the context of climate change: a review. Sustainability 10:2228. https://doi.org/10.3390/su10072228

Strassemeyer J, Daehmlow D, Dominic AR, Lorenz S, Golla B (2017) SYNOPS-WEB, an online tool for environmental risk assessment to evaluate pesticide strategies on field level. Crop Prot 97:28–44. https://doi.org/10.1016/J.CROPRO.2016.11.036

Stuth JW, Lyons BG (1993) Decision support systems for the management of grazing lands: emerging issues. Man and the Biosphere Series. Unesco, Paris; Parthenon Pub Group, Pearl River, N.Y., USA

Thirtle C, Piesse J (2007) Governance, agricultural productivity and poverty reduction in Africa, Asia and Latin America. Irrig Drain 56:165–177

Udias A, Pastori M, Dondeynaz C, Carmona Moreno C, Cattaneo L, Ali A, Cano J (2018) A decision support tool to enhance agricultural growth in the Mékrou River Basin (West Africa). Comput Electron Agric 154:467–481

van Ittersum MK, Cassman KG, Grassini P, Wolf J, Tittonell P, Hochman Z (2013) Yield gap analysis with local to global relevance – a review. Field Crops Res 143:4–17. https://doi.org/10.1016/j.fcr.2012.09.009

Van Keulen H, Breman H (1990) Agricultural development in the West African Sahelian region: a cure against land hunger? Agric Ecosyst Environ 32:177–197

Wang X, Williams JR, Gassman PW, Baffaut C, Izaurralde RC, Jeong J, Kiniry JR (2012) EPIC and APEX: model use, calibration, and validation. Trans ASABE. https://doi.org/10.13031/2013.42253

Weight D, Kelly VA (1998) Restoring soil fertility in Sub-Saharan Africa: Technical and Economic Issues. Food Security International Development Policy Syntheses 11374, Michigan State University, Department of Agricultural, Food, and Resource Economics, East Lansing, MI 48824, USA

Williams JR (1995) The EPIC model in: computer models of watershed hydrology. Water Resources Publications, Highlands Ranch

Xiong W, Holman I, Lin E, Conway D, Jiang J, Xu Y, Li Y (2010) Climate change, water availability and future cereal production in China. Agric Ecosyst Environ 135:58–69. https://doi.org/10.1016/j.agee.2009.08.015

Zougmoré R, Partey S, Ouédraogo M, Omitoyin B, Thomas T, Ayantunde A, Ericksen P, Said M, Jalloh A (2016) Toward climate-smart agriculture in West Africa: a review of climate change impacts, adaptation strategies and policy developments for the livestock, fishery and crop production sectors. Agric Food Secur 5:26. https://doi.org/10.1186/s40066-016-0075-3

Index

Y. Bamutaze et al. (eds.), *Agriculture and Ecosystem Resilience in Sub Saharan Africa*, Climate Change Management,
https://doi.org/10.1007/978-3-030-12974-3

S

Printed by Printforce, the Netherlands